INTRODUCTION
TO
DISCRETE EVENT SYSTEMS

THE KLUWER INTERNATIONAL SERIES ON DISCRETE EVENT DYNAMIC SYSTEMS

Series Editor

Yu-Chi Ho
Harvard University

INTRODUCTION
TO
DISCRETE EVENT SYSTEMS

by

Christos G. Cassandras
Boston University

Stéphane Lafortune
The University of Michigan

KLUWER ACADEMIC PUBLISHERS
Boston / Dordrecht / London

Distributors for North, Central and South America:
Kluwer Academic Publishers
101 Philip Drive
Assinippi Park
Norwell, Massachusetts 02061 USA
Telephone (781) 871-6600
Fax (781) 681-9045
E-Mail <kluwer@wkap.com>

Distributors for all other countries:
Kluwer Academic Publishers Group
Distribution Centre
Post Office Box 322
3300 AH Dordrecht, THE NETHERLANDS
Telephone 31 78 6392 392
Fax 31 78 6546 474
E-Mail <services@wkap.nl>

Electronic Services <http://www.wkap.nl>

Library of Congress Cataloging-in-Publication Data

Cassandras, Christos G.
      Introduction to discrete event systems / by Christos G.
Cassandras, Stéphane Lafortune.
            p.   cm.
      Includes bibliographical references.
      ISBN: 0-7923-8609-4 (alk. paper)
      1. System analysis. 2. Discrete-time systems. I. Lafortune,
Stéphane, 1958-   . II. Title.
      T57.6.C39 1999
      003--dc21   99-40716
                                                                              CIP

To Carol and Monica (C.G.C)

To Julien and Claire (S.L.)

Table of Contents

Preface

Though this be madness, yet there is method in't.

William Shakespeare, Hamlet

Over the past few decades, the rapid evolution of computing, communication, and sensor technologies has brought about the proliferation of "new" dynamic systems, mostly technological and often highly complex. Examples are all around us: computer and communication networks; automated manufacturing systems; air traffic control systems; highly integrated command, control, communication, and information (C^3I) systems; advanced monitoring and control systems in automobiles or large buildings; intelligent transportation systems; distributed software systems; and so forth. A significant portion of the "activity" in these systems, sometimes all of it, is governed by operational rules designed by humans; their dynamics are therefore characterized by asynchronous occurrences of discrete *events*, some controlled (like hitting a keyboard key, turning a piece of equipment "on", or sending a message packet) and some not (like a spontaneous equipment failure or a packet loss), some observed by sensors and some not. These features lend themselves to the term *discrete event system* for this class of dynamic systems.

The mathematical arsenal centered around differential and difference equations that has been employed in systems and control engineering to model and study the time-driven processes governed by the laws of nature is inadequate or simply inappropriate for discrete event systems. The challenge is to develop new modeling frameworks, analysis techniques, design tools, testing methods, and systematic control and optimization procedures for this new generation of highly complex systems. In order to face this challenge we need a multidisciplinary approach. First, we need to build on the concepts and techniques of system and control theory (for performance optimization via feedback control), computer science (for modeling and verification of event-driven processes), and operations research (for analysis and simulation of stochastic models of discrete event systems). Second, we need to develop new modeling frameworks, analysis techniques, and control procedures that are suited for discrete event systems.

Finally, we need to introduce new paradigms that combine mathematical techniques with processing of experimental data. The role of the computer itself as a tool for system design, analysis, and control is becoming critical in the development of these new techniques and paradigms.

The capabilities that discrete event systems have, or are intended to have, are extremely exciting. Their complexity, on the other hand, is overwhelming. Powerful methodologies are needed not only to enhance design procedures, but also to prevent failures, which can indeed be catastrophic at this level of complexity, and to deliver the full potential of these systems.

About this Book

A substantial portion of this book is a revised version of *Discrete Event Systems: Modeling and Performance Analysis*, written by the first author and published in 1993 (Irwin and Aksen Associates), which received the 1999 Harold Chestnut Prize, awarded by the International Federation of Automatic Control for best control engineering textbook. The present book includes additional material providing in-depth coverage of language and automata theory and new material on the supervisory control of discrete event systems; overall, it is intended to be a comprehensive introduction to the field of discrete event systems, emphasizing breadth of coverage and accessibility of the material to a large audience of readers with possibly different backgrounds. Its key feature is the emphasis placed on a unified modeling framework for the different facets of the study of discrete event systems. This modeling framework is centered on automata (and to a lesser extent on Petri nets) and is gradually refined: untimed models for logical properties concerned with the ordering of events, timed models for properties that involve timing considerations, and stochastic timed models for properties that involve a probabilistic setting. The unified modeling framework transcends specific application areas and allows linking of the following topics in a coherent manner for the study of discrete event systems: language and automata theory, supervisory control, Petri net theory, (max,+) algebra, Markov chains and queueing theory, discrete-event simulation, perturbation analysis, and concurrent estimation techniques. Until now, these topics had been treated in separate books or in the research literature only.

The book is written as a textbook for courses on discrete event systems at the senior undergraduate level or the first-year graduate level. It should be of interest to students in a variety of disciplines where the study of discrete event systems is relevant: control, communications, computer engineering, computer science, manufacturing engineering, operations research, and industrial engineering, to name a few. In this regard, examples throughout the book are drawn from many areas such as control engineering, networking, manufacturing, and software engineering.

We have attempted to make this book as self-contained as possible. It is assumed that the background of the reader includes set theory and elementary linear algebra and differential equations. A basic course in probability theory with some understanding of stochastic processes is essential for Chapters 6 through 11; a comprehensive review is provided in Appendix I. If readers have had an undergraduate course in systems and control, then the first part of Chapter 1 should be a refresher of fundamental modeling concepts. Some parts of Chapters 3, 9, and 11 are more advanced and appropriate for graduate courses.

A senior-level one-semester course taught at the University of Massachusetts at Amherst covered the material in Chapter 1, parts of Chapters 2 and 4, and most of Chapters 5, 6, 7, 8, and 10. A more advanced graduate-level course taught at Boston University, is based on Chapter 6 and Chapters 8 through 11, assuming knowledge of elementary random processes. At the University of Michigan, a first-year graduate course for students in electrical engineering and computer science covers Chapter 1 through 5; no prerequisite is required.

Organization of this Book

The basic road map and organizational philosophy of this book are as follows.

- **Chapter 1** introduces the defining characteristics of discrete event systems and places the different chapters of this book in perspective. This is preceded by an introduction to the fundamental concepts associated with the theory of systems and control engineering in Section 1.2. The motivation for starting with Section 1.2 is to help the reader appreciate the distinction between *continuous-variable (time-driven) dynamic systems* and *event-driven dynamic systems*. Readers with a background in linear systems and control are likely to be familiar with the material in Section 1.2 but not with that in Section 1.3 where discrete event systems (DES) are introduced. On the other hand, readers with a background in computing science and systems are likely to be familiar with the concept of event-driven system but not with the notions of state space model and feedback control covered in Section 1.2, which are important for the subsequent chapters.

The next 10 chapters are organized according to the level of abstraction (as defined in Section 1.3.3) chosen for modeling, analysis, control, performance optimization, and simulation of DES: *untimed* or *logical* (Chapters 2, 3, and 4), *timed* (Chapter 5), and *stochastic timed* (Chapters 6 through 11).

- **Chapter 2** introduces *language models* of DES and the representation of languages by *automata*. The automaton modeling formalism is discussed in detail, including composition operations on automata, observer automata, and diagnoser automata. While most of this chapter is "standard material" in formal languages and automata theory, the presentation is adapted to the needs and objectives of this book.

- **Chapter 3** presents an introduction to *supervisory control theory*. The goal is to study how to *control* a DES in order to satisfy a given set of logical (or qualitative) performance objectives on states and event ordering. The control paradigm is language-based and accounts for limited controllability of events and limited observability of events. This chapter is based on recent research and some parts are more advanced and suited for a graduate course. Chapter 2 is a necessary prerequisite for this chapter.

- **Chapter 4** presents the modeling formalism of *Petri nets* and discusses the analysis and control of untimed Petri net models. While many references are made to concepts introduced in Chapter 2 (and in Chapter 3 for Section 4.5), this chapter is to a large extent self-contained.

- In **Chapter 5**, the two classes of untimed models studied in Chapters 2 (automata) and 4 (Petri nets) are refined to include "time" by means of the *clock structure* mechanism, resulting in *timed automata* and *timed Petri nets*. The chapter also contains an introduction to the technique of the $(max, +)$ *algebra* for analyzing certain classes of timed DES, particularly timed marked graphs.

It is suggested that Chapters 1 through 5 may form the content of a course on modeling, analysis, diagnosis, and control of DES, with no consideration of probabilistic models.

- Starting with **Chapter 6**, the focus is on a *probabilistic setting* for the study of *stochastic timed* DES models. The timed automaton model of Chapter 4 is refined by the use of a *stochastic* clock structure, leading to *stochastic timed automata* and their associated *generalized semi-Markov stochastic processes* (GSMP). The *Poisson process* is then presented in depth as a "building block" for the stochastic clock structure of timed DES models. Using this building block (which is based on simple, physically plausible assumptions), the class of Markov chain models emerges rather naturally from the general GSMP model.

By the end of Chapter 6, two general directions emerge regarding analysis of stochastic timed DES models. The first direction is based on classical stochastic models for which *analytical techniques* have been developed based on probability theory (Chapters 7 through 9). In particular, Chapter 7 (Markov chains) and Chapter 8 (queueing theory) cover a limited class of stochastic DES models that can be handled through fairly traditional analysis. This material will look quite familiar to many readers.

The second direction relies on *computer simulation* and on some new techniques based on the analysis of sample paths of DES (Chapters 10 and 11). It should be pointed out that the reader can go directly from Chapter 6 to Chapter 10, completely bypassing the first direction if so desired.

- **Chapter 7** is concerned with the analysis of *Markov chain models*, introduced at the end of Chapter 6. Both discrete-time and continuous-time Markov chains are considered. The chapter also includes treatment of birth-death chains and uniformization of continuous-time chains.

- **Chapter 8** is an introduction to *queueing theory*. Queueing models are arguably the most well-known and studied class of stochastic DES models. The material in Chapter 8 includes the standard key results on simple Markovian queueing systems ($M/M/1$, $M/G/1$, etc.), as well as results on some special classes of queueing networks.

- While Chapters 7 and 8 cover the analysis part of Markov chains, **Chapter 9** covers the *control* part, based on the technique of *dynamic programming*. This chapter involves more advanced material and requires some additional mathematical maturity; it is, therefore, more suited to graduate students.

- **Chapter 10** brings the reader back to the "real world", where complex systems do not always conform to the "convenient" assumptions made in Chapters 7-9 in the analysis of stochastic timed DES models; hence, the need for *simulation*. The goal here is to help the reader become comfortable with building simulation models for DES, to introduce some basic techniques for analyzing the output data of a simulation for purposes such as estimating the performance of a complex DES, and to appreciate the advantages and limitations of such techniques.

  The stochastic timed automaton framework developed in Chapter 6 allows the introduction of discrete-event simulation in Chapter 10 to be particularly smooth and natural. This is because of the concept of a "clock structure" driving a DES considered in Chapters 5 (deterministic case) and 6 (stochastic case). When this clock structure is supplied through a computer random number generator, a "simulation" is simply a software implementation of a stochastic timed automaton.

- **Chapter 11** presents sensitivity analysis and concurrent estimation techniques for DES. It develops the theory behind perturbation analysis and the fundamental *sample path constructability problem*, based on which methodologies and concrete algorithms for "rapid learning" in the control and optimization of DES have been developed. This material is based exclusively on recent research. The emphasis is on presenting key ideas and basic results from which the reader can proceed to more detailed and advanced material.

It is suggested that Chapter 6 along with Chapters 8 through 11 may form the content of a more advanced graduate course on stochastic modeling, analysis, control, performance optimization, and simulation of DES.

Feedback to the Authors

The Web site http://vita.bu.edu/cgc/BOOK/ is being maintained for feedback between the authors and the readers. In particular, a list of errata is available there. Please take a look and send your comments! Throughout the book, we also refer readers to various relevant Web sites. These can also be accessed directly from http://vita.bu.edu/cgc/BOOK/. In particular, we urge readers to visit the interactive multimedia introduction to DES located at http://vita.bu.edu/cgc/MIDEDS/.

Acknowledgements

As mentioned earlier, a large part of this book is a revised version of *Discrete Event Systems: Modeling and Performance Analysis*, written by the first author and published in 1993. Therefore, several of the acknowledgements included in this earlier book are sill relevant here (see below). Regarding the present book as a joint effort by the two authors, special acknowledgements go to George Barrett, Rami Debouk, Feng Lin, and Karen Rudie, for carefully reviewing earlier versions of the new contributions in this book and making numerous suggestions for improvement. Special thanks also go to Jianyang Tai, whose help with the formatting of the book and with figure generation has been vital, and to Christos Panayiotou for his problem-solving contributions to some sticky formatting issues (he is also an instrumental contributor to some of the material of Chapter 11).

First Author:

A large part of the material included in this book was written while I was on sabbatical at the Division of Applied Sciences at Harvard University (September 1990 - January 1991). Professor Y.C. Ho was instrumental in providing me with a truly comfortable environment to do some serious writing during that period. I am also grateful to the Lilly Foundation for providing me with a Fellowship for the academic year 1991-92, during which another substantial portion of this book was written. In addition, I would like to acknowledge the support of the National Science Foundation, which, through a grant under its Combined Research and Curriculum Development Program, supported the creation of a course on discrete event systems at the Department of Electrical and Computer Engineering Department at the University of Massachusetts at Amherst, and, since 1997, at the Department of Manufacturing Engineering at Boston University as well. A similar acknowledgement goes toward several other funding organizations (the Air Force Office of Scientific Research and the Air Force Research Laboratory, the Office of Naval Research and the Naval Research Laboratory, DARPA, and United Technologies/OTIS) that have provided, over the past several years, support for my research work; some of this work has given rise to parts of Chapters 5, 6, 9, and 11.

A number of colleagues, friends, reviewers, former and current students have contributed in a variety of ways to the final form of this book. I am particularly grateful to Y.C. (Larry) Ho, because my continuing interaction with him (from Ph.D. thesis advisor to friend and colleague) has helped me realize that if there is such a thing as "joie de vivre", then there surely is something like "joie de recherche". And it is under the intoxicating influence of such a "joie de recherche" that books like this can come to be.

For many stimulating discussions which directly or indirectly have influenced me, many thanks to Wei-Bo Gong, Rajan Suri, Xiren Cao, R. (R.S) Sreenivas, Al Sisti, Jeff Wieselthier, Yorai Wardi, and Felisa Vázquez-Abad. For their constructive suggestions in reviewing my prior book (of which this is largely a revised version), I am grateful to Mark Andersland (University of Iowa), Ed Chong (Purdue University), Vijay K. Garg (University of Texas at Austin), P.R. Kumar (University of Illinois), Peter Ramadge (Princeton University), Mark Shayman (University of Maryland), and Pravin Varaiya (University of California at Berkeley). My co-author, Stéphane Lafortune, was originally also a reviewer; I am now delighted to have had the opportunity to collaborate with him on this followup project. His deep knowledge of supervisory control has contributed material in Chapters 2 through 4 which is a major asset of this new, truly comprehensive, book on DES.

For developing and testing the algorithm in Appendix II, I would like to thank Jie Pan and Wengang Zhai, former graduate students at the Control Of Discrete Event Systems (CODES) Laboratory at the University of Massachusetts at Amherst. All of my former and current graduate students at the University of Massachusetts and at Boston University have had a role to play in the body of knowledge this book encompasses. Special thanks must go to my very first Ph.D. student, Steve Strickland, who made important contributions to the material of Chapter 11, but all the rest are also gratefully acknowledged.

Finally, I was grateful to Carol during the writing of my last book, and am even more grateful now after completing this one...

Christos G. Cassandras

Second Author:

I wish to thank all the graduate students and colleagues with whom I have built my knowledge of discrete event systems. My contributions to this book have been shaped by my interactions with them. I thank Hyuck Yoo, Enke Chen, Sheng-Luen Chung, Nejib Ben Hadj-Alouane, Raja Sengupta, Meera Sampath, Yi-Liang Chen, Isaac Porche, George Barrett, and Rami Debouk; I also thank all the students who have taken "EECS 661" at Michigan over the last 11 years. I thank Feng Lin (Wayne State University), Karen Rudie (Queen's University), Kasim Sinnamohideen (Johnson Controls), Demosthenis Teneketzis (University of Michigan), and John Thistle (École Polytechnique de Montréal); doing research with them has been, and continues to be, inspiring and enjoyable. It is a pleasure to acknowledge the constant support that I have received since 1986 from the Department of Electrical Engineering and Computer Science at the University of Michigan regarding my research and teaching activities in discrete event systems. The National Science Foundation and the Army Research Office are also acknowledged for supporting my research in this area.

I extend very special thanks to Peter Ramadge (Princeton University), who first introduced me to the topic of discrete event systems and who has been a constant source of encouragement, and to W. Murray Wonham (University of Toronto), whose vision and seminal contributions have influenced greatly the field of discrete event systems. I am grateful to W.M. Wonham for kindly sharing with me his "Course Notes on Discrete Event Systems" for several years; these notes have been an important source of inspiration in the writing of Chapter 3.

Finally, I wish to thank Christos for asking me to contribute to the new version of this book. It has been a great experience.

Stéphane Lafortune

Chapter 1

Systems and Models

1.1 INTRODUCTION

As its title suggests, this book is about a special class of *systems* which in recent decades have become an integral part of our world. Before getting into the details of this particular class of systems, it is reasonable to start out by simply describing what we mean by a "system", and by presenting the fundamental concepts associated with system theory as it developed over the years. This defines the first objective of this chapter, which is for the benefit of readers with little or no prior exposure to introductory material on systems and control theory (Section 1.2). Readers who are already familiar with concepts such as "state spaces", "state equations", "sample paths", and "feedback" may immediately proceed to Section 1.3.

The second objective is to look at useful classifications of systems so as to reveal the features motivating our study of *discrete event* systems. Historically, scientists and engineers have concentrated on studying and harnessing natural phenomena which are well-modeled by the laws of gravity, classical and non-classical mechanics, physical chemistry, etc. In so doing, we typically deal with quantities such as the displacement, velocity, and acceleration of particles and rigid bodies, or the pressure, temperature, and flow rates of fluids and gases. These are "continuous variables" in the sense that they can take on any real value as *time* itself "continuously" evolves. Based on this fact, a vast body of mathematical tools and techniques has been developed to model, analyze, and control the systems around us. It is fair to say that the study of ordinary and partial differential equations currently provides the main infrastructure for system analysis and control.

But in the day-to-day life of our technological and increasingly computer-dependent world, we notice two things. First, many of the quantities we deal with are "discrete", typically involving counting integer numbers (how many parts are in an inventory, how many planes are in a runway, how many telephone calls are active). And second, what drives many of the processes we use and depend on are instantaneous "events" such as the pushing of a button, hitting a keyboard key, or a traffic light turning green. In fact, much of the technology we have invented and rely on (especially where digital computers are involved) is event-driven: Communication networks, manufacturing facilities, or the execution of a computer program are typical examples.

1.2 SYSTEM AND CONTROL BASICS

In this section, we will introduce carefully, but rather informally, the basic concepts of system theory. As we go along, we will identify fundamental criteria by which systems are distinguished and classified. These are summarized in Section 1.4, Fig. 1.31. Since this section may be viewed as a summary of system and control engineering basics, it may be skipped by readers with a background in this area, who can immediately proceed to Section 1.3 introducing the key elements of discrete event systems. By glancing at Fig. 1.31, these readers can also immediately place the class of discrete event systems in perspective.

1.2.1 The Concept of System

System is one of those primitive concepts (like *set* or *mapping*) whose understanding might best be left to intuition rather than an exact definition. Nonetheless, we can provide three representative definitions found in the literature:

- An aggregation or assemblage of things so combined by nature or man as to form an integral or complex whole [*Encyclopedia Americana*].

- A regularly interacting or interdependent group of items forming a unified whole [*Webster's Dictionary*].

- A combination of components that act together to perform a function not possible with any of the individual parts [*IEEE Standard Dictionary of Electrical and Electronic Terms*].

There are two salient features in these definitions. First, a system consists of interacting "components", and second a system is associated with a "function" it is presumably intended to perform. It is also worth pointing out that a system should not always be associated with physical objects and natural laws. For example, system theory has provided very convenient frameworks for describing economic mechanisms or modeling human behavior and population dynamics.

1.2.2 The Input-Output Modeling Process

As scientists and engineers, we are primarily concerned with the quantitative analysis of systems, and the development of techniques for design, control, and the explicit measurement of system performance based on well-defined criteria. Therefore, the purely qualitative definitions given above are inadequate. Instead, we seek a *model* of an actual system. Intuitively, we may think of a model as a device that simply duplicates the behavior of the system itself. To be more precise than that, we need to develop some mathematical means for describing this behavior.

To carry out the modeling process, we start out by defining a set of *measurable variables* associated with a given system. For example, particle positions and velocities, or voltages and currents in an electrical circuit, which are all real numbers. By measuring these variables over a period of time $[t_0, t_f]$ we may then collect *data*. Next, we select a subset of these variables and assume that we have the ability to vary them over time. This defines a set of time functions which we shall call the *input variables*

$$\{u_1(t), \ldots, u_p(t)\} \quad t_0 \leq t \leq t_f \tag{1.1}$$

Then, we select another set of variables which we assume we can directly measure while varying $u_1(t), \ldots, u_p(t)$, and thus define a set of *output variables*

$$\{y_1(t), \ldots, y_m(t)\} \quad t_0 \leq t \leq t_f \tag{1.2}$$

These may be thought of as describing the "response" to the "stimulus" provided by the selected input functions. Note that there may well be some variables which have not been associated with either the input or the output; these are sometimes referred to as *suppressed* output variables.

To simplify notation, we represent the input variables through a column vector $\mathbf{u}(t)$ and the output variables through another column vector $\mathbf{y}(t)$; for short, we refer to them as the *input* and *output* respectively. Thus, we will write

$$\mathbf{u}(t) = [u_1(t), \ldots, u_p(t)]^T$$

where $[\cdot]^T$ denotes the transpose of a vector, and, similarly,

$$\mathbf{y}(t) = [y_1(t), \ldots, y_m(t)]^T$$

To complete a model, it is reasonable to postulate that there exists some mathematical relationship between input and output. Thus, we assume we can define functions

$$y_1(t) = g_1(u_1(t), \dots, u_p(t)), \dots, y_m(t) = g_m(u_1(t), \dots, u_p(t))$$

and obtain the system model in the mathematical form

$$\mathbf{y} = \mathbf{g(u)} = [g_1(u_1(t), ..., u_p(t)), ..., g_m(u_1(t), ..., u_p(t))]^T \qquad (1.3)$$

where $\mathbf{g}(\cdot)$ denotes the column vector whose entries are the functions $g_1(\cdot), \dots, g_m(\cdot)$.

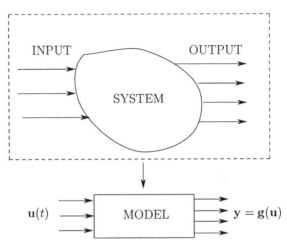

Figure 1.1: Simple modeling process.

   This is the simplest possible modeling process, and it is illustrated in Fig. 1.1. Strictly speaking, a system is "something real" (e.g., an amplifier, a car, a factory, a human body), whereas a model is an "abstraction" (a set of mathematical equations). Often, the model only approximates the true behavior of the system. However, once we are convinced we have obtained a "good" model, this distinction is usually dropped, and the terms *system* and *model* are used interchangeably. This is what we will be doing in the sequel. But, before doing so, it is worth making one final remark. For any given system, it is always possible (in principle) to obtain a model; the converse is not true, since mathematical equations do not always yield real solutions. For example, let $u(t) = -1$ for all t be a scalar input representing a constant voltage, and $y = \sqrt{u} = \sqrt{-1}$ be the output. Clearly, we cannot build any electrical system generating an imaginary voltage. In such cases, we say that a system is not physically *realizable*.

   It is important to emphasize the flexibility built into the modeling process, since no unique way to select input and output variables is imposed (see also Example 1.1). Thus, it is the modeler's task to identify these variables depending on a particular point of view or on the constraints imposed upon us by a particular application.

Example 1.1 (A voltage-divider system)
The voltage divider circuit of Fig. 1.2 constitutes a simple electrical system. Five variables are shown: the source voltage V, the current i, the two resistances r and R, and the voltage v across R. The simplest models we can construct make use of standard circuit theory relationships such as:

$$v = V\frac{R}{R+r} \tag{1.4}$$

$$v = iR \tag{1.5}$$

Assuming we can adjust the source voltage V (input) and are interested

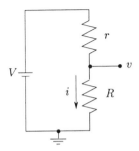

Figure 1.2: Simple electrical system for Example 1.1.

in regulating the voltage v (output), we can obtain the model shown in Fig. 1.3a. Alternatively, suppose the power source is fixed, but we have an adjustable resistance r, in which case our model might be that of Fig. 1.3b. Finally, if both V and r are adjustable and we are interested in regulating the current i, the model of Fig. 1.3c is appropriate. Thus, for the same underlying system, different models may be conceived. Of course, the functional relationships between variables do not change, but only the choice of input and output.

Example 1.2 (A spring-mass system)
Consider the spring-mass system of Fig. 1.4. Suppose that at time $t = 0$ we displace the mass from its rest position by an amount $u(0) = u_0 > 0$ (the positive direction is as shown in Fig. 1.4) and release it. Let the displacement at any time $t > 0$ be denoted by $y(t)$. We know, from simple mechanics, that the motion of the mass defines a harmonic oscillation described by the second-order differential equation

$$m\ddot{y} = -ky \tag{1.6}$$

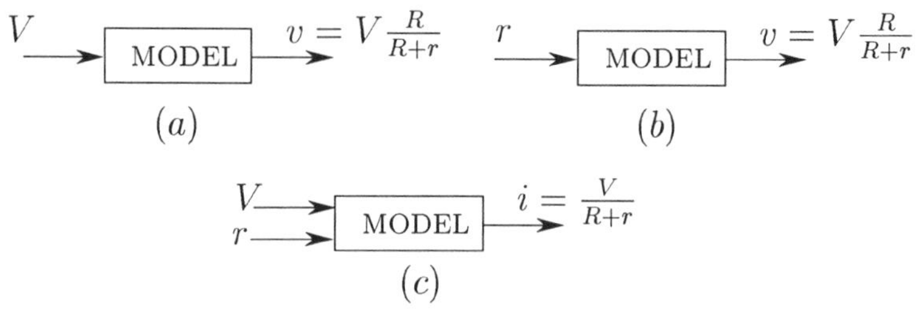

$V \xrightarrow{\hspace{1cm}} \boxed{\text{MODEL}} \xrightarrow{v = V \frac{R}{R+r}} \qquad r \xrightarrow{\hspace{1cm}} \boxed{\text{MODEL}} \xrightarrow{v = V \frac{R}{R+r}}$

(a)   (b)

$\begin{matrix} V \longrightarrow \\ r \longrightarrow \end{matrix} \boxed{\text{MODEL}} \xrightarrow{i = \frac{V}{R+r}}$

(c)

Figure 1.3: Three models for the system of Fig. 1.2.
Depending on what we view as controllable input and observable output, several alternate models of the same underlying system may be constructed.

with initial conditions $y(0) = u_0$, $\dot{y}(0) = 0$. If we are interested in controlling the initial displacement $u(0)$ and observing the position of the mass as a function of time, we can construct the model shown in Fig. 1.4, where the input is the function $u(t)$ defined so that

$$u(t) = \begin{cases} u_0 & t = 0 \\ 0 & \text{otherwise} \end{cases} \tag{1.7}$$

and the output $y(t)$ is the solution of the differential equation (1.6). Note that the variables k and m are assumed constant.

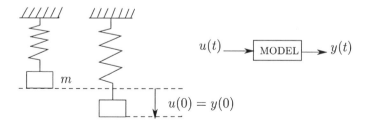

Figure 1.4: Simple mechanical system for Example 1.2 and corresponding model.

Static and Dynamic Systems

Comparing Examples 1.1 and 1.2 reveals one of the major ways to classify systems. In Example 1.1, all input-output functional relationships we can construct are simply described by *algebraic* equations. In Example 1.2, although $u(t)$ is constant for all $t > 0$, the output $y(t)$ is time-dependent (the solution of (1.6) is known to be a sinusoid), and is clearly affected by the value of $u(0)$. In this case, the output $y(t)$ at some $t > 0$ depends on past values of the input (in particular, $u(0)$), and the input-output functional relationship is a *differential* equation. Obtaining an explicit solution for $y(t)$ is not always an easy task.

We define a *static* system to be one where the output $y(t)$ is independent of past values of the input $u(\tau)$, $\tau < t$ for all t. A *dynamic* system is one where the output generally depends on past values of the input. Thus, determining the output of a static system requires no "memory" of the input history, which is not the case for a dynamic system. We will almost exclusively restrict attention to dynamic systems, which are much more interesting in practice.

Time-varying and Time-invariant Dynamic Systems

In considering the various types of input-output relationships in systems, it is reasonable to pose the following question: *Is the output always the same when the same input is applied?* The answer cannot always be "yes", and gives rise to another important way for classifying systems. In Example 1.2, for instance, suppose that the mass attached to the spring is actually a container full of some fluid, and that a leak develops immediately after time $t = 0$. Originally, the displacement u_0 is applied at time $t = 0$ and the system is observed for some interval, say $[0, t_1]$. At $t = t_1$ we bring back the mass to rest, displace it again by u_0, and repeat the process for the interval $[t_1, 2t_1]$. We would expect the system to behave quite differently, since by time t_1 the value of m will certainly have changed.

Returning to the input-output functional relationship (1.3), note that we have assumed $\mathbf{g}(\cdot)$ to be independent of time. However, a more general representation would be

$$\mathbf{y} = \mathbf{g}(\mathbf{u}, t) \tag{1.8}$$

where we allow $\mathbf{g}(\cdot)$ to explicitly depend on the time variable t. When this is not the case, we obtain the class of systems which are called *time-invariant*, as opposed to *time-varying*, or sometimes *stationary systems*. More precisely, a system is said to be time-invariant if it has the following property:

> if an input $\mathbf{u}(t)$ results in an output $\mathbf{y}(t)$, then the input $\mathbf{u}(t - \tau)$ results in the output $\mathbf{y}(t - \tau)$, for any τ.

In other words, if the input function is applied to the system τ units of time later than t, the resulting output function is identical to that obtained at t, translated by τ. This is illustrated in Fig. 1.5, where the input $u(t)$ applied at time $t = 0$ results in the output $y(t)$. When a replica of the function $u(t)$ is applied as input at time $t = \tau > 0$, the resulting output is an exact replica of the function $y(t)$.

1.2.3 The Concept of State

Roughly speaking, the *state* of a system at a time instant t should describe its behavior at that instant in some measurable way. In system theory, the term

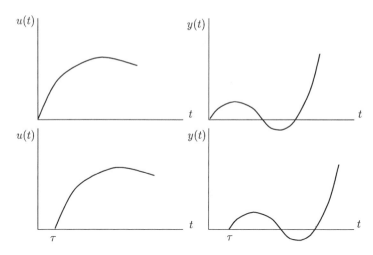

Figure 1.5: The time-invariance property.
The input $u(t)$ applied at time $t = 0$ results in the output $y(t)$. A replica of $u(t)$ applied at time $\tau > 0$ results in an exact replica of $y(t)$ shifted by τ.

state has a much more precise meaning and constitutes the cornerstone of the modeling process and many analytical techniques.

Returning to the input-output modeling process described in the previous section, suppose that the input vector $\mathbf{u}(t)$ is completely specified for all $t \geq t_0$. In addition, suppose that the output $\mathbf{y}(t)$ is observed at some time $t = t_1 \geq t_0$. The question we now raise is: *Is this information adequate to uniquely predict all future output $y(t), t > t_1$?* Surely, if our model is to be of any use, the ability to predict how the system behaves is critical.

Let us attempt to first answer this question for the system of Example 1.2. Let $t_1 > t_0$, and the mass displacement $y(t_1)$ be observed. Can we *uniquely* determine $y(t_1 + \tau)$ for some time increment $\tau > 0$? Before resorting to mathematics, a little thought shows that the answer is "no"; in fact, we can see that there are three distinct possibilities about $y(t_1 + \tau)$ compared to $y(t_1)$:

- $y(t_1 + \tau) > y(t_1)$ if the mass happens to be moving downwards at time t_1 (see Fig. 1.4).

- $y(t_1 + \tau) < y(t_1)$ if the mass happens to be moving upwards at time t_1.

- $y(t_1 + \tau) = y(t_1)$ if the mass happens to be at rest at time t_1.

How can we predict with certainty which of these three cases will arise? What is missing here is information regarding the *velocity* of the mass. Suppose we denote this velocity by $\dot{y}(t), t \geq t_0$. Then, $\dot{y}(t_1) > 0, \dot{y}(t_1) < 0$, and $\dot{y}(t_1) = 0$ correspond to the three cases given previously.

Mathematically, looking at the model we derived for Example 1.2 – in particular the second-order differential equation (1.6) – it should be clear that we cannot solve for $y(t_1 + \tau)$ with only one initial condition, i.e., $y(t_1)$; we also need information about the first derivative $\dot{y}(t_1)$, which is precisely the mass velocity.

Observe then that together $y(t_1)$ and $\dot{y}(t_1)$ provide the information we need which, along with full knowledge of the input function, allows us to obtain a unique solution of (1.6), and hence the value of $y(t_1 + \tau)$. This leads us to the following definition.

Definition. The *state* of a system at time t_0 is the information required at t_0 such that the output $\mathbf{y}(t)$, for all $t \geq t_0$, is uniquely determined from this information and from $\mathbf{u}(t), t \geq t_0$.   ◆

Like the input $\mathbf{u}(t)$ and the output $\mathbf{y}(t)$, the state is also generally a vector, which we shall denote by $\mathbf{x}(t)$. The components of this vector, $x_1(t), ..., x_n(t)$, are called *state variables*.

1.2.4   The State Space Modeling Process

With the idea of a system state in mind, we are now in a position to enhance the modeling process of Fig. 1.1. In addition to selecting input and output variables, we also identify state variables. The modeling process then consists of determining suitable mathematical relationships involving the input $\mathbf{u}(t)$, the output $\mathbf{y}(t)$, and the state $\mathbf{x}(t)$. These relationships are precisely what we refer to as the *dynamics* of a system. In particular, we are interested in obtaining expressions for $\mathbf{x}(t)$ given $\mathbf{x}(t_0)$ and the input function $\mathbf{u}(t)$, $t \geq t_0$. This leads to the following definition.

Definition. The set of equations required to specify the state $\mathbf{x}(t)$ for all $t \geq t_0$ given $\mathbf{x}(t_0)$ and the function $\mathbf{u}(t), t \geq t_0$, are called *state equations*.   ◆

Let us also introduce the term "state space".

Definition. The *state space* of a system, usually denoted by X, is the set of all possible values that the state may take.   ◆

The state equations could take several different forms. Most of system and control theory, however, is based on differential equations of the form

$$\dot{\mathbf{x}}(t) = \mathbf{f}(\mathbf{x}(t), \mathbf{u}(t), t) \tag{1.9}$$

We can then say that we have obtained a state space model of a system when we can completely specify the following set of equations:

$$\dot{\mathbf{x}}(t) = \mathbf{f}(\mathbf{x}(t), \mathbf{u}(t), t), \quad \mathbf{x}(t_0) = x_0 \tag{1.10}$$

$$\mathbf{y}(t) = \mathbf{g}(\mathbf{x}(t), \mathbf{u}(t), t) \tag{1.11}$$

where (1.10) is a set of *state equations* with initial conditions specified, and (1.11) is a set of *output equations*. Assuming there are n state variables, m output variables, and p input variables, these equations can be explicitly rewritten in scalar form as follows. There are n state equations

$$\dot{x}_1(t) = f_1(x_1(t), \ldots, x_n(t), u_1(t), \ldots, u_p(t), t), \qquad x_1(t_0) = x_{10}$$
$$\vdots$$
$$\dot{x}_n(t) = f_n(x_1(t), \ldots, x_n(t), u_1(t), \ldots, u_p(t), t), \qquad x_n(t_0) = x_{n0}$$

and m output equations

$$y_1(t) = g_1(x_1(t), \ldots, x_n(t), u_1(t), \ldots, u_p(t), t)$$
$$\vdots$$
$$y_m(t) = g_m(x_1(t), \ldots, x_n(t), u_1(t), \ldots, u_p(t), t)$$

Once again, the modeling process allows for considerable flexibility in the choice of state variables, as is the case in selecting input and output variables. In other words, there is no unique state representation for a given system. Usually, however, intuition, experience, and the presence of natural physical quantities that serve as state variables lead to good state space models.

Remark. For a static system, the state equation (1.10) reduces to $\dot{x}(t) = 0$ for all t, that is, the state remains fixed at all time, and a complete model is specified by the output equation (1.11). For a time-invariant system, the functions **f** and **g** do not explicitly depend on time, that is, we can write $\dot{\mathbf{x}}(t) = \mathbf{f}(\mathbf{x}(t), \mathbf{u}(t))$ and $\mathbf{y}(t) = \mathbf{g}(\mathbf{x}(t), \mathbf{u}(t))$.

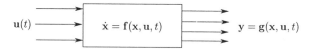

Figure 1.6: State space modeling process.

The input-output modeling process shown in Fig. 1.1 is sometimes also referred to as the *black-box* approach. This reflects the fact that what we know about the system is captured only by the output response $\mathbf{g}(\mathbf{u})$ to an input stimulus $\mathbf{u}(t)$, but the precise internal structure of the system is unspecified. Our task then is to deduce all we can about the "black box" by trying different kinds of input and observing the resulting output each time. In contrast, the state space modeling process contains some additional information captured by the state equation (1.10). This is shown in Fig. 1.6. Note that in this framework we can first focus on deriving the state equations, and hence the dynamics of a system. Then, we can choose output variables depending on the problem of interest, and observe them for different input functions.

Example 1.3

Consider once again the spring-mass system of Fig. 1.4. As in Example 1.2, let $y(t)$ be the output measuring the mass displacement at time t, and $u(t)$ the input function (1.7). Suppose we define the mass displacement as a state variable and denote it by $x_1(t)$. This immediately allows us to write down a trivial output equation

$$y(t) = x_1(t) \tag{1.12}$$

To obtain a state equation of the form (1.9), we have at our disposal the differential equation (1.6) which, since $y(t) = x_1(t)$, we rewrite as

$$m\ddot{x}_1 = -kx_1 \tag{1.13}$$

To obtain first-order differential equations, it is clear that we need to introduce an additional state variable, $x_2(t)$, which we define as

$$\dot{x}_1 = x_2 \tag{1.14}$$

We can then rewrite (1.13) in the form

$$\dot{x}_2 = -\frac{k}{m}x_1 \tag{1.15}$$

We now have a complete model with two state variables and a single output variable. In matrix form we can combine (1.12), (1.14), and (1.15) to write

$$\begin{bmatrix} \dot{x}_1 \\ \dot{x}_2 \end{bmatrix} = \begin{bmatrix} 0 & 1 \\ -\frac{k}{m} & 0 \end{bmatrix} \begin{bmatrix} x_1 \\ x_2 \end{bmatrix}, \qquad \begin{bmatrix} x_1(0) \\ x_2(0) \end{bmatrix} = \begin{bmatrix} u_0 \\ 0 \end{bmatrix} \tag{1.16}$$

$$y = \begin{bmatrix} 1 & 0 \end{bmatrix} \begin{bmatrix} x_1 \\ x_2 \end{bmatrix} \tag{1.17}$$

Note that this model uses both the mass displacement and velocity information as the state of the system, which is in agreement with our discussion in the previous section.

Example 1.4 (A simple RC circuit)

Consider the circuit of Fig. 1.7. We are interested in observing the voltage $v(t)$ across the resistance as a function of the source voltage $V(t)$. Thus, $V(t)$ and $v(t)$ are input and output variables respectively. To determine state variables, we begin by considering some simple relationships from circuit theory:

$$i = C\dot{v}_C, \quad v = iR, \quad V = v + v_C$$

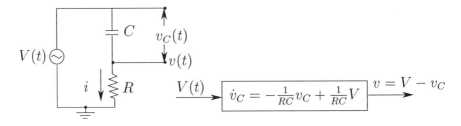

Figure 1.7: Electrical system for Example 1.4 and corresponding state-based model.

where v_C is the voltage across the capacitor. We then obtain

$$\dot{v}_C = \frac{1}{RC}(V - v_C)$$

and notice that v_C (which is neither an input nor an output variable) is a natural choice for a state variable. This leads to the following one-dimensional state variable model:

$$\dot{v}_C = -\frac{1}{RC}v_C + \frac{1}{RC}V \tag{1.18}$$

$$v = V - v_C \tag{1.19}$$

Example 1.5 (A simple flow system)
Consider a tank containing some fluid (see Fig. 1.8). The fluid flow into the tank is denoted by $\lambda(t)$ and the flow out is denoted by $\mu(t)$. These flows are regulated by valves which are typically used to maintain a "desirable" fluid level, $x(t)$. For instance, the tank may contain fuel at a gas station, the liquid component of an industrial chemical process, or it may represent the water reservoir for an entire city. Since the fluid level can never be negative, and the tank is limited by some storage capacity K, the variable $x(t)$ is constrained by $0 \le x(t) \le K$ for all time instants t. For this system, the fluid level $x(t)$ is an obvious natural state variable, whose rate of change can be described by

$$\dot{x}(t) = \lambda(t) - \mu(t)$$

subject to the constraint $0 \le x(t) \le K$. The effects of this constraint are the following:

1. If $x(t) = 0$, then the tank must remain empty and unchanged, that is $\dot{x}(t) = 0$, until the inflow exceeds the outflow, that is, $\lambda(t) > \mu(t)$.

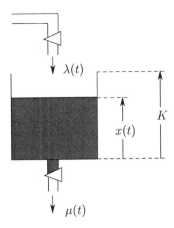

Figure 1.8: Flow system for Example 1.5.

2. Similarly, if $x(t) = K$, then the tank level must remain unchanged until $\lambda(t) < \mu(t)$.

Thus, the proper state equation is

$$\dot{x}(t) = \begin{cases} 0 & \text{if } [x(t) = 0 \text{ and } \lambda(t) \leq \mu(t)] \\ & \text{or } [x(t) = K \text{ and } \lambda(t) \geq \mu(t)] \\ \lambda(t) - \mu(t) & \text{otherwise} \end{cases} \tag{1.20}$$

To fully specify a model for this system, we shall consider $u_1(t) = \lambda(t)$ and $u_2(t) = \mu(t)$ to be input variables, and $y(t) = x(t)$ to be the output equation.

This system may appear deceptively simple, but the presence of the "boundary conditions" at $x(t) = 0$ and $x(t) = K$ makes the state equation (1.20) quite complicated to solve in general. The special case where $\lambda(t)$ and $\mu(t)$ are both constant is not particularly interesting, since:

1. If $\lambda - \mu > 0$, then the tank simply fills up at some time t_K and $x(t) = K$ for all $t > t_K$.

2. If $\lambda - \mu < 0$, then the tank becomes empty at some time t_0 and $x(t) = 0$ for all $t > t_0$.

Linear and Nonlinear Systems

The nature of the functions **f** and **g** in the model equations (1.10) and (1.11) serves to further classify systems into *linear* and *nonlinear* ones. The notion of linearity is fundamental in science and engineering, and is closely associated with the "principle of superposition", which is described by the following property: If a stimulus S_1 produces a response R_1, and a stimulus S_2 produces a response R_2, then the superposition of the two stimuli, $(S_1 + S_2)$, will produce the superposition of the two responses, $(R_1 + R_2)$. In its simplest form, i.e., $S_1 = S_2$, superposition amounts to proportionality; for example, doubling the input to a system results in doubling the output.

Mathematically, we use the notation $g : U \to Y$, to denote a function g mapping elements of a set U into elements of a set Y. For $u \in U$ and $y \in Y$, we write $y = g(u)$. We then have the following definition.

Definition. The function g is said to be *linear* if and only if

$$g(a_1 u_1 + a_2 u_2) = a_1 g(u_1) + a_2 g(u_2) \tag{1.21}$$

for any $u_1, u_2 \in U$, and any real numbers a_1, a_2. There is an obvious extension to the vector case: The function $\mathbf{g}$ is linear if and only if

$$\mathbf{g}(a_1 \mathbf{u}_1 + a_2 \mathbf{u}_2) = a_1 \mathbf{g}(\mathbf{u}_1) + a_2 \mathbf{g}(\mathbf{u}_2) \tag{1.22}$$

$\blacklozenge$

If we think of $\mathbf{u}_1$ and $\mathbf{u}_2$ as two input vectors to a system with output represented by $\mathbf{y} = \mathbf{g}(\mathbf{u})$, we may be led to a natural definition of a linear system based on (1.22). But there are two things to watch out for. First, $\mathbf{u}$ is actually a function of the time variable t, i.e., $\mathbf{u}(t) = [u_1(t), \dots, u_p(t)]^T$. Therefore, the set of inputs, U, consists of vectors of time functions, and the same is true for the output set, Y. However, mapping functions into functions is conceptually no harder than mapping points into points. Second, recall that the nature of the function $\mathbf{g}(\cdot)$ does not capture the complete dynamic behavior of the system, which was the reason that the state equation (1.10) became necessary. The functions $\mathbf{g}(\cdot)$ and $\mathbf{f}(\cdot)$ together tell the whole story.

Definition. A system modeled through (1.10) and (1.11) is said to be linear if and only if the functions $\mathbf{g}(\cdot)$ and $\mathbf{f}(\cdot)$ are both linear. $\blacklozenge$

In the case of linearity, the model (1.10) and (1.11) reduces to

$$\dot{\mathbf{x}}(t) = \mathbf{A}(t)\mathbf{x}(t) + \mathbf{B}(t)\mathbf{u}(t) \tag{1.23}$$
$$\mathbf{y}(t) = \mathbf{C}(t)\mathbf{x}(t) + \mathbf{D}(t)\mathbf{u}(t) \tag{1.24}$$

and if we assume n state variables, m output variables, and p input variables, then $\mathbf{A}(t)$ is an $n \times n$ matrix, $\mathbf{B}(t)$ is an $n \times p$ matrix, $\mathbf{C}(t)$ is an $m \times n$ matrix, and $\mathbf{D}(t)$ is an $m \times p$ matrix.

The class of linear systems is a small subset of all possible systems. Fortunately, it covers many cases of interest, or provides adequate approximations we can use for practical purposes. Much of system and control theory is in fact based on the analysis of linear systems (see selected references at the end of this chapter), and has led to plenty of success stories, from designing complex electromechanical structures to describing the behavior of economies and population growth.

We can further narrow down the class of linear systems by assuming the time-invariance property defined earlier. In this case, the matrices $\mathbf{A}(t)$, $\mathbf{B}(t)$, $\mathbf{C}(t)$, and $\mathbf{D}(t)$ are all constant, and we get

$$\dot{\mathbf{x}} = \mathbf{Ax} + \mathbf{Bu} \tag{1.25}$$

$$\mathbf{y} = \mathbf{Cx} + \mathbf{Du} \tag{1.26}$$

where, for simplicity, the time variable is dropped. The entries of the matrices $\mathbf{A}$, $\mathbf{B}$, $\mathbf{C}$, $\mathbf{D}$ are usually referred to as the *model parameters*. Parameters are assumed to remain fixed when one studies the dynamic behavior of such a system. However, part of many problems of interest consists of studying the solution of the state equation (1.25) for different parameter values, in order to select "desirable" ones in some sense. In an electrical circuit, resistances, inductances, and capacitances are typical parameters; the storage capacity of the tank in Example 1.5 is a parameter of the flow model (1.20).

It is fair to say that the model (1.25) and (1.26) constitutes the cornerstone of modern system theory. The solution and properties of these equations have been extensively studied using a variety of techniques, including the use of Laplace transforms. When the Laplace transform is applied to (1.25) and (1.26), one obtains a set of algebraic - rather than differential - equations, and replaces the time variable t by a new variable s. This variable may be given a frequency interpretation. Accordingly, the study of linear time-invariant systems has proceeded along two parallel paths: the *time domain*, where one uses the differential equation model (1.25) and (1.26), and the *frequency domain*, where one works with the algebraic equations resulting from the Laplace transforms of (1.25) and (1.26). Both approaches have their respective advantages and drawbacks, which we shall not discuss here. As it will become apparent later on, the time domain model above is most suited for our purposes.

Taking a look at the models obtained in Examples 1.3 and 1.4, we can see that they correspond to linear time-invariant systems. On the other hand, the system in Example 1.5 is time-invariant, but it is clearly nonlinear. In fact, we can see that $\dot{x}(t)$ in (1.20) is discontinuous at points in time such that $x(t) = 0$ or $x(t) = K$: the value of $\dot{x}(t)$ jumps to 0 at such points.

Remark. It is tempting to claim that all dynamic systems can be modeled through differential equations, no matter how nonlinear and complex they might be. Although it is certainly true that these models are immensely useful in system and control theory, we will soon see that for the discrete event systems we need to consider, differential equations simply do not capture the essential dynamic behavior, or they lead to design and control solutions that are not sufficiently accurate for many practical purposes.

1.2.5 Sample Paths of Dynamic Systems

Let us return to the state equation (1.10), and, for simplicity, consider the scalar case

$$\dot{x}(t) = f(x(t), u(t), t) \tag{1.27}$$

Suppose we select a particular input function $u(t) = u^1(t)$ and some initial condition at time $t = 0$, $x(0) = x_0$. Then, the solution to this differential equation is a particular function $x^1(t)$, as shown in Fig. 1.9. By changing $u(t)$ and forming a family of input functions $u^1(t), u^2(t), \ldots$, we can obtain a family of solutions $x^1(t), x^2(t), \ldots$, each one describing how the state evolves in time. Any member of this family of functions is referred to as a *sample path* of the system whose state equation is (1.27). The term *state trajectory* is also used.

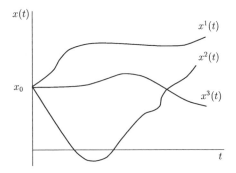

Figure 1.9: Sample paths of the state equation $\dot{x}(t) = f(x(t), u(t), t)$.

When there are more than one state variables in our model, there is a sample path corresponding to each one. There is, however, an alternative, perhaps a bit more abstract, way to think of a sample path in this case. All possible values that the n-dimensional state vector $\mathbf{x}(t)$ can take define an n-dimensional space. Suppose we fix t on a given sample path; this specifies a point (vector) in this space. As t varies, different points are visited, thus defining a curve, which we can also think of as a state trajectory of the system. This is best visualized in the 2-dimensional case shown in Fig. 1.10. An initial condition $(x_1(0), x_2(0))$ corresponds to a point in this space. For an input function $u(t) = u^1(t)$, the state is denoted by $(x_1^1(t), x_2^1(t))$. As t varies, $(x_1^1(t), x_2^1(t))$ represents a new point, giving rise to a trajectory describing how the state travels around this space. Different sample paths are shown in Fig. 1.10 for different inputs $u^1(t), u^2(t), \ldots$. Mathematically, this representation corresponds to eliminating the variable t from the functions $x_1(t), x_2(t)$ to obtain some function $h(x_1, x_2) = 0$. The graphical representation of this function in the 2-dimensional space defines a sample path.

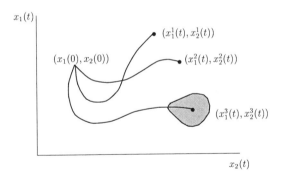

Figure 1.10: Sample paths in a two-dimensional case.

This representation is particularly useful in considering interesting questions such as: Can a particular state or set of states ever be reached from a given initial condition? In Fig. 1.10, such a desirable set is described by the shaded region in the (x_1, x_2) space, and we see that at least one sample path (i.e., some input function) allows the system to reach this region.

Example 1.6
Consider the system of Fig. 1.8 as in Example 1.5, whose state equation is (1.20). Let the storage capacity be $K = 2$ and the valve controlling the flow out of the tank be fixed so that $\mu(t) = 1$ for all $t \geq 0$. The initial state is set at $x(0) = 0$. We will derive two sample paths corresponding to two input functions defined by

$$\lambda_1(t) = \begin{cases} 3 & \text{for } 0 < t \leq 2 \\ 0 & \text{for } 2 < t \leq 3 \\ 1 & \text{for } t > 3 \end{cases} \qquad (1.28)$$

$$\lambda_2(t) = \begin{cases} 2 & \text{for } 0 < t \leq 1 \\ 1 & \text{for } 1 < t \leq 2 \\ 0 & \text{for } 2 < t \leq 4 \\ 11 & \text{for } t > 4 \end{cases} \qquad (1.29)$$

Using (1.20) with $\mu(t) = 1$, we can see that for $\lambda_1(t)$ in the interval $0 < t \leq 1$, the state satisfies the equation

$$\dot{x}_1(t) = 2, \qquad x_1(0) = 0 \qquad (1.30)$$

since $x_1(t) = 2$ at $t = 1$. From (1.20), we must then set $\dot{x}_1(t) = 0$ for all $1 < t \leq 2$. This corresponds to the tank overflowing at $t = 2$, and its level remaining at the maximum possible value (see Fig. 1.11).

In the interval $2 < t \leq 3$, the state equation becomes

$$\dot{x}_1(t) = -1, \qquad x_1(2) = 2 \tag{1.31}$$

In this case, the tank level decreases at unit rate, and at $t = 3$ we have $x_1(t) = 1$.

Finally, for $t > 3$, the state equation is simply

$$\dot{x}_1(t) = 0, \qquad x_1(3) = 1 \tag{1.32}$$

since $x_1(t)$ can no longer change (inflow = outflow). The complete sample path is shown in Fig. 1.11.

The case of $\lambda_2(t)$ is similarly handled and results in the sample path $x_2(t)$ shown in Fig. 1.11. In this case, the tank can be seen to overflow at time $t = 4.2$ and remain full thereafter.

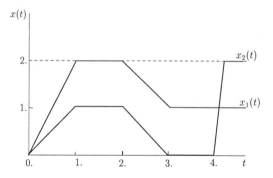

Figure 1.11: Sample paths for Example 1.6.

1.2.6 State Spaces

Thus far, the values of the state variables we have considered are real numbers. For instance, $x(t)$ in Example 1.5 can take any value in the interval $[0, K]$ which is a subset of the real number set $\mathbb{R}$. Real variables are of course very convenient when it comes to deriving models based on differential equations as in (1.10). However, there is nothing sacred about state variables always taking real number values, as opposed to integer values or just values from a given discrete set, such as {ON, OFF}, {HIGH, MEDIUM, LOW}, or {GREEN, RED, BLUE}. In fact, one should always keep in mind that the modeling process allows for substantial flexibility in defining the state, input, and output of a system depending on the application or problem of interest.

Continuous-State and Discrete-State Systems

Yet another way to classify systems is based on the nature of the state space selected for a model. In *continuous-state* models the state space X is a continuum consisting of all n-dimensional vectors of real (or sometimes complex) numbers. Normally, X is finite-dimensional (i.e., n is a finite number), although there are some exceptions where X is infinite-dimensional. This normally leads to differential equations as in (1.10) and associated techniques for analysis. In *discrete-state* models the state space is a discrete set. In this case, a typical sample path is a piecewise constant function, since state variables are only permitted to jump at discrete points in time from one discrete state value to another (see Example 1.8 below). Naturally, there are many situations in which a *hybrid* model may be appropriate, that is, some state variables are discrete and some are continuous.

The dynamic behavior of discrete-state systems is often simpler to visualize. This is because the state transition mechanism is normally based on simple logical statements of the form "if something specific happens and the current state is x, then the next state becomes x'." However, the mathematical machinery to formally express the state equations and to solve them may be considerably more complex, as we shall have occasion to find out. On the other hand, continuous-state models ultimately reduce to the analysis of differential equations, for which many mathematical tools and general solution techniques are available.

Example 1.7

In Example 1.3, we defined the 2-dimensional state vector $x = [x_1, \ x_2]^T$ for the mass-spring system. In this case, the state space X is the set $\mathbb{R}^2$, i.e., the real plane. This is an example of a continuous-state linear system. Of course, it is possible that constraints are imposed on the state variables so that not all points in the plane are part of X. For example, the displacement of the spring cannot exceed certain values defined by the physical limitations of the mass-spring setting. Once these limitations are specified we could better define X as a particular *subset* of $\mathbb{R}^2$.

Example 1.8 (A warehouse system)

This is a simple example of a natural discrete-state system. Consider a warehouse containing finished products manufactured in a factory. Whenever a new product is completed at the manufacturing process, it "arrives" at the warehouse and is stored there. A truck shows up periodically and loads up a certain number of products, which are thought of as "departures" from the warehouse (Fig. 1.12). Similar to the tank flow system of Example 1.5, we are interested in the level of inventory at this warehouse, that is, how many products are present at any given time. Thus, we define $x(t)$ to be the number of products at time t, and define an output equation for our model to be $y(t) = x(t)$. Since products are discrete

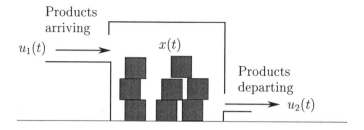

Figure 1.12: Warehouse system for Example 1.8.

entities, our obvious choice for a state space X is the set of non-negative integers $\{0, 1, 2, \ldots\}$.

Suppose we choose as input two functions of time defined as follows:

$$u_1(t) = \begin{cases} 1 & \text{if a product arrives at time } t \\ 0 & \text{otherwise} \end{cases} \qquad (1.33)$$

and

$$u_2(t) = \begin{cases} 1 & \text{if a truck arrives at time } t \\ 0 & \text{otherwise} \end{cases} \qquad (1.34)$$

The sequence of time instants when $u_1(t) = 1$ defines the "schedule" of product arrivals at the warehouse. Similarly, the sequence of time instants when $u_2(t) = 1$ defines the "schedule" of truck arrivals at the warehouse.

It is interesting to observe the similarity between the tank flow system of Fig. 1.8, and the warehouse system illustrated in Fig. 1.12. In both cases, we study the level of a storage container of some type, subject to specified inflow and outflow processes. The difference is in the nature of what is being stored: A fluid level is accurately measured through a real variable, whereas the level of an inventory of discrete entities requires an integer-valued state variable.

To simplify matters, let us assume that: (a) the warehouse is very large and its storage capacity is never reached, (b) loading a truck takes zero time, (c) the truck can only take away a single product at a time, and (d) a product arrival and a truck arrival never take place at exactly the same time instant, that is, there is no t such that $u_1(t) = u_2(t) = 1$.

In order to derive a state equation for this model, let us examine all possible state transitions we can think of:

1. $u_1(t) = 1, u_2(t) = 0$. This simply indicates an arrival into the warehouse. As a result, $x(t)$ should experience a jump of $+1$.

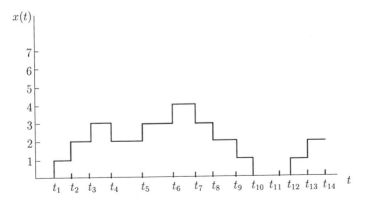

Figure 1.13: Typical sample path for Example 1.8.
When an arrival occurs, $x(t)$ increases by 1, and when a departure occurs it decreases by 1. Note that if $x(t) = 0$, only an arrival is possible.

2. $u_1(t) = 0, u_2(t) = 1$. This means a truck is present at time t, and we should reduce the warehouse content by 1. However, there are two subcases. If $x(t) > 0$, then the state change is indeed -1. But if $x(t) = 0$, the truck finds the warehouse empty and the state does not change.

3. $u_1(t) = 0, u_2(t) = 0$. Clearly, no change occurs at time t.

One problem we have here is that we obviously cannot define a derivative $\dot{x}(t)$ for $x(t)$. Instead, we use the notation t^+ to denote a time instant "just after" t. With this notation, based on the observations above we can write

$$
x(t^+) = \begin{cases} x(t) + 1 & \text{if } (u_1(t) = 1, u_2(t) = 0) \\ x(t) - 1 & \text{if } (u_1(t) = 0, u_2(t) = 1, x(t) > 0) \\ x(t) & \text{otherwise} \end{cases} \qquad (1.35)
$$

A typical sample path of this system is shown in Fig. 1.13. In this case, $u_1(t) = 1$ (i.e., finished products arrive) at time instants $t_1, t_2, t_3, t_5, t_6, t_{12}$, and t_{13}, and $u_2(t) = 1$ (i.e., a truck arrives) at time instants $t_4, t_7, t_8, t_9, t_{10}$, and t_{11}. Note that even though a truck arrival takes place at time t_{11}, the state $x(t) = 0$ does not change, in accordance with (1.35).

Remark. Equation (1.35) is representative of many situations we will be encountering in our study of discrete event systems. Although it certainly captures the simple dynamics of the warehouse in Example 1.8, it is not particularly elegant from a mathematical standpoint. The definition of

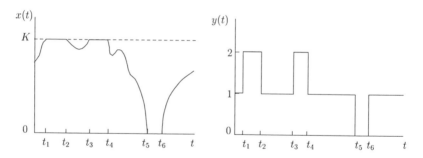

Figure 1.14: Sample paths for Example 1.9.

the input functions $u_1(t), u_2(t)$ is informal; the discontinuities due to the instantaneous nature of the state transitions require the use of the awkward notation t^+; and, more importantly, we can see that this is a rather inefficient way to describe the behavior of the system, since most of the time nothing interesting is happening. In fact, it is only at the few points when the *events* "truck arrival" and "product arrival" take place that we need to worry about state changes.

Example 1.9

Let us return to the tank flow system of Example 1.5. The state in this case was a scalar variable $x(t)$ with a state space $X = [0, K]$, a finite interval subset of $\mathbb{R}$. Therefore, this is a continuous-state system.

Suppose, that we are only interested in knowing three things about this system: whether the tank is empty (i.e., $x(t) = 0$), completely full or overflowing (i.e., $x(t) \geq K$), or anywhere in between (i.e., $0 < x(t) < K$). For simplicity, we associate "empty" with 0, "completely full or overflowing" with 2, and "anywhere in between" with 1. In this case, we could modify the model of Example 1.5 by defining a new output equation:

$$y(t) = \begin{cases} 0 & \text{if } x(t) = 0 \\ 1 & \text{if } 0 < x(t) < K \\ 2 & \text{otherwise} \end{cases}$$

Sample paths of $y(t)$ and $x(t)$ are shown in Fig. 1.14. Note that $x(t)$ is a continuous curve constrained by the empty ($x = 0$) and full ($x = K$) levels, whereas $y(t)$ simply jumps between the three allowed values whenever a boundary value (0 or K) is entered or left. Thus, the only "interesting" aspects of the $y(t)$ evolution are contained exclusively in these jumps at time instants $t_1, \ldots, t_6$ in Fig. 1.14.

In this example, even though we have a continuous-state system, the output is only defined over the discrete set $\{0, 1, 2\}$. A reasonable question might be: Why not choose our state space to be the set $S = \{0, 1,$

2} in the first place? Then, we might define a new state variable $s(t)$ which can only take values in S, and obtain a simple output equation $y(t) = s(t)$. To answer this question, we need to recall our definition of state in Section 1.2.3, and ask ourselves: *Is $s(t)$ a legitimate state?*

Suppose we know $s(t)$ at some time t, and assume the same input variables as before, that is, the inflow $\lambda(t)$ and outflow $\mu(t)$. Then, we can see that this information alone is not sufficient for predicting future values of $s(t)$. This is clear in looking at the sample path of Fig. 1.14. Suppose we know $s(t) = y(t) = 1$ at $t = 0$. To predict the transition shown at $t = t_1$ we need to evaluate t_1 and the tank level $x(t_1)$; but this is not possible without explicitly knowing $x(0)$ and solving the state equation (1.20) of the original model. Therefore, $x(t)$ must still be a state variable in this new model.

Deterministic and Stochastic Systems

Recall that our definition of state required the ability to predict at $t = t_0$ all future behavior of a system, assuming knowledge of the input $u(t)$ for all $t \geq t_0$. This latter assumption is not always reasonable, as it precludes the unpredictable effect of nature on the input to a given model. In the circuit of Example 1.4, the voltage $V(t)$ may be subject to random "noise" which we cannot account for with absolute certainty. A more obvious situation arises in the flow system of Example 1.5. Suppose the tank in Fig. 1.8 is a water reservoir whose input $\lambda(t)$ is actual rainfall; it would be rather naive to expect a precise model of when and how much rain will fall in the future. One, however, can develop realistic probabilistic models for rainfall and incorporate them into the state-based model of the system. Finally, in Example 1.8 our state equation (1.35) depends almost exclusively on the exact knowledge of the product arrival input function $u_1(t)$ and the truck arrival input function $u_2(t)$. In most practical problems of interest, such knowledge is not available. Common facts of life such as trucks delayed due to traffic or weather conditions, and production affected by random machine failures, worker absenteeism, or delays in the delivery of spare parts and raw materials continuously affect the state in ways that only probabilistic mechanisms can adequately capture. Of course, the state equation (1.35) still describes the system dynamics, but if the values of $u_1(t)$ and $u_2(t)$ are not deterministically known for all t, then neither is the value of the state $x(t)$.

We are thus led to yet another classification of systems. We define a system to be *stochastic* if at least one of its output variables is a random variable. Otherwise, the system is said to be *deterministic*. In general, the state of a stochastic dynamic system defines a random process, whose behavior can be described only probabilistically. Thus, in a deterministic system with the input $u(t)$ given for all $t \geq t_0$, the state $x(t)$ can be evaluated. In a stochastic system, the state at time t is a random vector, and it is only its probability distribution function that can be evaluated.

It is assumed that the reader is familiar with elementary probability theory and at least the notion of a random process. This material is also reviewed in Appendix I, and will be essential in later chapters. For the time being, the discussion of system fundamentals in the remainder of this chapter will be based on deterministic systems, unless otherwise noted.

1.2.7 The Concept of Control

Our discussion thus far has been limited to the basic issue: What happens to the system output for a given input? Systems, however, do not normally exist in a vacuum. In fact, we saw that the very definition of a system contains the idea of performing a particular function. In order for such a function to be performed, the system needs to be *controlled* by selecting the right input so as to achieve some "desired behavior". For example, in driving a car we control the position, speed, and acceleration of the vehicle by selecting the right input through the steering wheel and the accelerator and brake pedals. The desirable behavior in this case is keeping the car in the road at a reasonable speed.

Thus, the input to a system is often viewed as a *control signal* aimed at achieving a desired behavior. Conceptually, for a simple scalar case, we represent this desired behavior by a *reference signal* $r(t)$, and the control input to the actual system as

$$u(t) = \gamma(r(t), t) \tag{1.36}$$

This relationship is referred to as a *control law*, or just *control*. Thus, given the function $r(t)$ describing a "desired behavior" for the system at hand, our task as controllers is to select $u(t) = \gamma(r(t), t)$ to be the input function to the system. The extension to the vector case, where multiple reference signals are specified, leads to the control law

$$\mathbf{u}(t) = \boldsymbol{\gamma}(\mathbf{r}(t), t) \tag{1.37}$$

where $\boldsymbol{\gamma}(\cdot)$ denotes the column vector whose entries are the p scalar functions $u_1(t) = \gamma_1(\mathbf{r}(t), t), \ldots, u_p(t) = \gamma_p(\mathbf{r}(t), t)$.

As shown in Fig. 1.15, we can extend the diagram of Fig. 1.6 to include the control function in our modeling process, assuming that a desired behavior or reference $\mathbf{r}(t)$ is specified.

Example 1.10

Consider a linear time-invariant system whose state and output equations are

$$\begin{aligned} \dot{x}(t) &= u(t), \qquad x(0) = 0 \\ y(t) &= x(t) \end{aligned} \tag{1.38}$$

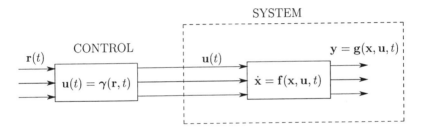

Figure 1.15: State space modeling with control input.

The desired behavior is to bring the state to the value $x(t) = 1$ and keep it there forever. Thus, we define a reference signal

$$r(t) = 1 \text{ for all } t \geq 0$$

The control input is limited to operate in the range $[-1, 1]$, that is, we have to choose our control law $u(t) = \gamma(r(t), t)$ from the set of all possible time functions constrained by $|u(t)| \leq 1$. Clearly, a simple control that works is

$$u_1(t) = \begin{cases} 1 & \text{for } t \leq 1 \\ 0 & \text{for } t > 1 \end{cases}$$

which results in achieving the desired state value at $t = 1$ (see Fig. 1.16) when (1.38) is solved. Intuitively, we can also see that $u_1(t)$ has the property of achieving the desired behavior in minimum time, since it makes use of the maximum allowable control subject to $|u(t)| \leq 1$. However, it is certainly not the only possible control law we can come up with. For example,

$$u_2(t) = \begin{cases} 1/2 & \text{for } t \leq 2 \\ 0 & \text{for } t > 2 \end{cases}$$

also achieves our objective.

1.2.8   The Concept of Feedback

The idea of *feedback* is intuitively simple: Use any available information about the system behavior in order to continuously adjust the control input. Feedback is used in our everyday life in a multitude of forms. In a conversation, we speak when the other party is silent, and switch to listening when the other party is beginning to talk. In driving, we monitor the car's position and speed in order to continuously make adjustments through our control of the steering wheel and accelerator and brake pedals. In heating a house, we use a thermostat which senses the actual temperature in order to turn a furnace on or off.

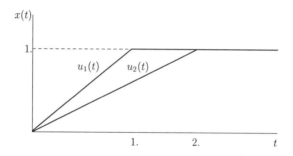

Figure 1.16: Sample paths corresponding to controls $u_1(t), u_2(t)$ in Example 1.10.

Returning to the example of driving a car, suppose we find a straight stretch of road, and set as our objective to drive the car from a given point A on this stretch to another point B, where A and B are one mile apart. One way to do this is the following: Set the wheels straight, start the car, and reach a constant speed of 60 miles per hour as we cross point A. We can then close our eyes and continue to drive straight at constant speed for one more minute, which should get the car to point B. This is clearly not a very efficient way to drive. A slight air gust or a deviation in holding the steering wheel straight - not to mention the possibility of an unexpected obstacle showing up in the middle of the road - and our control scheme fails miserably. This is because the input (the steering wheel, the accelerator, the brake) is constrained to remain fixed, and cannot respond to the observed system behavior. Yet, this problem is easily fixed as soon as we open our eyes and provide ourselves with simple *feedback* (the car position on the road and the speedometer value). Thus, the key property of feedback is in making corrections, especially in the presence of unexpected *disturbances* (air gusts, obstacles in the road, etc.).

Mathematically, using feedback implies that we should extend the control law in (1.37) to include, along with the reference $\mathbf{r}(t)$, the observed output $\mathbf{y}(t)$ or, more generally, the state $\mathbf{x}(t)$. We can then write a more general control law of the form

$$\mathbf{u}(t) = \boldsymbol{\gamma}(\mathbf{r}(t), \mathbf{x}(t), t) \qquad (1.39)$$

where $\boldsymbol{\gamma}(\cdot)$ denotes the column vector whose entries are the p scalar functions $u_1(t) = \gamma_1(\mathbf{r}(t), \mathbf{x}(t), t), \ldots, u_p(t) = \gamma_p(\mathbf{r}(t), \mathbf{x}(t), t)$.

Open-loop and Closed-loop Systems

The possibility of using feedback in the controlled system model of Fig. 1.15 leads to one more system classification. A system with a control input of the form $\mathbf{u}(t) = \boldsymbol{\gamma}(\mathbf{r}(t), t)$ is referred to as an *open-loop* system. On the other hand, a system with a control input of the form $\mathbf{u}(t) = \boldsymbol{\gamma}(\mathbf{r}(t), \mathbf{x}(t), t)$ is referred to as a *closed-loop* system. The distinction between these two forms of control is

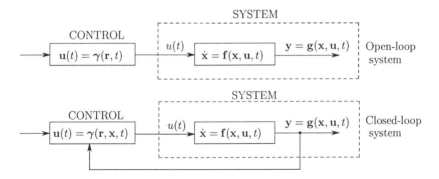

Figure 1.17: Open-loop and Closed-loop systems.

fundamental. In open-loop control, the input remains fixed regardless of the effect (good or bad) that it has on the observed output. In closed-loop control, the input depends on the effect it causes on the output. This distinction is illustrated in Fig. 1.17, which represents the most complete form of the modeling process we shall concern ourselves with. In the closed-loop case, it is assumed that the information fed back is some function of the state variables (not explicitly shown in the diagram), which is then included in the control law $\gamma(\cdot)$. Note that it may not be desirable (or in fact feasible) to include information about all the state variables in our model. The "loop" formed by the feedback process in the figure gives rise to the term "closed-loop" system.

Example 1.11 (Closed-loop control of a simple flow system)

A simple control problem for the tank system of Example 1.5 is to maintain the fluid level at its maximum possible value, which is given by the capacity K. To simplify the problem, suppose that an initial condition $x(0) = 0$ is given, and that the outflow rate is $\mu(t) = 0$ for all $t \geq 0$. In this case, our task is essentially the same as that of Example 1.10. We will control the inflow $\lambda(t)$ so as to achieve the desired behavior

$$r(t) = K \text{ for all } t \geq 0$$

with state and output equations

$$\dot{x}(t) = \begin{cases} \lambda(t) & \text{if } x(t) < K \\ 0 & \text{otherwise} \end{cases} \qquad x(0) = 0 \qquad (1.40)$$
$$y(t) = x(t)$$

We will also make the reasonable assumption that there is a maximum rate at which fluid can flow into the tank, denoted by ρ, that is, our control is constrained by $\lambda(t) \leq \rho$. Obviously, the inflow rate is also constrained to be non-negative, i.e., $\lambda(t) \geq 0$.

Let us first design an open-loop control system for this problem. As in Example 1.10, it is easy to see that we can choose

$$\lambda_{OL}(t) = \begin{cases} \rho & \text{for } t \le K/\rho \\ 0 & \text{for } t > K/\rho \end{cases} \tag{1.41}$$

since the solution of (1.40) with this control gives $x(t) = \rho t$, and the level $x(t) = K$ is therefore reached at time $t = K/\rho$.

We can immediately see two potential drawbacks of this open-loop control law:

1. The slightest error in measuring either of the two system parameters, K and ρ, will prevent us from meeting our desired behavior. Suppose, for instance, that we have underestimated the flow rate ρ. This means that the true value ρ_{actual} is such that $\rho_{actual} > \rho$. As a result, at time $t = K/\rho_{actual} \le K/\rho$ the tank level has already reached its maximum value. Yet, our open-loop control ignores this fact, and causes an overflow!

2. The slightest unexpected disturbance will lead to similar phenomena. For instance, the inflow may be unpredictably perturbed during the interval $0 \le t \le K/\rho$; or a small leak may develop in the tank that the open-loop control (1.41) is not capable of accounting for.

This provides the motivation for designing a *closed-loop* control system. This is easily accomplished by the following control law:

$$\lambda_{CL}(t) = \begin{cases} \rho & \text{if } x(t) < K \\ 0 & \text{if } x(t) = K \end{cases} \tag{1.42}$$

which keeps track of the state $x(t)$ itself. Note that small disturbances or errors in the flow rate do not adversely affect the function of this system. Essentially, the controller allows as much flow as possible into the tank until the tank is full; this of course requires some sort of sensing device (e.g., a floater capable of shutting off the inflow whenever $x(t) = K$). Such a mechanism is illustrated in Fig. 1.18.

This simple system represents the basic principle of the flush toilet, where the process of flushing is modeled through the state equation (1.40) with a closed-loop control given by (1.42).

There are certain obvious advantages to the use of feedback, some of which are clearly illustrated in Example 1.11. Briefly, without getting into details, we can point out the following:

1. The desired behavior of the system becomes less sensitive to unexpected disturbances.

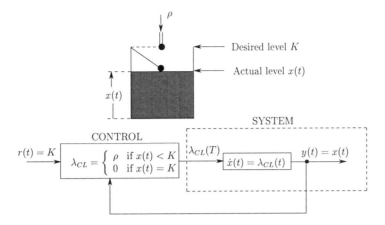

Figure 1.18: Flow system of Example 1.11 and closed-loop control model.

2. The desired behavior of the system becomes less sensitive to possible errors in the parameter values assumed in the model.

3. The output $\mathbf{y}(t)$ can *automatically* follow or track a desired reference signal $\mathbf{r}(t)$ by continuously seeking to minimize the difference $(\mathbf{y}(t) - \mathbf{r}(t))$ (which is sometimes referred to as the *error signal*).

On the other hand, feedback comes at some cost:

1. Sensors or other potentially complex equipment may be required to monitor the output and provide the necessary information to the controller.

2. Feedback requires some effort (measured in terms of the *gain* of the system), which may adversely affect the overall system performance.

3. Feedback could actually create some problems of undesirable system behavior, while correcting others.

As in many other areas of engineering, using feedback entails several tradeoffs. Control theory is to a large extent devoted to the study of the tradeoffs outlined above. Details may be found in many of the references provided at the end of this chapter.

1.2.9 Discrete-Time Systems

We have assumed thus far that time is a continuous variable. This certainly corresponds to our basic notion of time in the physical world. Moreover, it allows us to develop models of the form (1.10)-(1.11) based on differential equations, which are particularly attractive from a mathematical standpoint.

Suppose that we were to define the input and output variables of a system at discrete time instants only. As a result, we obtain what is called a *discrete-time* system, in contrast to the *continuous-time* systems considered up to this point. There are several good reasons why we might want to adopt such an approach.

1. Any digital computer we might use as a component of a system operates in discrete-time fashion, that is, it is equipped with an internal discrete-time clock. Whatever variables the computer recognizes or controls are only evaluated at those time instants corresponding to "clock ticks".

2. Many differential equations of interest in our continuous-time models can only be solved numerically through the use of a computer. Such computer-generated solutions are actually discrete-time versions of continuous-time functions. Therefore, starting out with discrete-time models is reasonable if the ultimate solutions are going to be in this form anyway.

3. Digital control techniques, which are based on discrete-time models, often provide considerable flexibility, speed, and low cost. This is because of advances in digital hardware and computer technology.

4. Some systems are inherently discrete-time, such as economic models based on data that is recorded only at regular discrete intervals (e.g., quarterly).

In discrete-time models, the time line is thought of as a sequence of intervals defined by a sequence of points $t_0 < t_1 < \ldots < t_k < \ldots$. It is assumed that all intervals are of equal length T, i.e., $t_{k+1} - t_k = T$ for all $k = 0, 1, 2, \ldots$. The constant T is sometimes referred to as the *sampling interval*. The real variable t is then replaced by an integer variable k, which counts the number of intervals elapsed since a given reference point, usually $k = 0$.

Figure 1.19 illustrates how a continuous-time sample path $x(t)$ can be sampled to give rise to a discrete-time sample path $x(k)$, with sampling period T. It is important to note that *discretization of time does not imply discretization of the state space*. In Fig. 1.19, the discrete-time sample path is piecewise constant, but the state can still take any value in the set of real numbers, as in the continuous-time case.

Virtually all of the framework developed thus far can be replicated with the input and output functions $\mathbf{u}(t)$ and $\mathbf{y}(t)$ replaced by sequences $\mathbf{u}(k)$ and $\mathbf{y}(k)$. Similarly, the state $\mathbf{x}(t)$ is replaced by $\mathbf{x}(k)$. Thus, the state-based model (1.10)-(1.11) becomes

$$\mathbf{x}(k+1) = \mathbf{f}(\mathbf{x}(k), \mathbf{u}(k), k), \quad \mathbf{x}(0) = \mathbf{x}_0 \qquad (1.43)$$
$$\mathbf{y}(k) = \mathbf{g}(\mathbf{x}(k), \mathbf{u}(k), k) \qquad (1.44)$$

with *difference equations* now replacing differential equations in describing the dynamic system behavior. For the case of linear discrete-time systems, the

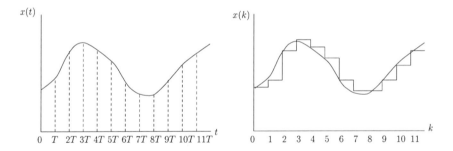

Figure 1.19: Continuous-time and discrete-time sample paths.

analog of (1.23)-(1.24) is

$$\mathbf{x}(k+1) = \mathbf{A}(k)\mathbf{x}(k) + \mathbf{B}(k)\mathbf{u}(k) \tag{1.45}$$
$$\mathbf{y}(k) = \mathbf{C}(k)\mathbf{x}(k) + \mathbf{D}(k)\mathbf{u}(k) \tag{1.46}$$

which in the time-invariant case reduces to

$$\mathbf{x}(k+1) = \mathbf{A}\mathbf{x}(k) + \mathbf{B}\mathbf{u}(k) \tag{1.47}$$
$$\mathbf{y}(k) = \mathbf{C}\mathbf{x}(k) + \mathbf{D}\mathbf{u}(k) \tag{1.48}$$

where **A, B, C,** and **D** are all constant matrices containing the system parameters.

1.3 DISCRETE EVENT SYSTEMS

When the state space of a system is naturally described by a discrete set like $\{0, 1, 2, \ldots\}$, and state transitions are only observed at discrete points in time, we associate these state transitions with "events" and talk about a "discrete event system". We begin our study of these systems by first identifying their fundamental characteristics, and looking at a few familiar examples of discrete event systems. For an interactive multimedia introduction to discrete event systems, the reader is referred to the Web site http://vita.bu.edu/cgc/MIDEDS/.

1.3.1 The Concept of Event

As with the term "system", we will not attempt to formally define what an "event" is. It is a primitive concept with a good intuitive basis. We only wish to emphasize that an event should be thought of as occurring instantaneously and causing transitions from one state value to another.

An event may be identified with a specific action taken (e.g., somebody presses a button). It may be viewed as a spontaneous occurrence dictated by nature (e.g., a computer goes down for whatever reason too complicated to

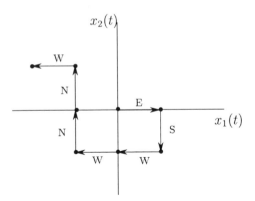

Figure 1.20: Random walk on a plane for Example 1.12.
Each event (N, S, W, or E) causes a state transition in the (x_1, x_2) state space. If an event does not occur, the state cannot change.

figure out). Or it may be the result of several conditions which are suddenly all met (e.g., the fluid level in the tank of Fig. 1.8 has just exceeded a given value).

For our purposes, we will use the symbol e to denote an event. When considering a system affected by different types of events, we will assume we can define an *event set* E whose elements are all these events. Clearly, E is a discrete set.

Example 1.12 (A random walk)

A "random walk" is a useful model for several interesting processes, including some games of chance. When the random walk takes place in two dimensions, we may visualize it as a particle which can be moved one unit of distance (a "step") at a time in one of four directions: north, south, west, or east. The direction is assumed to be chosen at random and independent of the present position. The state of this system is the position of the particle, (x_1, x_2), measured on a plane, with x_1, x_2 taking only integer values, that is, the state space is the discrete set $X = \{(i, j) : i, j = \ldots, -1, 0, 1, \ldots\}$.

In this case, a natural event set is

$$E = \{N, S, W, E\}$$

corresponding to the four events "one step north", "one step south", "one step west", and "one step east". Fig. 1.20 shows a sample path in the (x_1, x_2) space (as in Fig. 1.10) resulting from an initial state (0,0) and the sequence of events $\{E, S, W, W, N, N, W\}$.

Example 1.13
In the warehouse of Example 1.8, we saw that two events affect the state:
a "product arrival" and a "truck arrival". In this case, we may simply
define a set

$$E = \{P, T\}$$

where P denotes the "product arrival" event, and T denotes the "truck
arrival" event.

Time-driven and Event-driven Systems

In continuous-state systems the state generally changes as time changes.
This is particularly evident in discrete-time models: The "clock" is what drives
a typical sample path. With every "clock tick" the state is expected to change,
since *continuous* state variables *continuously* change with time. It is because
of this property that we refer to such systems as *time-driven systems*. In this
case, the time variable (t in continuous time or k in discrete time) is a natural
independent variable which appears as the argument of all input, state, and
output functions.

In discrete-state systems, we saw that the state changes only at certain
points in time through instantaneous transitions. With each such transition we
can associate an event. What we have not yet discussed is the timing mechanism
based on which events take place. Let us assume there exists a clock through
which we will measure time, and consider two possibilities:

1. At every clock tick an event e is to be selected from the event set E. If
   no event takes place, we can think of a "null event" as being a member
   of E, whose property is that it causes no state change.

2. At various time instants (not necessarily known in advance and not necessarily coinciding with clock ticks), some event e "announces" that it is
   occurring.

There is a fundamental difference between 1 and 2 above:

- In 1, state transitions are *synchronized* by the clock: There is a clock
  tick, an event (or no event) is selected, the state changes, and the process
  repeats. The clock alone is responsible for any possible state transition.

- In 2, every event $e \in E$ defines a distinct process through which the time
  instants when e occurs are determined. State transitions are the result of
  combining these *asynchronous* and *concurrent* event processes. Moreover,
  these processes need not be independent of each other.

The distinction between 1 and 2 gives rise to the terms *time-driven* and *event-
driven* systems respectively. As we already saw, continuous-state systems are

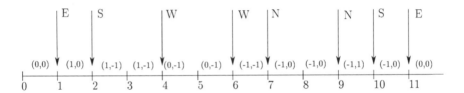

Figure 1.21: Event-driven random walk on a plane.

by their nature time-driven. However, in discrete-state systems this depends on whether state transitions are synchronized by a clock or occur asynchronously as in scheme 2 above. Clearly, event-driven systems are more complicated to model and analyze, since there are several asynchronous event-timing mechanisms to be specified as part of our understanding of the system.

It is worthwhile pointing out that the idea of event-driven state transitions corresponds to the familiar notion of an "interrupt" in computer systems. While many of the functions in a computer are synchronized by a clock, and are therefore time-driven, operating systems are designed to also respond to asynchronous calls that may occur at any time. For instance, an external user request or a timeout message may take place as a result of specific events, but completely independent of the computer clock.

Example 1.14 (Event-driven random walk)

The random walk described in Example 1.12 is a *time-driven* system. Given a clock, we can imagine that with every clock tick there is a single "player" who moves the particle, that is, the player selects an event from the set E.

However, there is an alternative view of the random walk, which comes from modifying the rules controlling the particle movements. Suppose there are four different players, each one responsible for moving the particle in a single direction (N, S, W, or E). Each player acts by occasionally issuing a signal to move the particle in his direction. This results in an *event-driven* system defined by these four asynchronously acting players.

As an example, suppose player N issues signals at discrete-time instants $\{7, 9\}$, S issues signals at $\{2, 10\}$, W issues signals at $\{4, 6\}$, and E issues signals at $\{1, 11\}$. Fig. 1.21 shows the resulting sample path in the form of a timing diagram, where state transitions are event-driven. It is assumed that the initial state is (0,0).

In Example 1.14, it is assumed that two events never take place at exactly the same time instant. If this were the case, then the resulting state transition should reflect the occurrence of both events. Suppose, for instance, that at time instant 1 both E and S issue a signal. We could easily assume that either one of the two events occurs before the other (but at the same time), since the resulting state in either way is ultimately $(1, -1)$. Of course, this is not always so. In general, the exact order in which two events affect the state can make all the difference in the world. For example, suppose the state is a bank account balance, which is currently zero. Let D denote the event "account owner makes a deposit of x dollars" and C denote the event "bank cashes a check of x dollars against this account." The events may occur at the same time, but one of them affects the account balance first. If D does first, the net effect is zero; but if C does, the account is penalized before the deposit, and the net effect is a negative balance corresponding to the service charge for a bounced check. In some cases, we must actually explicitly model the effect of two events occurring simultaneously as completely distinct from the occurrence of these two events in either order.

1.3.2 Characteristic Properties of Discrete Event Systems

We have pointed out that most of the successes in system and control engineering to date have relied on differential-equation-based models, such as (1.10)-(1.11) or its difference equation analog (1.43)-(1.44). To use these mathematically convenient models, there are two key properties that systems must satisfy:

1. They are *continuous-state* systems.

2. The state transition mechanism is *time-driven*.

The first property allows us to define the state by means of continuous variables, which can take on any real (or complex) values. It is because of this reason that we will refer to this class of systems as *Continuous-Variable Dynamic Systems* (CVDS). Common physical quantities such as position, velocity, acceleration, temperature, pressure, flow, etc. fall in this category. Since we can naturally define derivatives for these continuous variables, differential equation models like (1.10) can be used.

The second property points to the fact that the state generally changes as time changes. As a result, the time variable (t in continuous time or k in discrete time) is a natural independent variable for modeling such systems.

In contrast to CVDS, *Discrete Event Dynamic Systems* (DEDS) or, more broadly, *Discrete Event Systems* (DES), satisfy the following two properties:

1. The state space is a *discrete* set.

2. The state transition mechanism is *event-driven*.

In what follows, we will use the acronym DES. We present below an informal first definition of DES based on these properties. Formal definitions will be provided in the next chapter, as we attempt to develop detailed modeling procedures.

Definition. A *Discrete Event System* (DES) is a *discrete-state, event-driven* system, that is, its state evolution depends entirely on the occurrence of asynchronous discrete events over time. ♦

Many systems, particularly technological ones, are in fact discrete-state systems. Even if this is not the case, for many applications of interest a discrete-state view of a complex system may be necessary. Here are some simple examples of discrete-state systems:

- The state of a machine may be selected from a set such as {ON, OFF} or {BUSY, IDLE, DOWN}.

- A computer running a program may be viewed as being in one of three states: {WAITING FOR INPUT, RUNNING, DOWN}. Moreover, the RUNNING state may be broken down into several individual states depending, for example, on what line of code is being executed at any time.

- Any type of inventory consisting of discrete entities (e.g., products, monetary units, people) has a natural state space in the non-negative integers {0, 1, 2, ...}.

- Most games can be modeled as having a discrete state space. In chess, for instance, every possible board configuration defines a state; the resulting state space is enormous, but it is discrete.

The event-driven property of DES was discussed in the previous section. It refers to the fact that the state can only change at discrete points in time, which physically correspond to occurrences of asynchronously generated *discrete events*. From a modeling point of view, this has the following implication. If we can identify a set of "events" any one of which can cause a state transition, then time no longer serves the purpose of driving such a system and may no longer be an appropriate independent variable.

The two fundamental features that distinguish CVDS from DES are clearly revealed when comparing typical sample paths from each of these system classes, as in Fig. 1.22:

- For the CVDS shown, the state space X is the set of real numbers $\mathbb{R}$, and $x(t)$ can take any value from this set. The function $x(t)$ is the solution of a differential equation of the general form $\dot{x}(t) = f(x(t), u(t), t)$, where $u(t)$ is the input.

- For the DES, the state space is some discrete set $X = \{s_1, s_2, s_3, s_4, s_5, s_6\}$. The sample path can only jump from one state to another whenever an event occurs. Note that an event may take place, but not cause a state transition, as in the case of e_3. We have, at this point, no analog to $\dot{x}(t) = f(x(t), u(t), t)$, i.e., no mechanism is provided to specify how events

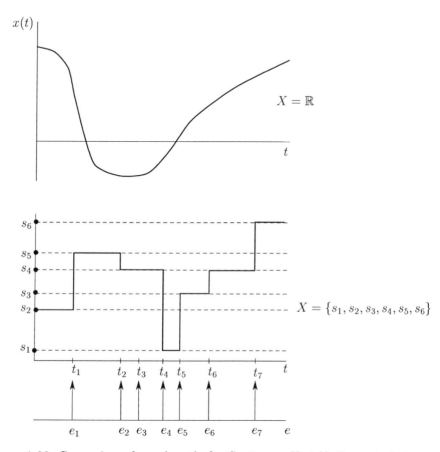

Figure 1.22: Comparison of sample paths for *Continuous-Variable Dynamic Systems* (CVDS) and *Discrete Event Systems* (DES).
In a CVDS, $x(t)$ is generally the solution of a differential equation $\dot{x}(t) = f(x(t), u(t), t)$. In a DES, $x(t)$ is a piecewise constant function, as the state jumps from one discrete value to another whenever an event takes place.

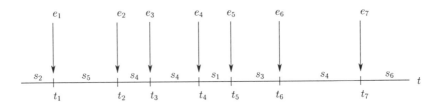

Figure 1.23: A DES sample path.

might interact over time or how their time of occurrence might be determined.

Much of our effort in subsequent chapters will be devoted to determining "state equations" for DES, in an attempt to parallel the model $\dot{x}(t) = f(x(t), u(t), t)$ for CVDS. In some cases, we also encounter systems with all the CVDS features, except for some discrete events that occasionally cause jumps in their sample paths. These events correspond to switches from one mode of operation (state equation) to another. Such systems are sometimes referred to as *Hybrid Systems*.

Remark. One should not confuse discrete event systems with *discrete-time* systems. The class of discrete-time systems contains both CVDS and DES. In other words, a DES may be modeled in continuous or in discrete time, just like a CVDS can. In Example 1.14, we saw a discrete-time model of a DES. If, however, players can issue signals at any real-valued time instant, we obtain a continuous-time model of the DES.

It is often convenient to represent a DES sample path as a timing diagram with events denoted by arrows at the times they occur, and states shown in between events. As an example, the DES sample path of Fig. 1.22 is redrawn in Fig. 1.23.

1.3.3 The Three Levels of Abstraction in the Study of Discrete Event Systems

Let us return to the DES sample path shown in Figs. 1.22 and 1.23. Instead of plotting sample paths of that form, it is often convenient to simply write the timed sequence of events

$$(e_1, t_1), (e_2, t_2), (e_3, t_3), (e_4, t_4), (e_5, t_5), (e_6, t_6), (e_7, t_7) \qquad (1.49)$$

which contains the same information as the sample path depicted in these figures. The first event is e_1 and it occurs at time $t = t_1$; the second event is e_2 and it occurs at time $t = t_2$, and so forth. When this notation is used, it is implicitly assumed that the initial state of the system, s_2 in this case, is known and that the system is "deterministic" in the sense that the next state after the occurrence of an event is unique. Thus, from the sequence of events in (1.49), we can recover the state of the system at any point in time and reconstruct the sample path depicted in Fig. 1.22.

Consider the set of all timed sequences of events that a given system can ever execute. We will call this set the *timed language* model of the system. The word "language" comes from the fact that we can think of the set of events E as an "alphabet" and of (finite) sequences of events as "words". We can further refine our model of the system if some statistical information is available about the set of sample paths of the system. Let us assume that probability distribution functions are available about the "lifetime" of each event, that is, the elapsed time between successive occurrences of each event. We will call a *stochastic timed language* a timed language together with associated probability distribution functions for the events. The stochastic timed language is then a model of the system that lists all possible sample paths together with relevant statistical information about them.

Stochastic timed language modeling is the most detailed, as it contains event information in the form of event occurrences and their orderings, information about the exact times at which the events occur (and not only their relative ordering), and statistical information about successive occurrences of events. If we omit the statistical information, then the corresponding timed language enumerates all the possible sample paths of the DES, with timing information. Finally, if we delete the timing information from a timed language we obtain an *untimed language*, or simply *language*, which is the set of all possible orderings of events that could happen in the given system. Deleting the timing information from a timed language means deleting the time of occurrence of each event in each timed sequence in the timed language. For example, the untimed sequence corresponding to the timed sequence of events in (1.49) is

$$e_1, e_2, e_3, e_4, e_5, e_6, e_7 \ .$$

Languages, timed languages, and stochastic timed languages represent the three levels of abstraction at which DES are modeled and studied: untimed (or logical), timed, and stochastic. The choice of the appropriate level of abstraction clearly depends on the objectives of the analysis. In many instances, we are interested in the "logical behavior" of the system, that is, in ensuring that a precise *ordering of events* takes place which satisfies a given set of specifications (e.g., first-come first-served in a job processing system). Or we may be interested in finding if a particular state (or set of states) of the system can be reached or not. For example, in Fig. 1.10 the shaded region represents a desirable set of states that may be reached by certain sample paths. In this context, the actual timing of events is not required, and it is sufficient to model only the untimed behavior of the system, that is, consider the language model of the system.

Next, we may become interested in *event timing* in order to answer questions such as: "How much time does the system spend at a particular state?" or "How soon can a particular state be reached?" or "Can this sequence of events be completed by a particular deadline?" These and related questions are often crucial parts of the design specifications. More generally, event timing is important in assessing the performance of a DES often measured through quantities such as *throughput* or *response time*. In these instances, we need to consider the timed language model of the system. As we shall see, the fact that different event processes are concurrent and often interdependent in complex ways presents great challenges both for modeling and analysis of timed DES.

Finally, we cannot ignore the fact that DES frequently operate in a stochastic setting, hence necessitating the development of probabilistic models and related analytical methodologies for design and performance analysis. In these cases, the stochastic timed language model of the system has to be considered.

These three levels of abstraction are complementary as they address different issues about the behavior of a DES. Indeed, the literature in DES is quite broad and varied as extensive research has been done on modeling, analysis, control, optimization, and simulation at all three levels of abstraction. One of the objectives of this book is to present important results about discrete event system and control theory at all three levels of abstraction in a manner that emphasizes the gradual refinement of models by going from untimed to timed and then to stochastic timed models.

As we shall see in the following chapters, this language-based approach to discrete event modeling is attractive to present modeling issues and discuss system-theoretic properties of DES. However, it is by itself not convenient to address verification, controller synthesis, or performance issues; what is also needed is a convenient way of *representing* languages, timed languages, and stochastic timed languages. If a language (or timed language or stochastic timed language) is finite, we could always list all its elements, that is, all the possible sample paths that the system can execute. Unfortunately, this is rarely practical. Preferably, we would like to use *discrete event modeling formalisms* that would allow us to represent languages in a manner that highlights structural information about the system behavior and that is convenient to manipulate when addressing analysis and controller synthesis issues. Discrete event modeling formalisms can be untimed, timed, or stochastic, according to the level of abstraction of interest.

In this book, two discrete event modeling formalisms will be presented and studied: *automata* and *Petri nets*. These formalisms have in common the fact that they represent languages by using a state transition structure, that is, by specifying what the possible events are in each state of the system. The formalisms differ by how they represent state information. They are also amenable to various composition operations, which allows building the discrete event model of a system from discrete event models of the system components. This makes automata and Petri nets convenient for model building. Analysis and synthesis issues are then typically addressed by making use of the structural properties of the transition structure in the model. This will become apparent in the subsequent chapters of this book.

This introduction to DES is intended to point out the main characteristics of these systems. Thus far, two elements emerge as essential in defining a DES: a discrete state space, which we shall denote by X, and a discrete event set, which we shall denote by E. In later chapters we will build on this basic understanding in order to develop formal models and techniques for analysis.

1.3.4   Examples of Discrete Event Systems

The next few subsections describe examples of DES drawn from the real world and common engineering experience. In order to present these systems effectively, we begin with a simple "building block" which will serve to represent many DES of interest.

Queueing Systems

The term "queueing" refers to a fact of life intrinsic in many of the most common systems we design and build: In order to use certain resources, we have to wait. For example, to use the resource "bank teller" in a bank, people form a line and wait for their turn. Sometimes, the waiting is not done by people, but by discrete objects or more abstract "entities". For example, to use the resource "truck", finished products wait in a warehouse, as in the example of Fig. 1.12. Similarly, to use the resource "CPU", various "tasks" wait somewhere in a computer until they are given access to the CPU through potentially complex mechanisms. There are three basic elements that comprise a queueing system:

- The *entities* that do the waiting in their quest for resources. We shall usually refer to these entities as *customers*.

- The *resources* for which the waiting is done. Since resources typically provide some form of service to the customers, we shall generically call them *servers*.

- The space where the waiting is done, which we shall call a *queue*.

Examples of customers are: people (e.g., waiting in a bank or at a bus stop); messages transmitted over some communication medium; tasks, jobs or transactions executed in a computer system; production parts in a manufacturing process; and cars using a road network.

Examples of servers are: people again (e.g., bank tellers or supermarket checkout tellers); communication channels responsible for the transmission of messages; computer processors or peripheral devices; various types of machines used in manufacturing; and traffic lights regulating the flow of cars.

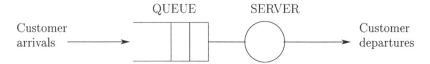

Figure 1.24: A simple queueing system.
Each arriving customer either immediately proceeds to the SERVER and is served, or must first wait in the QUEUE until the server is available. After receiving service, each customer departs.

Examples of visible queues are found in bank lobbies, bus stops, or warehouses. However, queues are also ever-present in communication networks or computer systems, where less tangible forms of customers, like telephone calls or executable tasks, are also allocated to waiting areas. In some cases, queues are also referred to as *buffers*.

The study of queueing systems is motivated by the simple fact that resources are not unlimited; if they were, no waiting would ever occur. This fact gives rise to obvious problems of resource allocation and related tradeoffs so that (a) customer needs are adequately satisfied, (b) resource access is provided in fair and efficient ways among different customers, and (c) the cost of designing and operating the system is maintained at acceptable levels.

We will utilize a queueing system as a basic building block for many of the more complex DES we will be considering. Graphically, we will represent a simple queueing system as shown in Fig. 1.24. A circle represents a server, and an open box represents a queue preceding this server. The slots in the queue are meant to indicate waiting customers. Customers are thought of as *arriving* at the queue, and *departing* from the server. It is also assumed that the process of serving customers normally takes a strictly positive amount of time (otherwise there would be no waiting). Thus, a server may be thought of as a "delay block" which holds a customer for some amount of service time.

In order to fully specify the characteristics of a queueing system such as the one shown in Fig. 1.24, we also need to define (a) the capacity of the queue, and (b) the queueing discipline. The *capacity* of a queue is the maximum number of customers that can be accommodated in the actual queueing space. In many models, it is assumed that the capacity is "infinite", in the sense that the queueing space is so large that the likelihood of it ever being full is negligible. The *queueing discipline* refers to the rule according to which the next customer to be served is selected from the queue. The simplest such rule is the First-In-First-Out (FIFO): Customers are served in the precise order in which they arrive.

Viewed as a DES, the queueing system of Fig. 1.24 has an event set $E = \{a, d\}$, where a denotes an "arrival" event, and d denotes a "departure" event. A natural state variable is the number of customers in queue, which we shall call *queue length*. By convention, the queue length at time t is allowed to

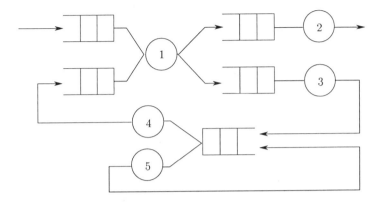

Figure 1.25: A queueing network example.

include a customer in service at time t. We shall adopt this convention, unless otherwise noted. Thus, the state space is the set of non-negative integers, $X = \{0, 1, 2, \ldots\}$.

Finally, note that a single queue may be connected to multiple servers, and several queues may be connected to a single server, or both. Clearly, arbitrary series and parallel connections of this building block are possible, as shown in Fig. 1.25. Thus, one can construct *queueing networks* much in the same way that one constructs networks consisting of electrical building blocks, such as resistors, inductors, and capacitors. In this case, customers are viewed as "flowing" through the components (queues and servers) of the network according to specific rules. For instance, customers departing from server 1 in Fig. 1.25 must follow certain rules in choosing one of the two queues connected to servers 2 and 3 respectively. Similarly, server 1 must adopt certain rules in selecting one of the two input queues from which a customer is served next. In fact, defining such rules and determining which rules are "better" than others is a large part of our responsibility as "control engineers" for queueing systems.

Computer Systems

In a typical computer system, *jobs*, *tasks*, or *transactions* are the customers competing for the attention of servers such as various *processors*, such as the CPU (central processing unit), or *peripheral devices* (e.g., printers, disks). When a server is busy at the time of job requests, the jobs are placed in queues, which are an integral part of the computer system.

It is often convenient to represent such a system through a queueing model, using the building block described above. An example of a common computer system configuration is shown in Fig. 1.26. In this system, jobs arrive at the CPU queue. Once served by the CPU, they either depart or request access to

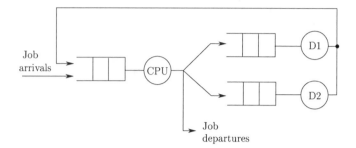

Figure 1.26: Queueing model of a computer system.

one of two disks. They then return for more service at the CPU.

The event set in the example of Fig. 1.26 normally consists of arrivals and departures at the various servers. In particular,

$$E = \{a, d, r_1, r_2, d_1, d_2\}$$

where

    a is an arrival from the outside world to the computer system

    d is a departure from the CPU to the outside world

  r_1, r_2 are departures from the CPU which are routed to disks 1 and 2, respectively

  d_1, d_2 are departures from disks 1, 2, respectively, which always return to the CPU queue.

One possible state representation of this system consists of a vector $\mathbf{x} = [x_{CPU}, x_1, x_2]^T$ corresponding to the three queue lengths at the CPU and disks 1, 2. In this case, the state space is

$$X = \{(x_{CPU}, x_1, x_2) : x_{CPU}, x_1, x_2 \geq 0\}$$

Communication Systems

The customers of a communication system are referred to as *messages, packets*, or *calls* (as in a telephone network). A message is typically generated by a user located at some "source" point and intended to reach some other user located at a "destination" point. Often, the connection between source and destination is not direct, but goes through one or more intermediate points. In order for messages to travel from source to destination, they must access various servers, which in this case are *switching equipment* such as simple telephone switches or fairly sophisticated computer processors, as well as *communication media* such as wires or radio. Queueing models, some quite similar to those of Fig. 1.26, are often used to describe the operation of such systems.

An important feature of communication systems is the need for control mechanisms which ensure that access to the servers is granted on a fair and efficient basis, and that the objectives of the communication process are met (i.e., a message is indeed successfully delivered to its destination). These control mechanisms, sometimes referred to as *protocols*, may be quite complex. Thus, their design and the process of verifying that they function as intended pose challenging problems.

As an example, consider two users, A and B, both sharing a common communication medium, which we shall call a *channel*. The channel is only capable of serving one user at a time. Thus, if B sends a message to the channel while A is using it, the result is an unintelligible scrambled signal, and therefore no message is successfully transmitted. This is referred to as a *collision*. It follows that the channel can be in one of three states:

> I idle
> T transmitting one message
> C transmitting two or more messages resulting in a collision.

Similarly, each user can be in one of three states:

> I idle
> T transmitting
> W waiting to transmit an existing message.

There are now two problems. First, each user does not know the state of the other, and, second, users may not know the state of the channel (or may only be able to detect one state). One can see that the possibility of continuous collisions is present in this setting, unless the users operate based on a set of rules guaranteeing at least some success for this process.

The events driving this system are arrivals of messages to be transmitted at A and B, the actions by A and B of sending a message to the channel for transmission, and the completion of a transmission by the channel. Thus, we can define a state space

$$X = \{(x_{CH}, x_A, x_B) : x_{CH} \in \{I, T, C\}, x_A \in \{I, T, W\}, x_B \in \{I, T, W\}\}$$

and an event set

$$E = \{a_A, a_B, \tau_A, \tau_B, \tau_{CH}\}$$

where

> a_A, a_B are arrivals of messages to be transmitted by A and B, respectively
> τ_A, τ_B are events of sending a message to the channel by A and B, respectively
> τ_{CH} is completion of a channel transmission (with one or more messages present).

Figure 1.27: Queueing model of a manufacturing system.

Manufacturing Systems

The customers of a manufacturing process are *production parts* or *pieces*. Parts compete for access at various servers, which in a typical factory are *machines* performing specific operations and *material handling devices* such as robots and conveyor belts. When parts are not being worked on they are stored in buffers until the server required for the next operation is available. Because of real physical limitations, buffers in a manufacturing system usually have finite capacities.

Once again, queueing models provide a convenient framework to describe manufacturing systems. A simple example is shown in Fig. 1.27. Parts go through two machines, where the buffer capacity in front of the second machine is limited to two. As a result, it is possible that a part completing service at machine 1 finds that machine 2 is busy and that both buffer slots are occupied. In this case, the part must remain at machine 1 even though it requires no more service there; moreover, other parts waiting to access machine 1 are forced to remain in queue. This situation is referred to as *blocking*.[1]

The event set for this example is

$$E = \{a, c_1, d_2\}$$

where

a is an arrival from the outside world to the first machine
c_1 is a completion of service at the first machine
d_2 is a departure from the second machine.

Note that a c_1 event does not imply movement of a part from machine 1 to the downstream buffer, since blocking is a possibility. The state of the system can be defined as a vector $x = [x_1, x_2]^T$ corresponding to the queue lengths at the two machines. In this case, x_2 is constrained to the values $\{0,1,2,3\}$. Note, however, that when $x_2 = 3$, machine 1 may either be blocked or still serving a part. To model the blocking phenomenon we introduce an additional value, B, that x_2 can take. Thus, $x_2 = B$ means that the queue length at the second machine is 3 and a part at the first machine is blocked. The state space then becomes the discrete set

$$X = \{(x_1, x_2) : x_1 \geq 0, x_2 \in \{0, 1, 2, 3, B\}\}$$

---

[1]Strictly speaking, we should refer to this situation as *time-blocking*, to differentiate it from logical (or untimed) blocking that will be considered in Chapters 2 and 3; however, the "time" qualifier is never used in the queueing literature.

To illustrate the flexibility of the modeling process (depending on the level of detail we wish to capture and the type of issues we need to address), observe that an alternative state space might be

$$X = \{(x_1, x_2) : x_1 \in \{I, W, B\}, x_2 \in \{I, W\}\}$$

where x_1 is the state of the first machine: idle (I), working (W), or blocked (B), and x_2 is the state of the second machine: idle (I), or working (W). In this model, we do not explicitly track queue lengths, but are only interested in the "logical" state of each machine.

Traffic Systems

In a traffic environment, *vehicles* make use of servers such as traffic lights, toll booths, and the physical space on roads. As an example, consider how a simple T-intersection (Fig. 1.28) may be viewed as a discrete event system. There are four types of vehicles:

(1,2) vehicles coming from point 1 and turning right towards 2,
(1,3) vehicles coming from 1 and turning left towards 3,
(2,3) vehicles going straight from 2 to 3, and
(3,2) vehicles going straight from 3 to 2.

The traffic light is set so that it either turns red for (1,2) and (1,3) vehicles (green for (2,3) and (3,2) vehicles), or it turns green for (1,2) and (1,3) vehicles (red for (2,3) and (3,2) vehicles).

In this case, the event set is

$$E = \{a_{12}, a_{13}, a_{23}, a_{32}, d_{12}, d_{13}, d_{23}, d_{32}, g, r\}$$

where

$a_{12}, a_{13}, a_{23}, a_{32}$ is a vehicle arrival for each of the four types,
$d_{12}, d_{13}, d_{23}, d_{32}$ is a vehicle departure upon clearing the intersection for each type,
g indicates that the light turns green for (1,2) and (1,3) vehicles,
r indicates that the light turns red for (1,2) and (1,3) vehicles.

A possible state space is defined by the queue lengths formed by the four vehicle types and the state of the traffic light itself, that is

$$X = \{(x_{12}, x_{13}, x_{23}, x_{32}, y) : x_{12}, x_{13}, x_{23}, x_{32} \geq 0, y \in \{G, R\}\}$$

where $x_{12}, x_{13}, x_{23}, x_{32}$ are the four queue lengths, and y is the state of the light (G denotes green for (1,2) and (1,3) vehicles, and R denotes red for (1,2) and (1,3) vehicles).

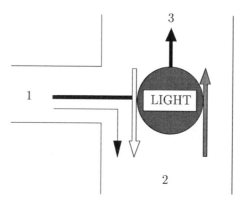

Figure 1.28: A simple intersection controlled by a traffic light.

Database Systems

A database management system (DBMS) must ensure that a database re-mains consistent despite the fact that many users may be doing retrievals and updates concurrently. The sequence of read and write operations of each user is called a *transaction*. When many transactions are being executed concurrently, the DBMS must schedule the interleaving of the operations of the transactions in order to preserve the consistency of the data. This is called the *concurrency control* problem. Concurrency control is a discrete event control problem. The events are the read and write operations of the individual transactions on the records in the database. Generic events are $r_1(a)$ and $w_2(b)$, denoting respec-tively a read operation on record a by transaction 1 and a write operation on record b by transaction 2. Given a set of transactions to be executed concur-rently on a database, a *schedule* is a sequence of events representing the actual order in which the operations of the transactions are executed on the database. The set of all possible interleavings of the operations of the given transactions, that is, the set of all schedules, is the language model of this system.

For example, let us consider a set of two banking transactions, labeled 1 and 2, where the first transaction is defined by the string of events

$$r_1(a)r_1(b)$$

and the second transaction is defined by the string of events

$$r_2(a)w_2(a)r_2(b)w_2(b) \ .$$

Assume that the purpose of transaction 1 is to obtain the total amount of money in records a and b, while the purpose of transaction 2 is to transfer \$100 from record a to record b. If the DBMS were to "blindly" let transactions 1 and 2 run concurrently, a possible schedule would be

$$S_x = r_2(a)w_2(a)r_1(a)r_1(b)r_2(b)w_2(b) \ .$$

We can see that with such an interleaving of the events of transactions 1 and 2, transaction 1 would return an incorrect result, namely it would return a total that is $100 less than it should be, since it is reading the contents of a and b after a has been decremented by $100 but before b has been incremented by $100. Schedule S_x shows the necessity for the DBMS to control the execution of a set of concurrent transactions.

The interesting aspect of concurrency control is that it is not necessary to completely prevent any interleaving of the events of individual transactions in order to obtain a correct result. For instance, schedule

$$S_y = r_2(a)w_2(a)r_1(a)r_2(b)w_2(b)r_1(b)$$

is admissible since transaction 1 would return the correct result. Thus, the DBMS must guarantee that it allows only admissible schedules, while at the same time allowing as many of these admissible schedules as possible, since allowing more interleavings of events from different transactions means less waiting on the part of the users who are submitting these transactions and thus better performance.

It is important to note that the DBMS must make sure that any partially-completed schedule can be completed to an admissible schedule. For example, if the DBMS allows the string of events

$$S_z = r_1(a)r_2(a)w_2(a)r_2(b)w_2(b)$$

to occur, then while nothing incorrect has been done so far – transaction 2 is performed correctly – the only possible completion of S_z with event $r_1(b)$ will lead to an incorrect result for transaction 1 – the amount returned will be $100 higher than it should be. Thus, S_z cannot be completed properly. We call this situation a *deadlock*; the term *blocking* will also be used in Chapters 2 and 3 for such situations. In practice, the deadlock will be resolved by aborting transaction 1 and starting it all over again.

In summary, we can model the concurrent execution of a set of database transactions as a DES whose language model is the set of all possible interleavings of the strings of events corresponding to the individual transactions. The control problem is to restrict the language of the controlled system to admissible schedules only. This is a problem that is posed at the logical level of abstraction, since the admissibility of a schedule only depends on the ordering of the events in it, and not on temporal or stochastic considerations.

Software Systems: Telephony

As a specific example of a complex software system, let us consider the software programs that are executed by (digital) switches in a telephone network. These programs have increased in complexity due to the introduction of new call processing features such as automatic call-back, call forwarding, call

screening, call waiting, multi-way calling, etc. At present, the software programs executed by a switch contain millions of lines of code. When a new call
processing feature has to be introduced, designers have to make sure that the
new feature will have no unanticipated side effects that could interfere with the
normal functionalities of existing features. For instance, when the call screening
feature was introduced after the call forwarding feature had been implemented,
the designers had to decide if a call that should normally be screened should
be forwarded or not when a user subscribes to and activates both features. In
fact, what we are describing here is a specific instance of the generic problem
of compatibility with prior versions when software is upgraded.

In the context of our telephone example, designers can make use of discrete
event models to study potential conflicts between existing features and new
ones, when the specifications for these new features are being developed. If the
specification of a new feature has been verified in the context of a formal model,
then the software code that will be written to implement the specification is
more likely to behave correctly (of course, bugs can still be introduced in the
process of writing the code). Examples of events to include in the development of a DES model for a call processing feature include: "phone i off-hook",
"request for connection from user i to user j", "establishment of connection
between users i and j", "forwarding of call from user i to user j to user k",
"connection cannot be established due to screening list of user j" (leading to
recorded message to calling party), "flash-hook by user i", "phone i on-hook",
and so forth. Events such as these can be used to describe how each feature
should operate at a switch. The language model of the joint operation of a
set of features would then be analyzed in order to identify strings that lead to
incorrect behavior. One form of incorrect behavior is the violation of the intent
of a feature, for instance, a connection is established for a call that should be
screened. Another form of incorrect behavior is a deadlock, e.g., when each of
two features assumes the other feature will make the decision about establishing
or not establishing a requested connection, and thus the calling party does not
get a ringing (or busy) signal or a special recording.

Monitoring and Control of Complex Systems

The control of complex automated systems (for example, assembly and manufacturing systems, automobile, jet, or locomotive engines, chemical processes,
heating, ventilation, and air-conditioning units in large buildings, semiconduc-
tor manufacturing, etc.) typically requires a multi-level hierarchy of controllers,
from servo-mechanisms at the lowest level of this hierarchy to software programs
implementing the necessary control logic at the highest level of the hierarchy.
The feedback loops at the lower levels of the hierarchy are based on CVDS representations of the system or component of interest. At the higher levels, a DES
view is often adequate for dealing with situations such as system start-up and
shut-down, change of mode of operation, exception handling, failure diagnosis,

failure recovery and system reconfiguration, and so forth.

Conceptually, one may view the complete monitoring and control architecture of a complex automated system as depicted in Fig. 1.29. The purpose of this simplified figure is to illustrate the "boundary" between feedback controllers designed based on CVDS models of the system and the control logic implemented by the supervisory controller. It is often the case that the system and associated continuous-variable controllers at the lower levels can be "abstracted" as a DES for the purpose of the higher levels. This abstraction occurs at the interface in the figure, where information from sensors and continuous-variable controllers is relayed to the supervisory controller in the form of events, while commands from this supervisory controller, which are also in the form of events, are translated by the interface to generate the appropriate input signals to the actuators or set points to the continuous-variable controllers.

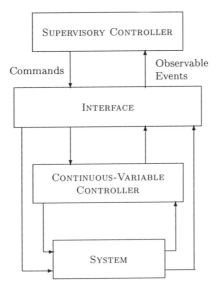

Figure 1.29: Conceptual monitoring and control system architecture.

To focus this discussion, let us assume that we are considering the heating system of a heating, ventilation, and air conditioning unit, along with its controller. This system is depicted in Fig. 1.30. A DES view of this controlled system for the purpose of diagnosing abrupt failures such as "valve stuck open", "valve stuck closed", "pump failed off", "controller failed on", etc., would associate with each system component a discrete set of states (for example, four states for the valve: open, closed, stuck open, and stuck closed) and could include as events (detected by the interface or commands by supervisory con-

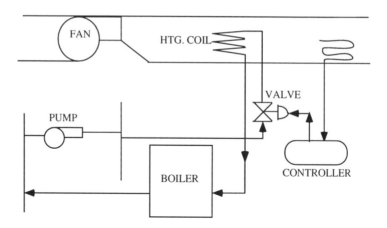

Figure 1.30: Heating system with controller.

troller): "set point increase", "fan on", "open valve", "pump on", "change of sensor reading from no-flow to flow", and so forth. If the system is started by opening the valve and turning on the pump, and no change in the flow sensor reading is observed within a certain time delay, then one would conclude that either the valve is stuck closed or the pump is failed off.

In conclusion, while many systems are inherently discrete and thus naturally modeled as DES, it may be convenient to model certain aspects of CVDS as DES when the level of detail contained in the discrete event model suffices for the problem at hand.

1.4 SUMMARY OF SYSTEM CLASSIFICATIONS

We briefly review the major system classifications presented throughout this chapter. These classifications are by no means exclusive. Still, they do serve the purpose of describing the scope of different aspects of system and control theory. More importantly, they help us identify the key features of the discrete event systems we will be studying in this book. Figure 1.31 provides a quick reference to the classifications we wish to emphasize.

- **Static and Dynamic Systems.** In static systems the output is always independent of past values of the input. In dynamic systems, the output does depend on past values of the input. Differential or difference equations are generally required to describe the behavior of dynamic systems.

- **Time-varying and Time-invariant Systems.** The behavior of time-invariant systems does not change with time. This property, also called *stationarity*, implies that we can apply a specific input to a system and

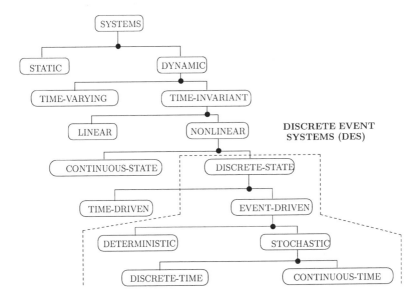

Figure 1.31: Major system classifications.

expect it to always respond in the same way.

- **Linear and Nonlinear Systems.** A linear system satisfies the condition $\mathbf{g}(a_1\mathbf{u}_1 + a_2\mathbf{u}_2) = a_1\mathbf{g}(\mathbf{u}_1) + a_2\mathbf{g}(\mathbf{u}_2)$, where $\mathbf{u}_1$, $\mathbf{u}_2$ are two input vectors, a_1, a_2 are two real numbers, and $\mathbf{g}(\cdot)$ is the resulting output. Linear time-invariant dynamic systems are described by the state space model (1.25) and (1.26), which has been studied in great detail in the system and control theory literature.

- **Continuous-State and Discrete-State Systems.** In continuous-state systems, the state variables can generally take on any real (or complex) value. In discrete-state systems, the state variables are elements of a discrete set (e.g., the non-negative integers).

- **Time-driven and Event-driven Systems.** In time-driven systems, the state continuously changes as time changes. In event-driven systems, it is only the occurrence of asynchronously generated discrete events that forces instantaneous state transitions. In between event occurrences the state remains unaffected.

- **Deterministic and Stochastic Systems.** A system becomes stochastic whenever one or more of its output variables is a random variable. In this case, the state of the system is described by a stochastic process, and a probabilistic framework is required to characterize the system behavior.

- **Discrete-time and Continuous-time Systems.** A continuous-time system is one where all input, state, and output variables are defined for all possible values of time. In discrete-time systems, one or more of these variables are defined at discrete points in time only, usually as the result of some sampling process.

As shown in Fig. 1.31, we will concentrate on dynamic, time-invariant, nonlinear, discrete-state, event-driven systems. The nonlinearity of DES is inherent in the discontinuities (jumps) of all state transitions resulting from event occurrences. Within the class of DES, we may consider either deterministic or stochastic and either discrete-time or continuous-time models.

1.5 THE GOALS OF SYSTEM THEORY

There is obviously a great deal of effort that goes into developing the notation and semantics used to classify and describe different types of systems. An even greater effort goes into modeling systems, as we have seen in preceding sections, and will be seeing further in subsequent chapters. Hopefully, this effort provides the necessary infrastructure for the solution of some real engineering problems and the building of practical systems that perform a set of desirable functions in efficient and economically feasible ways. In general terms, the goals of what we call "system theory" may be summarized as follows.

1. **Modeling and Analysis.** This is the first step toward understanding how an existing system actually works. We develop a model in order to see if we can reproduce the physical system (mathematically, in a laboratory, or through computer simulation). If the model is accurate, our goal becomes the study of the system behavior under different conditions, e.g., different parameter values or input functions.

2. **Design and Synthesis.** Once we have modeling techniques at our disposal which we believe to be accurate, we can address the issue of "how to build a system that behaves as we desire." We therefore attempt to bring together various components and to select parameter values which result in a "satisfactory" design.

3. **Control.** This is the next logical step to the basic design process. Referring to the diagrams of Fig. 1.17, we now attempt to select the input functions that will ensure the system will behave as desired under a variety of (possibly adverse) operating conditions. Once again, we need a model we can work with in testing and validating various control approaches we may come up with. In addition, we need techniques which make the process of selecting the right control as efficient as possible.

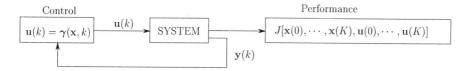

Figure 1.32: A closed-loop system with performance optimization.

4. **Performance evaluation.** After a system has been designed and controlled so as to ensure some proper behavior, we address issues such as "how well is the system really performing?" Measures of system performance may often be subjective or application-dependent. Note that several different controls may accomplish basic design objectives. In evaluating the performance of a system, we usually seek to refine or limit control schemes to the select few that satisfy our performance objectives.

5. **Optimization.** Given that a system can be designed and controlled to achieve some desirable performance, it is natural to ask "how can we control it so as to achieve the *best possible* performance?" This goal requires the development of additional analytical or experimental techniques for efficiently determining the optimal system behavior and the means for attaining it.

The processes of performance evaluation and optimization are illustrated in Fig. 1.32 for a discrete-time model. Here, the output of the system is used to evaluate some function of the form

$$J[\mathbf{x}(0), \mathbf{x}(1), \ldots, \mathbf{x}(K), \mathbf{u}(0), \mathbf{u}(1), \ldots, \mathbf{u}(K)] = \phi[\mathbf{x}(K)] + \sum_{k=0}^{K} L[\mathbf{x}(k), \mathbf{u}(k)]$$

$$(1.50)$$

which defines a *performance measure* for the system over a time interval [0, K]. It is also sometimes referred to as a *performance metric, performance index, cost function* or *cost criterion*. If $J(\cdot)$ represents a *cost*, then we can think of $L[\mathbf{x}(k), \mathbf{u}(k)]$ as the cost of operating the system with control $\mathbf{u}(k)$ when the state is $\mathbf{x}(k)$. Similarly, $\phi[\mathbf{x}(K)]$ is the cost of ending up at state $\mathbf{x}(K)$ after the whole process is complete. Note that $J(\cdot)$ depends not just on the current observed values of the state and input, but the entire history of the system over the interval [0, K]. The control sequence $\{u(0), u(1), \ldots, u(K)\}$ in this case is chosen so as to maximize (or minimize) the function $J(\cdot)$.

SUMMARY

- Historically, the systems that have been studied over the years involve quantities such as pressure, temperature, speed, and acceleration, which are continuous variables evolving over time. We refer to these systems as *Continuous-Variable Dynamic Systems* (CVDS). Modeling and analysis of CVDS heavily rely on the theory and techniques related to differential and difference equations.

■ The main system and control theory concepts one should feel comfortable with in order to proceed to the next chapter are *state*, *sample path*, and *feedback*. A basic state space model consists of a set of equations describing the evolution of state variables over time as a result of a given set of input functions.

■ *Discrete Event Dynamic Systems* (DEDS), or just *Discrete Event Systems* (DES), are systems whose state space is discrete and whose state can only change as a result of asynchronously occurring instantaneous events over time. Sample paths of DES are typically piecewise constant functions of time. Conventional differential equations are not suitable for describing such "discontinuous" behavior.

■ The main elements of a DES are (a) a discrete state space we will usually denote by X, and (b) a discrete event set we will usually denote by E.

■ In order to develop a mathematical model for a DES, we must first decide if we want to view its sample path as (a) merely a sequence of states, or (b) a sequence of states accompanied by the time instants at which state transitions take place. This distinction gives rise to the classes of *untimed* and *timed* models respectively. Timed models can include stochastic elements.

■ Examples of DES which we frequently encounter include computer, communication, manufacturing, software, and traffic systems. A "queueing system" is often used as a building block for modeling many types of DES. The higher-level behavior of complex automated systems with underlying continuous-variable dynamics is often modeled as a DES for the purpose of supervisory control, monitoring, and diagnostics.

■ For an interactive multimedia introduction to discrete event systems, the reader is referred to the Web site http://vita.bu.edu/cgc/MIDEDS/.

PROBLEMS

1.1 Several tasks are submitted to a computer system with two processors, P_1 and P_2, working in parallel. The process of submitting tasks can be described in discrete time by $a(t)$, $t = 0, 1, 2, \ldots$ where $a(t) = 1$ if a task is submitted at time t and $a(t) = 0$ otherwise (at most one task can be submitted in each time step). Suppose such a process is specified for the time interval $t = 0, 1, \ldots, 10$ as follows: $\{1, 1, 1, 0, 1, 0, 1, 1, 0, 0, 1\}$. When a task is seen by the computer system, the following rule for deciding which of the two processors to use is applied: Alternate between the two processors, with the first task going to P_1. It is assumed that if a task is sent to P_i, $i = 1, 2$, and that processor is busy, the task joins a queue of infinite capacity. The processing time of a task at P_1 alternates between 4 and 1 time units (starting with 4), whereas the processing time at P_2 is always 2 time units.

Let $y(t)$ be the total number of customers having departed from the system at time t, and $x_1(t)$ and $x_2(t)$ be the queue lengths at processors P_1 and P_2, respectively (including a task in process). If one or more events occur at time t, the values of these variables are taken to be *just after* the event occurrence(s).

(a) Draw a timing diagram with $t = 0, 1, \ldots, 10$ showing arrivals and departures (assume that $x_1(0) = x_2(0) = y(0) = 0$).

(b) Construct a table with the values of $x_1(t), x_2(t)$ and $y(t)$ for all $t = 0, 1, \ldots, 10$.

(c) Suppose we now work in continuous time. Arrivals occur at times 0.1, 0.7, 2.2, 5.2 and 9.9. The processing time at P_1 now alternates between 4.2 and 1.1, whereas that of P_2 is fixed at 2.0 time units. Consider an event-driven model with event set $E = \{a, d_1, d_2\}$, where $a =$ arrival, $d_i =$ departure from processor $P_i, i = 1, 2$. Construct a table with the values of $x_1(k), x_2(k), y(k), t(k)$, where $x_1(k), x_2(k), y(k)$ are the queue lengths and cumulative number of departures after the kth event occurs, $k = 1, 2, \ldots$, and $t(k)$ is the time instant of the kth event occurrence. If two events occur at the same time, assume that a departure always comes before an arrival. Compare the number of updates required in this model to a time-driven model with a time step of magnitude 0.1 time units.

1.2 Repeat Problem 1.1 with the following two decision rules regarding which processor to use for every arriving task (if two events occur at the same time, assume that a departure always comes before an arrival):

(a) Send the task to P_1 as long as it sees a queue length less than or equal to 2, otherwise send it to P_2.

(b) Send the task to the processor with the shortest queue seen. In case of a tie, send it to P_2.

1.3 A simple manufacturing process involves two machines M_1 and M_2 and a robot arm that removes a completed piece from M_1 and brings it to M_2. There is no buffering at either machine. Thus, if a piece is supplied to M_1 while the machine is busy, that piece is rejected. On the other hand, if the robot transports a piece to M_2 while it is busy, it waits there until M_2 can accept the piece. Note that after the robot arm places a piece in M_2 it still takes some time for it to return to its original position from which it can pick up a new piece from M_1. Thus, M_1 may occasionally be forced to hold a completed piece (and not accept any new arrivals) until the robot arm is available.

Let x_1 and x_2 describe the states of M_1 and M_2, respectively, and x_3 the state of the robot arm. Assume that the processing times at M_1, M_2 are 0.5 sec and 1.5 sec respectively, and that the robot arm requires 0.2 sec to transport a piece from M_1 to M_2, and 0.1 sec to return to its original position at M_1. Finally, suppose that arrivals of pieces are scheduled to arrive at M_1 at time instants: 0.1, 0.7, 1.1, 1.6, and 2.5 sec.

(a) Identify all possible values that each of x_1, x_2, and x_3 can take.

(b) Define an appropriate event set E (with the smallest possible number of elements) for this system.

(c) For the time interval [0.0, 3.0] sec, construct a table with the values of $x_1(k), x_2(k), x_3(k)$, and $t(k)$, where $x_1(k), x_2(k), x_3(k)$ are the states of M_1, M_2, and the robot arm after the kth event occurs, $k = 1, 2, \ldots$, and $t(k)$ is the time instant of the kth event occurrence. If two events occur at the same time, assume that a completion of processing at a machine always comes before the arrival of a new piece.

(d) Identify all states for which M_1 is forced to wait until the robot arm removes a completed piece.

SELECTED REFERENCES

■ *Background Material on Systems and Control Theory*

– Banks, S.P., *Control Systems Engineering*, Prentice-Hall, Englewood Cliffs, 1986.

– Brogan, W.L., *Modern Control Theory*, Prentice-Hall, Englewood Cliffs, 1985.

– Bryson, A.E., and Y.C. Ho, *Applied Optimal Control*, Hemisphere Publishing, Washington, 1975.

– Chen, C.T., *Linear System Theory and Design*, Holt, Rinehart and Winston, New York, 1984.

– D'Azzo J.J., and C.H. Houpis, *Linear Control System Analysis and Design*, McGraw-Hill, New York, 1988.

– Eyman, E., Modeling, *Simulation and Control*, West Publishing, St. Paul, 1988.

- Glisson, T.H., *Introduction to System Analysis*, McGraw-Hill, New York, 1985.

- Ogata, K., *State Space Analysis of Control Systems*, Prentice-Hall, Englewood Cliffs, 1967.

- Takahashi, Y., Rabins, M.J., and Auslander, D.M., *Control and Dynamic Systems*, Addison-Wesley, Reading, 1972.

- Zadeh, L.A., and C.A. Desoer, *Linear System Theory: The State Space Approach*, McGraw-Hill, New York, 1963.

■ *Discrete Event Systems*

- Baccelli, F., G. Cohen, G.J. Olsder, and J.-P. Quadrat, *Synchronization and Linearity: An Algebra for Discrete Event Systems*, Wiley, Chichester, 1992.

- Cassandras, C.G., and P.J. Ramadge (Eds.), "Special Section on Discrete Event Systems," *Control Systems Magazine*, Vol. 10, No. 4, pp. 66-112, 1990.

- Cassandras, C.G., and S. Lafortune, "Discrete Event Systems: The State of the Art and New Directions," in *Applied and Computational Control, Signals, and Circuits*, Vol. I, B. Datta, Ed., Birkhäuser, Boston, 1999.

- David, R., and H. Alla, *Petri Nets & Grafcet: Tools for Modelling Discrete Event Systems*, Prentice-Hall, New York, 1992.

- Glasserman, P., and D.D. Yao, *Monotone Structure in Discrete-Event Systems*, Wiley, New York, 1994.

- Ho, Y.C. (Ed.), *Discrete Event Dynamic Systems: Analyzing Complexity and Performance in the Modern World*, IEEE Press, New York, 1991.

- Ho, Y.C., and X. Cao, *Perturbation Analysis of Discrete Event Dynamic Systems*, Kluwer Academic Publishers, Boston, 1991.

- Kumar, R., and V.K. Garg, *Modeling and Control of Logical Discrete Event Systems*, Kluwer Academic Publishers, Boston, 1995.

- Zeigler, B.P., *Theory of Modeling and Simulation*, Wiley, New York, 1976.

Chapter 2

Languages and Automata

2.1 INTRODUCTION

We have seen how discrete-event systems (DES) differ from continuous-variable dynamic systems (CVDS) and why DES are not adequately modeled through differential or difference equations. Our first task, therefore, in studying DES is to develop appropriate models, which both adequately describe the behavior of these systems and provide a framework for analytical techniques to meet the goals of design, control, and performance evaluation.

When considering the state evolution of a DES, our first concern is with the sequence of states visited and the associated events causing these state transitions. To begin with, we will not concern ourselves with the issue of when the system enters a particular state or how long the system remains at that state. We will assume that the behavior of the DES is described in terms of event sequences of the form $e_1 e_2 \cdots e_n$. A sequence of that form specifies the order in which various events occur over time, but it does not provide the time instants associated with the occurrence of these events. This is the untimed or logical level of abstraction discussed in Section 1.3.3 in Chapter 1, where the behavior of the system is modeled by a *language*. Consequently, our first objective in this chapter is to discuss language models of DES and present operations on languages that will be used extensively in this and the next chapters.

As was mentioned in Section 1.3.3, the issue of representing languages using appropriate modeling formalisms is key for performing analysis and control of DES. The second objective of this chapter is to introduce and describe the first of the two untimed modeling formalisms for DES considered in this book to represent languages, *automata*. Automata form the most basic class of DES models. As we shall see in this chapter, they are intuitive, easy to use, amenable to composition operations, and amenable to analysis as well (in the finite-state case). On the other hand, they lack structure and for this reason may lead to very large state spaces when modeling complex systems. Nevertheless, any study of discrete event system and control theory must start with a study of automaton models. The second modeling formalism considered in this book, Petri nets, will be presented in Chapter 4. As we shall see in that chapter, Petri net models have more structure than automaton models, although they do not possess, in general, the same analytical power as automata. Other modeling formalisms for untimed DES have been proposed and considered. These models invariably have more complicated semantics than automata and Petri nets and for this reason, they are beyond the scope of this book; some relevant references are presented at the end of this chapter.

The third objective of this chapter is to present some of the fundamental logical behavior problems we encounter in our study of DES. Using the automaton formalism, we will present some solution techniques for a few of these problems. The following chapter will address the problem of *controlling* the behavior of a DES, in the sense of the feedback control loop presented in Section 1.2.8, in order to ensure that the logical behavior of the closed-loop system is satisfactory.

Finally, we emphasize that an important objective of this book is to study timed and stochastic models of DES; establishing untimed models constitutes the first stepping stone towards this goal.

2.2 THE CONCEPTS OF LANGUAGES AND AUTOMATA

2.2.1 Language Models of Discrete-Event Systems

One of the formal ways to study the logical behavior of DES is based on the theories of languages and automata. The starting point is the fact that any DES has an underlying event set E associated with it. The set E is thought of as the "alphabet" of a language and event sequences are thought of as "words" in that language. In this framework, we can pose questions such as "Can we build a system that speaks a given language?" or "What language does this system speak?"

To motivate our discussion of languages, let us consider a simple example. Suppose there is a machine we usually turn on once or twice a day (like a car, a photocopier, or a desktop computer), and we would like to design a simple system to perform the following basic task: When the machine is turned on, it should first issue a signal to tell us that it is in fact ON, then give us some simple status report (like, in the case of a car, "everything OK", "check oil", or "I need gas"), and conclude with another signal to inform us that "status report done". Each of these signals defines an event, and all of the possible signals the machine can issue define an alphabet (event set). Thus, our system has the makings of a DES driven by these events. This DES is responsible for recognizing events and giving the proper interpretation to any particular sequence received. For instance, the event sequence: "I'm ON", "everything is OK", "status report done", successfully completes our task. On the other hand, the event sequence: "I'm ON", "status report done", without some sort of actual status report in between, should be interpreted as an abnormal condition requiring special attention. We can therefore think of the combinations of signals issued by the machine as words belonging to the particular language spoken by this machine. In this particular example, the language of interest should be one with three-event words only, always beginning with "I'm ON" and ending with "status report done". When the DES we build sees such a word, it knows the task is done. When it sees any other word, it knows something is wrong. We will return to this type of system in Example 2.9 and see how we can build a simple DES to perform a "status check" task.

Language Notation and Definitions

We begin by viewing the event set E of a DES as an alphabet. We will assume that E is finite. A sequence of events taken out of this alphabet forms a "word" or "string" (short for "string of events"). We shall use the term "string" in this book; note that the term "trace" is also used in the literature. A string consisting of no events is called the *empty string* and is denoted by ε. (The symbol ε is not to be confused with the generic symbol e for an element of E.) The length of a string is the number of events contained in it, counting multiple occurrences of the same event. If s is a string, we will denote its length by $|s|$. By convention, the length of the empty string ε is taken to be zero.

Definition. (Language)
A *language* defined over an event set E is a set of finite-length strings formed from events in E.   ◆

As an example, let $E = \{a, b, g\}$ be the set of events. We may then define the language

$$L_1 = \{\varepsilon, a, abb\} \tag{2.1}$$

consisting of three strings only; or the language

$$L_2 = \{\text{all possible strings of length 3 starting with event } a\} \tag{2.2}$$

which contains nine strings; or the language

$$L_3 = \{\text{all possible strings of finite length which start with event } a\} \quad (2.3)$$

which contains an infinite number of strings.

The key operation involved in building strings, and thus languages, from a set of events E is *concatenation*. The string abb in L_1 above is the concatenation of the string ab with the event (or string of length one) b; ab is itself the concatenation of a and b. The concatenation uv of two strings u and v is the new string consisting of the events in u immediately followed by the events in v. The empty string ε is the *identity element* of concatenation: $u\varepsilon = \varepsilon u = u$ for any string u.

Let us denote by E^* the set of *all* finite strings of elements of E, including the empty string ε; the * operation is called the *Kleene-closure*. Observe that the set E^* is countably infinite since it contains strings of arbitrarily long length. For example, if $E = \{a, b, c\}$, then

$$E^* = \{\varepsilon, a, b, c, aa, ab, ac, ba, bb, bc, ca, cb, cc, aaa, \ldots\} \ .$$

A language over an event set E is therefore a *subset* of E^*. In particular, $\emptyset$, E, and E^* are languages.

We conclude this discussion with some terminology about strings. If $tuv = s$ with t, u, $v \in E^*$, then:

- t is called a *prefix* of s,

- u is called a *substring* of s, and

- v is called a *suffix* of s.

Observe that both ε and s are prefixes (substrings, suffixes) of s.

Operations on Languages

The usual set operations, such as union, intersection, difference, and complement with respect to E^*, are applicable to languages since languages are sets. In addition, we will also use the following operations:[1]

- *Concatenation*: Let $L_a, L_b \subseteq E^*$, then

$$L_a L_b := \{s \in E^* : (s = s_a s_b) \text{ and } (s_a \in L_a) \text{ and } (s_b \in L_b)\} \ .$$

  In words, a string is in $L_a L_b$ if it can be written as the concatenation of a string in L_a with a string in L_b.

---

[1] ":=" denotes "equal to by definition."

- *Prefix-closure*: Let $L \subseteq E^*$, then

$$\overline{L} := \{s \in E^* : \exists t \in E^* \ (st \in L)\} \,.$$

In words, the prefix closure of L is the language denoted by $\overline{L}$ and consisting of all the prefixes of all the strings in L. In general, $L \subseteq \overline{L}$.

L is said to be *prefix-closed* if $L = \overline{L}$. Thus language L is prefix-closed if any prefix of any string in L is also an element of L.

- *Kleene-closure*: Let $L \subseteq E^*$, then

$$L^* := \{\varepsilon\} \cup L \cup LL \cup LLL \cup \cdots$$

This is the same operation that we defined above for the set E, except that now it is applied to set L whose elements may be strings of length greater than one. An element of L^* is formed by the concatenation of a finite (but possibly arbitrarily large) number of elements of L; this includes the concatenation of "zero" elements, that is, the empty string ε. Note that the * operation is idempotent: $(L^*)^* = L^*$.

Observe that in expressions involving several operations on languages, prefix-closure and Kleene-closure should be applied first, and concatenation always precedes operations such as union, intersection, and set difference. (This was implicitly assumed in the above definition of L^*.)

Example 2.1 (Operations on languages)
Let $E = \{a, b, g\}$, and consider the two languages $L_1 = \{\varepsilon, a, abb\}$ and $L_4 = \{g\}$. Neither L_1 nor L_4 are prefix-closed, since $ab \notin L_1$ and $\varepsilon \notin L_4$. Then:

$$
\begin{aligned}
L_1 L_4 &= \{g, ag, abbg\} \\
\overline{L_1} &= \{\varepsilon, a, ab, abb\} \\
\overline{L_4} &= \{\varepsilon, g\} \\
L_1 \overline{L_4} &= \{\varepsilon, a, abb, g, ag, abbg\} \\
L_4^* &= \{\varepsilon, g, gg, ggg, \ldots\} \\
L_1^* &= \{\varepsilon, a, abb, aa, aabb, abba, abbabb, \ldots\} \,.
\end{aligned}
$$

We make the following observations for technical accuracy:

(i) $\varepsilon \notin \emptyset$;

(ii) $\{\varepsilon\}$ is a nonempty language containing only the empty string;

(iii) If $L = \emptyset$ then $\overline{L} = \emptyset$, and if $L \neq \emptyset$ then necessarily $\varepsilon \in \overline{L}$;

(iv) $\emptyset^* = \{\varepsilon\}$ and $\{\varepsilon\}^* = \{\varepsilon\}$.

Representation of Languages

A language may be thought of as a formal way of describing the behavior of a DES. It specifies all admissible sequences of events that the DES is capable of "processing" or "generating", while bypassing the need for any additional structure. Taking a closer look at the example languages L_1, L_2, and L_3 in equations 2.1-2.3 above, we can make the following observations. First, L_1 is easy to define by simple enumeration, since it consists of only three strings. Second, L_2 is defined descriptively, only because it is simpler to do so rather than writing down the nine strings it consists of; but we could also have easily enumerated these strings. Finally, in the case of L_3 we are limited to a descriptive definition, since full enumeration is not possible.

The difficulty here is that "simple" representations of languages are not always easy to specify or work with. In other words, we need a set of compact "structures" which define languages and which can be manipulated through well-defined operations so that we can construct, and subsequently manipulate and analyze, arbitrarily complex languages. In CVDS for instance, we can conveniently describe inputs we are interested in applying to a system by means of functional expressions of time such as $\sin(wt)$ or $(a + bt)^2$; the system itself is described by a differential or difference equation. Basic algebra and calculus provide the framework for manipulating such expressions and solving the problem of interest (for example, does the output trajectory meet certain requirements?). The next section will present the modeling formalism of automata as a framework for representing and manipulating languages and solving problems that pertain to the logical behavior of DES.

2.2.2 Automata

An automaton is a device that is capable of representing a language according to well-defined rules. This section focuses on the formal definition of automaton. The connection between languages and automata will be made in the next section. The simplest way to present the notion of automaton is to consider its directed graph representation, or state transition diagram. We use the following example for this purpose.

Example 2.2 (A simple automaton)
Consider the directed graph in Fig. 2.1, where nodes represent *states* and labeled arcs represent *transitions* between these states. This graph provides a complete description of an automaton. The set of nodes is the state set of the automation, $X = \{x, y, z\}$. The set of labels for the transitions is the event set (alphabet) of the automaton, $E = \{a, b, g\}$. The arcs in the graph provide a graphical representation of the *transition*

function of the automaton, which we denote as $f : X \times E \to X$:

$$f(x, a) = x \qquad\qquad f(x, g) = z$$
$$f(y, a) = x \qquad\qquad f(y, b) = y$$
$$f(z, b) = z \qquad\qquad f(z, a) = f(z, g) = y \ .$$

The notation $f(y, a) = x$ means that if the automaton is in state y, then upon the "occurrence" of event a, the automaton will make an instantaneous transition to state x. The cause of the occurrence of event a is irrelevant; the event could be an external input to the system modeled by the automaton, or it could be an event spontaneously "generated" by the system modeled by the automaton.

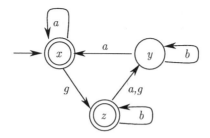

Figure 2.1: State transition diagram for Example 2.2.
The event set is $E = \{a, b, g\}$, and the state set is $X = \{x, y, z\}$. The initial state is x (marked by an arrow), and the set of marked states is $\{x, z\}$ (double circles).

Three observations are worth making regarding Example 2.2. First, an event may occur without changing the state, as in $f(x, a) = x$. Second, two distinct events may occur at a given state causing the exact same transition, as in $f(z, a) = f(z, g) = y$. What is interesting about the latter fact is that we may not be able to distinguish between events a and g by simply observing a transition from state z to state y. Third, the function f is a *partial function* on its domain $X \times E$, that is, there need not be a transition defined for each event in E at each state of X; for instance, $f(x, b)$ and $f(y, g)$ are not defined.

Two more ingredients are necessary to completely define an automaton: an initial state, denoted by x_0, and a subset X_m of X that represents the states of X that are *marked*. The role of the set X_m will become apparent in the remainder of this chapter as well as in Chapter 3. States are marked when it is desired to attach a special meaning to them. Marked states are also referred to as "accepting" states or "final" states. In the figures in this book, the initial state will be identified by an arrow pointing into it and states belonging to X_m will be identified by double circles.

We can now state the formal definition of an automaton. We begin with deterministic automata. Nondeterministic automata will be formally defined in Section 2.2.4.

Definition. (Deterministic automaton)

A *Deterministic Automaton*, denoted by G, is a six-tuple

$$G = (X, E, f, \Gamma, x_0, X_m)$$

where:

X is the set of *states*

E is the finite set of *events* associated with the transitions in G

$f : X \times E \to X$ is the *transition function*: $f(x, e) = y$ means that there is a transition labeled by event e from state x to state y; in general, f is a *partial* function on its domain

$\Gamma : X \to 2^E$ is the *active event function* (or feasible event function); $\Gamma(x)$ is the set of all events e for which $f(x, e)$ is defined and it is called the *active event set* (or feasible event set) of G at x

x_0 is the *initial* state

$X_m \subseteq X$ is the set of *marked states*.   ♦

(Given a set A, the notation 2^A means the power set of A, that is, the set of all subsets of A.)

We make the following remarks about this definition.

■ The words *state machine* and *generator* (which explains the notation G) are also often used to describe the above object.

■ If X is a finite set, we call G a *deterministic finite-state automaton*, often abbreviated as DFA in this book.

■ The automaton is said to be *deterministic* because f is a function over $X \times E$. In contrast, the transition structure of a *nondeterministic* automaton is defined by means of a relation over $X \times E \times X$ or, equivalently, a function from $X \times E$ to 2^X. Note that by default, the word automaton will refer to deterministic automaton in this book.

■ The fact that we allow the transition function f to be partially defined over its domain $X \times E$ is a variation over the "standard" definition of automaton in the computer science literature that is quite important in DES theory.

■ Formally speaking, the inclusion of Γ in the definition of G is superfluous in the sense that Γ is derived from f. For this reason, we will sometimes omit explicitly writing Γ when specifying an automaton when the active event function is not central to the discussion. One of the reasons why we care about the contents of $\Gamma(x)$ for state x is to help distinguish between events e that are feasible at x but cause no state transition, that is, $f(x, e) = x$, and events e' that are not feasible at x, that is, $f(x, e')$ is undefined.

- Proper selection of which states to mark is a modeling issue that depends on the problem of interest. By designating certain states as marked, we may for instance be recording that the system, upon entering these states, has completed some operation or task.

The automaton G operates as follows. It starts in the initial state x_0 and upon the occurrence of an event $e \in \Gamma(x_0) \subseteq E$ it will make a transition to state $f(x_0, e) \in X$. This process then continues based on the transitions for which f is defined.

For the sake of convenience, f is always extended from domain $X \times E$ to domain $X \times E^*$ in the following recursive manner:

$$
\begin{aligned}
f(x, \varepsilon) &:= x \\
f(x, se) &:= f(f(x, s), e) \text{ for } s \in E^* \text{ and } e \in E .
\end{aligned}
$$

Returning to the automaton in Example 2.2, we have for example that

$$
\begin{aligned}
f(y, \varepsilon) &= y \\
f(x, gba) &= f(f(x, gb), a) = f(f(f(x, g), b), a) = f(f(z, b), a) = f(z, a) = y \\
f(x, aagb) &= z \\
f(z, b^n) &= z \text{ for all } n \geq 0
\end{aligned}
$$

where b^n denotes n consecutive occurrences of event b. These results are easily seen by inspection of the state transition diagram in Fig. 2.1.

Remarks.

1. We emphasize that we do not wish to require at this point that the set X be finite. In particular, the concepts and operations introduced in the remainder of Section 2.2 and in Section 2.3 work for infinite-state automata. Of course, explicit representations of infinite-state automata would require infinite memory. Finite-state automata will be discussed in Section 2.4.

2. The automaton model defined in this section is also referred to as a *Generalized Semi-Markov Scheme* (abbreviated as GSMS) in the literature on stochastic processes. The term "semi-Markov" historically comes from the theory of Markov processes. We will cover this theory and use the GSMS terminology in later chapters in this book, in the context of our study of stochastic timed models of DES.

2.2.3   Languages Represented by Automata

The connection between languages and automata is easily made by inspecting the state transition diagram of an automaton. Consider all the directed paths that can be followed in the state transition diagram, starting at the initial state; consider among these all the paths that end in a marked state. This leads to the notions of the languages *generated* and *marked* by an automaton.

Definition. (Languages generated and marked)
The *language generated* by $G = (X, E, f, \Gamma, x_0, X_m)$ is

$$\mathcal{L}(G) := \{s \in E^* : f(x_0, s) \text{ is defined } \} .$$

The *language marked* by G is

$$\mathcal{L}_m(G) := \{s \in \mathcal{L}(G) : f(x_0, s) \in X_m\} . \qquad \blacklozenge$$

(The above definitions assume that we are working with the extended transition function $f : X \times E^* \to X$.)

The language $\mathcal{L}(G)$ represents all the directed paths that can be followed along the state transition diagram, starting at the initial state; the string corresponding to a path is the concatenation of the event labels of the transitions composing the path. Therefore, a string s is in $\mathcal{L}(G)$ if and only if it corresponds to an admissible path in the state transition diagram, equivalently, if and only if f is defined at (x_0, s). $\mathcal{L}(G)$ is prefix-closed by definition, since a path is only possible if all its prefixes are also possible. If f is a total function over its domain, then necessarily $\mathcal{L}(G) = E^*$.

The second language represented by G, $\mathcal{L}_m(G)$, is the subset of $\mathcal{L}(G)$ consisting only of the strings s for which $f(x_0, s) \in X_m$, that is, these strings correspond to paths that end at a marked state in the state transition diagram. Since not all states of X need be marked, the language marked by G, $\mathcal{L}_m(G)$, need not be prefix-closed in general. The language marked is also called the language *recognized* by the automaton, and we often say that the given automaton is a *recognizer* of the given language.

Example 2.3 (Marked language)
Let $E = \{a, b\}$ be an event set. Consider the language

$$L = \{a, aa, ba, aaa, aba, baa, bba, \ldots\}$$

consisting of all strings of a and b always followed by an event a. This language is marked by the finite-state automaton $G = (E, X, f, \Gamma, x_0, X_m)$ where $X = \{0, 1\}$, $x_0 = 0$, $X_m = \{1\}$, and f is defined as follows: $f(0, a) = 1$, $f(0, b) = 0$, $f(1, a) = 1$, $f(1, b) = 0$.

This can be seen as follows. With 0 as the initial state, the only way that the marked state 1 can be reached is if event a occurs at some point. Then, either the state remains forever unaffected or it eventually becomes 0 again if event b takes place. In the latter case, we are back where we started, and the process simply repeats. The state transition diagram of this automaton is shown in Fig. 2.2. We can see from the figure that $\mathcal{L}_m(G) = L$. Note that f is a total function and therefore the language generated by G is $\mathcal{L}(G) = E^*$.

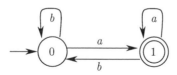

Figure 2.2: Automaton of Example 2.3.
This automaton marks the language $L = \{a, aa, ba, aaa, aba, baa, bba, \dots\}$ consisting of all strings of a and b followed by a, given the event set $E = \{a, b\}$.

Example 2.4 (Generated and marked languages)
If we modify the automaton in Example 2.3 by removing the self-loop due to event b at state 0 in Fig. 2.2, that is, by letting $f(0, b)$ be undefined, then $\mathcal{L}(G)$ now consists of those strings in E^* that start with event a and where there are no consecutive occurrences of event b. Any b in the string is either the last event of the string or it is immediately followed by an a. $\mathcal{L}_m(G)$ is the subset of $\mathcal{L}(G)$ consisting of those strings that end with event a.

Thus, an automaton G is a representation of *two* languages: $\mathcal{L}(G)$ and $\mathcal{L}_m(G)$. In the "standard" definition of automaton in automata theory, the function f is required to be a total function and the notion of language generated is not meaningful since it is always equal to E^*. In DES theory, allowing f to be partial is a consequence of the fact that a system may not be able to produce (or execute) all strings in E^*. Subsequent examples in this and the next chapters will illustrate this point.

Equivalence of Automata
It is clear that there are many ways to construct automata that generate, or mark, a given language. Two automata are said to be *equivalent* if they generate *and* mark the same languages. Formally:

Definition. (Equivalent automata)
Automata G_1 and G_2 are said to be *equivalent* if

$$\mathcal{L}(G_1) = \mathcal{L}(G_2) \qquad \text{and} \qquad \mathcal{L}_m(G_1) = \mathcal{L}_m(G_2) \ . \qquad \blacklozenge$$

Example 2.5 (Equivalent automata)
Let us return to the automaton described in Example 2.4. This automaton is shown in Fig. 2.3. The three automata shown in Fig. 2.3 are equivalent, as they all generate the same language and they all mark the same

language. Observe that third automaton in Fig. 2.3 has an infinite state set.

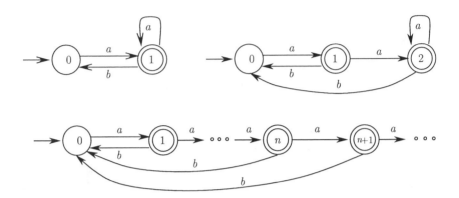

Figure 2.3: Three equivalent automata (Example 2.5).

Blocking

In general, we have from the definitions of G, $\mathcal{L}(G)$, and $\mathcal{L}_m(G)$ that

$$\mathcal{L}_m(G) \subseteq \overline{\mathcal{L}_m(G)} \subseteq \mathcal{L}(G) \ .$$

The first set inclusion is due to the fact that X_m may be a proper subset of X, while the second set inclusion is a consequence of the definition of $\mathcal{L}_m(G)$ and the fact that $\mathcal{L}(G)$ is prefix-closed by definition. It is worth examining this second set inclusion in more detail.

An automaton G could reach a state x where $\Gamma(x) = \emptyset$ but $x \notin X_m$. This is called a *deadlock* because no further event can be executed. Given our interpretation of marking, we say that the system "blocks" because it enters a deadlock state without having terminated the task at hand. If deadlock happens, then necessarily $\overline{\mathcal{L}_m(G)}$ will be a proper subset of $\mathcal{L}(G)$, since any string in $\mathcal{L}(G)$ that ends at state x cannot be a prefix of a string in $\mathcal{L}_m(G)$.

Another issue to consider is when there is a set of unmarked states in G that forms a strongly connected component (i.e., these states are reachable from one another), but with *no transition going out of the set*. If the system enters this set of states, then we get what is called a *livelock*. While the system is "live" in the sense that it can always execute an event, it can never complete the task started since no state in the set is marked and the system cannot leave this set of states. If livelock is possible, then again $\overline{\mathcal{L}_m(G)}$ will be a proper subset of $\mathcal{L}(G)$. Any string in $\mathcal{L}(G)$ that reaches the "absorbing" set of unmarked states cannot be a prefix of a string in $\mathcal{L}_m(G)$, since we assume that there is no way out of this set. Again, the system is "blocked" in the livelock.

The importance of deadlock and livelock in DES leads us to formulate the following definition.

Definition. (Blocking)
Automaton G is said to be *blocking* if

$$\overline{\mathcal{L}_m(G)} \subset \mathcal{L}(G)$$

where the set inclusion is proper[2], and *nonblocking* when

$$\overline{\mathcal{L}_m(G)} = \mathcal{L}(G) \ . \qquad\qquad\qquad \blacklozenge$$

Thus, if an automaton is blocking, this means that deadlock and/or livelock can happen.

The notion of marked states and the definitions that we have given for language generated, language marked, and blocking, provide an approach for considering deadlock and livelock that is useful in a wide variety of applications.

Example 2.6 (Blocking)
Let us examine the automaton G depicted in Fig. 2.4. Clearly, state 5 is a deadlock state. Moreover, states 3 and 4, with their associated a, b, and g transitions, form an absorbing strongly connected component; since neither 3 nor 4 is marked, any string that reaches state 3 will lead to a livelock. String $ag \in \mathcal{L}(G)$ but $ag \notin \overline{\mathcal{L}_m(G)}$; the same is true for any string in $\mathcal{L}(G)$ that starts with aa. Thus G is blocking since $\overline{\mathcal{L}_m(G)}$ is a proper subset of $\mathcal{L}(G)$.

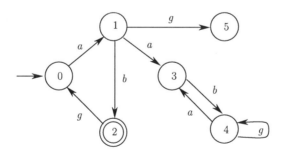

Figure 2.4: Blocking automaton of Example 2.6.
State 5 is a deadlock state, and states 3 and 4 are involved in a livelock.

We return to examples of DES presented in Chapter 1 in order to further illustrate blocking and livelock.

---

[2]We shall use the notation $\subset$ for "strictly contained in" and $\subseteq$ for "contained in or equal to."

Example 2.7 (Deadlock in database systems)

Recall our description of the concurrent execution of database transactions in Section 1.3.4. Let us assume that we want to build an automaton H_a whose language would exactly be the set of *admissible schedules* corresponding to the concurrent execution of transactions

$$r_1(a)r_1(b) \quad \text{and} \quad r_2(a)w_2(a)r_2(b)w_2(b) \; .$$

This H_a will have a single marked state, reached by all admissible schedules that contain all the events of transactions 1 and 2; that is, this marked state models the successful completion of the execution of both transactions, the desired goal in this problem. We will not completely build H_a here; this is done in Example 3.3 in Chapter 3. However, we have argued in Section 1.3.4 that the schedule

$$S_z = r_1(a)r_2(a)w_2(a)r_2(b)w_2(b)$$

is admissible but its only possible continuation, with event $r_1(b)$, leads to an inadmissible schedule. This means that the state reached in H_a after string S_z has to be a deadlock state, and consequently H_a will be a blocking automaton.

Example 2.8 (Livelock in telephony)

In Section 1.3.4, we discussed the modeling of telephone networks for the purpose of studying possible conflicts between call processing features. Let us suppose that we are building an automaton model of the behavior of three users, each subscribing to "call forwarding", and where: (i) user 1 forwards all his calls to user 2, (ii) user 2 forwards all his calls to user 3, and (iii) user 3 forwards all his calls to user 1. The marked state of this automaton could be the initial state, meaning that all call requests can be properly completed and the system eventually returns to the initial state. The automaton model of this system, under the above instructions for forwarding, would be blocking because it would contain a livelock. This livelock would occur after any request for connection by a user, since the resulting behavior would be an endless sequence of forwarding events with no actual connection of the call ever happening. (In practice, the calling party would hang up the phone out of frustration, but we assume here that such an event is not included in the model!)

Other Examples

We conclude this section with two more examples; the first one revisits the machine status check example discussed at the beginning of Section 2.2.1 while the second one presents an automaton model of the familiar queueing system.

Example 2.9 (Machine status check)

Let $E = \{a_1, a_2, b\}$ be the set of events. A task is defined as a three-event sequence which begins with event b, followed by event a_1 or a_2, and then event b, followed by any arbitrary event sequence. Thus, we would like to design a device that reads any string formed by this event set, but only recognizes, that is marks, strings of length 3 or more that satisfy the above definition of task. Each such string must begin with b and include a_1 or a_2 in the second position, followed by b in the third position. What follows after the third position is of no concern in this example.

A finite-state automaton that marks this language (and can therefore implement the desired task) is shown in Fig. 2.5. The state set $X = \{0, 1, 2, 3, 4, 5\}$ consists of arbitrarily labeled states with $x_0 = 0$ and $X_m = \{4\}$. Note that any string with less than 3 events always ends up in state 0, 1, 2, 3, or 5, and is therefore not marked. The only strings with 3 or more events that are marked are those that start with b, followed by either a_1 or a_2, and then reach state 4 through event b.

There is a simple interpretation for this automaton and the language it marks, which corresponds to the machine status check example introduced at the beginning of Section 2.2.1. When a machine is turned on, we consider it to be "Initializing" (state 0), and expect it to issue event b indicating that it is now on. This leads to state 1, whose meaning is simply "I am ON". At this point, the machine is supposed to perform a diagnostic test resulting in one of two events, a_1 or a_2. Event a_1 indicates "My status is OK", while event a_2 indicates "I have a problem". Finally, the machine must inform us that the initial checkup procedure is complete by issuing another event b, leading to state 4 whose meaning is "Status report done". Any sequence that is not of the above form leads to state 5, which should be interpreted as an "Error" state. As for any event occurring after the "Status report done" state is entered, it is simply ignored for the purposes of this task, whose marked (and final in this case) state has already been reached.

We observe that in this example, the focus is on the language marked by the automaton. Since the automaton is designed to accept input events issued by the system, all unexpected inputs, that is, those not in the set $\overline{\mathcal{L}_m(G)}$, send the automaton to state 5. This means that f is a total function; this ensures that the automaton is able to process any input, expected or not. In fact, state 5 is a state where livelock occurs, since it is impossible to reach marked state 4 from state 5. Therefore, our automaton model is a blocking automaton. Finally, we note that states 2 and 3 could be merged without affecting the languages generated and marked by the automaton. The issue of minimizing the number of states without affecting the language properties of an automaton will be discussed in Section 2.4.3.

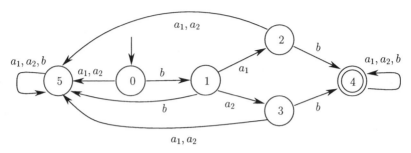

Figure 2.5: Automaton of Example 2.9.
This automaton recognizes event sequences of length 3 or more, which begin with b and contain a_1 or a_2 in the second position, followed by b in their third position. It can be used as a model of a "status check" system when a machine is turned on, if its states are given the following interpretation: 0 = "Initializing", 1= "I am ON", 2 = "My status is OK", 3 = "I have a problem", 4 = "Status report done", and 5 = "Error".

Example 2.10 (Queueing system)

Queueing systems form an important class of DES. A simple queueing system was introduced in the previous chapter and is shown again in Fig. 2.6. Customers arrive and request access to a server. If the server is already busy, they wait in the queue. When a customer completes service, he departs from the system and the next customer in queue (if any) immediately receives service. Thus, the events driving this system are:

  a: customer arrival

  d: customer departure.

We can define an infinite-state automaton model G for this system as follows:

$$
\begin{aligned}
E &= \{a, d\} \\
X &= \{0, 1, 2, \ldots\} \\
\Gamma(x) &= \{a, d\} \text{ for all } x > 0,\ \Gamma(0) = \{a\} \\
f(x, a) &= x + 1 \text{ for all } x \geq 0 \qquad\qquad (2.4) \\
f(x, d) &= x - 1 \text{ if } x > 0 . \qquad\qquad\qquad (2.5)
\end{aligned}
$$

Here, the state variable x represents the number of customers in the system (in service and in the queue, if any). The initial state x_0 would be chosen to be the initial number of customers of the system. The feasible event set $\Gamma(0)$ is limited to arrival (a) events, since no departures are possible when the queueing system is empty. Thus, $f(x, d)$ is not defined for $x = 0$. A state transition diagram for this system is shown in Fig. 2.6.

Note that the state space in this model is infinite, but countable; also, we have left X_m unspecified.

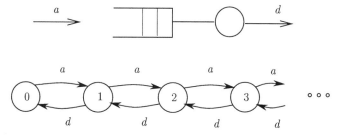

Figure 2.6: Simple queueing system and state transition diagram.
No d event is included at state 0, since departures are not feasible when $x = 0$.

Next, let us consider the same simple queueing system, except that now we focus on the state of the server rather than the whole queueing system. The server can be either Idle (denoted by I) or Busy (denoted by B). In addition, we will assume that the server occasionally breaks down. We will refer to this separate state as Down, and denote it by D. When the server breaks down, the customer in service is assumed to be lost; therefore, upon repair, the server is idle.

The events defining the input to this system are assumed to be:

α: service starts

β: service completes

λ: server breaks down

μ: server repaired.

The automaton model of this server is given by:

$$E = \{\alpha, \beta, \lambda, \mu\} \qquad X = \{I, B, D\}$$
$$\Gamma(I) = \{\alpha\} \qquad f(I, \alpha) = B$$
$$\Gamma(B) = \{\beta, \lambda\} \qquad f(B, \beta) = I \qquad f(B, \lambda) = D$$
$$\Gamma(D) = \{\mu\} \qquad f(D, \mu) = I.$$

A state transition diagram for this system is shown in Fig. 2.7. An observation worth making in this example is the following. Intuitively, assuming we do not want our server to remain unnecessarily idle, the event "start serving" should always occur immediately upon entering the I state. Of course, this is not possible when the queue is empty, but this model has no knowledge of the queue length. Thus, we limit ourselves to observations of α events by treating them as purely exogenous.

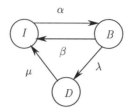

Figure 2.7: State transition diagram for a server with breakdowns.

2.2.4   Nondeterministic Automata

In our definition of automaton, state transitions describe how an event e causes a transition from some state x to a unique new state y. Suppose, however, that an event e at state x may cause transitions to more than one new states. The reason why we may want to allow for this possibility could simply be our own ignorance. Sometimes, we cannot say with certainty what the effect of an event might be. In this case, $f(x, e)$ should no longer represent a specific new state $x \in X$, but rather a *set* of possible new states. In addition, we may also want to allow the label ε in the state transition diagram of an automaton, that is, allow transitions between distinct states to have the empty string as label. Our motivation for including so-called "ε-transitions" is again our own ignorance. These transitions may represent events that cause a change in the internal state of a DES but are not "observable" by an outside observer – imagine that there is no sensor that records this state transition. Thus the outside observer cannot attach an event label to such a transition but it recognizes that the transition may occur by using the ε label.

These two changes, namely, that the event set is no longer E but $E \cup \{\varepsilon\}$ and that the transition function has different domain and co-domain, namely domain $X \times E \cup \{\varepsilon\}$ and co-domain 2^X, lead us to the notion of a *nondeterministic automaton*.

Definition. (Nondeterministic automaton)
A *Nondeterministic Automaton*, denoted by G_{nd}, is a six-tuple

$$G_{nd} = (X, E \cup \{\varepsilon\}, f_{nd}, \Gamma, x_0, X_m)$$

where these objects have the same interpretation as in the definition of deterministic automaton, with the two differences that:

1. f_{nd} is a function $f_{nd} : X \times E \cup \{\varepsilon\} \rightarrow 2^X$, that is, $f_{nd}(x, e) \subseteq X$ whenever it is defined.

2. The *initial* state may itself be a set of states, that is $x_0 \subseteq X$.   ◆

Example 2.11 (A simple nondeterministic automaton)
Consider the finite-state automaton of Fig. 2.8. Note that when event a occurs at state 0, the resulting transition is either to state 1 or back

to state 0. The state transition mappings are: $f_{nd}(0, a) = \{0, 1\}$ and $f_{nd}(1, b) = \{0\}$ where the values of f_{nd} are expressed as subsets of the state set X. The transitions $f_{nd}(0, b)$ and $f_{nd}(1, a)$ are undefined. This automaton marks any string of a events, as well as any string containing ab if b is immediately followed by a or ends the string.

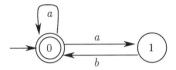

Figure 2.8: Nondeterministic automaton of Example 2.11.

We extend f_{nd} to apply to a string u (including the case $u = \varepsilon$), instead of only a single event e, just as we did for f in deterministic automata. In particular,

$$f_{nd}(x, ue) := \{z : z \in f_{nd}(y, e) \text{ for some state } y \in f_{nd}(x, u)\} .$$

In words, we first identify all states y that are reachable from x through string u, then we apply e to reach a state z in the set $f_{nd}(x, ue)$. As an example, $f_{nd}(0, ab) = \{0\}$ in the automaton of Fig. 2.8, since: (i) $f_{nd}(0, a) = \{0, 1\}$ and (ii) $f_{nd}(1, b) = \{0\}$ and $f_{nd}(0, b)$ is undefined.

Nondeterministic automata generate and mark languages similarly to automata. We define:

$$\begin{aligned} \mathcal{L}(G_{nd}) &= \{s \in E^* : \exists x \in x_0 \ (f_{nd}(x, s) \text{ is defined })\} \\ \mathcal{L}_m(G_{nd}) &= \{s \in \mathcal{L}(G_{nd}) : \exists x \in x_0 \ (f_{nd}(x, s) \cap X_m \neq \emptyset)\} . \end{aligned}$$

These definitions mean that a string is in the language generated by the nondeterministic automaton if there exists a path in the state transition diagram that is labeled by that string. If it is possible to follow a path that is labeled consistently with a given string and ends in a marked state, then that string is in the language marked by the automaton. For instance, the string aa is in the language marked by the automaton in Fig. 2.8 since we can do two self-loops and stay at state 0, which is marked; it does not matter that the same string can also take us to an unmarked state (1 in this case).

Example 2.12 (An equivalent deterministic automaton)
It is easily verified that the deterministic automaton (let us call it G) depicted in Fig. 2.9 is equivalent to the nondeterministic one in Fig. 2.8 (let us call it G_{nd}). Both automata generate and mark the same languages. In fact, we can think of state A of G as corresponding to state 0 of G_{nd} and of state B of G as corresponding to the set of states $\{0, 1\}$ of G_{nd}.

By "corresponds", we mean here that f of G and f_{nd} of G_{nd} match in the sense that:
(i) $f(A, a) = B$ and $f_{nd}(0, a) = \{0, 1\}$;
(ii) $f(A, b)$ and $f_{nd}(0, b)$ are undefined;
(iii) $f(B, a) = B$ and $f_{nd}(0, a) = \{0, 1\}$ with $f_{nd}(1, a)$ undefined; and
(iv) $f(B, b) = A$ and $f_{nd}(1, b) = \{0\}$ with $f_{nd}(0, b)$ undefined.

We will discuss the correspondence between nondeterministic and deterministic automata more formally in Section 2.3.3.

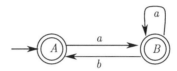

Figure 2.9: Deterministic automaton of Example 2.12.

2.3 OPERATIONS ON AUTOMATA

In order to analyze DES modeled by automata, we need to have a set of operations that allow us to combine, or compose, two or more automata, as well as operations on a single automaton in order to modify appropriately its state transition diagram. This is the first focus of this section. Then we will see how any nondeterministic automaton can be transformed into an equivalent deterministic one, that is, both will generate and mark the same languages. Finally, we will conclude with a brief discussion on automata with inputs and outputs. As in the preceding section, we will allow the state set of an automaton to be infinite.

2.3.1 Unary Operations

Accessible Part

From the definitions of $\mathcal{L}(G)$ and $\mathcal{L}_m(G)$, we see that we can delete from G all the states that are not *accessible* or *reachable* from x_0 by some string in $\mathcal{L}(G)$, without affecting the languages generated and marked by G. When we "delete" a state, we also delete all the transitions that are *attached* to that state. We will denote this operation by $Ac(G)$, where Ac stands for taking the "accessible" part. Formally,

$$
\begin{aligned}
Ac(G) &:= (X_{ac}, E, f_{ac}, x_0, X_{ac,m}) \quad \text{where} \\
X_{ac} &= \{x \in X : \exists s \in E^* \ (f(x_0, s) = x)\} \\
X_{ac,m} &= X_m \cap X_{ac} \\
f_{ac} &= f|_{X_{ac} \times E \to X_{ac}}.
\end{aligned}
$$

The notation $f|_{X_{ac} \times E \to X_{ac}}$ means that we are restricting f to the smaller domain of the accessible states X_{ac}.

Clearly, the Ac operation has no effect on $\mathcal{L}(G)$ and $\mathcal{L}_m(G)$. Thus, from now on, we will always assume, without loss of generality, that an automaton is *accessible*, that is, $G = Ac(G)$.

Coaccessible Part

A state x of G is said to be *coaccessible to X_m*, or simply coaccessible, if there is a string in $\mathcal{L}_m(G)$ that goes through x; this means that there is a path in the state transition diagram of G from state x to a marked state. We denote the operation of deleting all the states of G that are *not* coaccessible by $CoAc(G)$, where $CoAc$ stands for taking the "coaccessible" part.

Taking the coaccessible part of an automaton means building

$$
\begin{aligned}
CoAc(G) &:= (X_{coac}, E, f_{coac}, x_{0,coac}, X_m) \quad \text{where} \\
X_{coac} &= \{x \in X : \exists s \in E^* \ (f(x, s) \in X_m)\} \\
x_{0,coac} &= \begin{cases} x_0 & \text{if } x_0 \in X_{coac} \\ \text{undefined} & \text{otherwise} \end{cases} \\
f_{coac} &= f|_{X_{coac} \times E \to X_{coac}}.
\end{aligned}
$$

The $CoAc$ operation may shrink $\mathcal{L}(G)$, since we may be deleting states that are accessible from x_0; however, the $CoAc$ operation does not affect $\mathcal{L}_m(G)$, since any deleted state is not on any path from x_0 to X_m. If $G = CoAc(G)$, then G is said to be *coaccessible*; in this case, $\mathcal{L}(G) = \overline{\mathcal{L}_m(G)}$.

Coaccessibility is closely related to the concept of *blocking*; recall that an automaton is said to be blocking if $\mathcal{L}(G) \neq \overline{\mathcal{L}_m(G)}$. Therefore, blocking necessarily means that $\overline{\mathcal{L}_m(G)}$ is a proper subset of $\mathcal{L}(G)$ and consequently there are accessible states that are not coaccessible.

Note that if the $CoAc$ operation results in $X_{coac} = \emptyset$ (this would happen if $X_m = \emptyset$ for instance), then we obtain the *empty automaton*. The term "empty automaton" refers to an automaton whose state space is empty; an empty automaton necessarily generates and marks the empty set.

Trim Operation

An automaton that is both accessible and coaccessible is said to be *trim*. We define the *Trim* operation to be

$$Trim(G) := CoAc[Ac(G)] = Ac[CoAc(G)]$$

where the commutativity of Ac and $CoAc$ is easily verified.

Complement

Let us suppose that we have a trim automaton $G = (X, E, f, \Gamma, x_0, X_m)$ that marks the language $L \subseteq E^*$. Thus G generates the language $\overline{L}$. We can build another automaton, denoted by G^{comp}, that will mark the language $E^* \setminus L$. G^{comp} is built in two steps as follows.

The first step is to "complete" the transition function f of G and make it a total function; let us denote the new transition function by f_{tot}. This is done by adding a new state x_d to X, often called the "dead" or "dump" state. All undefined $f(x, e)$ in G are then assigned to x_d. Formally,

$$f_{tot}(x, e) = \begin{cases} f(x, e) & \text{if } e \in \Gamma(x) \\ x_d & \text{otherwise.} \end{cases}$$

Moreover, we set $f_{tot}(x_d, e) = x_d$ for all $e \in E$. Note also that the new state x_d is not marked. The new automaton

$$G_{tot} = (X \cup \{x_d\}, E, f_{tot}, x_0, X_m)$$

is such that $\mathcal{L}(G_{tot}) = E^*$ and $\mathcal{L}_m(G_{tot}) = L$.

The second step is to change the marking status of all states in G_{tot} by marking all unmarked states (including x_d) and unmarking all marked states. That is, we define

$$G^{comp} = (X \cup \{x_d\}, E, f_{tot}, x_0, (X \cup \{x_d\}) \setminus X_m) .$$

Clearly, $\mathcal{L}(G^{comp}) = E^*$ and $\mathcal{L}_m(G^{comp}) = E^* \setminus \mathcal{L}_m(G)$, as desired.

Example 2.13 (Unary operations)

Consider the automaton G depicted in Fig. 2.10. It is a slight variation of the automaton of Fig. 2.4. The new state 6 that has been added is clearly not accessible from state 0. To get $Ac(G)$, it suffices to delete state 6 and the two transitions attached to it; the result is as in Fig. 2.4.

In order to get $CoAc(G)$, we need to identify the states of G that are not coaccessible to the marked state 2. These are states 3, 4, and 5. Deleting these states and the transitions attached to them, we get $CoAc(G)$ depicted in Fig. 2.11 (a). Note that state 6 is not deleted since it can reach state 2. $Trim(G)$ is shown in Fig. 2.11 (b). We can see that the order in which the operations Ac and $CoAc$ are taken does not affect the final result.

Finally, let us take the complement of $Trim(G)$. First, we add a new state, labeled state d, and complete the transition function using this state. Note that $E = \{a, b, g\}$. Next, we reverse the marking of the states. The resulting automaton is shown in Fig. 2.11 (c).

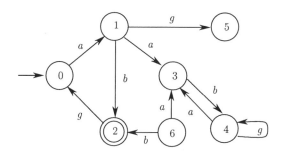

Figure 2.10: Automaton G of Example 2.13.

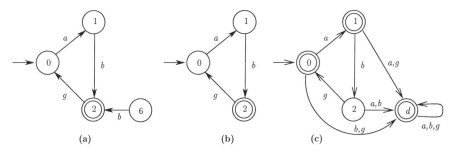

Figure 2.11: (a) $CoAc(G)$, (b) $Trim(G)$, and (c) Complement of $Trim(G)$, for G of Fig. 2.10 (Example 2.13).

2.3.2   Composition Operations

We define two operations on automata: the product, denoted by $\times$, and the parallel composition, denoted by $\|$. Parallel composition is often called synchronous composition and product is sometimes called completely synchronous composition. These operations model two forms of joint behavior of a set of automata that operate concurrently. For simplicity, we present these operations for two automata. We can think of $G_1 \times G_2$ and $G_1 \| G_2$ as two types of systems resulting from the interconnection of system components G_1 and G_2, as depicted in Fig. 2.12.

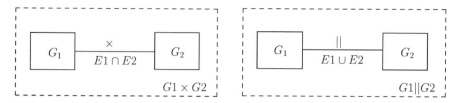

Figure 2.12: This figure illustrates the interconnection of two automata. *The × operation involves only events in $E_1 \cap E_2$ while the || operation involves all events in $E_1 \cup E_2$.*

Consider the two automata

$$G_1 = (X_1, E_1, f_1, \Gamma_1, x_{01}, X_{m1}) \text{ and } G_2 = (X_2, E_2, f_2, \Gamma_2, x_{02}, X_{m2}) .$$

As mentioned earlier, G_1 and G_2 are assumed to be accessible; however, they need not be coaccessible. No assumptions are made at this point about the two event sets E_1 and E_2.

Product

The *product* of G_1 and G_2 is the automaton

$$G_1 \times G_2 := Ac(X_1 \times X_2, E_1 \cap E_2, f, \Gamma_{1 \times 2}, (x_{01}, x_{02}), X_{m1} \times X_{m2})$$

where

$$f((x_1, x_2), e) := \begin{cases} (f_1(x_1, e), f_2(x_2, e)) & \text{if } e \in \Gamma_1(x_1) \cap \Gamma_2(x_2) \\ \text{undefined} & \text{otherwise} \end{cases}$$

and thus $\Gamma_{1 \times 2}(x_1, x_2) = \Gamma_1(x_1) \cap \Gamma_2(x_2)$. Observe that we take the accessible part on the right-hand side of the definition of $G_1 \times G_2$ since, as was mentioned earlier, we only care about the accessible part of an automaton.

In the product, the transitions of the two automata must always be synchronized on a common event, that is, an event in $E_1 \cap E_2$. $G_1 \times G_2$ thus represents the "lock-step" interconnection of G_1 and G_2, where an event occurs if and only if it occurs in both automata. The states of $G_1 \times G_2$ are denoted by pairs, where the first component is the (current) state of G_1 and the second component is the (current) state of G_2. It is easily verified that

$$\begin{aligned} \mathcal{L}(G_1 \times G_2) &= \mathcal{L}(G_1) \cap \mathcal{L}(G_2) \\ \mathcal{L}_m(G_1 \times G_2) &= \mathcal{L}_m(G_1) \cap \mathcal{L}_m(G_2) . \end{aligned}$$

This shows that the intersection of two languages can be "implemented" by doing the product of their automaton representations – an important result. If $E_1 \cap E_2 = \emptyset$, then $\mathcal{L}(G_1 \times G_2) = \{\varepsilon\}$; $\mathcal{L}_m(G_1 \times G_2)$ will be either $\emptyset$ or $\{\varepsilon\}$, depending on the marking status of the initial state (x_{01}, x_{02}).

Example 2.14 (Product of two automata)

Two automata are depicted in Fig. 2.13. The first one is the result of the product of the automata in Figs. 2.1 and 2.2. The set of common events is $\{a, b\}$. Note that we have denoted states of the product automaton by pairs, with the first component a state of the automaton of Fig. 2.1 and the second component a state of the automaton of Fig. 2.2. At the initial state $(x, 0)$, the only possible common event is a, which takes x to x and 0 to 1; thus the new state is $(x, 1)$. Comparing the active event sets of x and 1 in their respective automata, we see that the only possible common event is a again, which takes x to x and 1 to 1, that is, $(x, 1)$ again. After this, we are done constructing the product automaton. Only a transitions are possible since the automaton of Fig. 2.1 never reaches a state where event b is feasible. Observe that state $(x, 1)$ is marked since both x and 1 are marked in their respective automata.

The second automaton in Fig. 2.13 is the result of the product of the automata in Figs. 2.2 and 2.11 (b). Again, the set of common events is $\{a, b\}$. The only common behavior is string ab, which takes the product automaton to state $(0, 2)$, at which point this automaton deadlocks. Since there are no marked states, the language marked by the product automaton is empty, as expected if we compare the marked languages of the automata in Figs. 2.2 and 2.11 (b).

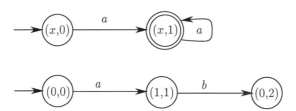

Figure 2.13: Product automata of Example 2.14.
The first automaton is the result of the product of the automata in Figs. 2.1 and 2.2. The second automaton is the result of the product of the automata in Figs. 2.2 and 2.11 (b).

Parallel Composition

The *parallel composition* of G_1 and G_2 is the automaton

$$G_1 \parallel G_2 := Ac(X_1 \times X_2, E_1 \cup E_2, f, \Gamma_{1\parallel 2}, (x_{01}, x_{02}), X_{m1} \times X_{m2})$$

where

$$f((x_1, x_2), e) := \begin{cases} (f_1(x_1, e), f_2(x_2, e)) & \text{if } e \in \Gamma_1(x_1) \cap \Gamma_2(x_2) \\ (f_1(x_1, e), x_2) & \text{if } e \in \Gamma_1(x_1) \setminus E_2 \\ (x_1, f_2(x_2, e)) & \text{if } e \in \Gamma_2(x_2) \setminus E_1 \\ \text{undefined} & \text{otherwise} \end{cases}$$

and thus $\Gamma_{1||2}(x_1, x_2) = [\Gamma_1(x_1) \cap \Gamma_2(x_2)] \cup [\Gamma_1(x_1) \setminus E_2] \cup [\Gamma_2(x_2) \setminus E_1]$.

In the parallel composition, a common event, that is, an event in $E_1 \cap E_2$, can only be executed if the two automata both execute it simultaneously. Thus, the two automata are "synchronized" on the common events. The other "private" events, that is, those in $(E_2 \setminus E_1) \cup (E_1 \setminus E_2)$, are not subject to such a constraint and can be executed whenever possible. In this kind of interconnection, a component can execute its private events without the participation of the other component; however, a common event can only happen if both components can execute it.

If $E_1 = E_2$, then the parallel composition reduces to the product, since all transitions are forced to be synchronized. If $E_1 \cap E_2 = \emptyset$, then there are no synchronized transitions and $G_1 || G_2$ is the *concurrent* behavior of G_1 and G_2. This is often termed the *shuffle* of G_1 and G_2.

It is not hard to verify the following commutativity and associativity results about $||$:

1. $G_1 || G_2 = G_2 || G_1$;

2. $G_1 || (G_2 || G_3) = (G_1 || G_2) || G_3$.

These results hold up to an appropriate renaming of the states, consistent with the commutativity and associativity operations.

In order to precisely characterize the languages generated and marked by $G_1 || G_2$ in terms of those of G_1 and G_2, let us define the *projections*

$$P_i : (E_1 \cup E_2)^* \rightarrow E_i^* \quad \text{for } i = 1, 2$$

as follows:

$$\begin{aligned} P_i(\varepsilon) &:= \varepsilon \\ P_i(e) &:= \begin{cases} e & \text{if } e \in E_i \\ \varepsilon & \text{if } e \notin E_i \end{cases} \\ P_i(se) &:= P_i(s)P_i(e) \quad \text{for } s \in (E_1 \cup E_2)^*, \ e \in (E_1 \cup E_2) . \end{aligned}$$

Given two event sets where one is a subset of the other, namely $E_1 \cup E_2$ and E_i in this case, this kind of projection *erases* events in a string formed from the larger event set $(E_1 \cup E_2)$ that do not belong to the smaller event set (either E_1 or E_2). This type of projection is called a *natural projection*.

We will also be working with the corresponding inverse maps

$$P_i^{-1} : E_i^* \to 2^{(E_1 \cup E_2)^*}$$

defined as follows:

$$P_i^{-1}(t) := \{s \in (E_1 \cup E_2)^* : P_i(s) = t\} \ .$$

Given a string in the smaller event set (E_i), the inverse projection returns the set of all strings in the larger event set $(E_1 \cup E_2)$ that project, with P_i, to the given string.

The projections P_i and their inverses P_i^{-1} are extended to languages by simply applying them to all the strings in the language. For $L \subseteq (E_1 \cup E_2)^*$,

$$P_i(L) := \{t \in E_i^* : \exists s \in L \ (P_i(s) = t)\}$$

and for $L_i \subseteq E_i^*$,

$$P_i^{-1}(L_i) := \{s \in (E_1 \cup E_2)^* : \exists t \in L_i \ (P_i(s) = t)\} \ .$$

Note that $P_i[P_i^{-1}(L)] = L$ but in general $L \subseteq P_i^{-1}[P_i(L)]$.

Example 2.15 (Projection)
Consider $E_1 = \{a, b\}$, $E_2 = \{b, c\}$, and

$$L = \{c, ccb, abc, cacb, cabcbbca\} \ .$$

We have that:

$$
\begin{aligned}
P_1(L) &= \{\varepsilon, b, ab, abbba\} \\
P_2(L) &= \{c, ccb, bc, cbcbb\} \\
P_1^{-1}(\{\varepsilon\}) &= \{c\}^* \\
P_1^{-1}(\{b\}) &= \{c\}^* \{b\} \{c\}^* \\
P_1^{-1}(\{ab\}) &= \{c\}^* \{a\} \{c\}^* \{b\} \{c\}^* \ .
\end{aligned}
$$

We can see that

$$P_1^{-1}[P_1(\{abc\})] = P_1^{-1}[\{ab\}] \supset \{abc\}$$

as we claimed above could happen.

Examining the expression of the inverse projections in the above example leads us to a simple implementation of this operation on automaton representations of languages. If $K_s = \mathcal{L}_m(G) \subseteq E_s^* \subset E_l^*$ and P is the natural projection from the larger event set E_l to the smaller event set E_s, then an automaton that marks $P^{-1}(K_s)$ can be obtained by adding self-loops for all the events in $E_l \setminus E_s$ at all the states of G.

Using the above-defined projections, we can characterize the languages resulting from a parallel composition:

1. $\mathcal{L}(G_1\|G_2) = P_1^{-1}[\mathcal{L}(G_1)] \cap P_2^{-1}[\mathcal{L}(G_2)]$

2. $\mathcal{L}_m(G_1\|G_2) = P_1^{-1}[\mathcal{L}_m(G_1)] \cap P_2^{-1}[\mathcal{L}_m(G_2)]$.

We will not present a formal proof of these results. However, they are fairly intuitive if we relate them to the implementation of inverse projection with self-loops and to the product of automata. The inverse projections on the right-hand side can be implemented by adding self-loops at all states of G_1 and G_2; these self-loops are for events in $E_2 \setminus E_1$ for G_1 and in $E_1 \setminus E_2$ for G_2. Then, we can see that the right-hand side corresponds to doing the product of the automata with self-loops. This is correct, as the self-loops guarantee that the private events of each automaton (before the addition of self-loops) will be occurring in the product without any constraint from the other automaton, since that other automaton will have a self-loop for this event at every state. Overall, the final result does indeed correspond to the definition of parallel composition.

A natural consequence of the above characterization of the behavior of automata under parallel composition using projections is to *define* a parallel composition for *languages*. With $L_i \subseteq E_i^*$ and P_i defined as above, the obvious definition is

$$L_1\|L_2 := P_1^{-1}(L_1) \cap P_2^{-1}(L_2) \ .$$

We present two examples to illustrate parallel composition.

Example 2.16 (Parallel composition)

The automaton in Fig. 2.14 is the result of the parallel composition of the automata in Figs. 2.1 and 2.2, referred to as G_1 and G_2, respectively, in this example. (Recall that the product of these automata is shown in Fig. 2.13.) The set of common events is $\{a, b\}$, and G_1 is the only one to have private events, event g in this case. As in the case of the product, the states of $G_1\|G_2$ are denoted by pairs. At the initial state $(x, 0)$, the only possible common event is a, which takes $(x, 0)$ to $(x, 1)$, a marked state since x is marked in G_1 and 1 is marked in G_2. In contrast to $G_1 \times G_2$, another transition is possible at $(x, 0)$ in $G_1\|G_2$. G_1 can execute event g without the participation of G_2 and take $G_1\|G_2$ to the new state $(z, 0)$; G_1 is in state z after event g and G_2 stays in state 0.

We repeat this process in a breadth-first manner and find all possible transitions at $(x, 1)$ and $(z, 0)$, then at all newly generated states, and so forth. We can see that in this example, all six states in $X_1 \times X_2$ are reachable from $(x, 0)$.

Example 2.16 shows that in general, we have the set inclusion

$$P_i[\mathcal{L}(G_1\|G_2)] \subseteq \mathcal{L}(G_i), \quad \text{for } i = 1, \ 2.$$

The fact that we have inclusion and not equality is because of the constraints imposed by the synchronization of the common events in the definition of parallel composition.

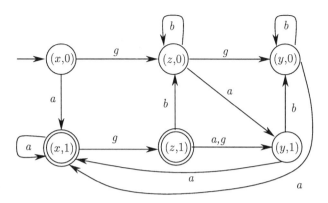

Figure 2.14: Automaton for Example 2.16.
This automaton is the result of the parallel composition of the automata in Figs. 2.1 and 2.2.

Example 2.17 (Two users of two common resources)

A common situation in DES is that of several users sharing a set of common resources. We can present this example in an amusing way by stating it in terms of the traditional story of the *dining philosophers*. For simplicity, we present our model in terms of two philosophers. These two philosophers are seated at a round table where there are two plates of food, one in front of each philosopher, and two forks, one on each side between the plates. The behavior of each philosopher is as follows. The philosopher may be thinking or he may want to eat. In order to go from the "thinking" state to the "eating" state, the philosopher needs to pick up both forks on the table, one at a time, in either order. After the philosopher is done eating, he places both forks back on the table and returns to the "thinking" state. Figure 2.15 shows the automaton models of the two philosophers, denoted by P_1 and P_2, where the events are denoted by *ifj* for "philosopher i picks up fork j" and *jf* for "philosopher j puts both forks down".

The automata P_1 and P_2 are incomplete to model our system. This is because we need to capture the fact that a fork can only be used by one philosopher at a time. If we build $P_1 \| P_2$, this will be a shuffle as these two automata have no common events, and thus we will have 16 states reachable from the initial state. In particular, we will have states where a given fork is being used simultaneously by the two philosophers, a situation that is not admissible. Therefore, we add two more automata, one for each fork, to capture the constraint on the resources in this system. These are automata F_1 and F_2 in Fig. 2.15. Fork 1 can be in state "available" or in state "in use". It will go from "available" to "in use" due to either event *1f1* or event *2f1*; it will return to "available" from "in use" upon the occurrence of either *1f* or *2f*.

The complete system behavior, which we denote by PF, is the parallel composition of all four automata, P_1, P_2, F_1, and F_2. This automaton is depicted in Fig. 2.16. Due to the common events between the fork automata and the philosopher automata, only nine states are reachable out of the 64 possible states. The automaton $PF = P_1||P_2||F_1||F_2$ is blocking as it contains two deadlock states. This happens when each philosopher is holding one fork; according to the models P_1 and P_2, each philosopher waits for the other fork to become available, and consequently they both starve to death in this deadlock. To avoid this deadlock, we would have to add a *controller* to our system.[3] The role of the controller would be to prevent a philosopher from picking up an available fork if the other philosopher is holding the other fork. As an exercise, the reader may build an automaton C for this purpose, so that

$$PF||C = CoAc(PF) .$$

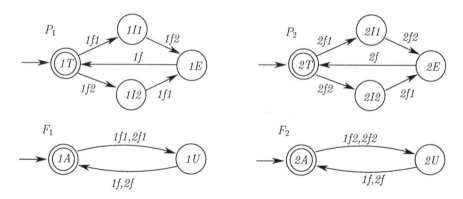

Figure 2.15: The four automata of the dining philosophers example (Example 2.17).

The above examples illustrate important observations about the computational complexity of analyzing "large" DES. Let us assume that we have a DES composed of 10 components, each modeled by a 5-state automaton. If the event sets of these 10 automata are distinct, then the model of the complete system has 5^{10} states (nearly 10 million), since it corresponds to the shuffle of the 10 automata. This is the origin of the "curse of dimensionality" in DES. This kind of exponential growth need not happen in a parallel composition when there are common events between the components, as we saw in Example 2.17. If we add an 11-th component that models how the above 10 components are

---

[3]The dining philosophers example is revisited in the context of supervisory control in Chapter 3, Example 3.16.

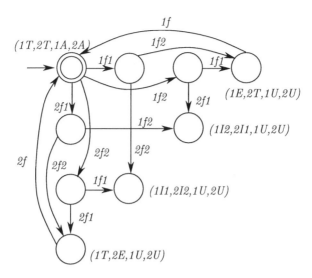

Figure 2.16: The parallel composition of the four automata in Fig. 2.15 (Example 2.17).
For the sake of readability, some state names have been omitted and those that are included are indicated next to the state.

interconnected and/or operate (e.g., a controller module), then all the events of that new component will belong to the union of the 10 event sets of the original components. The model of the system of 11 components may then have considerably less than the worst case of $5^{10} \times N_{11}$ states, where N_{11} is the number of states of the 11-th component, due to the synchronization requirements imposed by the common events.

We conclude this section with a technical remark that may be useful to remember.

Remark. Formally, given an automaton $G = (X, E, f, \Gamma, x_0, X_m)$, we can include in E events that do not appear in $\mathcal{L}(G)$, since E is a *parameter* in the definition of G. This can however lead to some confusion, as in such a case, the automaton is not entirely represented by its transition function f, or equivalently its state transition diagram, something that we find convenient. And we saw how the set E plays an important role in the parallel composition operation. Thus, from now on, unless explicitly stated otherwise, we will assume that E in the definition of an automaton G consists only of those events that appear in the strings in $\mathcal{L}(G)$.

2.3.3 Observer Automata

We introduced earlier the class of nondeterministic automata, which differ from deterministic automata by allowing the codomain of f to be 2^X, the power set of the state space of the automaton, and also allowing ε-transitions. It turns out that we can always transform a nondeterministic automaton, G_{nd}, into an equivalent deterministic one, that is, one that generates and marks the same languages as the original nondeterministic automaton. The state space of the equivalent deterministic automaton will be a subset of the power set of the state space of the nondeterministic one. This means that if the nondeterministic automaton is finite-state, then the equivalent deterministic one will also be finite-state. This latter statement has important implications that will be discussed in Section 2.4. The focus of this section is to present an algorithm for this language-preserving transformation from nondeterministic to deterministic automaton. We shall call the resulting equivalent deterministic automaton the *observer* corresponding to the nondeterministic automaton; we will denote the observer of G_{nd} by G_{obs}. This terminology is inspired from the concept of observer in system theory; it captures the fact that the equivalent deterministic automaton (the observer) keeps track of the estimate of the state of the nondeterministic automaton upon transitions labeled by events in E. (Recall that the event set of G_{nd} is $E \cup \{\varepsilon\}$.)

Before presenting the details of the algorithm to construct the observer, we illustrate the key points using a simple example. (The reader may also want to revisit Example 2.12 at this point.)

Example 2.18 (From nondeterministic to deterministic automata)
Consider the nondeterministic automaton G_{nd} in Fig. 2.17, where nondeterminism arises at states 1 and 2, since event b leads to two different states from state 1 and since we have ε-transitions in the active event sets of states 1 and 2. Let us start building the observer G_{obs} of G_{nd} by defining its initial state and calling it $\{0\}$. Since 0 is marked in G_{nd}, we mark $\{0\}$ in G_{obs} as well.

First, we analyze the state $\{0\}$ of G_{obs}.

- Event a is the only event defined at state 0 in G_{nd}. String a can take G_{nd} to states 1 (via a), 2 (via $a\varepsilon$), and 3 (via $a\varepsilon\varepsilon$) in G_{nd}, so we define a transition from $\{0\}$ to $\{1, 2, 3\}$, labeled a, in G_{obs}.

Next, we analyze the newly-created state $\{1, 2, 3\}$ of G_{obs}. We take the union of the active event sets of 1, 2, and 3 in G_{nd} and get events a and b, in addition to ε.

- Event a can only occur from state 2 and it takes G_{nd} to state 0. Thus, we add a transition from $\{1, 2, 3\}$ to $\{0\}$, labeled a, in G_{obs}.

- Event b can occur in states 1 and 3. From state 1, we can reach states 0 (via b), 1 (via b), 2 (via $b\varepsilon$), and 3 (via $b\varepsilon\varepsilon$). From state 3, we can reach state 0 (via b). Overall, the possible states that can be reached from $\{1,2,3\}$ with string b are 0, 1, 2, and 3. Thus, we add a transition from $\{1,2,3\}$ to $\{0,1,2,3\}$, labeled b, in G_{obs}.

Finally, we analyze the newly-created state $\{0,1,2,3\}$ of G_{obs}. Proceeding similarly as above, we identify the following transitions to be added to G_{obs}:

- A self-loop at $\{0,1,2,3\}$ labeled a;

- A self-loop at $\{0,1,2,3\}$ labeled b.

The first self-loop is due to the fact that from state 0, we can reach states 1, 2, and 3 under a, and from state 2, we can reach state 0 under a. Thus, all of $\{0,1,2,3\}$ is reachable under a from $\{0,1,2,3\}$. Similar reasoning explains the second self-loop. We mark state $\{0,1,2,3\}$ in G_{obs} since state 0 is marked in G_{nd}.

The process of building G_{obs} is completed since all created states have been examined. This observer is depicted in Fig. 2.17. It can be seen that G_{nd} and G_{obs} are indeed equivalent.

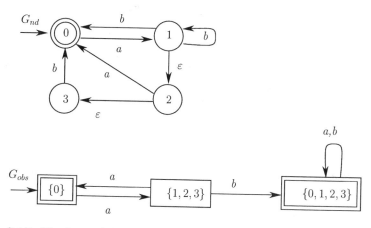

Figure 2.17: Nondeterministic automaton and its (deterministic) observer for Example 2.18.

With the intuition gained from this example, we present the formal steps of the algorithm to construct the observer.

Procedure for Building Observer G_{obs} of Nondeterministic Automaton G_{nd}

Let $G_{nd} = (X, E \cup \{\varepsilon\}, f_{nd}, x_0, X_m)$ be a nondeterministic automaton. Then $G_{obs} = (X_{obs}, E, f_{obs}, x_{0,obs}, X_{m,obs})$ and it is built as follows.

Step 1: Start with $X_{obs} = 2^X \setminus \emptyset$.

Step 2: For each state $x \in X$ define

$$UR(x) := f_{nd}(x, \varepsilon) \ .$$

Read UR as "unobservable reach" since ε transitions are not "observed". It is assumed here that we are working with the extension of function f_{nd} to strings in $(E \cup \{\varepsilon\})^*$, as described in Section 2.2.4.

For a set B, define

$$UR(B) = \bigcup_{x \in B} UR(x) \ .$$

Step 3: Define $x_{0,obs} = UR(x_0)$.

Step 4: For each $S \subseteq X$ and $e \in E$, define

$$
\begin{aligned}
f_{obs}(S, e) &= UR(\{x \in X : \exists x_e \in S \ [x \in f_{nd}(x_e, e)]\}) \\
&= \{x \in X : \exists x_e \in S \ [x \in f_{nd}(x_e, e)]\}
\end{aligned}
$$

by the definition of the extension of f_{nd} to strings.

Step 5: $X_{m,obs} = \{S \subseteq X : S \cap X_m \neq \emptyset\}$.

Step 6: In practice, the above is performed in a breadth-first manner so that only the accessible part of G_{obs} is constructed. The resulting state space X_{obs} is a subset of 2^X. Note that the empty subset of X need not be considered, since it is never an accessible state of X_{obs}.

We explain the key steps of this algorithm.

- The idea of this procedure is to start with $x_{0,obs}$ as the initial state of the observer. We then identify all the events in E that label all the transitions out of any state in the set $x_{0,obs}$; this results in the active event set of $x_{0,obs}$. For each event e in this active event set, we identify all the states in X that can be reached starting from a state in $x_{0,obs}$. We then extend this set of states to include its unobservable reach; this returns the state $f_{obs}(x_{0,obs}, e)$ of G_{obs}. This transition, namely event e taking state $x_{0,obs}$ to state $f_{obs}(x_{0,obs}, e)$, is then added to the state transition diagram of G_{obs}.

What the above means is that an "outside observer" that knows the system model G_{nd} but only observes the transitions of G_{nd} labeled by events in E will start with $x_{0,obs}$ as its estimate of the state of G_{nd}. Upon observing event $e \in E$, this outside observer will update its state estimate to $f_{obs}(x_{0,obs}, e)$, as this set represents all the states where G_{nd} could be after executing the string e, preceded and/or followed by ε.

- The procedure is repeated for each event in the active event set of $x_{0,obs}$ and then for each state that has been created as an immediate successor of $x_{0,obs}$, and so forth for each successor of $x_{0,obs}$. Clearly, the worst case for all states that could be created is no larger than the set of all non-empty subsets of X.

- Finally, any state of G_{obs} that contains a marked state of G_{nd} is considered to be marked from the viewpoint of G_{obs}. This is because this state is reachable from $x_{0,obs}$ by a string in $\mathcal{L}_m(G_{nd})$.

The important properties of G_{obs} are that:

1. G_{obs} is a deterministic automaton.

2. $\mathcal{L}(G_{obs}) = \mathcal{L}(G_{nd})$.

3. $\mathcal{L}_m(G_{obs}) = \mathcal{L}_m(G_{nd})$.

The first result is obvious; the other two results follow directly from the algorithm to construct G_{obs}.

2.3.4 State Space Refinement

Let us suppose that we are interested in "comparing" two languages L_1 and L_2, where $L_1 \subseteq L_2$, by comparing two automata representations of them, say G_1 and G_2, respectively. For instance, we may want to know what event(s), if any, are possible in L_2 but not in L_1 after string $t \in L_1 \cap L_2$. For simplicity, let us assume that L_1 and L_2 are prefix-closed languages; or, equivalently, we are interested in the languages generated by G_1 and G_2. In order to answer this question, we need to identify what states are reached in G_1 and G_2 after string t and then compare the active event sets of these two states.

In order to make such comparisons more computationally efficient if the above question has to be answered for a large set of strings t's, we would want to be able to "map" the states of G_1 to those of G_2. However, we know that even if $L_1 \subseteq L_2$, there need not be any relationship between the state transition diagram of G_1 and that of G_2. In particular, it could happen that state x of G_1 is reached by strings t_1 and t_2 of $\mathcal{L}(G_1)$ and in G_2, t_1 and t_2 lead to two different states, say y_1 and y_2. This means that state x_1 of G_1 should be mapped to state y_1 of G_2 if the string of interest is t_1, whereas x_1 of G_1 should be mapped to y_2

of G_2 if the string of interest is t_2. When this happens, it may be convenient to *refine* the state transition diagram of G_1, by adding new states and transitions but without changing the language properties of the automaton, in order to obtain a new automaton whose states could be mapped to those of G_2 by a function that is independent of the string used to reach a state. Intuitively, this means in the above example that we would want to split state x_1 into two states, one reached by t_1, which would then be mapped to y_1 of G_2, and one reached by t_2, which would then be mapped to y_2 of G_2.

This type of refinement, along with the desired function mapping the states of the refined automaton to those of G_2, is easily performed by the product operation as we now describe.

Refinement by Product

To refine the state space of G_1 in the manner described above, it suffices to build

$$G_{1,new} = G_1 \times G_2 .$$

By our earlier assumption that $L_1 = \mathcal{L}(G_1) \subseteq L_2 = \mathcal{L}(G_2)$, it is clear that $G_{1,new}$ will be equivalent to G_1. Moreover, since the states of $G_{1,new}$ are pairs (x, y), the second component y of a pair tells us the state that G_2 is in whenever $G_{1,new}$ is in state (x, y). Thus, the desired map from the state space of $G_{1,new}$ to that of G_2 consists of simply reading the second component of a state of $G_{1,new}$.

Example 2.19 (Refinement by product)
Figure 2.18 illustrates how the product operation can refine an automaton with respect to another one. State x_2 of the first automaton can correspond to either y_2 or y_3 of the second automaton, depending on whether it is reached by event a_1 or a_2, respectively. When the first automaton is replaced by the product of the two automata, state x_2 has been split into two states: (x_2, y_2) and (x_2, y_3).

Notion of Subautomaton

An alternative to refinement by product to map states of G_1 to states of G_2 is to require that the state transition diagram of G_1 be a *subgraph* of the state transition diagram of G_2. This idea of subgraph is formalized by the notion of *subautomaton*. We say that G_1 is a subautomaton of G_2, denoted by $G_1 \sqsubseteq G_2$, if

$$f_1(x_{01}, s) = f_2(x_{02}, s) \text{ for all } s \in \mathcal{L}(G_1) .$$

Note that this condition implies that $X_1 \subseteq X_2$, $x_{01} = x_{02}$, and $\mathcal{L}(G_1) \subseteq \mathcal{L}(G_2)$. This definition also implies that the state transition diagram of G_1 is a subgraph of that of G_2, as desired.

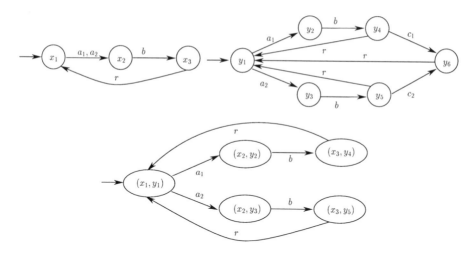

Figure 2.18: Two automata and their product (Example 2.19).

This form of correspondence between two automata is stronger than what refinement by product can achieve. In particular, we may have to modify both G_1 and G_2. On the other hand, the subgraph relationship makes it trivial to "match" the states of the two automata. It is not difficult to obtain a general procedure to build G_1' and G_2' such that $\mathcal{L}(G_i') = L_i$, $i = 1, 2$, and $G_1' \sqsubseteq G_2'$, given any G_i such that $\mathcal{L}(G_i) = L_i$, $i = 1, 2$:

1. Build $G_1' = G_1 \times G_2$;

2. (a) Examine each state x_1 of G_1 and add a self-loop for each event in $E_2 \setminus \Gamma_1(x_1)$ and call the result G_1^{sl};

   (b) Build $G_2' = G_1^{sl} \times G_2$.

The proof of the correctness of this procedure is left to the reader.

2.3.5   Automata with Inputs and Outputs

There are two variants to the definition of automaton given in Section 2.2.2 that are useful in system modeling: Moore automaton and Mealy automaton. (These kinds of automata are named after E. F. Moore and G. H. Mealy who defined them in 1956 and 1955, respectively.) The differences with the definition in Section 2.2.2 are quite simple and are depicted in Fig. 2.19.

- Moore automata are *automata with (state) outputs*. There is an output function that assigns an output to each state. The output associated with a state is shown in bold above the state in Fig. 2.19. This output is "emitted" by the automaton when it enters the state. We can think of "standard" automata as having two outputs: "unmarked" and "marked". Thus, we can view state outputs in Moore automata as a generalization of the notion of marking.

- Mealy automata are *input/output automata*. Transitions are labeled by "events" of the form *input event/output event*, as shown in Fig. 2.19. The set of output events, say E_{output}, need not be the same as the set of input events, E. The interpretation of a transition e_i/e_o from state x to state y is as follows. When the system is in state x, if the automaton "receives" input event e_i, it will make a transition to state y and in that process will "emit" the output event e_o.

One can see how the notions of state output and output event could be useful in building models of DES. Let us refer to systems composed of interacting electromechanical components (e.g., assembly and manufacturing systems, engines, process control systems, heating and air conditioning units, and so forth) as *physical systems*. These systems are usually equipped with a set of sensors that record the *physical state* of the system. For instance, an air handling system would be equipped with valve flow sensors, pump pressure sensors, thermostats, etc. Thus, Moore automata are a good class of models for such systems, where the output of a state is the set of readings of all sensors when the system is in that (physical) state. Mealy automata are also a convenient class of models since the notion of input-output mapping is central in system and control theory (cf. Section 1.2 in Chapter 1). In the modeling of communication protocols for instance, input/output events could model that upon reception of a certain message, the input event, a protocol entity issues a new message, the output event. The same viewpoint applies to software systems in general.

We claim that for the purposes of this book, we can always "transform" Mealy and Moore automata into "standard automata" (i.e., as defined in Section 2.2.2). For Mealy automata, this claim is based on the following interpretation. We can view a Mealy automaton as a standard one where events are of the form *input/output*. That is, we view the set E of events as the set of all input/output labels of the Mealy automaton. In this context, the language generated by the automaton will be the set of all *input/output strings* that can be generated by the Mealy automaton. Thus, the material presented in this chapter applies to Mealy automata, when those are interpreted in the above manner. For Moore automata, we can view the state output as the output event associated to all events that enter that state; refer to Fig. 2.20. This effectively transforms the Moore automaton into a Mealy automaton, which can then be interpreted as a standard automaton as we just described. We will revisit the issue of conversion from Moore to standard automata in Section 2.5.2 at the end of this chapter.

The above discussion is not meant to be rigorous but rather its purpose is to make the reader aware that for the material presented in this book, there is

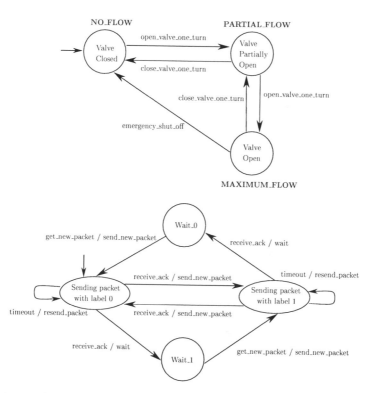

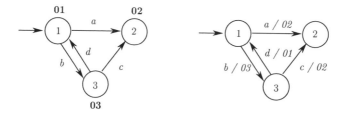

Figure 2.19: Automaton with output (Moore automaton) and input/output automaton (Mealy automaton).

The Moore automaton models a valve together with its flow sensor; the output of a state - indicated in bold next to the state - is the flow sensor reading for that state. The Mealy automaton models the sender in the context of the "Alternating Bit Protocol" for transmission of packets between two nodes in a communication network (half-duplex link in this case). Packets are labeled by a single bit in order to differentiate between subsequent transmissions. (See, e.g., Chapter 6 of Communication Networks. A First Course *by J. Walrand, McGraw-Hill, 1998, for further details on such protocols.)*

Figure 2.20: Conversion of automaton with output.

The output of each transition in the equivalent Mealy automaton on the right is the output of the state entered by the same transition in the Moore automaton on the left.

no "essential" lack of generality in considering standard automata as opposed to Moore or Mealy automata.

2.4 FINITE-STATE AUTOMATA

2.4.1 Definition and Properties of Regular Languages

As we mentioned at the beginning of this chapter, we are normally interested in specifying tasks as event sequences which define some language, and then in obtaining an automaton that can generate or mark (that is, represent) this language. The automaton representation is certainly likely to be more convenient to use than plain enumeration of all the strings in the language. Naturally, a crucial question is: Can we always do that? The answer is "yes", albeit it comes with a "practical" problem. Any language can be marked by an automaton: Simply build the automaton as a (possibly infinite) tree whose root is the initial state and where the nodes at layer n of the tree are entered by the strings of length n or the prefixes of length n of the longer strings. The state space is the set of nodes of the tree and a state is marked if and only if the string that reaches it from the root is an element of the language. Refer to Fig. 2.21. The tree automaton there represents the language

$$L = \{\varepsilon, ab, aabb, aaabbb, \ldots\} = \{a^n b^n : n \geq 0\}$$

over the event set $E = \{a, b\}$. The key point here is that any such tree automaton will have an infinite state space if the cardinality of the language is infinite. Of course, we know that there are infinite languages that can be represented by *finite-state* automata. For example, a single state suffices to represent the language E^*; it suffices to put a self-loop at that state for each event in E. Another example is provided by the automaton in Fig. 2.2.

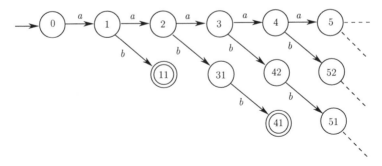

Figure 2.21: Tree automaton marking the language $L = \{a^n b^n : n \geq 0\}$.

The above discussion leads us to ask the question: Are there infinite languages that cannot be represented by finite-state automata? The answer is "yes" and a classic example is the language

$$L = \{a^n b^n : n \geq 0\}$$

mentioned above. To argue that this language cannot be represented by a finite-state automaton, we observe that a marked state must be reached after exactly the same number of b events as that of a events that started the string. Therefore, the automaton must have "memorized" how many a events occurred when it starts allowing b events, in order to allow the correct number of b events before entering a marked state. But the number of a events can be arbitrarily large, so any automaton with $2N - 1$ states would not be able to mark the string $a^N b^N$ without marking other strings not in L. Consequently, L cannot be marked by a *finite-state* automaton. This discussion leads us to the following definition.

Definition. (Regular language)

A language is said to be *regular* if it can be marked by a finite-state automaton. The class of regular languages is denoted by $\mathcal{R}$. ♦

We have established that $\mathcal{R}$ is a proper subset of 2^{E^*}. The class $\mathcal{R}$ is very important since it delimits the languages that possess automaton representations that require finite memory when stored in a computer. In other words, automata are a practical means of manipulating *regular* languages in analysis or controller synthesis problems. On the other hand, automata are not a practical means for representing *non-regular* languages, since they would require infinite memory. We will see in Chapter 4 that Petri nets, the other DES modeling formalism considered in this book, can represent some non-regular languages with a finite transition structure.

Nondeterministic Finite-State Automata

We presented in Section 2.3.3 a procedure to transform any nondeterministic automaton G_{nd} to an equivalent deterministic one G_{obs}, called the observer of G_{nd}. If we restrict attention to nondeterministic *finite-state* automata, then we can see immediately from the construction procedure for observers that the corresponding observers will be *finite-state* (deterministic) automata. As we noted earlier, the state space of an observer will be a subset of the power set of the state space of the original nondeterministic automaton; clearly, the power set of a finite set is also a finite set. In other words, for any nondeterministic finite-state automaton, there exists an equivalent deterministic finite-state one.

Theorem. (Regular languages and finite-state automata)

The class of languages representable by nondeterministic finite-state automata is exactly the same as the class of languages representable by deterministic finite-state automata: $\mathcal{R}$. ♦

This important result does not mean that nondeterministic automata are "useless". We commented earlier how nondeterminism may arise in system modeling. Another aspect is that nondeterministic automata may sometimes require fewer states than deterministic ones to describe certain languages, and this makes them quite useful.

Properties of Regular Languages

The following theorem about the properties of the class of regular languages, along with the proof that we present, illustrates how well-behaved the class $\mathcal{R}$ is and how the duality between regular languages and finite-state automata, be they deterministic or nondeterministic, can be exploited.

Theorem. (Properties of $\mathcal{R}$)
Let L_1 and L_2 be in $\mathcal{R}$. Then the following languages are also in $\mathcal{R}$:

1. $\overline{L_1}$
2. L_1^*
3. $L^c := E^* \setminus L_1$
4. $L_1 \cup L_2$
5. $L_1 L_2$
6. $L_1 \cap L_2$.

Proof. Let G_1 and G_2 be two finite-state automata that mark L_1 and L_2, respectively. We prove the first three properties by modifying G_1 in order to obtain finite-state automata that mark $\overline{L_1}$, L_1^*, and L_1^c, respectively. We prove the last three properties by building, from G_1 and G_2, a third finite-state automaton that marks the language obtained after the corresponding operation. Note that in all cases, it does not matter if a finite-state automaton is nondeterministic, as we know we can always build its (deterministic and finite-state) observer to mark the same language.

1. Take the trim of G_1 and then mark all of its states.
2. *Kleene-closure:* Add a new initial state, mark it, and connect it to the old initial state of G_1 by an ε-transition. Then add ε-transitions from every marked state of G_1 to the old initial state. The new finite-state automaton marks L_1^*.
3. *Complement:* This was proved when we considered the complement operation in Section 2.3.1; the automaton to build from G_1 has at most one more state than G_1.
4. *Union:* Create a new initial state and connect it, with two ε-transitions, to the initial states of G_1 and G_2. The result is a nondeterministic automaton that marks the union of L_1 and L_2.

5. *Concatenation:* Connect the marked states of G_1 to the initial state of G_2 by ε-transitions. Unmark all the states of G_1. The resulting nondeterministic automaton marks the language $L_1 L_2$.

6. *Intersection:* As was seen in Section 2.3.2, $G_1 \times G_2$ marks $L_1 \cap L_2$. **Q.E.D.**

2.4.2   Regular Expressions

Another way to describe regular languages is through compact expressions that are called *regular expressions*. We have already defined the operations of concatenation, Kleene-closure, and union on languages. We now establish some notational conventions. First, we adopt the symbol "+" rather than "∪" to remind us that this operation is equivalent to the logical "OR" function. Next, suppose we are interested in the Kleene-closure of $\{u\}$, where u is a string. This is given by $\{u\}^* = \{\varepsilon, u, uu, uuu, \dots\}$ as was seen earlier. Let us agree, however, to omit the braces and write u^* in place of $\{u\}^*$. By extension, let us agree to think of u as either a string or as the set $\{u\}$, depending on the context. Then, the concatenation of $\{u\}$ and $\{v\}$, $\{uv\}$, is written as uv, and the union of $\{u\}$ and $\{v\}$, $\{u,v\}$, is written as $(u+v)$. Expressions such as u^* or $(u+v)^*$ are therefore used to denote sets that may otherwise be too cumbersome to write down by enumerating their elements. These are what we refer to as regular expressions, which we can now define recursively as follows:

1. $\emptyset$ is a regular expression denoting the empty set, ε is a regular expression denoting the set $\{\varepsilon\}$, and e is a regular expression denoting the set $\{e\}$, for all $e \in E$.

2. If r and s are regular expressions, then rs, $(r+s)$, r^*, s^* are regular expressions.

3. There are no regular expressions other than those constructed by applying rules 1 and 2 above a finite number of times.

Regular expressions provide a compact *finite* representation for potentially complex languages with an infinite number of strings.

Example 2.20 (Regular expressions)

Let $E = \{a, b, g\}$ be the set of events. The regular expression $(a+b)g^*$ denotes the language

$$L = \{a, b, ag, bg, agg, bgg, aggg, bggg, \dots\}$$

which consists of all strings that start with either event a or event b and are followed by a repetition of event g. Note that even though L contains infinitely many elements, the corresponding regular expression provides a simple finite representation of L.

The regular expression $(ab)^* + g$ denotes the language

$$L = \{\varepsilon, g, ab, abab, ababab, \dots\}$$

which consists of the empty string, event g, and repetitions of the string ab any number of times.

The following theorem is an important result in automata theory. We shall state it without proof.

Theorem. (Regular expressions and regular languages)
Any language that can be denoted by a regular expression is a regular language; conversely, any regular language can be denoted by a regular expression. ♦

This theorem establishes the equivalence of regular expressions and finite-state automata in the representation of languages. This equivalence is known as Kleene's Theorem after S.C. Kleene who proved it in the 1950's.

2.4.3 State Space Minimization

There is no unique way to build an automaton that marks a given language. For $K \in \mathcal{R}$, define $\|K\|$ to be the minimum of $|X|$ (the cardinality of X) among all deterministic finite-state automata G that mark K. The automaton that achieves this minimum is called the *canonical recognizer* of K. Here we are only concerned with the language *marked* by an automaton, and not with the pair of languages $\mathcal{L}(G)$ and $\mathcal{L}_m(G)$. In order to simplify the discussion in this section, we will assume that the *automata under consideration have completely defined transition functions*, namely they all generate E^*. Under these conditions, the canonical recognizer of a language is unique, up to a renaming of the states.

For example, $\|\emptyset\| = \|E^*\| = 1$: in both cases, we have a single state with a self-loop for all events in E; this state is unmarked in the case of $\emptyset$ and marked in the case of E^*. If $E = \{a, b\}$ and $L = \{a\}^*$, then $\|L\| = 2$. (Recall that we require that the automata generate E^*.) It should be emphasized that $\| \cdot \|$ is not related to the cardinality of the language. For instance, $\|E^*\| = 1$ yet E^* is an infinite set. Thus a larger language may be (but need not be) representable with fewer states than a smaller language.

The usefulness of the canonical recognizer is that it provides a representation for a (regular) language that essentially minimizes the amount of memory required (in the automaton modeling formalism). We write "essentially" in the preceding sentence because the storage required is proportional to the number of transitions in the state transition diagram of the automaton. However, since we are dealing with deterministic automata, the active event set at any state of an automaton never exceeds $|E|$. In practical applications, the number of events is almost always much less than the number of states; recall our discussion about computational complexity at the end of Section 2.3.2. For this reason, it is customary to express the computational complexity of various manipulations of automata in terms of the cardinalities of the state spaces of the automata involved.

Obtaining the canonical recognizer for the language marked by a given automaton G means that we should look for states that are redundant, or *equivalent*, in the sense that they can be replaced by a single "aggregate" state. The notion of equivalence that matters here is the following:

two states x and y are equivalent if $\mathcal{L}_m(G(x)) = \mathcal{L}_m(G(y))$,

where $G(x)$ denotes the same automaton as G with the only difference that the initial state is state x. In other words, two states are equivalent if they have the same future behavior in terms of marked language - recall that by assumption, $\mathcal{L}(G(x)) = \mathcal{L}(G(y)) = E^*$ for the automata considered in this section. Clearly, two (or more) equivalent states can be "merged", that is, replaced by a single aggregate state.

There are some simple observations we can immediately make regarding the above definition of equivalence.

1. If we consider two states x, y such that $x \in X_m$ and $y \notin X_m$, then they can never be equivalent. For instance, $f(x, \varepsilon) \in X_m$, whereas $f(y, \varepsilon) \notin X_m$.

2. Suppose two states x, y are either both in X_m or neither one is in X_m. Therefore, they are eligible to be equivalent. If $f(x, e) = f(y, e)$ for any event $e \in E \cup \{\varepsilon\}$, then states x and y are always equivalent, since they both go through the exact same state sequence for any string applied to them.

3. The preceding property still holds if $f(x, e) = y$ and $f(y, e) = x$ for one or more events e, since the role of x and y is simply interchanged after e. Of course, for all remaining events e we must still have $f(x, e) = f(y, e)$.

4. More generally, let R be a set of states such that either $R \subseteq X_m$ or $R \cap X_m = \emptyset$. Then, R consists of equivalent states if $f(x, e) = z \notin R$ implies that $f(y, e) = z$, for any x, $y \in R$. In other words, all possible transitions from states in R to states outside R must be caused by the same event and must lead to the same next state.

Example 2.21 (A digit sequence detector)
In this example, we design an automaton model for a machine that reads digits from the set $\{1, 2, 3\}$ and detects (that is, marks) any string that ends with the substring 123. The event set is $E = \{1, 2, 3\}$, where event n means "the machine just read digit n," with $n = 1, 2, 3$. The automaton should generate the language E^*, since it should accept any input digit at any time, and it should mark the language

$$L = \{st \in E^* : s \in E^* \text{ and } t = 123\} .$$

Let us use simple logic to determine a state space X and build the corresponding transition function f. Let us begin by defining state x_0, the initial state, to represent that no digit has been read thus far. Since we are interested in the substring 123, let us also define states x_1 and x_{not1} to represent that the first event is 1, for x_1, or 2 or 3, for x_{not1}. Given the form of L, we can see that we need the automaton to "memorize" that the suffix of the string read thus far is either 1, 12, 123, or none of these. Therefore, the role of state x_1 is extended to represent that the most recent event is 1, corresponding to the suffix 1. In addition to x_1, we also need states to memorize suffixes 12 and 123; we call these states x_{12} and x_{123}, respectively. Finally, we know that any suffix other than 1, 12, and 123 need not be memorized. Thus, events that result in a suffix that is not 1, 12, or 123 will be sent to state x_{not1}, meaning that this state will represent that 1, 12, and 123 are not suffixes of the string read thus far.

Given $X = \{x_0, x_1, x_{not1}, x_{12}, x_{123}\}$, we construct the correct (completely defined) f as follows:

- From x_0: $f(x_0, 1) = x_1$ and $f(x_0, 2) = f(x_0, 3) = x_{not1}$, consistent with the above-described meaning of these three states.

- From x_1: $f(x_1, 1) = x_1$, $f(x_1, 2) = x_2$, and $f(x_1, 3) = x_{not1}$, in order for the automaton to be in the state corresponding to the current suffix, namely 1, 12, or none of 1, 12 and 123.

- From x_{12}: $f(x_{12}, 1) = x_1$, $f(x_{12}, 2) = x_{not1}$, and $f(x_{12}, 3) = x_{123}$, by the same reasoning as above; if suffix 12 is "broken" by a 1 event, then we have to return to x_1, whereas if it is broken by a 2 event, we must go to x_{not1}.

- From x_{123}: $f(x_{123}, 1) = x_1$, $f(x_{123}, 2) = x_{not1}$, and $f(x_{123}, 3) = x_{not1}$.

- From x_{not1}: $f(x_{not1}, 1) = x_1$, $f(x_{not1}, 2) = f(x_{not1}, 3) = x_{not1}$, by definition of these states.

These state transitions are shown in Fig. 2.22; the initial state is x_0 and the only marked state is x_{123}, since it is reached by any string that ends with the suffix 123. Consequently, the automaton in Fig. 2.22 marks L, as desired.

However, a simpler automaton, with four states instead of five, can be built to mark L (and generate E^*). This automaton is also shown in Fig. 2.22. It is obtained by applying the equivalence condition defined at the beginning of this section to the set of states $R = \{x_0, x_{not1}\}$. Specifically, when event 1 occurs we have $f(x_0, 1) = f(x_{not1}, 1) = x_1$. Moreover, $f(x_0, 2) = f(x_{not1}, 2) = x_{not1}$, and $f(x_0, 3) = f(x_{not1}, 3) = x_{not1}$. Thus, states x_0 and x_{not1} have indistinguishable future behaviors, which means that they can be merged into a single state, denoted by $x_{0,new}$ in the state transition diagram of the second automaton in Fig. 2.22. It is easily verified that there are no equivalent states in the four-state automaton and therefore this automaton is the canonical recognizer of L and $||L|| = 4$.

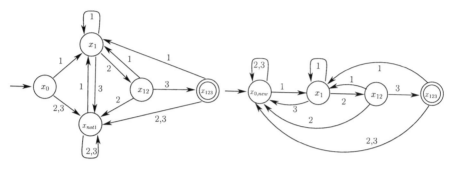

Figure 2.22: State transition diagrams for digit sequence detector in Example 2.21. *Both automata detect the event sequence 123 upon entering state x_{123}. However, in the first model (left), states x_0 and x_{not1} are equivalent, since $f(x_0, e) = f(x_{not1}, e)$ for any $e \in \{1, 2, 3\}$. These states are aggregated into a single state x_0 to give the second simpler model (right).*

The process of replacing a set of states by a single one is also referred to as state aggregation. This is an important process in many aspects of system theory, because it results in state space reduction, which in turn translates into a decrease of computational complexity for many analytical problems, as we mentioned earlier. Furthermore, *approximate* state aggregation is sometimes a good way to decrease complexity at the expense of some loss of information.

Obviously, it is of interest to develop a simple general procedure that can always give us the canonical recognizer. Let us assume that we are given automaton $G = (X, E, f, E_G, x_0, X_m)$ where $\mathcal{L}_m(G) = L$ and $\mathcal{L}(G) = E^*$. We want to build from G the canonical recognizer

$$G_{can} = (X_{can}, E, f_{can}, \Gamma_{can}, x_{0,can}, X_{m,can})$$

where $\mathcal{L}_m(G_{can}) = L$, $\mathcal{L}(G_{can}) = E^*$, and $|X_{can}|$ is minimum among all other automata that mark L and generate E^*. The algorithm below allows us to identify all sets of equivalent states. Once these sets are identified, we can form an aggregate state for each one, and thereby obtain the canonical recognizer. The state space X_{can} is the (smaller) set of states after aggregation, f_{can} is the resulting transition function after aggregation, $x_{0,can}$ is the state in X_{can} that contains x_0, and $X_{m,can}$ is the subset of X_{can} containing marked states; since equivalent states must share the same marking property (either marked or unmarked), there is no ambiguity in defining $X_{m,can}$.

The main idea in the algorithm is to begin by "flagging" all pairs of states (x, y) such that x is a marked state and y is not. Such flagged pairs cannot be equivalent. Here, of course, we are assuming that there is at least one state y that is not marked, that is, $X_m \neq X$. Next, every remaining pair of states (x, y) is considered, and we check whether some event $e \in E$ may lead to a new pair of states $(f(x, e), f(y, e))$ which is already flagged. If this is not the case, and $f(x, e) \neq f(y, e)$, then we create a list of states associated with $(f(x, e), f(y, e))$ and place (x, y) in it. As the process goes on, we will eventually determine if $(f(x, e), f(y, e))$ should be flagged; if it is, then the flagging propagates to (x, y) and all other pairs in its list.

Algorithm for Identifying Equivalent States

Step 1: Flag (x, y) for all $x \in X_m$, $y \notin X_m$.

Step 2: For every pair (x, y) not flagged in Step 1:

> **Step 2.1:** If $(f(x, e), f(y, e))$ is flagged for some $e \in E$, then:
>
>> **Step 2.1.1:** Flag (x, y).
>> **Step 2.1.2:** Flag all unflagged pairs (w, z) in the list of (x, y). Then, repeat this step for each (w, z) until no more flagging is possible.
>
> **Step 2.2:** Otherwise, that is, no $(f(x, e), f(y, e))$ is flagged, then for every $e \in E$:
>
>> **Step 2.2.1:** If $f(x, e) \neq f(y, e)$, then add (x, y) to the list of $(f(x, e), f(y, e))$.

At the end of this procedure, we examine all pairs that have remained unflagged. These pairs correspond to states that are equivalent. If two pairs have an element in common, we group them together into a set, and so forth, since equivalence is transitive. At the end, we get disjoint sets of equivalent states; each set is associated with a single aggregate state in the canonical recognizer.

The algorithm is easier to visualize through a table containing all possible state pairs, where an entry (x, y) is flagged to indicate that x and y are not equivalent. At the end of the algorithm, equivalent state pairs correspond to entries in the table that are not flagged. As an example, the table corresponding to the automaton of Fig. 2.22 before state aggregation is shown in Fig. 2.23. We begin by flagging (indicated by an "F" in the table) all entries in the first column, corresponding to pairs consisting of the marked state x_{123} and the remaining four states. Then, proceeding by column and applying the algorithm above, we obtain one pair of equivalent states, (x_0, x_{not1}), which are aggregated as in Fig. 2.22. More precisely, the step by step application of the algorithm proceeds as follows:

- Flag all first column entries.

- Flag (x_{12}, x_0), because $f(x_{12}, 3) = x_{123}$, $f(x_0, 3) = x_{not1}$, and (x_{123}, x_{not1}) is already flagged.

- Flag (x_{12}, x_1), because $f(x_{12}, 3) = x_{123}$, $f(x_1, 3) = x_{not1}$, and (x_{123}, x_{not1}) is already flagged.

- Flag (x_{12}, x_{not1}), because $f(x_{12}, 3) = x_{123}$, $f(x_{not1}, 3) = x_{not1}$, and (x_{123}, x_{not1}) is already flagged.

- Do not flag (x_{not1}, x_0), since $f(x_{not1}, e) = f(x_0, e)$ for all $e = 1, 2, 3$.

- Flag (x_{not1}, x_1), because $f(x_{not1}, 2) = x_{not1}$, $f(x_1, 2) = x_{12}$, and (x_{12}, x_{not1}) is already flagged.

- Flag (x_1, x_0), because $f(x_1, 2) = x_{12}$, $f(x_0, 2) = x_{not1}$, and (x_{12}, x_{not1}) is already flagged.

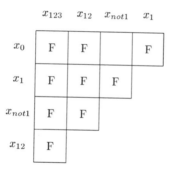

Figure 2.23: Identifying equivalent states in the automaton of Fig. 2.22. Flagged state pairs are identified by the letter F.

2.5   ANALYSIS OF DISCRETE-EVENT SYSTEMS

One of the key reasons for using finite-state automata to model DES is their amenability to analysis for answering various questions about the behavior of the system. Using efficient data structures, such as pointers and linked lists, we can easily navigate their state transition diagrams, forward and backward, and thus answer questions about the properties of the languages generated and marked by the automata. The computational complexity of navigating the state transition diagram of a deterministic automaton is linear in the size n of the state space of the automaton, that is, $O(n)$, unless iterations are necessary in which case the computational complexity will typically be $O(n^2)$. (Readers unfamiliar with the $O(\cdot)$ notation for complexity of algorithms may consult the book by Comer *et al.* listed in the end of chapter references.) Here, we have made the usual assumption that $|E| << n$. This approach generally works well for systems with up to about a million states (with today's computing power). Recently, researchers have developed symbolic techniques for representing and storing the state transition diagrams of automata; these symbolic techniques have allowed the analysis of systems with n of the order of 10^{20} (see the references at the end of this chapter).

We discuss in the next two subsections the most-often encountered analysis problems for DES. Section 2.5.1 considers deterministic automata where all the events are "observable". Section 2.5.2 considers the case of nondeterministic automata with "unobservable" events, where some form of state estimation is necessary; an unobservable event is either an ε-transition or some other event that cannot be seen by an outside observer.

2.5.1 Safety and Blocking Properties

Safety Properties

Safety properties are those properties that are concerned with the reachability of certain undesirable states (from the initial state or from some other state) in the automaton, the presence of certain undesirable strings or substrings in the language generated by the automaton, or more generally the inclusion of the language generated by the automaton in a given language, referred to as the "legal" or "admissible" language; we assume that an automaton model of the legal language is available. Observe that a DES model of a system is usually built in two steps. First, automaton models of the components of the system are built; then the complete system model is obtained by forming the parallel composition of the constituent automata. The safety questions are posed in terms of this "complete" automaton.

We first note that the unary operations discussed in Section 2.3.1, accessible, coaccessible, and complement, are easily implemented by algorithms of linear computational complexity in n, since they only involve a single traversal, either forward or backward, of the state transition diagram of the automaton. Composition operations such as product or parallel composition are implementable in $O(n_1 n_2)$, where n_i is the cardinality of the state space of automaton i, by simply building the accessible part of the composed automaton from its initial state, following the rules specific to product and parallel composition, respectively. In both cases, $n_1 n_2$ is the worst case of the size of the state space of the composed automaton.

The safety questions listed above are answered by rather straightforward algorithms:

- To determine if a given state y is reachable from another given state x, we take the accessible operation on the automaton, with x declared as the initial state, and look for state y in the result.

- To determine if a given substring is possible in the automaton, we try to "execute" the substring from all the accessible states in the automaton; this is easily done when the state transition diagram is represented as a linked list.

- Testing the inclusion $A \subseteq B$ is equivalent to testing $A \cap B^c = \emptyset$. We have shown earlier how to implement complement of languages in the automaton modeling formalism; the intersection is then implemented by taking the product of the corresponding automata.

- If other safety questions posed involve Kleene-closure, union, or concatenation of regular languages, then the techniques mentioned in the proof of the properties of $\mathcal{R}$ in Section 2.4.1 can be employed.

Blocking Properties

Blocking properties are those properties that are concerned with the coaccessibility of states to the set of marked states. Of course, we need only be concerned with states that are themselves accessible from the initial state. The most common blocking property is to determine if

$$\overline{\mathcal{L}_m(G)} = \mathcal{L}(G) \quad \text{or} \quad \overline{\mathcal{L}_m(G)} \subset \mathcal{L}(G) .$$

In the latter case, G is blocking. To determine if a given accessible G is blocking or not, we take the coaccessible operation. If any state is deleted, then G is blocking, otherwise it is nonblocking.

For a blocking G, we may wish to explicitly identify deadlock states and livelock cycles. For this purpose, we can start by finding all the states of G that are not coaccessible. Deadlock states are easily found by examining the active event sets of these states. Livelock cycles can be found by finding the strongly connected components of the part of G consisting of the non-coaccessible states of G and their associated transitions among themselves. There are algorithms that can find strongly connected components of a directed graph with n vertices in linear complexity in n.

2.5.2   State Estimation and Diagnostics

We briefly touched upon the concept of *unobservable event* when we introduced nondeterministic automata in Section 2.2.4. We motivated the notion of ε-transitions in a nondeterministic automaton as events that occur in the system modeled by the automaton but are not seen, or *observed*, by an outside observer of the system behavior. This lack of observability can be due to the absence of a sensor to record the occurrence of the event or to the fact that the event occurs at a remote location but is not communicated to the site being modeled - a typical situation in distributed information systems. *Fault events that do not cause any immediate change in sensor readings are an example of unobservable events.*

Instead of labeling all transitions due to unobservable events by ε and obtain a nondeterministic automaton model of a system, let us define "genuine" events for these transitions, but characterize these events as *unobservable*. In other words, our model for the system will be a deterministic automaton whose set of events E is partitioned into two disjoint subsets: E_o, the set of observable events, and E_{uo}, the set of unobservable events.

We can use the procedure presented in Section 2.3.3 to build an observer for a deterministic automaton with unobservable events. It suffices to treat all events in E_{uo} as if they were ε. The observer will therefore have event set E_o. Let G be the deterministic automaton with unobservable events. Let us denote by P the natural projection from E to E_o; natural projections were introduced in Section 2.3.2. Namely, we have the function $P : E^* \to E_o^*$ with

$$P(\varepsilon) := \varepsilon$$

$$P(e) := \begin{cases} e & \text{if } e \in E_o \\ \varepsilon & \text{if } e \notin E_o \end{cases}$$

$$P(se) := P(s)P(e) \text{ for } s \in E^*, e \in E .$$

Then by construction of the observer G_{obs} we have that:

- $\mathcal{L}(G_{obs}) = P[\mathcal{L}(G)]$.

- $\mathcal{L}_m(G_{obs}) = P[\mathcal{L}_m(G)]$.

- The state of G_{obs} that is reached after string $t \in P[\mathcal{L}(G)]$ will contain all states of G that can be reached after any of the strings in

$$P^{-1}(t) \cap \mathcal{L}(G) .$$

In other words, the state of G_{obs} is the union of all the states of G that are consistent with the observable events that have occurred so far (namely, string t). In this sense, the state of G_{obs} is an estimate of the current state of G.

Example 2.22 (Automaton with unobservable events)
Consider the automaton G shown in Fig. 2.24. The set of unobservable events of G is

$$E_{uo} = \{e_d, u, v\} .$$

The observer G_{obs} of G is shown in Fig. 2.25. State $\{8, 9, 10, 11, 12\}$ reached after string ab in G_{obs} records the fact that after observing ab, all the possible states where G may be are 8, 9, 10, 11, and 12.

We conclude this discussion about state estimation with observers with two remarks. First, the state of G_{obs} may be a singleton, meaning that at that time, the state of G is known exactly. However, uncertainty may arise again if unobservable events are possible in the future. Second, not all strings in $\mathcal{L}(G)$ whose projections enter a state in G_{obs} need to have the same projection. For instance, in Example 2.22, $s_1 = e_d uc$ has projection c which leads to state $\{5, 6, 7\}$ in G_{obs} of Fig. 2.25; string $s_2 = ae_d ug$, whose projection is ag, also leads to state $\{5, 6, 7\}$. But if the projection of $s \in \mathcal{L}(G)$ leads to state x of G_{obs}, then all strings $t \in \mathcal{L}(G)$ such that $P(t) = P(s)$ also lead, in their projected versions, to state x of G_{obs}, since G_{obs} is deterministic.

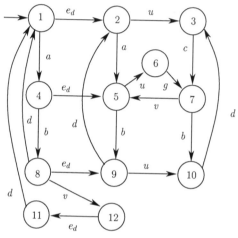

Figure 2.24: Automaton G of Example 2.22.
The set of unobservable events is $E_{uo} = \{e_d, u, v\}$.

Diagnostics

In many applications where the system model contains unobservable events, we may be interested in determining if certain unobservable events *could have occurred* or *have occurred with certainty*. This is the problem of *diagnostics*. If these unobservable event of interest model faults of system components, then knowing that one of these events has occurred is very important when monitoring the performance of the system. The key point here is that as we continue to make observations of the system behavior, we can reduce the uncertainty about prefixes of the string of events generated by the system. In Example 2.22 for instance, after observing $t = a$, we do not know if the system has executed unobservable event e_d or not. However, after observing $s = tg = ag$, we know with certainty that event e_d has occurred. Thus we have *diagnosed* the occurrence of unobservable event e_d after observing s.

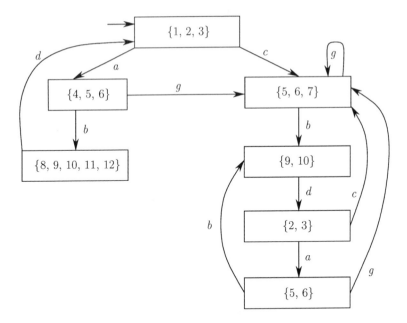

Figure 2.25: Observer G_{obs} of G in Fig. 2.24 (Example 2.22).

We can make this kind of inferencing about the past "systematic" if we make some simple modifications to the standard construction of the observer. To avoid confusion with G_{obs}, let us call this new kind of observer a *diagnoser* and denote it by G_{diag}. The idea is to attach, in the states of G_{diag}, "labels" to the states of G. For the sake of simplicity, let us assume that we are concerned with diagnosing a single unobservable event, say $e_d \in E_{uo}$. The procedure can be extended to diagnosing a set of unobservable events; see the references at the end of the chapter. We will use two kinds of labels: N for "No, e_d has not occurred yet" and Y for "Yes, e_d has occurred."

Recall the procedure for constructing G_{obs} presented in Section 2.3.3 and remember that each unobservable event in G is treated as ε in the context of this procedure. We shall not present a detailed algorithm for the construction of G_{diag} but only highlight the key modifications to the construction of G_{obs} for the purpose of building G_{diag}. These are:

M1. When building the unobservable reach of the initial state x_0 of G,

(a) Attach the label N to states that can be reached from x_0 by unobservable strings in $[E_{uo} \setminus \{e_d\}]^*$;

(b) Attach the label Y to states that can be reached from x_0 by unobservable strings that contain at least one occurrence of e_d;

(c) If state z can be reached both with and without executing e_d, then create two entries in the initial state set of G_{diag}: zN and zY.

M2. Build the subsequent states of G_{diag} by following the rules for G_{obs} (with the above modified way to build unobservable reaches) and by *propagating* the label Y. Namely, any state reachable from state zY should get the label Y, to indicate that e_d has occurred in the process of reaching z, and thus in the process of reaching the new state.

A consequence of modification M1.(c) is that G_{diag} may have more states than G_{obs}. The following example illustrates how and when the above two modifications come into play in building a diagnoser.

Example 2.23 (Diagnoser automaton)

Figure 2.26 shows the diagnoser G_{diag} for the automaton G in Fig. 2.24, where e_d is the event to be diagnosed. If we can compare G_{diag} with G_{obs} shown in Fig. 2.25, we can see that G_{diag} has more states than G_{obs}, due to the fact that some states of G may appear twice in a state of G_{diag}, once with label N and once with label Y. This is the case after the observed string abd, where both $1N$ and $1Y$ appear in the state of G_{diag}.

The state transition diagram of G_{diag} shows that after observing either event c or event g, we are sure of the past occurrence of e_d since all states of G that appear as entries in states of G_{diag} have label Y. While we may not know the exact state of G, we know for sure that event e_d has occurred. On the other hand, if the observed string does not contain any c or g events, we are in states of G_{diag} where there are entries with label N and entries with label Y. Thus, while it is possible that event e_d might have happened, we are uncertain about its occurrence.

This example shows how we can draw inferences about the occurrence of e_d by examination of the diagnoser:

- If all the states of G in the current state of G_{diag} have label N, then we are sure that the event e_d has not occurred yet.

- If all the states of G in the current state of G_{diag} have label Y, then we are sure that the event e_d has occurred at some point in the past.

- If the current state of G_{diag} contains at least one state of G with label N and at least one state of G with label Y, then the event e_d may or may not have occurred in the past.

Example 2.23 and Fig. 2.26 illustrate also an important issue in diagnostics, namely, is it possible to always remain uncertain about the occurrence of the event of interest, e_d, after this event happens in the system? This would mean that in such a system, the occurrence of e_d would not always be diagnosable. Let

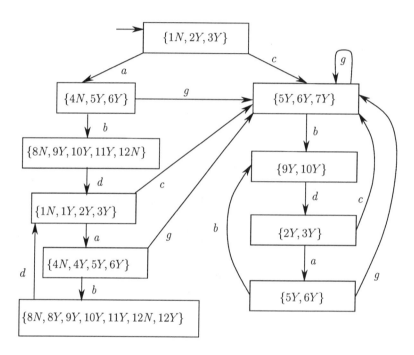

Figure 2.26: Diagnoser G_{diag} of G in Fig. 2.24 (Example 2.23).

us call a state of the diagnoser where both labels N and Y appear an *uncertain state*. When the diagnoser has a cycle of uncertain states, as is the case in G_{diag} of Fig. 2.26, we would like to know if the system can execute an arbitrarily long string containing e_d and whose projection would stay in the cycle of uncertain states in the diagnoser. If this happens, then we know that the occurrence of e_d along that string cannot be diagnosed. If we examine closely G and G_{diag} of Example 2.23, we will see that we can associate to the cycle of uncertain states in G_{diag} the cycle "$2 \rightarrow 5 \rightarrow 9 \rightarrow 2$" in G, and moreover any string that remains in this cycle in G must contain event e_d. Consequently, such a string will remain in the uncertain cycle in G_{diag}, meaning that this occurrence of e_d is never diagnosed with certainty.

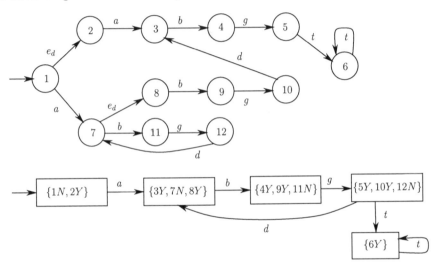

Figure 2.27: System and corresponding diagnoser for unobservable event e_d. *Observe that the diagnoser will exit its cycle of uncertain states, via event t, if event e_d occurs in the system. This will occur in at most six (observable) transitions after the occurrence of e_d, namely suffix bgdbgt.*

It is important to emphasize that presence of a cycle of uncertain states in a diagnoser does not necessarily imply inability to diagnose with certainty an occurrence of event e_d. Consider the system and diagnoser shown in Fig. 2.27 where e_d, the event to be diagnosed, is the only unobservable event in the system. This diagnoser has a cycle of uncertain states. However, we cannot form a cycle in the system from entries that appear in the uncertain states in the diagnoser and have the Y label. Indeed, the only system cycle that can cause the diagnoser to remain in its cycle of uncertain states is the cycle "$7 \rightarrow 11 \rightarrow 12 \rightarrow 7$", and these states all have the N label in the corresponding diagnoser states. Thus, if event e_d happens, the diagnoser will eventually enter state $\{6Y\}$. We can therefore say that the occurrence of event e_d in this system is always diagnosable. Here, the fact that the system may cycle in states 7, 11, and 12, causing the diagnoser to cycle in its cycle of uncertain states, is not interpreted as a lack of diagnosability, since strings causing such cycling do not contain e_d; in other words, it is admissible to remain uncertain for an arbitrarily long time, as long as the event of interest has not happened.

Model-Building for Estimation and Diagnostics

It is instructive to revisit at this point the issue of the "conversion" from Moore automata (automata with state outputs) to standard automata that we discussed in Section 2.3.5, but this time accounting for the presence of unobservable events in the system model. Recall that this conversion consisted in renaming a transition from state x to state y due to event a with a new event, (a, output of state y). Thus, the information about the output of a state is captured in the name of the events entering it. This works fine if event a is observable, as the interpretation of the new event is: event a happens and is immediately followed by an "output" (e.g., sensor reading) equal to the output of state y. However, this conversion is inadequate if a is an unobservable event, as the observability properties of the new event (a, output of state y) are not well defined: the a part is not observable, but the output is observable by definition of state outputs. A fix to this problem is to add a new state, say x_{new}, and have two transitions: unobservable event a from state x to state x_{new} and observable event output of state y from state x_{new} to state y. Equivalently, we could label the event from state x_{new} to state y as follows: output changed from "output of state x" to "output of state y".

The following example illustrates this conversion process.

Example 2.24 (Pump-valve-controller system)

Consider a portion of a heating system (as in Fig. 1.30 in Chapter 1) consisting of a pump, valve, and controller, together with two sensors, a valve flow sensor and a pump pressure sensor. Assume that the outputs of these two sensors have been discretized to two possible values (i.e., one bit) each: Flow (F) or No Flow (NF) for the valve flow sensor, and Positive Pressure (PP) or No Pressure (NP) for the pump pressure sensor.

The models of the three individual components are shown in Fig. 2.28. The model of the valve accounts for possible failure modes of this component, in states SC (for "Stuck Closed") and SO (for "Stuck Open"). The fault events STUCK_CLOSED and STUCK_OPEN that take the valve to the SC and SO states are assumed to be unobservable. All other events in Fig. 2.28 are assumed to be observable. It can be verified that the parallel composition of the three automata in Fig. 2.28 is an automaton with 12 reachable states. The (discretized) outputs of the sensors for these 12 states are shown in Table 2.1.

The conversion of the automaton Pump || Valve || Controller with its as-

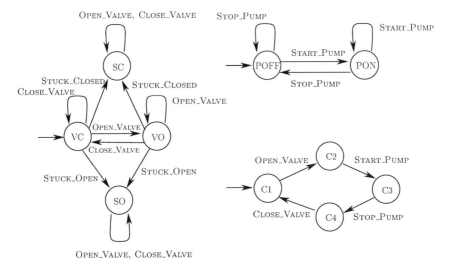

Figure 2.28: Component models of pump, valve, and controller (Example 2.24). *All events are assumed to be observable except STUCK_CLOSED and STUCK_OPEN in the model of the valve.*

sociated sensor outputs to a standard automaton requires the addition of one new state, denoted by x_{new}, in addition to the renaming of all observable events as described above. The addition of state x_{new} is necessary because unobservable event STUCK_CLOSED from state (PON, VO, C3) to state (PON, SC, C3) causes a change of flow sensor reading from F to NF. The event causing the transition from state x_{new} to state (PON, SC, C3) is observable and called $F \to NF$ to reflect that change in sensor reading. The (standard) automaton model for this system of three components and two sensors is shown in Fig. 2.29.

SUMMARY

- If we think of an event set E as an alphabet, then event sequences can form strings (or words), and a language is a set of such strings formed from this alphabet.

- An automaton G is a device capable of representing languages according to some rules. It is defined by event set E, state set X, transition function f, initial state x_0, and marked state set X_m. The events in E label the state transitions of G. All the strings of events that can be formed by sequences of state transitions from x_0 to any state of G constitute the language generated by the automaton; among these strings, those that

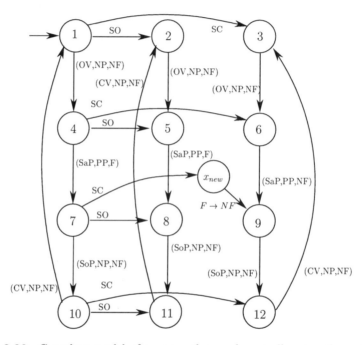

Figure 2.29: Complete model of pump, valve, and controller, together with flow sensor and pressure sensor (Example 2.24).

For ease of reading, states are denoted by their corresponding numbers in Table 2.1 and event names have been abbreviated: SC *for* STUCK_CLOSED, SO *for* STUCK_OPEN, OV *for* OPEN_VALVE, CV *for* CLOSE_VALVE, SaP *for* START_PUMP, *and* SoP *for* STOP_PUMP. *All events are observable except for* SO *and* SC.

State No.	State Name	Sensor Readings
1	(POFF, VC, C1)	NP, NF
2	(POFF, SO, C1)	NP, NF
3	(POFF, SC, C1)	NP, NF
4	(POFF, VO, C2)	NP, NF
5	(POFF, SO, C2)	NP, NF
6	(POFF, SC, C2)	NP, NF
7	(PON, VO, C3)	PP, F
8	(PON, SO, C3)	PP, F
9	(PON, SC, C3)	PP, NF
10	(POFF, VO, C4)	NP, NF
11	(POFF, SO, C4)	NP, NF
12	(POFF, SC, C4)	NP, NF

Table 2.1: Sensor map for the pump-valve-controller example (Example 2.24).

take G from x_0 to a state in X_m constitute the language marked by G.

- An automaton that has states that are not coaccessible to states in X_m, that is, for which the prefix-closure of the language marked is a proper subset of the language generated, is said to be blocking.

- Automata can be composed by product and parallel composition.

- Observer automata are deterministic automata that can be built from nondeterministic automata (with ε-transitions and/or multiple transitions with same event label out of a state) and from deterministic automata with unobservable events. At any time, their state is the best estimate of the state of the automaton from which they are built, based on the events observed so far.

- Diagnoser automata are a variation of observer automata where labels are appended to state estimates in order to track the occurrence of one or more unobservable events of interest.

- The class of regular languages is the class of languages that can be represented by finite-state (deterministic or nondeterministic) automata. Not all languages over a given event set are regular. The class of regular languages is closed under prefix-closure, Kleene-closure, complement, concatenation, union, and intersection. Observer and diagnoser automata of finite-state automata also have a finite state space.

- Regular expressions are another mechanism for representing regular languages.

- The state space of an automaton can be minimized as well as refined without changing the language properties of the automaton.

- Automaton models of DES can be used to study many properties of DES, among them: safety, blocking, state estimation, and diagnostics.

Software for Manipulation of Automata

Readers interested in software implementations of many of the operations on automata described in this chapter may download the library of routines UMDES-LIB available at the Web site www.eecs.umich.edu/umdes. Some of the problems that follow in the next section require the use of such software as the state spaces involved are too large for hand calculations.

PROBLEMS

2.1 Answer true or false. (The notation e^* is shorthand for $\{e\}^*$.)

(a) $baa \in a^*b^*a^*b^*$

(b) $b^*a^* \cap a^*b^* = a^* \cup b^*$

(c) $a^*b^* \cap c^*d^* = \emptyset$

(d) $abcd \in (a(cd)^*b)^*$

2.2 Consider three languages L_1, L_2, and L_3, and assume that L_1 does not contain the empty string. Show that if $L_2 = L_1 L_2 \cup L_3$, then $L_2 = L_1^* L_3$.

2.3 Two languages L_1 and L_2 are said to be *nonconflicting* if

$$\overline{L_1 \cap L_2} = \overline{L_1} \cap \overline{L_2} .$$

Give an example of two *conflicting* languages.

2.4 A language $L \subseteq M = \overline{M}$ is said to be M-closed if $L = \overline{L} \cap M$.

(a) Show that if L_i is M-closed for $i = 1, 2$ then so is $L_1 \cap L_2$.

(b) Show that if $L_a = \overline{L_a}$ and we define $L_{am} = L_a \cap M$, then L_{am} is M-closed.

2.5 Consider the English language alphabet and construct a deterministic finite-state automaton that marks the words *man* and *woman*.

2.6 Construct an automaton that marks the following language:

$\{s \in \{a, b\}^* : \text{each } a \text{ in } s \text{ is immediately preceded and followed by a } b\}$.

2.7 A combination lock for a safe is designed based on two symbols only, 0 and 1. The combination that opens the safe consists of exactly four symbols such that the last two are the same (either two 0's or two 1's). If the combination is not correct, an alarm sounds. Construct a finite-state automaton to describe the process of trying to open the safe.

2.8 Build automaton models for M_1, M_2, and the robot arm described in Problem 1.3 of Chapter 1. (Choose physically meaningful states for system components and an event set that captures the behavior of individual components as well as their coupling in the system.)

2.9 A computer processor is preceded by a queue that can accommodate one job only, so that the total capacity of this system is two jobs. Two different job types request service from the processor, and type 1 has higher priority than type 2. Thus, if a type 1 job arrives and finds a type 2 job in service, it preempts it; the type 2 job is returned to the queue and the type 1 job receives service. If any job arrives and finds the system full, it is rejected and lost. Build an automaton model of this system.

2.10 A computer system operates with two parallel processors P_1 and P_2. The total capacity (queue and server included) of P_1 is $K_1 = 1$, and that of P_2 is $K_2 = 2$. The system receives two types of jobs, labeled J_1 and J_2. Jobs of type J_1 must be processed at P_1, and jobs of type J_2 must be processed at P_2. When a job is processed, it leaves the system. If a job finds a full queue upon arrival, then the job is simply rejected. Build an automaton model of this system.

2.11 A system consists of three servers as shown in Fig. 2.30. Arriving customers first enter the queue preceding server 1, and after completing service they are routed to either server 2 or server 3. The routing policy is to always send a customer to the server with the shortest queue, and in case of a tie to send to server 2. Customers completing service at servers 2 and 3 leave the system. A customer arriving from the outside is accepted into this system only as long as the total number of customers present is less than or equal to 2; otherwise, customers are rejected. Build an automaton model of this system.

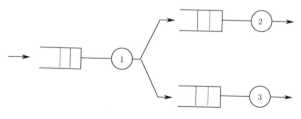

Figure 2.30: Problem 2.11.

2.12 Build an automaton that models the concurrent execution of the two database transactions considered in the "Database Systems" example in Section 1.3.4 of Chapter 1. Assume no concurrency control is done, that is, the language generated by your automaton should include all possible schedules, admissible or not. Explain your choice of marked state(s).

2.13 A useful control mechanism in many DES is a timeout, where some state transition is forced to take place when a "timeout event" occurs. Suppose a computer processor operates as follows. When a job arrives, it is broken down into two tasks. First, task 1 is executed, and, if execution is successfully completed, then task 2 is executed. If the execution of task 2 is also successfully completed, then the job is done and it leaves the system. If, however, task 1 takes too long, a timeout event is generated, and the entire job leaves at once. Similarly, if task 2 takes too long, another timeout event is generated, and the job leaves. Assume the system has a total capacity of one job. Four events are required to model this process: JOB_ARRIVAL, SUCCESSFUL_COMPLETION_OF_TASK_1, SUCCESS-FUL_COMPLETION_OF_TASK_2, TIMEOUT. Based on this set of events, build an automaton to represent this process (three states should suffice).

2.14 For each of the state transition diagrams shown in Fig. 2.31, construct a simple queueing model and provide a physical description of the process it represents. Be sure to define what each event stands for in your queueing models.

(a) $E = \{e_1, e_2, e_3\}$.

(b) $E = \{e_1, e_2\}$.

(c) $E = \{e_1, e_2, e_3\}$.

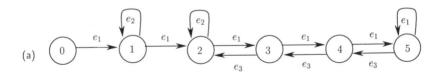

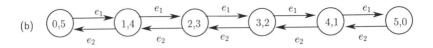

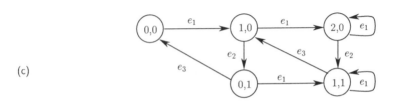

Figure 2.31: Problem 2.14.

2.15 Consider the automaton in Fig. 2.2 and denote it as G. Find another automaton H such that the product automaton $H \times G$: (i) is trim, (ii) generates the same language as G, and (iii) marks only those traces in $\mathcal{L}_m(G)$ that end after three or more consecutive occurrences of a (as opposed to one or more in G).

2.16 Let $E = \{a, b\}$, $E_{uc} = \{b\}$, $M = \overline{\{a\}\{b\}^*}$ and $L = \{ab\}$. Build automata that mark L and M. Use these automata to build an automaton that generates the language

$$\overline{L}E_{uc}^* \cap M$$

using the operations on automata presented in Section 2.3.

2.17 Consider the problem of the dining philosophers presented in Example 2.17.

(a) Build an automaton C such that

$$C||(P_1||P_2||F_1||F_2) = CoAc(P_1||P_2||F_1||F_2) \ .$$

(b) Assume that there are five philosophers and five forks. How many deadlock states are there in this case? Are there any livelocks that cause blocking?

2.18 A workcell consists of two machines M_1 and M_2 and an automated guided vehicle AGV.[4] The automaton models of these three components are shown in Fig. 2.32. The complete system is $G = M_1||M_2||AGV$.

(a) Find G.

(b) Is G blocking or nonblocking?

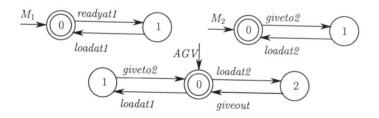

Figure 2.32: Problem 2.18.

2.19 Consider two event sets E_o and E, with $E_o \subseteq E$, and the corresponding natural projection $P : E^* \rightarrow E_o^*$. A language $K \subseteq M$, where $M = \overline{M} \subseteq E^*$, is said to be *normal* with respect to M and P if

$$\overline{K} = P^{-1}[P(\overline{K})] \cap M \ .$$

---

[4]From W. M. Wonham, "Notes on Control of Discrete-Event Systems," Dept. of Elec. and Comp. Eng., Univ. of Toronto, 1994.

In words, $\overline{K}$ can be reconstructed from its projection and M. Give an example of a normal language and an example of a language that is not normal.

2.20 Build the observer of the nondeterministic automaton shown in Fig. 2.33.

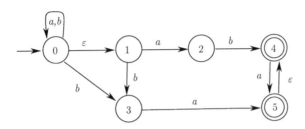

Figure 2.33: Problem 2.20.

2.21 Construct a nondeterministic automaton whose observer is not minimum-state, that is, it has equivalent states.

2.22 Consider the event set $E = \{a, b, g\}$. Find:

(a) A regular expression for the language with each string containing at least one b.

(b) A regular expression for the language with each string containing exactly two b's.

(c) A regular expression for the language with each string containing at least two b's.

(d) Two more different regular expressions for (c).

2.23 Consider the event set $E = \{a, b\}$. For each of the following pairs of regular expressions, either show that they denote the same language, or provide a counterexample to this fact:

(a) $(a + b)^*$ and $(a + b)^* ab(a + b)^* + b^* a^*$.

(b) $(a^* + b)^*$ and $(a + b)^*$.

(c) $(ab)^* ab$ and $b(a + b)^* ab$.

(d) $(a^* b)^* a^*$ and $a^* (ba^*)^*$.

2.24 Given the event set $E = \{a, b, g\}$, construct a finite-state automaton that marks:

(a) The language $\{abg, abb, bgg\}$.

(b) The language denoted by the regular expression $(ab)^* g$.

(c) The language denoted by the regular expression $a(ba)^*a^*$.

(d) The language denoted by the regular expression $(a+g)(ba)^*+a(ba)^*a^*$.

2.25 Prove or disprove the following statements:

(a) Every subset of a regular language is regular.

(b) Every regular language has a proper subset that is also a regular language.

(c) If language L is regular, then so is the language $\{xy|x \in L$ and $y \notin L\}$.

(d) The language defined by $\{w|w = w^R\}$, where w^R denotes the reverse of string w, is regular.

2.26 Find a regular expression for the language marked by:

(a) The automaton shown in Fig. 2.34 (a).

(b) The automaton shown in Fig. 2.34 (b).

(c) The automaton shown in Fig. 2.34 (c).

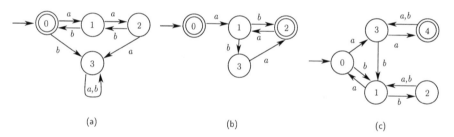

Figure 2.34: Problem 2.26.

2.27 Find a minimum-state automaton (with completely defined transition function) that marks the same language as the automaton in Fig. 2.35.

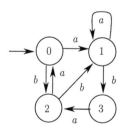

Figure 2.35: Problem 2.27.

2.28 Construct an automaton that marks the complement of the language represented by the regular expression $a(ba)^*a^*$; the set of events is $E = \{a, b, c\}$.

2.29 Consider the automaton H depicted in Fig. 2.36. The set of events is $E = \{a, b, c, d\}$ and the set of observable events is $E_o = \{b, c, d\}$. Build H_{obs}, the observer of H. Compare the blocking properties of H with those of H_{obs}.

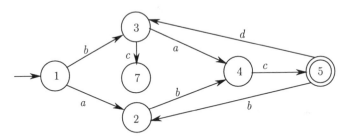

Figure 2.36: Problem 2.29.

2.30 Build the diagnoser of unobservable event e for the nondeterministic automaton shown in Fig. 2.37. (All the events are observable but e.)

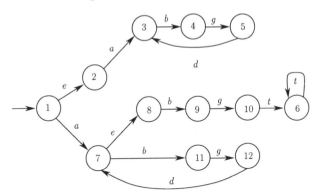

Figure 2.37: Problem 2.30.

2.31 Consider the pump-valve-controller system discussed in Example 2.24. Build the diagnosers of the complete system model depicted in Fig. 2.29, for the two faults STUCK_CLOSED and STUCK_OPEN. Are these faults diagnosable?

2.32 Consider the pump-valve-controller system discussed in Example 2.24. Modify the controller by interchanging the order of the command events: START_PUMP followed by OPEN_VALVE to go from state C1 to state C3, and CLOSE_VALVE followed by STOP_PUMP to return to state C1.

(a) Calculate the parallel composition of the pump, valve, and modified controller.

(b) Identify the output of the flow and pressure sensors for each state of the composed system.

(c) Transform the above Moore automaton into a "standard" automaton, by renaming events and adding new states, as necessary, as described in Section 2.5.2.

(d) Build the diagnosers of the complete system obtained in (c), for the two faults STUCK_CLOSED and STUCK_OPEN. Are these faults diagnosable?

2.33 Show that if G_{diag} has a cycle of uncertain states, then you can always construct a corresponding cycle in G by using only entries with the N label in the states of G_{diag}. (Recall that Fig. 2.27 shows that it is *not* always possible to construct a corresponding cycle in G using only entries with the Y label in the states of G_{diag}.)

2.34 Consider the sequential transmission of packets between two nodes, denoted A and B, in a data network. The transmission channel is noisy and thus packets can "get lost" (namely, they can be corrupted by errors or dropped if memory buffers are full). One class of *protocols* (or distributed algorithms) that control the transmission between A and B, is the class of STOP and WAIT protocols:

1. A sends packet to B then starts timer;

2.a. A receives acknowledgement (ACK) from B: Go back to 1;

2.b. A times out: A resends packet to B and starts timer. Go back to 2.

A little thought shows that node B needs to know if a packet received is a new packet or a retransmission of a previous packet by node A; this is because the transmission delay is not known exactly and also because ACKs can get lost too. A widely-used strategy to address this problem is to add one bit to each packet header, resulting in "0-packets" and "1-packets". This header bit is flipped by node A each time A sends a new packet (as opposed to a retransmission). This is called the *Alternating Bit Protocol* (abbreviated as ABP). We wish to model a simple version of ABP, for transmission from A to B only, and with a half-duplex channel. We assume that the channel can only contain one packet at a time. Based on the above discussion, we define the following list of nine events for this system: [5]

Event 0: contents of channel lost

---

[5] Our model follows that in *Computer Networks - 3rd Ed.*, by A. Tanenbaum, Prentice-Hall, 1996, Sections 3.3.3 and 3.5.1.

Event 1: 0-packet received by B; ACK sent by B to A; 0-packet delivered by node B

Event 2: ACK received by A; 1-packet sent by A to B

Event 3: 1-packet received by B; ACK sent by B to A; 1-packet delivered by node B

Event 4: ACK received by A; 0-packet sent by A to B

Event 5: 0-packet received by B; ACK sent by B to A

Event 6: 1-packet received by B; ACK sent by B to A

Event 7: timeout at A; 0-packet retransmitted to B

Event 8: timeout at A; 1-packet retransmitted to B

(A packet is "delivered by node B" when B forwards the packet to the application program running at that node.)

(a) Using the above events, build automaton models for the sender at node A, the receiver at node B, and the channel. The models for the sender and the receiver automata, denoted by S and R, respectively, should have two states each, namely one state for sending (receiving) 0-packets and one state for sending (receiving) 1-packets. The channel automaton, denoted by C, should have four states, for the four possible channel contents: empty, 1-packet, 0-packet, and ACK.

(b) Assume that the initial state of the system is that where S is sending a 0-packet, R is expecting a 0-packet, and C contains a 0-packet. Build $S||R||C$ for this choice of initial state. (Your answer should have 10 reachable states.)

(c) Verify that $S||R||C$ is nonblocking.

(d) To verify the correctness of ABP, we build a *monitor automaton*, denoted by M, that captures the two specifications:

- Spec-1: events 1 and 3 should alternate;
- Spec-2: no packet should be lost, that is, there should be no 2 events between events 3 and 1 and no 4 events between events 1 and 3.

Construct the desired M and then verify that $\mathcal{L}(S||R||C) \subseteq \mathcal{L}(M)$.

2.35 A card game is played by three players.[6] Each player uses cards of two colors, red (R) and black (B). At any point in time, each player has a card on the table in front of him with its color showing. The player then uses a specific rule (known only to this player) for replacing this card by another one. We assume that the card replacement for all three players is done simultaneously.

---

[6]Problem due I. Aleksander, *Designing Intelligent Systems*, Unipub, New York, 1984.

(a) Suppose the players use the following rules:
Player 1: Plays R if and only if both he and Player 2 show R;
Player 2: Plays the color opposite to that shown by Player 1;
Player 3: Plays R if and only if both he and Player 2 show B.
Model the game as an automaton, where the only event is "everybody plays." What is the state space X? Can any of the states be visited more than once during the game?

(b) Suppose Player 2 changes his strategy as follows. Play the color opposite to that shown by Player 1, unless BRR is showing on the table. How are the state space and state transition function of the automaton affected? Are any of the states revisited in this case?

(c) Now suppose further that Player 3 also decides to modify his strategy as follows. He determines the next card to play by tossing a fair coin. The outcome of the coin now represents an event. What are X (state space) and E (event set) in this case? Construct the new state transition function, and identify which states can be revisited?

(d) Suppose you are an outside observer and in each of the above three cases you know the players' strategies. You are allowed to look at the cards initially showing on the table and then bet on what the next state will be. Can you guarantee yourself a continuous win in each case (a), (b) and (c)? In case (c), if initially RRR is showing, are there any states you should never bet on?

2.36 A simple manufacturing process involves two machines, M_1 and M_2, and a buffer B in between.[7] There is an infinite supply of parts to M_1. When a part is processed at M_1, it is placed in B, which has a capacity of one part only. The part is subsequently processed by M_2. Let us suppose that we build the uncontrolled model of M_i, $i = 1, 2$, as follows. Each machine has three states: Idle (the initial state), Processing, and Down. Each machine has four transitions: event START from Idle to Processing, event END from Processing to Idle, event BREAKDOWN from Processing to Down, and event REPAIR from Down to Idle. The behavior of the system need to be restricted to satisfy the following rules:

(i) M_1 can only begin processing if the buffer is empty;

(ii) M_2 can only begin processing if the buffer is full;

(iii) M_1 cannot begin processing if M_2 is down;

(iv) If both machines are down, then M_2 gets repaired first.

Answer the following questions.

---

[7] From P. J. Ramadge & W. M. Wonham, "Supervisory Control of a Class of Discrete-Event Processes," *SIAM J. Contr. Opt.*, May 1987.

(a) Construct an automaton that represents the admissible behavior, as captured by (i) to (iv) above. This automaton should generate a sublanguage of $\mathcal{L}(M_1||M_2)$.

(b) Suppose that the events START and REPAIR of each machine can be controlled (that is, enabled or disabled) by a controller. For each state of your automaton in (a) above, identify which *feasible* events in M_1 and M_2 the controller should disable.

2.37 Consider a DES G with event set E, along with its observer G_{obs} corresponding to projection $P : E^* \rightarrow E_o^*$. Define the inverse projection operation on automata, $P^{-1}(G_{obs})$, to consist of adding self-loops for all events in $E_{uo} = E \setminus E_o$ at all states of G_{obs}. Show that

$$[G \times P^{-1}(G_{obs})]_{obs} = G_{obs}$$

where the equality is up to a renaming of the states (i.e., the two automata are isomorphic).

SELECTED REFERENCES

■ *Languages and Automata Theory*[8]

  - Hopcroft, J.E., and J.D. Ullman, *Introduction to Automata Theory, Languages, and Computation*, Addison-Wesley, Reading, MA, 1979.

  - Savage, J.E., *Models of Computation*, Addison-Wesley, Reading, MA, 1998.

  - Sipser, M., *Introduction to the Theory of Computation*, PWS Publishing Company, Boston, 1997.

■ *Automata and Related Modeling Formalisms*

  - Arnold, A., *Finite Transition Systems*, Prentice-Hall, Englewood Cliffs, NJ, 1994.

  - Harel, D., and M. Politi, *Modeling Reactive Systems with Statecharts: The Statemate Approach*, Wiley, New York, 1998.

  - Kurshan, R.P., *Computer-Aided Verification of Coordinating Processes: The Automata-Theoretic Approach*, Princeton University Press, NJ, 1994.

---

[8]From the viewpoint of the theory of computation.

- *Some Other (Untimed) Modeling Formalisms for Discrete Event Systems*[9]

  - Baeten, J.C.M., and W.P. Weijland, *Process Algebra*, Volume 18 of *Cambridge Tracts in Theoretical Computer Science.* Cambridge University Press, Great Britain, 1990.

  - Hoare, C.A.R., *Communicating Sequential Processes*, Prentice-Hall, Englewood Cliffs, NJ, 1985.

  - Inan, K., and P.P. Varaiya, "Algebras of Discrete Event Models," *Proceedings of the IEEE*, Vol. 77, pp. 24-38, January 1989.

  - Milner, R., *Communication and Concurrency*, Prentice-Hall, New York, 1989.

- *Miscellaneous*[10]

  - Burch, J.R., E.M. Clarke, K.L. McMillan, D.L. Dill, and L.J. Hwang. "Symbolic Model Checking: 10^{20} States and Beyond," *Information and Computation*, Vol. 98, No. 2, pp. 142–170, 1992.

  - Comer, T.H., C. Leiserson, and R.L. Rivest, *Introduction to Algorithms*, MIT Press and McGraw-Hill, New York, 1990.

  - Dini, P., R. Boutaba, and L. Logrippo (Eds.), *Feature Interactions in Telecommunication Networks IV*, IOS Press, Amsterdam, The Netherlands, 1997.

  - Holzmann, G.J., *Design and Validation of Computer Protocols*, Prentice-Hall, Englewood Cliffs, NJ, 1991.

  - McMillan, K.L., *Symbolic Model Checking*, Kluwer Academic Publishers, Boston, 1993.

  - Papadimitriou, C., *The Theory of Database Concurrency Control*, Computer Science Press, Rockville, MD, 1986.

  - Sampath, M., R. Sengupta, S. Lafortune, K. Sinnamohideen, and D. Teneketzis, "Failure Diagnosis Using Discrete Event Models," *IEEE Transactions on Control Systems Technology*, Vol. 4, pp. 105–124, March 1996.

---

[9]This brief list excludes Petri Nets, which are considered in Chapter 4.

[10]These references provide further details on some issues mentioned in this chapter.

Chapter 3

Supervisory Control

3.1 INTRODUCTION

The situation under consideration in this chapter is that of a given DES, modeled at the untimed (or logical) level of abstraction, and whose behavior must be modified by feedback control in order to achieve a given set of *specifications*. This is reminiscent of the feedback control loop introduced in Chapter 1, in Section 1.2.8. Let us assume that the given DES is modeled by automaton G, where the state space of G need not be finite. Let E be the event set of G. Automaton G models the "uncontrolled behavior" of the DES. The premise is that this behavior is not satisfactory and must be "modified" by control; modifying the behavior is to be understood as *restricting the behavior to a subset of* $\mathcal{L}(G)$. In order to alter the behavior of G we introduce a *supervisor*; supervisors will be denoted by S. Note that we separate the "plant" G from the "controller" (or supervisor) S, as is customary in control theory. This raises two questions: (a) What do we mean by specifications? and (b) How does S modify the behavior of G?

Specifications are interpreted as follows. The language $\mathcal{L}(G)$ contains strings that are not acceptable because they violate some condition that we wish to impose on the system. It could be that certain states of G are undesirable and should be avoided. These could be states where G blocks, via deadlock or livelock; or they could be states that are physically inadmissible, for example, a collision of a robot with an automated guided vehicle or an attempt to place a part in a full buffer in an automated manufacturing system. Moreover, it could be that some strings in $\mathcal{L}(G)$ contain substrings that are not allowed. These substrings may violate a desired ordering of certain events, for example, requests for the use of a common resource should be granted in a "first-come first-served" manner. Thus, we will be considering *sublanguages* of $\mathcal{L}(G)$ that represent the "legal" or "admissible" behavior for the controlled system. In some cases, we may be interested in a range of sublanguages of $\mathcal{L}(G)$, of the form

$$L_r \subset L_a \subseteq \mathcal{L}(G)$$

where the objective is to restrict the behavior of the system to within the range delimited by L_r and L_a; here, L_a is interpreted as the maximal admissible behavior and L_r as the minimal required behavior.

We will consider a very general control paradigm for how S interacts with G. Our objective is to understand key system- and control-theoretic properties that arise in the control of DES rather than to develop intricate control mechanisms. This is also consistent with our choice of automata as the primary modeling formalism for DES in this textbook. In our paradigm, S sees (or observes) some, possibly all, of the events that G executes. Then, S tells G which events in the current active event set of G are allowed next. More precisely, S has the capability of disabling some, but not necessarily all, feasible events of G. The decision about which events to disable will be allowed to change whenever S observes the execution of a new event by G. In this manner, S exerts *dynamic* feedback control on G. The two key considerations here are that S is limited in terms of observing the events executed by G and that S is limited in terms of disabling feasible events of G. Thus, we will be talking of the *observable events* in E – those that S can observe – and the *controllable events* in E – those that S can disable.

The organization of this chapter is as follows. We will treat the two questions posed above in Sections 3.2 (Question (b): Feedback loop of G and S) and 3.3 (Question (a): Admissible behavior). The issues of limited controllability and limited observability of S with respect to G will be studied in Sections 3.4 and 3.7, respectively. The issue of avoiding or mitigating blocking in the controlled system will be considered in Section 3.5. Finally, modular and decentralized control architectures will be treated in Sections 3.6 and 3.8, respectively.

Our approach in presenting the results in this chapter will be to characterize existence results for supervisors and associated key discrete-event system-theoretic properties in the framework of languages; these properties include controllability (Section 3.4), nonconflicting (Section 3.6), observability (Section 3.7), and co-observability (Section 3.8). This means that these properties will be defined in a manner that is independent of any particular DES modeling formalism chosen to represent languages, thus allowing for complete generality.

Algorithmic issues regarding verification of these properties and synthesis of supervisors will be presented in the framework of finite-state automata, under the assumption that we are dealing with regular languages. Consequently, the results in this chapter build extensively on the material presented in Chapter 2.

The examples presented in this chapter are fairly simple and often academic. Their purpose is to illustrate, by hand calculations, the concepts and algorithms that we introduce. More realistic examples from various application areas almost always involve automata containing more than "a few" states; such examples are tedious if not impossible to solve by hand and moreover it is impractical to draw their state transition diagrams. We refer the reader to the literature (see references at the end of this chapter) for such examples.

Finally, we note that the body of work presented in this chapter is part of what is generally known as "Supervisory Control Theory" in the field of DES. The key results of Supervisory Control Theory were developed by W. M. Wonham and P. J. Ramadge, and their co-workers, in the 1980's. Since then, many other researchers have also made contributions to the development and application of this discrete event system theory.

3.2   FEEDBACK CONTROL WITH SUPERVISORS

3.2.1   Controlled Discrete Event Systems

The formulation of the control problem considered in this chapter proceeds as follows. Consider a DES modeled by a pair of languages, L and L_m, where L is the set of all strings that the DES can generate and $L_m \subseteq L$ is the language of *marked* strings that is used to represent the completion of some operations or tasks; the definition of L_m is a modeling issue. L and L_m are defined over the event set E. L is always prefix-closed, that is, $L = \overline{L}$, while L_m need not be prefix-closed. Without loss of generality and for the sake of convenience, assume that L and L_m are the languages generated and marked by automaton

$$G = (X, E, f, \Gamma, x_0, X_m)$$

where X need not be finite; that is,

$$\mathcal{L}(G) = L \quad \text{and} \quad \mathcal{L}_m(G) = L_m .$$

Thus, we will talk of the "DES G."

We wish to adjoin a supervisor, denoted by S, to interact with G in a feedback manner, as depicted in Fig. 3.1. In this regard, let E be partitioned into two disjoint subsets

$$E = E_c \cup E_{uc}$$

where

- E_c is the set of *controllable* events: these are the events that can be prevented from happening, or disabled, by supervisor S;

- E_{uc} is the set of *uncontrollable* events: these events cannot be prevented from happening by supervisor S.

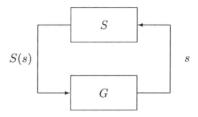

Figure 3.1: The feedback loop of supervisory control, with G representing the uncontrolled system and S the supervisor.

There are many reasons why an event would be modeled as uncontrollable: it is inherently unpreventable (for example, a fault event); it models a change of sensor readings not due to a command; it cannot be prevented due to hardware or actuation limitations; or it is modeled as uncontrollable by choice, as for example when the event has high priority and thus should not be disabled or when the event represents the tick of a clock.

Assume that all the events in E executed by G are observed by supervisor S. Thus, in Fig. 3.1, s is the string of all events executed so far by G and s is entirely seen by S. Control under partial event observation will be presented in Section 3.2.2 and then studied in Sections 3.7 and 3.8.

The control paradigm is as follows. The transition function of G can be controlled by S in the sense that the controllable events of G can be dynamically enabled or disabled by S. Formally, a supervisor S is a *function* from the language generated by G to the power set of E:

$$S : \mathcal{L}(G) \rightarrow 2^E .$$

For each $s \in \mathcal{L}(G)$ generated so far by G (under the control of S),

$$S(s) \cap \Gamma(f(x_0, s))$$

is the set of *enabled events* that G can execute at its current state $f(x_0, s)$. In other words, G cannot execute an event that is in its current active event set, $\Gamma(f(x_0, s))$, if that event is not also contained in $S(s)$. In view of the partition of E into controllable and uncontrollable events, we will say that supervisor S is *admissible* if for all $s \in \mathcal{L}(G)$

$$E_{uc} \cap \Gamma(f(x_0, s)) \subseteq S(s)$$

which means that S is not allowed to ever disable a feasible uncontrollable event. From now on, we will only consider admissible supervisors.

We call $S(s)$ the *control action* at s. S is the *control policy*. This feedback loop is an instance of dynamic feedback in the sense that the domain of S is $\mathcal{L}(G)$ and not X; thus the control action may change on subsequent visits to a state. We do not wish to specify at this point how the function S, whose domain may be an infinite set, is to be "realized" for implementation purposes. This will be discussed later when we present existence and synthesis results for supervisors. Given G and admissible S, the resulting closed-loop system is denoted by S/G (read as "S controlling G"). The controlled system S/G *is* a DES, and we can characterize its generated and marked languages. These two languages are simply the subsets of $\mathcal{L}(G)$ and $\mathcal{L}_m(G)$ containing the strings that remain feasible in the presence of S. This is formalized in the following definition.

Definition. (Languages generated and marked by S/G)
The *language generated* by S/G is defined recursively as follows:

1. $\varepsilon \in \mathcal{L}(S/G)$
2. $[(s \in \mathcal{L}(S/G))$ and $(s\sigma \in \mathcal{L}(G))$ and $(\sigma \in S(s))] \Leftrightarrow [s\sigma \in \mathcal{L}(S/G)]$.

The *language marked* by S/G is defined as follows:

$$\mathcal{L}_m(S/G) := \mathcal{L}(S/G) \cap \mathcal{L}_m(G) .$$ ♦

Clearly, $\mathcal{L}(S/G) \subseteq \mathcal{L}(G)$ and it is prefix-closed by definition. As for $\mathcal{L}_m(S/G)$, it consists exactly of the marked strings of G that survive under the control of S. Overall, we have the set inclusions

$$\emptyset \subseteq \mathcal{L}_m(S/G) \subseteq \overline{\mathcal{L}_m(S/G)} \subseteq \mathcal{L}(S/G) \subseteq \mathcal{L}(G) .$$

The empty string ε is always in $\mathcal{L}(S/G)$ since it is always contained in $\mathcal{L}(G)$; here, we have excluded the degenerate case where G is the empty automaton.[1] Thus, when we adjoin S to G, G is to start in its initial state at which time its possible first transition will be constrained by the control action $S(\varepsilon)$.

The notion of blocking defined in Chapter 2 for automata is also meaningful for the DES S/G, since this DES has a generated language and a marked language associated with it. The "obvious" way to state this definition is as follows.

Definition. (Blocking in controlled system)
The DES S/G is *blocking* if

$$\mathcal{L}(S/G) \neq \overline{\mathcal{L}_m(S/G)}$$

---

[1] Recall from Chapter 2, Section 2.3.1, that the term "empty automaton" refers to an automaton whose state space is empty and consequently whose generated and marked languages are empty.

and *nonblocking* when

$$\mathcal{L}(S/G) = \overline{\mathcal{L}_m(S/G)} \ . \qquad \blacklozenge$$

For the sake of illustrating the above concepts, let us consider a very simple example.

Example 3.1 (Controlled system S/G)
Assume that $\mathcal{L}(G) = \overline{\{abc\}}$ and $\mathcal{L}_m(G) = \{ab\}$. This system G is blocking since string $abc \in \mathcal{L}(G)$ is not a prefix of a string in $\mathcal{L}_m(G)$. If we adjoin to G supervisor S such that $S(\varepsilon) = \{a\}$, $S(a) = \{b\}$, and $S(ab) = \emptyset$, then the controlled system S/G is nonblocking since

$$\mathcal{L}(S/G) = \overline{\mathcal{L}_m(S/G)} = \overline{\{ab\}} \ .$$

However, if event c were uncontrollable, then we could not by control remove the blocking strings. If we keep the same $S(\varepsilon)$ and $S(a)$, then by admissibility $S(ab) = \{c\}$, and thus S/G is the same as G. If we set $S(a) = \emptyset$, then the controlled system will block after string a, while if we set $S(\varepsilon) = \emptyset$, then the controlled system will block after string ε.

Since the blocking properties of S/G are as much the result of S as of the structure of G, we shall say from now on that *supervisor S controlling DES G is blocking if S/G is blocking*, and *supervisor S is nonblocking if S/G is nonblocking*. Since marked strings represent completed tasks or record the completion of some particular operation (by choice at modeling), a blocking supervisor results in a controlled system that cannot terminate the execution of the task at hand. As was seen in Chapter 2, the notions of marked strings and blocking allow to model *deadlock* and *livelock* and thus they are very useful.

3.2.2 Control under Partial Observation

We now consider the situation where the supervisor does not "see" or "observe" all the events that G executes. In this case, the event set E is partitioned into two disjoint subsets

$$E = E_o \cup E_{uo}$$

where

- E_o is the set of *observable* events: these are the events that can be seen by the supervisor;

- E_{uo} is the set of *unobservable* events: these are the events that cannot be seen by the supervisor.

The principal causes of unobservability of events are the limitations of the sensors attached to the system and the distributed nature of some systems where events at some locations are not seen at other locations.

The feedback loop for control under partial observation is depicted in Fig. 3.2 and includes a natural projection P between G and the supervisor; to reflect the presence of P, we will denote partial-observation supervisors by S_P. Natural projections were defined in Chapter 2, in Section 2.3.2; see also Section 2.5.2. Here, the natural projection P is defined from domain E^* to codomain E_o^*, $P : E^* \to E_o^*$, as in Section 2.5.2.

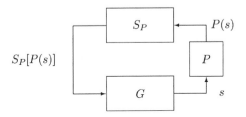

Figure 3.2: The feedback loop of supervisory control in the case of partial observation.
The projection $P : E^* \to E_o^*$ hides the unobservable events executed by G from supervisor S_P.

Due to the presence of P, the supervisor cannot distinguish between two strings s_1 and s_2 that have the same projection, that is, $P(s_1) = P(s_2)$; for such $s_1, s_2 \in \mathcal{L}(G)$, the supervisor will necessarily issue the same control action, $S_P[P(s_1)]$. In order to capture this fact, we define a partial-observation supervisor as a function

$$S_P : P[\mathcal{L}(G)] \to 2^E$$

and call S_P a *P-supervisor*. This means that the control action can change only after the occurrence of an observable event, that is, when $P(s)$ changes. When an (enabled) observable event occurs, the control action is *instantaneously* updated. We shall assume that this update occurs before any unobservable event occurs. This is an important point, and it is necessary as we may wish to update the control action for some of these unobservable events. In this regard, observe that we are not making any specific assumptions about the relation between the controllability and observability properties of an event; an unobservable event could be controllable, an uncontrollable event could be observable, and so forth.

As in the full-observation case, we need to enforce the constraint that P-supervisors be admissible, that is, they should not disable feasible uncontrollable events. This admissibility condition is more complicated to state in the partial-observation case. The control action $S_P(t)$, for $t \in P[\mathcal{L}(G)]$, is applied by S_P immediately after the execution by G of the last (observable) event of t and remains in effect until the next observable event is executed by G. Let us take $t = t'\sigma$ (where $\sigma \in E_o$). $S_P(t)$ is the control action that applies to all strings in $\mathcal{L}(G)$ that belong to $P^{-1}(t')\{\sigma\}$ as well as to the unobservable continuations of these strings. However, $S_P(t)$ may disable unobservable events and thus prevent some of these unobservable continuations. In view of these observations, we define

$$L_t = P^{-1}(t')\{\sigma\}(S_P(t) \cap E_{uo})^* \cap \mathcal{L}(G) .$$

L_t contains all the strings in $\mathcal{L}(G)$ that are effectively subject to the control action $S_P(t)$, when S_P controls G. Since a supervisor is admissible if it does not disable a feasible uncontrollable event, we conclude that S_P is admissible if for all $t = t'\sigma \in P[\mathcal{L}(G)]$,

$$E_{uc} \cap \Big[\bigcup_{s \in L_t} \Gamma(f(x_0, s)) \Big] \subseteq S_P(t) .$$

The term in brackets represents all feasible continuations in $\mathcal{L}(G)$ of all strings that $S_P(t)$ applies to. (We note that if control actions had been defined in terms of *disabled events* instead of enabled events, then admissibility would be stated much more simply: $S_P^{disable}(t) \cap E_{uc} = \emptyset$. However, the above characterization of admissibility in terms of L_t does help in understanding the effect of S_P on G.) From now on, we will only consider admissible P-supervisors.

Example 3.2 (Admissibility of P-supervisors)
In Example 3.1, if we assume that $E_{uo} = \{b\}$ and $E_{uc} = \{c\}$, then if $b \in S_P(a)$, we must also have that $c \in S_P(a)$, since c is in the active event set of G after string ab.

The closed-loop behavior when S_P is controlling G is defined analogously to the case of full observation.

Definition. (Languages generated and marked by S_P/G)
The *language generated* by S_P/G is defined recursively as follows:

1. $\varepsilon \in \mathcal{L}(S_P/G)$
2. $[(s \in \mathcal{L}(S_P/G))$ and $(s\sigma \in \mathcal{L}(G))$ and $(\sigma \in S_P[P(s)])] \Leftrightarrow [s\sigma \in \mathcal{L}(S_P/G)] .$

The *language marked* by S_P/G is defined as follows:

$$\mathcal{L}_m(S_P/G) := \mathcal{L}(S_P/G) \cap \mathcal{L}_m(G) . \qquad \blacklozenge$$

It is important to note that the languages $\mathcal{L}(S_P/G)$ and $\mathcal{L}_m(S_P/G)$ are defined over E, and not E_o, that is, they correspond to the closed-loop behavior of G before the effect of projection P, namely, as seen at the "output" of G in Fig. 3.2.

3.3   SPECIFICATIONS ON CONTROLLED SYSTEM

As was mentioned in Section 3.1, the supervisor S (or S_P in the partial-observation case) is introduced because the behavior of the uncontrolled DES G is assumed to be unsatisfactory and must be restricted to within a subset of $\mathcal{L}(G)$,

$$\mathcal{L}(S/G) \subseteq L_a \subset \mathcal{L}(G) \text{ and/or } \mathcal{L}_m(S/G) \subseteq L_{am} \subset \mathcal{L}_m(G)$$

or within a range of sublanguages of $\mathcal{L}(G)$,

$$L_r \subseteq \mathcal{L}(S/G) \subseteq L_a \subset \mathcal{L}(G) \text{ and/or } L_{rm} \subseteq \mathcal{L}_m(S/G) \subseteq L_{am} \subset \mathcal{L}_m(G) .$$

For partial-observation problems, S is replaced by S_P in the above equations. We adopt the notation L_a for "admissible sublanguage of the generated language" and L_{am} for "admissible sublanguage of the marked language." We shall assume that L_a is prefix-closed, that is, $L_a = \overline{L_a}$. In the case of the range problem, L_r stands for "required sublanguage" and L_{rm} for "required marked sublanguage."

In problems where blocking is not of concern, the admissible behavior will be described by L_a and we will be focusing on ensuring that

$$\mathcal{L}(S/G) \subseteq L_a .$$

In problems where blocking is an issue, the admissible behavior will be described by L_{am} and we will be focusing on ensuring that

$$\mathcal{L}_m(S/G) \subseteq L_{am}$$

as well as mitigating blocking in the controlled system.

Let us consider the situation where the automaton model G of the system is finite-state and the language requirements L_a, L_{am}, L_r, or L_{rm}, are regular. As we shall see in the rest of this chapter, in order to synthesize supervisors that are provably correct in terms of the applicable language requirements, we will need automata that represent (namely, mark) these languages, in addition to the automaton model G of the system. If automaton H marks language K, we often say that H is a *recognizer* of K. In practice, we must usually construct these recognizer automata since the specifications on the behavior of G are typically given as natural language statements. Examples of such statements are: avoid a list of illegal states of G, enforce a first-come first-served policy, alternate events a and b, execute a set of events in a certain priority order, do not allow event a to occur more than n times between two occurrences of event b, execute event c only if event a precedes event b in the string generated so far, and so forth. The task of building the appropriate automata from G and statements of the above type is by no means trivial and requires experience in DES modeling. Section 3.3.1 below presents techniques for capturing in automata some natural language specifications that are common in practice. We note that building the system model G itself is another potentially difficult task. Typically, this is done by building models of individual components first and then composing them by parallel composition or product.

3.3.1 Automaton Models of Specifications

We present several natural language specifications that arise often in applications and show how to enforce them in automaton models. In most of these cases, the construction of the automaton that generates/marks the applicable language requirement, say L_a for the sake of discussion, is preceded by the construction of a simple automaton that captures the essence of the natural language specification. Let us call this automaton H_{spec}. We then combine H_{spec} with G, using either product or parallel composition, as appropriate, to obtain H_a where

$$\mathcal{L}(H_a) = L_a .$$

In the case of several natural language specifications, there will be many $H_{spec,i}$, $i = 1, \ldots, n$, that will be combined with G.

The choice of product or parallel composition to compose H_{spec} with G is based on the respective event sets of these automata:

- If the events that appear in G but not in H_{spec} are irrelevant to the specification that H_{spec} implements, then we use parallel composition.

- On the other hand, if these events are absent from H_{spec} because they should not happen in the admissible behavior L_a, then product is the right composition operation.

In most cases, all the states of H_{spec} will be marked so that marking in H_a will be solely determined by G.

The cases that follow illustrate the preceding discussion.

Illegal States

If a specification identifies certain states of G as illegal, then it suffices to delete these states from G, that is, remove these states and all the transitions attached to them, and then do the Ac operation to obtain H_a such that $\mathcal{L}(H_a) = L_a$. If the specification also requires nonblocking behavior after the illegal states have been removed, then we do the *Trim* operation instead of the Ac operation and obtain H_a such that $\mathcal{L}_m(H_a) = L_{am}$ and $\mathcal{L}(H_a) = \overline{L_{am}}$.

State Splitting

If a specification requires remembering how a particular state of G was reached in order to determine what future behavior is admissible, then that state must be split into as many states as necessary. The active event set of each newly introduced state is adjusted according to the respective admissible continuations.

Example 3.3 (Database concurrency control)

Let us return to the problem of database concurrency control discussed in Section 1.3.4 and Example 2.7. Consider the two transactions

$$T_1 = a_1 b_1 \qquad \text{and} \qquad T_2 = a_2 b_2$$

where, to simplify the notation and subsequent discussion, we have omitted specifying if a given operation is a read or a write. Thus, x_i means some operation by transaction i on database record x. The uncontrolled concurrent execution of T_1 and T_2 is modeled by automaton G depicted in Fig. 3.3, where the only marked state is state 8 corresponding to the completion of T_1 and T_2. The language $\mathcal{L}(G)$ contains inadmissible strings (or schedules, in database terminology). From the theory of database concurrency control, it can be shown that the only admissible strings are those where

$$a_1 \text{ precedes } a_2 \text{ if and only if } b_1 \text{ precedes } b_2.$$

Since marking and blocking are issues of concern here (recall the discussion in Section 1.3.4 and Example 2.7), we want to build *trim* automaton H_a such that $\mathcal{L}_m(H_a)$ contains only those strings in $\mathcal{L}_m(G)$ that satisfy the above ordering constraint. In order to do this, we need to remember how state 4 of G is reached, that is, we need to split this state in H_a; let us number as 4 and 9 the two resulting states of H_a, where 4 is the state reached by string $a_2 a_1$ and 9 is the state reached by string $a_1 a_2$. Then event b_1 should not occur in state 4 and event b_2 should not occur in state 9. The desired H_a is depicted in Fig. 3.3. As compared with G, H_a marks only strings where a_1 precedes a_2 if and only if b_1 precedes b_2; H_a is also nonblocking.

Event Alternance

If a specification requires the alternance of two events, say a and b with a the first event to occur, then we can build a two-state automaton H_{spec} that captures this alternance; H_{spec} is depicted in Fig. 3.4. The desired automaton H_a is obtained by

$$H_a = H_{spec} \| G .$$

Here, the rules of parallel composition will prevent a second consecutive occurrence of a until b is executed by G, and vice-versa. The other events of G are not affected since they do not appear in H_{spec}. We marked both states of H_{spec} since the specification on event alternance does not involve blocking; therefore, marking in H_a is consistent with marking in G.

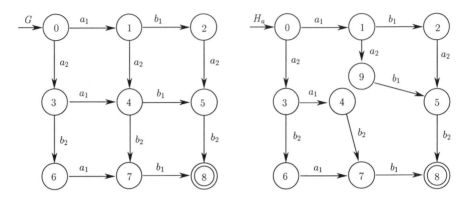

Figure 3.3: Automata G and H_a for Example 3.3.
G corresponds to the uncontrolled execution of two database transactions. The first transaction is $T_1 = a_1 b_1$ and the second transaction is $T_2 = a_2 b_2$, where x and y are two records in the database. H_a is nonblocking and models the admissible concurrent execution of transactions T_1 and T_2, namely the marked strings in $\mathcal{L}(G)$ where a_1 precedes a_2 if and only if b_1 precedes b_2.

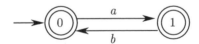

Figure 3.4: Two-state automaton that enforces the alternance of events a and b.

Illegal Substring

If a specification identifies as illegal all strings of $\mathcal{L}(G)$ that contain substring

$$s_f = \sigma_1 \cdots \sigma_n \in E^*$$

where f stands for *forbidden*, then we account for this specification by building automaton

$$H_{spec} = (X, E, f, x_0, X)$$

as follows.

1. $X = \{\varepsilon, \sigma_1, \sigma_1\sigma_2, \ldots, \sigma_1 \cdots \sigma_{n-1}\}$
   that is, we associate a state of H_{spec} to every proper prefix of s_f.

2. The transition function f is constructed in two steps:

   (a) $f(\sigma_1 \cdots \sigma_i, \sigma_{i+1}) = \sigma_1 \cdots \sigma_{i+1}$, for $i = 0, \ldots, n-2$.

   (b) Complete f to E as follows for all the states in X, except for state $\sigma_1 \cdots \sigma_{n-1}$ which is completed to $E \setminus \{\sigma_n\}$ (since that last event is illegal in that state):

   $f(\sigma_1 \cdots \sigma_i, \gamma)$
   $= $ state in X corresponding to the longest suffix of $\sigma_1 \cdots \sigma_i\gamma$.

3. Take $x_0 = \varepsilon$.

It is not difficult to show that

$$\mathcal{L}(H_{spec}) = \mathcal{L}_m(H_{spec}) = E^* \setminus E^*\{s_f\}E^* \ .$$

Consequently, the desired H_a is obtained by

$$H_a = H_{spec} \times G \ .$$

Observe that we marked all the states of H_{spec} so that state marking in H_a is solely determined by G; this is because the specification regarding s_f does not involve blocking.

Example 3.4 (Illegal substring)
Given event set $E = \{a, b, c, d\}$ and illegal substring $s_f = abcd$, H_{spec} such that

$$\mathcal{L}(H_{spec}) = E^* \setminus E^*\{s_f\}E^*$$

is depicted in Fig. 3.5.

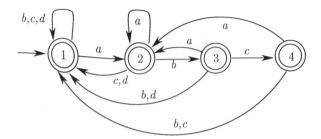

Figure 3.5: Automaton that generates all strings in $\{a, b, c, d\}^*$ except those that contain $abcd$ as a substring.

3.3.2 The Need for Formal Methods

At this stage of our discussion, it is useful to examine intuitively how one can deal with issues of uncontrollability and unobservability in supervisor design. For this purpose, we will use the database concurrency control example presented in Example 3.3; the uncontrolled system model is represented by the automaton G in Fig. 3.3, while the admissible behavior is represented by the automaton H_a in Fig. 3.3.

First, let us assume that all events are controllable and observable. Intuitively, we can see that we can use H_a to implement supervisor S_1 that perfectly achieves the desired objectives

$$\mathcal{L}(S_1/G) = \mathcal{L}(H_a) \quad \text{and} \quad \mathcal{L}_m(S_1/G) = \mathcal{L}_m(H_a)$$

if we proceed as follows. S_1 should use H_a to track the behavior of the system, namely, every event executed by G should cause S_1 to execute the same event in H_a and thus update the state of H_a. When H_a enters state 9, S should disable event b_2, and when H_a enters state 4, S_1 should disable event b_1; otherwise, S_1 enables all feasible events.

Second, to complicate the problem a little, let us assume that $E_{uc} = \{a_2, b_2\}$. What should a supervisor do in such a case? The supervisor for this case, denoted by S_2, can again use H_a to track the behavior of G. If S_2 allows reaching state 9 of H_a, then it is too late to prevent inadmissible strings from being generated, since we know that S_2 must disable event b_2 when H_a is in state 9, but b_2 is uncontrollable and thus cannot be disabled. The predecessor of state 9 in H_a is state 1, and state 1 has a transition to state 9 with uncontrollable event a_2. This means that S_2 should not allow reaching state 1 of H_a either. Consequently, S_2 must disable event a_1 when H_a is in state 0, that is, when the system is "started". After observing event a_2, S_2 can enable a_1. Finally, S_2 should disable b_1 when H_a enters state 4. We see that the presence of uncontrollable events forces restricting the behavior of the system to a proper subset of $\mathcal{L}(H_a)$, even though all of $\mathcal{L}(H_a)$ is admissible.

Third, let us assume that all events are controllable, but that $E_{uo} = \{a_2\}$.

We denote by S_3 the supervisor to design for this case. We claim that if S_3 enables both a_1 and a_2 at the beginning, when H_a is in state 0, then S_3 will run into trouble, that is, S_3 will cause blocking or will allow illegal strings. The reason is that if both a_1 and a_2 are enabled at the outset, then when S_3 observes event a_1, it will assume that right at that moment (i.e., before any other event happens) H_a is in state 1 or state 4, since S_3 does not see event a_2 and a_2 could have occurred before a_1. Let us assume for the moment that S_3 keeps a_2 enabled after seeing a_1 (we will see below what happens if a_2 is disabled immediately after observing a_1). Then the system can make an unobservable transition that takes H_a from state 1 to state 9. Therefore, S_3 does not know if H_a is in state 1, 4, or 9, until the next observable event occurs. But for state 9 of H_a, S_3 wants to enable b_1, while for state 4 it wants to disable b_1. Consequently, S_3 faces a "conflict" and is unable to achieve all of $\mathcal{L}(H_a)$ under control. In fact, S_3 must disable b_1 to avoid illegal strings from being generated. In this case however, we could get into a deadlock at state 9 of H_a since the only admissible event at state 9 has been disabled. To avoid this potential deadlock at state 9 of H_a, S_3 would want to disable a_2 in addition to b_1 immediately after observing a_1 (this is the option we said we would return to). But then the controlled system would deadlock when H_a is in state 1, since the two possible events out of that state have been disabled (b_1, since S_3 must consider that H_a could be in state 4, and a_2, to avoid blocking at state 9). We conclude that in order to avoid blocking while still guaranteeing legality, S_3 can only enable one of a_1 and a_2 at the beginning, but not both.

- If S_3 chooses to enable only a_2, then it enables also b_2 at the beginning, waits until it sees event b_2, then enables both a_1 and b_1. Thus the only marked string allowed under control in this case is $a_2 b_2 a_1 b_1$. Let us call this solution *Solution A*.

- If S_3 chooses to enable only a_1 at first, then it does so and waits until it observes a_1. After seeing a_1, S_3 knows that H_a is in state 1. It therefore enables a_2 and b_1, but disables b_2 since the system could execute a_2 causing H_a to transition to state 9 without S_3 being aware of it. Finally, after seeing b_1, S_3 can safely enable b_2. Two marked strings are allowed by this solution, called *Solution B*: $a_1 a_2 b_1 b_2$ and $a_1 b_1 a_2 b_2$.

Fourth and last, let us discuss the case when $E_{uo} = \{a_2\}$ and in addition not all events are controllable. If $E_{uc} = \{a_1\}$, then we must resort to Solution B above, as Solution A requires disabling a_1. Similarly, if $E_{uc} = \{a_2\}$, then only Solution A can be chosen. If $E_{uc} = \{b_2\}$, then we must also pick Solution A as Solution B needs to disable event b_2.

We could continue this discussion with different choices of E_{uc} and E_{uo}. But the point that we want to illustrate here is that even for this simple example, with only a few marked strings and at most 10 states, uncontrollability and unobservability can very rapidly complicate finding a supervisor that guarantees legality and nonblocking of the controlled system. Our intuition may still allow us to design a correct supervisor for this example, but obviously, *formal methods* are required if one is to solve any problem with more than a few dozen states in G and in the automaton representation of the admissible language, in the presence of uncontrollable and/or unobservable events. The purpose of the remainder of this chapter is to present formal approaches, with associated algorithms, for "dealing with" uncontrollability, blocking, and unobservability in supervisory control.

3.4 DEALING WITH UNCONTROLLABILITY

The purpose of this section is to study what controlled behaviors can be achieved when controlling a DES G with a full-observation supervisor S. Our focus will be on how to deal with the presence of uncontrollable events. We will be concerned with the generated language $\mathcal{L}(S/G)$ only. In the next section, we will bring the marked language $\mathcal{L}_m(S/G)$ into the picture in order to address blocking issues.

3.4.1 Controllability Theorem

We start by stating the key existence result for supervisors in the presence of uncontrollable events. We shall refer to this result as the *Controllability Theorem* (abbreviated as CT hereafter).

Theorem. [Controllability Theorem (CT)]
Consider DES $G = (X, E, f, \Gamma, x_0)$ where $E_{uc} \subseteq E$ is the set of uncontrollable events. Let $K \subseteq \mathcal{L}(G)$, where $K \neq \emptyset$. Then there exists supervisor S such that $\mathcal{L}(S/G) = \overline{K}$ if and only if

$$\overline{K} E_{uc} \cap \mathcal{L}(G) \subseteq \overline{K} .$$

This condition on K is called the *controllability condition*.

Proof. (IF) The proof of this part is constructive. For $s \in \mathcal{L}(G)$, define $S(s)$ according to

$$S(s) = [E_{uc} \cap \Gamma(f(x_0, s))] \cup \{\sigma \in E_c : s\sigma \in \overline{K}\} .$$

This supervisor enables after string s: (i) all uncontrollable events that are feasible in G after string s and (ii) all controllable events that extend s inside of $\overline{K}$. Part (i) ensures that S is admissible, that is, S never disables a feasible uncontrollable event.

We now prove that with this S, $\mathcal{L}(S/G) = \overline{K}$. The proof is by induction on the length of the strings in the two languages.

- The base case is for strings of length 0. But $\varepsilon \in \mathcal{L}(S/G)$ by definition and $\varepsilon \in \overline{K}$ since $K \neq \emptyset$ by assumption. Thus the base case holds.

- The induction hypothesis is that for all strings s such that $|s| \leq n$, $s \in \mathcal{L}(S/G)$ if and only if $s \in \overline{K}$. We now prove the same for strings of the form $s\sigma$.

  - Let $s\sigma \in \mathcal{L}(S/G)$. By definition of $\mathcal{L}(S/G)$, this implies that

$$[s \in \mathcal{L}(S/G)] \text{ and } [\sigma \in S(s)] \text{ and } [s\sigma \in \mathcal{L}(G)]$$

  which in turn implies that

$$[s \in \overline{K}] \text{ and } [\sigma \in S(s)] \text{ and } [s\sigma \in \mathcal{L}(G)]$$

  using the induction hypothesis.
    * If $\sigma \in E_{uc}$, then the controllability condition immediately yields $s\sigma \in \overline{K}$.
    * On the other hand, if $\sigma \in E_c$, then by the definition of S, we also obtain that $s\sigma \in \overline{K}$.
  - For the other direction of the induction step, let $s\sigma \in \overline{K}$. Then $s\sigma \in \mathcal{L}(G)$ since by assumption $\overline{K} \subseteq \mathcal{L}(G)$.
    * If $\sigma \in E_{uc}$, then $\sigma \in S(s)$ by the admissibility of $S(s)$.
    * On the other hand, if $\sigma \in E_c$, then by the above definition of $S(s)$, we also obtain that $\sigma \in S(s)$.
  Overall, we have that

$$[s \in \overline{K}] \text{ and } [\sigma \in S(s)] \text{ and } [s\sigma \in \mathcal{L}(G)]$$

  which in turns implies that

$$[s \in \mathcal{L}(S/G)] \text{ and } [\sigma \in S(s)] \text{ and } [s\sigma \in \mathcal{L}(G)]$$

  using the induction hypothesis. It then immediately follows that $s\sigma \in \mathcal{L}(S/G)$.
  - This completes the proof of the induction step.

(ONLY IF) Let there exist an admissible S such that $\mathcal{L}(S/G) = \overline{K}$. Let $s \in \overline{K}$, $\sigma \in E_{uc}$, and $s\sigma \in \mathcal{L}(G)$.

- Then $\sigma \in S(s)$ since any admissible supervisor is not allowed to disable a feasible uncontrollable event.

- But by definition of $\mathcal{L}(S/G)$, we have that

$$[s \in \overline{K} = \mathcal{L}(S/G)] \text{ and } [s\sigma \in \mathcal{L}(G)] \text{ and } [\sigma \in S(s)] \Rightarrow s\sigma \in \mathcal{L}(S/G) = \overline{K} \,.$$

- Thus we have shown that

$$[s \in \overline{K}] \text{ and } [\sigma \in E_{uc}] \text{ and } [s\sigma \in \mathcal{L}(G)] \Rightarrow s\sigma \in \overline{K}$$

or, in terms of languages,

$$\overline{K} E_{uc} \cap \mathcal{L}(G) \subseteq \overline{K}$$

which is the controllability condition. **Q.E.D.**

It is important to note that the proof of CT is constructive. If the controllability condition is satisfied, then the supervisor that achieves exactly the required behavior, $\overline{K}$, is:

$$S(s) = [E_{uc} \cap \Gamma(f(x_0, s))] \cup \{\sigma \in E_c : s\sigma \in \overline{K}\} .$$

The reason for not allowing $K = \emptyset$ in the statement of CT is because we always have $\varepsilon \in \mathcal{L}(S/G)$ (by definition). There is a difference between $\mathcal{L}(S/G) = \{\varepsilon\}$ and $\mathcal{L}(S/G) = \emptyset$. While $\mathcal{L}(S/G) = \emptyset$ is not a possible controlled behavior, $\mathcal{L}(S/G) = \{\varepsilon\}$ is possible and means that the controlled system stays in its initial state and does not execute any events (they are all disabled by the supervisor). One can think of $\mathcal{L}(S/G) = \{\varepsilon\}$ as a controlled system that "is turned on but does nothing." Of course, this is only allowed if the active event set of the initial state of G does not contain any uncontrollable events.

The controllability condition in CT is intuitive and a central concept in supervisory control. It can be paraphrased by:

"If you cannot prevent it, then it should be legal."

We state a general definition of this notion.

Definition. (Controllability)
Let K and $M = \overline{M}$ be languages over event set E. Let E_{uc} be a designated subset of E. K is said to be *controllable* with respect to M and E_{uc} if

$$\overline{K} E_{uc} \cap M \subseteq \overline{K} . \qquad \blacklozenge$$

By definition, controllability is a property of the prefix-closure of a language. Thus K is controllable if and only if $\overline{K}$ is controllable. The language expression for the controllability condition can be rewritten as follows:

for all $s \in \overline{K}$, for all $e \in E_{uc}$, $se \in M \Rightarrow se \in \overline{K}$.

In the case of regular languages K and M, the results of Chapter 2 can be used to obtain an implementable test for verifying the controllability of K with respect to M. Let H be an automaton that generates $\overline{K}$ and G be an automaton that generates M. Controllability can be checked by comparing the active event set of each state of $H \times G$ with the active event set of the corresponding state of G (given by the second component of the state of $H \times G$). If there is an uncontrollable event in the latter that does not appear in the former, then K is not controllable. The worst-case computational complexity of this test is $O(|E|mn)$, where m is the number of states of H and n is the number of states of G.

Example 3.5 (Controllability of languages)
Consider the two automata G and H_a of Example 3.3, where G and H_a are shown in Fig. 3.3. Take $M = \mathcal{L}(G)$ and $K = \mathcal{L}_m(H_a)$. If we form the product $H_a \times G$, then the resulting automaton will be identical to H_a except for a renaming of the states. States 0 to 8 of H_a will be renamed $(0,0)$ to $(8,8)$, respectively, while state 9 will be renamed $(9,4)$. Comparing the active event sets of $H_a \times G$ and G, we can see that these differ in only two states of $H_a \times G$: $(9,4)$ and $(4,4)$.

1. In state $(9,4)$, event b_2 is feasible in G but not in $H_a \times G$.

2. In state $(4,4)$, event b_1 is feasible in G but not in $H_a \times G$.

This means that for any E_{uc} that contains either b_1 or b_2, K is not controllable, while K is controllable for any choice of E_{uc} that does not contain b_1 and b_2. For instance, if $E_{uc} = \{b_1\}$, then string $s = a_2a_1 \in \overline{K}$ can be extended in M by uncontrollable event b_1, and since $a_2a_1b_1$ is not in $\overline{K}$, K is not controllable.

We will see later what can be done when the given language K is not controllable. But before that, we discuss the issue of building "realizations" of supervisors.

3.4.2 Realization of Supervisors

Let us assume that language $K \subseteq \mathcal{L}(G)$ is controllable. From CT, we know that supervisor S defined by

$$S(s) = [E_{uc} \cap \Gamma(f(x_0, s))] \cup \{\sigma \in E_c : s\sigma \in \overline{K}\}$$

results in

$$\mathcal{L}(S/G) = \overline{K} .$$

We rule out the two cases $\overline{K} = \mathcal{L}(G)$ and $\overline{K} = \emptyset$. In the first case, S plays no role, so it might as well be omitted. The second case is not allowed (see statement of CT).

The issue at hand is that for implementation purposes, we need to build a convenient *representation* of the function S; this is because it would be impractical to list $S(s)$ for all $s \in \mathcal{L}(S/G)$, as was done when we initially defined the notion of supervisor in Section 3.2.1. Observe that the domain of S can be restricted to $\mathcal{L}(S/G) = \overline{K}$ without loss of generality. Given that we are using an automaton to represent the system, let us also use an automaton to represent the supervisor S. In this case, when we will be dealing with *regular* languages, namely, the languages $\mathcal{L}(G)$ and K, the required representations will be *finite* and thus implementable. We will call an automaton representation of supervisor S a *realization* of S.

It is important to emphasize that we are concerned with building *off-line* a complete realization of S for all possible behaviors of the controlled system $\mathcal{L}(S/G)$. This realization will then be stored and at run time it will suffice to "read" the desired control action that is, the control action for the string of events observed up to now; this is often referred to as *table lookup*.

In order to build an automaton realization of S, it suffices to build an automaton that marks the language $\overline{K}$. Let R be such an automaton, that is, let

$$R = (Y, E, g, \Gamma_R, y_0, Y)$$

where R is trim and

$$\mathcal{L}_m(R) = \mathcal{L}(R) = \overline{K} \ .$$

If we "connect" R to G by the product operation, the result $R \times G$ is exactly the behavior that we desire for the closed-loop system S/G:

$$
\begin{aligned}
\mathcal{L}(R \times G) &= \mathcal{L}(R) \cap \mathcal{L}(G) \\
&= \overline{K} \cap \mathcal{L}(G) \\
&= \overline{K} = \mathcal{L}(S/G) \\
\mathcal{L}_m(R \times G) &= \mathcal{L}_m(R) \cap \mathcal{L}_m(G) \\
&= \overline{K} \cap \mathcal{L}_m(G) \\
&= \mathcal{L}(S/G) \cap \mathcal{L}_m(G) = \mathcal{L}_m(S/G) \ .
\end{aligned}
$$

Note that since R is defined to have the same event set as G (namely E), then $R \parallel G = R \times G$. What the above means is that the control action $S(s)$ is "encoded" into the transition structure of R. Namely,

$$
\begin{aligned}
S(s) &= [E_{uc} \cap \Gamma(f(x_0, s))] \cup \{\sigma \in E_c : s\sigma \in \overline{K}\} \\
&= \Gamma_R(g(y_0, s)) \\
&= \Gamma_{R \times G}(g \times f((y_0, x_0), s))
\end{aligned}
$$

where the first equality follows from the controllability of K and the second equality follows from the fact that $\overline{K} \subseteq \mathcal{L}(G)$. Here, $\Gamma_{R \times G}$ and $g \times f$ denote the active event set and transition function of $R \times G$, respectively.

Of course, $R \times G$ is a composition of two automata that is defined without reference to a control mechanism *à la* S/G. The interpretation with our control paradigm is as follows:

"*Let G be in state x and R be in state y following the execution of string $s \in \mathcal{L}(S/G)$. G generates an event σ that is currently enabled. This means that this event is also present in the active event set of R at y. Thus R also executes the event, as a passive observer of G. Let x' and y' be the new states of G and R after the execution of σ. The set of enabled events of G after string $s\sigma$ is now given by the active event set of R at y'.*"

Thus, we have built a representation of S that in the case of a regular K will only require finite memory. We will call the R derived by the above process the *standard realization* of S.

Example 3.6 (Standard realization of supervisors)

Consider the automaton G shown in Fig. 3.3 and the language $L_{am} = \mathcal{L}_m(H_a)$ where H_a is also depicted in Fig. 3.3. As was argued in Example 3.5, this L_{am} is controllable when $E_{uc} = \{a_1, a_2\}$. This means that there exists S such that $\mathcal{L}(S/G) = \overline{L_{am}}$. This supervisor S can be realized by automaton R where R is the same as H_a of Fig. 3.3 with the exception that all the states of R are marked, not only state 8. The active event set of R at state $g(y_0, s)$ gives the control action $S(s)$, for all $s \in \overline{L_{am}}$.

Induced Supervisors

The standard realization of S by automaton R raises the reverse question. If we are given automaton C and form the product $C \times G$, can that be interpreted as controlling G by some supervisor? The answer is: Not always! It depends on the controllability of $\mathcal{L}(C)$.

Let $C = (Y, E, h, \Gamma_C, y_0, Y)$. Let us define S_i^C, the supervisor for G induced by C, as follows. For all $s \in \mathcal{L}(G)$,

$$S_i^C(s) = \begin{cases} [E_{uc} \cap \Gamma(f(x_0, s))] \cup \{\sigma \in E_c : s\sigma \in \mathcal{L}(C)\} & \text{if } s \in \mathcal{L}(G) \cap \mathcal{L}(C) \\ E_{uc} & \text{otherwise.} \end{cases}$$

Note that we need to add $E_{uc} \cap \Gamma(f(x_0, s))$ to make sure that S_i^C is an admissible supervisor, namely, that it does not disable a feasible uncontrollable event of G. Precisely for that reason, we get the following fact, whose proof is omitted:

"$\mathcal{L}(S_i^C/G) = \mathcal{L}(C \times G)$ if and only if $\mathcal{L}(C)$ is controllable with respect to $\mathcal{L}(G)$ and E_{uc}."

In other words, if $\mathcal{L}(C)$ is controllable with respect to $\mathcal{L}(G)$, then when doing $C \times G$, C can never prevent G from executing an uncontrollable event, because all such transitions will always be defined in C. Thus the product $C \times G$ can indeed be viewed as the control of G by S_i^C, since only controllable events are "disabled" by the product. In this case, the resulting closed-loop behavior is

$$
\begin{aligned}
\mathcal{L}(S_i^C/G) &= \mathcal{L}(C \times G) \\
&= \mathcal{L}(C) \cap \mathcal{L}(G) \\
\mathcal{L}_m(S_i^C/G) &= \mathcal{L}_m(C \times G) \\
&= \mathcal{L}_m(C) \cap \mathcal{L}_m(G) \\
&= \mathcal{L}(C) \cap \mathcal{L}_m(G) \\
&= \mathcal{L}(C) \cap \mathcal{L}(G) \cap \mathcal{L}_m(G) \\
&= \mathcal{L}(S_i^C/G) \cap \mathcal{L}_m(G) \ .
\end{aligned}
$$

Reduced-State Realization

When $K \subseteq \mathcal{L}(G)$ is a *regular* controllable language, the standard realization R of supervisor S presented earlier guarantees that $\mathcal{L}(S/G) = \overline{K}$, but it may not be the most economical way of representing the function S in terms of memory requirements (for storing the state transition structure of R), even if we pick R to be the *canonical recognizer*[2] of $\overline{K}$. It may be possible to connect a different automaton than R with G, in the same manner as R is connected with G in the standard realization, and obtain the same closed-loop behavior $\mathcal{L}(S/G)$. The key is to relax the requirement that $\mathcal{L}(R) = K$ to the weaker condition

$$
\mathcal{L}(R_{rs}) \supseteq K
$$

where we are using the notation R_{rs} - for reduced-state - to differentiate this realization from the standard one. We can still obtain

$$
\mathcal{L}(R_{rs} \times G) = \mathcal{L}(S/G)
$$

in this case if the "extra enabled events" in R_{rs}, that is, the events in the active event set of R_{rs} that do not appear in the active event set of the corresponding state of R, are not feasible in the corresponding state(s) of G. In essence, there is no harm in enabling events that are not feasible. One can take advantage of this "degree of freedom" to design a *reduced-state realization* R_{rs} of S that has fewer states than R, even when R is the canonical recognizer of $\overline{K}$. This is illustrated in the following example. The process of finding the *minimum-state* R_{rs} for a given K and G is somewhat involved and it will not be discussed here.

Example 3.7 (Reduced-state realization)

Let us return to the supervisor S of Example 3.6, whose standard realization R is the same as the automaton H_a depicted in Fig. 3.3 but with all states marked. The uncontrolled system is the automaton G of Fig. 3.3. By examining R and G, we can see that the only disabling actions that S performs are to disable event b_2 in state 9 and to disable event b_1 in state

---

[2]See Section 2.4.3 in Chapter 2.

4. Consequently, we can merge states 2, 5, 6, 7, and 8 of R into a single state, say state 10, and obtain automaton R_{rs} shown in Fig. 3.6. The active event set of state 10 of R_{rs} is $\{a_2, b_2, a_1, b_1\}$, and all these events self-loop at state 10. The rest of R_{rs} is the same as in R. The fact that the active event set of state 10 of R_{rs} is larger than that of each individual corresponding state of G is of no concern as enabling an infeasible event has no effect. Overall, R_{rs} has all the necessary information to perform the desired disabling actions since it unambiguously recognizes states 4 and 9 of R.

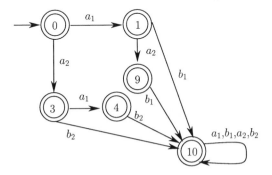

Figure 3.6: Reduced-state realization of supervisor in Examples 3.7 and 3.6.

3.4.3   The Property of Controllability

The goal of this section is to present several results about the property of controllability and in particular to discuss controllable sublanguages and superlanguages of an uncontrollable language. Suppose that a given language K is not controllable with respect to given $M = \overline{M} \subseteq E^*$ and $E_{uc} \subseteq E$, that is,

$$\overline{K} E_{uc} \cap M \not\subseteq \overline{K} .$$

For the sake of generality, the language K under consideration will *not* be assumed to be prefix-closed; however, we will assume that $K \subseteq M$. Unless otherwise specified, controllability will always be with respect to M and E_{uc}.

We will consider the following two languages derived from K:

- $K^{\uparrow C}$: the *supremal controllable sublanguage of K*;

- $K^{\downarrow C}$: the *infimal prefix-closed and controllable superlanguage of K*.

Overall, we have the following inequalities:

$$\emptyset \subseteq K^{\uparrow C} \subseteq K \subseteq \overline{K} \subseteq K^{\downarrow C} \subseteq M .$$

We now prove some properties that guarantee the existence of these two languages.

Proposition. (Properties of controllability)

1. If K_1 and K_2 are controllable, then $K_1 \cup K_2$ is controllable.

2. If K_1 and K_2 are controllable, then $K_1 \cap K_2$ need not be controllable.

3. If $\overline{K_1} \cap \overline{K_2} = \overline{(K_1 \cap K_2)}$ and K_1 and K_2 are controllable, then $K_1 \cap K_2$ is controllable.

4. If K_1 and K_2 are prefix-closed and controllable, then $K_1 \cap K_2$ is prefix-closed and controllable.

Proof.   1. The result is proved using the definition of controllability and properties of prefix-closure:

$$
\begin{aligned}
\overline{(K_1 \cup K_2)} E_{uc} \cap M &= (\overline{K_1} \cup \overline{K_2}) E_{uc} \cap M \\
&= (\overline{K_1} E_{uc} \cap M) \cup (\overline{K_2} E_{uc} \cap M) \\
&\subseteq \overline{K_1} \cup \overline{K_2} \\
&= \overline{(K_1 \cup K_2)} \ .
\end{aligned}
$$

2. Consider the following counter-example:

$$
\begin{aligned}
M &= \{\varepsilon, \alpha, \alpha\beta, \alpha\gamma\} & E &= \{\alpha, \beta, \gamma\} & E_{uc} &= \{\alpha\} \\
K_1 &= \{\varepsilon, \alpha\beta\} & K_2 &= \{\varepsilon, \alpha\gamma\} \ .
\end{aligned}
$$

3. We have that

$$
\begin{aligned}
\overline{(K_1 \cap K_2)} E_{uc} \cap M &\subseteq (\overline{K_1} \cap \overline{K_2}) E_{uc} \cap M \\
&= (\overline{K_1} E_{uc} \cap M) \cap (\overline{K_2} E_{uc} \cap M) \\
&\subseteq \overline{K_1} \cap \overline{K_2} \\
&= \overline{(K_1 \cap K_2)}
\end{aligned}
$$

where the last subset inclusion holds by controllability of K_1 and K_2, and the last equality holds by the assumption regarding prefix-closure and intersection.[3]

4. Immediate from above since prefix-closed languages satisfy the condition $\overline{K_1} \cap \overline{K_2} = \overline{(K_1 \cap K_2)}$.   **Q.E.D.**

The condition in part 3 of the above proposition will be encountered many times in this chapter, so we restate it in the context of a definition.

---

[3]We could also have used this assumption to replace the first subset inclusion by an equality, but this is not necessary as $\overline{(K_1 \cap K_2)} \subseteq \overline{K_1} \cap \overline{K_2}$ is always true.

Definition. (Nonconflicting languages)

Languages K_1 and K_2 are said to be *nonconflicting* if they satisfy the condition

$$\overline{K_1} \cap \overline{K_2} = \overline{(K_1 \cap K_2)} .$$

In words, if K_1 and K_2 share a prefix, then they must share a string containing that prefix.

◆

Observe that the inequality $\overline{K_1} \cap \overline{K_2} \supseteq \overline{(K_1 \cap K_2)}$ always holds; it is the other direction that may not hold in general, as is the case in the counter-example in part 2 of the above proof, where K_1 and K_2 share prefix α but do not share a string containing α as a prefix.

Let us define the class of controllable sublanguages of K and the class of prefix-closed and controllable superlanguages of K:

$$\mathcal{C}_{in}(K) \quad := \quad \{L \subseteq K : \overline{L}E_{uc} \cap M \subseteq \overline{L}\}$$
$$\mathcal{CC}_{out}(K) \quad := \quad \{L \subseteq E^* : (K \subseteq L \subseteq M) \text{ and } (\overline{L} = L) \text{ and } (\overline{L}E_{uc} \cap M \subseteq \overline{L})\} .$$

These classes are not empty since $\emptyset \in \mathcal{C}_{in}(K)$ and $M \in \mathcal{CC}_{out}(K)$. We discuss distinguished elements of these two classes in the next two subsections.

Supremal Controllable Sublanguage: Existence and Properties

If a given language K is not controllable, we would like to find the "largest" sublanguage of K that is controllable, where "largest" is in terms of set inclusion. The question is: Does such a sublanguage of K exist? Let us take the union of *all* the elements of the class $\mathcal{C}_{in}(K)$ and denote the result by $K^{\uparrow C}$:

$$K^{\uparrow C} := \bigcup_{L \in \mathcal{C}_{in}(K)} L .$$

The result is certainly a well-defined sublanguage of K; this is why we denote it by the superscript $\uparrow$, to be read as "inside K", while the superscript C refers to the fact that we are concerned with the controllability property. But is $K^{\uparrow C}$ controllable?

From part 1 of the above proposition, we know that "the class $\mathcal{C}_{in}(K)$ is closed under union"; by this we mean that if K_1 and K_2 are two elements of $\mathcal{C}_{in}(K)$, then $K_1 \cup K_2$ is also in $\mathcal{C}_{in}(K)$. We can use an induction argument to argue that $\mathcal{C}_{in}(K)$ is closed under *finite* unions. However, since the class $\mathcal{C}_{in}(K)$ may have infinite cardinality, we must verify if the proof holds for *infinite* unions too. The key step in proving closure of $\mathcal{C}_{in}(K)$ under union is the equality $\overline{K_1 \cup K_2} = \overline{K_1} \cup \overline{K_2}$ (cf. part 1 of the proof of the preceding proposition). But this result holds for infinite unions too, since

$$\overline{\cup_{i \in I} K_i} = \cup_{i \in I} \overline{K_i}$$

when I is an infinite index set. We conclude that controllability is preserved under *arbitrary unions* and consequently

$$K^{\uparrow C} \in \mathcal{C}_{in}(K) \ .$$

Since by definition $L \subseteq K^{\uparrow C}$ for any $L \in \mathcal{C}_{in}(K)$, we call $K^{\uparrow C}$ the *supremal controllable sublanguage of K*.

- In the "worst" case, $K^{\uparrow C} = \emptyset$, since $\emptyset \in \mathcal{C}_{in}(K)$.

- If K is controllable, then $K^{\uparrow C} = K$.

We will refer to "$\uparrow C$" as the operation of obtaining the supremal controllable sublanguage. It is immediate from the definition of $\uparrow C$ that this operation is *monotone*, that is,

$$K_1 \subseteq K_2 \Rightarrow K_1^{\uparrow C} \subseteq K_2^{\uparrow C} \ .$$

Moreover, the property of prefix-closure is preserved under the $\uparrow C$ operation.

Proposition. If K is prefix-closed, then so is $K^{\uparrow}$.

Proof. Since $K^{\uparrow C}$ is controllable, then so is its prefix-closure. But $\overline{K^{\uparrow C}} \subseteq \overline{K} = K$, which implies that $\overline{K^{\uparrow C}} \subseteq K^{\uparrow C}$. This suffices to prove the result. **Q.E.D.**

Example 3.8 (Supremal controllable sublanguages)

Part 1. For the first part of this example, we consider again the two automata G and H_a of Example 3.3, where G and H_a are shown in Fig. 3.3. Take $M = \mathcal{L}(G)$ and $K = \mathcal{L}_m(H_a)$. Explicitly, we have

$$K = \{a_2 b_2 a_1 b_1, a_2 a_1 b_2 b_1, a_1 a_2 b_1 b_2, a_1 b_1 a_2 b_2\} \ .$$

Let the set of uncontrollable events be $E_{uc} = \{a_2, b_2\}$, that is, the events of transaction 2 cannot be disabled. We can see that K is not controllable, as the string $a_1 a_2 \in \overline{K}$ can be extended in M by the uncontrollable event b_2, and $a_1 a_2 b_2$ is not in $\overline{K}$. This means that any controllable sublanguage of K cannot contain a string that contains $a_1 a_2$ as a prefix. If we remove from K all strings that contain $a_1 a_2$ as a prefix, we get the language

$$K_1 = \{a_2 b_2 a_1 b_1, a_2 a_1 b_2 b_1, a_1 b_1 a_2 b_2\} \ .$$

However, K_1 is not controllable! It contains as a prefix a_1 which can be extended within M by uncontrollable event a_2, and $a_1 a_2 \notin \overline{K_1}$. This means that we need to remove from K_1 all strings that contain a_1 as a prefix. The result is

$$K_2 = \{a_2 b_2 a_1 b_1, a_2 a_1 b_2 b_1\}$$

which is controllable. Thus, $K^{\uparrow C} = K_2$.

Let us repeat this example with the new larger set of uncontrollable events $E_{uc} = \{a_2, b_2, a_1\}$. In this case, the above K_2 is not controllable as $\varepsilon \in \overline{K_2}$, $a_1 \in M \cap \overline{K_2} E_{uc}$, but $a_1 \notin \overline{K_2}$. Therefore, we must remove from K_2 all strings that contain the empty string ε as a prefix, namely, all of K_2. This means that in this case $K^{\uparrow C} = \emptyset$. Observe that prefix ε does not have to be removed at the outset, when we first compare K and M. Yet, in the end, the uncontrollability of a_1 forces $K^{\uparrow C}$ to be empty.

Part 2. For the second part of this example, we pick new K, M, and E_{uc}:

$$M = \overline{\{abababa\}} \qquad K = \{a, aba, ababa\} \qquad E_{uc} = \{b\} \ .$$

Proceeding similarly as in the first part of this example, it is straightforward to verify that

$$K^{\uparrow C} = \emptyset \qquad (\overline{K})^{\uparrow C} = \{\varepsilon, a, ab, aba, abab\} \ .$$

Note that $\overline{K^{\uparrow C}}$ is a proper subset of $(\overline{K})^{\uparrow C}$, which means that prefix-closure and $\uparrow C$ do not commute. The implications of this result will be discussed in Section 3.5.2.

Other useful properties of the $\uparrow C$ operation are presented in the following proposition. These results can be proved from the definition of controllability and the supremality of $K^\uparrow$; recall that controllability need not preserved under intersection, but will be preserved if the languages of interest are nonconflicting.

Proposition. (Properties of $\uparrow C$ operation)

1. $(K_1 \cap K_2)^{\uparrow C} \subseteq K_1^{\uparrow C} \cap K_2^{\uparrow C}$
2. $(K_1 \cap K_2)^{\uparrow C} = (K_1^{\uparrow C} \cap K_2^{\uparrow C})^{\uparrow C}$
3. If $K_1^{\uparrow C}$ and $K_2^{\uparrow C}$ are nonconflicting, then $(K_1 \cap K_2)^{\uparrow C} = K_1^{\uparrow C} \cap K_2^{\uparrow C}$.
4. $(K_1 \cup K_2)^{\uparrow C} \supseteq K_1^{\uparrow C} \cup K_2^{\uparrow C}$. ◆

Algorithms that implement the $\uparrow C$ operation with automata in the case of regular languages are presented later in this chapter; see Section 3.4.5 for prefix-closed languages and Section 3.5.3 for the general case.

Infimal Prefix-Closed Controllable Superlanguage: Existence and Properties

We saw earlier that the property of controllability need not be preserved under intersection, but will be preserved if the languages involved are prefix-closed. Therefore, the class of prefix-closed and controllable superlanguages of

K, $\mathcal{CC}_{out}(K)$, is closed under intersection. Let us take the intersection of *all* the elements of $\mathcal{CC}_{out}(K)$ and denote the result by $K^{\downarrow C}$:

$$K^{\downarrow C} := \bigcap_{L \in \mathcal{CC}_{out}(K)} L .$$

The superscript $\downarrow$ denotes the fact that we are "outside of K". As was the case with $K^{\uparrow C}$ and $\mathcal{C}_{in}(K)$, we must make sure that $K^{\downarrow C}$ is indeed an element of $\mathcal{CC}_{out}(K)$. $K^{\downarrow C}$ is certainly prefix-closed and in the range between K and M; but is it controllable? The earlier proof that $\mathcal{CC}_{out}(K)$ is closed under finite intersections extends to the intersection of an infinite number of elements since

$$\overline{\cap_{i \in I} K_i} = \cap_{i \in I} \overline{K_i}$$

when all the K_i's are prefix-closed, where I is an infinite index set. Therefore, we conclude that

$$K^{\downarrow C} \in \mathcal{CC}_{out}(K) .$$

Since by definition $K^{\downarrow C} \subseteq L$ for any $L \in \mathcal{CC}_{out}(K)$, we call $K^{\downarrow C}$ the *infimal prefix-closed and controllable superlanguage of K*.

- In the "worst" case, $K^{\downarrow C} = M$, since $M \in \mathcal{CC}_{out}(K)$.

- If K is controllable, then $K^{\downarrow C} = \overline{K}$.

We will refer to "$\downarrow C$" as the operation of obtaining the infimal prefix-closed and controllable superlanguage. Using the definition of the infimal prefix-closed controllable superlanguage and the properties of controllability, we can prove the following useful properties of the $\downarrow C$ operation.

Proposition. (Properties of $\downarrow C$ operation)

1. $K_1 \subseteq K_2 \Rightarrow K_1^{\downarrow C} \subseteq K_2^{\downarrow C}$
2. $(K_1 \cap K_2)^{\downarrow C} \subseteq K_1^{\downarrow C} \cap K_2^{\downarrow C}$
3. If K_1 and K_2 are nonconflicting, then $(K_1 \cap K_2)^{\downarrow C} = K_1^{\downarrow C} \cap K_2^{\downarrow C}$
4. $(K_1 \cup K_2)^{\downarrow C} = K_1^{\downarrow C} \cup K_2^{\downarrow C}$.   ◆

Example 3.9 (Infimal prefix-closed controllable superlanguage)
If we consider the languages M and K of part 1 of Example 3.8, together with $E_{uc} = \{a_2, b_2\}$, then all we need to do to make $\overline{K}$ controllable is to extend string $a_1 a_2$ with a string of uncontrollable events of length one, namely b_2, as there are no other possible continuations of strings of $\overline{K}$ with uncontrollable events in M. Therefore,

$$K^{\downarrow C} = \overline{K} \cup \{a_1 a_2 b_2\} .$$

An algorithm that implements the $\downarrow C$ operation using automata in the case of regular languages is presented in Section 3.4.6.

3.4.4 Some Supervisory Control Problems and Their Solutions

As stated earlier, we have the following inequalities relating the controllable languages that we have presented in the preceding section:

$$\emptyset \subseteq K^{\uparrow C} \subseteq K \subseteq \overline{K} \subseteq K^{\downarrow C} \subseteq M \ .$$

With the concepts of *supremal controllable sublanguage* and *infimal prefix-closed controllable superlanguage* defined and some of their properties stated, we are ready to formulate a set of supervisory control problems and present their solutions in the presence of uncontrollable events.

BSCP: Basic Supervisory Control Problem

Given DES G with event set E, uncontrollable event set $E_{uc} \subseteq E$, and admissible language $L_a = \overline{L_a} \subseteq \mathcal{L}(G)$, find a supervisor S such that:

1. $\mathcal{L}(S/G) \subseteq L_a$

2. $\mathcal{L}(S/G)$ is "the largest it can be," that is, for any other supervisor S_{other} such that $\mathcal{L}(S_{other}/G) \subseteq L_a$,

$$\mathcal{L}(S_{other}/G) \subseteq \mathcal{L}(S/G) \ .$$

Solution of BSCP:

Requirement 2 means that we wish the solution S to be "optimal" with *set inclusion* as the criterion of optimality for this logical control problem: the optimal solution *contains* all other solutions. Such a solution is said to be *minimally restrictive*. Using the results of the preceding section, the solution is to choose S such that

$$\mathcal{L}(S/G) = L_a^{\uparrow C}$$

as long as $L_a^{\uparrow C} \neq \emptyset$. S can be realized by building a recognizer of $L_a^{\uparrow C}$, that is, an automaton whose marked language is $L_a^{\uparrow C}$; this is the standard realization presented in Section 3.4.2. We will see later how to build this recognizer from the recognizer of L_a. Note that the size of the realization of S is not a consideration in the requirements of BSCP.

If we obtain $L_a^{\uparrow} = \emptyset$, then we know that this is not an allowed controlled behavior. This means that there exists a string of uncontrollable events from the initial state of G that does not belong to L_a. This can be paraphrased as "the controlled system cannot be turned on."

Comments:

BSCP is indeed the "basic" problem in supervisory control: Restrict the behavior of the system G in order to stay inside the admissible behavior, but do not restrict G more than necessary. L_a is obtained from $\mathcal{L}(G)$ by removing illegal states in G and illegal strings in $\mathcal{L}(G)$ according to the given specifications. Blocking is not a consideration here. The essence of the control problem is how to deal with the presence of uncontrollable events; this is done by an application of the $\uparrow C$ operation on L_a.

Often, we have a "range problem" where it is required that

$$L_r \subseteq \mathcal{L}(S/G) \subseteq L_a$$

for two given prefix-closed languages L_r and L_a, both subsets of $\mathcal{L}(G)$. In this case we can compute $L_a^{\uparrow C}$ and check whether or not

$$L_r \subseteq L_a^{\uparrow C} .$$

If this subset inclusion does not hold, then the range problem does not have a solution as $L_a^{\uparrow C}$ is the largest solution inside L_a; if it does not contain L_r, then no other solution inside L_a will. If this subset inclusion holds, then the range problem has at least one solution. In this case, there may be other solutions, that is, other controllable languages in the range from L_r to $L_a^{\uparrow C}$. This leads us to the question: What is the smallest solution inside this range (if any exists)? This is the motivation for the supervisory control problem that we address next.

DuSCP: Dual Version of BSCP

Given DES G with event set E, uncontrollable event set $E_{uc} \subseteq E$, and minimum required language $L_r \subseteq \mathcal{L}(G)$, find a supervisor S such that:

1. $\mathcal{L}(S/G) \supseteq L_r$

2. $\mathcal{L}(S/G)$ is "the smallest it can be," that is, for any other supervisor S_{other} such that $\mathcal{L}(S_{other}/G) \supseteq L_r$,

$$\mathcal{L}(S_{other}/G) \supseteq \mathcal{L}(S/G) .$$

Note that L_r need not be prefix-closed and could be given as a subset of $\mathcal{L}_m(G)$. However, in DuSCP, we are concerned with $\mathcal{L}(S/G)$ and not with $\mathcal{L}_m(S/G)$.

Solution of DuSCP:

The desired solution is to take S such that

$$\mathcal{L}(S/G) = L_r^{\downarrow C}$$

which clearly meets requirements 1 and 2 by definition of the $\downarrow C$ operation. S can be realized by building a recognizer of $L_r^{\downarrow C}$. We will see later how to build such a recognizer from a recognizer of L_r.

As in the case of BSCP, set inclusion is the criterion used to compare candidate solutions in DuSCP and $L_r^{\downarrow C}$ is "optimal" with respect to that criterion since it is contained in any solution that satisfies requirement 1.

Comments:

As in BSCP, the essence of the control problem in DuSCP is how to deal with the presence of uncontrollable events; this is done by an application of the $\downarrow C$ operation on L_r. The notion of minimum required behavior arises in many supervisory control problems, such as the range problem discussed after the presentation of BSCP. In this range problem, the two languages $L_r^{\downarrow C}$ and $L_a^{\uparrow C}$ delimit the set of possible solutions, assuming that the first language is a subset of the second one.

The last supervisory control problem discussed in this section generalizes the notion of range by considering "desired" and "tolerated" behaviors.

SCPT: Supervisory Control Problem with Tolerance

Finally, consider the following supervisory control problem where we are given a *desired* language, L_{des}, and a larger *tolerated* language, L_{tol}. The idea is to achieve as much as possible of the desired language without ever exceeding the tolerated language. In contrast to L_r in the range problem, we allow not achieving all of L_{des}, as long as we achieve as much of it as possible. We can think of L_{des} as the solution we would adopt if all events were controllable. Due to the presence of uncontrollable events, we allow "deviations" from L_{des}. However, such deviations must keep in mind all the "hard" specifications imposed on G, which are captured by L_{tol}. Thus, as for L_a considered above, L_{tol} should never be exceeded. The precise formulation of SCPT is as follows.

Given DES G with event set E, uncontrollable event set $E_{uc} \subseteq E$, desired language $L_{des} \subseteq \mathcal{L}(G)$ and tolerated language $L_{tol} = \overline{L_{tol}} \subseteq \mathcal{L}(G)$, where $\overline{L_{des}} \subseteq L_{tol}$, find a supervisor S such that:

1. $\mathcal{L}(S/G) \subseteq L_{tol}$

   (this means that S/G can never exceed the tolerated language)

2. For all prefix-closed and controllable $K \subseteq L_{tol}$,

$$K \cap L_{des} \subseteq \mathcal{L}(S/G) \cap L_{des}$$

   (this means that we want S/G to achieve as much of the desired language as possible)

3. For all prefix-closed and controllable $K \subseteq L_{tol}$,

$$K \cap L_{des} = \mathcal{L}(S/G) \cap L_{des} \Rightarrow \mathcal{L}(S/G) \subseteq K$$

   (this means that requirement 2 is achieved with the smallest possible controlled behavior $\mathcal{L}(S/G)$).

Solution of SCPT:

A little thought shows that the solution of SCPT is obtained by taking S such that

$$\mathcal{L}(S/G) = (L_{tol}^{\uparrow C} \cap L_{des})^{\downarrow C} .$$

First, we identify the supremal controllable sublanguage of L_{tol}, since we cannot exceed that language by requirement 1. The intersection of that supremal with L_{des} gives us the largest part of L_{des} that can be achieved, which is what we need to obtain by requirement 2. Finally, taking the infimal prefix-closed controllable superlanguage of the resulting intersection yields the smallest solution that will achieve this intersection, in agreement with requirement 3.

The solutions of the above-formulated problems, BSCP, DuSCP, and SCPT, lead us to study the implementation of the $\uparrow C$ and $\downarrow C$ operations.

3.4.5 Computation of $K^{\uparrow C}$: Prefix-Closed Case

In this section, we discuss the calculation of $K^{\uparrow C}$ for a given *prefix-closed* language K that is not controllable. The general case for K's that need not be prefix-closed will be considered in Section 3.5.3. As before, controllability is with respect to prefix-closed M and uncontrollable event set E_{uc}.

When K is prefix-closed, it can be shown that $K^{\uparrow C}$ is given by a relatively simple formula on languages.

Theorem. (Formula for $\uparrow C$)
If $K = \overline{K}$, then
$$K^{\uparrow C} = K \setminus [(M \setminus K)/E_{uc}^*]E^*$$

where the "/" symbol denotes the *quotient* operation on languages. ◆

Definition. (Quotient of languages)
The quotient operation for languages L_1, $L_2 \subseteq E^*$ is defined as follows:
$$L_1/L_2 := \{s \in E^* : \exists t \in L_2 \ (st \in L_1)\} .$$
◆

In the special case of interest to us, this definition reduces to
$$(M \setminus K)/E_{uc}^* = \{s \in E^* : \exists t \in E_{uc}^* \ (st \in M \setminus K)\}$$

or, in words, the resulting set contains the longest prefix of string v whose last event is controllable, plus all its uncontrollable continuations in $M \setminus K$, for all $v \in M \setminus K$.

We shall not present a formal proof of the above formula for $K^{\downarrow C}$, but rather we explain it intuitively.

- If K contains string s that belongs to $(M \setminus K)/E_{uc}^*$, then s *must* be removed from K because it violates the controllability condition; namely, a suffix composed only of uncontrollable events can extend s to a string in $M \setminus K$, that is, outside of K. This explains the set difference in the formula. In other words, the "boundary" between $K^{\uparrow C}$ and its complement in M should consist of strings that are extended by *controllable* events only when leaving $K^{\uparrow C}$, otherwise the controllability condition will be violated.

- If string s must be removed from K to make it controllable, then any string in K that contains s as a prefix *must also* be removed from K. This is a consequence of the controllability condition for a language L: $\overline{L}E_{uc} \cap M \subseteq \overline{L}$; namely, if string u contains s as a prefix and s must be removed for the above reason, then u must be removed as well since u also leads to a violation of the controllability condition. This explains the concatenation of $(M \setminus K)/E_{uc}^*$ with E^* in the formula.

- Nothing else needs to be removed from K to make it controllable, so the result is indeed the supremal controllable sublanguage of K.

Having a formula to characterize the $\uparrow C$ operation is very useful. Results in automata theory tell us that the class of regular languages is closed under the quotient operation, that is, if L_1 and L_2 are regular languages, then so is L_1/L_2. This can be seen more easily in the special case of interest to us, where $L_2 = E_{uc}^*$. Observe that $L_1/E_{uc}^* \subseteq \overline{L_1}$. To build an automaton that marks L_1/E_{uc}^*, we can start from the marked states of an automaton that marks L_1 and "backtrack" along paths composed of uncontrollable events only in order to identify the new set of marked states that will allow us to mark the language L_1/E_{uc}^*. In fact, with an appropriate data structure that includes forward and backward pointers, this can be done in linear complexity in the state space of the automaton that marks L_1. We have also seen in Chapter 2 that the class of regular languages is closed under concatenation and set difference. This leads to the following important result.

Theorem. (Regularity and $\uparrow C$)
If prefix-closed language K is regular and M is regular, then $K^{\uparrow C}$ is regular. ♦

The importance of this result lies in the fact that for finite-state DES and (prefix-closed) admissible languages that can be represented by finite-state automata, supervisors that solve BSCP can be realized by *finite-state* automata. (We will see in the next section that this result remains true for admissible languages that are not prefix-closed.) In order to build an automaton that generates $K^{\uparrow C}$, we can get insight from the procedures to implement concatenation and set difference of regular languages on finite-state automata along with the above discussion on the implementation of quotient with E_{uc}^*. It can be shown that a "clever" implementation of the $\uparrow C$ operation for prefix-closed languages with worst-case computational complexity of $O(nm|E|)$ exists, where n and m are the numbers of states of the recognizers of K and M, respectively. Instead of going into the details of such an implementation here, we illustrate by means of an example a "less clever" algorithm that has quadratic complexity but is easier to explain and convenient to use for hand calculations. (This is in fact the same algorithm as the one that will be presented in Section 3.5.3 for the computation of the $\uparrow C$ operation in the general case.)

Example 3.10 (Calculation of supremal controllable sublanguage)
Consider the two automata G and H in Fig. 3.7. For the moment, ignore the second component in the labeling of the states of H and consider only the first component. We wish to compute $\mathcal{L}(H)^{\uparrow C}$ with respect to $\mathcal{L}(G)$, when the set of uncontrollable events is $E_{uc} = \{a_1, b_1\}$. We can interpret G as the parallel composition of two processes, each generating the prefix-closure of $(a_i b_i)^*$, $i = 1, 2$, and H as implementing on G the specification that "after an occurrence of b_1, b_1 cannot occur again until b_2 occurs at least once."

Our first step is to form the product $H \times G$ in order to map the states of H to those of G. The result is isomorphic to H with a renaming of the states. In fact, we obtain the H of Fig. 3.7 with the labeling of states as shown in that figure, where the second component refers to the corresponding states of G.

Intuitively, we can immediately see that $\mathcal{L}(H)^{\uparrow C} = \emptyset$. Since string $a_1 b_1 a_1 b_1 \in \mathcal{L}(G) \setminus \mathcal{L}(H)$ and all events in that string are uncontrollable, then string $\varepsilon \in \mathcal{L}(H)$ can be uncontrollably extended to $\mathcal{L}(G) \setminus \mathcal{L}(H)$; therefore, all strings in $\mathcal{L}(H)$ that contain ε as a prefix must be removed, resulting in the empty set.

Algorithmically, we can proceed as follows to obtain an automaton that generates the desired answer - the empty automaton in this case. First, comparing the active event sets of the states of H with the active event sets of the corresponding states of G, we see that any string ending at state $(D, 1)$ or $(H, 3)$ of H violates the controllability condition, due to the possible continuation in G with uncontrollable event b_1 in both cases. Thus, these two states must be deleted from H; this is the automaton implementation of the removal from $\mathcal{L}(H)$ of all strings that contain as a prefix any of the "problem strings" that end at $(D, 1)$ or $(H, 3)$. Before the removal of $(D, 1)$ and $(H, 3)$, the other states of H were safe in regard to uncontrollable continuations in G. After the removal of $(D, 1)$ and $(H, 3)$ however, states $(C, 0)$ and $(G, 2)$ now pose a problem since the modified H does not contain the continuation with a_1 that is possible in states 0 and 2 of G. Consequently, the second iteration of our procedure forces the deletion of $(C, 0)$ and $(G, 2)$ from H. However, after this iteration, states $(B, 1)$ and $(F, 3)$ now pose a problem due to uncontrollable event b_1. Therefore, at the third iteration we delete from H states $(B, 1)$ and $(F, 3)$. Finally, at the fourth and last iteration, we see that since $(B, 1)$ has been deleted, the controllability condition is violated at the initial state $(A, 0)$ and consequently the answer returned by our algorithm is the empty automaton, as expected.

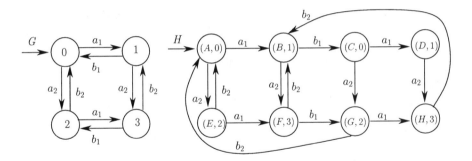

Figure 3.7: Automata G and H for Example 3.10.

3.4.6 Computation of $K^{\downarrow C}$

In this section, we consider the computation of $K^{\downarrow C}$ for a given language K that is not controllable. Again, controllability is with respect to prefix-closed M and uncontrollable event set E_{uc}.

Theorem. (Formula for $\downarrow C$)

1. $K^{\downarrow C} = \overline{K} E_{uc}^* \cap M$.

2. If K and M are regular, then $K^{\downarrow C}$ is regular.

Proof. 1. Let $K' := \overline{K} E_{uc}^* \cap M$. The inclusion $K^{\downarrow C} \subseteq K'$ follows by observing that $K' \supseteq K$ and that it is prefix-closed and by verifying that K' is controllable:

$$
\begin{aligned}
K' E_{uc} \cap M &= \overline{K} E_{uc}^* E_{uc} \cap M E_{uc} \cap M \\
&\subseteq \overline{K} E_{uc}^* \cap M = K' .
\end{aligned}
$$

For the reverse inclusion, take any $L \in \mathcal{CC}_{out}(K)$. Using the definition of $\mathcal{CC}_{out}$, we have that:

$$
\begin{aligned}
\overline{K} \cap M &\subseteq L \\
\overline{K} E_{uc} \cap M &\subseteq L E_{uc} \cap M \subseteq L \\
\overline{K} E_{uc}^2 \cap M &\subseteq L E_{uc} \cap M \subseteq L \\
&\cdots \\
\overline{K} E_{uc}^r \cap M &\subseteq L E_{uc} \cap M \subseteq L \quad \text{for all } r \geq 0 .
\end{aligned}
$$

Thus $K' \subseteq L$. Since L was arbitrary, we get that

$$
K' \subseteq \bigcap_{L \in \mathcal{CC}_{out}(K)} L =: K^{\downarrow C} .
$$

2. The proof of this part follows from the fact that the class of regular languages is closed under concatenation and intersection. **Q.E.D.**

The formula for $\downarrow C$ leads to an algorithm for building an automaton that generates $K^{\downarrow C}$ from automata H and G that generate $\overline{K}$ and M, respectively.

1. Build a deterministic automaton that generates $\overline{K}E^*_{uc}$ as follows. Add a new state to H and "partially complete to E_{uc}" the transition function of H by adding all the missing uncontrollable transitions (in the active event set of each state of H) to the new state. Also, add self-loop transitions for all uncontrollable events at the new state. Call the resulting automaton H_{aug}.

2. The intersection in the formula for $K^{\downarrow C}$ is implemented by doing

$$H_{aug} \times G =: H_{\downarrow C}$$

and consequently

$$\mathcal{L}(H_{\downarrow C}) = K^{\downarrow C} .$$

Example 3.11 (Infimal prefix-closed controllable superlanguage)
Let us use the same H and G as in Example 3.10 and Fig. 3.7. As we did in Example 3.10 in order to avoid drawing two isomorphic copies of the same automaton, when considering the original automaton H, we disregard the second component in the state names in Fig. 3.7. We take $E_{uc} = \{a_1, b_1\}$ and wish to compute $\mathcal{L}(H)^{\downarrow C}$. Let us denote as N the new state to be added in H at step 1 of the algorithmic implementation of $\downarrow C$. It is straightforward to verify that the automaton $H_{\downarrow C}$ obtained by this algorithm will consist of H in Fig. 3.7, this time with state names as indicated in this figure, with four extra states, $(N, 0)$, $(N, 1)$, $(N, 2)$, and $(N, 3)$, and the following six additional transitions, listed in the form $(from\_state, event, to\_state)$: $[(D, 1), b_1, (N, 0)]$, $[(H, 3), b_1, (N, 2)]$, $[(N, 0), a_1, (N, 1)]$, $[(N, 1), b_1, (N, 0)]$, $[(N, 2), a_1, (N, 3)]$, $[(N, 3), b_1, (N, 2)]$.

3.5 DEALING WITH BLOCKING

In this section, we consider supervisory control problems where we are concerned with both $\mathcal{L}(S/G)$ and $\mathcal{L}_m(S/G)$. Typically, the specifications on the controlled system will be given as a sublanguage of $\mathcal{L}_m(G)$ and it will be required that the supervisor S be nonblocking, that is,

$$\overline{\mathcal{L}_m(S/G)} = \mathcal{L}(S/G) .$$

Our development will parallel and complement that in the previous section, where the focus was on $\mathcal{L}(S/G)$ only. The key underlying result in Section 3.4

was the Controllability Theorem (CT) presented in Section 3.4.1. The extension of CT to nonblocking supervisors, termed "Nonblocking Controllability Theorem" and abbreviated as NCT, is presented next.

3.5.1 Nonblocking Controllability Theorem

Theorem. [Nonblocking Controllability Theorem (NCT)]
Consider DES $G = (X, E, f, \Gamma, x_0, X_m)$ where $E_{uc} \subseteq E$ is the set of uncontrollable events. Consider the language $K \subseteq \mathcal{L}_m(G)$, where $K \neq \emptyset$. There exists a *nonblocking* supervisor S for G such that

$$\mathcal{L}_m(S/G) = K \quad \text{and} \quad \mathcal{L}(S/G) = \overline{K}$$

if and only if the two following conditions hold:

1. Controllability: $\overline{K} E_{uc} \cap \mathcal{L}(G) \subseteq \overline{K}$
2. $\mathcal{L}_m(G)$-closure: $K = \overline{K} \cap \mathcal{L}_m(G)$.

Proof. The proof of the necessity and sufficiency of the controllability condition is identical to the proof of CT. In particular, the desired supervisor for the "IF" part of the proof is the same as in the proof of CT, namely

$$S(s) = [E_{uc} \cap \Gamma(f(x_0, s))] \cup \{\sigma \in E_c : s\sigma \in \overline{K}\} .$$

Given that $\mathcal{L}(S/G) = \overline{K}$ with this S, we have that

$$\begin{aligned} \mathcal{L}_m(S/G) &= \overline{K} \cap \mathcal{L}_m(G) \\ &= K \end{aligned}$$

where the second equality follows by applying the $\mathcal{L}_m(G)$-closure condition. Thus S is nonblocking.

For the "ONLY IF" part of the proof, let there exist an admissible nonblocking S such that

$$\mathcal{L}(S/G) = \overline{K} \quad \text{and} \quad \mathcal{L}_m(S/G) = K .$$

Then by definition of $\mathcal{L}_m(S/G)$, we have that

$$K = \overline{K} \cap \mathcal{L}_m(G)$$

which is the $\mathcal{L}_m(G)$-closure condition.   **Q.E.D.**

We make the important observation that the nonblocking supervisor constructed in the proof of NCT is the same as that in the proof of CT:

$$S(s) = [E_{uc} \cap \Gamma(f(x_0, s))] \cup \{\sigma \in E_c : s\sigma \in \overline{K}\} .$$

This means that this supervisor can be realized in the same manner as presented in Section 3.4.2. The only difference between CT and NCT is the so-called $\mathcal{L}_m(G)$-closure condition

$$K = \overline{K} \cap \mathcal{L}_m(G)$$

which, together with the controllability condition, is necessary and sufficient to guarantee that the above-defined supervisor works as desired. Observe that the inclusion

$$K \subseteq \overline{K} \cap \mathcal{L}_m(G)$$

always holds, since we assume that $K \subseteq \mathcal{L}_m(G)$. It is the other inclusion

$$K \supseteq \overline{K} \cap \mathcal{L}_m(G)$$

that need not hold in general. For instance, take

$$
\begin{aligned}
\mathcal{L}_m(G) &= \{\alpha_1, \alpha_1\beta_1\alpha_1, \alpha_1\beta_1\alpha_1\beta_1\alpha_1\} \\
K &= \{\alpha_1\beta_1\alpha_1\} \, .
\end{aligned}
$$

This K violates the $\mathcal{L}_m(G)$-closure condition since it does not contain string α_1.

3.5.2 Nonblocking Supervisory Control

It can be argued that unlike the controllability condition, which truly depends on the properties of the system under consideration, the $\mathcal{L}_m(G)$-closure condition is a technical condition that will typically hold *by construction of K*, when K is interpreted as "admissible marked behavior." The following reasons support this argument.

1. In practice, marking is a property of the uncontrolled system G, modeled by proper construction of X_m.

2. Specifications are usually stated in terms of prefix-closed languages, say $\overline{K_{spec}}$; recall the discussion in Section 3.3.

3. The admissible marked language is obtained by forming $K = \overline{K_{spec}} \cap \mathcal{L}_m(G)$.

4. Such a K is guaranteed to be $\mathcal{L}_m(G)$-closed.

In view of these observations, we will assume that any given admissible marked behavior satisfies the $\mathcal{L}_m(G)$-closure condition. Therefore, controllability remains the issue at hand. Our treatment of the property of controllability in Section 3.4.3, most particularly the existence of the supremal controllable sublanguage, was general and included the case of languages that are not prefix-closed. However, the computation of supremal controllable sublanguages covered in Section 3.4.5 was restricted to prefix-closed languages. Consequently, we need to address two issues:

1. What is the effect of the $\uparrow C$ operation on the $\mathcal{L}_m(G)$-closure properties of a language?

2. How do we perform the $\uparrow C$ operation on regular languages that are not prefix-closed?

The answer to the second question will be presented in Section 3.5.3. The answer to the first question is quite nice, as seen from the first part of the following proposition. Here, we assume that controllability is with respect to $\mathcal{L}(G)$ and E_{uc}.

Proposition. (Further properties of the $\uparrow C$ operation)

1. If $K \subseteq \mathcal{L}_m(G)$ is $\mathcal{L}_m(G)$-closed, then so is $K^{\uparrow C}$.

2. In general, $\overline{K^{\uparrow C}} \subseteq (\overline{K})^{\uparrow C}$.

Proof.   1. We need to show that

$$K^{\uparrow C} = \overline{K^{\uparrow C}} \cap \mathcal{L}_m(G) .$$

The inclusion ($\subseteq$) is immediate. For ($\supseteq$), define $K' = \overline{K^{\uparrow C}} \cap \mathcal{L}_m(G)$.

- Then

$$
\begin{aligned}
K^{\uparrow C} &\subseteq \overline{K^{\uparrow C}} \cap \mathcal{L}_m(G) \\
\Rightarrow \overline{K^{\uparrow C}} &\subseteq \overline{\overline{K^{\uparrow C}} \cap \mathcal{L}_m(G)} = \overline{K'} .
\end{aligned}
$$

- Also

$$
\begin{aligned}
\overline{K^{\uparrow C}} \cap \mathcal{L}_m(G) &\subseteq \overline{K^{\uparrow C}} \\
\Rightarrow \quad \overline{\overline{K^{\uparrow C}} \cap \mathcal{L}_m(G)} &\subseteq \overline{K^{\uparrow C}} \\
\Rightarrow \quad \overline{K'} &\subseteq \overline{K^{\uparrow C}} .
\end{aligned}
$$

- Therefore, $\overline{K'} = \overline{K^{\uparrow C}}$. Since $\overline{K^{\uparrow C}}$ in controllable, then so is K', by definition of controllability.
- But $K' \subseteq \overline{K} \cap \mathcal{L}_m(G) = K$ (by hypothesis). It follows that $K' \subseteq K^{\uparrow C}$, that is,

$$\overline{K^{\uparrow C}} \cap \mathcal{L}_m(G) \subseteq K^{\uparrow C} .$$

2. The inclusion $\overline{K^{\uparrow C}} \subseteq (\overline{K})^{\uparrow C}$ is easily proved from the definition and monotonicity of $\uparrow C$. Part 2 of Example 3.8 in Section 3.4.3 shows that the inclusion can be strict.   **Q.E.D.**

This proposition allows us to state the answer to the nonblocking version of BSCP considered in Section 3.4.4.

BSCP-NB: Basic Supervisory Control Problem – Nonblocking Case

Given DES G with event set E, uncontrollable event set $E_{uc} \subseteq E$, and admissible marked language $L_{am} \subseteq L_m(G)$, with L_{am} assumed to be $L_m(G)$-closed, find a *nonblocking* supervisor S such that:

1. $L_m(S/G) \subseteq L_{am}$

2. $L_m(S/G)$ is "the largest it can be," that is, for any other nonblocking supervisor S_{other} such that $L_m(S_{other}/G) \subseteq L_{am}$,

$$L_m(S_{other}/G) \subseteq L_m(S/G) .$$

Solution of BSCP-NB:

Due to requirement 2, we call the desired solution S the *minimally restrictive nonblocking solution*. Using the results presented above and in Section 3.4.3, the solution is to choose S such that

$$\mathcal{L}(S/G) = \overline{L_{am}^{\uparrow C}} \quad \text{and} \quad L_m(S/G) = L_{am}^{\uparrow C}$$

as long as $L_{am}^{\uparrow C} \neq \emptyset$. It is important to note that since L_{am} is assumed to be $L_m(G)$-closed, then $L_{am}^{\uparrow C}$ is also $L_m(G)$-closed, which guarantees that

$$L_m(S/G) = L_{am}^{\uparrow C}$$

whenever

$$\mathcal{L}(S/G) = \overline{L_{am}^{\uparrow C}} .$$

S can be realized by building a recognizer of $\overline{L_{am}^{\uparrow C}}$. We will see in Section 3.5.3 how to build this recognizer from a recognizer of L_{am}.

Example 3.12 (Solution to BSCP-NB)

The database concurrency control example presented in Example 3.3 and used in many examples in this chapter is an instance of BSCP-NB. Namely, the admissible marked language L_{am} is the language marked by the automaton H_a of Fig. 3.3. The goal is to obtain a nonblocking supervisor. We solved an instance of this problem for $E_{uc} = \{a_2, b_2\}$ in part 1 of Example 3.8. Any automaton that generates and marks language

$$\overline{L_{am}^{\uparrow C}} = \overline{\{a_2 b_2 a_1 b_1, a_2 a_1 b_2 b_1\}}$$

realizes the desired nonblocking supervisor.

Important Observation about Blocking:

Choosing S_{alt} such that

$$\mathcal{L}(S_{alt}/G) = (\overline{L_{am}})^{\uparrow C}$$

will satisfy requirement 1, *but S_{alt} may be blocking*. This will happen when the set inclusion

$$\overline{L_{am}^{\uparrow C}} \subseteq (\overline{L_{am}})^{\uparrow C}$$

is strict, which is possible as was seen earlier. Consider the following example.

Example 3.13 (Incorrect solution to BSCP-NB)
Take

$$
\begin{aligned}
\mathcal{L}(G) &= \overline{\{abababa\}} \\
\mathcal{L}_m(G) &= \{a, aba, ababa, abababa\} \\
L_{am} &= \{a, aba, ababa\} \\
E_{uc} &= \{b\} \ .
\end{aligned}
$$

These languages were encountered in part 2 of Example 3.8 where we obtained

$$
\begin{aligned}
L_{am}^{\uparrow C} &= \emptyset \\
(\overline{L_{am}})^{\uparrow C} &= \{\varepsilon, a, ab, aba, abab\} \ .
\end{aligned}
$$

Using a supervisor that enforces

$$\mathcal{L}(S/G) = (\overline{L_{am}})^{\uparrow C}$$

we obtain the marked language under control

$$\mathcal{L}_m(S/G) = \{a, aba\} \subset L_{am} \ .$$

Thus, while we guarantee that $\mathcal{L}_m(S/G) \subseteq L_{am}$, the controlled system blocks (in this case, deadlocks) after string $abab \in \mathcal{L}(S/G)$, which means that such a supervisor does not solve BSCP-NB. In fact, BSCP-NB does not have a solution here as $L_{am}^{\uparrow C} = \emptyset$.

3.5.3   Computation of $K^{\uparrow C}$: General Case

In this section, we generalize the results of Section 3.4.5 and discuss how to implement the $\uparrow C$ operation in the case of regular languages that are not prefix-closed. More precisely, we are concerned with computing $K^{\uparrow C}$ for a regular language K that need not be prefix-closed; the controllability condition is with respect to regular and prefix-closed language M and uncontrollable event set E_{uc}. Unlike the situation in Section 3.4.5, there is no closed-form expression for $K^{\uparrow C}$ in the present case. However, we do have the following important result.

Theorem. (Regularity of supremal controllable sublanguage)
Let K be a regular language that is marked by a trim automaton with n states. Let M be a regular language that is generated by an automaton with m states. Then $K^{\uparrow C}$ is also regular and it can be represented (i.e., marked) by a trim automaton with at most nm states.   ◆

The proof of this result is somewhat involved technically and it will not be presented here. Instead, we present an intuitive algorithm for the implementation of the $\uparrow C$ operation. This algorithm will "show" that a recognizer for $K^{\uparrow C}$ has in the worst case nm states. The worst-case complexity of this algorithm is $O(n^2 m^2 |E|)$. Thus, the complexity of calculating $K^{\uparrow C}$ is quadratic in the product nm, in contrast to the prefix-closed case where it is linear in nm. We shall call this algorithm the "standard algorithm for $\uparrow C$." The reason for quadratic complexity is the fact that we may have to iterate in the computation of the recognizer of $K^{\uparrow C}$, as we did in Example 3.10.

Let us give some intuition before formally stating the standard algorithm. Assume that we are given trim automaton

$$H = (Y, E, g, \Gamma_H, y_0, Y_m) \ ,$$

with n states, such that

$$\mathcal{L}_m(H) = K$$

and automaton

$$G = (X, E, f, \Gamma, x_0) \ ,$$

with m states, such that

$$\mathcal{L}(G) = M \ .$$

Observe that marking of the states of G is disregarded here, since we are only concerned with the language M. We know that we need to be able to map the states of H to the states of G in order test the controllability condition for languages, $\overline{K} E_{uc} \cap M \subseteq \overline{K}$, on the automaton representations of these languages. To do this, we use the technique of refinement by product presented in Chapter 2, Section 2.3.4, and form the new automaton

$$H_0 := (Y_0, E, g_0, \Gamma_{H_0}, (y_0, x_0), Y_{0,m}) = H \times G$$

where $Y_0 \subseteq Y \times X$ and $Y_{0,m} \subseteq Y_m \times X$. Marking of the states of H_0 is solely determined by the states of H, since as mentioned above marking of G is not considered in this algorithm. Automaton H_0 has at most nm states and is clearly equivalent to H.

In order to check which strings in K, if any, violate the controllability condition, we need to check if

$$\Gamma_{H_0}[(y, x)] \supseteq \Gamma(x) \cap E_{uc}$$

at every state $(y, x) \in Y_0$. We call this condition on states the *active event set constraint* imposed by the controllability condition. Now, from a language viewpoint, all strings in K that possess a prefix that violates the controllability condition must be deleted from K in order to make this language controllable. From an automaton viewpoint, all states where the active event set constraint is violated must be deleted from H_0 and since we are interested in the language *marked* by H_0, these deletions *must be followed by taking the trim operation.* Let us call the resulting automaton H_1.

The key observation here is that the process of performing the trim may result in new violations of the active event set constraint at states that did not violate this condition prior to the trim. From a language viewpoint, removing one or more strings from K because they violate the controllability condition may lead to new violations of the controllability condition by other strings, thus forcing the process of iterative removal of strings until the resulting sublanguage of K is controllable. Recall Example 3.8 where such iterations were performed. In the context of automaton representations, the same iterative process must be performed, that is, we need to examine the states of H_1 for possible violations of the active event set constraint, remove states that violate it, take the trim operation to obtain H_2, and repeat. (This is in fact what we did in Example 3.10, although there the trim operation was reduced to the accessible operation since we were dealing with prefix-closed languages.) Clearly, the maximum number of iterations is nm since H_0 has nm states and at least one state is removed at each iteration. This explains the complexity result stated above.

We are now ready to formally state the standard algorithm.

Standard Algorithm for $\uparrow C$

Step 0: Let $G = (X, E, f, \Gamma, x_0)$ be an automaton that generates M, i.e., $\mathcal{L}(G) = M$.

Let $H = (Y, E, g, \Gamma_H, y_0, Y_m)$ be such that $\mathcal{L}_m(H) = K$ and $\mathcal{L}(H) = \overline{K}$, where it is assumed that $K \subseteq \mathcal{L}(G)$.

Step 1: Let
$$H_0 := (Y_0, E, g_0, \Gamma_{H_0}, (y_0, x_0), Y_{0,m}) = H \times G$$

where $Y_0 \subseteq Y \times X$. Treat all states of G as marked for the purpose of determining $Y_{0,m}$.

By assumption, $\mathcal{L}_m(H_0) = K$ and $\mathcal{L}(H_0) = \overline{K}$.

States of H_0 will be denoted by pairs (y, x).

Set $i = 0$.

Step 2: Calculate

Step 2.1:

$$
\begin{aligned}
Y_i' &= \{(y, x) \in Y_i : \Gamma(x) \cap \Sigma_{uc} \subseteq \Gamma_{H_i}((y, x))\} \\
g_i' &= g_i | Y_i' \\
Y_{i,m}' &= Y_{i,m} \cap Y_i'
\end{aligned}
$$

where the notation "|" stands for "restricted to."

Step 2.2:

$$H_{i+1} = Trim(Y_i', E, g_i', (y_0, x_0), Y_{i,m}') .$$

If H_{i+1} is the empty automaton, i.e, (y_0, x_0) is deleted in the above calculation, then $K^{\uparrow C} = \emptyset$ and STOP.

Otherwise, set

$$H_{i+1} =: (Y_{i+1}, E, g_{i+1}, (y_0, x_0), Y_{i+1,m}) .$$

Step 3: If $H_{i+1} = H_i$, then

$$\mathcal{L}_m(H_{i+1}) = K^{\uparrow C} \quad \text{and} \quad \mathcal{L}(H_{i+1}) = \overline{K^{\uparrow C}}$$

and STOP.

Otherwise, set $i \leftarrow i + 1$ and go to Step 2.

We make the following comment about step 1 above. By definition, a state of H_0 is marked if and only if the corresponding state of H is marked. This is because we want H_0 to be equivalent to H and therefore state marking in G should not affect H_0. If the given K happens to be a subset of $\mathcal{L}_m(G)$, then the second component of all marked states of H_0 will be marked in G.

We present next a simple example that illustrates all the steps of this algorithm.

Example 3.14 (Iterative calculation of $K^{\uparrow C}$)
We consider a variation of the languages used in part 2 of Example 3.8. The language M is generated by automaton G shown in Fig. 3.8. The language $K \subseteq M$ is marked by trim automaton H shown in Fig. 3.8. We take $E_{uc} = \{b\}$. The goal is to calculate $K^{\uparrow C}$ with respect to M and E_{uc}.

The refined automaton for K, H_0 of step 1, is shown in Fig. 3.9. We shall not draw the figures for the intermediate automata H_i obtained after each iteration of step 2, as they are easily derived from H_0. We instead describe what states are deleted in the iterative procedure.

1. At the first iteration of step 2, states $(F, 5)$, $(H, 5)$, $(G, 5)$, $(H, 3)$, and $(G, 3)$ are deleted since they violate the active event set constraint imposed by the controllability condition. Uncontrollable event b is not defined at any of these states of H_0, yet b can occur at states 5 and 3 of G.

   Then state $(E, 4)$ is deleted in the trim operation that produces H_1.

2. At the second iteration of step 2, state $(D, 3)$ is deleted since it violates the active event set constraint.

   Then state $(C, 2)$ is deleted in the trim operation that produces H_2.

3. At the third iteration of step 2, state $(B, 1)$ is deleted since it violates the active event set constraint.

Then state $(A, 0)$ is deleted in the trim operation that produces H_3. Since H_3 is the empty automaton, the algorithm terminates and $K^{\uparrow C} = \emptyset$.

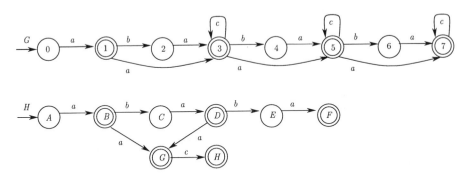

Figure 3.8: Automata G and H for Example 3.14.

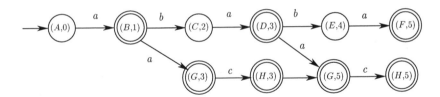

Figure 3.9: Automaton H_0 for Example 3.14.

3.5.4 Dealing with Blocking Supervisors

In the formulation of BSCP-NB, we require nonblocking supervisors as the only admissible solutions. In many applications however, it may be that nonblocking solutions are too restrictive in the sense that they constrain too much the behavior of the system. This is because we account for all possible behaviors of the system in the synthesis of the nonblocking supervisor. If blocking of the controlled system can be "easily" detected and resolved (by some mechanism not modeled here), then it may be of interest to relax the nonblocking requirement and consider the synthesis of blocking supervisors. In database applications for instance, blocking (deadlock) is detected by keeping track of the "waiting-for" relations between transactions and is resolved by aborting one or more of the transactions involved in a deadlock. (Aborts are feasible in this application because it is easy to "undo" read and write operations; a log is kept for that purpose.) The motivation for considering blocking solutions is that by allowing a certain amount of blocking (which is deemed acceptable in the given application), we can increase, in the sense of set inclusion, the part of L_{am} that can be achieved under control, thus potentially allowing for better "performance" of the controlled system. A complete characterization of this performance would require a richer modeling formalism, where one accounts for the probability of blocking and the cost to recover from a blocking solution. This is beyond the scope of this chapter. However, the above considerations motivate the study of the properties of blocking solutions in the context of our untimed modeling formalism.

In order to formally pose the issues raised in the above discussion, we consider the following problem formulation whose role is to serve as a framework for the study of blocking supervisors.

SCPB: Supervisory Control Problem with Blocking

Given DES G with event set E, uncontrollable event set $E_{uc} \subseteq E$, admissible language $L_a = \overline{L}_a \subseteq \mathcal{L}(G)$, and admissible marked language $L_{am} \subseteq \mathcal{L}_m(G)$, where the two admissible languages satisfy the condition

$$L_a \cap \mathcal{L}_m(G) = L_{am} \, ,$$

find a supervisor S, possibly blocking, such that:

1. $\mathcal{L}(S/G) \subseteq L_a$, which implies that $\mathcal{L}_m(S/G) \subseteq L_{am}$.

2. $\mathcal{L}(S/G)$ is *Pareto optimal* with respect to the two following sets which serve as measures of performance:

   (a) The *Blocking Measure* of S, $BM(S)$, defined as

   $$BM(S) := \mathcal{L}(S/G) \setminus \overline{\mathcal{L}_m(S/G)} \, ,$$

   which is the set of all strings of the controlled system that lead to blocking.

   (b) The *Satisficing Measure* of S, $SM(S)$, defined as

   $$SM(S) := \mathcal{L}_m(S/G) \cap L_{am} \, ,$$

   which is the set of admissible marked strings allowed by the controlled system. In view of requirement 1, we have that $SM(S) = \mathcal{L}_m(S/G)$.

   Pareto optimality means that any possible improvement of $BM(S)$ by shrinking this set is necessarily accompanied by a deterioration of $SM(S)$, that is, this set also shrinks. Similarly, any possible improvement of $SM(S)$ by enlarging this set is necessarily accompanied by a deterioration of $BM(S)$, that is, this set is also enlarged.

Solution of SCPB:

As formulated above, SCPB does not have a unique solution, but rather a set of Pareto optimal solutions that are incomparable in our logical framework where set inclusion is the mechanism for comparing solutions. Thus, we will focus our attention on determining if a given, possibly blocking, supervisor S satisfies or does not satisfy the condition of Pareto optimality for BM and SM in SCPB. Before presenting algorithms that answer this question, we make the following observations about SCPB. All of these observations follow from results presented earlier about the $\uparrow C$ and $\downarrow C$ operations and by comparing SCPB with BSCP, DuSCP, and BSCP-NB.

- There is only one *nonblocking* solution S_{nb} to SCPB, that is, a solution for which $BM(S_{nb}) = \emptyset$: It is the minimally restrictive nonblocking solution where
$$\mathcal{L}(S_{nb}/G) = \overline{L_{am}^{\uparrow C}} \quad \text{and} \quad \mathcal{L}_m(S_{nb}/G) = L_{am}^{\uparrow C} .$$

  Thus $SM(S_{nb}) = L_{am}^{\uparrow C}$. Observe that it is clear from earlier results that S_{nb} is Pareto optimal.

- From requirement 1 in the problem statement, the *largest* solution of interest is
$$\mathcal{L}(S_{mrs}/G) = L_a^{\uparrow C} \cap L_{am}^{\downarrow C} .$$

  We call it the *minimally restrictive solution*. It can be shown that this solution is Pareto optimal. Observe that $SM(S_{mrs}) = L_a^{\uparrow C} \cap L_{am}$.

- From the above observations, the set of *Candidate Solutions* of SCPB is
$$\mathcal{L}_{cand} = \{K : (\overline{L_{am}^{\uparrow C}} \subseteq K \subseteq L_a^{\uparrow C} \cap L_{am}^{\downarrow C}) \text{ and } (K = \overline{K}) \text{ and } (K = K^{\uparrow C}\} .$$

  Observe that this set is closed under unions and intersections. Not all candidate solutions to SCPB are Pareto optimal, although the Pareto optimal ones are those of interest.

- If $L_{am}^{\downarrow C} \subseteq L_a$, then there exists a *completely satisficing solution* S_{css} in the sense that $SM(S_{css}) = L_{am}$; namely, take
$$\mathcal{L}(S_{css}/G) = L_{am}^{\downarrow C} .$$

  It can be shown that CSS is Pareto optimal. Observe that $BM(S_{css}) = L_{am}^{\downarrow C} \setminus \overline{L_{am}}$.

Example 3.15 (Extremal solutions of SCPB)

The database concurrency control problem presented in Example 3.3 can be used to illustrate SCPB. Given the requirement that the only admissible strings are those where "a_1 precedes a_2 if and only if b_1 precedes b_2," we can choose

$$
\begin{aligned}
L_a &= \mathcal{L}(G) \setminus \{a_1 a_2 b_2 b_1, a_2 a_1 b_1 b_2\} \\
L_{am} &= \mathcal{L}(H_a)
\end{aligned}
$$

with automata G and H_a depicted in Fig. 3.3. Note here that we include more strings in L_a than contained in $\overline{L_{am}}$. The only strings excluded from L_a are those that violate the precedence requirement; this means that L_a will contain strings that do not violate the precedence requirement but lead to blocking as all their feasible continuations do violate the precedence requirement.

In the case where $E_{uc} = \{a_2, b_2\}$ (as in Section 3.3.2 and part 1 of Example 3.8):

- The minimally restrictive nonblocking solution of SCPB is

$$
\mathcal{L}(S_{nb}/G) = \overline{L_{am}^{\uparrow C}}
$$

  found in part 1 of Example 3.8.

- Since $L_{am}^{\downarrow C}$ found in Example 3.9 is contained in L_a, the minimally restrictive solution is

$$
\mathcal{L}(S_{mrs}/G) = L_{am}^{\downarrow C}
$$

  and it is completely satisficing.

Improving the Blocking Measure

The first problem that we address is: Given S such that $\mathcal{L}(S/G) \in \mathcal{L}_{cand}$, can we find S' such that

1. $BM(S') \subset BM(S)$

2. $SM(S') \supseteq SM(S)$.

In such a case, we say that S' is *BM-better* than S.

The solution to this problem is straightforward and obtained by a single application of the $\downarrow C$ operation. It is given by the operator

$$
\mathcal{A}_{BM} : \mathcal{L}_{cand} \to \mathcal{L}_{cand}
$$

defined as follows:

$$
\mathcal{A}_{BM}(L_p) := (L_p \cap L_{am})^{\downarrow C} .
$$

If $\mathcal{A}_{BM}(\mathcal{L}(S/G)) \subset \mathcal{L}(S/G)$, then it can be verified that S' defined by

$$
\mathcal{L}(S'/G) = \mathcal{A}_{BM}(\mathcal{L}(S/G))
$$

is *BM-better* than S. This is a consequence of the infimality property of the $\downarrow C$ operation. Moreover, it is not difficult to prove that the $\mathcal{A}_{BM}$ operator is *idempotent*, namely,

$$\mathcal{A}_{BM}(\mathcal{A}_{BM}(K)) = \mathcal{A}_{BM}(K) \, .$$

Improving the Satisficing Measure

The second problem that we address is: Given S such that $\mathcal{L}(S/G) \in \mathcal{L}_{cand}$, can we find S' such that

1. $BM(S') \subseteq BM(S)$

2. $SM(S') \supset SM(S)$.

In such a case, we say that S' is *SM-better* than S.

The solution to this problem is more intricate than the solution to the first problem considered above. Basically, we need to try to enlarge $SM(S)$ by adding strings that will not cause any additional blocking. The solution is given by the operator

$$\mathcal{A}_{SM} : \mathcal{L}_{cand} \rightarrow \mathcal{L}_{cand}$$

defined as follows:

$$
\begin{aligned}
\mathcal{A}_{SM}(L_p) \;&:=\; L_p \cup K_{max} \\
&=\; L_p \cup K_{max}^{\downarrow C}
\end{aligned}
$$

where $L_p \cap K_{max} = \emptyset$ and where

$$K_{max} := sup\{K : (K \subseteq L_{am} \setminus L_p) \text{ and } (K^{\downarrow C} \subseteq L_p \cup L_{am})\} \, .$$

(The notation *sup* means the largest language satisfying the conditions in the braces.) If L_p denotes the language generated by S/G, then the first clause ensures that adding K_{max} does enlarge the satisficing measure and the second clause guarantees that this increase in the satisficing measure is not accompanied by an increase in the blocking measure. Consequently, if $K_{max} \neq \emptyset$, then it can be shown that S' defined as

$$\mathcal{L}(S'/G) = \mathcal{A}_{SM}(\mathcal{L}(S/G))$$

is *SM-better* than S.

We shall not go into further details on SCPB. Our objective was to illustrate how the $\uparrow C$ and $\downarrow C$ operations can be used as "building blocks" for solving more complicated supervisory control problems.

3.6 MODULAR CONTROL

By modular control, we refer to the situation where the control action of supervisor S is given by some combination of the control actions of two or more supervisors. For simplicity, we consider the case of two supervisors and discuss their conjunction. Given admissible supervisors S_1 and S_2 each defined for DES G, we define the *modular* supervisor $S_{mod12} : \mathcal{L}(G) \to 2^E$ corresponding to the *conjunction* of S_1 and S_2 as follows:

$$S_{mod12}(s) := S_1(s) \cap S_2(s) \ .$$

This means that an event is enabled by S_{mod12} if and only if it is enabled by both S_1 and S_2, or, in other words, it suffices for *one* supervisor to disable an event for the event to be effectively disabled in this modular control architecture which is depicted in Fig. 3.10. It is straightforward to verify that the closed-loop behavior under modular control is

$$\begin{aligned} \mathcal{L}(S_{mod12}/G) &= \mathcal{L}(S_1/G) \cap \mathcal{L}(S_2/G) \\ \mathcal{L}_m(S_{mod12}/G) &= \mathcal{L}_m(S_1/G) \cap \mathcal{L}_m(S_2/G) \ . \end{aligned}$$

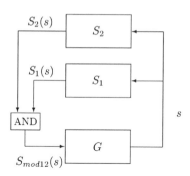

Figure 3.10: Modular control architecture with two supervisors.
The resulting control action on G is the intersection of the control actions of the individual supervisors, which means that an event is enabled if and only if it is enabled by both supervisors.

Given standard realizations R_1 and R_2 of S_1 and S_2, respectively, the standard realization of S_{mod12} could be obtained by building $R = R_1 \times R_2$. However, the point here is precisely *not* to build this realization, but rather to use the existing R_1 and R_2 and realize the control action $S_{mod12}(s)$ by taking the *intersection of the active event sets of R_1 and R_2* at their respective states after the execution of s, in accordance with the block diagram in Fig. 3.10. We call this the *modular realization* of modular supervisor S_{mod12}. Significant memory savings may be possible by realizing S_{mod12} in this manner. If R_1 has n_1 states and R_2 has n_2 states, then we need only to store a total of $n_1 + n_2$ states for this modular realization instead of possibly as many as $n_1 n_2$ states if $R = R_1 \times R_2$ were built. Note that we can *interpret* the supervision of G by S_{mod12} as the product $R_1 \times R_2 \times G$.

It is a similar complexity argument that motivates the *synthesis* of a supervisor in modular form. If the admissible language L_a for BSCP is given as (or can be decomposed as) the intersection of two prefix-closed languages

$$L_a = L_{a1} \cap L_{a2}$$

then we would like to synthesize S_i for $L_{ai}^{\uparrow C}$, $i = 1, 2$, and use these two supervisors in conjunction instead of doing the full calculation $L_a^{\uparrow C}$. Using this modular approach, the total (worst case) computational complexity for supervisor synthesis is reduced from $O(n_1 n_2 m)$ to $O(\max(n_1, n_2)m)$. This modular synthesis is a correct approach because in the case of prefix-closed languages,

$$(K_1 \cap K_2)^{\uparrow C} = K_1^{\uparrow C} \cap K_2^{\uparrow C}$$

as was seen in Section 3.4.3; therefore, we do get the desired answer.

This discussion is formalized in the following modular version of BCSP, which we denote by MSCP.

MSCP: Modular Supervisory Control Problem

Given DES G with event set E, uncontrollable event set $E_{uc} \subseteq E$, and admissible language

$$L_a = L_{a1} \cap L_{a2}$$

where $L_{ai} = \overline{L_{ai}} \subseteq \mathcal{L}(G)$ for $i = 1, 2$, find a *modular* supervisor S_{mod} (according to the architecture in Fig. 3.10) such that

$$\mathcal{L}(S_{mod}/G) = L_a^{\uparrow C} ,$$

which is the same as what can be achieved by a "monolithic" approach (cf. BSCP).

Solution of MSCP:

From the above discussion, we build the standard realizations R_i of S_i such that

$$\mathcal{L}(S_i/G) = L_{ai}^{\uparrow C}$$

for $i = 1, 2$, and then take S_{mod} to be the modular supervisor

$$S_{mod}(s) := S_{mod12}(s) = S_1(s) \cap S_2(s).$$

With this choice of modular supervisor S_{mod} we get

$$
\begin{aligned}
\mathcal{L}(S_{mod}/G) &= L_{a1}^{\uparrow C} \cap L_{a2}^{\uparrow C} \\
&= (L_{a1} \cap L_{a2})^{\uparrow C} \\
&= L_{a}^{\uparrow C}
\end{aligned}
$$

which is the desired solution.

It is unfortunate that the same simple approach to synthesizing and realizing a modular supervisor will not, in general, work for the *nonblocking* version of BSCP, BSCP-NB. The problem is that the conjunction of two nonblocking supervisors *need not* be a nonblocking supervisor. Consider the following result, whose proof is straightforward from the definitions of the nonblocking and nonconflicting properties.

Proposition. (Nonblocking modular supervisors)
Let S_i, $i = 1, 2$, be individual nonblocking supervisors for G. Then S_{mod12} is nonblocking if and only if $\mathcal{L}_m(S_1/G)$ and $\mathcal{L}_m(S_2/G)$ are nonconflicting languages, that is, if and only if

$$
\overline{\mathcal{L}_m(S_1/G) \cap \mathcal{L}_m(S_2/G)} = \overline{\mathcal{L}_m(S_1/G)} \cap \overline{\mathcal{L}_m(S_2/G)} . \qquad \blacklozenge
$$

This result has the following implication. If we consider the "modular version" of BSCP-NB where

$$
L_{am} = L_{am1} \cap L_{am2}
$$

and where $L_{ami} \subseteq \mathcal{L}_m(G)$ and is $\mathcal{L}_m(G)$-closed, for $i = 1, 2$ (which implies that L_{am} itself is $\mathcal{L}_m(G)$-closed), the intuitive approach of first synthesizing S_i such that

$$
\mathcal{L}(S_i/G) = \overline{L_{ami}^{\uparrow C}}
$$

for $i = 1, 2$, and then forming the modular supervisor S_{mod12} yields:

$$
\begin{aligned}
\mathcal{L}(S_{mod12}/G) &= \overline{L_{am1}^{\uparrow C}} \cap \overline{L_{am2}^{\uparrow}} \\
\mathcal{L}_m(S_{mod12}/G) &= \overline{L_{am1}^{\uparrow C}} \cap \overline{L_{am2}^{\uparrow C}} \cap \mathcal{L}_m(G) \\
&= L_{am1}^{\uparrow C} \cap L_{am2}^{\uparrow C} \\
&\supseteq (L_{am1} \cap L_{am2})^{\uparrow C} \\
&= L_{am}^{\uparrow C}
\end{aligned}
$$

where we have used results from Section 3.4.3. This means that the modular supervisor S_{mod12} could be blocking, even though it satisfies the legality requirement of BSCP-NB in the sense that

$$
\mathcal{L}_m(S_{mod12}/G) \subseteq L_{am} .
$$

By the above proposition, the above approach for finding a modular solution to BSCP-NB will result in a nonblocking modular solution if and only if $L^{\uparrow C}_{am1}$ and $L^{\uparrow C}_{am2}$ are nonconflicting languages. It is important to emphasize that this condition cannot be verified *before* doing the $\uparrow C$ calculations. Moreover, to verify this condition, we have to examine $L^{\uparrow C}_{am1}$ and $L^{\uparrow C}_{am2}$ "together" (i.e., form their product). In contrast, the monolithic approach to BSCP-NB used in Section 3.5.2 requires taking the intersection $L_{am1} \cap L_{am2}$ and then applying the $\uparrow C$ operation on the result, a calculation that has essentially the same (worst case) computational complexity as the verification of the above nonconflicting condition. In addition, the monolithic approach *guarantees* a nonblocking (global) supervisor, namely,

$$\mathcal{L}(S/G) = \overline{L^{\uparrow C}_{am}} \, ,$$

the minimally restrictive nonblocking solution of BSCP-NB. Yet, if the above modular solution is indeed nonblocking, then, as was mentioned earlier, it may still be advantageous from an implementation viewpoint as it can lead to memory savings.

Example 3.16 (Two users of two common resources)

Let us revisit the dining philosophers example presented in Chapter 2, Example 2.17. This is the classic example of two users of two resources. The uncontrolled system G is the parallel composition of automata P_1 and P_2 in Fig. 2.15 and it contains 16 states. Let us assume that the events "philosopher i picks up fork j" (abbreviated as ifj in Fig. 2.15) are controllable but that the events "philosopher i puts down both forks" (abbreviated as jf in Fig. 2.15) are uncontrollable. The control specification is to ensure that each fork is being used by at most one philosopher at any time, as well as to ensure that the resulting behavior is nonblocking. Since the initial state of G is the only marked state of G, a nonblocking solution means that each philosopher is able to eventually return to his thinking state after eating.

The above-described modular approach to this control problem would synthesize two supervisors, S_1 and S_2, where S_1 would take care of fork 1 and S_2 of fork 2. To design S_i, we identify in G all illegal states where fork i is being used by both philosophers. We delete all such illegal states and take the trim operation. The resulting automaton marks the admissible marked language L_{ami} for fork i; this language is $\mathcal{L}_m(G)$-closed by construction. Consequently, S_i is realized by automaton R_i, where R_i generates (and marks, since all its states are marked by construction of the standard realization) the language

$$\overline{L^{\uparrow C}_{ami}} \, .$$

The controlled system S_i/G, which is formally identical to $R_i \times G$, is depicted in Fig. 3.11, where Fig. 3.11 (a) shows S_1/G and Fig. 3.11 (b) shows S_2/G. Clearly, each controlled system is nonblocking and legal from the point of view of the fork under consideration.

Now, if we form the modular supervisor S_{mod12}, we get the behavior shown in Fig. 3.12, which is formally identical to $R_1 \times R_2 \times G$. We can see that the controlled system under such modular supervision is blocking, even though it ensures that each fork is never in use by two philosophers at any given time. (In fact, Fig. 3.12 is identical to Fig. 2.16 in Chapter 2, although Fig. 2.16 was obtained by an intuitive approach to ensuring mutual exclusion of fork usage and not by a formal supervisor synthesis procedure, as was followed here.)

The blocking problem arises here because languages $L_{am1}^{\uparrow C}$ and $L_{am2}^{\uparrow C}$ are conflicting, as is clearly seen from an examination of Fig. 3.11. The monolithic solution to this problem would be to form the language

$$L_{am} := L_{am1} \cap L_{am2}$$

which would be marked by the automaton obtained by deleting from G all illegal states for fork 1 *and* fork 2 and then taking the trim operation. The desired S would be realized by the automaton generating (and marking)

$$\overline{L_{am}^{\uparrow C}} .$$

It can be verified that in this case the controlled system S/G is equal to the trim of the automaton in Fig. 3.12. (This is not true in general; it is a consequence of our choice of E_{uc} in this example.)

3.7 DEALING WITH UNOBSERVABILITY

We presented the feedback loop of supervisory control in the case of partial event observation in Section 3.2.2; see Fig. 3.2. In this case, we have to deal with the presence of *unobservable events* in addition to the presence of uncontrollable events. The purpose of this section is to study what controlled behaviors can be achieved when controlling DES G with partial-observation supervisor S_P. Clearly, unobservable events impose further limitations on the controlled behaviors that can be achieved by P-supervisors. Therefore, we wish to identify the precise technical condition on languages that is key for the existence result of P-supervisors, in addition to the condition of controllability encountered in the Controllability Theorem of Section 3.4.1 and the condition of $\mathcal{L}_m(G)$-closure encountered in the Nonblocking Controllability Theorem of Section 3.5.1. This additional condition will be called *observability*.

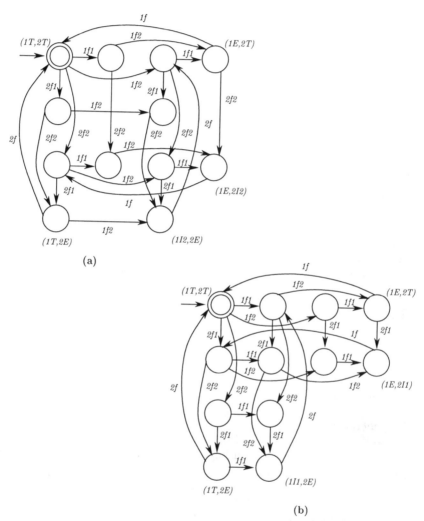

Figure 3.11: Supervisors for **(a)** fork 1 and **(b)** fork 2 for the dining philosophers example (Example 3.16).
Each figure depicts the controlled behavior under the supervisor for one fork only.

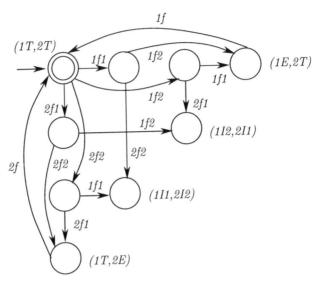

Figure 3.12: Modular supervision for the dining philosophers example (Example 3.16).
This figure depicts the controlled behavior under modular supervision with the supervisors for fork 1 and fork 2. This resulting behavior is blocking.

The main goal of this section is to present and understand the property of observability. While we shall roughly follow the same organization as in Section 3.4, we shall not go into as much detail as we did in our study of controllability.

3.7.1 Controllability and Observability Theorem

The generalization of the Controllability Theorem (CT) and the Nonblocking Controllability Theorem (NCT) to control under partial observation necessarily requires another condition beyond controllability and $\mathcal{L}_m(G)$-closure. Intuitively, observability means:

> *If you cannot differentiate between two strings, then these strings should require the same control action."*

Another way to phrase this, from the point of view of event disablement, is:

> *"If you must disable an event after observing a string, then by doing so you should not disable any string that appears in the desired behavior."*

This intuition is formalized in the following definition.

Definition. (Observability)

Let K and $M = \overline{M}$ be languages over event set E. Let E_c be a designated subset of E. Let E_o be another designated subset of E with P as the corresponding natural projection from E^* to E_o^*.

K is said to be *observable* with respect to M, P, and E_c if for all $s \in \overline{K}$ and for all $\sigma \in E_c$,

$$(s\sigma \notin \overline{K}) \text{ and } (s\sigma \in M) \;\Rightarrow\; P^{-1}[P(s)]\sigma \cap \overline{K} = \emptyset \,. \qquad \blacklozenge$$

(Note the slight abuse of notation in the definition: $P^{-1}[P(s)]\sigma$ stands for $P^{-1}[P(s)]\{\sigma\}$.)

Comments:

1. The term $P^{-1}[P(s)]\sigma \cap \overline{K}$ on the right-hand side of the implication in the definition of observability identifies all strings in $\overline{K}$ that have the same projection as s and can be continued within $\overline{K}$ with event σ. If this set is not empty, namely if K is not observable, this means that $\overline{K}$ contains two strings, s and s', such that $P(s) = P(s')$, and where $s\sigma \notin \overline{K}$ while $s'\sigma \in \overline{K}$. If this happens, then clearly no P-supervisor can exactly achieve language $\overline{K}$, since a P-supervisor cannot differentiate between s and s', yet, these strings require a different control action regarding event σ.

2. If the parameter E_c is omitted from the definition of observability, then it will be understood to be equal to E.

   The parameter E_c is included in the definition in order for the property of observability not to "overlap" with the property of controllability. (See the proof of the Controllability and Observability Theorem below.) Examining the definition of observability, we can see that when $s \in \overline{K}$, $s\sigma \in M$, and $\sigma \in E_{uc}$, then controllability implies that $s\sigma \in \overline{K}$, that is, there is no need to "worry" about observability issues for uncontrollable events, as long as K is controllable with respect to M and E_{uc}.

3. As in the case of controllability, the observability of a language depends only on the prefix-closure of this language. Thus K is observable if and only if $\overline{K}$ is observable.

4. We shall use interchangeably E_o and its corresponding natural projection P when talking about the observability of a language.

Example 3.17 (Observability of languages)

Let $E = \{u, b\}$ and consider the language $M = \overline{\{ub, bu\}}$. Let $E_{uo} = \{u\}$ and $E_{uc} = \{b\}$. Is the language $K_1 = \{bu\}$ observable with respect to M, E_o, and E_c? Looking at the definition of observability, we see that we need to first examine the string $s = \varepsilon \in \overline{K_1}$. This string can be extended outside of $\overline{K_1}$ by controllable event $\sigma = u$, and $u \in M$. On the other hand, there is no other string in $\overline{K_1}$ that starts with a string that has the same projection as ε and ends with u. This means that for this s and σ, $P^{-1}[P(s)]\sigma \cap \overline{K_1} = \emptyset$, which is consistent with the property of observability. Next, we need to examine $s = b$; however, there is no controllable event that extends this string s in M but not in $\overline{K_1}$, and consequently the observability condition is necessarily satisfied. The same happens when considering $s = bu$. We conclude that K_1 is observable.

As a second example, consider the language $K_2 = \{ub\}$ and modify E_c to $E_c = \{b\}$. In this case, if we take $s = \varepsilon$ and $\sigma = b \in E_c$, we have that $s\sigma = b \in M \setminus \overline{K_2}$. But $s' = u \in P^{-1}[P(s)]$ since $u \in E_{uo}$, and $s'\sigma = ub \in \overline{K_2}$, which contradicts the property of observability. Consequently, with this choice of E_c, K_2 is not observable. Note that if event b were uncontrollable (as considered above with K_1), then K_2 would be observable but would not be controllable.

We are now ready to present the generalization of CT and NCT to control under partial observation.

Theorem. [Controllability and Observability Theorem (COT)]
Consider DES $G = (X, E, f, \Gamma, x_0, X_m)$, where $E_{uc} \subseteq E$ is the set of uncontrollable events and $E_o \subseteq E$ is the set of observable events. Let P be the natural projection from E^* to E_o^*. Consider also the language $K \subseteq \mathcal{L}_m(G)$, where $K \neq \emptyset$. There exists a nonblocking P-supervisor S_P for G such that

$$\mathcal{L}_m(S_P/G) = K \quad \text{and} \quad \mathcal{L}(S_P/G) = \overline{K}$$

if and only if the three following conditions hold:

1. K is controllable with respect to $\mathcal{L}(G)$ and E_{uc}.

2. K is observable with respect to $\mathcal{L}(G)$, P, and E_c.

3. K is $\mathcal{L}_m(G)$-closed.

Proof. (IF) For $t \in P[\mathcal{L}(G)]$, define $S_P(t)$ according to

$$S_P(t) = E_{uc} \cup \{\sigma \in E_c : \exists s'\sigma \in \overline{K} \ (P(s') = t)\} \ .$$

This supervisor enables after string $t \in P[\mathcal{L}(G)]$:

(i) all uncontrollable events, and

(ii) all controllable events that extend any string s', that projects to t, inside of $\overline{K}$.

Part (i) ensures that S_P is admissible, namely, that it never disables a feasible uncontrollable event, since all uncontrollable events are always enabled.

(Strictly speaking, part (i) need only enable *feasible* uncontrollable events – compare with the admissibility condition for P-supervisors stated in Section 3.2.2 – but enabling all uncontrollable events has no adverse effect and simplifies the notation.)

We now prove that with this S_P, $\mathcal{L}(S_P/G) = \overline{K}$. Then, as in the case of NCT, the $\mathcal{L}_m(G)$-closure condition will imply that $\mathcal{L}_m(S_P/G) = K$ and S_P is nonblocking. The proof is by induction on the length of the strings in the two languages.

- The base case is for strings of length 0. But $\varepsilon \in \mathcal{L}(S_P/G)$ by definition and $\varepsilon \in \overline{K}$ since $K \neq \emptyset$ by hypothesis. Thus the base case holds.

- The induction hypothesis is that for all strings s such that $|s| \leq n$, $s \in \mathcal{L}(S_P/G)$ if and only if $s \in \overline{K}$. We now prove the same for strings of the form $s\sigma$ where $|s| = n$.

  - Let $s\sigma \in \mathcal{L}(S_P/G)$. By definition of $\mathcal{L}(S_P/G)$, this implies that

  $$[s \in \mathcal{L}(S_P/G)] \text{ and } [\sigma \in S_P[P(s)]] \text{ and } [s\sigma \in \mathcal{L}(G)]$$

  which in turn implies that

  $$[s \in \overline{K}] \text{ and } [\sigma \in S_P[P(s)]] \text{ and } [s\sigma \in \mathcal{L}(G)]$$

  using the induction hypothesis.

    * Now, if $\sigma \in E_{uc}$, then the controllability condition immediately yields $s\sigma \in \overline{K}$.
    * On the other hand, if $\sigma \in E_c$, then by the above definition of S_P and with $t = P(s)$, we obtain that there exists $s'\sigma \in \overline{K}$ such that $P(s') = t = P(s)$, that is, we have that

  $$P^{-1}[P(s)]\sigma \cap \overline{K} \neq \emptyset \ .$$

  But since $s\sigma \in \mathcal{L}(G)$, we must have that $s\sigma \in \overline{K}$, otherwise observability would be contradicted.

  This completes the proof that $s\sigma \in \overline{K}$.

  - For the other direction of the induction step, let $s\sigma \in \overline{K}$. Then $s\sigma \in \mathcal{L}(G)$ since by assumption $\overline{K} \subseteq \overline{\mathcal{L}_m(G)} \subseteq \mathcal{L}(G)$.
    * Now, if $\sigma \in E_{uc}$, then $\sigma \in S_P[P(s)]$ since by definition S_P is guaranteed to be admissible.
    * On the other hand, if $\sigma \in E_c$, then by the above definition of $S_P[P(s)]$, we also obtain that $\sigma \in S_P[P(s)]$.

  Overall, we have that

  $$[s \in \overline{K}] \text{ and } [\sigma \in S_P[P(s)]] \text{ and } [s\sigma \in \mathcal{L}(G)]$$

which in turns implies that

$$[s \in \mathcal{L}(S_P/G)] \text{ and } [\sigma \in S_P[P(s)]] \text{ and } [s\sigma \in \mathcal{L}(G)]$$

using the induction hypothesis. It then immediately follows that $s\sigma \in \mathcal{L}(S_P/G)$.

  – This completes the proof of the induction step.

(ONLY IF) Let S_P be an admissible P-supervisor such that $\mathcal{L}(S_P/G) = \overline{K}$ and $\mathcal{L}_m(S_P/G) = K$.

- The proof that controllability and $\mathcal{L}_m(G)$-closure hold is the same as in CT for controllability and NCT for $\mathcal{L}_m(G)$-closure; the only modification is that $S(s)$ there is now replaced by $S_P[P(s)]$.
- To prove that observability must also hold, take $s \in \overline{K}$ and $\sigma \in E_c$ such that $s\sigma \notin \overline{K}$ and $s\sigma \in \mathcal{L}(G)$.

  – From $s \in \overline{K}$, $\sigma \in E_c$, $s\sigma \notin \overline{K}$, and $\mathcal{L}(S_P/G) = \overline{K}$, we must have that $\sigma \notin S_P[P(s)]$, otherwise $\mathcal{L}(S_P/G) \neq \overline{K}$.
  – But this means that there cannot exist $s'\sigma \in \overline{K}$ such that $P(s') = P(s)$, otherwise $\mathcal{L}(S_P/G) \neq \overline{K}$ since S_P does not distinguish between s and s'.
  – That is, we must have that

  $$P^{-1}[P(s)]\sigma \cap \overline{K} = \emptyset .$$

This proves the observability condition.   Q.E.D.

As in the case of CT and NCT, we note that the proof of COT is *constructive* in the sense that if the controllability, observability, and $\mathcal{L}_m(G)$-closure conditions are satisfied, it gives us a supervisor that will achieve the required behavior. That supervisor is:

$$S_P(t) = E_{uc} \cup \{\sigma \in E_c : \exists s'\sigma \in \overline{K} \ (P(s') = t)\} ,$$

which is necessarily admissible according to Section 3.2.2.

We present a useful corollary that explicitly states a special case of COT when only $\mathcal{L}(S_P/G)$ is of concern.

Corollary. Consider DES $G = (X, E, f, \Gamma, x_0)$ where $E_{uc} \subseteq E$ is the set of uncontrollable events and $E_o \subseteq E$ is the set of observable events. Let P be the natural projection from E^* to E_o^*. Let $K \subseteq \mathcal{L}(G)$, $K \neq \emptyset$. Then there exists S_P such that

$$\mathcal{L}(S_P/G) = \overline{K}$$

if and only if K is controllable with respect to $\mathcal{L}(G)$ and E_{uc} and observable with respect to $\mathcal{L}(G)$, P, and E_c.   ♦

In view of the key role played by the notion of observability in COT, we would like to know if observability of a regular language K with respect to a regular language M can be tested efficiently; this is not obvious from the way the definition of observability is stated. Results in the literature show that observability can be tested with worst-case complexity of $O(m^2 n)$, when m and n are the numbers of states of automata that generate $\overline{K}$ and M, respectively; note that we have absorbed $|E|$ in the constants of $O(\cdot)$. We shall not present the details of this test here. Instead, we present a test that is less computationally efficient but insightful regarding control under partial observation. This test requires building the observer of the language under consideration. Observer automata were presented in Chapter 2, in Sections 2.3.3 and 2.5.2. In the present context, we are interested in building the observer of an automaton with unobservable events, as discussed in Section 2.5.2 and presented in Example 2.22. We present our test by means of an example. This example is very simple, yet it illustrates all the steps of the testing procedure that we wish to present.

Example 3.18 (Test for observability)

Take $E = \{u, b\}$, $M = \overline{\{ub, bu\}}$, $E_{uo} = \{u\}$, $K = \{ub\}$, and $E_{uc} = \{u\}$. (This is the same data as in the second part of Example 3.17.)

Instead of exhaustively testing all strings in $\overline{K}$ according to the definition of observability, let us build automata H and G to represent K and M, respectively. We want to build H and G such that each state of H can be mapped to a corresponding state of G; refinement by product can be used for this purpose (see Section 2.3.4). This way, we are able to list for each state of H which controllable events must be enabled and disabled; namely, all controllable events in the active event set of H must be enabled, and all controllable events in the active event set of G but not in that of H must be disabled. Automata H and G that meet this condition are shown in Fig. 3.13.

The next step is to build H_{obs}, the observer of H for the given set E_o; H_{obs} is also shown in Fig. 3.13. We then examine each state of H_{obs} and determine the sets of controllable events that "must be enabled" and "must be disabled" by a P-supervisor that is based on H_{obs}. This is done by taking the union of these respective sets over all states of H appearing in the state of H_{obs}. At the initial state of H_{obs}, we have that uncontrollable event b is in the "must-be-disabled" list of state 1 of H and is also in the "must-be-enabled" list of state 2 of H. Therefore, the P-supervisor does not know what control action to take initially regarding controllable event b; we call this situation a *control conflict*. On the one hand, the P-supervisor needs to disable b in order to prevent the string $b \notin \overline{K}$; on the other hand, it needs to enable b in order to allow string $ub \in \overline{K}$, since it does not see event u.

In general, if at some state of H_{obs} the two sets "must-be-enabled" and "must-be-disabled" have an non-empty intersection, then there is a control conflict in H_{obs}. This means that there exists two strings in $\overline{K}$, say s and s', that have the same projection (since they lead to two different states in H but to the same state in H_{obs}), but have different continuations in $\overline{K}$ with respect to some controllable event, say σ; in other words, $s\sigma \in M \setminus \overline{K}$ but $s'\sigma \in \overline{K}$, which contradicts observability. Consequently, the presence of a control conflict in H_{obs} implies that the language generated by H is not observable. Conversely, the absence of any control conflict in H_{obs} means that a P-supervisor based on H_{obs} will always be able to make correct decisions regarding the control actions needed for the controllable events, despite its limited observability. A correct decision is given by the "must-be-enabled" list at each state of H_{obs}; since none of these events needs to be disabled, the controlled system will achieve all of $\overline{K}$, and nothing else.

In conclusion, the above K is not observable with respect to the above M, E_o, and E_c. However, K is controllable. If instead we had $E_{uc} = \{b\}$, then K would not be controllable, but it would be observable. The control conflict at the initial state of H_{obs} would not appear since the test only pertains to controllable events.

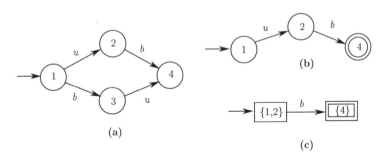

Figure 3.13: (a) G, (b) H, and (c) H_{obs} for Example 3.18. H_{obs} is used to test the observability of $\mathcal{L}_m(H)$ with respect to $\mathcal{L}(G)$ and $E_o = E_c = \{b\}$.

Example 3.19 (Test for observability: Section 3.3.2 revisited)

As a second illustration of our test for observability, we return to the database concurrency control example discussed in Section 3.3.2. The system of interest is the automaton G of Fig. 3.3 and the admissible language K is the language marked by automaton H_a of Fig. 3.3. We argued in Section 3.3.2 that K cannot be achieved by supervisory control when $E_{uo} = \{a_2\}$, even if all events are controllable. To formally verify this, we construct the observer of H_a, denoted by $H_{a,obs}$ and depicted in

Fig. 3.14. We can see that a control conflict arises at state $\{1,4,9\}$ of $H_{a,obs}$, as the control action at state 4 of H_a dictates that event b_2 must be enabled and event b_1 must be disabled. This is exactly the opposite of the control action action required at state 9 of $H_{a,obs}$, where b_2 must be disabled and b_1 enabled. The presence of this control conflict in $H_{a,obs}$ indicates that K is not an observable language.

We presented in Section 3.3.2 two possible solutions, termed Solutions A and B, that each achieved a sublanguage of K under control. We can formally verify the intuitive development of that section by building the observers of the two languages achieved by these two solutions. These observers are shown in Fig. 3.15. We can verify that no control conflict arises in these two observers, confirming that the two solutions do generate observable languages.

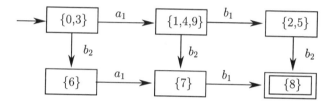

Figure 3.14: Observer of automaton H_a of Fig. 3.3, when $E_{uo} = \{a_2\}$.

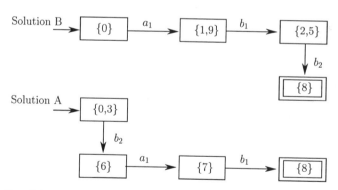

Figure 3.15: Observers for the two languages generated by Solutions A and B described in Section 3.3.2.
The set of unobservable events is $E_{uo} = \{a_2\}$.

Before we discuss how to deal with unobservable languages in supervisory control problems, we present how to realize P-supervisors by automata for desired languages that are observable and controllable.

3.7.2 Realization of P-supervisors

We presented in Section 3.4.2 the standard realization of a supervisor that achieves a given controllable language; this realization is based on an automaton representation of the given language and therefore requires finite memory when this language is regular. We will similarly realize P-supervisors using automata, which again will ensure finite realizations in the case of regular languages. However, the process of building a realization of a P-supervisor that achieves a given observable and controllable language K is slightly more involved, due to the presence of unobservable events (which may or may not be controllable).

Based on the construction of S_P in the proof of the Controllability and Observability Theorem, consider the following initial steps for constructing a realization of S_P, where $\mathcal{L}(S_P/G) = \overline{K}$:

1. Build a trim automaton R that generates and marks $\overline{K}$; the event set of R is E and E_o is the subset of observable events.

2. Build R_{obs}, the observer for R corresponding to the set E_o of observable events; R_{obs} is built using the procedure presented in Chapter 2, Section 2.5.2 (where unobservable events are treated as if they were ε-transitions).

At this point, unlike the standard realization in the case of full observation, we cannot write that the active event set of R_{obs} encodes the set of events enabled by the function S_P, since the event set of R_{obs} is E_o and thus contains no information about the desired control action regarding the unobservable events. That information is contained in R; however, since each state of R_{obs} is a set of states of R, we can recover the required information. Here is how to proceed to recover the control action:

3. Let t be the current string of *observable* events and let $x_{obs,current}$ be the state of R_{obs} after the execution of t. This means that after the last (observable) event in t but before the next observable event, automaton R could be in anyone of the states in the set $x_{obs,current}$.

4. Then we have that

$$S_P^{real}(t) = \bigcup_{x \in x_{obs,current}} [\Gamma_R(x)]$$

where Γ_R is the active event function of R. We use the notation S_P^{real}, where *real* stands for "realized", to distinguish this supervisor from the supervisor S_P in the proof of COT: S_P enables *all* uncontrollable events (a design choice for the sake of simpler notation) while S_P^{real} only enables *feasible* uncontrollable events in K. (Recall the discussion about admissibility of P-supervisors in Section 3.2.2.) S_P^{real} is admissible (since $\overline{K}$ is controllable) and $S_P^{real}(t) \cap E_c = S_P(t) \cap E_c$. Note that we can precompute all the enabled events for each state of R_{obs} so that it is not necessary to store R itself.

Here, the interpretation of the realization with R_{obs} (and indirectly R) is one where R_{obs} is a "passive observer" that follows the observable (only) transitions of G; the desired control action, namely $S_P^{real}(t)$, is obtained by looking at the current state of R_{obs}, which is a set of states of R, and by considering the corresponding active event sets in R of all the states in this set. When K is regular, all this information requires finite memory, which was our goal.

We call this realization with R_{obs} and the associated set of enabled events at each state of R_{obs} the *standard realization* of S_P.

Example 3.20 (Standard realization of P-supervisors)

The observers shown in Fig. 3.15 can be used as standard realizations of Solutions A and B presented in Section 3.3.2 for our familiar database concurrency control example. For Solution A, we adjoin to the observer in Fig. 3.15 the following control actions:

- At state $\{0,3\}$, enable a_2 and b_2 (thus a_1 is disabled).
- At state $\{6\}$, enable a_1.
- At state $\{7\}$, enable b_1.

For Solution B, the control actions of the observer in Fig. 3.15 are:

- At state $\{0\}$, enable a_1 (thus a_2 is disabled).
- At state $\{1,9\}$, enable a_2 and b_1 (thus b_2 is disabled).
- At state $\{2,5\}$, enable a_2 and b_2.

3.7.3 The Property of Observability

Observability and Union

The property of observability is more difficult to deal with than the property of controllability in the context of supervisory control because it is *not* preserved under *union*, as shown by the following example.

Example 3.21 (Observability and union)

Let $E = E_c = \{\alpha, \beta\}$ and $E_o = \{\beta\}$ and consider the languages

$$M = \{\epsilon, \alpha, \beta, \alpha\beta\} \qquad K_1 = \{\alpha\} \qquad K_2 = \{\beta\} .$$

Then

1. K_1 and K_2 are observable with respect to M, E_o, and E_c. To see this, consider P-supervisors that enable only α (for K_1) or β (for K_2) at the outset.

2. $K = K_1 \cup K_2$ is not observable. To see this, take $s = \alpha$, $s' = \epsilon$ and $\sigma = \beta$. Then, $s\sigma \notin \overline{K}$, $s\sigma \in M$, $s'\sigma \in \overline{K}$, and $s'\sigma \in P^{-1}[P(s)]\sigma$ since $P(s) = P(s')$. But this is a violation of the definition of observability. Intuitively, in order to exactly achieve $K_1 \cup K_2$, a P-supervisor should disable β only after α has occurred, but α is not observable so the language $K_1 \cup K_2$ is not achievable by control. Note that taking the prefix-closure of K_1 and K_2 would not help here.

Observability and Intersection

Observability does possess a useful property. Similarly to controllability, it is closed under *intersection* in the case of *prefix-closed* languages.

Proposition. (Observability and intersection)
If K_1 and K_2 are both prefix-closed and observable with respect to M, E_o, and E_c, then so is $K_1 \cap K_2$.

Proof. Clearly, $K := K_1 \cap K_2$ is prefix-closed. By contradiction, suppose that K is not observable. Then there exists s and s', with $P(s) = P(s')$, and $\sigma \in E_c$ such that

$$(s'\sigma \in \overline{K}) \text{ and } (s \in \overline{K}) \text{ and } (s\sigma \in M \setminus \overline{K}) .$$

But

$$s\sigma \in M \setminus \overline{K} \Leftrightarrow (s\sigma \in M) \text{ and } (s\sigma \notin \overline{K_1} \text{ or } s\sigma \notin \overline{K_2}) .$$

If $s\sigma \notin \overline{K_1}$, then we have that

$$(s'\sigma \in \overline{K_1}) \text{ and } (s \in \overline{K_1}) \text{ and } (s\sigma \in M \setminus \overline{K_1})$$

which contradicts the assumption that K_1 is observable. If $s\sigma \notin \overline{K_2}$, then a similar argument contradicts the observability of K_2. Thus no such s, s', and σ exist and we conclude that K is observable. **Q.E.D.**

It can be verified that the above fact and proof hold for arbitrary intersections as well. (Observe where the prefix-closure assumption for K_1 and K_2 is used in this proof.) This suggests that we proceed as we did for the class $CC_{out}(K)$ in our study of controllability and define the class of languages:

$$CO_{out}(K) := \{L \subseteq E^* : (K \subseteq L \subseteq M) \text{ and } (\overline{L} = L) \text{ and } (L \text{ is observable})\} .$$

Here, observability is with respect to fixed M, E_o (equivalently, the corresponding projection P), and E_c. This class is not empty because $M \in CO_{out}(K)$. In view of the above result, this class possesses a unique *infimal* element. Namely,

$$K^{\downarrow O} := \bigcap_{L \in CO_{out}(K)} L$$

is a well-defined element of $CO_{out}(K)$. We call $K^{\downarrow O}$ the *infimal prefix-closed and observable superlanguage of K*. In the "worst" case, $K^{\downarrow O} = M$. If K is observable, then $K^{\downarrow O} = \overline{K}$. We refer to $\downarrow O$ as the operation of obtaining the infimal prefix-closed and observable superlanguage.

Observability, Controllability, and Intersection

We can combine the results about $\downarrow O$ with those about $\downarrow C$ for controllability and conclude that the *infimal prefix-closed observable and controllable superlanguage* of a given language does exist. To see this, consider the class of languages

$$CCO_{out}(K) := CC_{out}(K) \cap CO_{out}(K) .$$

$CCO_{out}(K)$ contains the superlanguages of K that are prefix-closed, controllable, and observable. $CCO_{out}(K)$ is closed under arbitrary intersections, and thus its infimal element exists and is the infimal prefix-closed observable and controllable superlanguage of K. We denote this infimal language by $K^{\downarrow(CO)}$. In the "worst" case, $K^{\downarrow(CO)} = M$.

3.7.4   Supervisory Control Problems under Partial Observation

Let us denote the partial observation versions of BSCP and BSCP-NB as BSCOP and BSCOP-NB, respectively, meaning "Basic Supervisory Control and Observation Problem." The statements of BSCOP and BSCOP-NB are as follows.

BSCOP: Basic Supervisory Control and Observation Problem

Given DES G with event set E, uncontrollable event set $E_{uc} \subseteq E$, observable event set $E_o \subseteq E$ with corresponding natural projection $P : E^* \to E_o^*$, and admissible language $L_a = \overline{L_a} \subseteq \mathcal{L}(G)$, find a P-supervisor S_P such that:

1. $\mathcal{L}(S_P/G) \subseteq L_a$

2. $\mathcal{L}(S_P/G)$ is "the largest it can be," that is, for any other P-supervisor S_P' such that $\mathcal{L}(S_P'/G) \subseteq L_a$,

$$\mathcal{L}(S_P'/G) \subseteq \mathcal{L}(S_P/G) .$$

BSCOP-NB: Nonblocking Version of BSCOP

Given DES G with event set E, uncontrollable event set $E_{uc} \subseteq E$, observable event set $E_o \subseteq E$ with corresponding natural projection $P : E^* \to E_o^*$, and admissible marked language $L_{am} \subseteq \mathcal{L}_m(G)$, with L_{am} assumed to be $\mathcal{L}_m(G)$-closed, find a *nonblocking* P-supervisor S_P such that:

1. $\mathcal{L}_m(S_P/G) \subseteq L_{am}$

2. $\mathcal{L}_m(S_P/G)$ is "the largest it can be," that is, for any other nonblocking P-supervisor S'_P such that $\mathcal{L}_m(S'_P/G) \subseteq L_{am}$,

$$\mathcal{L}_m(S'_P/G) \subseteq \mathcal{L}_m(S_P/G) .$$

Solving BSCOP and BSCOP-NB:

In view of the results of Section 3.7.3 about observability and union, the supremal observable sublanguage of a given language need not exist. Consequently, the supremal observable and controllable sublanguage of a given language need not exist. This implies that BSCOP and BSCOP-NB do *not* possess, *in general*, solutions that satisfy requirement 2 in the formulation of these problems.

Various approaches have been considered to deal with this difficulty. We mention three of them.

1. Calculate sublanguages of L_a or L_{am}, as appropriate, that are *maximal* (with respect to set inclusion) observable and controllable sublanguages of L_a or L_{am} (and also $\mathcal{L}_m(G)$-closed, if necessary). By "maximal" we mean that there is no other observable and controllable sublanguage that is *strictly larger* than the maximal one; on the other hand, there may be many other *incomparable* maximals.

This means that requirement 2 in the statement of BSCOP is replaced by the weaker condition:

2'. $\mathcal{L}(S_P/G)$ is *maximal*, that is, for any other P-supervisor S'_P,

$$\mathcal{L}(S'_P/G) \subseteq L_a \Rightarrow \mathcal{L}(S'_P/G) \not\supseteq \mathcal{L}(S_P/G)$$

and similarly for BSCOP-NB.

The computation of maximal observable (and controllable) sublanguages of a given language has been considered in the literature; however, we will not discuss this issue in this book.

2. Identify a property of languages that is *stronger* than observability (in the sense that it *implies* observability) and that *is* closed under union. The property of *normality* discussed in Section 3.7.5 hereafter satisfies these requirements. In this case however, there is no guarantee (in general) that the solution is maximal and consequently requirement 2' above need not hold; only requirement 1 of BSCOP and BSCOP-NB will hold in general.

3. Identify special situations where the supremal observable and controllable sublanguage of a given language does exist. As we show in Section 3.7.5, one such situation is when $E_c \subseteq E_o$, that is, when all the controllable events are observable or, stated differently, when all the unobservable events are uncontrollable.

Example 3.22 (Maximal solutions of BSCOP-NB)

Solutions A and B presented in Section 3.3.2 and realized as described in Example 3.20 are examples of maximal solutions of BSCOP-NB. In this case, these two solutions are disjoint, but that need not be the case in general. Maximality of these solutions is verified by arguing that at each state of their realizations, it is unsafe - in terms of the admissible behavior or in terms of blocking - to enable any of the disabled events (examine the control actions listed in Example 3.20 and revisit the discussion in Section 3.3.2).

While BSCOP and BSCOP-NB cannot be solved "optimally" in general (in the sense of a supremal solution), the existence of $K^{\downarrow(CO)}$ means that the partial observation version of DuSCP does have a solution.

DuSCOP: Dual Version of BSCOP

Given DES G with event set E, uncontrollable event set $E_{uc} \subseteq E$, observable event set $E_o \subseteq E$ with corresponding projection $P : E^* \to E_o^*$, and minimum required language $L_r \subseteq \mathcal{L}(G)$, find a P-supervisor S_P such that:

1. $\mathcal{L}(S_P/G) \supseteq L_r$

2. $\mathcal{L}(S_P/G)$ is "the smallest it can be," that is, for any other S_P' such that $\mathcal{L}(S_P'/G) \supseteq L_r$,

$$\mathcal{L}(S_P'/G) \supseteq \mathcal{L}(S_P/G) .$$

(Note that L_r need not be prefix-closed and could be given as a subset of $\mathcal{L}_m(G)$.)

Solution of DuSCOP:

The solution is to take S_P such that

$$\mathcal{L}(S_P/G) = L_r^{\downarrow(CO)}.$$

(Here, the $\downarrow(CO)$ operation is with respect to $M = \mathcal{L}(G)$.)

Some Comments on $\downarrow O$ and $\downarrow(CO)$

We shall not discuss in detail the $\downarrow O$ and $\downarrow(CO)$ operations. We mention without proof useful results from the literature in order to give some insight into these operations.

- $K^{\downarrow O}$ and $K^{\downarrow(CO)}$ can be expressed in terms of formulas on languages involving set operations, concatenation, projection, and inverse projection.

These formulas are:

$$K^{\downarrow O} = M \setminus [E^* E \setminus \bigcup_{\sigma \in E_c} [P^{-1}(P(\overline{K}\sigma \cap \overline{K})) \cap E^*\sigma]] E^*$$

$$K^{\downarrow(CO)} = M \setminus [E^* E_c \setminus \bigcup_{\sigma \in E_c} [P^{-1}(P(\overline{K}\sigma \cap \overline{K})) \cap E^*\sigma]] E^* .$$

- These formulas prove that if K and M are regular, then $K^{\downarrow O}$ and $K^{\downarrow(CO)}$ are regular. The formulas also suggest algorithms for computing these languages.

3.7.5 The Property of Normality

Definition. (Normality)
Consider $M = \overline{M} \subseteq E^*$ and natural projection $P : E^* \to E_o^*$. A language $K \subseteq M$ is said to be *normal* with respect to M and P if

$$\overline{K} = P^{-1}[P(\overline{K})] \cap M . \qquad \blacklozenge$$

In words, normality means that

$\overline{K}$ *can be exactly recovered from its projection* $P(\overline{K})$ *and from* M.

Observe that $\emptyset$ and M are both normal, so the property is not vacuous. As was done for controllability and observability, the property of normality is defined on the *prefix-closure* of a language. Thus K is normal if and only if $\overline{K}$ is normal. The inequality

$$\overline{K} \subseteq P^{-1}[P(\overline{K})] \cap M$$

is always true. If furthermore

$$P^{-1}[P(\overline{K})] \cap M \subseteq \overline{K}$$

holds, then K is normal (with respect to M and P).

Example 3.23 (Normal languages)
Let us consider again Solutions A and B of Section 3.3.2. The set of observable events is $E_{uo} = \{a_2\}$ and M is the language generated by automaton G in Fig. 3.3.

The language marked by Solution B is

$$L_B = \{a_1 b_1 a_2 b_2, a_1 a_2 b_1 b_2\} .$$

We have that $P(\overline{L_B}) = \overline{\{a_1 b_1 b_2\}}$, which in fact is the language generated by the observer used to realize Solution B; this observer is shown in Fig. 3.15. It is straightforward to verify that

$$a_2 \in P^{-1}[P(\overline{L_B})] \cap M$$

which leads us to conclude that L_B is not a normal language.

On the other hand, the language marked by Solution A,

$$L_A = \{a_2 b_2 a_1 b_1\} \ ,$$

can be verified to be normal.

There are essentially three reasons motivating our study of normality: (a) normality implies observability; (b) normality is preserved under union; (c) under certain conditions, normality and observability are equivalent. We present next two propositions that formalize reasons (a) and (b); reason (c) is discussed in the subsection that follows the two propositions.

Proposition. (Normality and observability)
If $K \subseteq M$ is normal with respect to M and P, then K is observable with respect to M, P, and E_c for all $E_c \subseteq E$. However, the converse statement is not true in general.

Proof. It suffices to prove the result for $E_c = E$. By contradiction, suppose that K is normal but not observable. Then, there exists s, σ, and s', such that $s \in \overline{K}$, $\sigma \in E_c = E$, $s\sigma \notin \overline{K}$, $s\sigma \in M$, and $s'\sigma \in \overline{K}$, with $P(s) = P(s')$. Clearly, $P(s\sigma) = P(s'\sigma) \in P(\overline{K})$. But then $P(s\sigma) \in P(\overline{K})$ and $s\sigma \in M$ imply that

$$s\sigma \in P^{-1}[P(\overline{K})] \cap M \ .$$

Thus, by normality, $s\sigma \in \overline{K}$, a contradiction.

To show that normality is stronger than observability, consider the following example. Take $M = \overline{\{\beta\alpha\}}$, $K = \{\beta\}$, and $E_o = \{\beta\}$. Then it can be verified that K is observable for any $E_c \subseteq E$ (intuition: disable α immediately after observing β) but not normal since

$$P^{-1}[P(\overline{K})] \cap M = M \supset \overline{K} \ . \qquad \textbf{Q.E.D.}$$

Proposition. (Normality and union)
If K_1, $K_2 \subseteq M$ are normal with respect to M and P, then so is $K_1 \cup K_2$.

Proof.

$$
\begin{aligned}
P^{-1}[P(\overline{K_1 \cup K_2})] \cap M &= P^{-1}[P(\overline{K_1} \cup \overline{K_2})] \cap M \\
&= P^{-1}[P(\overline{K_1}) \cup P(\overline{K_2})] \cap M \\
&= (P^{-1}[P(\overline{K_1})] \cup P^{-1}[P(\overline{K_2})]) \cap M \\
&= (P^{-1}[P(\overline{K_1})] \cap M) \cup (P^{-1}[P(\overline{K_2})] \cap M) \\
&= \overline{K_1} \cup \overline{K_2} \\
&= \overline{K_1 \cup K_2}
\end{aligned}
$$

where the above equalities follow from set operations, properties of prefix-closure, and the normality assumptions. $\qquad \textbf{Q.E.D.}$

The above proposition and proof remain true for arbitrary unions. Consequently, we can proceed as we did before for other extremal sub(super)languages and establish the existence of

- $K^{\uparrow N}$, the *supremal normal sublanguage of K*

- $K^{\uparrow(CN)}$, the *supremal controllable and normal sublanguage of K.*

As for $K^{\uparrow C}$, these languages always exist although they may be empty.

$K^{\uparrow(CN)}$ is of interest because it provides a "sub-optimal" solution to BSCOP and BSCOP-NB in the sense that it meets requirement 1 of these problems but not necessarily requirement 2. It should be emphasized though that this solution may not in general be maximal, that is, there may be controllable and observable sublanguages that are strictly larger than the supremal controllable and normal sublanguage.

Controllability, Normality, and Observability

There is a very interesting feature about the supremal controllable and normal sublanguage and about the relation between normality, observability, and controllability. Consider the following result.

Theorem. (Equivalence of normality and observability)
Assume that $E_c \subseteq E_o$. If K is controllable (with respect to M and E_{uc}) and observable (with respect to M, P, and E_c), then K is normal (with respect to P and M).

Proof. The result is trivial if $K = \emptyset$, so assume that K is not empty. By contradiction, assume that K is not normal. Then there exists $t \in M$ such that

$$(t \notin \overline{K}) \text{ and } (P(t) \in P(\overline{K})) .$$

But $t \neq \epsilon$ (since $K \neq \emptyset$). Let $s\sigma$ be the *shortest* such t. Therefore

$$(s \in \overline{K}) \text{ and } (s\sigma \in M) \text{ and } (s\sigma \notin \overline{K}) \text{ and } (P(s\sigma) \in P(\overline{K})) .$$

Since K is controllable, this implies that $\sigma \in E_c \subseteq E_o$. Now,

$$(\sigma \in E_o) \text{ and } (P(s\sigma) \in P(\overline{K})) \Rightarrow \exists s'\sigma \in \overline{K} \ (P(s') = P(s)) .$$

For this $s'\sigma$, we have that

$$s'\sigma \in P^{-1}[P(s)]\sigma \cap \overline{K}$$

which implies that K is not observable, a contradiction. We therefore conclude that K is normal.   **Q.E.D.**

The importance of this theorem lies in the fact that when the assumption $E_c \subseteq E_o$ holds, the theorem implies that *the supremal observable and controllable sublanguage does exist* and consequently BSCOP and BSCOP-NB do have "optimal" solutions. These solutions can be obtained by calculating $L_a^{\uparrow(CN)}$ or $L_{am}^{\uparrow(CN)}$, as appropriate, or by calculating any maximal controllable and observable sublanguage of L_a or L_{am} (since there can only be one maximal, namely the supremal!).

The intuition behind the above theorem is that when all the controllable events are observable, or equivalently when all the unobservable events are uncontrollable, the intrinsic difficulties associated with observability and in particular the lack of existence of a supremal observable sublanguage are alleviated if controllability enters the picture. This is because controllability will "take care of" some of the effect of unobservable events and "reduce" observability to normality, a property that is better behaved.

Some Comments on $\uparrow (CN)$

In view of the preceding theorem, we are interested in the properties and implementation of the $\uparrow N$ and $\uparrow (CN)$ operations. A thorough discussion of these operations is somewhat involved and beyond the scope of this book. We shall only state without proof some relevant results that can be found in the literature.

- When K and M are regular, then so are $K^{\uparrow N}$ and $K^{\uparrow(CN)}$.

- If K is prefix-closed, then so are $K^{\uparrow N}$ and $K^{\uparrow(CN)}$.

- If K is $\mathcal{L}_m(G)$-closed, then so are $K^{\uparrow N}$ and $K^{\uparrow(CN)}$. (This result is invoked when solving BSCOP-NB by the $\uparrow (CN)$ operation, when $E_c \subseteq E_o$.)

- If K is prefix-closed, $K^{\uparrow N}$ is given by the formula

$$K^{\uparrow N} = K \setminus (P^{-1}[P(M \setminus K)])E^* .$$

In the general case, there exists an iterative procedure for the computation of $K^{\uparrow N}$. The idea is essentially to iterate the above formula until convergence; namely, one can proceed as follows:

$$K_{i+1} = (\overline{K_i})^{\uparrow N} \cap \mathcal{L}_m(G)$$

with $K_0 = K$. In the case of regular languages, convergence occurs in finite steps.

- The computation of $\uparrow (CN)$ can be done by iteratively performing the $\uparrow C$ and $\uparrow N$ operations until convergence; iterations are necessary as each operation need not preserve the other property. The key point is that in the case of regular languages, only a finite number of iterations needs to be performed.

■ The computational complexity of performing $\uparrow N$ and $\uparrow (CN)$ is exponential (in the worst case) in the number of states of the automaton representation of the language of interest; this is due to the fact that an observer must be built in the procedure.

Example 3.24 (Supremal normal sublanguage)
We conclude this section by revisiting one last time Solutions A and B of Section 3.3.2, whose properties were analyzed in many examples in this section.

■ We saw in Example 3.23 that the language

$$L_B = \{a_1 b_1 a_2 b_2, a_1 a_2 b_1 b_2\}$$

marked by Solution B is not normal when $E_{uo} = \{a_2\}$, due to unobservable string a_2 possible in $M = \mathcal{L}(G)$ (where G appears in Fig. 3.3) but not in $\overline{L_B}$. Due to this string, we conclude that $L_B^{\uparrow N} = \emptyset$ and $(\overline{L_B})^{\uparrow N} = \emptyset$.

■ The admissible marked language in the discussion in Section 3.3.2 is the language

$$\mathcal{L}_m(H_a) = \{a_1 b_1 a_2 b_2, a_1 a_2 b_1 b_2, a_2 a_1 b_2 b_1, a_2 b_2 a_1 b_1\}$$

where H_a appears in Fig. 3.3. What is $\mathcal{L}_m(H_a)^{\uparrow N}$ when $E_{uo} = \{a_2\}$? Examining Fig. 3.3, we can see that we must remove from $\mathcal{L}_m(H_a)$ any string that contains a prefix that projects to $a_1 b_2$, since illegal string $a_1 a_2 b_2$ also projects to $a_1 b_2$; this forces the removal of $a_2 a_1 b_2 b_1$. By a similar argument, we must remove from $\mathcal{L}_m(H_a)$ any string that contains a prefix that projects to $a_1 b_1$, since $a_2 a_1 b_1$ is illegal; this forces the removal of $a_1 b_1 a_2 b_2$ and $a_1 a_2 b_1 b_2$. What we are left with after these removals is the language marked by Solution A,

$$L_A = \{a_2 b_2 a_1 b_1\} .$$

Therefore, $\mathcal{L}_m(H_a)^{\uparrow N} = L_A$.

■ When $E_{uc} = \{a_2\}$, $\mathcal{L}_m(H_a)^{\uparrow (CN)} = \mathcal{L}_m(H_a)^{\uparrow (CO)} = L_A$.

■ If $E_{uc} = \{a_1\}$, then $\mathcal{L}_m(H_a)^{\uparrow (CN)} = \emptyset$ since $L_A^{\uparrow C} = \emptyset$. However, in that case Solution B is a maximal solution, namely, L_B is a maximal observable and controllable sublanguage of $\mathcal{L}_m(H_a)$ as was mentioned in Example 3.22. This illustrates that the $\uparrow (CN)$ operation need not produce a maximal solution of BSCOP and BSCOP-NB.

3.8   DECENTRALIZED CONTROL

We wish to conclude this chapter on supervisory control with a brief discussion of *decentralized* supervisory control. The feedback loop depicted in Fig. 3.2 for supervisory control under partial observation is an instance of centralized control, since there is a single supervisor S_P. In many technological systems, we have several "processing nodes" that are jointly controlling a given system that is inherently distributed. This is, for instance, the situation in complex automated manufacturing systems composed of several workstations interconnected by conveyors or automated guided vehicles. Another example is a platoon of vehicles in an automated highway system. Communication and computer networks are also an example of distributed systems that are controlled in a decentralized manner by several controllers located at different nodes. What distinguishes these control architectures from the modular control architecture depicted in Fig. 3.10 and studied in Section 3.6 is the fact that the individual supervisors may be partial-observation supervisors and moreover their respective sets of observable and controllable events need not all be the same. In other words, due to the distributed nature of the system, supervisors at different "sites" in the distributed system may see the effect of different (possibly overlapping) sets of sensors and may control different (again, possibly overlapping) sets of actuators. This leads us to consider the decentralized control architecture shown in Fig. 3.16, which in some sense is a hybrid of Figs. 3.2 and 3.10. For simplicity, only two supervisors are depicted in Fig. 3.16; it is straightforward to extend the diagram to the case of n supervisors.

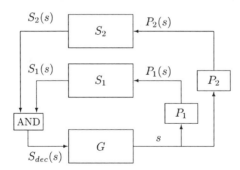

Figure 3.16: Decentralized control architecture.

Let us formulate more precisely the decentralized supervisory control problem that we wish to study in this section. We have a set of n partial-observation supervisors, each associated with a different projection P_i, $i = 1, \ldots, n$, jointly controlling the given system G with event set E. It is convenient to extend the domain of partial-observation supervisor i from $P_i[\mathcal{L}(G)]$ to $\mathcal{L}(G)$. Let us denote the extended-domain supervisor by S_i (this explains the notation in Fig. 3.16); clearly, the right way to define $S_i(s)$ for $s \in \mathcal{L}(G)$ is by

$$S_i(s) = S_{P_i}[P_i(s)] .$$

As in modular control, the net control action on G will be the *intersection* of the sets of events enabled by each supervisor, that is, we define the combined control policy $S_{dec} : \mathcal{L}(G) \to 2^E$ acting on G as follows:

$$S_{dec}(s) = \bigcap_{i=1}^{n} S_i(s) .$$

The resulting controlled behavior is described by the languages $\mathcal{L}(S_{dec}/G)$ and $\mathcal{L}_m(S_{dec}/G)$. We often say that $\mathcal{L}(S_{dec}/G)$ is the "global" behavior, as opposed to the "local" behaviors $P_i[\mathcal{L}(S_{dec}/G)]$, $i = 1, \ldots, n$, seen by the individual supervisors.

Associated with G are the four usual sets E_c, E_{uc}, E_o, and E_{uo}. Corresponding to supervisor S_i at site i, $i = 1, \ldots, n$, we have:

- the set of controllable events $E_{i,c} \subseteq E_c$, where $\cup_{i=1}^n E_{i,c} = E_c$;

- the set of observable events $E_{i,o} \subseteq E_o$, where $\cup_{i=1}^n E_{i,o} = E_o$;

- the (natural) projection $P_i : E^* \to E_{i,o}^*$ corresponding to $E_{i,o}$.

Given the above description of the situation under consideration, let us ask the following question. If we are concerned with restricting the global behavior to within a subset K of $\mathcal{L}(G)$, what is the necessary and sufficient condition on K, beyond controllability, that will ensure the existence of S_i, $i = 1, \ldots, n$, such that $\mathcal{L}(S_{dec}/G) = K$? In other words, what is the *decentralized version of the (centralized) observability condition*? Intuitively, we know that the required condition will have to be *stronger than*

K is observable with respect to $\mathcal{L}(G)$, E_o, and E_c

since if the centralized problem cannot be solved, then the decentralized problem surely cannot be solved either. However, the required condition should be *weaker than*

K is observable with respect to $\mathcal{L}(G)$, P_i, and $E_{i,c}$, for each $i = 1, \ldots, n$

since there may be events that can be controlled by more than one supervisors, and therefore we may not need full "local" observability at all sites as the supervisors may somehow be able to "share the work" on the *common* controllable events. The supervisors certainly "share the work" in the sense that each one is entirely responsible for the events that are controllable at its site only. But the supervisors may also be able to "share the work" for the common controllable events, in the sense that no single supervisor is uniquely responsible for disabling these events; rather, which supervisor disables a common event (when the event must be disabled) could depend on the string of events executed so far by the system.

Let us consider an example to strengthen the above intuitive comments.

Example 3.25 (Decentralized solution)

Consider the following decentralized problem involving two sites.

$$L(G) = \overline{\{bg, bbg, ag, abg\}} \qquad E_{1,o} = \{a\} \qquad E_{1,c} = \{a, g\}$$
$$K = \overline{\{bg, bb, ab\}} \qquad E_{2,o} = \{b\} \qquad E_{2,c} = \{b, g\} \, .$$

Clearly, a centralized supervisor that observes $E_{1,o} \cup E_{2,o}$ and controls $E_{1,c} \cup E_{2,c}$ can achieve K as follows:

1. initially enable events a and b;

2. enable event b after observing string a;

3. enable events b and g after observing string b;

4. do not enable any events after strings ab and bb.

However, we claim that neither S_1 nor S_2 can achieve K if acting *alone*. This is not a controllability problem, since K is controllable with respect to $L(G)$ and $E_{i,c}$, $i = 1, 2$ (note that both S_1 and S_2 can control g). The problem lies with observability. Concerning S_1, since it does not see event b, it will not be able to allow string bg and disallow string bbg. Concerning S_2, since it does not see event a, it will not be able to allow string bg and disallow string abg. What happens here is that on the one hand S_1 has enough information to decide about its controllable event a (enable until it occurs) and S_2 has enough information to decide about its controllable event b (enable the first two occurrences), but on the other hand neither S_1 nor S_2 knows, individually, what to do about event g. In other words, K is not observable with respect to $L(G)$, $E_{i,o}$, and $E_{i,c}$, $i = 1, 2$.

The interesting point here is that S_1 and S_2 *together* have enough information to properly control their common controllable event g. Namely, they can proceed as follows:

1. S_1 can start by enabling events a, b (uncontrollable to S_1), and g, and then enable only b after it sees string a; in other words, S_1 disables event g only after it sees string a;

2. S_2 can start by enabling events a (uncontrollable to S_2) and b, then enable a, b, and g after it sees string b, and finally enable event a

only after it sees string bb (i.e., g is disabled at the beginning and after bb).

Since at any time the control action on G is the intersection of the current sets of events enabled by S_1 and S_2, string bbg that poses a problem to S_1 when S_1 initially enables g (in order to allow bg) is taken care of by S_2 which disables g after string bb. On the other hand, the problem that S_2 faces by enabling event g after string b and until it sees string bb (in order to allow bg), namely the possible occurrence of string abg, is taken care of by S_1 which disables event g after string a. Supervisors S_1 and S_2 are indeed "sharing the work" regarding event g.

The key feature of the control policies used in Example 3.25 to exactly achieve the desired language K is that each supervisor enables an event when it needs to enable this event in order to allow some string in $\overline{K}$. The fact that this enabling may also allow a string not in $\overline{K}$ to occur (as far as this supervisor can tell from its own observations) is "disregarded" by this supervisor. When this happens, we say that the supervisor *passes the buck.* Of course, there is no guarantee that "passing the buck" will always work. Consider the following example.

Example 3.26 ("Passing the buck" does not always work)
Consider the following variation of the decentralized problem of Example 3.25, where $\mathcal{L}(G)$ and K have been modified.

$$\mathcal{L}(G) = \overline{\{aag, abg, bag, bbg\}} \qquad E_{1,o} = \{a\} \qquad E_{1,c} = \{a, g\}$$
$$K = \overline{\{aa, bb, ba, abg\}} \qquad E_{2,o} = \{b\} \qquad E_{2,c} = \{b, g\} \ .$$

In this problem, $\overline{K}$ contains the strings aa, bb, ab, and ba as well as abg, but it does not contain string bag, which is possible in $\mathcal{L}(G)$. This means that the correct control action for event g depends on the order in which events a and b occur. There is no controllability problem here as both S_1 and S_2 control event g, the only event that needs to be disabled. However, neither S_1 nor S_2 observes both a and b, and consequently no supervisor knows the order in which these events occur. If we follow the strategy employed in Example 3.25 and let both supervisors adopt the "pass the buck" policy, then S_1 will enable event g after it sees string a, in order to allow abg, and S_2 will enable event g after it sees string b, also in order to allow abg; overall, this will result in both abg and bag being possible in the controlled language, a violation of K.

Based on the above discussion and examples, the decentralized version of the property of observability that we are seeking to identify can roughly be stated as a property where having each supervisor follow the "pass the buck" policy results exactly in the desired language K. This concept can be formalized in a manner closely related to the way we defined observability in Section 3.7.1. The name of this new property is *co-observability*.

Definition. (Co-Observability)
Let K and $M = \overline{M}$ be languages over event set E. Let $E_{i,c}$ and $E_{i,o}$ be sets of controllable and observable events, respectively, for $i = 1, \ldots, n$. Let P_i be the natural projection corresponding to $E_{i,o}$, with $P_i : E^* \to E_{i,o}^*$.
K is said to be *co-observable* with respect to M, P_i, and $E_{i,c}$, $i = 1, \ldots, n$, if for all $s \in \overline{K}$ and for all $\sigma \in E_c = \cup_{i=1}^n E_{i,c}$,

$$(s\sigma \notin \overline{K}) \text{ and } (s\sigma \in M) \Rightarrow$$

there exists $i \in \{1, \ldots, n\}$ such that $P_i^{-1}[P_i(s)]\sigma \cap \overline{K} = \emptyset$ and $\sigma \in E_{i,c}$. ◆

We can paraphrase co-observability as follows:

> *If event σ needs to be disabled, then at least one of the supervisors that can control σ must unambiguously know that it must disable σ, that is, from this supervisor's viewpoint, disabling σ does not prevent any string in $\overline{K}$; consequently, each supervisor can still follow the "pass the buck" policy.*

Comments:

1. If we set $E_{i,o} = E_o$, $E_{i,c} = E_c$, and $E_{j,o} = E_{j,c} = \emptyset$ for $j = 1, \ldots, n$, $j \neq i$, then co-observability reduces to observability, as expected.

2. If $E_{i,c} \cap E_{j,c} = \emptyset$ for all $i, j = 1, \ldots, n$, then since each controllable event can only be controlled by one supervisor, the notion of "passing the buck" does not apply. Consequently, each supervisor must unambiguously know when to disable all its controllable events. This means that in this case co-observability of language K is equivalent to

   "K is observable with respect to $\mathcal{L}(G)$, P_i, and $E_{i,c}$, for each $i = 1, \ldots, n$."

   This result is easily seen by comparing the respective definitions of observability and co-observability.

The notion of co-observability leads us to stating the decentralized version of the Controllability and Observability Theorem (COT) presented in Section 3.7.1. We call this new result the "Controllability and Co-Observability Theorem" (abbreviated as CC-OT).

Theorem. [Controllability and Co-Observability Theorem (CC-OT)]
Consider DES $G = (X, E, f, \Gamma, x_0, X_m)$, where $E_{uc} \subseteq E$ is the set of uncontrollable events, $E_c = E \setminus E_{uc}$ is the set of controllable events, and $E_o \subseteq E$

is the set of observable events. For each site i, $i = 1, \dots, n$, consider the set of controllable events $E_{i,c}$ and the set of observable events $E_{i,o}$; overall, $\cup_{i=1}^n E_{i,c} = E_c$ and $\cup_{i=1}^n E_{i,o} = E_o$. Let P_i be the natural projection from E^* to $E_{i,o}^*$, $i = 1, \dots, n$. Consider also the language $K \subseteq \mathcal{L}_m(G)$, where $K \neq \emptyset$. There exists a nonblocking decentralized supervisor S_{dec} for G (according to the control architecture in Fig. 3.16) such that

$$\mathcal{L}_m(S_{dec}/G) = K \quad \text{and} \quad \mathcal{L}(S_{dec}/G) = \overline{K}$$

if and only if the three following conditions hold:

1. K is controllable with respect to $\mathcal{L}(G)$ and E_{uc};
2. K is co-observable with respect to $\mathcal{L}(G)$, P_i, and $E_{i,c}$, $i = 1, \dots, n$;
3. K is $\mathcal{L}_m(G)$-closed. ◆

We shall not present the proof of CC-OT, as it follows closely the steps of the proof of COT. We can however write the expression of the desired supervisors, as this expression is very similar to that of S_P in COT. Namely, the desired language K is achieved by the decentralized supervisor S_{dec} consisting of the conjunction of the extended-domain versions S_i of the local partial-observation supervisors S_{P_i}, $i = 1, \dots, n$, where for $s \in \mathcal{L}(G)$ and $P_i(s) = s_i$,

$$S_i(s) = S_{P_i}(s_i) = (E_{i,uc} \cup \{\sigma \in E_{i,c} : \exists s'\sigma \in \overline{K} \ (P_i(s') = s_i)\} .$$

It is important to note that $E_{i,uc} = E \setminus E_{i,c}$ and therefore all events that S_i cannot control are by default enabled. (For the sake of a simpler expression for $S_i(s)$, infeasible uncontrollable events are enabled.) As far as the events that S_i can control are concerned, then an event is enabled if, as far as S_i knows, the event needs to be enabled in order to achieve all of K; this is precisely the "pass the buck" policy, which is guaranteed to work by the co-observability of K. (Looking at it in terms of disablement of events, then S_i disables an event only when it is "certain" that the event needs to be disabled.)

It is quite interesting to remark that for regular languages $\mathcal{L}(G)$ and K, the property of co-observability can be tested in polynomial time. We refer the reader to the literature for the details of this test. We note however that in the special case of mutually disjoint controllable event sets, our test based on observers presented in Example 3.18 can be used for testing observability at each site; in view of Comment 2 above, if all sites pass the test, then K is co-observable.

SUMMARY

- Supervisory control consists of restricting the behavior of an uncontrolled DES G by means of a supervisor S in order to satisfy given specifications on event ordering and legality of states. Supervisors see the observable events executed by G and can disable the controllable events of G; the control action is allowed to change instantaneously after the execution of an observable event by G. Supervisors can be realized by finite-state automata when the language generated by the controlled system is regular. When there are unobservable events, observer automata are used to realize supervisors.

- The three fundamental issues in supervisory control are: (a) how to deal with uncontrollable events; (b) how to deal with blocking in the controlled system; (c) how to deal with unobservable events.

- The property of controllability delineates the class of languages that can be achieved by supervisory control in the presence of uncontrollable events. Controllable sublanguages that are supremal with respect to set inclusion (i.e., they contain all other controllable sublanguages) can be synthesized by using the $\uparrow C$ operation; for regular languages, this operation can be implemented using finite-state automata with algorithms of quadratic complexity in the state spaces of the automata.

- Blocking in the controlled system can be completely eliminated if the specification language possesses the $\mathcal{L}_m(G)$-closure property. This property is preserved under the $\uparrow C$ operation. If completely eliminating blocking turns out to be too restrictive, there are techniques for synthesizing blocking solutions that are "locally optimal." In the case of modular control architectures, the complete elimination of blocking also requires a nonconflicting condition between the component solutions.

- The property of observability delineates the class of languages that can be achieved by supervisory control in the presence of unobservable events. In general, there is no notion of supremal (with respect to set inclusion) observable sublanguage when there are unobservable events. However, if all unobservable events are uncontrollable, then there is a supremal controllable and observable sublanguage. If some unobservable events are controllable, then there may be many incomparable maximal observable and controllable sublanguages.

- The property of co-observability delineates the class of languages that can be achieved by decentralized supervisory control in the presence of unobservable events. Co-observability guarantees that it is safe for the supervisors to enable an event when they are not sure if an event should be disabled ("pass the buck" policy).

Software for Supervisory Control Operations

The library of routines UMDES-LIB mentioned at the end of Chapter 2 [4] contains routines that implement many of the supervisory control operations presented in this chapter. Another software package that implements supervisory control operations is CTCT.[5] UMDES-LIB and CTCT can be used to solve the problems in the next section that are too tedious to solve by hand calculations.

PROBLEMS

3.1 Let G be a trim automaton with $\mathcal{L}_m(G) = (\alpha\gamma^*\beta)^*$. Let $K = (\alpha\beta)^*$ and $E_{uc} = \{\alpha, \gamma\}$.

   (a) Find a supervisor S such that $\mathcal{L}_m(S/G) = K$. Is your supervisor blocking?

   (b) Can you find a nonblocking supervisor S' such that $\mathcal{L}_m(S'/G) = K$? Explain your reasoning.

3.2 Let $\mathcal{L}(G) = \overline{\{u_1\alpha\gamma, u_1\alpha\beta, u_2\alpha\gamma\}}$, where $E_c = E$ and $E_{uo} = \{u_1, u_2\}$.

   (a) Let $K_1 = \overline{\{u_1\alpha\gamma, u_1\alpha\beta\}}$. Can you find a P-supervisor such that $\mathcal{L}(S_P/G) = K_1$?

   (b) Let $K_2 = \overline{\{u_1\alpha\gamma, u_1\alpha\beta, u_2\alpha\}}$. Can you find a P-supervisor such that $\mathcal{L}(S_P/G) = K_2$?

3.3 Consider $E = \{\alpha_1, \beta_1, \gamma_1, \alpha_2, \beta_2, \gamma_2\}$ and the string $f = \alpha_1\gamma_1\alpha_2\gamma_1$. Build an R such that $\mathcal{L}(R) = E^* \setminus E^*\{f\}E^*$.

3.4 You are given automaton G with event set $E = \{a_1, a_2, b_1, b_2, g_1, g_2\}$. You wish to build another automaton that will generate the sublanguage of $\mathcal{L}(G)$ where all strings in $\mathcal{L}(G)$ that contain the substrings $a_1a_2b_2$ and $a_1a_2g_2$ have been removed. Show how to obtain such an automaton by taking the product of G with an appropriately built automaton.

3.5 Consider the system G such that $\mathcal{L}(G) = a^*b^*$ and the prefix-closed admissible language is

$$L_a = \{a^n b^m : n \geq m \geq 0\}.$$

   Let the set of uncontrollable events be $E_{uc} = \{a\}$. Is L_a controllable? Is L_a regular?

---

[4] UMDES-LIB is available at the Web site www.eecs.umich.edu/umdes.
[5] CTCT is available at the Web site www.control.utoronto.ca/DES.

3.6 Show that the controllability condition can be restated as:

$$\overline{K}E_{uc}^* \cap M \subseteq \overline{K} \ .$$

3.7 Can you *always* construct a supervisor S for a given G such that:

(a) $\mathcal{L}(S/G) = \mathcal{L}(G)$

(b) $\mathcal{L}(S/G) = \mathcal{L}_m(G)$

(c) $\mathcal{L}(S/G) = \emptyset$

(d) $\mathcal{L}(S/G) = \{\epsilon\}$

(e) $\mathcal{L}_m(S/G) = \emptyset$.

Briefly explain your answers. If your answer is no, can you give sufficient conditions that guarantee the desired result?

3.8 Prove the following results (stated without proof in Section 3.4.3) using the definition of the $\uparrow C$ operation:

(a) $(K_1 \cap K_2)^{\uparrow C} = (K_1^{\uparrow C} \cap K_2^{\uparrow C})^{\uparrow C}$

(b) If $K_1^{\uparrow C}$ and $K_2^{\uparrow C}$ are nonconflicting, then $(K_1 \cap K_2)^{\uparrow C} = K_1^{\uparrow C} \cap K_2^{\uparrow C}$.

3.9 A simple manufacturing process involves two machines, denoted by M_1 and M_2, and a buffer, denoted by B, in between (cf. Problem 2.36 in Chapter 2). There is an infinite supply of parts to M_1. When a part is processed at M_1, it is placed in B, which has a capacity of one part only. The part is subsequently processed by M_2. Let us suppose that we build the uncontrolled model of M_i, $i = 1, 2$, as follows. Each machine has three states: Idle (the initial state), Processing, and Down. The only marked state is Idle. Each machine has four transitions: event START from Idle to Processing, event END from Processing to Idle, event BREAKDOWN from Processing to Down, and event REPAIR from Down to Idle. Let $G = M_1 || M_2$. Let the only controllable events of G be the START and REPAIR events of each machine. Let K_1 be the admissible language corresponding to the specification that the buffer should not overflow or underflow. Let K_2 be the admissible language corresponding to the specification that if both machines are down, then M_2 should be repaired first.

(a) Is K_1 controllable? If your answer is no, find $K_1^{\uparrow C}$.

(b) Is K_2 controllable? If your answer is no, find $K_2^{\uparrow C}$.

(c) Repeat (a) but this time assuming that the initial state and only marked state of G is the state where both machines are down.

(d) Repeat (c) but for the language $\overline{K_1}$ instead of K_1.

3.10 Consider the dining philospher example discussed in Example 3.16.

(a) Define $G = P_1||P_2||F_1||F_2$. Let the set of uncontrollable events be $E_{uc} = \{2f1, 2f, 1f\}$. Find the standard realization of the *nonblocking* supervisor that achieves the largest sublanguage of $\mathcal{L}_m(G)$ possible.

(b) Define $G = P_1||P_2$ and $L_{am} = \mathcal{L}_m(P_1||P_2||F_1||F_2)$. Find the smallest (with respect to set inclusion) set E_{uc} such that $L_{am}^{\downarrow C} = \mathcal{L}(G)$.

3.11 Consider the simple database example discussed in Section 3.3.2; we have that $G = G_1||G_2$ where $\mathcal{L}_m(G_i) = \{a_i b_i\}$ for $i = 1, 2$ and each G_i is the canonical recognizer of the corresponding language.

(a) Consider the admissible marked language

$$L_{am} = \{s \in \mathcal{L}_m(G) \mid a_1 \text{ precedes } a_2 \text{ in } s \text{ iff } b_1 \text{ precedes } b_2 \text{ in } s\}.$$

Let the set of uncontrollable events be $E_{uc} = \{a_1, b_1, b_2\}$. Build the standard realization of the supervisor S that solves BSCP-NB for this system. (Indicate the events that are effectively disabled by S.)

(b) Let the minimum required language be

$$L_r = \mathcal{L}_m(G_1)\mathcal{L}_m(G_2) \cup \mathcal{L}_m(G_2)\mathcal{L}_m(G_1)$$

which corresponds to the two possible serial (i.e., with no interleaving) executions of the database transactions. Let the set of uncontrollable events be $E_{uc} = \{b_1, a_2\}$. Build the standard realization of the supervisor S that solves DuSCP for this system.

3.12 Consider the discrete event system G where G is depicted in Fig. 3.17. We have that

$$E = \{a_1, a_2, b_1, b_2\} \quad \text{and} \quad E_{uc} = \{b_1, b_2\} .$$

Consider the admissible marked language $L_{am} \subseteq \mathcal{L}_m(G)$ where L_{am} is marked by the machine H depicted in Fig. 3.17.

(a) Solve BSCP-NB for these G and L_{am}. Provide the standard realization of the desired supervisor.

(b) Let S_1 be the nonblocking supervisor you obtained in part (a). Does there exist a (possibly blocking) supervisor S_2 such that

$$\mathcal{L}(S_2/G) \subseteq \overline{L_{am}} \quad \text{and} \quad \mathcal{L}(S_2/G) \cap L_{am} \supset \mathcal{L}(S_1/G) \cap L_{am}$$

(where the set inclusion in the second condition is proper)? Clearly explain your reasoning.

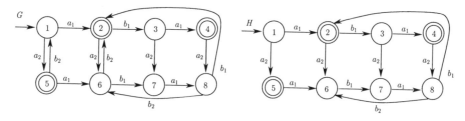

Figure 3.17: Problem 3.12: G and H.

(c) Interpreting L_{am} as a *minimum required behavior*, solve DuSCP for these G and $L_r = L_{am}$. Provide the standard realization of the desired supervisor.

3.13 Recall the formulation of SCPT. Define

$$L_1 = (L_{des} \cap L_{tol}^{\uparrow})^{\downarrow}$$
$$L_2 = L_{des}^{\downarrow} \cap L_{tol}^{\uparrow} .$$

How are L_1 and L_2 related, with respect to set inclusion?

3.14 Suppose that we are given the admissible behavior $L_{am} \subseteq \mathcal{L}_m(G)$. The goal is to ensure that

$$\mathcal{L}_m(S/G) \subseteq L_{am} .$$

Is it possible to be forced into blocking after string $t \in \overline{L_{am}}$, that is, blocking is unavoidable even though the string generated so far is a prefix of an admissible marked string? If your answer is yes, how should S be designed to avoid such blocking?

3.15 Show that if $A \subseteq B$, then A and B are nonconflicting.

3.16 Consider the class of languages

$$\mathcal{CNC}_{in}(K) := \{ L \subseteq K : (\overline{L \cap F} = \overline{L} \cap \overline{F}) \text{ and } (\overline{L}E_{uc} \cap M \subseteq \overline{L}) \}$$

where K, F, and M are given languages over the event set E, $E_{uc} \subseteq E$ is the set of uncontrollable events, $M = \overline{M}$, and K and F are subsets of M. In words, $\mathcal{CNC}_{in}(K)$ is the class of sublanguages of K that (1) are nonconflicting with respect to F and (2) are controllable with respect M and E_{uc}.

It has been shown in the literature that $\mathcal{CNC}_{in}(K)$ is closed under arbitrary unions and thus it possesses a supremal element, denoted by $K^{\uparrow(CNC)}$. Thus $K^{\uparrow(CNC)}$ is the *supremal nonconflicting (with respect to F) and controllable sublanguage* of K.

Recall the discussion on the modular version of BSCP-NB in Section 3.6. Show that

$$L_{am1}^{\uparrow C} \cap L_{am2}^{\uparrow(CNC)} = (L_{am1} \cap L_{am2})^{\uparrow C}$$

where in the $\uparrow (CNC)$ operation the nonconflicting condition on the sublanguage of L_{am2} is with respect to $L_{am1}^{\uparrow C}$. Discuss the implications of this result in modular supervisory control.

3.17 Let $E = \{\alpha, \beta, \lambda, \mu\}$, $E_o = \{\alpha, \beta\}$, $\mathcal{L}(G) = \overline{\{\alpha, \beta, \lambda\alpha, \mu\alpha\}}$, $K = \overline{\{\alpha, \lambda\alpha\}}$. Verify that K is observable with respect to $\mathcal{L}(G)$, P (corresponding to this E_o), and $E_c = E$.

3.18 Give an example of fixed M, K, P, and two different E_c's such that K is observable with respect to the first choice of E_c but not with respect to the second one.

3.19 Prove or disprove the following conjecture:

$$K^{\downarrow(CO)} \supseteq (K^{\downarrow O})^{\downarrow C} .$$

3.20 Consider the languages $M = \overline{M} \subseteq E^*$ and $L = \overline{L} \subseteq E_o^*$, where $P : E \to E_o$ is a natural projection. Let $E_{uc} \subseteq E$ be the set of uncontrollable events. Assume that $L \subseteq P(M)$. Define the language

$$K = P^{-1}(L) \cap M .$$

(a) Prove that K is observable with respect to M, P (E_o), and E.

(b) Show that "K is controllable with respect to M and E_{uc}" if and only if "L is controllable with respect to $P(M)$ and $E_{uc} \cap E_o$."

3.21 Consider the languages $K \subseteq M = \overline{M} \subseteq E^*$ and the natural projection $P : E \to E_o$. Show that the language

$$P^{-1}[P(K)] \cap M$$

is observable with respect to M, E_o, and *all choices* of $E_c \subseteq E$.

3.22 Build an example that illustrates that the property of controllability is not always preserved under the $\uparrow N$ operation.

3.23 Consider the system G and specification automaton H depicted in Fig. 3.18. A decentralized control architecture with two supervisors, S_1 and S_2, is to be used to control G. The admissible behavior is the language $L_a = \overline{L_a} = \mathcal{L}(H)$. The local sets of observable and controllable events are:

$$E_{1,o} = \{a, b, c\} \quad E_{2,o} = \{b, c, d\} \quad E_{1,c} = \{a, c\} \quad E_{2,c} = \{b, d\} .$$

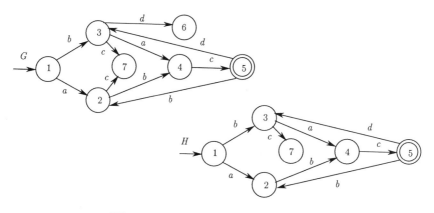

Figure 3.18: Problem 3.23: G and H.

(a) Show that L_a is co-observable with respect to the above information by constructing the standard realizations of S_1 and S_2 and verifying that the controlled behavior is indeed equal to L_a.

(b) Modify one or more of the four sets $E_{i,j}$ until L_a is no longer co-observable.

3.24 Recall Example 3.25. Build realizations of S_1 and S_2 that result in $\mathcal{L}(S_{dec}/G) = K$; this will show that K is co-observable.

3.25 Show that the language K in Example 3.26 is not co-observable with respect to the given sets.

3.26 The discussion on decentralized control in Section 3.8 assumes that an event can be disabled if any *single* supervisor decides to disable that event (cf. the "AND" block for enabled events in Fig. 3.16). Consider a different "decision fusion" rule that requires k supervisors, $1 < k \leq n$, to simultaneously disable an event in order for the event to actually be disabled in the system G. Prove that the class of controlled behaviors achievable by this new decision fusion rule is smaller than the class of co-observable languages, for fixed sets $E_{i,o}$ and $E_{i,c}$, $i = 1, \ldots, n$. (Assume that controllability and $\mathcal{L}_m(G)$-closure hold; cf. CC-OT.)

SELECTED REFERENCES

- *Book or survey papers on supervisory control*[6]

    - Kumar, R., and V.K. Garg, *Modeling and Control of Logical Discrete Event Systems*, Kluwer Academic Publishers, Boston, 1995.

[6]These references contain extensive bibliographies.

- Ramadge, P.J., and W.M. Wonham, "The control of discrete event systems," *Proceedings of the IEEE*, Vol. 77, No. 1, pp. 81–98, January 1989.

- Thistle, J.G., "Supervisory control of discrete event systems," *Mathematical and Computer Modelling*, Vol. 23, No. 11/12, pp. 25–53, 1996.

■ *Original references on main concepts presented in this chapter*[7]

- Chen, E., and S. Lafortune. "Dealing with blocking in supervisory control of discrete event systems," *IEEE Transactions on Automatic Control*, Vol. 36, No. 6, pp. 724–735, June 1991.

- Cieslak, R., C. Desclaux, A. Fawaz, and P. Varaiya, "Supervisory control of discrete-event processes with partial observations," *IEEE Transactions on Automatic Control*, Vol. 33, No. 3, pp. 249–260, March 1988.

- Lafortune, S., and E. Chen, "The infimal closed controllable superlanguage and its application in supervisory control," *IEEE Transactions on Automatic Control*, Vol. 35, No. 4, pp. 398–405, April 1990.

- Lin, F., and W.M. Wonham, "On observability of discrete-event systems," *Information Sciences*, Vol. 44, pp. 173-198, 1988.

- Ramadge, P.J., and W.M. Wonham, "Supervisory control of a class of discrete event processes," *SIAM Journal on Control and Optimization*, Vol. 25, No. 1, pp. 206–230, January 1987.

- Rudie, K., and W. M. Wonham, "Think globally, act locally: Decentralized supervisory control," *IEEE Transactions on Automatic Control*, Vol. 37, No. 11, pp. 1692–1708, November 1992.

- Wonham, W.M. , and P.J. Ramadge, "On the supremal controllable sublanguage of a given language," *SIAM Journal on Control and Optimization*, Vol. 25, No. 3, pp. 637–659, May 1987.

- Wonham, W.M., and P.J. Ramadge, "Modular supervisory control of discrete-event systems," *Mathematics of Control, Signals, and Systems*, Vol. 1, No. 1, pp. 13-30, 1988.

■ *For further reading on some topics only briefly covered in this chapter*

- Ben Hadj-Alouane, N., S. Lafortune, and F. Lin, "Centralized and distributed algorithms for on-line synthesis of maximal control policies under partial observation," *Discrete Event Dynamic Systems: Theory and Applications*, Vol. 6, No. 4, pp. 379-427, October 1996.

---

[7]Most of the theorems and propositions as well as some of the examples and problems in this chapter are from (or inspired by) these references.

– Brandt, R. D., V. Garg, R. Kumar, F. Lin, S. I. Marcus, and W. M. Wonham, "Formulas for calculating supremal controllable and normal sublanguages," *Systems & Control Letters*, Vo. 15, No. 2, pp. 111-117, August 1990.

– Cho, H., and S.I. Marcus, "On supremal languages of classes of sublanguages that arise in supervisor synthesis problems with partial observation," *Mathematics of Control, Signals, and Systems*, Vol. 2, No. 2, pp. 47–69, 1989.

– Inan, K., "An algebraic approach to supervisory control," *Mathematics of Control, Signals, and Systems*, Vol. 5, pp. 151–164, 1992.

– Kumar, R., and M. Shayman, "Formulas relating controllability, observability, and co-observability," *Automatica*, Vol. 34, No. 2, pp. 211-215, 1998.

– Lafortune, S., and F. Lin, "On tolerable and desirable behaviors in supervisory control of discrete event systems," *Discrete Event Dynamic Systems: Theory and Applications*, Vol. 1, No. 1, pp. 61–92, May 1991.

– Lin, F., A. Vaz, and W.M. Wonham, "Supervisor specification and synthesis for discrete event systems," *International Journal of Control*, Vol. 48, No. 1, pp. 321-332, 1988.

– Rudie, K., and J.C. Willems, "The computational complexity of decentralized discrete-event control problems," *IEEE Transactions on Automatic Control*, Vol. 40, No. 7, pp. 1313-1319, July 1995.

– Rudie, K., and W.M. Wonham, "The infimal prefix-closed and observable superlanguage of a given language," *Systems & Control Letters*, Vol. 15, pp. 361-371, 1990.

– Tsitsiklis, J.N., "On the control of discrete-event dynamical systems," *Mathematics of Control, Signals, and Systems*, Vol. 2, No. 1, pp. 95-107, 1989.

– Vaz, A. , and W.M. Wonham, "On supervisor reduction in discrete-event systems," *International Journal of Control*, Vol. 44, No. 2, pp. 475-491, 1986.

■ *Some applications of supervisory control*

– Balemi, S., G. J. Hoffmann, P. Gyugyi, H. Wong-Toi, and G. F. Franklin, "Supervisory control of a rapid thermal multiprocessor," *IEEE Transactions on Automatic Control*, Vol. 38, No. 7, pp. 1040–1059, July 1993.

– Brandin, B.A., "The Real-Time Supervisory Control of an Experimental Manufacturing Cell," *IEEE Transactions on Robotics and Automation*, Vol. 12, No. 1, pp. 1-14, February 1996.

– Lafortune, S., "Modeling and analysis of transaction execution in database systems," *IEEE Transactions on Automatic Control*, Vol. 33, No. 5, pp. 439–447, May 1988.

– Rudie, K. , W.M. Wonham, "Supervisory control of communicating processes," in *Protocol Specification, Testing, and Verification, X*, L. Logrippo, R. L. Probert, and H. Ural (Eds.), Elsevier Science Publishers, 1990, pp. 243–257.

– Thistle, J.G., R.P. Malhamé, H.-H. Hoang, and S. Lafortune, "Feature interaction modelling, detection and resolution: A supervisory control approach," in *Feature Interactions in Telecommunication Networks IV*, P. Dini, R. Boutaba, and L. Logrippo, Eds., IOS Press, 1997, pp. 93–107.

■ *Other paradigms for control of untimed DES*[8]

– Caines, P.E., and S. Wang," "COCOLOG: A conditional observer and controller logic for finite machines," *SIAM Journal on Control and Optimization*, Vol. 33, No. 6, pp. 1687–1715, November 1995.

■ *Notes*

1. The "Notes on Control of Discrete-Event Systems," by W.M. Wonham, Dept. of Electrical and Computer Engineering, University of Toronto (available at Web site www.control.utoronto.ca/DES) were an important source of inspiration in the writing of this chapter, most particularly regarding our presentation of the main theorems CT, NCT, and COT, the associated topics of supervisor realization and induced supervisors, and Example 3.16.

2. Our definitions of the properties of observability (Section 3.7.1) and co-observability (Section 3.8) are slightly different from those in the published literature in the sense that we excluded the "marking condition" from these definitions and we included E_c as a parameter in the case of observability. The rest of these definitions is equivalent to what appears in the published literature, except that it is stated in a different manner; this restatement, which in particular allows easy comparison of observability and co-observability, is due to G. Barrett.

3. The proof of the theorem on the equivalence between normality and observability in Section 3.7.5 does not appear in the published literature and is due to F. Lin.

4. The result of Problem 3.16 does not appear in the published literature and is due to J. Thistle.

---

[8]This excludes DES modeled by Petri nets, which are covered in the next chapter.

Chapter 4
Petri Nets

4.1 INTRODUCTION

An alternative to automata for untimed models of DES is provided by Petri nets. These models were first developed by C. A. Petri in the early 1960's. As we will see, Petri nets are related to automata in the sense that they also explicitly represent the transition function of DES. Like an automaton, a Petri net is a device that manipulates events according to certain rules. One of its features is that it includes explicit conditions under which an event can be enabled; this allows the representation of very general DES whose operation depends on potentially complex control schemes. This representation is conveniently described graphically, at least for small systems, resulting in *Petri net graphs*; Petri net graphs are intuitive and capture a lot of structural information about the system. We will see that an automaton can always be represented as a Petri net; on the other hand, not all Petri nets can be represented as *finite-state* automata. Consequently, Petri nets can represent a larger class of languages than the class of regular languages, $\mathcal{R}$. Another motivation for considering Petri net models of DES is the body of analysis techniques that have been developed for studying them. Such techniques cover not only untimed Petri net models but timed Petri net models as well; in this regard, we will see in the next chapter that there is a well-developed theory, called the "max-plus algebra," for a certain class of timed Petri nets (cf. Section 5.4). Finally, we mention that *Grafcet*, the widely-used programming language for programmable logic controllers (or PLCs) used in industrial automation, is inspired by Petri nets.

4.2 PETRI NET BASICS

The process of defining a Petri net involves two steps. First, we define the *Petri net graph*, also called *Petri net structure*, which is analogous to the state transition diagram of an automaton; this is the focus of Section 4.2.1. Then we adjoin to this graph an initial state, a set of marked states, and a transition labeling function, resulting in the complete Petri net model, its associated dynamics, and the languages that it generates and marks; this is treated in Sections 4.2.2 to 4.2.4.

4.2.1 Petri Net Notation and Definitions

In Petri nets, events are associated with *transitions*. In order for a transition to occur, several conditions may have to be satisfied. Information related to these conditions is contained in *places*. Some such places are viewed as the "input" to a transition; they are associated with the conditions required for this transition to occur. Other places are viewed as the output of a transition; they are associated with conditions that are affected by the occurrence of this transition. Transitions, places, and certain relationships between them define the basic components of a *Petri net graph*. A Petri net graph has two types of nodes, places and transitions, and arcs connecting these. It is a *bipartite graph* in the sense that arcs cannot directly connect nodes of the same type; rather, arcs connect place nodes to transition nodes and transition nodes to place nodes. The precise definition of Petri net graph is as follows.

Definition. (Petri net graph)
A *Petri net graph* (or *Petri net structure*) is a weighted bipartite graph

$$(P, T, A, w)$$

where

P is the finite set of *places* (one type of node in the graph)

T is the finite set of *transitions* (the other type of node in the graph)

$A \subseteq (P \times T) \cup (T \times P)$ is the set of arcs from places to transitions and from transitions to places in the graph

$w : A \to \{1, 2, 3, \dots\}$ is the *weight function* on the arcs.

We assume that (P, T, A, w) has no isolated places or transitions. ◆

We will normally represent the set of places by $P = \{p_1, p_2, \ldots, p_n\}$, and the set of transitions by $T = \{t_1, t_2, \ldots, t_m\}$. A typical arc is of the form (p_i, t_j) or (t_j, p_i), and the weight related to an arc is a positive integer. Note that we could allow P and T to be countable, rather than finite sets, as is the case for the state set in automata. It turns out, however, that a finite number of transitions and places is almost always perfectly adequate in modeling DES of interest.

We can see that a Petri net graph is somewhat more elaborate than the state transition diagram of an automaton. First, the nodes of a state transition diagram correspond to states selected from a single set X. In a Petri net graph, the nodes are either places, selected from the set P, or transitions, selected from the set T. Second, in a state transition diagram there is a single arc for each event causing a state transition. In a Petri net graph, we allow multiple arcs to connect two nodes, or, equivalently, we assign a weight to each arc representing the number of arcs. This is why we call this structure a *multigraph*.

In describing a Petri net graph, it is convenient to use $I(t_j)$ to represent the set of input places to transition t_j. Similarly, $O(t_j)$ represents the set of output places from transition t_j. Thus, we have

$$I(t_j) = \{p_i \in P : (p_i, t_j) \in A\}, \qquad O(t_j) = \{p_i \in P : (t_j, p_i) \in A\} .$$

Similar notation can be used to describe input and output transitions for a given place p_i: $I(p_i)$ and $O(p_i)$.

When drawing Petri net graphs, we need to differentiate between the two types of nodes, places and transitions. The convention is to use circles to represent places and bars to represent transitions. The arcs connecting places and transitions represent elements of the arc set A. Thus, an arc directed from place p_i to transition t_j means that $p_i \in I(t_j)$. Moreover, if $w(p_i, t_j) = k$, then there are k arcs from p_i to t_j or, equivalently, a single arc accompanied by its weight k. Similarly, if there are k arcs directed from transition t_j to place p_i, this means that $p_i \in O(t_j)$ and $w(t_j, p_i) = k$. We will generally represent weights through multiple arcs on a graph. However, when large weights are involved in a Petri net, writing the weight on the arc is a much more efficient representation. If no weight is shown on an arc of a Petri net graph, we will assume it to be 1. Finally, we mention that it is convenient to extend the domain and co-domain of the weight function w and write

$$w(p_i, t_j) = 0 \quad \text{when} \quad p_i \notin I(t_j) \qquad \text{and} \qquad w(t_j, p_i) = 0 \quad \text{when} \quad p_i \notin O(t_j) .$$

Example 4.1 (Simple Petri net graph)
Consider the Petri net graph defined by

$$P = \{p_1, p_2\} \qquad T = \{t_1\} \qquad A = \{(p_1, t_1), (t_1, p_2)\}$$

$$w(p_1, t_1) = 2 \qquad w(t_1, p_2) = 1$$

In this case, $I(t_1) = \{p_1\}$ and $O(t_1) = \{p_2\}$. The corresponding Petri net graph is shown in Fig. 4.1. The fact that $w(p_1, t_1) = 2$ is indicated by the presence of two input arcs from place p_1 to transition t_1.

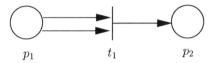

Figure 4.1: Petri net graph for Example 4.1.

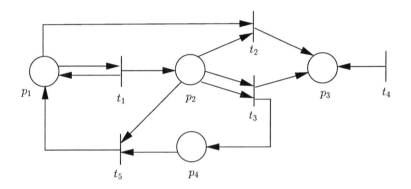

Figure 4.2: Petri net graph for Example 4.2.

Example 4.2

Consider the Petri net graph shown in Fig.4.2. The Petri net it represents is specified by

$$P = \{p_1, p_2, p_3, p_4\} \qquad T = \{t_1, t_2, t_3, t_4, t_5\}$$

$$A = \{(p_1, t_1), (p_1, t_2), (p_2, t_2), (p_2, t_3), (p_2, t_5), (p_4, t_5), (t_1, p_1),$$
$$(t_1, p_2), (t_2, p_3), (t_3, p_3), (t_3, p_4), (t_4, p_3), (t_5, p_1)\}$$

$$
\begin{array}{llll}
w(p_1, t_1) = 1 & w(p_1, t_2) = 1 & w(p_2, t_2) = 1 & w(p_2, t_3) = 2 \\
w(p_2, t_5) = 1 & w(p_4, t_5) = 1 & w(t_1, p_1) = 1 & w(t_1, p_2) = 1 \\
w(t_2, p_3) = 1 & w(t_3, p_3) = 1 & w(t_3, p_4) = 1 & w(t_4, p_3) = 1 \\
w(t_5, p_1) = 1 \; .
\end{array}
$$

It is worth commenting on the fact that transition t_4 has no input places. If we think of transitions as events and places as conditions related to event occurrences, then the event corresponding to t_4 takes place unconditionally. In contrast, the event corresponding to transition t_2, for example, depends on certain conditions related to places p_1 and p_2.

4.2.2   Petri Net Markings and State Spaces

Let us return to the idea that transitions in a Petri net graph represent the events driving a DES, and that places describe the conditions under which these events can occur. In this framework, we need a mechanism indicating whether these conditions are in fact met or not. This mechanism is provided by assigning *tokens* to places. A token is something we "put in a place" essentially to indicate the fact that the condition described by that place is satisfied. The way in which tokens are assigned to a Petri net graph defines a *marking*. Formally, a *marking* x of a Petri net graph (P, T, A, w) is a function $x : P \rightarrow \mathbb{N} = \{0, 1, 2, \ldots\}$. Thus, marking x defines row vector $\mathbf{x} = [x(p_1), x(p_2), \ldots, x(p_n)]$, where n is the number of places in the Petri net. The ith entry of this vector indicates the (non-negative integer) number of tokens in place p_i, $x(p_i) \in \mathbb{N}$. In Petri net graphs, a token is indicated by a dark dot positioned in the appropriate place.

Definition. (Marked Petri net)
A *marked Petri net* is a five-tuple (P, T, A, w, x) where (P, T, A, w) is a Petri net graph and x is a marking of the set of places P; $\mathbf{x} = [x(p_1), x(p_2), \ldots, x(p_n)] \in \mathbb{N}^n$ is the row vector associated with x.   ♦

Example 4.3
Consider the Petri net of Fig. 4.1. In Fig. 4.3, we show two possible markings, namely the row vectors $\mathbf{x}_1 = [1, 0]$ and $\mathbf{x}_2 = [2, 1]$.

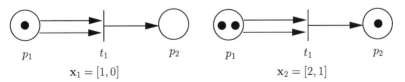

p_1 t_1 p_2 p_1 t_1 p_2
$\mathbf{x}_1 = [1, 0]$ $\mathbf{x}_2 = [2, 1]$

Figure 4.3: Two markings, $\mathbf{x}_1$ and $\mathbf{x}_2$, for the Petri net graph in Fig. 4.1 (Example 4.3).

For simplicity, we shall henceforth refer to a marked Petri net as just a Petri net. Now, since our system modeling efforts have always relied on the concept of state, in the case of a Petri net we identify place marking with the *state* of the Petri net. That is, we define the *state* of a Petri net to be its marking row vector $\mathbf{x} = [x(p_1), x(p_2), \ldots, x(p_n)]$. Note that the number of tokens assigned to a place is an arbitrary non-negative integer, not necessarily bounded. It follows that the number of states we can have is, in general, infinite. Thus, the state space X of a Petri net with n places is defined by all n-dimensional vectors whose entries are non-negative integers, that is, $X = \mathbb{N}^n$. While the term "marking" is more common than "state" in the Petri net literature, the term state is consistent with the role of state in system dynamics, as we shall now see; moreover, the term state avoids the potential confusion between marking in Petri net graphs and marking in the sense of *marked states* in automata (and Petri nets too, cf. Section 4.2.4).

The above definitions do not explicitly describe the state transition mechanism of Petri nets. This is clearly a crucial point since we want to use Petri nets to model *dynamic* DES. It turns out that the state transition mechanism is captured by the structure of the Petri net graph. In order to define the state transition mechanism, we first need to introduce the notion of *enabled transition*. Basically, for a transition $t \in T$ to "happen" or to "be enabled," we require a token to be present in each place (i.e., condition) which is input to the transition. Since, however, we allow weighted arcs from places to transitions, we use a slightly more general definition.

Definition. (Enabled transition)
A transition $t_j \in T$ in a Petri net is said to be *enabled* if

$$x(p_i) \geq w(p_i, t_j) \quad \text{for all } p_i \in I(t_j) . \qquad \blacklozenge$$

In words, transition t_j in the Petri net is enabled when the number of tokens in p_i is at least as large as the weight of the arc connecting p_i to t_j, for all places p_i that are input to transition t_j. In Fig. 4.3 with state $\mathbf{x}_1$, $x(p_1) = 1 < w(p_1, t_1) = 2$, and therefore t_1 is not enabled. But with state $\mathbf{x}_2$, we have $x(p_1) = 2 = w(p_1, t_1)$, and t_1 is enabled.

As was mentioned above, since places are associated with conditions regarding the occurrence of a transition, then a transition is enabled when all the conditions required for its occurrence are satisfied; tokens are the mechanism used to determine satisfaction of conditions. The set of enabled transitions in a given state of the Petri net is equivalent to the active event set in a given state of an automaton. We are now ready to describe the dynamical evolution of Petri nets.

4.2.3 Petri Net Dynamics

In automata, the state transition mechanism is directly captured by the arcs connecting the nodes (states) in the state transition diagram, equivalently by the transition function f. The state transition mechanism in Petri nets is provided by *moving tokens* through the net and hence changing the state of the Petri net. When a transition is enabled, we say that it can *fire* or that it can occur (the term "firing" is standard in the Petri net literature). The state transition function of a Petri net is defined through the change in the state of the Petri net due to the firing of an enabled transition.

Definition. (Petri net dynamics)
The *state transition function*, $f : \mathbb{N}^n \times T \to \mathbb{N}^n$, of Petri net (P, T, A, w, x) is defined for transition $t_j \in T$ if and only if

$$x(p_i) \geq w(p_i, t_j) \quad \text{for all } p_i \in I(t_j) . \qquad (4.1)$$

If $f(\mathbf{x}, t_j)$ is defined, then we set $\mathbf{x}' = f(\mathbf{x}, t_j)$ where

$$x'(p_i) = x(p_i) - w(p_i, t_j) + w(t_j, p_i), \quad i = 1, \ldots, n . \tag{4.2}$$

♦

Condition 4.1 ensures that the state transition function is defined only for transitions that are enabled; an "enabled transition" is therefore equivalent to a "feasible event" in an automaton, as we mentioned above. But whereas in automata the state transition function was quite arbitrary, here the state transition function is based on the structure of the Petri net. Thus, the next state defined by equation 4.2 explicitly depends on the input and output places of a transition and on the weights of the arcs connecting these places to the transition.

According to equation 4.2, if p_i is an input place of t_j, it loses as many tokens as the weight of the arc from p_i to t_j; if it is an output place of t_j, it gains as many tokens as the weight of the arc from t_j to p_i. Clearly, it is possible that p_i is both an input and output place of t_j, in which case equation 4.2 removes $w(p_i, t_j)$ tokens from p_i, and then immediately places $w(t_j, p_i)$ new tokens back in it.

It is important to remark that the number of tokens need not be conserved upon the firing of a transition in a Petri net. This is immediately clear from equation 4.2, since it is entirely possible that

$$\sum_{p_i \in P} w(t_j, p_i) > \sum_{p_i \in P} w(p_i, t_j) \quad \text{or} \quad \sum_{p_i \in P} w(t_j, p_i) < \sum_{p_i \in P} w(p_i, t_j)$$

in which case either $\mathbf{x}' = f(\mathbf{x}, t_j)$ contains more tokens than $\mathbf{x}$ or less tokens than $\mathbf{x}$. In general, it is entirely possible that after several transition firings, the resulting state is $\mathbf{x} = [0, \ldots, 0]$, or that the number of tokens in one or more places grows arbitrarily large after an arbitrarily large number of transition firings. The latter phenomenon is a key difference with automata, where finite-state automata have only a finite number of states, by definition. In contrast, a finite Petri net graph may result in a Petri net with an unbounded number of states. We will comment further on this issue later in this section.

Example 4.4 (Firing of transitions)

To illustrate the process of firing transitions and changing the state of a Petri net, consider the Petri net of Fig. 4.4 (a), where the "initial" state is $\mathbf{x}_0 = [2, 0, 0, 1]$. We can see that the only transition enabled is t_1, since it requires a single token from place p_1 and we have $x_0(p_1) = 2$. In other words, $x_0(p_1) \geq w(p_1, t_1)$, and condition 4.1 is satisfied for transition t_1. When t_1 fires, one token is removed from p_1, and one token is placed in each of places p_2 and p_3, as can be seen from the Petri net graph. We can also directly apply equation 4.2 to obtain the new state $\mathbf{x}_1 = [1, 1, 1, 1]$, as shown in Fig. 4.4 (b). In this state, all three transitions t_1, t_2, and t_3 are enabled.

Next, suppose transition t_2 fires. One token is removed from each of the input places, p_2 and p_3. The output places are p_2 and p_4. Therefore, a token is immediately placed back in p_2, since $p_2 \in I(t_2) \cap O(t_2)$. In addition, a token is added to p_4. The new state is $\mathbf{x}_2 = [1, 1, 0, 2]$, as shown in Fig. 4.4 (c). At this state, t_2 and t_3 are no longer enabled, but t_1 still is.

Let us now return to state $\mathbf{x}_1$ of Fig. 4.4 (b), where all three transitions are enabled. Instead of firing t_2, let us fire t_3. We remove a token from each of the input places, p_1, p_3, and p_4. Note that there are no output places. The new state is denoted by $\mathbf{x}_2'$ and is given by $\mathbf{x}_2' = [0, 1, 0, 0]$, as shown in Fig. 4.4 (d). We see that no transition is now enabled. No further state changes are possible, and $[0, 1, 0, 0]$ is a "deadlock" state of this Petri net.

The number of tokens is not conserved in this Petri net since $\mathbf{x}_0$ contains three tokens while $\mathbf{x}_1$ and $\mathbf{x}_2$ each contain four tokens and $\mathbf{x}_2'$ contains one token.

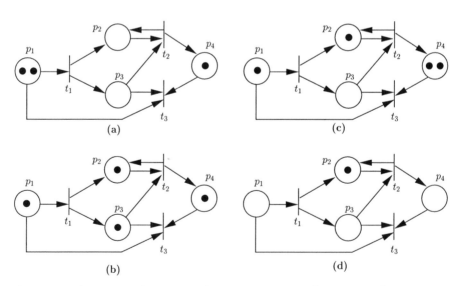

Figure 4.4: Sequences of transition firings in a Petri net (Example 4.4).
In (a), only transition t_1 is enabled in initial state $\mathbf{x}_0$. The new state resulting from the firing of t_1 in (a) is denoted by $\mathbf{x}_1$ and is shown in (b). The new state resulting from the firing of t_2 in (b) is denoted by $\mathbf{x}_2$ and is shown in (c). The new state resulting from the firing of t_3 in state $\mathbf{x}_1$ in (b) is denoted by $\mathbf{x}_2'$ and is shown in (d).

The preceding example shows that the sequence in which transitions fire is not pre-specified in a Petri net. At state $\mathbf{x}_1$, any one of the three transitions could fire next. This is the same as having three events in the active event set of a state in an automaton model of a DES. In the study of untimed Petri net models of DES, we must examine each possible sequence of transitions (events), as we did for automata in Chapters 2 and 3.

Another important observation about the dynamic behavior of Petri nets is that not all states in $\mathbb{N}^n$ can necessarily be reached from a Petri net graph with a given initial state. For instance, if we examine the Petri net graph in Fig. 4.3 together with the initial state $\mathbf{x}_2 = [2, 1]$, then we see that the only state that can be reached from $\mathbf{x}_2$ is $[0, 3]$. This leads us to defining the set of reachable states, $R[(P, T, A, w, x)]$, of Petri net (P, T, A, w, x). In this regard, we first need to extend the state transition function f from domain $\mathbb{N}^n \times T$ to domain $\mathbb{N}^n \times T^*$, in the same manner as we extended the transition function of automata in Section 2.2.2:

$$
\begin{aligned}
f(\mathbf{x}, \varepsilon) &:= \mathbf{x} \\
f(\mathbf{x}, st) &:= f(f(\mathbf{x}, s), t) \text{ for } s \in T^* \text{ and } t \in T .
\end{aligned}
$$

(Here the symbol ε is to be interpreted as the absence of transition firing.)

Definition. (Reachable states)
The set of *reachable states* of Petri net (P, T, A, w, x) is

$$
R[(P, T, A, w, x)] := \{\mathbf{y} \in \mathbb{N}^n : \exists s \in T^*(f(\mathbf{x}, s) = \mathbf{y})\} . \qquad \blacklozenge
$$

The above definitions of the extended form of the state transition function and of the set of reachable states assume that enabled transitions fire *one at a time*. Consider the Petri net in Fig. 4.4 (b), where all three transitions t_1, t_2, and t_3 are enabled. We can see that transitions t_1 and t_2 could fire simultaneously, since they "consume" tokens from disjoint sets of places: $\{p_1\}$ for t_1 and $\{p_2, p_3\}$ for t_2. Since we are interested in all possible states that can be reached, and since later on we will be labeling transitions with event names and considering the languages represented by Petri nets, we shall henceforth exclude such simultaneous firings of transitions and assume that transitions fire one at a time.

State Equations

Let us return to equation 4.2 describing how the state value of an individual place changes when a transition fires. It is not difficult to see how we can generate a vector equation from equation 4.2, in order to specify the next Petri net state $\mathbf{x}' = [x'(p_1), x'(p_2), \dots, x'(p_n)]$ given the current one $\mathbf{x} = [x(p_1), x(p_2), \dots, x(p_n)]$ and the fact that a particular transition, t_j, has fired. To do so, let us first define the *firing vector* $\mathbf{u}$, an m-dimensional row vector of the form

$$\mathbf{u} = [0, \ldots, 0, 1, 0, \ldots, 0] \tag{4.3}$$

where the only 1 appears in the jth position, $j \in \{1, \ldots, m\}$, to indicate the fact that the jth transition is currently firing. In addition, we define the *incidence matrix* of a Petri net, $\mathbf{A}$, an $m \times n$ matrix whose (j, i) entry is of the form

$$a_{ji} = w(t_j, p_i) - w(p_i, t_j) . \tag{4.4}$$

This matches the weight difference that appears in equation 4.2 in updating $x(p_i)$. Using the incidence matrix $\mathbf{A}$, we can now write a vector state equation

$$\mathbf{x}' = \mathbf{x} + \mathbf{u}\mathbf{A} \tag{4.5}$$

which describes the state transition process as a result of an "input" $\mathbf{u}$, that is, a particular transition firing. The ith equation in 4.5 is precisely equation 4.2. We see, therefore, that $f(\mathbf{x}, t_j) = \mathbf{x} + \mathbf{u}\mathbf{A}$, where $f(\mathbf{x}, t_j)$ is the transition function we defined earlier. The argument t_j in this function indicates that it is the jth entry in $\mathbf{u}$ which is nonzero. The state equation provides a convenient algebraic tool and an alternative to purely graphical means for describing the process of firing transitions and changing the state of a Petri net.

Example 4.5 (State equation)

Let us reconsider the Petri net of Fig. 4.4 (a), with the initial state $\mathbf{x}_0 = [2, 0, 0, 1]$. We can first write down the incidence matrix by inspection of the Petri net graph, which in this case is

$$\mathbf{A} = \begin{bmatrix} -1 & 1 & 1 & 0 \\ 0 & 0 & -1 & 1 \\ -1 & 0 & -1 & -1 \end{bmatrix} .$$

The (1, 2) entry, for example, is given by $w(t_1, p_2) - w(p_2, t_1) = 1 - 0$. Using equation 4.5, the state equation when transition t_1 fires at state $\mathbf{x}_0$ is

$$\mathbf{x}_1 = \begin{bmatrix} 2 & 0 & 0 & 1 \end{bmatrix} + \begin{bmatrix} 1 & 0 & 0 \end{bmatrix} \begin{bmatrix} -1 & 1 & 1 & 0 \\ 0 & 0 & -1 & 1 \\ -1 & 0 & -1 & -1 \end{bmatrix}$$

$$= \begin{bmatrix} 2 & 0 & 0 & 1 \end{bmatrix} + \begin{bmatrix} -1 & 1 & 1 & 0 \end{bmatrix} = \begin{bmatrix} 1 & 1 & 1 & 1 \end{bmatrix}$$

which is precisely what we obtained in Example 4.4. Similarly, we can determine $\mathbf{x}_2$ as a result of firing t_2 next, as in Fig. 4.4 (c):

$$\mathbf{x}_2 = \begin{bmatrix} 1 & 1 & 1 & 1 \end{bmatrix} + \begin{bmatrix} 0 & 1 & 0 \end{bmatrix} \begin{bmatrix} -1 & 1 & 1 & 0 \\ 0 & 0 & -1 & 1 \\ -1 & 0 & -1 & -1 \end{bmatrix}$$

$$= \begin{bmatrix} 1 & 1 & 0 & 2 \end{bmatrix} .$$

Viewed as a dynamic system, a Petri net gives rise to sample paths similar to those of automata. Specifically, the sample path of a Petri net is a sequence of states $\{\mathbf{x}_0, \mathbf{x}_1, \dots\}$ resulting from a transition firing sequence input $\{t^1, t^2, \dots\}$, where t^k is the kth transition fired. Given an initial state $\mathbf{x}_0$, the entire state sequence can be generated through

$$\mathbf{x}_{k+1} = f(\mathbf{x}_k, t_k) = \mathbf{x}_k + \mathbf{u}_k \mathbf{A}$$

where $\mathbf{u}_k$ contains the information regarding the kth transition fired. If a state is reached such that no more transitions can fire, we say that the execution of the Petri net deadlocks at that state (as in Fig. 4.4 (d)).

4.2.4 Petri Net Languages

So far, we have focused on the state dynamics of Petri nets, where transitions are enumerated as elements of the set T. We have assumed that transitions correspond to events, but we have not made precise statements about this correspondence. If we wish to look at Petri nets as a modeling formalism for representing *languages*, as was the focus for automata in Chapter 2, then we need to specify precisely what event each transition corresponds to; this will allow us to specify the language(s) represented (generated and marked) by a Petri net.

Let E be the set of events of the DES under consideration and whose language is to be modeled by a Petri net. Of course, we could require that the Petri net model of the system be such that each transition in T corresponds to a distinct event from the set of events E, and vice-versa. But this would be unnecessarily restrictive; in automaton models, we allow having two different arcs (originating from two different states) labeled with the same event. This leads us to the definition of a *labeled Petri net*.

Definition. (Labeled Petri net)
A *labeled Petri net* N is an eight-tuple

$$N = (P, T, A, w, E, \ell, \mathbf{x}_0, \mathbf{X}_m)$$

where

(P, T, A, w) is a Petri net graph

E is the event set for transition labeling

$\ell : T \to E$ is the transition labeling function

$\mathbf{x}_0 \in \mathbb{N}^n$ is the initial state of the net (i.e., the initial number of tokens in each place)

$\mathbf{X}_m \subseteq \mathbb{N}^n$ is the set of *marked states* of the net. ♦

In Petri net graphs, the label of a transition is indicated next to the transition. The notion of "marked states" in this definition is completely analogous to the notion of marked states in the definition of automaton in Chapter 2. We introduce marked states in order to define the language *marked* by a labeled Petri net.

Definition. (Languages generated and marked)
The language generated by labeled Petri net $N = (P, T, A, w, E, \ell, \mathbf{x}_0, \mathbf{X}_m)$ is

$$\mathcal{L}(N) := \{\ell(s) \in E^* : s \in T^* \text{ and } f(\mathbf{x}_0, s) \text{ is defined } \} .$$

The language marked by N is

$$\mathcal{L}_m(N) := \{\ell(s) \in \mathcal{L}(N) : s \in T^* \text{ and } f(\mathbf{x}_0, s) \in \mathbf{X}_m)\} . \qquad \blacklozenge$$

(This definition uses the extended form of the transition labeling function $\ell : T^* \to E^*$; this extension is done in the usual manner.) We can see that these definitions are completely consistent with the corresponding definitions for automata. The language $\mathcal{L}(N)$ represents all strings of transition labels that are obtained by all possible (finite) sequences of transition firings in N, starting in the initial state $\mathbf{x}_0$ of N; the marked language $\mathcal{L}_m(N)$ is the subset of these strings that leave the Petri net in a state that is a member of the set of marked states given in the definition of N.

The class of languages that can be represented by labeled Petri nets is

$$\mathcal{PNL} := \{K \subseteq E^* : \exists N = (P, T, A, w, E, \ell, \mathbf{x}_0, \mathbf{X}_m) \ [\mathcal{L}_m(N) = K]\} .$$

This is a general definition and the properties of $\mathcal{PNL}$ depend heavily on the specific assumptions that are made about ℓ (e.g., injective or not) and $\mathbf{X}_m$ (e.g., finite or infinite). We shall restrict ourselves to the following situation: ℓ is not necessarily injective and $\mathbf{X}_m$ need not be finite. Section 4.3 compares Petri nets with automata, and in particular $\mathcal{R}$ with $\mathcal{PNL}$. Finally, we note that we shall often refer to "labeled Petri nets" as "Petri nets"; this will not create confusion with "unlabeled" Petri nets, as it will be clear from the context if we are interested in state properties (in which case event labeling of transitions is irrelevant) or in language properties.

4.2.5 Petri Net Models for Queueing Systems

In Example 2.10 in Section 2.2.3, we saw how automata can be used to represent the dynamic behavior of a simple queueing system. We now repeat this process using a Petri net structure. We begin by considering three events (transitions) driving the system:

a: customer arrives

s: service starts

c: service completes and customer departs.

We form the transition set $T = \{a, s, c\}$. (In this example, we shall not need to consider labeled Petri nets; equivalently, we can assume that $E = T$ and that ℓ is a one-to-one mapping between these two sets.)

Transition a is spontaneous and requires no conditions (input places). On the other hand, transition s depends on two conditions: the presence of customers in the queue, and the server being idle. We represent these conditions through two input places for this transition, place Q (queue) and place I (idle server). Finally, transition c requires that the server be busy, so we introduce an input place B (busy server) for it. Thus, our place set is $P = \{Q, I, B\}$.

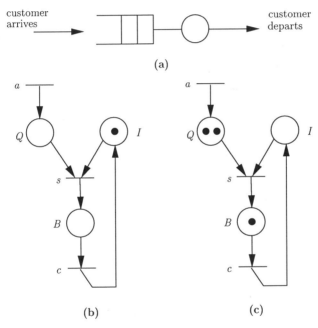

Figure 4.5: (a) Simple queueing system. (b) Petri net model of simple queueing system in initial state $[0, 1, 0]$. (c) Petri net model of simple queueing system with initial state $[0, 1, 0]$ after firing sequence $\{a, s, a, a, c, s, a\}$.

The complete Petri net graph, along with the simple queueing system it models, are shown in Fig. 4.5 (a) and (b). No tokens are placed in Q, indicating that the queue is empty, and a token is placed in I, indicating that the server is idle. This defines the initial state $\mathbf{x_0} = [0, 1, 0]$. Since transition a is always enabled, we can generate several possible sample paths. As an example, Fig. 4.5 (c) shows state $[2, 0, 1]$ resulting from the transition firing sequence $\{a, s, a, a, c, s, a\}$. This state corresponds to two customers waiting in queue, while a third is in service (the first arrival in the sequence has already departed after transition c).

Note that a c transition always enables an s transition as long as there is a token in Q. If we assume that the enabled s transition then immediately fires, this is equivalent to the automaton model using only two events (arrivals a and departures d), but specifying a feasible event set $\Gamma(0) = \{a\}$.

A slightly more detailed model of the same queueing system could include the additional transition

d: customer departs

which requires condition F (finished customer). In this case, transition c only means that "service completes". In addition, the external arrival process can also be represented by an input place to transition a, denoted by A, as long as a token is always maintained in A to keep transition a enabled. Thus, in this alternative model we have

$$T = \{a, s, c, d\} \quad \text{and} \quad P = \{A, Q, I, B, F\} \, .$$

The resulting model is shown at state $[1, 0, 1, 0, 0]$ in Fig. 4.6 (a). Comparing this model with the automaton using only events $\{a, d\}$, we can see that events c and d are combined into one, since transition c is always enabling transition d. This is yet another indication of the flexibility of any modeling process; using the additional transition d may not be needed in general, but there are applications where differentiating between "service completion" of a customer and "customer departure" is useful, if not necessary.

The model of Fig. 4.6 (a) can be further modified if we allow the server to break down (as in the automaton model of Fig. 2.7). In this case, we introduce two new transitions

b: server breaks down

r: server repaired

and place D (server down) which is input to transition r. Thus,

$$T = \{a, s, c, d, b, r\} \quad \text{and} \quad P = \{A, Q, I, B, F, D\}.$$

The Petri net model is shown at state $[1, 0, 1, 0, 0, 0]$ in Fig. 4.6 (b).

4.3 COMPARISON OF PETRI NETS AND AUTOMATA

Automata and Petri nets can both be used to represent the behavior of a DES. We have seen that both formalisms explicitly represent the transition structure of the DES. In automata, this is done by explicitly enumerating all possible states and then "connecting" these states with the possible transitions

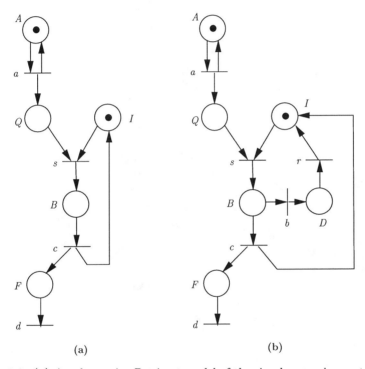

(a)   (b)

Figure 4.6: **(a)** An alternative Petri net model of the simple queueing system. **(b)** Petri net model of the simple queueing system with server breakdowns.

between them, resulting in the transition function of the automaton. This is not particularly elegant, yet, automaton models are easily combined by operations such as product and parallel composition and therefore models of complex systems can be built from models of individual components in a systematic manner. Petri nets on the other hand have more structure in their representation of the transition function. States are not enumerated; rather, state information is "distributed" among a set of places that capture key conditions that govern the operation of the system. Of course, some work needs to be done at modeling time to identify all relevant "conditions" that should be captured by places, and then to properly connect these places to the transitions.

It is reasonable to ask the question: Which is a better model of a given DES, an automaton or a Petri net? There is no obvious answer to such a question, as modeling is always subject to personal biases and very frequently depends on the particular application considered. However, if we reformulate the above question more precisely in the context of specific criteria for comparison, it is possible to draw some conclusions.

Language Expressiveness

As a first criterion for comparing automata and Petri nets, let us consider the class of languages that can be represented by each formalism, when constrained to models that require *finite memory*, an obvious practical consideration. We claim that the class $\mathcal{PNL}$ is strictly larger than the class $\mathcal{R}$, meaning that Petri nets with finite sets of places and transitions can represent (i.e., mark) more languages in E^* than finite-state automata. In order to prove this result, let us first see how any *finite-state* automaton can always be "transformed" into a Petri net that generates and marks the same languages. Then we will complete the proof by presenting a non-regular language that can be marked by a Petri net.

Suppose we are given a finite-state automaton $G = (X, E, f_G, \Gamma, x_0, X_m)$, where X (and necessarily E) is a finite set. To construct a (labeled) Petri net $N = (P, T, A, w, E, \ell, \mathbf{x}_0, \mathbf{X}_m)$ such that

$$\mathcal{L}(N) = \mathcal{L}(G) \quad \text{and} \quad \mathcal{L}_m(N) = \mathcal{L}_m(G)$$

we can proceed as follows.

1. We first view each state in X as defining a unique place in P, that is, $P = X$. This also immediately specifies:

    - the initial state $\mathbf{x}_0$ of N: it is the row vector $[0, \ldots, 0, 1, 0, \ldots, 0]$ where the only non-zero entry is for the place in P that corresponds to $x_0 \in X$;

    - the set of marked states $\mathbf{X}_m$ of N: it is the set of all row vectors $[0, \ldots, 0, 1, 0, \ldots, 0]$ where the non-zero entry is for a place that corresponds to a marked state in X_m.

2. Next, we associate each triple (x, e, x') in G, where $x' = f_G(x, e)$ for some $e \in \Gamma(x)$, with a transition $t_{(x,e,x')}$ in T of N. In other words, T has the same cardinality as the set of arcs in the state transition diagram of G. We then:

- label transition $t_{(x,e,x')}$ in T by the event $e \in E$;

- define two arcs in A for each (x, e, x') triple in G: arc $(x, t_{(x,e,x')})$ and arc $(t_{(x,e,x')}, x')$. All these arcs have weight equal to 1. This mapping of states and arcs in G to places, transitions, and arcs in N is illustrated in Fig. 4.7.

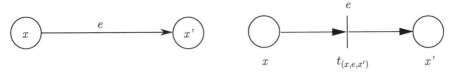

Figure 4.7: Mapping of a transition triple (x, e, x') in G to two places, one labeled transition, and two arcs in the graph of Petri net N.

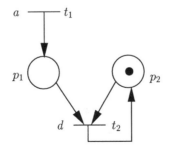

Figure 4.8: Modified Petri net model for queueing system in Fig. 4.5 (a). *This Petri net generates the same language as the automaton in Fig. 2.6, where we take $x_0 = 0$.*

We should point out that this transformation technique for building a Petri net out of a given finite-state automaton is merely for the sake of substantiating our above claim relating $\mathcal{PNL}$ and $\mathcal{R}$ and may not result in the most intuitive Petri net model. Indeed, the physical system itself is likely to suggest much better ways for deriving the Petri net model by capturing the structural information in the system.

To prove that $\mathcal{PNL}$ is strictly larger than $\mathcal{R}$, let us return to the queueing system modeled in Fig. 4.5 and slightly modify it by removing place B and transition c. Let us relabel the transition "service starts" by event d. The resulting Petri net is shown in Fig. 4.8. If we take $\mathbf{x}_0 = [0, 1]$ and $\mathbf{X}_m = \{[x, y] :$

$x \in \mathbb{N}$ and $y \in \{0, 1\}\}$, then we can see that this Petri net marks the same *non-regular* language as the infinite-state automaton in Fig. 2.6 in Chapter 2, when $x_0 = 0$ and $X_m = \mathbb{N}$ for that automaton: the set of all strings in $\{a, d\}^*$ where any prefix of any string contains no more d events than a events. Intuitively, the memory required to ensure that any string does not contain more d events than a events and forces the automaton to have an arbitrarily large number of states (since there can be arbitrarily more a events than d events in a string) is taken care of by the tokens in place p_1. Of course, the number of tokens in place p_1 can be arbitrarily large. But this is not a problem in the sense that the Petri net graph requires finite memory (unlike explicit storage of the state transition diagram in Fig. 2.6); a variable can be used to store $x(p_1)$ as it is not necessary to explicitly show all possible values that $x(p_1)$ can take.

While not all Petri nets can be transformed into equivalent finite-state automata, it is clear that any Petri net N whose reachable state set $R(N)$ is finite can be transformed into an equivalent finite-state automaton. It suffices to create an automaton state for each $\mathbf{x} \in R(N)$ and connect these automaton states by appropriate arcs according to the enabled transitions in $\mathbf{x}$, the states they reach, and their labels.

We conclude that the larger expressive power of Petri nets for languages as compared with finite-state automata occurs when the behavior of any Petri net representation of the given language results in an unbounded number of tokens in one or more places, leading to an infinite set of reachable Petri net states.

Modular Model-Building

Despite the potential complexity of Petri net graphs required to model even relatively simple DES, the Petri net structure possesses some inherent advantages. One such advantage is its ability to decompose or modularize a potentially complex system. Suppose we have two interacting systems with state spaces X_1 and X_2 modeled as automata. If we combine these two systems into one, its state space, X, can be as large as all of the states in $X_1 \times X_2$, as was seen in Chapter 2; in particular, this upper bound will be achieved if the two systems have no common events. This means that combining multiple systems rapidly increases the complexity of automaton models. On the other hand, if the systems are modeled through Petri nets, the combined system is often easier to obtain by leaving the original nets as they are and simply adding a few places and/or transitions (or merging some places) representing the coupling effects between the two. Moreover, by looking at such a Petri net graph, one can conveniently see the individual components, discern the level of their interaction, and ultimately decompose a system into logical distinct modules. Example 4.6 below illustrates these claims and shows an instance where the model of the complete system grows "linearly" in the models of the system components. Of course, such a "linear" combination of sub-models into a complete model is a *modeling issue* and not a systematic procedure like combination by product and parallel composition in the case of automata. In fact, we can define product and parallel composition of Petri nets (although we shall not do so in this book) in order to systematically compose sub-models into a complete model; however, these composition operations will not in general result in linear growth.

Along the same lines, Petri nets are oriented towards capturing the concurrent nature of separate processes forming a DES. The combination of two such asynchronous processes in automaton models tends to become complex and hide some of the intuitive structure involved in this combination. Petri nets form a much more natural framework for these situations, and make it easier to visualize this structure.

Example 4.6 (Dining philosophers)

Let us consider the example of two "dining philosophers" that we described in Example 2.17 in Chapter 2. Figures 2.15 and 2.16 show an automaton model of this system; the complete model is obtained by doing the parallel composition of four component models, one for each philosopher and one for each fork. Figure 4.9 shows a Petri net model of the same system, with the same set of events. In part (a) of Fig. 4.9, we have the model of one philosopher. There is a place corresponding to the condition "philosopher eating" and a place corresponding to the condition "philosopher thinking". This model also accounts for the availability or unavailability of each fork since it includes two "fork" places. In this sense, this model is more elaborate than automaton P_1 in Fig. 2.15. The reason for using the two fork places is to allow easy interconnection of the two philosophers. If we create a second copy of Fig. 4.9 (a) for the second philosopher, then a little thought shows that the complete system model is obtained by combining the Petri net models of the two philosophers, where the fork places are simply *overlapped* while all other places, all transitions, and all arcs are preserved. This complete model is shown in Fig. 4.9 (b). This merger of the fork places works because the two forks are shared resources, and thus the conditions "fork available" and "fork unavailable" are common to the two component models. The initial state of the Petri net is obtained by placing four tokens, namely one in each of the places corresponding to: philosopher 1 thinking, philosopher 2 thinking, fork 1 available, and fork 2 available.

This example illustrates that while we could have constructed the complete Petri net model by transforming the finite-state automaton of the complete system model (two philosophers and two forks) shown in Fig. 2.16 into a Petri net using the method described earlier in this section, a model derived component by component from the original system description and associating places with conditions governing the operation of the system (as opposed to associating places with system states) is much more intuitive and modular.

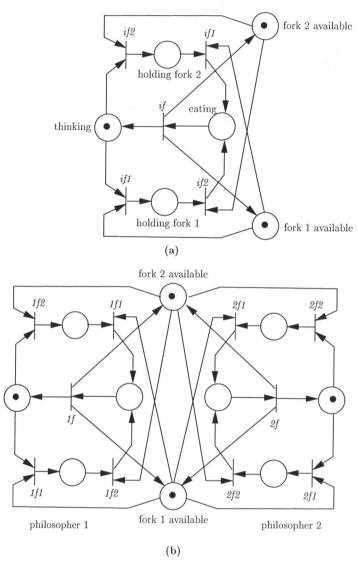

Figure 4.9: Petri net model of the dining philosophers example, with two philosophers and two forks. (Example 4.6).
Part **(a)** *shows the model of one philosopher interacting with two forks. Part* **(b)** *shows the complete model for two philosophers and two forks.*

Decidability

Another issue in comparing automata and Petri nets is that of *decidability*. Suppose we pose a question such as: Is this event sequence recognized by this finite-state automaton? We say that the problem is "decidable" if there is a specific ("algorithmic") procedure we can apply to come up with a "yes" or "no" answer for it. An attractive feature of finite-state automata is the fact that all such questions are indeed decidable (precisely because the state space is finite). Unfortunately, this is not always true in dealing with Petri nets, reflecting a natural tradeoff between decidability and model-richness.

Overall, it is probably most helpful to think of Petri nets and automata as complementary modeling approaches, rather than competing ones. As already pointed out, it is often the specific application that suggests which approach might be better suited.

4.4   ANALYSIS OF PETRI NETS

We begin by categorizing some of the most common problems related to the analysis of (untimed) Petri net models in Section 4.4.1. These problems are essentially related to the issues of safety and blocking that we considered in the case of automaton models in Section 2.5.1. However, the structural information contained in Petri net models is often used to pose more specific versions of these problems, such as boundedness, conservation, coverability, and persistence presented below. In Sections 4.4.2 and 4.4.3 we will present the technique of the *coverability tree* to answer some the problems posed in Section 4.4.1.

4.4.1   Problem Classification

The following are some of the key issues related to the logical behavior of Petri nets. These issues relate primarily to desirable properties that often have their direct analogs in CVDS. Many of these properties are motivated by the fact that Petri net models are often used in resource sharing environments where we would like to ensure efficient and fair usage of the resources.

Boundedness

In many instances, tokens represent customers in a resource sharing system. For example, the tokens in place Q of the Petri net in Fig. 4.5 represent customers entering a queue. Clearly, allowing queues to grow to infinity is undesirable, since it means that customers wait forever to access a server. In classical system theory, a state variable that is allowed to grow to infinity is generally an indicator of instability in the system. Similarly here, unbounded growth in state components (markings) leads to some form of instability.

Boundedness refers to the property of a place to maintain a number of tokens that never exceeds a given positive integer.

Definition. (Boundedness)
Place $p_i \in P$ in Petri net N with initial state $\mathbf{x}_0$ is said to be *k-bounded*, or *k-safe*, if $x(p_i) \leq k$ for all states $\mathbf{x} \in R(N)$, that is, for all reachable states. ◆

If a place is 1-bounded, it is simply called *safe*. If we can find some integer k, not necessarily prespecified, such that a place is k-bounded, then the place is said to be *bounded*. If all places in a Petri net are bounded, then the net is called bounded. The Petri net in Fig. 4.5 is not bounded since the number of tokens in place Q can become arbitrarily large.

Given a DES modeled as a Petri net, a boundedness problem consists of checking if the net is bounded and determining a bound. If boundedness is not satisfied, then our task may be to alter the model so as to ensure boundedness. If the Petri net is bounded, then as we mentioned previously it can be transformed into an equivalent finite-state automaton, allowing the application of the analysis techniques for finite-state automata, if so desired.

Safety and Blocking

These are the same issues that were posed in Section 2.5.1 in the context of automaton models. They can of course be posed for Petri net models as well. These issues can be posed in terms of states or in terms of languages, as in the case of automata. We could be concerned with the reachability of certain states or of certain substrings; we could ask whether there are states where the net deadlocks or whether the net is blocking from a language viewpoint; we could ask whether the behavior of a net is "contained" in the behavior of another net, either from the viewpoint of $R(N)$ or from the viewpoint of $\mathcal{L}(N)$; and so forth. If the Petri net of interest is bounded, then safety and blocking properties can be determined algorithmically, as we shall see in Section 4.4.3. (In essence, we build the equivalent finite-state automaton and answer the desired questions on this equivalent model.)

State Coverability

The concept of *state coverability* is a generalization of the concept of state reachability. It is also related to the concept of eventually being able to fire a particular transition. In order to enable a transition, it is often required that a certain number of tokens be present in some places. Consider, for instance, some state $\mathbf{y} = [y(p_1), y(p_2), \ldots, y(p_n)]$ which includes in each place the minimum number of tokens required to enable some transition t_j. Suppose we are currently in state $\mathbf{x}_0$, and we would like to see if t_j could eventually be enabled. It is, therefore, essential to know whether we can reach a state $\mathbf{x}$ from the current state $\mathbf{x}_0$ such that $x(p_i) \geq y(p_i)$ for all $i = 1, \ldots, n$. If this is the case, we say that state $\mathbf{x}$ *covers* state $\mathbf{y}$.

Definition. (State coverability)
Given Petri net N with initial state $\mathbf{x}_0$, state $\mathbf{y}$ is said to be *coverable* if there exists $\mathbf{x} \in R(N)$ such that $x(p_i) \geq y(p_i)$ for all $i = 1, \ldots, n$.   ◆

Conservation

Sometimes tokens represent resources, rather than customers, in a resource sharing system. For example, the token in place I of the Petri net in Fig. 4.5 represents the lone server of a simple queueing system. This token is passed on to place B and subsequently returned to I in order to denote the changing state of the server (idle or busy). However, it can never be lost nor can the number of tokens in I or B increase, since there is only one server in our system.

Conservation is a property of Petri nets to maintain a fixed number of tokens for all states reached in a sample path. However, this may be too constraining a property. For instance, consider the Petri net model of the dining philosophers in Fig. 4.9. Initially, there are two "fork" tokens. When a philosopher is eating, these two tokens are replaced by a single token representing an eating philosopher. When the philosopher is done eating, the two fork tokens reappear. Thus, tokens representing resources are not exactly conserved all the time.

We therefore introduce a weighting vector $\gamma = [\gamma_1, \gamma_2, \ldots, \gamma_n]$, with $\gamma_i \geq 0$ for all $i = 1, \ldots, n$, corresponding to places $p_1, p_2, \ldots, p_n$, and define conservation relative to these *weights* (not to be confused with the arc weights of the Petri net). We will also take a weight γ_i to always be an integer number; this is not absolutely necessary, and we do so for simplicity only.

Definition. (Conservation)
Petri net N with initial state $\mathbf{x}_0$ is said to be *conservative with respect to* $\gamma = [\gamma_1, \gamma_2, \ldots, \gamma_n]$ if

$$\sum_{i=1}^{n} \gamma_i x(p_i) = \text{constant} \tag{4.6}$$

for all states $\mathbf{x} \in R(N)$.   ◆

Given a Petri net model of a DES, we are often required to ensure conservation with respect to certain weights representing the fact that resources are not lost or gained. More generally, the loss or gain of tokens must always reflect physical "conservation" properties of the DES we are modeling.

Liveness

A complement to the properties of deadlock and blocking is the notion of *live* transitions. Here, instead of being concerned with being unable to fire any transition or unable to eventually reach a marked state, we are concerned with being able to *eventually* fire a given transition.

Definition. (Liveness)

Petri net N with initial state x_0 is said to be *live* if there always exists some sample path such that any transition can eventually fire from any state reached from x_0. ◆

Clearly, this is a very stringent condition on the behavior of the system. Moreover, checking for liveness as defined above is an extremely tedious process, often practically infeasible for many systems of interest. This has motivated a further classification of liveness into four levels. Thus, given an initial state x_0, a *transition* in a Petri net may be:

- *Dead* or *L0-live*, if the transition can never fire from this state;

- *L1-live*, if there is some firing sequence from x_0 such that the transition can fire at least once;

- *L2-live*, if the transition can fire at least k times for some given positive integer k;

- *L3-live*, if there exists some infinite firing sequence in which the transition appears infinitely often;

- *Live* or *L4-live*, if the transition is L1-live for every possible state reached from x_0.

The concept of coverability is closely related to that of L1-liveness. If y is the state that includes in each place the minimum number of tokens required to enable some transition t_j, then if y is not coverable from the current state, transition t_j is dead. Thus, it is possible to identify dead transitions by checking for coverability.

Example 4.7 (Liveness)

We illustrate some of the different liveness levels through the Petri net of Fig. 4.10. Here, transition t_2 is dead. This is because the only way it can fire is if a token is present in both p_1 and p_2, which can never happen (if t_1 fires, p_2 receives a token, but p_1 loses its token). Transition t_1 is L1-live, since it can fire, but only once. In fact, if t_1 fires, then all transitions become dead in the new state. Finally, transition t_3 is L3-live, since it is possible for it to fire infinitely often. It is not, however, L4-live, since it can become dead in the state resulting from t_1 firing.

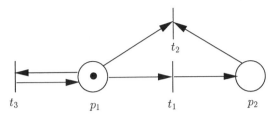

Figure 4.10: Petri net for Example 4.7.
Transition t_2 is dead, because it can never fire. Transition t_1 is L1-live, since it can fire once. Transition t_3 is L3-live, because it can fire an infinite number of times, but it is not L4-live, since it becomes dead if t_1 fires.

Persistence

It is sometimes the case that two different transitions are enabled by the same set of conditions. If one fires, does the other one remain enabled? In general, there is no guarantee that this is so. Actually, the two transitions do not have to depend on the exact same conditions, but only have one condition in common. Persistence is the property of a Petri net not to disable any enabled transition because of the firing of another enabled transition.

Definition. (Persistence)
A Petri net is said to be *persistent* if, for any two enabled transitions, the firing of one cannot disable the other. ◆

An example of a Petri net that is not persistent is the net of Fig. 4.10. While both t_1 and t_3 are enabled in the state shown, if t_1 occurs, then t_3 is no longer enabled. Similarly, the Petri net of Fig. 4.9 is not persistent. On the other hand, the Petri net model of a simple queueing system (Fig. 4.5) can easily be seen to be persistent.

It turns out that persistence is equivalent to a property referred to as *non-interruptedness* in the study of timed DES models. To better understand the choice of the term "non-interruptedness", think of each enabled transition as having to go through some process before it can actually fire, hence incurring some delay. If the firing of a transition disables another enabled transition, it effectively "interrupts" this process. In many ways the absence of the persistence property introduces technical complications which we can roughly compare to nonlinear phenomena in the study of CVDS. We will have the chance to return to this point in Chapter 11 and further elaborate on non-interruptedness.

4.4.2 The Coverability Tree

We shall limit ourselves to one analysis technique, the *coverability tree*, which may be used to solve some, but not all, of the problems described above. This technique is based on the construction of a tree where nodes are Petri net states and arcs represent transitions. The key idea is quite simple and is best illustrated through some examples.

Example 4.8 (Simple reachability tree)

Consider the Petri net of Fig. 4.11, with initial state $[1,1,0]$. Also shown in the same figure is a tree whose root node is $[1,1,0]$. We then examine all transitions that can fire from this state, define new nodes in the tree, and repeat until all possible reachable states are identified. In this simple case, the only transition that can initially fire is t_1, and the next state is $[0,0,1]$. Now t_2 can fire, and the next state is $[1,1,0]$. Since this state already exists (it is the root node), we stop there. In essence, we have constructed an equivalent automaton to the Petri net, although we duplicated state $[1,1,0]$ and stopped there; in the equivalent finite-state automaton, the arc labeled t_2 out of state $[0,0,1]$ goes back to the root node.

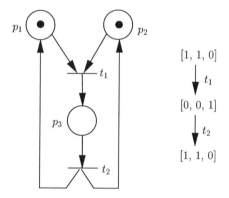

Figure 4.11: Reachability tree for Example 4.8.
The construction of the reachability tree stops at state (node) $[1,1,0]$, since this state is a duplicate of the initial state of the tree.

Example 4.9 (Infinite reachability tree)

Consider the Petri net of Fig. 4.12, with initial state $[1,0,0,0]$. The reachability tree starts out with this state, from which only transition t_1 can fire. As shown in Fig. 4.12, the next state is $[0,1,1,0]$. Now either t_2 or t_3 can fire, so we create two branches, with corresponding next states $[1,0,1,0]$ and $[0,0,1,1]$, respectively. Note that no transition can fire at state $[0,0,1,1]$, that is, this state is a deadlock or *terminal* state. (We will use the term "terminal" instead of "deadlock" since the former is more frequently used in the literature when discussing the coverability tree.) The left branch allows transition t_1 to fire, resulting in new state

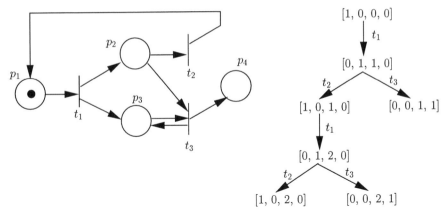

Figure 4.12: Part of the reachability tree for Example 4.9.
The construction of the tree stops at states that are terminal (deadlock) states or that dominate states encountered along the same branch from the root of the tree.

$[0, 1, 2, 0]$. Once again, either t_2 or t_3 can fire, and the corresponding next states are $[1, 0, 2, 0]$ and $[1, 0, 2, 1]$. This is sufficient to detect the pattern of this tree. Every right branch eventually leads to a terminal state, but with one additional token in place p_3. Every left branch repeats itself, but with one extra token in place p_3. In other words, we see that place p_3 is unbounded, meaning that the reachability tree is infinite, since the set of reachable states is infinite. However, we do not build this infinite tree. Rather, upon observing that state $[1, 0, 2, 0]$ is larger, component-wise, than state $[1, 0, 1, 0]$ that precedes it in the path from the root to $[1, 0, 2, 0]$, we stop at state $[1, 0, 2, 0]$. We say that state $[1, 0, 2, 0]$ *dominates* state $[1, 0, 1, 0]$.

As Examples 4.8 and 4.9 suggest, the reachability tree is conceptually easy to construct. It may, however, be infinite. Our task, therefore, is to seek a finite representation of this tree. This is possible, but at the expense of losing some information. The finite version of an infinite reachability tree will be called a *coverability tree*.

We will present an algorithm for constructing a finite coverability tree. To do so, we first introduce some notation:

1. *Root node.* This is the first node of the tree, corresponding to the initial state of the given Petri net. For example, $[1, 0, 0, 0]$ is the root node in the tree of Fig. 4.12.

2. *Terminal node.* This is any node from which no transition can fire. For example, in the tree of Fig. 4.12, $[0, 0, 1, 1]$ is a terminal node.

3. *Duplicate node.* This is a node that is identical to a node already in the tree. For example, in the tree of Fig. 4.11, node $[1, 1, 0]$ resulting from the sequence $t_1 t_2$ is a duplicate node of the root node.

4. *Node dominance.* Let $\mathbf{x} = [x(p_1), \dots, x(p_n)]$ and $\mathbf{y} = [y(p_1), \dots, y(p_n)]$ be two states, i.e., nodes in the coverability tree. We will say that "$\mathbf{x}$ *dominates* $\mathbf{y}$," denoted by $\mathbf{x} >_d \mathbf{y}$, if the following two conditions hold:

   (a) $x(p_i) \geq y(p_i)$, for all $i = 1, \dots, n$

   (b) $x(p_i) > y(p_i)$, for at least some $i = 1, \dots, n$.

   For example, in the tree of Fig. 4.12 we have $[1, 0, 2, 0] >_d [1, 0, 1, 0]$. But $[1, 0, 2, 0]$ does not dominate $[0, 1, 1, 0]$. Note that condition (a) above is the definition of coverability for states $\mathbf{x}$, $\mathbf{y}$; however, dominance requires the additional condition (b). Thus, dominance corresponds to "strict" covering.

5. *The symbol ω.* This may be thought of as "infinity" in representing the marking (state component) of an unbounded place. We use ω when we identify a node dominance relationship in the coverability tree. In particular, if $\mathbf{x} >_d \mathbf{y}$, then for all i such that $x(p_i) > y(p_i)$ we replace the value of $x(p_i)$ by ω. Note that adding tokens to a place which is already marked by ω, does not have any effect, that is, $\omega + k = \omega$ for any $k = 0, 1, 2, \dots$. As an example, in Fig. 4.12 we have $[1, 0, 1, 0] >_d [1, 0, 0, 0]$. We can then replace $[1, 0, 1, 0]$ by $[1, 0, \omega, 0]$.

Coverability Tree Construction Algorithm

Step 1: Initialize $\mathbf{x} = \mathbf{x}_0$ (initial state).

Step 2: For each new node, $\mathbf{x}$, evaluate the transition function $f(\mathbf{x}, t_j)$ for all $t_j \in T$:

   Step 2.1: If $f(\mathbf{x}, t_j)$ is undefined for all $t_j \in T$ (i.e., no transition is enabled at state x), then $\mathbf{x}$ is a terminal node.

   Step 2.2: If $f(\mathbf{x}, t_j)$ is defined for some $t_j \in T$, create a new node $\mathbf{x}' = f(\mathbf{x}, t_j)$. Then:

   Step 2.2.1: If $x(p_i) = \omega$ for some p_i, set $x'(pi) = \omega$.

   Step 2.2.2: If there exists a node $\mathbf{y}$ in the path from the root node $\mathbf{x}_0$ (included) to $\mathbf{x}$ such that $\mathbf{x}' >_d \mathbf{y}$, set $x'(p_i) = \omega$ for all p_i such that $x'(p_i) > y(p_i)$.

   Step 2.2.3: Otherwise, set $x'(p_i) = f(\mathbf{x}, t_j)$.

Step 3: If all new nodes are either terminal or duplicate nodes, stop.

It can be shown that the coverability tree constructed by this algorithm is indeed finite. The proof (which is beyond the scope of this book) is based on a straightforward contradiction argument making use of some elementary properties of trees and of sequences of non-negative integers.

Example 4.10 (Coverability tree of infinite reachability tree)

Consider once again the Petri net of Fig. 4.12, with initial state $x_0 = [1, 0, 0, 0]$. We construct its coverability tree by formally following the steps of the algorithm above, as shown in Fig. 4.13.

The first transition fired results in $x' = f(x_0, t_1) = [0, 1, 1, 0]$. If transition t_3 fires next, then we obtain the terminal node $[0, 0, 1, 1]$. If transition t_2 fires, we obtain the new node $[1, 0, 1, 0]$. Going through Step 2.2.2 of the algorithm, we observe that $[1, 0, 1, 0] >_d [1, 0, 0, 0]$, and therefore we replace $[1, 0, 1, 0]$ by $[1, 0, \omega, 0]$. When transition t_1 fires next, by Step 2.2.1, the symbol ω remains unchanged. The new node, $[0, 1, \omega, 0]$, dominates node $[0, 1, 1, 0]$ in the path from the root to $[1, 0, \omega, 0]$. However, the symbol ω has already been used in place p_3. We continue with transition t_3, which leads to the terminal node $[0, 0, \omega, 1]$, and with t_2, which leads to the duplicate node $[1, 0, \omega, 0]$. This completes the construction of the coverability tree.

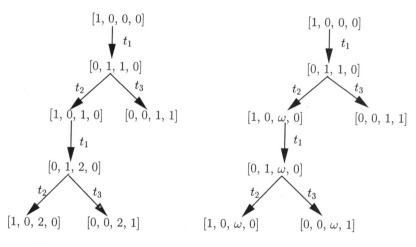

Figure 4.13: Coverability tree construction for Example 4.10.

Example 4.11 (Coverability tree of a simple queueing system)

Let us construct the coverability tree of the Petri net model of the simple queueing system shown in Fig. 4.5 and redrawn in Fig. 4.14. Let us assume that we start at the initial state $x_0 = [0, 1, 0]$ (no customer in system), where the elements of x_0 correspond to places Q, I, and B,

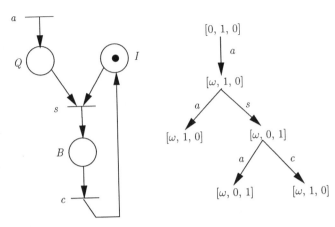

Figure 4.14: Coverability tree construction for a simple queueing system (Example 4.11).

respectively. Transition a fires and the next state is $[1, 1, 0]$. Observing that $[1, 1, 0] >_d [0, 1, 0]$, we replace $[1, 1, 0]$ by $[\omega, 1, 0]$ (see Fig. 4.14). Two transitions are now enabled. If a fires, the next state is the duplicate node $[\omega, 1, 0]$. If s fires, the next state is $[\omega, 0, 1]$, which is neither a duplicate nor a terminal node. If transition a fires next, the new state is the duplicate node $[\omega, 0, 1]$. If c fires, the new state is the duplicate node $[\omega, 1, 0]$, and the coverability tree construction is complete.

4.4.3 Applications of the Coverability Tree

In Section 4.4.1, we presented several problems related to the analysis of Petri net models of untimed DES. Let us now see how some of these problems can be solved by using the coverability tree. In the process, we will also identify the limitations of this approach.

Boundedness, Safety, and Blocking Problems

The problem of boundedness is easily solved using a coverability tree. A necessary and sufficient condition for a Petri net to be bounded is that the symbol ω never appears in its coverability tree. Since ω represents an infinite number of tokens in some place, if ω appears in place p_i, then p_i is unbounded. For example, in Fig. 4.14, place Q is unbounded; this is to be expected, since there is no limit to the number of customers that may reside in the queue at any time instant.

If ω never appears in the coverability tree, then we are guaranteed that the state space of the DES we are modeling is finite. The coverability tree is then the reachability tree, since it contains all the reachable states. Analysis of such systems becomes much easier to handle, since we can examine all possible states and transitions between them. For instance, to answer safety questions concerning states (including blocking questions), it suffices to examine the finite reachability tree to determine if any of the illegal states is reachable. For safety questions that are posed in language terms (i.e., illegal substrings) one would want to indicate the labels of the transitions in the reachability tree in order to answer these questions; since the finite reachability tree is essentially a finite-state automaton, the discussion in Section 2.5.1 applies here as well.

Finally, note that if ω does not appear in place p_i, then the largest value of $x(p_i)$ for any state encountered in the tree specifies a bound for the number of tokens in p_i. For example, $x(I) \leq 1$ in Fig. 4.14. Thus, place I is 1-bounded (or safe). If the coverability (reachability) tree of a Petri net contains states with 0 and 1 as the only place markings, then all places are guaranteed to be safe, and the Petri net is safe.

Coverability Problems

By simple inspection of the coverability tree, it is always possible to determine whether some state $\mathbf{y}$ of a given Petri net with initial state $\mathbf{x}_0$ is coverable or not. We construct the coverability tree starting with $\mathbf{x}_0$. If we can find a node $\mathbf{x}$ in the tree such that $x(p_i) \geq y(p_i)$ for all $i = 1, \dots, n$, then $\mathbf{y}$ is coverable. The path from the root $\mathbf{x}_0$ to $\mathbf{x}$ describes the firing sequence that leads to the covering state $\mathbf{x}$. However, if $\mathbf{x}$ contains ω in one or more places, we should realize that this path must include a loop. Depending on the particular values of $y(p_i)$, $i = 1, \dots, n$, that need to be covered, we can then determine the number of loops involved in this path until $\mathbf{y}$ is covered.

As an example, state $[3, 1, 0]$ in the Petri net of Fig. 4.14 is coverable, since $[\omega, 1, 0]$ is reachable from the root node $[0, 1, 0]$; in fact, $[\omega, 1, 0]$ appears in three separate positions in the tree. Upon closer inspection, we can see that there are several firing sequences that could lead to this state, such as $\{a, a, a\}$ or $\{a, s, a, c, s, a, c, a, a\}$.

Conservation Problems

This is also a problem that can be solved using the coverability tree. Recall that conservation with respect to a weighting vector $\gamma = [\gamma_1, \gamma_2, \dots, \gamma_n]$, with γ_i a non-negative integer, is defined by condition 4.6, which must be checked for all reachable states. First, we need to observe that if $x(p_i) = \omega$ for some p_i, then we must have $\gamma_i = 0$ if the Petri net is to be conservative. This is because ω does not represent a fixed integer value, but rather an infinite set of values. Suppose we consider two states $\mathbf{x}$ and $\mathbf{y}$ such that $x(p_i) = y(p_i)$ for all p_i, and $x(p_k) = y(p_k) = \omega$ for some p_k. The *actual* values of $x(p_k)$ and $y(p_k)$ may be different in general, since ω simply denotes infinitely many integer values (cf.

Example 4.12 hereafter). As a result, if $\omega_k > 0$, we cannot satisfy

$$\sum_{i=1}^{n} \gamma_i x(p_i) = \sum_{i=1}^{n} \gamma_i y(p_i) \; .$$

With this observation in mind, if a vector γ is specified such that $\gamma_i = 0$ for all i with $x(p_i) = \omega$ somewhere in the coverability tree, then we can check for conservation by evaluating the weighted sum above for each state in the coverability tree. Conservation requires that this sum be fixed.

Conversely, we can pose the following problem. Given a Petri net, is there a weighting vector γ such that the conservation condition 4.6 is satisfied? To solve this problem, we first set $\gamma_i = 0$ for all unbounded places p_i. Next, suppose there are $b \leq n$ bounded places, and let the coverability tree consist of r nodes. Finally, let the value of the constant sum in condition 4.6 be C. We then have a set of r equations of the form

$$\sum_{i=1}^{b} \gamma_i x(p_i) = C \qquad \text{for each state } \mathbf{x} \text{ in the coverability tree}$$

with $(b+1)$ unknowns, namely, the b strictly positive weights for the bounded places and the constant C. Thus, this conservation problem reduces to a standard algebraic problem for solving a set of linear equations. It is not necessary for a solution to exist, but if one or more solutions do exist, then it is possible to determine them by standard techniques.

As an example, consider the coverability tree of Fig. 4.14, where we set $\gamma_1 = 0$ for unbounded place Q. Then we are left with $r = 6$ equations of the form

$$\gamma_2 x(p_2) + \gamma_3 x(p_3) = C$$

and the three unknowns: γ_2, γ_3, and C. It is easy to see that setting $\gamma_2 = \gamma_3 = 1$ and $C = 1$ yields a conservative Petri net with respect to $\gamma = [0, 1, 1]$. This merely reflects the fact that the token representing the server in this queueing system is always conserved.

Coverability Tree Limitations

Safety properties are in general not solvable using the coverability tree for unbounded Petri nets. The reason is that the symbol ω represents a set of reachable states, but it does not explicitly specify them. In other words, ω is used to aggregate a set of states so that a convenient *finite* representation of the *infinite* set of reachable states can be obtained; however, in the process of obtaining this finite representation, some information is lost. This "loss of information" is illustrated by the following example.

Example 4.12 (Loss of information in coverability tree)
This example illustrates the loss of information regarding specific reachable states when a coverability tree is constructed for an unbounded Petri net. In Fig. 4.15 (a) we show a simple Petri net and its coverability tree. Observe that the first time transition t_2 fires the new state is $[1, 0, 1]$. The next time transition t_2 fires the new state becomes $[1, 0, 2]$, and so on. Thus, the reachable markings for place p_3 are $x(p_3) = 1, 2, \ldots$. All those are aggregated under the symbol ω.

In Fig. 4.15 (b), we show a similar - but different - Petri net along with its coverability tree. Note that in the Petri net of Fig. 4.15 (b), the first time transition t_2 fires the new state is $[1, 0, 2]$. The next time transition t_2 fires the new state becomes $[1, 0, 4]$, and so on. The reachable markings for p_3 in this case are $x(p_3) = 2, 4, \ldots$, and not all positive integers as in the Petri net of Fig. 4.15 (a). Despite this difference, the coverability trees in Figs. 4.15 (a) and (b) are identical. The reason is the inherent vagueness of ω.

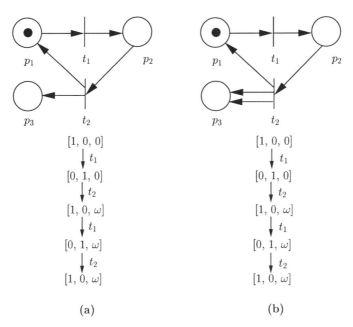

(a)   (b)

Figure 4.15: Two unbounded Petri nets and their coverability trees (Example 4.12). In (a), the symbol ω for place p_3 represents the set $\{1, 2, \ldots\}$, while in (b), ω for the same place represents the set $\{2, 4, \ldots\}$.

As Example 4.12 illustrates, state reachability issues cannot be dealt with, in general, using a coverability tree approach. These limitations of the coverability tree also affect the analysis of blocking properties. (Further discussion on blocking and trimming of blocking Petri nets can be found in Section 4.5.1.)

It is sometimes possible that the coverability tree does contain sufficient information to solve certain problems. It is often possible, for instance, to determine that some state is not reachable, while being unable to check if some other state is reachable. We can also limit general Petri nets to certain classes which are easier to analyze, yet still rich enough to model several interesting types of DES we encounter in practice. The class of bounded Petri nets is an example. Several other interesting classes have been identified in the literature, based on the structural properties of the Petri net graph. One such class is the class of *marked graphs*, which are Petri nets where all arc weights are 1 and where each place has a single input arc and a single output arc (thus the place has only one input transition and only one output transition). Marked graphs are also called *event graphs* or *decision-free Petri nets*. We will consider this important class in the next chapter, when we are faced with analysis problems related to event timing; this will lead us to the (max,+) algebra as an analytical tool for certain timed DES. Several analytical results are also available for untimed marked graphs. We shall not cover these results in this book; the interested reader may consult the references mentioned at the end of this chapter.

4.4.4 Linear-Algebraic Techniques

We conclude our discussion on the analysis of Petri nets by mentioning that the state equation 4.5 that we derived in Section 4.2.3 can be used to address problems such as state reachability and conservation. It provides an algebraic alternative to the graphical methodology based on the coverability tree, and it can be quite powerful in identifying structural properties that are mostly dependent on the topology of the Petri net graph captured in the incidence matrix $\mathbf{A}$.

With regard to reachability for instance, by looking at equation 4.5, we can see that a necessary condition for state $\mathbf{x}$ to be reachable from initial state $\mathbf{x}_0$ is for the equation

$$\mathbf{vA} = \mathbf{x} - \mathbf{x}_0$$

to have a solution $\mathbf{v}$ where all the entries of $\mathbf{v}$ are non-negative integers. Such a vector $\mathbf{v}$ is called a *firing count vector*, since its components specify the number of times that each transition fires in "attempting" to reach state $\mathbf{x}$ from $\mathbf{x}_0$. However, this necessary condition is not in general sufficient, as the existence of a non-negative integer solution $\mathbf{v}$ does not guarantee that the entries in the firing count vector $\mathbf{v}$ can be mapped to an actual feasible ordering of individual transition firings.[1] Consider the following example.

---

[1] This explains the use of the word "attempting" in the preceding sentence.

Example 4.13

The Petri net in Fig. 4.16 has incidence matrix

$$A = \begin{bmatrix} -1 & 1 & -1 & 0 \\ 0 & -1 & 1 & 1 \\ 1 & 0 & 0 & -1 \end{bmatrix}.$$

and initial state $x_0 = [1, 0, 0, 0]$. Consider state $x = [0, 0, 0, 1]$. We can see that the system of equations

$$vA = x - x_0 = [-1, 0, 0, 1]$$

has a solution $v = [1, 1, 0]$. While the entries of v are indeed non-negative integers, neither t_1 nor t_2 are enabled in state x_0. Thus none of the two possible orderings of transition firings consistent with v, $t_1 t_2$ and $t_2 t_1$, are feasible from x_0.

On the "positive" side, if we are interested in the reachability of $x' = [0, 1, 0, 0]$ from x_0, then the fact that the system of equations

$$vA = x' - x_0 = [-1, 1, 0, 0]$$

does not have a solution allows us to conclude that x' is not reachable from x_0.

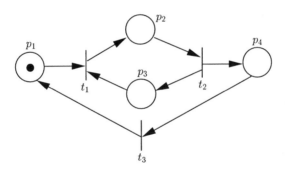

Figure 4.16: Petri net for Example 4.13.

Examination of the state equation 4.5 can also yield some information regarding conservation properties. Let A be the incidence matrix of a Petri net. If there exists a vector $\gamma = [\gamma_1, \ldots, \gamma_n]$, whose entries satisfy $\gamma_i \geq 0$ for $i = 1, \ldots, n$, and such that

$$A\gamma^T = 0$$

(where T denotes transposed), then we can write

$$\begin{aligned} x &= x_0 + vA \\ x\gamma^T &= x_0\gamma^T + vA\gamma^T \\ x\gamma^T &= x_0\gamma^T . \end{aligned}$$

Since this last equality holds for all states $\mathbf{x}$ that are reachable from $\mathbf{x}_0$, this means that the Petri net is conservative with respect to this γ, *for any choice* of $\mathbf{x}_0$.

Many more results are available in the literature regarding structural analysis of Petri nets and subclasses of Petri nets, such as marked graphs, using linear-algebraic techniques. We shall not, however, go into further details on this approach in this book.

4.5 ON THE CONTROL OF PETRI NETS

There is large body of literature on the control of Petri nets. We shall limit ourselves to a discussion of this topic. Since in Chapter 3 we presented a comprehensive control theory for DES, known as "supervisory control theory," the first question that one may ask is: Can Petri nets be used instead of automata as the modeling formalism in the context of supervisory control theory? The key system-theoretic properties (among them, controllability, nonconflicting, observability, co-observability) studied in Chapter 3 are language properties and therefore they are independent of the particular modeling formalism used to represent languages. The algorithms presented in Chapter 3 for verification of these properties and for computation of supervisors assume that the languages of interest (uncontrolled system language and admissible language) are regular and are based on finite-state automaton representations of these languages. As was seen in that chapter, the class of regular languages possesses nice properties in the context of supervisory control; in particular, this class is closed under the supervisor synthesis operations of interest: $\uparrow C$, $\downarrow C$, $\downarrow O$, $\uparrow N$, and so forth. Moreover, we saw that automaton models were appropriate for the implementation of these synthesis operations. These closure properties as well as the fact that finitely-convergent algorithms exist for the principal supervisor synthesis operations are results of great practical importance.

Clearly, it is also of interest to consider supervisory control problems where Petri nets are used to model the uncontrolled system language, the admissible language, and/or to realize the supervisor. That is, using the notation of Chapter 3, G could be given as a (labeled) Petri net, L_{am} could be represented by a (labeled) Petri net, and/or the supervisor S may be realized by a Petri net N (as opposed to being realized by an automaton R). An important motivation for using Petri net models is to extend the applicability of the results of supervisory control theory beyond the class of regular languages. Unfortunately, it turns out that several difficulties arise in this endeavor. For this reason, no comprehensive set of results has been developed yet in this regard. We will highlight some of these difficulties in Section 4.5.1 below.

Of course, if $\mathcal{L}_m(G)$ and L_{am} are both regular languages, then it is not necessary to use Petri net models, even if these languages are given in terms of their Petri net representations. These Petri nets will necessarily be bounded and thus we can always build equivalent finite-state automata and then apply the algorithms of Chapter 3 to solve the supervisory control problems. However, it may still be useful to work with the Petri net representations for computational complexity reasons, since these Petri nets may be more compact than the corresponding finite-state automata (because of the system structure that they capture).

In the same vein, the structural information captured by Petri net models has motivated the development of several other approaches to the control of Petri nets. Unlike the language-based approach of supervisory control theory, these results are state-based; Section 4.5.2 concludes this chapter with a very brief discussion on this topic.

4.5.1   Petri Nets and Supervisory Control Theory

It is instructive to begin this section with some comments on infinite-state systems and/or specifications.[2]

Infinite-State Systems or Specifications

It should be noted that if $\mathcal{L}_m(G)$ is not regular (say, G is an unbounded Petri net) but L_{am} is *regular and controllable* (with respect to $\mathcal{L}(G)$ and E_{uc}), then the non-regularity of $\mathcal{L}_m(G)$ is of no consequence since the desired supervisor S can be realized by a finite-state automaton (as was done in Chapter 3, Section 3.4.2). The "catch" here is the verification of the controllability properties of L_{am}, which in general requires manipulating G. In this regard, it should be observed that if L_{am} is controllable with respect to a *superset* of $\mathcal{L}(G)$, then L_{am} will necessarily be controllable with respect to $\mathcal{L}(G)$. Thus if one conjectures that L_{am} is controllable, then one can attempt to formally verify that conjecture by proving the controllability of L_{am} with respect to a *regular* superlanguage of $\mathcal{L}(G)$.

On the other hand, if $\mathcal{L}_m(G)$ is regular but L_{am} is *controllable but not regular*, then S cannot be realized by a finite-state automaton. (The regularity of $\mathcal{L}_m(G)$ is of no help.) But one may be able to realize S by a Petri net. Consider the following example.

Example 4.14 (Petri net realization of supervisor)

Consider the system G where $\mathcal{L}(G) = a^*b^*$ together with the prefix-closed admissible language L_a of the form

$$L_a = \overline{\{a^n b^m : n \geq m \geq 0\}} \ .$$

Note that while $\mathcal{L}(G)$ is regular, L_a is not regular. The set of uncontrollable events is $E_{uc} = \{a\}$.

---

[2]This section assumes that the reader is familiar with the material presented in Chapter 3, principally Sections 3.2, 3.4, and 3.5.

Since event b is controllable, it is clear that L_a is controllable. Therefore, there exists S such that $\mathcal{L}(S/G) = L_a$, but since L_a is not regular, S cannot be realized by a finite-state automaton. So instead, we wish to realize S with a Petri net. We claim that the labeled Petri net N in Fig. 4.17, shown in its initial state $\mathbf{x}_0$, can be used for this purpose.

Let f denote the state transition function of N. Then S can be realized as follows; for $s \in \mathcal{L}(G)$ and $\mathbf{x} = f(\mathbf{x}_0, s)$, define

$$S(s) = \begin{cases} \{a,b\} & \text{if } x(p_3) > 0 \\ \{a\} & \text{if } x(p_3) = 0 \ . \end{cases}$$

Consequently, we do have a realization of S that requires finite memory, namely storing N and the above rule for $S(s)$. It suffices to update the state of N in response to events executed by G (by firing the currently enabled transition with same event label), and read the contents of place p_3 to determine if event b should be enabled or not. In other words, event b is enabled by S whenever a transition labeled by b is enabled in N. Note that $\mathcal{L}(N) = L_a$. This is consistent with the standard realization of supervisors presented in Section 3.4.2 in Chapter 3, except that a Petri net is used instead of a finite-state automaton. Note that while two transitions of N share event label b, only one of these two is enabled at any given time; consequently, there is no ambiguity regarding which transition to fire in N if G executes event b.

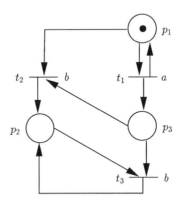

Figure 4.17: Petri net N of Example 4.14.

Some Difficulties that Arise

We claim that the extension of the applicability of the supervisor synthesis results of supervisory control theory to non-regular languages by the use of

Petri net representations of languages leads to several difficulties that are not easily circumvented; we substantiate this claim by highlighting some of these difficulties.[3]

First, as was seen in Chapter 3, many supervisor synthesis procedures require performing the trim of a (blocking) automaton. The trim operation will most certainly be invoked in any supervisory control problem where blocking is an issue and where it is required that the controlled system be nonblocking. We saw in Chapter 2 that the trim operation is straightforward to implement on automata. Suppose that we have a Petri net N such that $\mathcal{L}(N) \supset \overline{\mathcal{L}_m(N)}$ and we wish to "transform" N into N' such that $\mathcal{L}_m(N') = \mathcal{L}_m(N)$ and $\mathcal{L}(N') = \overline{\mathcal{L}_m(N')}$. One notices immediately that trimming a blocking Petri net is by no means straightforward as the state of the Petri net is a "distributed object" (number of tokens in places) and thus we cannot simply "delete" blocking states. In fact, the structure of N, namely the set of places and transitions, may have to be modified in order to obtain the desired N'.

It turns out that a fundamental difficulty may arise in trimming blocking Petri nets. Consider the Petri net graph in Fig. 4.18, and take $\mathbf{x}_0 = [1, 0, 0, 0]$ and $\mathbf{X}_m = \{[0, 0, 0, 1]\}$; call the resulting net N. It can be verified that the languages generated and marked by N are:

$$\mathcal{L}(N) = \overline{\{a^m b a^n b : m \geq n \geq 0\}}$$
$$\mathcal{L}_m(N) = \{a^m b a^m b : m \geq 0\} \ .$$

N is blocking since $\mathcal{L}(N) \supset \mathcal{L}_m(N)$; for example, N deadlocks after string $aabab$ since the resulting state, $[0, 1, 0, 1]$, is a deadlock state and it is not in $\mathbf{X}_m$.

In order to trim N, we need to construct N' such that

$$\mathcal{L}(N') = \overline{\mathcal{L}_m(N')} = \overline{\mathcal{L}_m(N)} \quad \text{and} \quad \mathcal{L}_m(N') = \mathcal{L}_m(N) \ .$$

The problem that arises here is that $\overline{\mathcal{L}_m(N)} \notin \mathcal{PNL}$, that is, it is not a Petri net language! Thus the desired N' *does not exist*. A formal proof of this result is beyond the scope of this book. Intuitively, in order to trim N, we essentially require that place p_2 be empty before transition t_4 can be allowed to fire. We also note that place p_2 is unbounded. It can be shown that it is not possible to transform N in a manner that enforces the condition "p_2 empty before t_4 fires" without using an infinite number of places, an outcome that defeats the purpose of using a Petri net model![4]

The second difficulty that we wish to mention concerns the use of Petri nets as realizations of supervisors. Given the semantics of the feedback loop

---

[3]This section is included for readers interested in the connection between the results of Chapter 3 and Petri net models. Its material is more specialized than that in the rest of this chapter and it may be skipped without loss of continuity.

[4]An alternative is to use an *inhibitor arc* – see the references listed at the end of the chapter – but this is undesirable as Petri nets with inhibitor arcs are not tractable analytically.

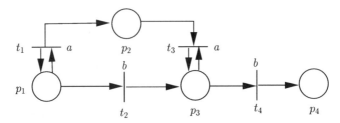

Figure 4.18: Petri net graph that is blocking with $\mathbf{x}_0 = [1, 0, 0, 0]$ and $\mathbf{X}_m = \{[0, 0, 0, 1]\}$.

of supervisory control, it is necessary that the Petri net realization of a supervisor be "deterministic" in the sense that there is never any ambiguity about which transition to fire in the Petri net realization in response to the execution of an event by the system. In Chapter 3, we built deterministic automaton realizations of supervisors, and consequently this issue did not arise. But the transition labeling function of Petri nets need not be injective (cf. Section 4.2.4) and therefore this issue needs to be addressed. In the Petri net in Fig. 4.17 considered in Example 4.14 for instance, two transitions have the event label b. If both of these transitions were enabled and the system executed event b, then which transition should the Petri net supervisor fire? This issue did not arise in the controlled system considered in Example 4.14, but it may arise in other examples.

This leads us to consider the subclass of Petri nets that possess the "deterministic property" of never having two distinct transitions with the same event label simultaneously enabled in any of their reachable states. Let us call Petri nets that possess this property *deterministic Petri nets* and let us denote by $\mathcal{PNL}^D$ the class of languages that can be represented by deterministic Petri nets. Clearly, $\mathcal{PNL}^D \subset \mathcal{PNL}$. In view of the difficulty about trimming first mentioned in this section and of the technique of standard realization presented in Chapter 3, an appropriate subclass of $\mathcal{PNL}$ appears to be

$$\mathcal{PNL}^{DP} := \{K \in \mathcal{PNL} : \overline{K} \in \mathcal{PNL}^D\} .$$

That is, we are interested in those languages K whose prefix-closure can be generated by a deterministic Petri net; note that we say nothing about K being a deterministic Petri net language or not. If a given language $K \in \mathcal{PNL}^{DP}$ is controllable and $\mathcal{L}_m(G)$-closed, then the results of Chapter 3 and the above definition of the class $\mathcal{PNL}^{DP}$ imply that there exists a nonblocking supervisor S such that $\mathcal{L}_m(S/N) = K$ and moreover S *can be realized by a deterministic Petri net*, which was our original objective.

A key difficulty that arises in using the above result for solving supervisory control problems such as BSCP and BSCP-NB for admissible languages that are not regular is: How do we know if the supremal controllable sublanguage of the admissible language, namely $L_a^{\uparrow C}$ for BSCP and $L_{am}^{\uparrow C}$ for BSCP-NB, will be a member of the class $\mathcal{PNL}^{DP}$? To complicate matters further, it has been shown that the class $\mathcal{PNL}^{DP}$ is not closed under the $\uparrow C$ operation, that is, $K^{\uparrow C}$ need not be in $\mathcal{PNL}^{DP}$ even if $K \in \mathcal{PNL}^{DP}$.

We conclude from the discussion in this section that while Petri nets are a more powerful discrete event modeling formalism than finite-state automata, they do not appear to be a good candidate formalism for a general extension of the *algorithmic* component of supervisory control theory to *non-regular* languages.

4.5.2   State-Based Control of Petri Nets

The negative flavor of the results discussed in the preceding section by no means implies that there is no control theory for DES modeled by Petri nets. In fact, there is a large body of results on the control of Petri nets when the control paradigm is changed from the "language viewpoint" of supervisory control theory to a "state viewpoint" for a given (uncontrolled) system modeled by a Petri net. By this we mean that the specifications on the uncontrolled system are stated in terms of *forbidden states* and the control mechanism is to decide which of the controllable and enabled transitions in the system should be allowed to fire by the "controller". In the same spirit as the notion of supremal sublanguages in supervisory control, the controller should guarantee that none of the forbidden states is reached, and should do so with "minimal intervention" on the behavior of the system. The synthesis of such a controller should exploit the system structure captured by the Petri net model.

Coverage of this important theory is beyond the scope of this book; we refer the reader to the references at the end of this chapter. We shall limit ourselves to presenting a simple example, where a Petri net with undesirable behavior can be modified by the addition of places and arcs, resulting in a "controlled Petri net" with satisfactory behavior. This example highlights the fact that exploiting the structural properties of the system captured by the Petri net model is key to successfully controlling the system modeled by the Petri net.

Example 4.15

Consider the Petri net N shown in Fig. 4.19; assume that N models the uncontrolled system behavior. Suppose that the specification on the states is given in terms of the linear inequality

$$x(p_2) + x(p_3) \leq 2$$

where a state is deemed admissible if and only if it satisfies this inequality. It has been shown in the literature that when the admissible behavior is described by such linear inequalities, the linear-algebraic approach for describing the behavior of a Petri net can be used to synthesize an "augmented" Petri net, with additional places and arcs, that will satisfy the given linear inequality. In this simple example, the modified net is N_{aug} in Fig. 4.19; the additional place p_c is connected to transitions t_4, t_1, and t_5. Place p_c can be thought of as a control place realizing a given control policy for N; given the structure of "solution" N_{aug}, it is implicitly assumed in this solution that transitions t_1 and t_5 are controllable. The initial number of tokens in new place p_c is determined by "augmenting" the linear inequality to a linear equality with p_c as the new variable (often termed "slack variable"):

$$x(p_2) + x(p_3) + x(p_c) = 2$$

leading to $x_0(p_c) = 2$. Examination of the behavior of N_{aug} shows that the original linear inequality holds for all reachable states.

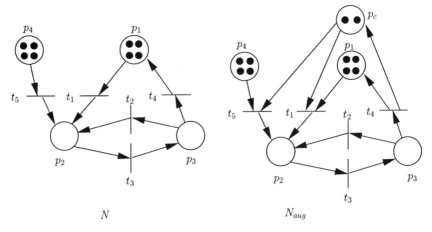

Figure 4.19: Petri net N for uncontrolled system and augmented Petri net N_{aug} where the constraint $x(p_2) + x(p_3) \leq 2$ is enforced (Example 4.15).

SUMMARY

- Petri nets are another modeling formalism for DES. A Petri net is defined by a set of places P, a set of transitions T, a set of arcs A, and a weight function w applied to arcs. Transitions correspond to events, and the places input to a transition are associated with conditions required for an event to occur.

- The dynamic behavior of a Petri net is described by using tokens in places to enable transitions, which then cause tokens to move around. The state of a Petri net is defined by a vector $[x(p_1), x(p_2), \ldots, x(p_n)]$ where $x(p_i)$

is the number of tokens present in place p_i. When transition t_j fires, the state components (markings) of its input and output places change according to $x'(p_i) = x(p_i) - w(p_i, t_j) + w(t_j, p_i)$.

- It is always possible to obtain a Petri net from a finite-state automaton, but the converse need not be true if the set of reachable states of the Petri net is infinite, which occurs when the number of tokens in a place can become arbitrarily large. This means that Petri nets can represent a larger class of DES than finite-state automata, or in other words the class of languages that can be marked by Petri nets is strictly larger than the class $\mathcal{R}$. The drawback is that analyzing Petri nets with unbounded places is much more challenging; in fact, many questions are undecidable.

- The coverability tree can be used to solve some, but not all, of the basic problems related to the behavior of Petri nets, such as boundedness, coverability, and conservation. For Petri nets with finite state spaces, the coverability tree becomes the reachability tree and is essentially a finite-state automaton that is equivalent to the original Petri net.

- The extension of the scope of supervisory control theory to non-regular languages by the use of Petri net models leads to several technical difficulties. Control paradigms that exploit directly the system structure captured in the Petri net model have proven more appropriate.

PROBLEMS

4.1 Consider the Petri net defined by:

$$P = \{p_1, p_2, p_3\} \qquad T = \{t_1, t_2, t_3\}$$
$$A = \{(p_1, t_1), (p_1, t_3), (p_2, t_1), (p_2, t_2), (p_3, t_3), (t_1, p_2), (t_1, p_3), (t_2, p_3),$$
$$(t_3, p_1), (t_3, p_2)\}$$

with all arc weights equal to 1 except for $w(p_1, t_1) = 2$.

(a) Draw the corresponding Petri net graph.

(b) Let $\mathbf{x}_0 = [1, 0, 1]$ be the initial state. Show that in any subsequent operation of the Petri net, transition t_1 can never be enabled.

(c) Let $\mathbf{x}_0 = [2, 1, 1]$ be another initial state. Show that in any subsequent operation of the Petri net, either a deadlock occurs (no transition can be enabled) or a return to $\mathbf{x}_0$ results.

4.2 Consider the state transition diagram in part (b) of Problem 2.14 (Fig. 2.31 (b)). Draw the corresponding Petri net graph. Then:

(a) Show one possible Petri net state that corresponds to one of the six states.

(b) Are there any states such that no transition is enabled?

4.3 For the system described in Problem 2.36, draw a Petri net graph and make sure to describe what each place and transition represents physically, keeping in mind the operating rules specified. For simplicity, assume that machine M_1 never breaks down. (There may be several Petri net models you can come up with, but the simplest one should consist of eight places.) Construct the coverability tree for the Petri net model you obtain.

4.4 Construct a Petri net model of the DES described in Problem 2.13. Show a state where a timeout may occur while task 1 is being processed.

4.5 Consider the Petri net shown in Fig. 4.20, with initial state $x_0 = [1, 1, 0, 2]$.

(a) After the Petri net fires twice, find a state where all transitions are dead.

(b) Suppose we want to apply the firing sequence $(t_3, t_1, t_3, t_1, ...)$. Show that this is not possible for all future times.

(c) Find the state x_s resulting from the firing sequence $(t_1, t_2, t_3, t_3, t_3)$.

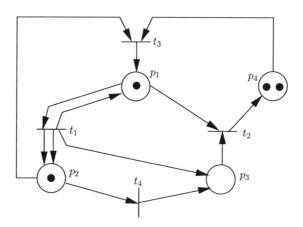

Figure 4.20: Problem 4.5.

4.6 Consider the Petri net shown in Fig. 4.21, with initial state $x_0 = [1, 1, 0, 0]$. Construct its coverability tree and use it to show that the empty state $[0, 0, 0, 0]$ is not reachable. Is this Petri net bounded?

4.7 Petri nets are often very useful in modeling the flow of data as they are processed for computational purposes. Transitions may represent operations (e.g. addition, multiplication), and places are used to represent the

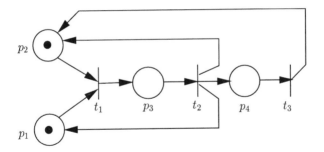

Figure 4.21: Problem 4.6.

storage of data (e.g. placing a token in a place called $X + Y$ represents the result of an operation where two numbers X and Y are added). Construct a Petri net graph to model the following computation:

$$(X + Y)(X - Z) + \frac{XY}{X - Y}$$

the result of which is denoted by R. (Make sure your model includes checks to prevent illegal operations from taking place.)

4.8 A Petri net in which there are no closed loops (directed circuits) formed is called *acyclic*. For this class of Petri nets, a necessary and sufficient condition for state $\mathbf{x}$ to be reachable from initial state $\mathbf{x_0}$ is that there exists a non-negative integer solution $\mathbf{z}$ to the equation:

$$\mathbf{x} = \mathbf{x_0} + \mathbf{zA}$$

where $\mathbf{A}$ is the incidence matrix (defined in Section 4.2.3).

(a) Use this result to show that $\mathbf{x} = [0, 0, 1, 2]$ is reachable from $\mathbf{x_0} = [3, 1, 0, 0]$ for the Petri net:

$$P = \{p_1, p_2, p_3, p_4\} \qquad T = \{t_1, t_2\}$$
$$A = \{(p_1, t_1), (p_2, t_2), (p_3, t_2), (t_1, p_2), (t_1, p_3), (t_2, p_4)\}$$

with all arc weights equal to 1 except for $w(p_2, t_2) = 2$.

(b) What is the interpretation of the vector $\mathbf{z}$?

(c) Prove the sufficiency of the result above (i.e., the existence of a non-negative integer solution $\mathbf{z}$ implies reachability).

4.9 Construct a labeled Petri net N such that

$$\mathcal{L}_m(N) = \{a^n b^n c^n, n \geq 0\} \ .$$

4.10 Consider the Petri net graph in Fig. 4.22.

(a) Determine if this Petri net is conservative for any choice of $\mathbf{x}_0$.

(b) Find an initial state such that this Petri net is live.

(c) Use the state equation of the Petri net to verify that state $\mathbf{x} = [0, 1, 1, 1, 1]$ is reachable from initial state $\mathbf{x}_0 = [2, 0, 0, 0, 0]$.

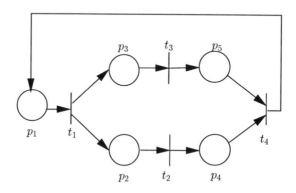

Figure 4.22: Problem 4.10.

4.11 Prove or disprove:

(a) The class of Petri net languages $\mathcal{PNL}^{DP}$ is strictly larger than the class of regular languages $\mathcal{R}$.

(b) The class of marked graph languages (i.e., languages recognized by marked graphs) is incomparable with $\mathcal{R}$.

4.12 Consider the Petri net N in Fig. 4.23 where arc weight $r > 1$. Take $\mathbf{X}_m = \{[0, 0, 1]\}$. Verify that N is blocking. Find Petri net N' such that $N' = Trim(N)$. (*Hint:* Add $r - 1$ transitions between places p_1 and p_2.)

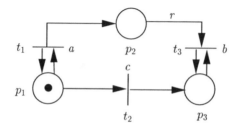

Figure 4.23: Problem 4.12.

4.13 Consider the DES G where $\mathcal{L}(G) = \overline{a^*ba^*}$ and $\mathcal{L}_m(G) = a^*ba^*$. Let $E_{uc} = \{b\}$. Take the admissible language to be

$$L_{am} = \{a^m ba^n : m \geq n \geq 0\} .$$

(a) Is $\overline{L_{am}}$ a regular language?

(b) Verify that there exists a nonblocking supervisor S such that

$$\mathcal{L}_m(S/G) = L_{am} .$$

(c) Build a Petri net realization of S.

4.14 In many DES modeled by Petri nets, places often have a maximum capacity in terms of the number of tokens that they can hold (e.g., the place may model a finite buffer). Imposing a maximum capacity on a place is a change in the Petri net dynamics described in Section 4.2.3, since an enabled transition cannot fire when its firing would lead to a violation of the capacity of one of its output places. Show how a finite-capacity place can be modeled without changing the operational rules of a Petri net (i.e., as described in Section 4.2.3). (*Hint:* Add a "complementary place" to the finite-capacity place and connect this new place appropriately with the existing transitions.)

4.15 Consider the Petri net N depicted in Fig. 4.24. Modify this Petri net in order to get a "controlled net" N_{aug} that satisfies the constraint $x(p_2) \leq x(p_1)$ for all reachable states from the initial state.

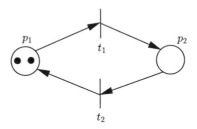

Figure 4.24: Problem 4.15.

4.16 Consider the Petri net N in Fig. 4.25. Note that all arcs have weight 1 except the arcs from p_3 to t_2 and from t_4 to p_3, which have weight k.

(a) Is N bounded?

(b) Is N live?

(c) Is N conservative?

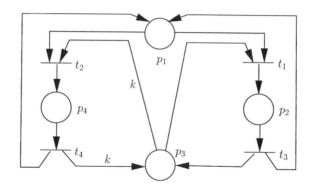

Figure 4.25: Problem 4.16.

SELECTED REFERENCES

- *Introductory books or survey papers on Petri nets*

    - David, R., and H. Alla, *Petri Nets & Grafcet: Tools for Modelling Discrete Event Systems*, Prentice-Hall, New York, 1992.

    - Murata, T., "Petri Nets: Properties, Analysis and Applications," *Proceedings of the IEEE*, Vol. 77, No. 4, pp. 541-580, 1989.

    - Peterson, J.L., *Petri Net Theory and the Modeling of Systems*, Prentice Hall, Englewood Cliffs, 1981.

- *Control of Petri nets*

    - Holloway, L.E., B.H. Krogh, and A. Giua, "A Survey of Petri Net Methods for Controlled Discrete Event Systems," *Journal of Discrete Event Dynamic Systems: Theory and Applications*, Vol. 7, No. 2, pp. 151–190, April 1997.

    - Moody, J.O., and P.J. Antsaklis, *Supervisory Control of Discrete Event Systems Using Petri Nets*, Kluwer Academic Publishers, Boston, 1998.

- *Miscellaneous*

    - Desrochers, A.A., and R.Y. Al-Jaar, *Applications of Petri Nets in Automated Manufacturing Systems: Modeling, Control, and Performance Analysis*, IEEE Press, Piscataway, NJ, 1995.

    - Giua, A., and F. DiCesare, "Blocking and Controllability of Petri Nets in Supervisory Control," *IEEE Transactions on Automatic Control*, Vol. 39, No. 4, pp. 818-823, April 1994.

- Sreenivas, R.S., and B.H. Krogh, "On Petri Net Models of Infinite State Supervisors," *IEEE Transactions on Automatic Control*, Vol. 37, No. 2, pp. 274–277, February 1992.

- Zhou, M.C., and F. DiCesare, *Petri Net Synthesis for Discrete Event Control of Manufacturing Systems*, Kluwer Academic Publishers, Boston, 1993.

■ *Notes*

1. The above survey paper by Murata contains a wealth of results and an extensive bibliography on Petri net models and their analysis (in particular, the linear-algebraic approach).

2. Readers interested in more details on Petri net languages and decidability issues (cf. discussion in Section 4.3) may consult the above book by Peterson.

3. The above survey paper by Holloway, Krogh, and Giua contains an insightful discussion and a detailed bibliography on state-based and language-based control methods for Petri nets.

4. The discussion and examples in Section 4.5.1 are primarily based on the work of Giua & DiCesare and Sreenivas & Krogh listed above.

5. The above book by Moody & Antsaklis covers in detail the state-based approach for control of Petri nets mentioned in Example 4.15.

6. The Web site `www.daimi.au.dk/PetriNets/` ("World of Petri nets") is a very useful source of information on Petri nets; in particular, this site contains bibliographical data on Petri nets and information about software tools for analyzing and simulating Petri net models.

Chapter 5
Timed Models

5.1 INTRODUCTION

From this point on we will concentrate on timed models of DES. This means that the sample paths we consider can no longer be specified as event sequences $\{e_1, e_2, \dots\}$ or state sequences $\{x_0, x_1, \dots\}$, but they must include some form of timing information. For example, let t_k, $k = 1, 2, \dots$, denote the time instant when the kth state transition occurs (with t_0 given); then a timed sample path of a DES may be described by the sequence $\{(x_0, t_0), (x_1, t_1), \dots\}$. Creating a framework for timed DES models will enable us to address questions such as "how many events of a particular type can occur in a given time interval?", or "how long does the system spend in a given state?" These issues are of critical importance in analyzing the behavior of DES such as computer systems and communication networks, since they provide us with particularly useful measures of system performance.

The first objective of this chapter is to describe timing mechanisms for DES. Using these mechanisms, we will find that we can directly extend automaton models, which were introduced in previous chapters, to create a modeling framework for DES that includes event timing. Establishing the framework of timed automata, on which most of the analytical techniques in subsequent chapters will be based, is our second objective. Finally, some alternative models will also be presented. In particular, the Petri net framework introduced in Chapter 4 can be readily extended to model DES as *timed Petri nets*. Another approach we will briefly cover uses a special kind of algebra, referred to as a *dioid algebra*, and attempts to parallel some of the analytical techniques used for linear CVDS. Finally, as in the case of untimed DES models, we should point out that there are several other possible modeling approaches. These, however, go beyond our purposes; we include a brief discussion in Section 5.5 and relevant references at the end of the chapter.

Throughout this chapter, we will assume that all input information required for the models we develop is available. This means that we know in advance exactly when certain events will occur in the future. Of course, it is unrealistic to expect this to be the case. A computer system user does not announce the jobs he will submit and their times of submission for the next week. Nor does a common person plan days in advance when he is going to place various telephone calls. In short, we will ultimately have to resort to probabilistic models in order to carry out some analysis of practical value; this is going to be the goal of the next chapter. We limit ourselves here to an input description which is fully specified in advance, so as to gain an understanding of the basic event timing dynamics of a DES, independent of the probabilistic characterization of its input.

5.2 TIMED AUTOMATA

We begin with an automaton model (X, E, f, Γ, x_0), where

X	is a countable *state space*
E	is a countable *event set*
$f : X \times E \to X$	is a *state transition function* and is generally a *partial* function on its domain
$\Gamma : X \to 2^E$	is the *active event function* (or feasible event function); $\Gamma(x)$ is the set of all events e for which $f(x, e)$ is defined and it is called the *active event set* (or feasible event set)
x_0	is the *initial* state

This is the same as the automaton defined in Chapter 2 with only a few minor changes. First, we allow for generally countable (as opposed to finite) sets X and E. We also leave out of the definition any consideration for marked states, since we will not be considering blocking issues.

5.2.1 The Clock Structure

In order to introduce the key ideas of the timing mechanism we need, we start out by discussing the simplest possible DES we can think of. We then gradually proceed with more complicated cases.

A DES with a single event. Let $E = \{\alpha\}$, and $\Gamma(x) = \{\alpha\}$ for all $x \in X$. A sample path of this simple system on the time line is shown in Fig. 5.1. The event sequence associated with this sample path is denoted by $\{e_1, e_2, \ldots, \}$, where $e_k = \alpha$ for all $k = 1, 2, \ldots$ The time instant associated with the kth occurrence of the event is denoted by t_k, $k = 1, 2, \ldots$ The length of the time interval defined by two successive occurrences of the event is called a *lifetime*. Thus, we define

$$v_k = t_k - t_{k-1} \qquad k = 1, 2, \ldots \tag{5.1}$$

to be the kth lifetime of the event. Clearly, this is a nonnegative real number, i.e., $v_k \in \mathbb{R}^+$.

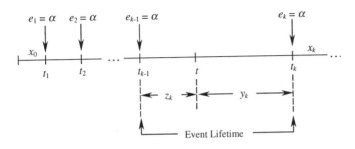

Figure 5.1: Sample path of a DES with $E = \{\alpha\}$.

The evolution of this system over time can be described as follows. At time t_{k-1}, the kth event is said to be *activated* or *enabled*, and is given a lifetime v_k. A clock associated with the event is immediately set at the value specified by v_k, and then it starts ticking down to 0. During this time interval, the kth event is said to be *active*. The clock reaches zero when the lifetime expires at time $t_k = t_{k-1} + v_k$. At this point, the event has to occur. This will cause a state transition. The process then repeats with the $(k+1)$th event becoming active.

Observe the difference between the event being "active" and the event "occurring". The event α is active as long as it is feasible in the current state, that is, $\alpha \in \Gamma(x)$. The event actually occurs when its clock runs down to 0, and a state transition takes place. This is similar to the distinction between "enabling" a transition and "firing" it in Petri nets, as discussed in the last chapter. In Fig. 5.1, the event α is in fact continuously active, but it only occurs at the time instants $t_1, t_2, \ldots$ Every time it occurs, it is immediately activated anew, since it is always feasible in this example.

To introduce some further notation, let t be any time instant, not necessarily associated with an event occurrence. Suppose $t_{k-1} \le t \le t_k$. Then t divides the interval $[t_{k-1}, t_k]$ into two parts (see Fig. 5.1) such that

$$y_k = t_k - t \tag{5.2}$$

is called the *clock* or *residual lifetime* of the kth event, and

$$z_k = t - t_{k-1} \tag{5.3}$$

is called the *age* of the kth event. It is immediately obvious that

$$v_k = z_k + y_k \tag{5.4}$$

It should be clear that a sample path of this DES is completely specified by the lifetime sequence $\{v_1, v_2, \ldots\}$. This is also referred to as the *clock sequence* of event α.

A DES with two permanently active events. Things become more interesting if we consider a DES with $E = \{\alpha, \beta\}$. For simplicity, we first assume that $\Gamma(x) = \{\alpha, \beta\}$ for all $x \in X$, so that both events are always active. Suppose that a clock sequence for each event is specified, that is, there are two known sequences of lifetimes:

$$\mathbf{v}_\alpha = \{v_{\alpha,1}, v_{\alpha,2}, \ldots\}, \quad \mathbf{v}_\beta = \{v_{\beta,1}, v_{\beta,2}, \ldots\}$$

Starting with a given time t_0, the first question we address is: *Which event occurs next?* It is reasonable to answer this question by comparing $v_{\alpha,1}$ to $v_{\beta,1}$ and selecting the event with the shortest lifetime. Thus, if

$$v_{\alpha,1} < v_{\beta,1}$$

α is the first event to occur at time $t_1 = t_0 + v_{\alpha,1}$, as illustrated in Fig. 5.2.

We are now at time t_1, and we may pose the same question once again: *Which event occurs next?* Since β is still active, it has a clock value given by

$$y_{\beta,1} = v_{\beta,1} - v_{\alpha,1}$$

On the other hand, α is also active. Since α just took place, its next occurrence is defined by the new lifetime $v_{\alpha,2}$, taken out of the given clock sequence. We now compare $v_{\alpha,2}$ to $y_{\beta,1}$ and select the smaller of the two. Thus, if

$$y_{\beta,1} < v_{\alpha,2}$$

β is the second event to occur at time $t_2 = t_1 + y_{\beta,1}$, as illustrated in Fig. 5.2.

The mechanism for selecting the "next event" at any time is therefore based on *comparing clock values and selecting the smallest one*. If an event has just occurred, then its clock is actually reset to a new lifetime value supplied by its clock sequence.

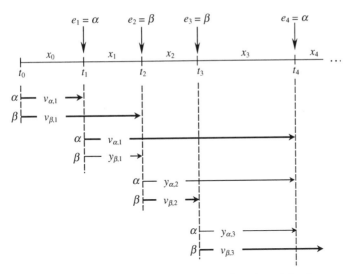

Figure 5.2: Sample path of a DES with $E = \{\alpha, \beta\}$ and both events permanently active.

The next event to occur is always the one with the smallest clock value. When an event occurs, it is immediately activated again, and its clock is reset to the next available lifetime value in its assigned clock sequence.

A DES with two not permanently active events. Our modeling framework would not be interesting if no events could ever be made inactive. Suppose that in the previous DES, $\Gamma(x) = \{\alpha, \beta\}$ for some $x \in X$, and $\Gamma(x) = \{\beta\}$ for all remaining $x \in X$. For a given initial state x_0, let $\Gamma(x_0) = \{\alpha, \beta\}$. Thus, as before, the first event to occur is α (see Fig. 5.3). Next, let us assume that the new state, x_1, is such that $\Gamma(x_1) = \{\beta\}$. In this case, the only remaining feasible event is β, and no comparison with any other clock is required. Next, when β occurs at time t_2, it causes a state transition into x_2. Suppose both α and β are feasible once again. In this case, both events are activated. We compare their clock values (the new lifetimes assigned to them at this point), and find that $v_{\beta,2} < v_{\alpha,2}$, so event β occurs next (see Fig. 5.3). Now suppose that the new state, x_3, is such that $\Gamma(x_3) = \{\beta\}$. Since α is no longer feasible, it is *deactivated* or *disabled*. Its clock becomes irrelevant in the determination of the next event, and is therefore ignored. Finally, at time t_4, β occurs and causes a transition to a new state x_4. Assuming $\Gamma(x_4) = \{\alpha, \beta\}$, both events are once again activated. Note that a completely new lifetime, $v_{\alpha,3}$, is selected for event α, since $v_{\alpha,2}$ was discarded when α was deactivated at t_3.

Therefore, in general, the mechanism for selecting the "next event" at any time is based on three simple rules:

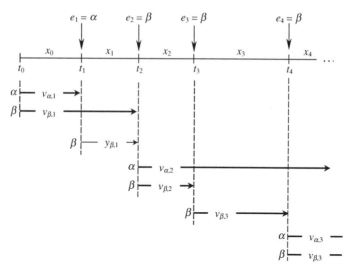

Figure 5.3: Sample path of a DES with $E = \{\alpha, \beta\}$ and α **not** feasible in some states.

In this example, α is not feasible in states x_1 and x_3. At t_1, α is not activated again after it occurs. At t_3, α is explicitly deactivated when β occurs. The clock value of α at this point is discarded. When it is activated again at t_4, a new lifetime is assigned to the clock of α.

Rule 1. To determine the next event, compare clock values of all events feasible at the current state and select the smallest one.

Rule 2. An event e is activated when

- e has just occurred and it remains feasible in the new state (e.g., event β at time t_2 in Fig. 5.3), or

- a different event occurs while e was not feasible, causing a transition to a new state where e is feasible (e.g., event α when β occurs at time t_2 in Fig. 5.3).

Rule 3. An event e is deactivated when a different event occurs causing a transition to a new state where e is not feasible (e.g., event α when β occurs at time t_3 in Fig. 5.3).

Note that activating an event is equivalent to setting its clock to a new lifetime value selected from the corresponding assigned clock sequence. If an event is deactivated, its clock value at the time is discarded. Finally, it is important to observe that determination of the next event requires (a) knowledge of the current state x_k, so that $\Gamma(x_k)$ can be evaluated, and (b) knowledge of all current clock values $y_{i,k}$, with $i \in \Gamma(x_k)$.

This discussion, which will be formalized in the next section, provides the fundamental principles for the event timing mechanism we will use. The new feature which we have introduced here, and which serves to define a timed model of a DES, is a *clock structure*:

Definition. The *Clock Structure* (or *Timing Structure*) associated with an event set E is a set

$$\mathbf{V} = \{\mathbf{v}_i \ : \ i \in E\}$$

of clock (or lifetime) sequences

$$\mathbf{v}_i = \{v_{i,1}, v_{i,2}, \dots\}, \quad i \in E, \quad v_{i,k} \in \mathbb{R}^+, \quad k = 1, 2, \dots \qquad \blacklozenge$$

The clock structure will be viewed as the input to a DES. The timing mechanism then translates this information into an actual event sequence that drives the underlying state automaton. We will assume that $\mathbf{V}$ is completely specified in advance. In some cases, a lifetime $v_{i,k}$ may actually have to be transformed depending on the current state before it is assigned to the clock of event i, but this only adds technical complications which do not affect the principles of these models.

Another useful concept is that of an event score.

Definition. The *score*, $N_{i,k}$, of an event $i \in E$ after the kth state transition on a given sample path is the number of times that i has been activated in the interval $[t_0, t_k]$. $\qquad \blacklozenge$

For example, in Fig. 5.2 the scores are initialized to $N_{\alpha,0} = N_{\beta,0} = 1$, since both events are activated at time t_0. At time t_1 we have $N_{\alpha,1} = 2$ and $N_{\beta,1} = 1$. At time t_2 we have $N_{\alpha,2} = N_{\beta,1} = 2$. Note that with every event occurrence we must check what new activations take place, and increase the corresponding event scores by 1. The score of event i serves as a pointer to its clock sequence $\mathbf{v}_i$, which specifies the next lifetime to be assigned to its clock when i is activated.

5.2.2   Event Timing Dynamics

Consider an automaton (X, E, f, Γ, x_0). For simplicity, we will assume that the event set E is finite and consists of m elements. We have seen that the system dynamics are captured by the state transition equation

$$x_{k+1} = f(x_k, e_{k+1}), \quad k = 0, 1, \dots \qquad (5.5)$$

where x_k is the current state and x_{k+1} is the next state obtained when event e_{k+1} occurs. In this framework, a given sequence of events $\{e_1, e_2, \dots, e_{k+1}, \dots\}$ constitutes the input to the system. Recalling the precise definition of "state"

(Section 1.4 of Chapter 1), we can see that knowledge of this input event sequence along with x_k are sufficient to predict the entire future state evolution through (5.5).

Let us now equip the automaton (X, E, f, Γ, x_0) with a clock structure $\mathbf{V} = \{\mathbf{v}_i : i = 1, \ldots, m\}$. This clock structure is to be viewed as the input to the automaton. The event sequence $\{e_1, e_2, \ldots, e_{k+1}, \ldots\}$ is no longer given. In fact, it has to be determined based on some timing mechanism which we will explicitly define. In other words, we now seek a relationship of the form

$$e_{k+1} = h(x_k, \mathbf{v}_1, \ldots, \mathbf{v}_m)$$

so that we can replace (5.5) by a state equation

$$x_{k+1} = f(x_k, \mathbf{v}_1, \ldots, \mathbf{v}_m)$$

This would imply that x_k is still the only state variable required to describe the dynamic behavior of a DES. This is too much to expect. The discussion of the previous section actually suggests that determining the next event, e_{k+1}, involves a mechanism of clock comparisons and updates, as well as tracking of event scores. Thus, our objective becomes to specify the next event through an internal timing mechanism captured by a relationship of the form

$$e_{k+1} = h(x_k, \mathbf{v}_1, \ldots, \mathbf{v}_m, \cdot) \tag{5.6}$$

where "·" represents as yet undetermined additional state variables. Once this relationship is fully specified, (5.5) and (5.6) together will describe the dynamics of our system.

We will now present in detail the event timing mechanism informally described in the previous section. To avoid excessive use of subscripts, we will omit the event counting index k, and adopt the following notation:

x is the current state
e is the most recent event (causing transition into state x)
t is the most recent event time (corresponding to event e)

In addition, we define two more variables associated with each event $i \in E$:

N_i is the current *score* of event i, $N_i \in \{0, 1, \ldots\}$
y_i is the current *clock* value of event i, $y_i \in \mathbb{R}^+$

The definitions of N_i and y_i were informally given earlier, but they will be made mathematically precise in what follows.

The "next" state, event, and event time are denoted using the prime (′) notation. Thus,

e' is the next event, also called the *triggering event*; clearly, $e' \in \Gamma(x)$
t' is the next event time (corresponding to event e')
x' is the next state, given by $x' = f(x, e')$

Similarly, we set:

N_i' is the next score of event i (after event e' occurs)
y_i' is the next clock value of event i (after event e' occurs)

Remark. It should be clear that in this notation x corresponds to x_k, and x' to x_{k+1}, $k = 0, 1, \ldots$, and similarly for all other variables defined previously.

Given x, our first task is to determine the triggering event e'. We proceed as follows:

Step 1. Since x is known, we can evaluate the feasible event set $\Gamma(x)$.

Step 2. Associated with each event $i \in \Gamma(x)$ is a clock value y_i. We can then determine the smallest clock value among those, denoted by y^*:

$$y^* = \min_{i \in \Gamma(x)} \{y_i\} \qquad (5.7)$$

Step 3. Determine the triggering event, e', as the value of i in (5.7) that defines y^*. We express this as

$$e' = \arg \min_{i \in \Gamma(x)} \{y_i\} \qquad (5.8)$$

Step 4. With e' defined by (5.8), determine the next state:

$$x' = f(x, e') \qquad (5.9)$$

Step 5. With y^* defined by (5.7), determine the next event time:

$$t' = t + y^* \qquad (5.10)$$

This process then repeats with x', e', and t' specified. Step 2, however, requires the new clock values y_i'. Therefore, at least one more step is needed:

Step 6. Determine the new clock values for all new feasible events $i \in \Gamma(x')$:

$$y_i' = \begin{cases} y_i - y^* & \text{if } i \neq e' \text{ and } i \in \Gamma(x) \\ v_{i,N_i+1} & \text{if } i = e' \text{ or } i \notin \Gamma(x) \end{cases} \qquad i \in \Gamma(x') \qquad (5.11)$$

To explain the two cases in (5.11), observe the following:
(a) The first case deals with all events that remain active in the next state x' except the triggering event. Their clock values are reduced by y^* (the amount of time elapsed since event e occurred).
(b) The second case deals with the triggering event e' and all events which were inactive in state x, but are activated in the new state x' (this is precisely Rule 2

of the previous section). All these events require a new lifetime, which is the next available in their assigned clock sequence. For event i, the first N_i lifetimes have already been used, so the next available one is the $(N_i + 1)$th element in the sequence $\mathbf{v}_i$.

(c) Note that y_i' is undefined for all $i \notin \Gamma(x')$. Thus, if i was active in x and is deactivated when e' occurs, its clock value is discarded (recall Rule 3 of the previous section).

When the process repeats, Step 6 requires new score values N_i'. Therefore, one last step is needed:

Step 7. Determine the new scores for all new feasible events $i \in \Gamma(x')$:

$$N_i' = \begin{cases} N_i + 1 & \text{if } i = e' \text{ or } i \notin \Gamma(x) \\ N_i & \text{otherwise} \end{cases} \quad i \in \Gamma(x') \qquad (5.12)$$

Here, the score of event i is incremented by 1 whenever the event is activated. This happens when $i \in \Gamma(x')$ and i happens to be the triggering event, or it was inactive and is activated upon entering the new state x'. This condition is identical to the second case of (5.11).

The smallest event clock value, y^*, determined through (5.7), also specifies the length of the time interval defined by two successive events. Thus, we shall refer to y^* as the *interevent* time.

Remark. It is conceivable that $y_i = y_j$ for two distinct events $i, j \in \Gamma(x)$. To resolve this conflict, we have to impose some priority rules over the event set E, that is, if two event occurrences are simultaneous, which of the two events should be considered as affecting the state first? Unless otherwise specified, we assume that events are ordered in E by priority, that is, event i has higher priority than $j > i$. In practice, when working in continuous time, this problem does not normally arise, unless there are explicit control rules that force two or more events to occur simultaneously.

We can now give a formal definition of a timed automaton.

Definition. A *Timed Automaton* is a six-tuple

$$(X, E, f, \Gamma, x_0, \mathbf{V})$$

where $\mathbf{V} = \{\mathbf{v}_i : i \in E\}$ is a clock structure, and (X, E, f, Γ, x_0) is an automaton. The automaton generates a state sequence $x' = f(x, e')$ driven by an event sequence $\{e_1, e_2, \ldots\}$ generated through

$$e' = \arg \min_{i \in \Gamma(x)} \{y_i\}$$

with the *clock values* y_i, $i \in E$, defined by

$$y_i' = \begin{cases} y_i - y^* & \text{if } i \neq e' \text{ and } i \in \Gamma(x) \\ v_{i, N_i + 1} & \text{if } i = e' \text{ or } i \notin \Gamma(x) \end{cases} \quad i \in \Gamma(x')$$

where the *interevent time* y^* is defined as

$$y^* = \min_{i \in \Gamma(x)} \{y_i\}$$

and the *event scores* N_i, $i \in E$, are defined by

$$N_i' = \begin{cases} N_i + 1 & \text{if } i = e' \text{ or } i \notin \Gamma(x) \\ N_i & \text{otherwise} \end{cases} \qquad i \in \Gamma(x')$$

In addition, initial conditions are: $y_i = v_{i,1}$ and $N_i = 1$ for all $i \in \Gamma(x_0)$. If $i \notin \Gamma(x_0)$, then y_i is undefined and $N_i = 0$. ♦

This is admittedly a rather intricate definition for what is, hopefully, much easier to grasp in an intuitive way. The ensuing discussion and examples serve to further illustrate the key features of timed automata viewed as dynamic systems.

5.2.3   A State Space Model

Our objective here is to emphasize the fact that the definition of a timed automaton is essentially the description of a state space model for a dynamic system.

Beginning with the automaton state transition mechanism described by (5.9):

$$x' = f(x, e')$$

we see that we need to specify e' through (5.8):

$$e' = \arg \min_{i \in \Gamma(x)} \{y_i\}$$

Since y_i is defined through the recursive equation (5.11), we must adopt it as a state variable of our model. Moreover, (5.11) depends on N_i, which is also recursively defined through (5.12); this too is a state variable. We now recognize a complete state space model where $x, y_1, \ldots, y_m, N_1, \ldots, N_m$ are *state variables* and $\mathbf{v}_1, \ldots, \mathbf{v}_m$ are the *input* sequences driving the system. To provide a more compact description of the model, define:

$$\begin{aligned} f^x(\cdot) : & \quad \text{the } \textit{next state function} \text{ in (5.9)} \\ f_i^y(\cdot) : & \quad \text{the } i\text{th } \textit{next clock function} \text{ in (5.11), } i \in \Gamma(x') \\ f_i^N(\cdot) : & \quad \text{the } i\text{th } \textit{next score function} \text{ in (5.12), } i \in \Gamma(x') \end{aligned}$$

and let $\mathbf{y}$ and $\mathbf{N}$ be the current clock and score (column) vectors

$$\mathbf{y} = [y_1, \ldots, y_m]^T, \qquad \mathbf{N} = [N_1, \ldots, N_m]^T$$

It is worth pointing out that the state variables $N_1, \ldots, N_m$ are only used to update themselves and to help us count event occurrences. They do not, however, affect the automaton state x or the clock values $y_1, \ldots, y_m$. Although we retain them here as part of our state description for the sake of completeness, we will see later on that they are often superfluous.

We can now write the state equations for our model as follows:

$$x' = f^x(x, \mathbf{y}, \mathbf{N}, \mathbf{v}_1, \ldots, \mathbf{v}_m) \tag{5.13}$$

$$\mathbf{y}' = \mathbf{f}^{\mathbf{y}}(x, \mathbf{y}, \mathbf{N}, \mathbf{v}_1, \ldots, \mathbf{v}_m) \tag{5.14}$$

$$\mathbf{N}' = \mathbf{f}^{\mathbf{N}}(x, \mathbf{y}, \mathbf{N}, \mathbf{v}_1, \ldots, \mathbf{v}_m) \tag{5.15}$$

The model is summarized in Fig. 5.4. The input consists of the clock structure $\mathbf{V} = \{\mathbf{v}_i \; : \; i = 1, \ldots, m\}$. The output is viewed as a timed event sequence, which defines the sample path of the DES. Note that the actual time variable, updated through (5.10), is not an essential part of the dynamics in the model. It is convenient, however, in defining the output sequence $\{(e_1, t_1), (e_2, t_2), \ldots\}$.

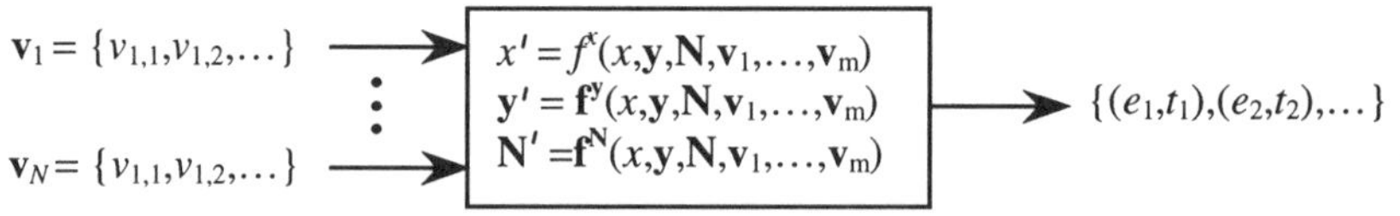

$$\mathbf{v}_1 = \{v_{1,1}, v_{1,2}, \ldots\} \quad \longrightarrow \quad \boxed{\begin{array}{l} x' = f^x(x, \mathbf{y}, \mathbf{N}, \mathbf{v}_1, \ldots, \mathbf{v}_m) \\ \mathbf{y}' = \mathbf{f}^{\mathbf{y}}(x, \mathbf{y}, \mathbf{N}, \mathbf{v}_1, \ldots, \mathbf{v}_m) \\ \mathbf{N}' = \mathbf{f}^{\mathbf{N}}(x, \mathbf{y}, \mathbf{N}, \mathbf{v}_1, \ldots, \mathbf{v}_m) \end{array}} \quad \longrightarrow \quad \{(e_1, t_1), (e_2, t_2), \ldots\}$$

$$\mathbf{v}_N = \{v_{1,1}, v_{1,2}, \ldots\}$$

Figure 5.4: Timed state automaton model of a DES: The three state equations are given by (5.9), (5.11), and (5.12), and time is updated through $t' = t + y^*$ where y^* is given by (5.7).

The initial conditions for the model are as specified in the definition of a timed automaton in the previous section. Given x_0, we set $N_{i,0} = 1$ for all initially active events $i \in \Gamma(x_0)$, and $N_{i,0} = 0$ for all $i \notin \Gamma(x_0)$. We also set $y_{i,0} = v_{i,1}$ for all $i \in \Gamma(x_0)$; $y_{i,0}$ is undefined for all $i \notin \Gamma(x_0)$.

Remark. There is a semantic difficulty with this model in the use of the term "state". Strictly speaking, the "state" of the system consists of the variables $(x, y_1, \ldots, y_m, N_1, \ldots, N_m)$. All these variables, along with the clock structure $\mathbf{V}$, are needed to predict the future behavior of the DES. On the other hand, we also refer to x as the "state", even though x alone only partially describes the DES. The underlying automaton state x is sometimes referred to as the *physical* or *external* state of the system, to contrast it to the variables $(y_1, \ldots, y_m, N_1, \ldots, N_m)$ which form the *internal* state. When the distinction absolutely has to be made clear, we will refer to x as the *process state* and to $(x, y_1, \ldots, y_m, N_1, \ldots, N_m)$ as the *system state*. We will learn to live with this inconsistency by usually referring to x as just the state, while always keeping in mind that the mathematically correct system state is defined by the process state x along with the clock and score vectors $\mathbf{y}$ and $\mathbf{N}$. As a result, the actual state space for the system is given by $X \times \mathbb{R}^{+m} \times \mathbb{Z}^{+m}$, where $x \in X$, $\mathbf{y} \in \mathbb{R}^{+m}$, and $\mathbf{N} \in \mathbb{Z}^{+m}$. Here, $\mathbb{Z}^+$ is the set of nonnegative integers $\{0, 1, 2, \ldots\}$.

An alternative state space model is based on the use of the *event ages* z_i, $i = 1, \ldots, m$, instead of the clocks y_i, $i = 1, \ldots, m$. It is easy to obtain this model by making use of (5.4). The state equation (5.11) is replaced by

$$z_i' = \begin{cases} z_i + y^* & \text{if } i \neq e' \text{ and } i \in \Gamma(x) \\ 0 & \text{if } i = e' \text{ or } i \notin \Gamma(x) \end{cases} \qquad i \in \Gamma(x') \tag{5.16}$$

where

$$y^* = \min_{i \in \Gamma(x)} \{(v_{i, N_i} - z_i)\} \tag{5.17}$$

$$e' = \arg \min_{i \in \Gamma(x)} \{(v_{i, N_i} - z_i)\} \tag{5.18}$$

and the remaining state equations (5.9) and (5.12) are unchanged. In (5.16), we increase the age of all events that remain active in the new state x', except the triggering event e'; we set the age to 0 for all newly activated events, including the triggering event e', provided $e' \in \Gamma(x')$.

Initial conditions in this model are also easily specified once x_0 is given. As before, we set $N_{i,0} = 1$ for all initially active events $i \in \Gamma(x_0)$, and $N_{i,0} = 0$ for all $i \notin \Gamma(x_0)$. In addition, we set $z_{i,0} = 0$ for all $i \in \Gamma(x_0)$; $z_{i,0}$ is undefined for all $i \notin \Gamma(x_0)$.

The notational complexity of the state space model is mainly due to the discontinuous nature of the clock and score state equations. However, the model reflects a simple intuitive way for generating timed DES sample paths, as initially described in Section 5.2.1. Putting aside the issue of mathematical representation, one should observe that the actual sample path generation relies on simple comparisons of real numbers (to determine the smallest clock value, and hence the triggering event), and elementary arithmetic (addition/subtraction) to update the event clocks and scores.

Example 5.1

Consider an automaton with $E = \{\alpha, \beta\}$ and $X = \{0, 1, 2\}$. A state transition diagram is shown in Fig. 5.5. Note that $\Gamma(0) = \Gamma(2) = \{\alpha, \beta\}$, but $\Gamma(1) = \{\alpha\}$. Let the initial state be $x_0 = 0$ at time $t_0 = 0.0$.

A clock structure $\mathbf{V}$ is supplied as follows (only the first few lifetimes are shown):

$$\mathbf{v}_\alpha = \{1.0, 1.5, 1.5, \ldots\}, \qquad \mathbf{v}_\beta = \{2.0, 0.5, 1.5, \ldots\}$$

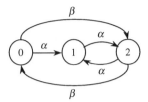

Figure 5.5: State transition diagram for timed automaton of Example 5.1.

The sample path of this timed DES for the first four event occurrences is shown in Fig. 5.6. Initially, both events are active and their clock and score values are set to

$$y_{\alpha,0} = v_{\alpha,1} = 1.0, \qquad y_{\beta,0} = v_{\beta,1} = 2.0$$
$$N_{\alpha,0} = 1, \qquad\qquad N_{\beta,0} = 1$$

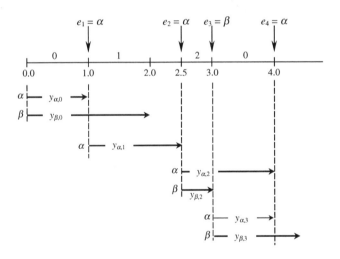

Figure 5.6: Sample path for Example 5.1.

We can now determine the entire sample path by inspection, without really having to resort to the model equations (5.7) through (5.12). First,

$$y_0^* = \min\{y_{\alpha,0}, y_{\beta,0}\} = y_{\alpha,0} \implies e_1 = \alpha$$

Hence, by looking at the state transition diagram, the next state is

$$x_1 = f^x(0, \alpha) = 1$$

Also, the event time is simply given by $t_1 = y_0^* = 1.0$.

We repeat the process with $\Gamma(1) = \{\alpha\}$. Note that the clock value of β is now irrelevant, since β is not feasible in this state. Since α was the triggering event, it is activated again with a score incremented by 1. Its new clock value is

$$y_{\alpha,1} = v_{\alpha,2} = 1.5$$

Obviously, the next triggering event is $e_2 = \alpha$. Hence the next state is

$$x_2 = f^x(1, \alpha) = 2$$

and time is moved forward to $t_2 = t_1 + y_1^* = 2.5$.

Next, we have $\Gamma(2) = \{\alpha, \beta\}$, and both events are activated anew. Their scores are also incremented. Thus, we set

$$y_{\alpha,2} = v_{\alpha,3} = 1.5, \quad y_{\beta,2} = v_{\beta,2} = 0.5$$

and we get

$$y_2^* = \min\{y_{\alpha,2}, y_{\beta,2}\} = y_{\beta,2} \quad \Longrightarrow \quad e_3 = \beta$$

Hence, the next state is

$$x_3 = f^x(2, \beta) = 0$$

The event time is $t_3 = t_2 + y_2^* = 2.5 + y_{\beta,2} = 3.0$.

The last transition shown in Fig. 5.6 occurs at state 0. Event α is still active, whereas β was the triggering event. Only the score of β is incremented. We update the clock values so that

$$y_{\alpha,3} = y_{\alpha,2} - y_2^* = 1.0, \quad y_{\beta,3} = v_{\beta,3} = 1.5$$

and we get

$$y_3^* = \min\{y_{\alpha,3}, y_{\beta,3}\} = y_{\alpha,3} \quad \Longrightarrow \quad e_4 = \alpha$$

The next state is

$$x_4 = f^x(0, \alpha) = 1$$

The event time is $t_4 = t_3 + y_3^* = 3.0 + y_{\alpha,3} = 4.0$.

Example 5.2
Many processes of interest consist of various tasks to be performed sequentially or concurrently or a combination of both. Each task requires some amount of time in order to be performed, which we may model

through various event lifetimes. This is true in manufacturing a product, executing an algorithm on a computer, or simply carrying out a "mission" or "project" (for example, turn engine on to warm up to a given temperature; while this is going on, check gas tank; fill it, if below some level; then, check oil pressure; if too low, turn engine off and "abort" mission; otherwise, check engine temperature and "go" if at desired level, otherwise wait until this level is reached and then "go").

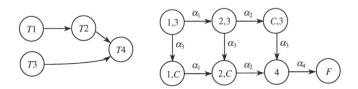

Figure 5.7: Process for Example 5.2 and state transition diagram.

In Fig. 5.7, we show a simple process consisting of two sequential tasks, $T1$ and $T2$, which may be performed in parallel with a third task, $T3$. We may think of these tasks as computer operations performed on two parallel processors. The output of $T1$ and $T3$ must then be combined before $T4$ can start on a single processor.

We can describe this process through an automaton whose state transition diagram is also shown in Fig. 5.7. There are four events driving this process, $E = \{a_i : i = 1, \dots, 4\}$, where a_i denotes completion of task i. The state space X consists of seven states, as shown. State $(1, 3)$ indicates that $T1$ and $T3$ are currently ongoing. Similarly, for state $(2, 3)$. On the other hand, $(1, C)$ indicates that $T1$ is ongoing, while $T3$ has "completed". Similarly for $(2, C)$. State $(C, 3)$ indicates that $T3$ is ongoing, while the sequence of tasks $T1, T2$ has completed. Finally, state 4 indicates that $T4$ is ongoing, and state F that the process is "finished". It is clear that event a_i is feasible only when task i is ongoing.

In this example, every sample path begins at state $(1, 3)$ and ends with state F. Each event can only occur once, so the clock structure is simply a set of four numbers, one for each event lifetime. A typical sample path is shown in Fig. 5.8. In this particular case, note that the time interval $[t_2, t_3]$ represents a synchronization delay in the start of $T4$, due to the requirement that the output of $T2$ and $T3$ be both available.

Finally, it is important to note that the automaton (X, E, f, Γ, x_0) and the clock structure $\mathbf{V} = \{v_i : i \in E\}$ are distinctly separate. What we mean by this is that (X, E, f, Γ, x_0) is simply a system in need of an event sequence to drive it, and $\mathbf{V}$ is just a set of independently generated sequences. Thus, $\mathbf{V}$ could be

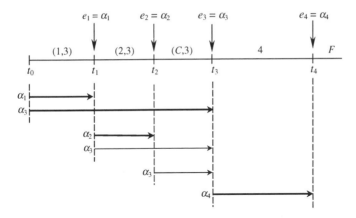

Figure 5.8: A typical sample path for the process of Example 5.2.

used as an input to several different timed automata (as long as their events sets are the same). The connection between the automaton and its clock structure manifests itself only through the feasible event set $\Gamma(x)$. This is evident in the state equations (5.11) and (5.12), where x appears only through $\Gamma(x)$. Thus, the timing mechanism is captured by a relationship of the form

$$e' = h[\mathbf{y}, \mathbf{N}, \Gamma(x)]$$

where $\mathbf{y}$ and $\mathbf{N}$ may be viewed as auxiliary state variables, required to update the automaton state through

$$x' = f^x(x, e')$$

5.2.4   Queueing Systems as Timed Automata

We already developed an automaton model for a simple queueing system in Example 2.10 in Section 2.2.3 of Chapter 2. The model is as follows:

$$E = \{a, d\}, \quad X = \{0, 1, 2, \dots\}$$
$$\Gamma(x) = \{a, d\} \text{ for all } x > 0, \quad \Gamma(0) = \{a\} \tag{5.19}$$
$$f(x, e') = \begin{cases} x + 1 & \text{if } e' = a \\ x - 1 & \text{if } e' = d \text{ and } x > 0 \end{cases}$$

where a is a customer arrival event, and d is a customer departure event. The automaton state x represents the queue length (including a customer in service). Note that $f(x, e')$ is not defined for $x = 0$, $e' = d$, since $d \notin \Gamma(0)$.

Given a clock structure $\mathbf{V}$, note that the triggering event is always obtained by comparing the clock values of the a and d events, except for the case where

$x = 0$, when only an a event is feasible. Thus, we have

$$e' = \begin{cases} d & \text{if } y_d < y_a \text{ and } x > 0 \\ a & \text{otherwise} \end{cases}$$

Once e' is determined, the event clock and score state equations (5.11) and (5.12) can be written as follows:

$$y'_a = \begin{cases} y_a - y_d & \text{if } e' = d \\ v_{a,N_a+1} & \text{otherwise} \end{cases} \tag{5.20}$$

$$y'_d = \begin{cases} y_d - y_a & \text{if } e' = a \text{ and } x > 0 \\ v_{d,N_d+1} & \text{otherwise} \end{cases} \tag{5.21}$$

$$N'_a = \begin{cases} N_a + 1 & \text{if } e' = a \\ N_a & \text{otherwise} \end{cases} \tag{5.22}$$

$$N'_d = \begin{cases} N_d + 1 & \text{if } [e' = a \text{ and } x = 0] \text{ or } [e' = d \text{ and } x > 1] \\ N_d & \text{otherwise} \end{cases} \tag{5.23}$$

Because of the special structure of this model, it is customary to represent sample paths in terms of the queue length evolution over time. An example is shown in Fig. 5.9, for the two specific clock sequences of a and d shown. All a events cause a positive unit jump, and all d events cause a negative unit jump. It is important to note that the first d event is only activated at t_1. This is because initially $x = 0$, and $\Gamma(0) = \{a\}$. A similar observation applies to the fifth d event at t_{11}, which is activated at t_9 following the end of the empty queue time interval $[t_8, t_9]$.

It is worthwhile at this point to emphasize once again the differences between the event-driven model (5.19) through (5.23) and a time-driven model for the same queueing system. First, observe that it is difficult to set up a continuous-time model for this system using the traditional CVDS differential equation formulation we discussed in Chapter 1. This is because $x(t)$ is defined over the discrete set of nonnegative integers, and, therefore, the derivative $\dot{x}(t)$ does not, strictly speaking, exist. We can, however, set up a discrete-time model as follows.

Let $k = 1, 2, \ldots$ be the time index. The time line is partitioned into slots $\tau, 2\tau, \ldots, k\tau, \ldots$ such that at most one arrival and one departure event can occur in a slot of length τ. Thus, we may define an input sequence $a(k)$ such that:

$$a(k) = \begin{cases} 1 & \text{if an arrival occurs in the } k\text{th time slot} \\ 0 & \text{otherwise} \end{cases} \quad k = 1, 2, \ldots$$

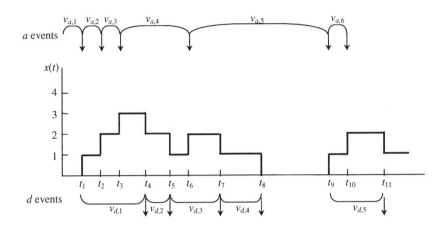

Figure 5.9: Sample path of a simple queueing system.
The queue length $x(t)$ increases by 1 with every arrival (a event), and decreases by 1 with every departure (d event). Activations of a take place with every a occurrence. Activations of d take place only when d occurs and the queue remains nonempty, or when a occurs when the queue is empty.

Then, the state equation for $x(k)$ is of the form

$$
x(k+1) = \begin{cases}
x(k)+1 & \text{if } a(k) = 1 \text{ and} \\
& \text{no departure occurs in the } k\text{th time slot} \\
x(k)-1 & \text{if } a(k) = 0 \text{ and} \\
& \text{a departure occurs in the } k\text{th time slot} \\
x(k) & \text{otherwise}
\end{cases}
$$

where the condition "departure occurs in the kth time slot" requires an analog to the clock state variable y_d in (5.21). Establishing this analog involves tedious details which we omit, since the important point to make here is different. Specifically, observe that this time-driven model requires $x(k)$ to remain unchanged if no event occurs in the kth slot. If slots are chosen so that their lengths are small, most slots are likely to contain no event. This is a potentially inefficient model, since *time is frequently updated when no change in the system state takes place.*

In addition, observe that $a(k)$ represents an *approximation of the actual arrival clock sequence* $\mathbf{v}_a$. Therefore, we may want to use a very small value of τ in order to satisfy the requirement that no two arrivals occur in the same slot. This may make the approximation better, but it also amplifies the inefficiency mentioned above.

5.2.5 The Event Scheduling Scheme

In this section we describe a variation of the automaton model summarized in Fig. 5.4, which turns out to be simpler from a computer implementation standpoint. In fact, it constitutes the cornerstone of discrete-event computer simulation to be further discussed in Chapter 10.

The modification we make is the following. Whenever an event i is activated at some event time t_n with score N_i, we *schedule* its next occurrence to be at time $(t_n + v_{i,N_i+1})$. Thus, instead of tracking clock values y_i, we maintain a *scheduled event list*

$$L = \{(e_k, t_k)\}, \quad k = 1, 2, \ldots, m_L$$

where $m_L \leq m$ is the number of feasible events in the current state. Moreover, the scheduled event list is *always ordered on a smallest-scheduled-time-first basis*. This implies that the first event in the list, e_1, is always the triggering event.

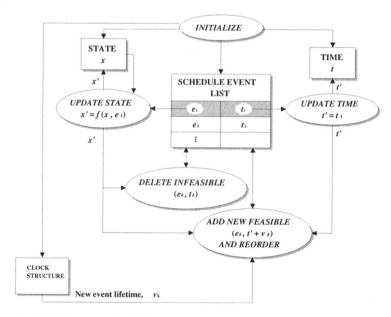

Figure 5.10: The event scheduling scheme.
The SCHEDULED EVENT LIST contains all feasible events at the current STATE, sorted on a smallest-scheduled-time-first. The next event is always e_1 and it occurs at time t_1. It causes updates to STATE and TIME. Then, based on the new value of STATE, some events are activated and some are deactivated. Events that are activated are entered in the SCHEDULED EVENT LIST with their scheduled occurrence times, maintaining the right sorting order.

This "event scheduling" scheme simply addresses the issue of how one wants to do the required bookkeeping for generating the DES sample path. There is nothing conceptually new or different here compared to the model of Fig. 5.4. However, the scheduled event list (sometimes referred to as the *event calendar*) does help to capture the essence of the event timing mechanism central in our discussion thus far.

The event scheduling scheme should be thought of as a procedure for generating sample paths based on the model we have presented and driven by a given clock structure. This is illustrated in Fig. 5.10. The STATE is initialized to some given value x_0, and TIME is usually set to zero. The CLOCK STRUCTURE is the set of clock sequences, one for each event. The SCHEDULED EVENT LIST is initialized so as to contain all events feasible in x_0 with scheduled times provided by the appropriate clock sequence. It is also sorted in increasing order of scheduled times. The procedure is essentially identical to the seven steps described in Section 5.2.2. Specifically:

Step 1. Remove the first entry (e_1, t_1) from the SCHEDULED EVENT LIST.

Step 2. Update TIME to the new event time t_1.

Step 3. Update STATE according to the state transition function:
$x' = f(x, e_1)$.

Step 4. Delete from the SCHEDULED EVENT LIST any entries corresponding to infeasible events in the new state, i.e., delete all $(e_k, t_k) \in L$ such that $e_k \notin \Gamma(x')$.

Step 5. Add to the SCHEDULED EVENT LIST any feasible event which is not already scheduled (possibly including the triggering event removed in Step 1). The scheduled event time for some such i is obtained by adding to the current time a new lifetime from the ith sequence in the CLOCK STRUCTURE.

Step 6. Reorder the updated SCHEDULED EVENT LIST on a *smallest-scheduled-time-first basis*.

The procedure then repeats with Step 1 for the new ordered list. As already mentioned, we will encounter this scheme again when we study simulation techniques for DES.

5.3   TIMED PETRI NETS

In the case of untimed DES models, we saw that Petri nets provide a very general framework. This framework can be extended to timed models by equipping a Petri net with a clock structure and turning it into a *timed Petri net*.

Timed automata can then always be obtained from timed Petri nets. A discussion of the relative merits of Petri nets and automata was provided in the previous chapter; the points made there still apply to the timed versions of these two modeling frameworks.

Let (P, T, A, w, x) be a marked Petri net, as defined in Chapter 4. Recall that T is a set of transitions (events), and P is a set of places (conditions required to enable various events). The set A contains all arcs (p_i, t_j) such that p_i is an input place of t_j, and (t_j, p_i) such that p_i is an output place of t_j. Each arc is assigned a weight $w(p_i, t_j)$ or $w(t_j, p_i)$, which is a positive integer. Also, recall that we use the notation $I(t_j)$ and $O(t_j)$ to denote the sets of input and output places respectively of transition t_j. Finally, x is the marking function of the Petri net, where a marking represents the number of tokens present in each place in the net.

Let us now introduce a clock structure similar to the one defined for automata. The only difference is that a clock sequence v_j is now associated with a transition t_j. A positive real number, $v_{j,k}$, assigned to t_j has the following meaning: When transition t_j is enabled for the kth time, it does not fire immediately, but incurs a firing delay given by $v_{j,k}$; during this delay, tokens are kept in the input places of t_j.

Not all transitions are required to have firing delays. Some transitions may always fire as soon as they are enabled. Thus, we partition T into subsets T_0 and T_D, such that $T = T_0 \cup T_D$. T_0 is the set of transitions always incurring zero firing delay, and T_D is the set of transitions that generally incur some firing delay. The latter are called *timed transitions*.

Definition. The *clock structure* (or *timing structure*) associated with the a set of timed transitions $T_D \subseteq T$ of a marked Petri net (P, T, A, w, x) is a set

$$\mathbf{V} = \{\mathbf{v}_j \ : \ t_j \in T_D\}$$

of *lifetime sequences*

$$\mathbf{v}_j = \{v_{j,1}, v_{j,2}, \dots\}, \quad t_j \in T_D, \quad v_{j,k} \in \mathbb{R}^+, \quad k = 1, 2, \dots \qquad \blacklozenge$$

Graphically, transitions with no firing delay are still represented by bars, whereas timed transitions are represented by rectangles. The clock sequence associated with a timed transition is normally written next to the rectangle.

Definition. A *Timed Petri Net* is a six-tuple

$$(P, T, A, w, x, \mathbf{V})$$

where (P, T, A, w, x) is a marked Petri net, and $\mathbf{V} = \{\mathbf{v}_j : t_j \in T_D\}$ is a clock structure. $\qquad \blacklozenge$

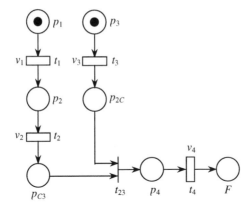

Figure 5.11: Timed Petri net for Example 5.3.

Remark. We emphasize once again the importance of distinguishing between *enabling* a transition and *firing* it. A timed transition is enabled as soon as all input places have the necessary number of tokens; but it fires only after a firing delay (supplied by an element of its clock sequence).

Example 5.3

Let us return to the process of Example 5.2, where a sequence of two tasks T_1 and T_2 is performed concurrently with a third task T_3, and then a fourth task T_4 is executed by combining the output of T_2 and T_3. One possible Petri net model of this process is shown in Fig. 5.11.

The timed transitions $t_1, \ldots, t_4$ correspond to the task completion events, and they are indicated by the rectangles accompanied by the corresponding firing delays $v_1, \ldots, v_4$. When task 2 is done, it adds a token to place p_{C3} (indicating that T_2 is complete, but T_3 may still be ongoing). Similarly, for place p_{2C}. Thus, the untimed transition t_{23} is enabled when both tasks 2 and 3 are complete (note that places p_{C3} and p_{2C} could directly enable transition t_4, omitting t_{23} and p_4). The process is finished when a token is added to place F. In Fig. 5.11 we show the Petri net in an initial state such that tasks 1 and 3 are ongoing.

5.3.1   Timed Petri Net Dynamics

A model based on the transition firing time dynamics of a Petri net can be derived by proceeding exactly as in the development of the model (5.7) through (5.12), with events replaced by transitions. More importantly, however, we can exploit the structure of the Petri net by decomposing the model into the dynamics of individual transitions (events). In other words, the state equations in such a model generate transition firing sequences of the form

$$\{\tau_{j,1}, \tau_{j,2}, \dots\}, \quad j = 1, \dots, m$$

where $\tau_{j,k}$ is the kth firing time of transition t_j, $k = 1, 2, \dots$ This is illustrated in Fig. 5.12. The m transition firing sequences output should be contrasted to the timed automaton model of Fig. 5.4, where the output is viewed as a single timed event sequence $\{(e_1, t_1), (e_2, t_2), \dots\}$.

Figure 5.12: Timed Petri net model of a DES.

Unfortunately, for arbitrary Petri nets, the task of deriving general-purpose state equations for all transition firing times is quite complex. But for certain special classes of Petri nets (for instance, the familiar simple queueing system considered in the next section), it is relatively simple to obtain a model for the transition firing time dynamics. To show how such models may be derived, let us limit ourselves to a restricted type of Petri nets.

Let us consider Petri nets with the following property: Every place has only one input and only one output transition. In addition, all arc weights are 1. Such a Petri net is known as a *marked graph*. Let us discuss some implications of the definition of a marked graph.

We first focus on the fact that a place p_i has only one input transition t_r. Define

$$\pi_{i,k} : \quad \text{time instant when place } p_i \text{ receives its } k\text{th token}, \ k = 1, 2, \dots$$

Suppose that $x(p_i) = 0$ initially. Then, the kth time a token is deposited in p_i is precisely the kth firing time of t_r, which we have denoted by $\tau_{r,k}$. If, on the other hand, p_i initially contains x_{i0} tokens, then the kth firing time of t_r is the time when p_i receives its $(x_{i0} + k)$th token. We therefore have the following relationship:

$$\pi_{i,k+x_{i0}} = \tau_{r,k}, \quad p_i \in O(t_r), \quad k = 1, 2, \dots \qquad (5.24)$$

where x_{i0} is the initial marking of place p_i and $\pi_{i,k} = 0$ for all $k = 1, \dots, x_{i0}$. Equivalently,

$$\pi_{i,k} = \tau_{r,k-x_{i0}}, \quad p_i \in O(t_r), \quad k = x_{i0} + 1, x_{i0} + 2, \dots \qquad (5.25)$$

Next, we focus on the fact that a place p_i has only one output transition t_j. Suppose p_i is the only input place of t_j. If t_j is untimed, then the kth firing time of t_j is precisely the time when p_i receives its kth token, and hence enables t_j. Then, we have the simple relationship:

$$\tau_{j,k} = \pi_{i,k}, \qquad k = 1, 2, \ldots$$

If, on the other hand, t_j is timed with a clock sequence $\mathbf{v}_j$, then this relationship becomes

$$\tau_{j,k} = \pi_{i,k} + v_{j,k}, \qquad k = 1, 2, \ldots$$

Now if p_i is not the only input place of t_j, then t_j is enabled for the kth time whenever the last input place in the set $I(t_j)$ acquires its kth token, that is, at some time $\pi_{s,k}$, $p_s \in I(t_j)$, such that

$$\pi_{s,k} \geq \pi_{i,k} \quad \text{for all } p_i \in I(t_j)$$

We can express this simple fact as follows:

$$\tau_{j,k} = \max_{p_i \in I(t_j)} \{\pi_{i,k}\} + v_{j,k}, \qquad k = 1, 2, \ldots \tag{5.26}$$

In summary, the combination of (5.24) through (5.26) provides a set of recursive equations for determining transition firing times for this class of Petri nets.

Example 5.4
Consider the Petri net of Fig. 5.13. Note that it is a marked graph, since every place has one input and one output arc of weight 1. The initial state is $[1, 1, 0]$. Thus, a set of recursive equations of the form (5.24) through (5.26) may be derived.

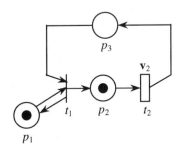

Figure 5.13: Timed Petri net for Example 5.4.

By inspection, we can immediately write the following equations for $\tau_{1,k}$ and $\tau_{2,k}$, $k = 1, 2, \ldots$:

$$\tau_{1,k} = \max\{\pi_{1,k}, \pi_{3,k}\} \tag{5.27}$$

$$\tau_{2,k} = \pi_{2,k} + v_{2,k} \tag{5.28}$$

since transition t_1 is untimed and fires when the last of the input places $\{p_1, p_3\}$ acquires a token; and transition t_2 is timed and fires following a delay $v_{2,k}$ after p_2 acquires a token.

Also, by inspection we can see that:

$$\pi_{1,k} = \tau_{1,k-1}, \quad k = 2, 3, \ldots, \quad \pi_{1,1} = 0 \qquad (5.29)$$

since p_1 initially contains a token, and must await the next firing of t_1 to reacquire it. Similarly,

$$\pi_{2,k} = \tau_{1,k-1}, \quad k = 2, 3, \ldots, \quad \pi_{2,1} = 0 \qquad (5.30)$$

Finally, p_3 receives a token whenever t_2 fires, and since it started out with no tokens we have

$$\pi_{3,k} = \tau_{2,k}, \quad k = 1, 2, \ldots \qquad (5.31)$$

Combining (5.27) through (5.31), we can eliminate $\pi_{1,k}$, $\pi_{2,k}$, and $\pi_{3,k}$ and obtain for $k = 1, 2, \ldots$,

$$
\begin{aligned}
\tau_{1,k} &= \max\{\tau_{1,k-1}, \tau_{1,k-1} + v_{2,k}\} \\
&= \tau_{1,k-1} + v_{2,k}, \qquad \tau_{1,0} = 0 \qquad (5.32) \\
\tau_{2,k} &= \tau_{1,k-1} + v_{2,k} \qquad\qquad\qquad\qquad (5.33)
\end{aligned}
$$

These recursive equations correspond to the intuitively obvious fact that both transitions repeatedly wait for the firing delay at t_2. Also, $\tau_{1,k} = \tau_{2,k}$ for all $k = 1, 2, \ldots$, since t_1 always fires as soon as t_2 fires.

For general Petri nets, it is still possible to derive equations of the form (5.24) through (5.26) for the transition firing times and the place enabling times. However, the recursive relationships become significantly more complex.

5.3.2 Queueing Systems as Timed Petri Nets

Recall the Petri net model of a simple queueing system presented in Chapter 4, with a place set $P = \{Q, I, B\}$. A timed version of this Petri net model is shown in Fig. 5.14 at state $x = [0, 1, 0]$, that is, when the queue is empty and the server is idle. In this case, the timed transition set is $T_D = \{a, d\}$, corresponding to customer arrivals and to departures from the server. Transition s, on the other hand, incurs no firing delay: Service starts as soon as the server is idle and a customer is in the queue.

The clock structure for this model consists of $\mathbf{v}_a = \{v_{a,1}, v_{a,2}, \ldots\}$ and $\mathbf{v}_d = \{v_{d,1}, v_{d,2}, \ldots\}$. This is the same as the clock structure for the timed automaton model presented earlier. Moreover, if we let

$$x(t) = x(Q) + x(B)$$

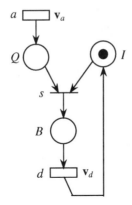

Figure 5.14: Timed Petri net model of a simple queueing system.

denote the total number of tokens in places Q and B at time t, note that Fig. 5.9 represents a typical sample path of this variable in the timed Petri net framework.

Looking at Fig. 5.14, we can see that our model satisfies the requirements of a marked graph. Therefore, we can immediately derive equations of the form (5.24) through (5.26) to describe the transition firing dynamics of a simple queueing system. In particular, let

$$a_k : \quad k\text{th arrival time}$$
$$d_k : \quad k\text{th departure time}$$
$$s_k : \quad k\text{th service start time}$$

and

$$\pi_{Q,k} : \quad \text{time when } Q \text{ receives its } k\text{th token}$$
$$\pi_{I,k} : \quad \text{time when } I \text{ receives its } k\text{th token}$$
$$\pi_{B,k} : \quad \text{time when } B \text{ receives its } k\text{th token}$$

with initial condition $\pi_{I,1} = 0$.

From (5.26), or directly by inspection, we have

$$a_k = a_{k-1} + v_{a,k}, \qquad k = 1, 2, \ldots, \qquad a_0 = 0 \tag{5.34}$$
$$s_k = \max\{\pi_{Q,k}, \pi_{I,k}\}, \qquad k = 1, 2, \ldots \tag{5.35}$$
$$d_k = \pi_{B,k} + v_{d,k}, \qquad k = 1, 2, \ldots \tag{5.36}$$

and

$$\pi_{Q,k} = a_k, \qquad k = 1, 2, \ldots \tag{5.37}$$
$$\pi_{I,k} = d_{k-1}, \qquad k = 2, 3, \ldots, \qquad \pi_{I,1} = 0 \tag{5.38}$$
$$\pi_{B,k} = s_k, \qquad k = 1, 2, \ldots \tag{5.39}$$

Combining (5.35) through (5.39) to eliminate $\pi_{Q,k}$, $\pi_{I,k}$, and $\pi_{B,k}$, we get

$$s_k = \max\{a_k, d_{k-1}\}, \quad k = 1, 2, \ldots, \quad d_0 = 0 \tag{5.40}$$

$$d_k = s_k + v_{d,k}, \quad k = 1, 2, \ldots \tag{5.41}$$

We can further eliminate s_k from (5.41) to obtain the following important fundamental relationship for this simple queueing system:

$$d_k = \max\{a_k, d_{k-1}\} + v_{d,k}, \quad k = 1, 2, \ldots, \quad d_0 = 0 \tag{5.42}$$

This is a simple recursive relationship characterizing customer departure times. It captures the fact that the kth departure occurs $v_{d,k}$ time units after the $(k-1)$th departure, except when $a_k > d_{k-1}$. The latter case occurs when the departure at d_{k-1} leaves the queue empty; the server must then await the next arrival at time a_k, and generate the next departure at time $a_k + v_{d,k}$.

Rewriting (5.34) and (5.42) for $k = 2, 3, \ldots$ as follows:

$$a_k = a_{k-1} + v_{a,k}, \quad a_0 = 0 \tag{5.43}$$

$$d_k = \max\{a_{k-1} + v_{a,k}, d_{k-1}\} + v_{d,k}, \quad d_0 = 0 \tag{5.44}$$

we obtain a state space model in the framework of Fig. 5.12. The queueing system model is driven by the clock sequences $\mathbf{v}_a$ and $\mathbf{v}_d$, and its output consists of the arrival and departure time sequences $\{a_1, a_2, \ldots\}$ and $\{d_1, d_2, \ldots\}$ generated through the state equations (5.34) and (5.42).

5.4 DIOID ALGEBRAS

Another modeling framework we will briefly describe in what follows is based on developing an algebra using two operations: $\min\{a, b\}$ (or $\max\{a, b\}$) for any real numbers a and b, and addition $(a + b)$. The motivation comes from the observation that the operations "min" and "+" are the only ones required to develop the timed automaton model summarized in Fig. 5.4. Similarly, the operations "max" and "+" are the only ones used in developing the timed Petri net models described in the previous section.

The term "dioid" (meaning "two") refers to the fact that this algebra is based on two operations. The operations are formally named *addition* and *multiplication* and denoted by $\oplus$ and $\otimes$ respectively. However, their actual meaning (in terms of regular algebra) is different. For any two real numbers a and b, we define

$$\text{Addition}: \quad a \oplus b \equiv \max\{a, b\} \tag{5.45}$$

$$\text{Multiplication}: \quad a \otimes b \equiv a + b \tag{5.46}$$

This is admittedly somewhat confusing, since $\otimes$ really means $+$, and $\oplus$ really means max. This dioid algebra is also called a (max, +) algebra.

Remark. The (max, +) algebra could easily be replaced by a (min, −) algebra. This is simply because

$$\max\{a, b\} = -\min\{(-a), (-b)\}$$

and

$$a + b = a - (-b)$$

As already mentioned, the motivation for introducing this algebra may be found directly in the model (5.7) through (5.12) which we derived for timed automata. But the motivation goes even further. If we consider a standard linear discrete-time CVDS, its state equation is of the form

$$\mathbf{x}(k + 1) = \mathbf{A}\mathbf{x}(k) + \mathbf{B}\mathbf{u}(k)$$

which involves (regular) multiplication (×) and addition (+). A question one can raise, therefore, is the following: Can we use a (max, +) algebra with DES, replacing the (+, ×) algebra of CVDS, in order to come up with a representation similar to the one above? Moreover, can we parallel the analysis of linear CVDS using the (max, +) algebra in DES? To a considerable extent, the answer to these questions is "yes."

The study of dioid algebras (sometimes also called min-max algebras) has appeared in the literature since the 1970s. Their use in the modeling and analysis of timed DES, however, was pioneered by Cohen et al. in the mid-1980s. We shall limit ourselves here to a brief description of how the (max, +) algebra can be used to analyze a simple timed DES (a simple queueing system). We will therefore be able to get the main ideas across, as well as identify the key advantages and limitations of this modeling framework.

5.4.1   Basic Properties of the (max, +) Algebra

Using the definitions of $\oplus$ and $\otimes$ in (5.45) and (5.46), we can verify the following properties.

Commutativity of $\oplus$, $\otimes$:

$$
\begin{aligned}
a \oplus b &= \max\{a, b\} = \max\{b, a\} = b \oplus a \\
a \otimes b &= (a + b) = (b + a) = b \otimes a
\end{aligned}
$$

Associativity of $\oplus$, $\otimes$:

$$
\begin{aligned}
(a \oplus b) \oplus c &= \max\{\max\{a, b\}, c\} = \max\{a, \max\{b, c\}\} = a \oplus (b \oplus c) \\
(a \otimes b) \otimes c &= (a + b) + c = a + (b + c) = a \otimes (b \otimes c)
\end{aligned}
$$

Distributivity of $\otimes$ over $\oplus$:

$$
\begin{aligned}
(a \oplus b) \otimes c &= \max\{a, b\} + c = \max\{(a + c), (b + c)\} \\
&= (a \otimes c) \oplus (b \otimes c)
\end{aligned}
$$

Null element in $\oplus$:

$$a \oplus \eta = \max\{a, \eta\} = a \quad \text{for any } a \in \mathbb{R} \text{ and some } \eta \in \mathbb{R}$$

In particular, we can select $\eta = -\infty$, since $\max\{a, -\infty\} = a$ for any $a \in \mathbb{R}$. Alternatively, if we are only dealing with a subset of $\mathbb{R}$, say A, we can select any negative real number $-L$ that satisfies $-L < a$ for any $a \in A$.

Absorbing null element in $\otimes$:

$$a \otimes \eta = a + (-\infty) = \eta$$

Idempotency in $\oplus$:

$$a \oplus a = \max\{a, a\} = a$$

We will subsequently drop the explicit use of the product symbol $\otimes$, as is customary with regular multiplication. We must keep in mind, however, that ab means $a + b$. This is particularly important in using matrix notation.

Example 5.5

This is an example of matrix multiplication in the (max, +) algebra:

$$\begin{bmatrix} 1 & 0 \\ 2 & -2 \end{bmatrix} \begin{bmatrix} 2 & -1 \\ 3 & 1 \end{bmatrix}$$
$$= \begin{bmatrix} \max(1+2, 0+3) & \max(1-1, 0+1) \\ \max(2+2, -2+3) & \max(2-1, -2+1) \end{bmatrix} = \begin{bmatrix} 3 & 1 \\ 4 & 1 \end{bmatrix}$$

Suppose we now multiply the resulting matrix with a scalar a:

$$a \begin{bmatrix} 3 & 1 \\ 4 & 1 \end{bmatrix} = \begin{bmatrix} 3+a & 1+a \\ 4+a & 1+a \end{bmatrix}$$

Note that if $a = \eta$, all elements of the resulting matrix are η.

5.4.2   Modeling Queueing Systems in the (max, +) Algebra

We will use the (max, +) algebra to develop a model of the simple queueing system we have already considered in previous sections. Our starting point is the set of equations (5.43) and (5.44) derived using a timed Petri net model:

$$a_k = a_{k-1} + v_{a,k}, \quad a_0 = 0$$
$$d_k = \max\{a_{k-1} + v_{a,k}, d_{k-1}\} + v_{d,k}, \quad d_0 = 0$$

These equations capture the fundamental dynamic behavior of a queueing system in terms of customer arrival and departure times. For simplicity, we will assume that

$$v_{a,k} = C_a, \quad v_{d,k} = C_d \quad \text{for all } k = 1, 2, \ldots$$

where C_a and C_d are given constants. Note that C_a represents a constant interarrival time, and C_d represents a constant service time for all customers. We will assume that

$$C_a > C_d$$

which is reasonable, since otherwise customers arrive faster than the service rate.

We now rewrite the preceding equations as follows:

$$a_{k+1} \;=\; a_k + C_a, \qquad\qquad a_1 = C_a$$
$$d_k \;=\; \max\{a_k, d_{k-1}\} + C_d, \qquad d_0 = 0$$

Using the $(\max, +)$ algebra, we rewrite these equations as

$$a_{k+1} \;=\; (a_k \otimes C_a) \oplus (d_{k-1} \otimes -L)$$
$$d_k \;=\; (a_k \otimes C_d) \oplus (d_{k-1} \otimes C_d)$$

where $-L$ is any sufficiently small negative number so that $\max\{a_k + C_a, d_{k-1} - L\} = a_k + C_a$.

In matrix notation, we have

$$\begin{bmatrix} a_{k+1} \\ d_k \end{bmatrix} = \begin{bmatrix} C_a & -L \\ C_d & C_d \end{bmatrix} \begin{bmatrix} a_k \\ d_{k-1} \end{bmatrix} \tag{5.47}$$

Defining

$$\mathbf{x}_k = \begin{bmatrix} a_{k+1} \\ d_k \end{bmatrix}, \qquad \mathbf{A} = \begin{bmatrix} C_a & -L \\ C_d & C_d \end{bmatrix}$$

we get

$$\mathbf{x}_{k+1} = \mathbf{A}\mathbf{x}_k, \qquad \mathbf{x}_0 = \begin{bmatrix} C_a \\ 0 \end{bmatrix} \tag{5.48}$$

which in fact looks like a linear CVDS model, except that the addition and multiplication operations are in the $(\max, +)$ algebra. As in CVDS, the matrix $\mathbf{A}$ contains all the information relevant to the system dynamics. Some of its properties are described below.

Critical Circuit of the System Matrix

We can associate with the matrix $\mathbf{A}$ a simple directed graph as shown in Fig. 5.15. The number of nodes is equal to the dimension of the (square) matrix $\mathbf{A}$. Each arc corresponds to an entry of the matrix, and is given a weight specified by that entry. In our case, $\mathbf{A}_{11} = C_a$, which is the weight of the $(1, 1)$ arc. Similarly, the weight of the $(1, 2)$ arc is given by $\mathbf{A}_{12} = -L$.

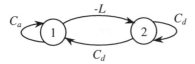

Figure 5.15: Circuits for matrix **A**.
*Each arc is associated with an entry of the **A** matrix, and is assigned a weight specified by that entry. A circuit is formed by every closed loop in the diagram. In this example, the circuits are $(1, 1)$, $(2, 2)$, and $(1, 2, 1)$.*

Every closed loop in the graph forms a *circuit*. We are only interested in simple circuits, where each node is visited only once. In Fig. 5.15, there are three such circuits: $(1, 1)$, $(1, 2, 1)$, and $(2, 2)$. The *length* of a circuit is the number of arcs forming the circuit. The weight of a circuit is the sum of arc weights in the circuit. The *average weight* of a circuit is its weight divided by its length. In Fig. 5.15, the average weights are: C_a for the $(1, 1)$ circuit; $(-L + C_d)/2$ for the $(1, 2, 1)$ circuit; and C_d for the $(2, 2)$ circuit.

Definition. The *critical circuit* of a matrix **A** is the circuit with the maximum average weight. ♦

In Fig. 5.15, the critical circuit is $(1, 1)$, since $C_a > C_d > (-L + C_d)/2$, under our assumption that $C_a > C_d$. Note that its average weight is C_a. Also note that the length of the critical circuit is 1.

Eigenvalues and Eigenvectors

In classical system theory, some of the most basic analytical techniques rely on the determination of the eigenvalues and eigenvectors of the system matrix. In regular algebra, $\lambda \in \mathbb{R}$ is an eigenvalue of a matrix **A**, and $\mathbf{u} \neq \mathbf{0}$ is an eigenvector, if λ and $\mathbf{u}$ satisfy the following relationship:

$$\lambda \mathbf{u} = \mathbf{A} \mathbf{u}$$

Therefore, a reasonable question to raise is whether there exist λ and $\mathbf{u}$ that satisfy a similar relationship in the (max, +) algebra, with the vector $\mathbf{u}$ such that its entries are not all η (i.e., $-\infty$).

Let $\mathbf{u} = [u_1, u_2]^T$, and $\lambda \in \mathbb{R}$. We have

$$\lambda \mathbf{u} = \begin{bmatrix} u_1 + \lambda \\ u_2 + \lambda \end{bmatrix}, \quad \mathbf{A} \mathbf{u} = \begin{bmatrix} C_a & -L \\ C_d & C_d \end{bmatrix} \begin{bmatrix} u_1 \\ u_2 \end{bmatrix} = \begin{bmatrix} u_1 + C_a \\ \max(u_1, u_2) + C_d \end{bmatrix}$$

If $\mathbf{u}$ is an eigenvector of **A**, and λ is an eigenvalue, we must have

$$\lambda = C_a$$
$$u_2 + \lambda = \max(u_1, u_2) + C_d$$

Setting $u_1 = u_2 + C_a - C_d$, where $C_a > C_d$, we find that for any $u \in \mathbb{R}$ the vector

$$\mathbf{u} = \begin{bmatrix} u + (C_a - C_d) \\ u \end{bmatrix}$$

is an eigenvector, with eigenvalue $\lambda = C_a$. Note that λ is precisely the average weight of the critical circuit in Fig. 5.15. In fact, it can be shown that this is always the case in the $(\max, +)$ algebra.

Periodicity of the System Matrix

Returning to the state equation (5.48), note that

$$\mathbf{x}_k = \mathbf{A}\mathbf{x}_{k-1} = \mathbf{A}^2 \mathbf{x}_{k-2} = \cdots = \mathbf{A}^k \mathbf{x}_0, \qquad \mathbf{x}_0 = \begin{bmatrix} C_a \\ 0 \end{bmatrix} \qquad (5.49)$$

where $\mathbf{A}^k = \mathbf{A} \otimes \cdots \otimes \mathbf{A}$ (k times). In order to calculate $\mathbf{A}^k$ (and hence solve for $\mathbf{x}_k$), let us first rewrite the matrix $\mathbf{A}$ as follows:

$$\mathbf{A} = \begin{bmatrix} C_a & -L \\ C_d & C_d \end{bmatrix} = C_a \begin{bmatrix} 0 & -L' \\ -(C_a - C_d) & -(C_a - C_d) \end{bmatrix} \qquad (5.50)$$

where $-L' = -L + C_a$, and $-(C_a - C_d) < 0$. Setting

$$\mathbf{B} = \begin{bmatrix} 0 & -L' \\ -(C_a - C_d) & -(C_a - C_d) \end{bmatrix}$$

we now calculate $\mathbf{B}^k$, $k = 2, 3, \ldots$,

$$\begin{aligned} \mathbf{B}^2 &= \begin{bmatrix} 0 & -L' \\ -(C_a - C_d) & -(C_a - C_d) \end{bmatrix} \begin{bmatrix} 0 & -L' \\ -(C_a - C_d) & -(C_a - C_d) \end{bmatrix} \\ &= \begin{bmatrix} 0 & -L' \\ -(C_a - C_d) & b_2 \end{bmatrix} \end{aligned}$$

where

$$b_2 = \max\{-L' - (C_a - C_d), \; -2(C_a - C_d)\}$$

Proceeding in the same manner, we can obtain

$$\mathbf{B}^k = \begin{bmatrix} 0 & -L' \\ -(C_a - C_d) & b_k \end{bmatrix} \qquad (5.51)$$

where

$$b_k = \max\{-L' - (C_a - C_d), \; -k(C_a - C_d)\} \qquad (5.52)$$

Observe that all entries of $\mathbf{B}^k$, $k = 2, 3, \ldots$, are fixed, except for b_k. However, looking at (5.52), note that as k increases there will be some value, say k^*, such that $-k(C_a - C_d)$ becomes sufficiently small to ensure that

$$
\begin{aligned}
b_k &= \max\{-L' - (C_a - C_d), \ -k(C_a - C_d)\} \\
&= -L' - (C_a - C_d) \qquad \text{for all } k > k^*
\end{aligned}
$$

It follows that, for sufficiently large k, we have

$$
\mathbf{B}^k = \begin{bmatrix} 0 & -L' \\ -(C_a - C_d) & -L' - (C_a - C_d) \end{bmatrix} = \mathbf{P}
$$

where $\mathbf{P}$ is a fixed matrix. Of course, the value of k^* depends on our choice of $-L'$, so the closer to 0 the value of this negative number can be made, the smaller k^* needs to be.

Definition. A matrix $\mathbf{M}$ is said to be n-periodic in the $(\max, +)$ algebra if and only if there exists an integer k^* such that

$$
\mathbf{M}^{k+n} = \mathbf{M}^k \qquad \text{for all } k > k^* \qquad \blacklozenge
$$

Thus, the matrix $\mathbf{B}$ defined earlier is 1-periodic, since for sufficiently large k we have $\mathbf{B}^{k+1} = \mathbf{B}^k = \mathbf{P}$. Recall that the length of the critical circuit of $\mathbf{A}$ is also 1. In fact, it can be shown that if there is a unique critical circuit whose length is n, then the matrix $\mathbf{B} = \mathbf{A}/w$ is n-periodic, where w is the average weight of this critical circuit. The division $\mathbf{A}/w$ is defined in the $(\max, +)$ algebra, so that $w\mathbf{B} = \mathbf{A}$. In our case, $w = C_a$, and $w\mathbf{B} = \mathbf{A}$ is given in (5.52).

We can now write (5.49) in terms of $\mathbf{P}$. Then, for sufficiently large k (i.e., after a sufficiently large number of events), we have

$$
\begin{aligned}
\mathbf{x}_k &= \begin{bmatrix} a_{k+1} \\ d_k \end{bmatrix} = (kC_a)\mathbf{P}\mathbf{x}_0 \\
&= (kC_a)\begin{bmatrix} 0 & -L' \\ -(C_a - C_d) & -L' - (C_a - C_d) \end{bmatrix}\begin{bmatrix} C_a \\ 0 \end{bmatrix}
\end{aligned}
$$

This gives us the "steady state" values of the arrival and departure times:

$$
\begin{aligned}
a_{k+1} &= (kC_a) \otimes C_a = (k+1)C_a & (5.53) \\
d_k &= (kC_a) \otimes C_d = kC_a + C_d & (5.54)
\end{aligned}
$$

In particular, (5.54) indicates that the kth departure occurs after k arrivals and a service delay C_d. This departure time pattern is a reflection of the periodicity of $\mathbf{B}$.

In summary, the $(\max, +)$ algebra provides a framework for both modeling and analysis of DES, which is analogous to the framework used for linear

CVDS (once the obstacle of replacing regular algebra operations by $\oplus$ and $\otimes$ is overcome). The determination of the periodicity properties of the system matrix and the calculation of its eigenvalues are useful tools for understanding the dynamic behavior of a DES. One limitation of the $(\max, +)$ algebra lies in the fact that the inverse of the max operation does not exist. Also note that the types of models, similar to (5.47), that we can derive and analyze in the $(\max, +)$ framework are limited to the class of marked graphs, which we discussed in Section 5.3.1. Another drawback is that extensions to stochastic timed DES models are not easily obtained. Thus, when event times are not deterministically specified, the type of analysis carried out in this setting cannot be applied.

It is worthwhile to point out, however, that the $(\max, +)$ algebra approach clearly reveals a salient feature of timed DES models, apparent in state automata and Petri net models as well: The only operations required in the analysis of DES are

- addition, to increment/decrement clocks, and

- max (or min), to compare clock values used in determining the events causing state transitions.

5.5 CONCLUDING COMMENTS

As the reader may have noticed, the addition of a *deterministic* clock structure $\mathbf{V}$ to an untimed model (automaton or Petri net) means that the language generated by the timed model becomes a *single string* of events for a given initial state, if there are no ties between residual lifetimes or if ties are resolved in a deterministic manner. The language generated by the untimed model contains, in general, a set of strings of events, namely all possible strings of events from the initial state *for all possible clock structures*; this set reduces to a single string if a specific clock structure is selected.

An important objective of the subsequent chapters of this book is to study stochastic DES, in particular stochastic timed automata. When the clock structure becomes stochastic - the topic of the next chapter - the resulting stochastic timed model will no longer generate a single string of events for a given initial state but rather a set of strings corresponding to all possible realizations of the underlying random variables associated with the clock structure. While in Chapters 2 and 3 we were interested in the "logical" properties of this set of strings, with no reference to the probability of a string, in the chapters that follow we will be interested in quantitative properties associated with this set of strings of events, or with the set of states visited by these strings, leading us to define and study stochastic processes that are members of the class of *generalized semi-Markov processes*. It is in this context that the approach adopted in this chapter for introducing time into untimed models should be interpreted.

One could imagine, however, introducing time into an untimed model in a manner that exhibits the randomness in the system but without the explicit use of probability distributions for the clock structure. By this we mean that the timed model would allow for more than one string to be generated by the system without attaching probabilities to these strings. A simple example would be to attach time intervals to the transitions in an automaton, with no explicit probabilistic information other than the event can only occur during the given interval. A transition in the state transition diagram would be labeled by an event name and a time interval during which this event could happen from the current state. This time interval would be with respect to a specific clock (or timer) and the semantics of the model would determine when the clock would have to be reset to zero. In this manner, a timed model would account for relevant timing information while still allowing for several possible behaviors (strings of events) of the system, but without explicit consideration of probabilities of strings. Such models could be used for instance to determine if some timing constraints are satisfied by all possible strings of events; an example is the constraint "state x is always reached from state x_0 before time T." One could also formulate control problems for such timed models; in the same vein, a control objective could be to dynamically enable/disable events in order to guarantee that state x is always reached from state x_0 before time T.

A presentation of timed models of the above form is beyond the scope of this book. Some relevant references are mentioned at the end of the chapter. In particular, we mention three specific timed models that have received consideration and for which results are available for analysis and/or controller synthesis: the "timed automaton model" of Alur & Dill, the "timed DES model" of Brandin & Wonham, and the "timed transition model" of Ostroff; roughly speaking, these three models are based on the idea of associating time intervals to the transitions, although they have some differences in their semantics.

As our final remark, we note that timing considerations could be included into an untimed model without explicitly resorting to (continuous or discrete) quantification of time and without using time intervals and timers, but rather by the use of *logical operators* such as "*eventually* [(···) will be true]", "[(···) is true] *until* [(···) becomes true]", or "*henceforth* [(···) is true]", where (···) means some logical expression on the states of the system. Various types of timing specifications on the system behavior can be stated as logical clauses involving such operators. This is the realm of *temporal logic*. Again, we refer the reader to the references at the end of the chapter for coverage of this topic.

Clearly, the choice of an appropriate modeling formalism among the different ones mentioned above for timed and stochastic DES should be dictated by the salient features of the analysis/control/optimization problem at hand.

SUMMARY

- A *timed automaton* is an automaton equipped with a *clock structure*. The clock structure is a set of sequences, one for each event, defining *event lifetimes*, and acting as the input to our model. This information is used to generate the event sequence that drives the underlying automaton.

- In a timed automaton, an event e is *activated* if it just occurred and it remains feasible in the new state; or if a different event occurs while e was not feasible, causing a transition to a new state where e is feasible. Whenever an event is activated, its *clock* is assigned a lifetime from the associated clock sequence.

- An event e is *deactivated* when a different event occurs causing a transition to a new state where e is not feasible. If an event is deactivated, its clock value is discarded.

- The *triggering event* at a given state is the one with the smallest clock value among all events that are active at that state. When the triggering event occurs, all other event clocks are reduced by the amount of time elapsed since the previous event occurrence, unless an event is deactivated.

- The *age* of a currently active event is the time elapsed since its most recent activation. Its *residual lifetime* is the time remaining until its next occurrence. The clock value of an active event is always its residual lifetime.

- In computer-generated timed sample paths of DES, the *event scheduling scheme* is used to simulate the operation of a timed automaton with a given a clock structure as input. The event scheduling scheme shown in Fig. 5.10 captures the main idea behind discrete event simulation, which we will further discuss in Chapter 10.

- A *timed Petri net* is a Petri net equipped with a clock structure, where each clock sequence is associated with a transition.

- The $(\max, +)$ algebra provides an algebraic modeling framework which parallels that of linear CVDS. In the $(\max, +)$ algebra, we can study the periodicity properties of a system matrix and obtain its eigenvalues and eigenvectors, in order to characterize its steady state behavior.

- There are a number of timed modeling frameworks for DES. Timed automata contain all the elements we need to proceed with our study of stochastic DES in the next chapter. Timed Petri nets provide added generality at the expense of greater modeling complexity. The $(\max, +)$ approach provides more detailed algebraic tools, but it is limited to certain classes of DES. Choosing an appropriate model is part of one's study of a DES, but one should also remember: There may be differences between various models, but the underlying DES is still the same.

PROBLEMS

5.1 Consider a DES with state space $X = \{x_1, x_2, x_3\}$, event set $E = \{\alpha, \beta, \gamma\}$, and state transition diagram as shown. Suppose α events are scheduled to occur at times $(0., 1., 2.3, 5.6)$ and similarly for β events at $(0.2, 3.1, 6.4, 9.7)$ and for γ events at $(2.2, 5.3)$.

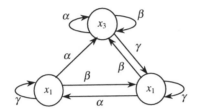

(a) Given that the initial state is x_1, construct a timing diagram for the resulting sample path over the interval $[0., 10.]$ (similar to Fig. 5.3 or Fig. 5.6), making sure you include times when events occur, their type, and the system state in every interevent interval (do not explicitly show all updated clock values at all states, unless it helps you in the construction).

(b) Suppose we are interested in estimating the *probability of finding the system at state x_2*. Based on this short finite sample path, what is your estimate of this probability?

5.2 Customers arrive at a bank where they either form a single queue waiting for one of two ATMs and taking the first available one, or they join a queue waiting to be served by a teller. Every customer prefers to join the ATM queue, but if there are already four people in that area (two in queue and two using the ATMs), the customer is forced to wait for the teller. The teller always takes twice as long as an ATM for any kind of transaction. Transactions are always classified as T_1, T_2, or T_3 and they require 1.0 min, 0.5 min, and 0.2 min respectively on an ATM. The schedule of customer arrival times and their transaction types over the interval $[0.0, 2.0]$ min is as follows:
$(0.0, T_1)$, $(0.2, T_2)$, $(0.25, T_1)$, $(0.3, T_1)$, $(0.4, T_1)$, $(0.6, T_3)$, $(0.9, T_3)$, $(1.1, T_2)$, $(1.15, T_1)$, $(1.4, T_2)$, $(1.45, T_1)$, $(1.7, T_3)$, $(1.8, T_3)$, $(2.0, T_2)$
Assume that the system is initially empty, and that if two events occur at the same time a completion of processing precedes an arrival.

(a) Define a timed automaton model for this DES using three event types and two state variables.

(b) Construct a timing diagram for the sample path (as in Problem 5.1) over the interval $[0.0, 2.5]$ min.

5.3 A transmitter with capacity 1024 bits/sec handles two types of data packets: long (L) and short (S) ones. The L packets are 1024 bits long and the S packets are 128 bits long. When packets arrive and the transmitter is busy, they wait in a queue of infinite capacity. In deciding which type of packet to process next, the transmitter works as follows: It always favors L packets over S packets, unless there are no L packets waiting or there are ≤ 1 L packets waiting and ≥ 2 S packets waiting. In addition, if an S packet has waited for more than 0.5 sec in queue, it is immediately discarded and lost. The schedule of packet arrival times and their types over the interval $[0.0, 2.0]$ sec is as follows:
$(0.0, L)$, $(0.2, S)$, $(0.3, S)$, $(0.55, S)$, $(0.57, S)$, $(0.8, L)$, $(1.1, S)$, $(1.15, L)$, $(1.25, S)$, $(1.28, S)$, $(1.3, L)$, $(1.5, L)$, $(1.6, S)$, $(1.65, S)$, $(1.75, L)$
Assume that the system is initially empty, and that if two events occur at the same time a completion of transmission precedes an arrival.

(a) Define a timed automaton model for this DES using three event types and two state variables.

(b) Construct a timing diagram for the sample path (as in Problems 5.1 and 5.2) over the interval $[0.0, 2.5]$ sec.

(c) Based on your observations over this sample path, what is your estimate of the probability that an S packet is lost?

5.4 Similar to Problem 2.13, consider a processor that breaks down a job into two tasks, first executes task 1, then executes task 2, and then the job leaves the system. Now, however, let the system have a total capacity of 3 jobs (not tasks). Note that the processor must finish *both* tasks for one job before proceeding to the next job. There is only one timeout mechanism for each job, and it operates as follows: When a job enters the system, it is given 1.0 sec to complete task 1; if the job has resided in the system for 1.0 sec and task 1 is not done, then the job is immediately thrown out and the next job in queue (if one is present) is processed. Note that there are six possible event types in this system: job arrival, completion of task 1, completion of task 2, timeout of the first job present, timeout of the second job present, and timeout of the third job present.

(a) With the system initially empty, suppose we are given a job arrival schedule at times $(0., 0.2, 0.9, 1.6, 2.0)$. We are also given the precise task execution times for each of the five jobs:
Job 1: 0.8 sec for task 1 and 0.7 sec for task 2

Job 2: 0.9 sec for task 1 and 0.7 sec for task 2
Job 3: 0.3 sec for task 1 and 0.6 sec for task 2
Job 4: 0.9 sec for task 1 and 0.5 sec for task 2
Job 5: 0.1 sec for task 1 and 1.2 sec for task 2
Show a timing diagram for this sample path, including event occurrence times and event types. Based on your diagram, determine which jobs successfully complete and which are thrown out for exceeding the 1.0 sec. deadline for task 1.

(b) Construct a state transition diagram for this DES using *only one* state variable (you should require seven states).

5.5 For each one of Problems 5.1 through 5.4, use the event scheduling scheme discussed in Section 5.2.5 to construct the corresponding sample paths. In particular, show the scheduled event list (see Fig. 5.6) upon initialization and after every event occurrence in the sample path.

5.6 Consider the Petri net in Fig. 5.11 (Example 5.3) which models the concurrent execution of task T_3 with the sequence of tasks T_1 and T_2, and then the execution of task T_4 combining the output of T_2 and T_3. Suppose we attempt to simplify the model by eliminating transition t_{23}, having transitions t_2 and t_3 directly deposit a token in place p_4, and assigning a weight of 2 to the arc (p_4, t_4).

(a) Will this work? Justify your answer.

(b) For the original Petri net in Fig. 5.11, derive a set of recursive equations for the transition firing times.

(c) Repeat (b) for the modified model. What is the difficulty in this case?

5.7 Consider the queueing model shown. We may think of this as a simple manufacturing process, where an infinite supply of parts is provided to machine 1, there is an infinite-capacity buffer between machines 1 and 2, and no buffer at all between machines 2 and 3. Thus, when a part completes processing at machine 2, it can only depart if machine 3 is idle; otherwise, the part remains at machine 2 until machine 3 becomes idle. This is referred to as *blocking*[1]: Machine 2 is "blocked" by such parts and cannot be used even though there is nothing wrong with it.

---

[1] As we noted in Chapter 1 (see "Manufacturing Systems" example in Section 1.3.4), the term "blocking" here refers to *time-blocking*, which is different from the logical notion of blocking that was considered in Chapters 2 and 3.

Let the processing times at the three machines be given constants v_1, v_2, and v_3 respectively, and let $d_{i,k}$ be the time when the kth part leaves machine i, $i = 1, 2, 3$. Assume the system is initially empty. Construct a timed Petri net graph, and derive a set of recursive equations, similar to (5.43) and (5.44), for $d_{i,k}$.

5.8 Derive a max-plus model for the system of Problem 5.7. Determine the circuits of the system matrix and its eigenvalues and eigenvectors.

SELECTED REFERENCES

■ *Timed Automata and Timed Petri Nets*

– Alur, R., and D.L. Dill, "A Theory of Timed Automata," *Theoretical Computer Science*, No. 126, pp. 183-235, 1994.

– Brandin, B., and W.M. Wonham, "Supervisory Control of Timed Discrete-Event Systems," *IEEE Transactions on Automatic Control*, Vol. 39, No. 2, pp. 329-342, February 1994.

– David, R., and H. Alla, *Petri Nets & Grafcet: Tools for Modelling Discrete Event Systems*, Prentice-Hall, New York, 1992.

– Murata, T., "Petri Nets: Properties, Analysis and Applications," *Proceedings of the IEEE*, Vol. 77, 4, pp. 541-580, 1989.

– Ostroff, J.S., *Temporal Logic for Real-Time Systems*, Research Studies Press and John Wiley & Sons, New York, 1989.

■ *Max-Plus Algebra*

– Baccelli, F., G. Cohen, G.J. Olsder, and J.P. Quadrat, *Synchronization and Linearity*, Wiley, 1992.

– Cohen, G., P. Moller, J.P. Quadrat, and M. Viot, "Algebraic Tools for the Performance Evaluation of Discrete Event Systems," *Proceedings of the IEEE*, Vol. 77, 1, pp. 39-58, 1989.

– Cunningham-Greene, R.A., *Minimax Algebra*, Springer-Verlag, Berlin, 1979.

■ *Miscellaneous*

– Benveniste, A., and P. Le Guernic, "Hybrid Dynamical Systems Theory and the SIGNAL Language," *IEEE Transactions on Automatic Control*, Vol. 35, No. 5, pp. 535-546, May 1990.

– Holloway, L.E., B.H. Krogh, and A. Giua, "A Survey of Petri Net Methods for Controlled Discrete Event Systems," *Journal of Discrete Event Dynamic Systems: Theory and Applications*, Vol. 7, No. 2, pp. 151–190, April 1997.

- Levi, S.-T., and A.K. Agrawala, *Real-Time System Design*, McGraw-Hill, New York, 1990.

- Manna, Z., and A. Pnueli, *The Temporal Logic of Reactive and Concurrent Systems: Specification*, Springer-Verlag, New York, 1992.

- Manna, Z., and A. Pnueli, *Temporal Verification of Reactive Systems: Safety*, Springer-Verlag, New York, 1995.

- Ostroff, J.S., "Formal Methods for the Specification and Design of Real-Time Safety-Critical Systems," *Journal of Systems and Software*, Vol. 18, pp. 33-60, April 1992.

Chapter 6
Stochastic Timed Automata

6.1 INTRODUCTION

In practice, systems always operate in environments which are constantly plagued by uncertainty. This is especially true in dealing with DES, which, by their nature, often involve unpredictable human actions and machine failures. The process of resource sharing (which provides an important motivation for studying DES) is inherently characterized by such unpredictability: changing user demand, computer breakdowns, inconsistencies in human decision making, etc. While the untimed (or logical) models considered in Chapters 2 to 4 do account for "all possible behaviors" of the system, their use is limited to *logical* (or qualitative) performance objectives. Deterministic timed models of the type considered in Chapter 5 certainly contribute to our basic understanding of some *quantitative* properties of the dynamic behavior of a system (for instance, the periodic behavior of systems that can be modeled as marked graphs). But their use is limited since models with deterministic clock structures only capture a single timed string of events (or states), or, in other words, a single sample path of the system. If we are to develop either descriptive or prescriptive techniques for evaluating performance and for "optimally" controlling timed DES with respect to quantitative performance measures and in the presence of uncertainty, more refined models that incorporate stochastic elements are required.

In this chapter we introduce some basic stochastic models which will allow us to extend the framework developed thus far, and to derive quantitative means of analysis. In particular, we concentrate on timed automata models. We saw that the input for these models is a clock structure, which determines the actual event sequence driving the underlying automaton. Our main effort will be to develop stochastic models for the clock structure, and to study the properties of the resulting DES.

The first objective here is to review the fundamental ideas involved in defining and categorizing stochastic processes. This will allow us to establish the framework we will be using for most of the remainder of this book in studying DES. As is often the case, we proceed by narrowing down a general framework to a class of simpler, but useful, models. These are based on the introduction of the *Poisson counting process* as a basic stochastic building block. Understanding the properties of this process and its use in DES modeling is the second major objective of this chapter.

The following sections are based on the premise that readers have a fundamental understanding of probability theory and related techniques. Appendix I provides a review of this material, which should be well-digested before attempting to launch into stochastic modeling for DES.

6.2 STOCHASTIC PROCESS BASICS

We briefly review the definition of a stochastic process (see also Appendix I) before discussing various means of classification. A *stochastic* or *random process* is simply a collection of random variables indexed through some parameter (which is normally thought of as "time"). Here is a simple example. Let Ω = {UP, DOWN} be the sample space for describing the random state of a machine. By mapping UP into 1 and DOWN into 0 we may define a random variable $X(\omega)$, $\omega \in \Omega$, which takes the values 0 or 1. Thus, $X(\omega) = 1$ means the machine is UP, and $X(\omega) = 0$ means the machine is DOWN. Next, suppose we wish to describe the state of the machine once a day every day, and let $k = 1, 2, \ldots$ index the days. Thus, $X(\omega, k)$ is a random variable describing the state of the machine on the kth day. The collection of random variables $\{X(\omega, 1), X(\omega, 2), \ldots, X(\omega, k), \ldots\}$ defines a stochastic process.

Note that a random variable $X(\omega)$ is defined through a mapping from some sample space to the set of real numbers (a point-to-point mapping), $X : \Omega \to \mathbb{R}$. A random process $X(\omega, t)$, on the other hand, is defined through a mapping from a sample space to the set of time functions (a point-to-function mapping).

A stochastic process $X(\omega, t)$ is often used to characterize the state of a system. If ω is fixed, the resulting deterministic time function is a *sample path* or *realization* of the stochastic process. A sample path may be thought of as the outcome of a particular random experiment describing the state of the system over time. On the other hand, if t is fixed, we obtain a random variable. This random variable may be thought of as describing the (random) state of the system at time t.

Definition. A *stochastic* or *random process* $X(\omega, t)$ is a collection of random variables indexed by t. The random variables are defined over a common probability space $(\Omega, \mathbb{E}, P)$ with $\omega \in \Omega$. The variable t ranges over some given set $\mathcal{T} \subseteq \mathbb{R}$.   ♦

The precise definition of a probability space is given in Appendix I; the reader is reminded that $\mathbb{E}$ is an event space (not to be confused with "events" which cause state transitions in DES), and P is a probability measure. To simplify notation, we will normally denote a stochastic process by $\{X(t)\}$. When there is no ambiguity, we also use $X(t)$ to denote the process. We will also use upper-case letters to denote random variables, and reserve lower-case letters to denote values that these random variables may take. For example, x is a value that the random variable X may take.

To classify stochastic processes, we look at (a) possible state spaces, (b) the nature of the index set $\mathcal{T}$, and (c) the statistical characteristics of the random variables defining the process.

6.2.1 Continuous-state and Discrete-state Stochastic Processes

As in the classification of systems, we are interested in the state space of a stochastic process, that is, the set of allowable values for the random variables $X(t)$, $t \in \mathcal{T}$. If $\{X(t)\}$ is defined over a finite or countable set (e.g., the set of non-negative integers $\{0, 1, 2, \dots\}$), then it is called a *discrete-state process*. Usually, a discrete-state process is referred to as a *chain*. In all other cases, $\{X(t)\}$ is called a *continuous-state process*.

As we have seen, DES are characterized by discrete state spaces. Thus, of particular interest to our study is the use of chains for modeling stochastic DES.

6.2.2 Continuous-time and Discrete-time Stochastic Processes

If the set $\mathcal{T}$, which specifies the allowable values of the index variable t, is finite or countable, then $\{X(t)\}$ is called a *discrete-time* process. In this case, $\mathcal{T}$ is normally taken to be the set of non-negative integers $\{0, 1, 2, \dots\}$, and we use the term *stochastic* or *random sequence* rather than process. We also use the notation $\{X_k\}$, $k = 0, 1, 2, \dots$ In all other cases, we use the notation $\{X(t)\}$ and refer to it as a *continuous-time process*.

6.2.3 Some Important Classes of Stochastic Processes

In order to fully characterize a real variable x, it suffices to assign it a real number. To fully characterize a random variable X, we normally specify a cumulative distribution function (cdf) $F_X(x) = P[X \leq x]$ for it. To fully characterize a stochastic process $\{X(t)\}$ we specify the joint cdf that describes the interdependence of *all* random variables that define the process. Thus, we define a random vector

$$\mathbf{X} = [X(t_0), X(t_1), \ldots, X(t_n)]$$

which can take values

$$\mathbf{x} = [x_0, x_1, \ldots, x_n]$$

Then, for all possible $(x_0, x_1, \ldots, x_n)$ and $(t_0, t_1, \ldots, t_n)$, it is necessary to specify a joint cdf

$$F_X(x_0, \ldots, x_n; t_0, \ldots, t_n) = P[X(t_0) \leq x_0, \ldots, X(t_n) \leq x_n] \quad (6.1)$$

For an arbitrary stochastic process this is an extremely tedious task. Fortunately, for many of the stochastic processes which turn out to be useful models of systems we encounter in practice, there are relatively simple means to accomplish this task. In what follows, we identify some of the most important classes of stochastic processes we will use in modeling DES.

Stationary Processes

The notion of stationarity was discussed in the context of systems in Chapter 1. Stationarity is the property of a system to maintain its dynamic behavior invariant to time shifts. The same idea applies to stochastic processes. Formally, $\{X(t)\}$ is said to be a *stationary process* if and only if

$$F_X(x_0, \ldots, x_n; t_0 + \tau, \ldots, t_n + \tau) = F_X(x_0, \ldots, x_n; t_0, \ldots, t_n)$$
$$\text{for any } \tau \in \mathbb{R} \quad (6.2)$$

In simple terms, we can choose to study a stationary process at any point in time and be certain that its stochastic behavior will always be the same. This is particularly important, since it implies that an experiment performed now on a stationary process is statistically indistinguishable from the same experiment repeated τ time units later (or earlier).

The property defined through (6.2) is sometimes referred to as *strict-sense stationarity*, to differentiate it from a weaker requirement called *wide-sense stationarity*. $\{X(t)\}$ is said to be a *wide-sense stationary process* if and only if

$$E[X(t)] = C \quad \text{and} \quad E[X(t) \cdot X(t + \tau)] = g(\tau) \quad (6.3)$$

where C is a constant independent of t, and $g(\tau)$ is a function of τ but not of t. Thus, wide-sense stationarity requires the first two moments only (as opposed to *all* moments) to be independent of t.

Independent Processes

The simplest possible stochastic process we can consider is just a sequence of random variables $\{X_0, X_1, \ldots, X_n\}$ which are all mutually independent. We say that $\{X_k\}$, $k = 0, 1, \ldots, n$, is an independent process if and only if

$$F_X(x_0, \ldots, x_n; t_0, \ldots, t_n) = F_{X_0}(x_0; t_1) \cdots F_{X_n}(x_n; t_n) \qquad (6.4)$$

If, in addition, all random variables are drawn from the same distribution, then the entire stochastic behavior of the process is captured by a single cdf, $F_{X_k}(x_k; t_k)$, for any $k = 0, 1, \ldots, n$. In this case, we refer to $\{X_k\}$ as a sequence of *independent and identically distributed* random variables, or just an *iid sequence*.

Markov Processes

Suppose we observe a chain from time t_0 up to time t_k, and let $t_0 \leq t_1 \leq \cdots \leq t_{k-1} \leq t_k$. Let us think of the *observed* value x_k at t_k as the "present state" of the chain, and of $\{x_0, x_1, \ldots, x_{k-1}\}$ as its observed "past history". Then, $\{X_{k+1}, X_{k+2}, \ldots\}$, for time instants $t_{k+1} \leq t_{k+2} \leq \cdots$, represents the unknown "future". In an independent chain, this future is completely independent of the past. This is a very strong property, which is partly relaxed in a *Markov chain*. In this type of chain, the future is *conditionally* independent of the past history, given the present state. In other words, the entire past history is summarized in the present state. This fact is often referred to as the *memoryless property*, since we need no memory of the past history (only the present) to probabilistically predict the future.

Discrete-state or continuous-state processes may possess this property. Formally, we define $\{X(t)\}$ to be a *Markov process* if

$$\begin{aligned} P[X(t_{k+1}) \leq x_{k+1} \mid X(t_k) = x_k, X(t_{k-1}) = x_{k-1}, \ldots, X(t_0) = x_0] \\ = P[X(t_{k+1}) \leq x_{k+1} \mid X(t_k) = x_k] \end{aligned} \qquad (6.5)$$

for any $t_0 \leq t_1 \leq \cdots \leq t_k \leq t_{k+1}$. This memoryless property is also referred to as the *Markov property*, after A.A. Markov who first described and studied this class of processes in the early 1900's.

In the case of a *Markov chain* (discrete state space), (6.5) is written as

$$\begin{aligned} P[X(t_{k+1}) = x_{k+1} \mid X(t_k) = x_k, X(t_{k-1}) = x_{k-1}, \ldots, X(t_0) = x_0] \\ = P[X(t_{k+1}) = x_{k+1} \mid X(t_k) = x_k] \end{aligned} \qquad (6.6)$$

In the case of a discrete-time Markov chain, state transitions are constrained to occur at time instants $0, 1, 2, \ldots, k, \ldots$, and (6.6) may be written as

$$\begin{aligned} P[X_{k+1} = x_{k+1} \mid X_k = x_k, X_{k-1} = x_{k-1}, \ldots, X_0 = x_0] \\ = P[X_{k+1} = x_{k+1} \mid X_k = x_k] \end{aligned} \qquad (6.7)$$

Conceptually, the idea of summarizing the past history of a system in its present state may not be overly constraining. For instance, the chain describing a process as complex as a game of chess (where each possible next move may be associated with some probability), possesses the Markov property: The current state of the chess board summarizes the history of the game, and the probability of making a transition to any new state depends only on what this current state is (except for the castle/king exchange which needs some extra memory regarding the king's prior moves...). The constraining nature of the Markov property is revealed when one thinks as follows. Suppose that at time t_1 the process has just made a transition into state x. At time $t_2 > t_1$ let the state still be x, that is, no state transition has yet occurred. Based on the Markov property, if at t_2 we try to predict the future given the information "the state is x at time t_2" we should come up with the same result as having the information "the state is x at time t_2 and has been x over the whole interval $[t_1, t_2]$". Therefore, the amount of time that has elapsed since the chain entered state x (the "age" of the current state) must be irrelevant.

We now see that the "memoryless property" has two aspects:

(M1) All past state information is irrelevant (no state memory needed), and

(M2) How long the process has been in the current state is irrelevant (no state age memory needed).

(M2) in particular is responsible for imposing serious constraints on the nature of the random variable that specifies the time interval between consecutive state transitions. In a DES, state transitions are caused by event occurrences, so the random variable directly constrained by the memoryless property is the interevent time. Even so, there are many real-life processes we encounter in the study of DES for which this property holds or at least can be justified to a large extent. Thus, Markov chains provide a very powerful framework for the analysis of many practical DES.

Example 6.1

Consider a discrete-time chain with state space $\{\ldots, -2, -1, 0, 1, 2, \ldots\}$ defined as follows:

$$X_{k+1} = X_k - X_{k-1} \tag{6.8}$$

with initial conditions:

$$P[X_0 = 0] = 0.5, \quad P[X_0 = 1] = 0.5$$
$$P[X_1 = 0] = 0.5, \quad P[X_1 = 1] = 0.5$$

Thus, the randomness in this simple process comes only from the uncertainty in the initial conditions. Two possible sample paths are:

$$\omega_1 : \quad \{0, 0, 0, 0, 0, 0, 0, 0, \ldots\}$$
$$\omega_2 : \quad \{1, 0, -1, -1, 0, 1, 1, 0, \ldots\}$$

This chain does not possess the Markov property. This should be immediately obvious from (6.8), where we see that X_{k+1} depends not only on X_k, but also X_{k-1}. To see this more explicitly, note that

$$P[X_2 = 0 \mid X_1 = 0, X_0 = 0] = 1$$
$$P[X_2 = 0 \mid X_1 = 0, X_0 = 1] = 0 \neq P[X_2 = 0 \mid X_1 = 0, X_0 = 0]$$

which clearly shows the dependence of the values X_2 can take on the observed value for X_0.

Next, suppose we define a random vector $Z_k = [X_k, Y_k]^T$, where $Y_k = X_{k-1}$. In other words, we use the variable Y_k to "store" the $(k-1)$th value of the original chain. As a result, we generate a vector chain

$$Z_{k+1} = \begin{bmatrix} X_{k+1} \\ Y_{k+1} \end{bmatrix} = \begin{bmatrix} 1 & -1 \\ 1 & 0 \end{bmatrix} \begin{bmatrix} X_k \\ Y_k \end{bmatrix} = \begin{bmatrix} 1 & -1 \\ 1 & 0 \end{bmatrix} Z_k$$

which does possess the Markov property, since information about the current state value, $z_k = [x_k, y_k]^T$, is all that is needed to determine the value of the next state $z_{k+1} = [x_k - y_k, x_k]^T$, that is,

$$P\left[Z_{k+1} = [x_k - y_k, x_k]^T \mid Z_k = [x_k, x_k]^T\right] = 1$$

which is fixed and independent of any past history $Z_{k-1}, \ldots, Z_0$.

Semi-Markov Processes

We saw that the Markov property implies facts **(M1)** and **(M2)** in the previous section. As we will later see, a consequence of these facts is that the cdf of interevent times of a Markov chain is the exponential distribution function. A *semi-Markov process* is an extension of a Markov process where constraint **(M2)** is relaxed. As a result, the interevent times are no longer constrained to be exponentially distributed. A state transition may now occur at any time, and interevent times can have arbitrary probability distributions. However, when an event does occur, the process behaves like a normal Markov chain and obeys **(M1)**: The probability of making a transition to any new state depends only on the current value of the state (and not on any past states).

Renewal Processes

A *renewal process* is a chain $\{N(t)\}$ with state space $\{0, 1, 2, \ldots\}$ whose purpose is to count state transitions. The time intervals defined by successive state transitions are assumed to be independent and characterized by a *common* distribution. This is the only constraint imposed on the process, as this distribution may be arbitrary. Normally, we set an initial condition $N(0) = 0$. In the context of DES, a renewal process counts the number of events that occur in the time interval $(0, t]$. It should be clear that in such a counting process

$$N(0) \leq N(t_1) \leq \cdots \leq N(t_k)$$

for any $0 \leq t_1 \leq \cdots \leq t_k$.

Renewal processes are particularly important in the study of timed DES models, since the process of "counting events" is obviously of interest. In fact, as we will see, an event-counting process can serve as a building block for developing a complete model structure based on which we can analyze a large class of DES.

6.3 STOCHASTIC CLOCK STRUCTURES

Recall from Chapter 5 that timed models for DES require the specification of a clock structure $\mathbf{V} = \{\mathbf{v}_i : i \in E\}$, where $\mathbf{v}_i = \{v_{i,1}, v_{i,2}, \dots\}$ is a clock sequence for event $i \in E$. The event timing mechanism summarized in Fig. 5.4 (or equivalently the event scheduling scheme of Fig. 5.10) is then used to generate sample paths of the underlying system. Our goal here is to extend these models by incorporating stochastic features that are often encountered in DES.

Thus far, the clock structure $\mathbf{V}$ was assumed to be fully specified in a deterministic sense. Let us now assume that the clock sequences v_i, $i \in E$, are specified only as stochastic sequences. This means that we no longer have at our disposal real numbers $\{v_{i,1}, v_{i,2}, \dots\}$ for each event i, but rather a distribution function, denoted by G_i, which describes the *random* clock sequence $\{V_{i,k}\} = \{V_{i,1}, V_{i,2}, \dots\}$. As we will see, by focusing on clock sequences with particular properties, we will be able to analyze certain interesting and useful classes of timed DES models.

Definition. The *Stochastic Clock Structure* (or *Timing Structure*) associated with an event set E is a set of distribution functions

$$G = \{G_i : i \in E\} \tag{6.9}$$

characterizing the *stochastic clock sequences*

$$\{V_{i,k}\} = \{V_{i,1}, V_{i,2}, \dots\}, \quad i \in E, \quad V_{i,k} \in \mathbb{R}^+, k = 1, 2, \dots \quad \blacklozenge$$

We will limit ourselves to clock sequences which are iid and independent of each other. Thus, each $\{V_{i,k}\}$ is completely characterized by a distribution function $G_i(t) = P[V_i \leq t]$. This assumption is made to keep notation simple and avoid complications which might obscure the essential features of the models we want to discuss. There are, however, several ways in which a clock structure can be extended to include situations where elements of a sequence $\{V_{i,k}\}$ are correlated or two clock sequences are dependent on each other.

We may now revisit the timed models we considered in the previous chapter (namely, timed automata and timed Petri nets) with the understanding that the clock sequences, which are the input to these models, are stochastic processes rather than sequences of real numbers. We will choose to concentrate on automata, because they provide the simplest and most convenient setting for introducing the basic stochastic building blocks we will be using in the next few chapters. We observe, however, that the discussion on *stochastic* timed automata can be readily applied to *stochastic* timed Petri nets.

In order to avoid notational confusion between random variables and sets (both usually represented by upper-case letters), we will use $\mathcal{E}$ and $\mathcal{X}$ to denote the *event set* and the *state space* of the underlying automaton respectively.

6.4   STOCHASTIC TIMED AUTOMATA

We begin by adopting a random variable notation paralleling the one used for timed automata:

X is the current state
E is the most recent event (causing the transition into state X)
T is the most recent event time (corresponding to event E)
N_i is the current score of event i
Y_i is the current clock value of event i

As in the deterministic case, we use the prime ($'$) notation to denote the next state X', triggering event E', next event time T', next score of i, N_i', and next clock of i, Y_i'.

Besides the stochastic clock structure specification, there are two additional probabilistic features we can include in our modeling framework:

1. The initial automaton state x_0 may not be deterministically known. In general, we assume that the initial state is a random variable X_0. What we need to specify then is the *probability mass function (pmf) of the initial state*

$$p_0(x) = P[X_0 = x], \quad x \in \mathcal{X} \tag{6.10}$$

2. The state transition function f may not be deterministic. In general, we assume that if the current state is x and the triggering event is e', the next state x' is probabilistically specified through a *transition probability*

$$p(x'; x, e') = P[X' = x' \mid X = x, E' = e'], \quad x, x' \in \mathcal{X}, \ e' \in \mathcal{E} \tag{6.11}$$

Note that if $e' \notin \Gamma(x)$, then $p(x'; x, e') = 0$ for all $x' \in \mathcal{X}$.

A timed automaton equipped with a stochastic clock structure, an initial state cdf, and state transition probabilities defines a *Stochastic Timed Automaton*. The input to this system consists of the random sequences $\{V_{i,k}\}$ generated through G.

Definition. A *Stochastic Timed Automaton* is a six-tuple

$$(\mathcal{E}, \mathcal{X}, \Gamma, p, p_0, G)$$

where

$\mathcal{E}$ is a countable *event set*

$\mathcal{X}$ is a countable *state space*

$\Gamma(x)$ is a set of *feasible* or *enabled* events, defined for all $x \in \mathcal{X}$ with $\Gamma(x) \subseteq \mathcal{E}$

$p(x'; x, e')$ is a *state transition probability*, defined for all $x, x' \in \mathcal{X}, e' \in \mathcal{E}$, and such that $p(x'; x, e') = 0$ for all $e' \notin \Gamma(x)$

$p_0(x)$ is the pmf $P[X_0 = x]$, $x \in \mathcal{X}$, of the initial state X_0

and $G = \{G_i : i \in \mathcal{E}\}$ is a *stochastic clock structure*.

The automaton generates a stochastic state sequence $\{X_0, X_1, \dots\}$ through a transition mechanism (based on observations $X = x, E' = e'$):

$$X' = x' \text{ with probability } p(x'; x, e') \tag{6.12}$$

and it is driven by a stochastic event sequence $\{E_1, E_2, \dots\}$ generated through

$$E' = arg \min_{i \in \Gamma(X)} \{Y_i\} \tag{6.13}$$

with the stochastic *clock values* $Y_i, i \in \mathcal{E}$, defined by

$$Y_i' = \begin{cases} Y_i - Y^* & \text{if } i \neq E' \text{ and } i \in \Gamma(X) \\ V_{i,N_i+1} & \text{if } i = E' \text{ or } i \notin \Gamma(X) \end{cases} \quad i \in \Gamma(X') \tag{6.14}$$

where the *interevent time* Y^* is defined as

$$Y^* = \min_{i \in \Gamma(X)} \{Y_i\} \tag{6.15}$$

the *event scores* $N_i, i \in \mathcal{E}$, are defined through

$$N_i' = \begin{cases} N_i + 1 & \text{if } i = E' \text{ or } i \notin \Gamma(X) \\ N_i & \text{otherwise} \end{cases} \quad i \in \Gamma(X') \tag{6.16}$$

and

$$\{V_{i,k}\} \sim G_i \tag{6.17}$$

where the tilde ($\sim$) notation denotes "with distribution." In addition, initial conditions are: $X_0 \sim p_0(x)$, and $Y_i = V_{i,1}$ and $N_i = 1$ if $i \in \Gamma(X_0)$. If $i \notin \Gamma(X_0)$, Y_i is undefined and $N_i = 0$.   ◆

The intricacy of this definition is even scarier than that of the (deterministic) timed automaton given in Chapter 5. Again, however, the complexity is mostly in the notation rather than the basic idea, which is no different than the deterministic case:

- Compare clock values for feasible events i.

- Determine the triggering event.

- Update the process state.

- Update the clock values and scores.

In addition, event times are updated through

$$T' = T + Y^* \tag{6.18}$$

The detailed interpretation of (6.13) through (6.16) is the same as in the case of timed automata with deterministic clock structures. The next event E' is the one with the smallest clock value among all feasible events $i \in \Gamma(X)$. The corresponding clock value, Y^*, is the interevent time between the occurrence of E and E', and it provides the amount by which time moves forward in (6.18). Clock values for all events that remain active in state X' are decremented by Y^*, except for the triggering event E' and all newly activated events, which are assigned a new lifetime V_{i,N_i+1}. Event scores are incremented whenever a new lifetime is assigned to them.

Remark. As in the case of (deterministic) timed automata, it is conceivable for two events to occur at the same time, in which case we need a priority scheme over all events in $\mathcal{E}$. Normally, we will assume that every cdf G_i in the clock structure is absolutely continuous over $[0, \infty)$ (so that its density function exists) and has a finite mean. This implies that two events can occur at exactly the same time only with probability 0.

6.5   THE GENERALIZED SEMI-MARKOV PROCESS

A stochastic timed automaton is used to generate the stochastic process $\{X(t)\}$. This stochastic process is referred to as a *Generalized Semi-Markov Process* (GSMP).

Definition. A *Generalized Semi-Markov Process* (GSMP) is a stochastic process $\{X(t)\}$ with state space $\mathcal{X}$, generated by a stochastic timed automaton $(\mathcal{E}, \mathcal{X}, \Gamma, p, p_0, G)$. ♦

The Markovian aspect of the GSMP stems from the fact that the process state behaves like a Markov chain at state transition points, as (6.12) indicates:

The value that the next state X' can take depends on the current state only and not the past history. However, the interevent times, specified through (6.15), have, in general, arbitrary distributions. Recall that these were the defining characteristics of a semi-Markov process. In a semi-Markov process, however, the interevent time distribution is externally specified if the model is to be complete. In other words, no $\mathcal{E}, \Gamma$, or G information is required; instead, a distribution for Y^* in (6.15) is supplied. In a *generalized* Semi-Markov process, on the other hand, the distribution of Y^* is not given; rather, it depends on the distributions G_i, $i \in \mathcal{E}$, and the clock and score updating mechanisms defined through (6.14) and (6.16). In fact, suppose we think of another stochastic process whose "state" consists of both $X(t)$ and all the clock values Y_i such that $i \in \Gamma(X)$. This is a semi-Markov process. It is a continuous-time process, since part of its state is described by the real-valued clocks.

The GSMP is a model for the state behavior of a wide class of stochastic DES. The distributions G_i are quite general (except for the technical constraints we impose to avoid pathological problems such as two events occurring at precisely the same time). One issue that arises in this model is whether an event lifetime $V_{i,k}$ may depend on the current state x. Since this introduces complications we do not wish to consider at this point, we will assume that this is not the case. At the end of this chapter, however, we discuss means by which this limitation can be overcome.

A sample path of a GSMP is generated much like a sample path of the state of a standard timed automaton, provided the stochastic clock structure and state transition probabilities are specified. Then we proceed by applying the mechanism (6.12) through (6.18) and *sampling* from the appropriate distributions as required. By "sampling" we mean that we await for nature to determine a value of the random variable involved; alternatively, we assume the existence of an artificial "random number" generating process, as in computer simulation (to be discussed in Chapter 10). Note that the event scheduling scheme outlined in Fig. 5.10 is still applicable. The only fundamental difference is that the CLOCK STRUCTURE component is replaced by "nature", which generates the event lifetimes required (based on G) when an event is activated; alternatively, it is replaced by a "random number generator" which a computer employs for simulation purposes.

Remark. As we will see, understanding the mechanism by which stochastic timed automata generate a GSMP is tantamount to understanding the basic principles of discrete-event computer simulation, which we will discuss in Chapter 10. This point is brought up to call attention to the fact that simulation is perhaps the most general experimental tool available to us for studying complex DES, and it serves as an "electronic laboratory" in which we may not only gain insight about how specific DES behave, but also experiment with various ways for controlling them and ensuring their satisfactory performance.

Example 6.2

We repeat Example 5.1 in order to emphasize the distinction between the deterministic and stochastic aspects of the same underlying automaton, whose state transition diagram is redrawn in Fig. 6.1. A GSMP is generated by the following stochastic timed automaton model:

$$\mathcal{E} = \{\alpha, \beta\}, \quad \mathcal{X} = \{0, 1, 2\}$$
$$\Gamma(0) = \Gamma(2) = \{\alpha, \beta\}, \quad \Gamma(1) = \{\alpha\}$$
$$p(1; 0, \alpha) = 1, \quad p(2; 0, \beta) = 1$$
$$p(2; 1, \alpha) = 1$$
$$p(0; 2, \beta) = 1, \quad p(1; 2, \alpha) = 1$$

with all other transition probabilities equal to 0. This is a probabilistic formulation of the deterministic state transition structure shown in Fig. 6.1. The stochastic clock structure is completely specified through two cdf's, G_α and G_β. In addition, an initial state cdf p_0 is also provided.

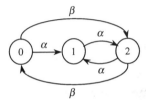

Figure 6.1: State transition diagram for Example 6.2.

The first step in constructing a sample path is to sample from p_0. In other words, we perform a simple random experiment whose outcome is characterized by $p_0 = P[X_0 = x]$, $x = 0, 1, 2$. Suppose we obtain the outcome $x_0 = 0$. This begins to define a sample path for the process state at time $t = 0.0$, as shown in Fig. 6.2.

Since $\Gamma(0) = \{\alpha, \beta\}$, both events are active and their score values are initially set to 1. In order to set their clock values, we need to sample from the two cdf's G_α and G_β. Let us assume we do so and obtain the outcomes $v_{\alpha,1} = 1.0$ and $v_{\beta,1} = 2.0$. Hence, we set

$$y_{\alpha,0} = 1.0, \qquad y_{\beta,0} = 2.0$$

as shown in Fig. 6.2.

We now have at our disposal real numbers $y_{\alpha,0}$ and $y_{\beta,0}$ to work with, just as in Example 5.1. Using (6.13) and (6.14) we easily see that

$$y_0^* = \min\{y_{\alpha,0}, y_{\beta,0}\} = y_{\alpha,0} \Rightarrow e_1 = \alpha$$

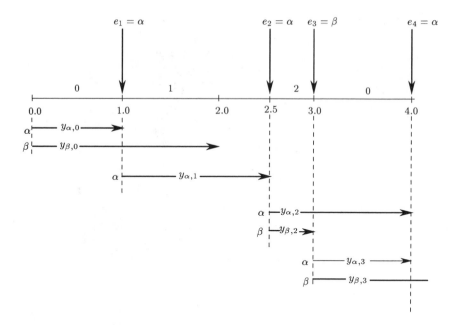

Figure 6.2: Sample path for Example 6.2.

and, therefore, the next state is determined from the fact that

$$p(1;0,\alpha) = 1 \Rightarrow x_1 = 1$$

Also, the event time is simply given by $t_1 = y_0^* = 1.0$.

We repeat the process with $\Gamma(1) = \{\alpha\}$. The event β is disabled in this state. Since α was the triggering event, it is activated again with a score incremented by 1. To determine a new clock value, we sample again from G_α to assign the random variable $V_{\alpha,2}$ in (6.14) some value $v_{\alpha,2}$. Suppose we obtain $v_{\alpha,2} = 1.5$. Then, we set $y_{\alpha,1} = 1.5$.

Obviously, the next triggering event is $e_2 = \alpha$. Since $p(2;1,\alpha) = 1$, the next state is $x_2 = 2$, and time is moved forward to $t_2 = t_1 + y_1^* = 1.0 + y_{\alpha,1} = 2.5$.

The process continues with $\Gamma(2) = \{\alpha,\beta\}$, and two new samples $v_{\alpha,3} = 1.5, v_{\beta,2} = 0.5$, as shown in Fig. 6.2. We get $e_3 = \beta, x_3 = 0$, and $t_3 = 3.0$. Finally, we sample from G_β to obtain the next clock value $v_{\beta,3} = 1.5$. The last transition shown in Fig. 6.2 occurs with $e_4 = \alpha$ at $t_4 = 4.0$.

Remark. More often than not, the state transition structures we will be dealing with are deterministic. In other words, given a state x and triggering event e', state transition probabilities are of the form $p(x'; x, e') = 1$ for some x'. In such cases, we will avoid the cumbersome probabilistic notation by specifying our model through the deterministic transition function $x' = f(x, e')$, and writing $(\mathcal{E}, \mathcal{X}, \Gamma, f, p_0, G)$ instead of $(\mathcal{E}, \mathcal{X}, \Gamma, p, p_0, G)$.

It is also worth pointing out, as in the case of deterministic timed automata models, that the automaton structure and the stochastic clock structure are quite separate. This is made clear in Example 6.2 above: Except for the sampling process (to specify event lifetimes for the clock values and the initial state), all steps for constructing the sample path are identical to the deterministic case. This observation has significant ramifications in the study of several structural properties of timed DES models, which we will examine in later chapters.

6.5.1   Queueing Systems as Stochastic Timed Automata

We return to the simple queueing system we have already considered several times in the past.[1] The system is shown in Fig. 6.3 along with its state transition diagram, and a timed automaton model with a stochastic clock structure defined through G_a and G_d. This automaton gives rise to a GSMP $\{X(t)\}$, where $X(t)$ represents the queue length, that is, the state of a simple queueing system.

Example 6.3
We construct a sample path of the queueing system in Fig. 6.3, where we are given a deterministic initial condition $x_0 = 0$, and the following clock structure:

$$V_{a,k} = 1.0 \quad \text{for all } k = 1, 2, \ldots$$
$$V_{d,k} \sim G_d \quad \text{for all } k = 1, 2, \ldots$$

where G_d is some given cdf.

Thus, arrivals occur deterministically every 1.0 time unit, whereas service times are iid random variables drawn from a common distribution G_d. Suppose that we sample from this distribution and obtain a sequence of service times $\{s_1, s_2, \ldots\}$. The resulting sample path (for the first five departures) is shown in Fig. 6.4. Note that the first d event is activated at time $t = 1.0$ following the first arrival. Similarly, the fourth d event is activated at time $t = 4.0$, when the queue length becomes positive again after the third departure.

6.5.2   GSMP Analysis

Given a particular stochastic clock structure G, we have seen how a timed automaton generates a GSMP $\{X(t)\}$ to model the state of a stochastic DES. Our goal then is to deduce the characteristics of this process.

---

[1]For an interactive simulation of a queueing system, the reader is referred to the Web site http://vita.bu.edu/cgc/MIDEDS/.

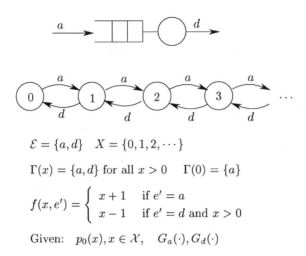

$$\mathcal{E} = \{a, d\} \quad X = \{0, 1, 2, \cdots\}$$

$$\Gamma(x) = \{a, d\} \text{ for all } x > 0 \quad \Gamma(0) = \{a\}$$

$$f(x, e') = \begin{cases} x + 1 & \text{if } e' = a \\ x - 1 & \text{if } e' = d \text{ and } x > 0 \end{cases}$$

Given: $p_0(x), x \in \mathcal{X}, \quad G_a(\cdot), G_d(\cdot)$

Figure 6.3: GSMP model for a simple queueing system.

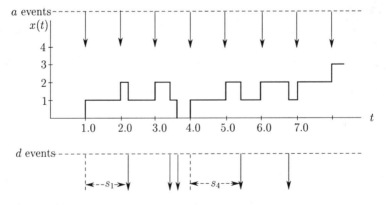

Figure 6.4: A sample path of the queueing system in Example 6.3.

Perhaps our most fundamental task is to calculate the probability of finding the DES at state x at time t, that is, $P[X(t) = x]$. In this way, we can probabilistically assess which states are more likely than others. Note that embedded in this process is the state sequence $\{X_k\} = \{X_0, X_1, \ldots\}$, where $k = 1, 2, \ldots$ counts event occurrences. In addition, however, there are various other processes of interest which are related to $X(t)$. Such processes are:

- The *interevent time sequence* $\{Y_k^*\} = \{Y_0^*, Y_1^*, \ldots\}$, from which we can characterize the time elapsed between successive event occurrences. Since event occurrences correspond to state transitions (including the case $x' = x$ where an event occurrence leaves the state unchanged), this sequence also provides information about *state holding times*. This information is particularly important, as it leads to the evaluation of useful performance measures for many practical DES.

- The event sequence $\{E_k\} = \{E_1, E_2, \ldots\}$, from which we can probabilistically assess which events are more likely than others, and compute probabilities of the form $P[E_k = i], i \in \mathcal{E}$, or $P[E_k = i \mid X_k = x]$.

- The score process $N_i(t)$ for each event $i \in \mathcal{E}$, which is a counting process for event i occurrences in the interval $(0, t]$.

Our starting point is the development of simple useful models for event clock sequences. We will base this development on a fundamental event counting process, referred to as the *Poisson process*.

6.6   THE POISSON COUNTING PROCESS

Consider a DES with a single event. Let $\{N(t)\}$ be the process that counts the number of event occurrences in $(0, t]$. Thus, the state space of the process is the set of nonnegative integers $\{0, 1, 2, \ldots\}$. By the simple counting nature of the process, we have

$$N(0) \leq N(t_1) \leq \cdots \leq N(t_k)$$

for any $0 \leq t_1 \leq \cdots \leq t_k$. A typical sample path $n(t)$ of a counting process $\{N(t)\}$ is shown in Fig. 6.5.

Let us partition the time line into an arbitrary number of intervals $(t_{k-1}, t_k]$, $k = 1, 2, \ldots$, each of arbitrary length, as illustrated in Fig. 6.6. We also set $t_0 = 0$, and assume an initial event count of 0, i.e., $N(0) = 0$. Finally, define

$$N(t_{k-1}, t_k) = N(t_k) - N(t_{k-1}), \quad k = 1, 2, \ldots \tag{6.19}$$

to count the number of events occurring in the interval $(t_{k-1}, t_k]$ only (which includes the point t_k).

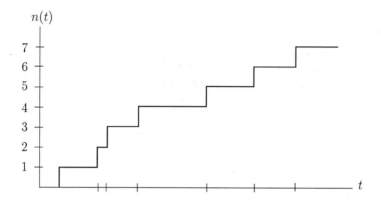

Figure 6.5: Typical sample path of a counting process $\{N(t)\}$.

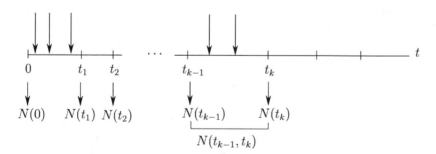

Figure 6.6: Time line partitioning for an event counting process, $N(t_{k-1}, t_k) = N(t_k) - N(t_{k-1})$.

We will now make three assumptions that will allow us to narrow down this general counting process to a simpler one, which we can fully characterize by means of a single distribution. In particular, under these assumptions, we will derive an explicit expression for the probability that n events occur in $(0, t]$:

$$P_n(t) \equiv P[N(t) = n], \qquad n = 0, 1, \ldots \qquad (6.20)$$

(A1) At most one event can occur at any time instant.

This is simply an assertion that two or more events cannot occur simultaneously.

(A2) The random variables $N(t), N(t, t_1), N(t_1, t_2), \ldots, N(t_{k-1}, t_k), \ldots$ are *mutually independent* for any $0 \leq t \leq t_1 \cdots \leq t_k$ and any $k = 1, 2, \ldots$

The assumption states that the event count in some time interval $(t_{k-1}, t_k]$ should not be affected by the event count in other intervals $(t_{j-1}, t_j], j \neq k$. Thus, **(A2)** implies that future events in some time interval $(t_{k-1}, t_k]$ are not affected by what has taken place in the past, that is, in intervals prior to time t_{k-1}.

(A3) $P[N(t_{k-1}, t_k) = n], n = 0, 1, \ldots$ may depend on the interval length $(t_k - t_{k-1})$, but it is independent of the time instants t_{k-1}, t_k for any interval $(t_{k-1}, t_k], k = 1, 2, \ldots$

It is natural to expect that $N(t_{k-1}, t_k)$ depends on $(t_k - t_{k-1})$; intuitively, the longer the interval is the more events we are likely to get. On the other hand, independence from t_{k-1}, t_k is a form of stationarity; the stochastic behavior of the counting process is the same regardless of where exactly on the time line an interval of length $(t_k - t_{k-1})$ is placed. Thus, beginning a count at time $t_{k-1} = t$ or at $t_{k-1} = t' \neq t$ should yield the same stochastic behavior over an interval $(t_{k-1}, t_k]$ of *fixed* length.

Definition. A process $\{N(t)\}$ satisfying **(A2)** is said to be a process with *independent increments*. If, in addition, **(A3)** is satisfied, $\{N(t)\}$ is a process with *stationary independent increments*.   ◆

A direct implication of **(A3)**, which we shall make ample use of, is the following. Since the distribution of $N(t_{k-1}, t_k)$ does not depend on t_{k-1}, t_k, but only on $s = t_k - t_{k-1}$, we can reposition the interval $(t_{k-1}, t_k]$ at $(0, s]$. Then, by **(A3)**:

$$P[N(t_{k-1}, t_k) = n] = P[N(s) = n], \qquad s = t_k - t_{k-1}$$

for any $n = 0, 1, \ldots$ Setting $t_{k-1} = t$ and $t_k = t + s$, $s > 0$, we can rewrite this relationship as

$$P[N(t, t + s) = n] = P[N(s) = n] \qquad (6.21)$$

We now proceed with the derivation of the probability $P[N(t) = n]$. We do so in several steps which we identify below.

Step 1. Determine $P[N(t) = 0]$.

Consider two arbitrary points t and $t+s$ on the time line such that $t+s \geq t \geq 0$. Observe that

$$P[N(t+s) = 0] = P[N(t) = 0 \ \text{ and } \ N(t, t+s) = 0]$$

since the event $[N(t+s) = 0]$ necessarily implies: (a) that $N(t) = 0$, and (b) that no event occurs in the interval $(t, t+s]$. By assumption **(A2)**, $N(t, t+s)$ and $N(t)$ are independent. It follows that

$$P[N(t+s) = 0] = P[N(t) = 0] \cdot P[N(t, t+s) = 0]$$

Moreover, from (6.21), $P[N(t, t+s) = 0] = P[N(s) = 0]$, and we get

$$P[N(t+s) = 0] = P[N(t) = 0] \cdot P[N(s) = 0]$$

Recalling the definition of $P_n(t)$ in (6.20), we rewrite this equation as follows:

$$P_0(t+s) = P_0(t) \cdot P_0(s) \tag{6.22}$$

This relationship, which holds for arbitrary $s, t \geq 0$, imposes a constraint on $P_0(t)$. The following auxiliary result sheds some light on the nature of this constraint.

Lemma 6.1 Let $g(t)$ be a differentiable function over all $t \geq 0$ such that

$$g(0) = 1 \text{ and } g(t) \leq 1 \text{ for all } t \geq 0$$

Then, for any $s, t \geq 0$,

$$g(t+s) = g(t) \cdot g(s) \tag{6.23}$$

if and only if

$$g(t) = e^{-\lambda t}, \quad \text{for some } \lambda > 0 \tag{6.24}$$

Proof. To show that (6.23) implies (6.24), differentiate (6.23) with respect to s to get

$$\frac{d}{ds} g(t+s) = g(t) \frac{dg(s)}{ds}$$

Let $g'(s) = dg(s)/ds$, and set $s = 0$ in the preceding equation:

$$\frac{d}{ds} g(t) = g(t) \cdot g'(0)$$

Let $c = g'(0)$. The solution of this differential equation, with an initial condition given as $g(0) = 1$, is

$$g(t) = e^{ct}$$

To satisfy $g(t) \leq 1$ for all $t \geq 0$, we must choose $c < 0$. Let $c = -\lambda$ for some $\lambda > 0$ and (6.24) is obtained.

Finally, the fact that (6.24) implies (6.23) is easy to see, since $e^{-\lambda(t+s)} = e^{-\lambda} \cdot e^{-\lambda t}$.   **Q.E.D.**

Returning to (6.22), note that $P_0(0) = 1$, since we have assumed that the initial event count is $N(0) = 0$. Moreover, $P_0(t) \leq 1$ for all $t \geq 0$. Assuming $P_0(t)$ is also differentiable, it follows from Lemma 6.1 that

$$P_0(t) = P[N(t) = 0] = e^{-\lambda t} \tag{6.25}$$

which is the probability that no event occurs in any interval of length t.

Step 2. Determine $P[N(\Delta t) = 0]$ for a "small" interval of length Δt.

Consider an interval of "small" length Δt. The reason we do so is to call attention to the following fact. If we set $t = \Delta t$ in (6.25) and write a Taylor series for the exponential function, we get

$$P[N(\Delta t) = 0] = e^{-\lambda \Delta t} = 1 - \lambda \Delta t + \lambda^2 \frac{(\Delta t)^2}{2!} - \lambda^3 \frac{(\Delta t)^3}{3!} + \ldots \tag{6.26}$$

Terms of second and higher order in Δt in this expression become negligible as $\Delta t \to 0$. Mathematically, we express this fact by lumping all these *small order* Δt terms into a single term denoted by $o(\Delta t)$. More generally, this notation is used to define any function $o(\Delta t)$ satisfying:

$$\frac{o(\Delta t)}{\Delta t} \to 0 \text{ as } \Delta t \to 0$$

We can therefore write (6.26) as follows:

$$P[N(\Delta t) = 0] = 1 - \lambda \Delta t + o(\Delta t) \tag{6.27}$$

Step 3. Determine $P[N(\Delta t) = n], n = 1, 2, \ldots$ for a "small" interval of length Δt.

By assumption (**A1**), two or more events cannot occur at the same precise time instant. It follows that we can always make an interval length Δt so small as to fit at most one event in the interval. In other words, as $\Delta t \to 0$, the probabilities $P[N(\Delta t) = n], n = 2, 3, \ldots$ become negligible. Mathematically, using once again the notation $o(\Delta t)$, we write

$$P[N(\Delta t) = n] = o(\Delta t) \quad \text{for all } n \geq 2 \tag{6.28}$$

Combining (6.27) and (6.28) and recalling that $\sum_{n=0}^{\infty} P[N(\Delta t) = n] = 1$, we see that

$$P[N(\Delta t) = 1] = \lambda \Delta t + o(\Delta t) \tag{6.29}$$

In other words, as $\Delta t \to 0$, the only possible events are $[N(\Delta t) = 1]$ and $[N(\Delta t) = 0]$. And since by (6.27) we have $P[N(\Delta t) = 0] = 1 - \lambda \Delta t + o(\Delta t)$, (6.29) follows from the requirement that the sum of $P[N(\Delta t) = 1]$ and $P[N(\Delta t) = 0]$ is 1 as $\Delta t \to 0$ (it is possible to derive Step 3 more rigorously, but the argument used is unnecessarily involved for our purposes).

Step 4. Derive an expression for $P[N(t + \Delta t) = n]$ in terms of $P[N(t) = n]$.

Consider the intervals $(0, t]$ and $(t, t + \Delta t]$, $\Delta t > 0$. Suppose that the event $[N(t + \Delta t) = n]$ with $n > 0$ occurs. There are n mutually exclusive events that could have taken place to result in this event:

$$[N(t) = n \quad and \quad N(t, t + \Delta t) = 0],$$
$$[N(t) = n - 1 \quad and \quad N(t, t + \Delta t) = 1], \ldots ,$$
$$[N(t) = 0 \quad and \quad N(t, t + \Delta t) = n]$$

Therefore,

$$P[N(t + \Delta t) = n] = \sum_{j=0}^{n} P[N(t, t + \Delta t) = n - j \; and \; N(t) = j] \tag{6.30}$$

By assumption **(A2)**, $N(t, t + \Delta t)$ and $N(t)$ are independent; therefore,

$$P[N(t, t + \Delta t) = n - j \; and \; N(t) = j]$$
$$= P[N(t, t + \Delta t) = n - j] \cdot P[N(t) = j]$$

Moreover, by (6.21), $P[N(t, t + \Delta t) = n - j] = P[N(\Delta t) = n - j]$. We can then rewrite (6.30) as

$$P[N(t + \Delta t) = n] = \sum_{j=0}^{n} P[N(\Delta t) = n - j] \cdot P[N(t) = j] \tag{6.31}$$

We now make use of the probabilities derived in (6.27) through (6.29). In the sum above, the only events of the form $[N(\Delta t) = n - j]$ contributing probabilities other than $o(\Delta t)$ are $[N(\Delta t) = 0]$ (and hence $[N(t) = n]$), and $[N(\Delta t) = 1]$ (and hence $[N(t) = n - 1]$). Thus, (6.31) becomes

$$P[N(t + \Delta t) = n] = P[N(\Delta t) = 0] \cdot P[N(t) = n]$$
$$+ P[N(\Delta t) = 1] \cdot P[N(t) = n - 1] + o(\Delta t)$$
$$= [1 - \lambda \Delta t + o(\Delta t)] \cdot P[N(t) = n]$$
$$+ [\lambda \Delta t + o(\Delta t)] \cdot P[N(t) = n - 1] + o(\Delta t)$$

Grouping together all $o(\Delta t)$ terms into a single term, and recalling the definition of $P_n(t)$ in equation (6.20), we get

$$P_n(t + \Delta t) = (1 - \lambda \Delta t)P_n(t) + \lambda \Delta t P_{n-1}(t) + o(\Delta t), \qquad n > 0 \qquad (6.32)$$

Step 5. Determine $P[N(t) = n], n > 0$, by letting $\Delta t \to 0$.

Subtracting $P_n(t)$ from both sides of equation (6.32), and dividing by Δt, we obtain

$$\frac{P_n(t + \Delta t) - P_n(t)}{\Delta t} = -\lambda P_n(t) + \lambda P_{n-1}(t) + \frac{o(\Delta t)}{\Delta t}$$

and by taking the limit as $\Delta t \to 0$, we get the difference-differential equation

$$\frac{dP_n(t)}{dt} = -\lambda P_n(t) + \lambda P_{n-1}(t), \qquad n > 0 \qquad (6.33)$$

Step 6. Solve (6.33) for $n = 1, 2, \ldots$

We already know that $P_0(t) = e^{-\lambda t}$ from (6.25). Thus, (6.33) for $n = 1$ becomes:

$$\frac{dP_1(t)}{dt} = -\lambda P_1(t) + \lambda e^{-\lambda t}$$

One can easily check that the solution to this differential equation is

$$P_1(t) = (\lambda t)e^{-\lambda t}$$

Proceeding in similar fashion for $n = 2, 3, \ldots$ we obtain the general solution

$$P_n(t) = \frac{(\lambda t)^n}{n!} e^{-\lambda t}, \qquad t \geq 0, \quad n = 0, 1, 2, \ldots \qquad (6.34)$$

which includes the case $n = 0$ in (6.25).

This completes our derivation. The expression in (6.34) is known as the *Poisson distribution*. It fully characterizes the stochastic process $\{N(t)\}$, which counts event occurrences in $(0, t]$ under assumptions **(A1)** through **(A3)**.

There are several reasons why the Poisson counting process is an essential building block in the stochastic modeling and analysis of DES. First of all, as we shall see in the next section, it possesses several properties that permit us to simplify GSMP models of DES to obtain Markov chains that are much easier to analyze. However, the Poisson process is not just an abstraction we have invented for our convenience; there is abundant experimental evidence (ranging from telephone traffic engineering to particle physics) that this simple counting process accurately models many practically interesting processes. Roughly speaking, the Poisson process counts events that occur in a very random but time-invariant way, so that assumptions **(A1)** through **(A3)** are indeed satisfied. Finally, (6.34) often serves as an approximation (like the normal distribution) for the superposition of multiple event processes of fairly general nature; the approximation is exact in the limit as the number of such combined processes grows to infinity.

Before exploring the properties of the Poisson counting process, we will calculate the mean and variance of the random variable $N(t)$. This may be thought of as the event score at time t of a simple DES with a single event.

Mean and Variance of the Poisson distribution

A direct way to evaluate $E[N(t)]$ is to proceed from the definition of expectation:

$$E[N(t)] = \sum_{n=0}^{\infty} n P_n(t) = e^{-\lambda t} \sum_{n=0}^{\infty} n \frac{(\lambda t)^n}{n!} = e^{-\lambda t} \sum_{n=1}^{\infty} \frac{(\lambda t)^{n-1}}{(n-1)!} (\lambda t)$$

Changing the summation index to $m = n - 1$ gives

$$E[N(t)] = (\lambda t) e^{-\lambda t} \sum_{m=0}^{\infty} \frac{(\lambda t)^m}{m!}$$

Recognizing that this sum is the definition of the exponential function $e^{\lambda t} = 1 + (\lambda t) + (\lambda t)^2/2! + \ldots$, we get

$$E[N(t)] = \lambda t \tag{6.35}$$

We are therefore able to give a simple intuitive interpretation to the parameter $\lambda = E[N(t)]/t$: It is the average rate at which events occur per unit time. We sometimes refer to λ as the *intensity* or the *rate* of the Poisson counting process.

A similar calculation gives us the second moment of $N(t)$:

$$E[N^2(t)] = \sum_{n=0}^{\infty} n^2 P_n(t) = e^{-\lambda t} \sum_{n=0}^{\infty} n^2 \frac{(\lambda t)^n}{n!} = e^{-\lambda t} \sum_{n=1}^{\infty} n \frac{(\lambda t)^n}{(n-1)!}$$

Adding and subtracting the term $[e^{-\lambda t} \sum_{n=1}^{\infty} (\lambda t)^n/(n-1)!]$ we get

$$E[N^2(t)] = e^{-\lambda t} \sum_{n=1}^{\infty} (n-1) \frac{(\lambda t)^n}{(n-1)!} + e^{-\lambda t} \sum_{n=1}^{\infty} \frac{(\lambda t)^n}{(n-1)!}$$

$$= e^{-\lambda t} \sum_{n=2}^{\infty} \frac{(\lambda t)^{n-2}}{(n-2)!} (\lambda t)^2 + e^{-\lambda t} \sum_{n=1}^{\infty} \frac{(\lambda t)^{n-1}}{(n-1)!} (\lambda t)$$

Changing the summation index to $m = n-2$ for the first sum, and to $m = n-1$ for the second gives

$$E[N_2(t)] = (\lambda t)^2 e^{-\lambda t} \sum_{m=0}^{\infty} \frac{(\lambda t)^m}{m!} + (\lambda t) e^{-\lambda t} \sum_{m=0}^{\infty} \frac{(\lambda t)^m}{m!}$$

As before, each of the two sums defines $e^{\lambda t}$. Therefore,

$$E[N^2(t)] = (\lambda t)^2 + (\lambda t) \tag{6.36}$$

This allows us to obtain the variance of $N(t)$, denoted by $\mathrm{Var}[N(t)]$, as follows. Since

$$\mathrm{Var}[N(t)] = E[N^2(t)] - (E[N(t)])^2$$

we use (6.36) and $E[N(t)] = \lambda t$ from (6.35) to get

$$\mathrm{Var}[N(t)] = \lambda t \tag{6.37}$$

Note that the mean and variance of $N(t)$ are identical, both given by λt. Also note that they are both unbounded as $t \to \infty$.

6.7 PROPERTIES OF THE POISSON PROCESS

As already mentioned, the Poisson process possesses several remarkable properties. In what follows, we identify and establish some of the most important ones that are particularly useful for our purposes. Some additional properties of the Poisson process are considered in Problems 6.6, 6.7, 6.8 at the end of this chapter.

6.7.1 Exponentially Distributed Interevent Times

A very natural first question is: *How are interevent times distributed in a Poisson process?* The answer turns out to be easy to obtain by arguing as follows.

Suppose the $(k-1)$th event takes place at some random time T_{k-1}. Let V_k denote the interevent time random variable following this event, and let $G_k(t)$ be its cdf, so that

$$G_k(t) = P[V_k \le t] = 1 - P[V_k > t] \tag{6.38}$$

Now suppose we observe the occurrence time of this event, so that $T_{k-1} = t_{k-1}$. Conditioning on this observation, let us consider the event $[V_k > t \mid T_{k-1} = t_{k-1}]$. The crucial observation is that this event is identical to the event $[N(t_{k-1}, t_{k-1}+t) = 0]$, as illustrated in Fig. 6.7. Clearly, if no event takes place in a given interval $(t_{k-1}, t_{k-1} + t]$ for some $t > 0$, then the kth interevent time must be longer than t. We can therefore write

$$P[V_k > t \mid T_{k-1} = t_{k-1}] = P[N(t_{k-1}, t_{k-1} + t) = 0]$$

By (6.21), we have $P[N(t_{k-1}, t_{k-1} + t) = 0] = P[N(t) = 0] = P_0(t)$. Since this is independent of t_{k-1}, it follows that the left-hand-side expression in the

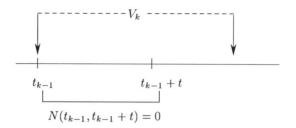

Figure 6.7: Illustrating the fact that $P[V_k > t \mid T_{k-1} = t_{k-1}] = P[N(t_{k-1}, t_{k-1}+t) = 0]$.

If $N(t_{k-1}, t_{k-1}+t) = 0$, the next event following t_{k-1} must occur after $t_{k-1}+t$. This means that the interevent time V_k must be longer than t.

previous equation must be independent of the value of T_{k-1}. Therefore,

$$P[V_k > t \mid T_{k-1} = t_{k-1}] = P[V_k > t] = P_0(t)$$

and we can rewrite (6.38) as

$$G_k(t) = 1 - P_0(t)$$

Finally, we know from (6.35) that $P_0(t) = e^{-\lambda t}$. Therefore,

$$G_k(t) = 1 - e^{-\lambda t} \tag{6.39}$$

which is independent of k. We conclude from this derivation that

(P1) *the interevent process* $\{V_k\}, k = 1, 2, \ldots$, *of a Poisson process is an iid stochastic sequence characterized by the cdf:*

$$G(t) = P[V_k \leq t] = 1 - e^{-\lambda t}, \quad t \geq 0$$

This is the *exponential distribution*. It is a cdf with a single parameter, $\lambda > 0$. The corresponding pdf is obtained by differentiating $G(t)$ in (6.39) to give

$$g(t) = \lambda e^{-\lambda t}, \quad t \geq 0 \tag{6.40}$$

The mean and variance of this distribution are $1/\lambda$ and $1/\lambda^2$ respectively (see Appendix I for derivation). The functions $G(t)$ and $g(t)$ are plotted in Fig. 6.8.

Recalling the definition of a renewal process, note also that the Poisson process is a renewal process with interevent times characterized by an exponential distribution.

In the following sections, we explore some more properties of the Poisson process. We have already seen in **(P1)**, however, that we can always consider a "typical" interevent time V, instead of explicitly specifying its index by writing V_k, since the interevent time distribution is fixed. This allows us to drop the subscript k in what follows (unless we explicitly need it), and study a typical interevent interval.

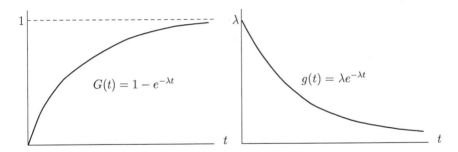

Figure 6.8: The exponential distribution $G(t) = 1 - e^{-\lambda t}$ and pdf $g(t) = \lambda e^{-\lambda t}$.

6.7.2 The Memoryless Property

Another natural question regarding the Poisson process is: *How are residual interevent times distributed?* The answer to this question reveals a unique fact about the exponential distribution that we refer to as the *memoryless property*.

Consider an event taking place at time T, and let V be the next interevent time. Now suppose that at some time instant $T + z > T$ the next event has not yet occurred (in other words, z is the age of the current event at that time). This observation corresponds to the fact that the event $[V > z]$ has already taken place. In general, we expect the information $[V > z]$ to tell us something about the distribution of the random variable V. For one thing, it tells us that the event $[V \leq z]$ cannot occur. So, we attempt to calculate the probability of the event $[V \leq z + t]$ for some $t > 0$, given that we already know $[V > z]$ (see Fig. 6.9), that is, the conditional probability distribution $P[V \leq z + t \mid V > z]$. The calculation of this distribution now proceeds as follows:

$$P[V \leq z + t \mid V > z] = \frac{P[V \leq z + t \text{ and } V > z]}{P[V > z]}$$
$$= \frac{P[z < V \leq z + t]}{1 - P[V \leq z]}$$

Since we know from property **(P1)** of the Poisson process that $V \sim (1 - e^{-\lambda t})$, it follows that

$$P[V \leq z + t \mid V > z] = \frac{(1 - e^{-\lambda(z+t)}) - (1 - e^{-\lambda z})}{e^{-\lambda z}}$$
$$= \frac{-e^{-\lambda z} \cdot e^{-\lambda t} + e^{-\lambda z}}{e^{-\lambda z}}$$

which finally gives

$$P[V \leq z + t \mid V > z] = 1 - e^{-\lambda t} \tag{6.41}$$

Not only is this distribution independent of the event age z, but it is also

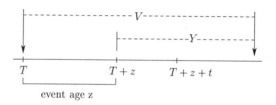

event age z

Figure 6.9: Calculation of $P[V \le z + t \mid V > z]$.

identical to the original interevent time distribution $G(t) = 1 - e^{-\lambda t}$! This is what we refer to as the *memoryless property of the exponential distribution*.

The result we have obtained in (6.41) readily translates into the distribution of the residual lifetime Y following the current event (see Fig. 6.9). Clearly, the residual lifetime needs to be specified with respect to some given event age z, that is, conditioned on the event $[V > z]$. Thus, we define

$$H(t, z) = P[Y \le t \mid V > z] \tag{6.42}$$

Now, by definition, $V = Y + z$, so (6.42) can be rewritten as

$$H(t, z) = P[V \le z + t \mid V > z]$$

This is precisely the probability obtained in (6.41). Therefore,

$$H(t, z) = 1 - e^{-\lambda t} \tag{6.43}$$

No matter how long it has been since the last occurrence of an event in a Poisson process, the distribution of the time remaining until the next event occurrence remains fixed and is given by $G(t) = 1 - e^{-\lambda t}$. The term *memoryless* is justified by the fact that no event age memory is required. In summary,

(P2) *every residual lifetime (clock) in a Poisson process is characterized by the exact same exponential distribution as the original lifetime:*

$$H(t, z) = P[Y \le t \mid V > z] = G(t) = 1 - e^{-\lambda t}$$

regardless of the event age z.

One cannot overemphasize the importance of property **(P2)**. This importance is amplified by the fact that the memoryless property is *unique* to the exponential distribution, as established through the following theorem.

Theorem 6.1 Let V be a positive random variable with a differentiable distribution function $G(t) = P[V \le t]$. Then

$$H(t, z) = P[V \le z + t \mid V > z]$$

is independent of z (i.e., it is *memoryless*) if and only if

$$G(t) = 1 - e^{-\lambda t} \quad \text{for some } \lambda > 0$$

Moreover, if $G(t) = 1 - e^{-\lambda t}$, then $H(t, z) = 1 - e^{-\lambda t}$.

Proof. First, we show that the memoryless property implies that $G(t) = 1 - e^{-\lambda t}$. Since, by assumption, $P[V \leq z + t \mid V > z]$ is independent of z, we have

$$P[V \leq z + t \mid V > z] = h(t)$$

where $h(t)$ is some function of t *only*. Consider the left-hand side above, and observe that

$$
\begin{aligned}
P[V \leq z + t \mid V > z] &= 1 - P[V > z + t \mid V > z] \\
&= 1 - \frac{P[V > z + t \text{ and } V > z]}{P[V > z]} \qquad (6.44) \\
&= 1 - \frac{P[V > z + t]}{P[V > z]}
\end{aligned}
$$

since $z + t \geq z$. It follows that

$$1 - \frac{P[V > z + t]}{P[V > z]} = h(t)$$

or, setting $[1 - h(t)] = \tilde{h}(t)$,

$$P[V > z + t] = P[V > z] \cdot \tilde{h}(t) \qquad (6.45)$$

For $z = 0$, we get

$$P[V > t] = P[V > 0] \cdot \tilde{h}(t) = \tilde{h}(t) \qquad (6.46)$$

since V is a positive random variable, therefore $P[V > 0] = 1$. Combining (6.45) and (6.46) to eliminate $\tilde{h}(t)$, we obtain the relationship:

$$P[V > z + t] = P[V > z] \cdot P[V > t]$$

We now invoke Lemma 6.1, which was used in the derivation of the Poisson distribution. Set

$$g(t) = P[V > t]$$

and observe that condition (6.23) in Lemma 6.1 holds, that is, $g(t + z) = g(t) \cdot g(z)$ for all $t, z \geq 0$. Moreover, the remaining conditions of the lemma are satisfied, that is, $P[V > t]$ is differentiable, $g(0) = P[V > 0] = 1$, and $g(t) \leq 1$ for all $t \geq 0$. Thus, by Lemma 6.1,

$$g(t) = P[V > t] = e^{-\lambda t}$$

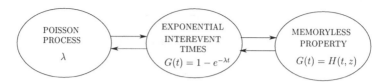

Figure 6.10: Properties of the Poisson process.

for some $\lambda > 0$. This implies that V is exponentially distributed:

$$G(t) = P[V \le t] = 1 - e^{-\lambda t}$$

The converse statement of the theorem is easily established, since from (6.44):

$$P[V \le z + t \mid V > z] = 1 - \frac{P[V > z + t]}{P[V > z]}$$

and with $P[V \le t] = 1 - e^{-\lambda t}$ this becomes

$$P[V \le z + t \mid V > z] = 1 - \frac{e^{-\lambda(z+t)}}{e^{-\lambda z}} = 1 - e^{-\lambda t}$$

which is independent of z. This also establishes the fact that if $G(t) = 1 - e^{-\lambda t}$, then $P[V \le z + t \mid V > z] = 1 - e^{-\lambda t}$. **Q.E.D.**

This result helps us reveal the intrinsic connection between the Poisson process, the memoryless property, and the exponential distribution, as summarized in Fig. 6.10. In particular, the Poisson process is characterized by exponentially distributed interevent times and possesses the memoryless property. Conversely, the memoryless property implies exponentially distributed interevent times, which in turn implies a Poisson process (we have not explicitly established this latter fact; it is, however, straightforward and is left as an exercise).

Recall also that the memoryless property is the defining feature of a Markov chain, as discussed in Section 6.2.3. Not surprisingly, there is a close connection between Poisson counting processes and Markov chains, which we will further explore in subsequent sections.

6.7.3 Superposition of Poisson Processes

Thus far, we have limited ourselves to a DES with a single event, which we have modeled as a Poisson process. We now wish to consider a DES consisting of m events, $m > 1$. We assume that each event sequence is modeled as a Poisson process with parameter λ_i, $i = 1, \ldots, m$. We also assume that these m Poisson processes are mutually *independent*.

The resulting stochastic process is a superposition of m Poisson processes, each satisfying properties **(P1)** and **(P2)** derived in the previous sections. The question we pose is: *Are properties* **(P1)** *and* **(P2)** *preserved in the resulting stochastic process?* And if so, *in what form?*

Suppose an event j in the compound process occurs at time T. Let Y_i be the residual lifetime (clock) of the ith event at that time. For event j, the clock Y_j is simply a lifetime from the corresponding Poisson process. Clearly, the next interevent time is the smallest residual lifetime:

$$Y^* = \min_{i=1,\dots,m} \{Y_i\}$$

To determine the distribution of each Y_i, we require, in general, knowledge of the corresponding event age. However, by the memoryless property **(P2)** of each Poisson process, we know that

$$Y_i \sim 1 - e^{-\lambda_i t} \tag{6.47}$$

We can therefore obtain the distribution of the interevent time Y^* as follows:

$$P[Y^* \le t] = 1 - P[Y^* > t] = 1 - P\left[\min_{i=1,\dots,m} \{Y_i\} > t\right]$$

Observe that the event $[\min_{i=1,\dots,m}\{Y_i\} > t]$ is identical to the event $[Y_1 > t \ \text{and}\ Y_2 > t \cdots \ \text{and}\ Y_m > t]$. Moreover, since the m Poisson processes are independent, the random variables $Y_1, Y_2, \dots, Y_m$ are independent. It follows that

$$P[Y^* \le t] = 1 - \prod_{i=1}^{m} P[Y_i > t]$$

and, by (6.47),

$$P[Y^* \le t] = 1 - \prod_{i=1}^{m} e^{-\lambda_i t} \tag{6.48}$$

Let Λ be the sum of all Poisson parameters:

$$\Lambda = \sum_{i=1}^{m} \lambda_i \tag{6.49}$$

Then, (6.48) becomes

$$P[Y^* \le t] = 1 - e^{-\Lambda t} \tag{6.50}$$

We therefore reach the interesting conclusion that

(P3) *the superposition of m independent Poisson processes with parameters $\lambda_i, i = 1, \dots, m$, is also a Poisson process with parameter*

$$\Lambda = \sum_{i=1}^{m} \lambda_i$$

6.7.4 The Residual Lifetime Paradox

Consider a Poisson process as a model of bus arrivals at some bus station. Thus, by **(P1)**, bus interarrival times at the station are exponentially distributed. Since the mean of the exponential distribution $G(t) = 1 - e^{-\lambda t}$ is $1/\lambda$, a bus arrives every $1/\lambda$ minutes on the average. Now suppose a traveler shows up at the bus station at some random time instant and waits for the next bus. The question we pose is: *How long does the traveler have to wait, on the average, until the next bus?*

Answer 1. One can present the following reasonable argument. The time interval defined by two successive bus arrivals is on the average $1/\lambda$ minutes long. The traveler shows up at a random point equally likely to be anywhere in this time interval. Therefore, he should wait for half that interval on the average, that is, $1/2\lambda$ minutes. The problem with this answer is that it violates the memoryless property **(P2)**: Since the residual lifetime for bus arrival events is also exponentially distributed with mean $1/\lambda$, the traveler should wait, on the average, for $1/\lambda$ minutes, not $1/2\lambda$!

Answer 2. Here is another reasonable argument, which is based precisely on the memoryless property **(P2)**. Because of this property, as explained above, the traveler should wait for $1/\lambda$ minutes on the average. Of course, the same is true for the time since the last bus arrived (by simply reversing the direction of time on the time axis, since the traveler shows up at random). It follows that the interarrival time is, on the average, $1/\lambda + 1/\lambda = 2/\lambda$ minutes, which contradicts the fact that expected interarrival times are $1/\lambda$ minutes!

We see that either answer leads to a contradiction, hence the "residual lifetime paradox" (also called the "renewal paradox", because the same apparent contradiction can be reached for any renewal process modeling bus arrivals, not just a Poisson process).

The correct answer is Answer 2, even though it seems to imply that the interarrival time seen by the traveler is not "correctly" distributed. The intuitive explanation is the following. As interarrival times are lined up on the time axis, some are longer than the mean, and some are shorter. The traveler is more likely to land on a longer interval than a shorter one, since *longer intervals occupy more space on the time line than shorter ones*. As an extreme case, suppose there are two interarrival times, $V_1 = 9$ minutes and $V_2 = 1$ minute, and the traveler arrives some time in the interval $[0, 10]$ minutes; it should be clear that he is nine times more likely to show up in the first interval.

In the case of a Poisson process, it turns out that the traveler sees an interval twice as long as a "typical" one. Referring to Fig. 6.11, let $T_1, T_2, \ldots$ denote bus

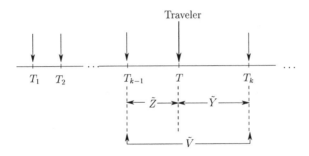

Figure 6.11: Traveler sees an interarrival time $\tilde{V}$ with residual lifetime $\tilde{Y}$.

arrival instants, and let T be the time when the traveler comes to the station, where $T \in (T_{k-1}, T_k]$. Let $\tilde{V} = T_k - T_{k-1}$ be the interarrival time of interest, and $\tilde{Z}$, $\tilde{Y}$ the age and residual lifetime of the kth bus arrival respectively. The argument leading to Answer 2 can be extended to show that in fact $\tilde{V}$ is not even exponentially distributed.

To see this, we proceed as follows (based on the discussion of the renewal paradox in Kleinrock, 1975, and Walrand, 1988). Let $G(t)$ denote the cdf of interarrival times, with a corresponding pdf denoted by $g(t)$. In addition, let

$$\tilde{G}(t) = P[\tilde{V} \leq t] \tag{6.51}$$

be the cdf of the "special" interarrival time which the traveler sees, and let $\tilde{g}(t)$ be the corresponding pdf. We have argued that the traveler is more likely to see a longer interarrival time than a shorter one. In other words, the probability of seeing an interarrival time of length t is proportional to t. This probability is also proportional to the probability that a typical interarrival time is of length t. We are therefore led to the relationship:

$$\tilde{g}(t)dt = c \cdot tg(t)dt \tag{6.52}$$

where $\tilde{g}(t)dt$ is the probability that the traveler sees an interarrival time of length t, and $g(t)dt$ is the probability that a typical interarrival time is of length t. The constant of proportionality, c, is determined so as to properly normalize the pdf $\tilde{g}(t)$. In particular, from (6.52) we get:

$$1 = \int_0^\infty \tilde{g}(t)dt = c \int_0^\infty tg(t)dt$$

where the integral on the right hand side is the expected interarrival time $E[V]$. It follows that $c = 1/E[V]$, and (6.52) implies that

$$\tilde{g}(t) = \frac{t}{E[V]}g(t) \tag{6.53}$$

This result holds for any interarrival time pdf $g(t)$. For our particular case, where $g(t) = \lambda e^{-\lambda t}$ and $E[V] = 1/\lambda$, we get

$$\tilde{g}(t) = (\lambda^2 t)e^{-\lambda t} \tag{6.54}$$

We can now obtain $E[\tilde{V}]$ as the mean of this distribution:

$$E[\tilde{V}] = \int_0^\infty t \cdot (\lambda^2 t)e^{-\lambda t}dt = -\lambda \int_0^\infty t^2 de^{-\lambda t}$$

Integration by parts yields

$$E[\tilde{V}] = 2 \int_0^\infty (\lambda t)e^{-\lambda t}dt$$

where the integral on the right-hand side is the mean of the exponential pdf $\lambda e^{-\lambda t}$, which we know is $1/\lambda$. It follows that

$$E[\tilde{V}] = \frac{2}{\lambda} = 2E[V] \tag{6.55}$$

as originally claimed.

Remark. The residual lifetime paradox illustrates the fact that there are two ways one can look at the bus arrival process (or indeed any service-providing process): (a) As an outside observer, one sees "typical" interarrival times whose average is $E[V]$, and (b) As a traveler interested in getting bus service, one actually selects a particular interarrival time which is generally a "longer than typical" one with mean $E[\tilde{V}]$. As we have seen, distributions of V intervals are generally different than those of $\tilde{V}$ intervals. The latter are also referred to as *Palm distributions.*

The analysis above can be carried out further to derive the pdf of the residual lifetime of a bus arrival event in an interarrival time $\tilde{V}$ seen by the randomly arriving traveler. Denoting by $\tilde{h}(t)$ this pdf, it can be shown that

$$\tilde{h}(t) = \frac{1 - G(t)}{E[V]} \tag{6.56}$$

In the Poisson case where $G(t) = 1 - e^{-\lambda t}$ and $E[V] = 1/\lambda$, we then get $\tilde{h}(t) = \lambda e^{-\lambda t}$, which is the exponential pdf, as expected.

Remark. Suppose the bus arrival process is deterministic, that is, buses always arrive every A minutes, where A is a constant. Thus, $G(t) = 0$ for all $0 \le t < A$, and $G(t) = 1$ for all $t \ge A$. Then, using (6.56) we get $\tilde{h}(t) = 1/A$ for all $0 \le t < A$, and $\tilde{h}(t) = 0$ for all $t \ge A$. In other words, the residual lifetime is uniformly distributed in $[0, A]$. It follows that the traveler waits for $A/2$ minutes on the average. If we choose $A = 1/\lambda$, the average waiting time is therefore $1/2\lambda$. In contrast, when the arrival process is Poisson, the average waiting time is $1/\lambda$, twice as long as the deterministic case with the same expected bus interarrival time. This observation serves to illustrate a fundamental fact: *Randomness always increases waiting*. This fact is of great importance in the performance analysis of DES, where keeping delays as low as possible is one of the main design and control objectives.

6.8   AUTOMATA WITH POISSON CLOCK STRUCTURE

Now that we have established the basic properties of a Poisson process, we will use it as a building block for more complicated DES. Already, in Section 6.7.3 we considered a DES consisting of m events which are permanently active, each with a Poisson clock sequence. We will extend this to cases where events may become infeasible at various states, and analyze the properties of the resulting models.

As a starting point, it is helpful to adopt a GSMP view of a simple Poisson process with parameter λ. Thus, a Poisson process may be viewed as a GSMP generated by a stochastic timed automaton $(\mathcal{E}, \mathcal{X}, \Gamma, f, p_0, G)$, where $\mathcal{E} = \{e\}$ consists of a single event, $\mathcal{X} = \{0, 1, \ldots\}$, $\Gamma(x) = \mathcal{E}$ for all $x, p_0(0) = 1$, and $f(x, e) = x + 1$ for all x, since the event count is simply incremented with every occurrence of e. Finally, the clock structure G is specified by $G(t) = 1 - e^{-\lambda t}$, a single cdf characterizing all event lifetimes for this model.

Next, let us consider an event set $\mathcal{E}$ consisting of m events. We impose no constraints on $\Gamma(x)$, the function p (which could be a deterministic transition function f), or $p_0(x)$. Thus, we generate a general GSMP, except for the clock structure which we constrain to be:

$$G = \{G_i, i = 1, \ldots, m\}, \quad \text{with } G_i(t) = 1 - e^{-\lambda_i t}$$

In other words, we consider each clock sequence to be a Poisson process with parameter $\lambda_i > 0$. The only other assumption we will make is that the m Poisson processes are independent.

6.8.1   Distribution of Interevent Times

We can derive the interevent time distribution for this GSMP by proceeding as in Section 6.7.3 where we combined m Poisson processes without ever disabling any event. Suppose an event occurs and state x is entered. By referring to equation (6.15) of the stochastic timed automaton definition, the next interevent time is given by

$$Y^* = \min_{i \in \Gamma(x)} \{Y_i\}$$

Proceeding exactly as in Section 6.7.3, we write

$$P[Y^* \leq t] = 1 - P[Y^* > t] = 1 - P[\min_{i \in \Gamma(x)} \{Y_i\} > t]$$

where the latter event is identical to the event $[Y_i > t$ for all $i \in \Gamma(x)]$. Exploiting the independence of $Y_i, i \in \Gamma(x)$, and since by the memoryless property **(P2)** of each Poisson process, we have $Y_i \sim 1 - e^{-\lambda_i t}$, we obtain the analog of (6.48):

$$P[Y^* \leq t] = 1 - \prod_{i \in \Gamma(x)} e^{-\lambda_i t} \tag{6.57}$$

Define the sum of Poisson parameters in a given state x to be

$$\Lambda(x) = \sum_{i \in \Gamma(x)} \lambda_i \tag{6.58}$$

and the distribution of the interevent time when the state is x:

$$G(t, x) = P\left[\min_{i \in \Gamma(x)} \{Y_i\} \leq t\right] \tag{6.59}$$

Then, from (6.57) we obtain the interevent time distribution at state x:

$$G(t, x) = P[Y^*(x) \leq t] = 1 - e^{-\Lambda(x)t} \tag{6.60}$$

which is still exponential, although it does generally depend on the process state x through the feasible event set $\Gamma(x)$. Note that (6.60) is a generalization of property **(P3)**, in which we assumed $\Gamma(x) = \mathcal{E}$ for all states x.

Once again, we see the presence of the exponential distribution, and take the opportunity to reemphasize its importance. Observe that the memoryless property still holds for $G(t, x)$, since this is the exponential cdf. This follows from Theorem 6.1, where the fixed Poisson rate λ is now replaced by the state-dependent parameter $\Lambda(x)$. Thus, for a GSMP with a Poisson clock structure we have, for any state x,

$$P[Y^*(x) \leq z + t \mid Y^*(x) > z] = P[Y^*(x) \leq t] = 1 - e^{-\Lambda(x)t}$$

Therefore, the memoryless property of the Poisson process extends to a GSMP formed by superimposing m Poisson processes subject to event enabling and disabling through $\Gamma(x)$. Intuitively, if some event i is occasionally disabled, at the time i is reenabled its new lifetime is simply taken to be some residual lifetime from the original Poisson clock sequence; by property **(P2)**, this residual lifetime has the same distribution as a lifetime. The only effect of enabling and disabling is the changing Poisson parameter $\Lambda(x)$ for the GSMP interevent times. Therefore, interevent times are no longer identically distributed.

6.8.2 Distribution of Events

The triggering event distribution for a given state x is defined by

$$p(i, x) = P[E' = i \mid X = x], \qquad i = 1, \ldots, m \tag{6.61}$$

where E' is the next event to occur when the state is x. Clearly, $p(i, x) = 0$ for all $i \notin \Gamma(x)$. For all other events $i \in \Gamma(x)$, we can derive $p(i, x)$ as follows.

Suppose the process state is x and the event $[E' = i]$ occurs. Observe that this event is identical to the event

$$\left[Y_i \leq \min_{\substack{j \in \Gamma(x) \\ j \neq i}} \{Y_j\} \right] \tag{6.62}$$

since event i must have the shortest residual lifetime among all events feasible in state x. To avoid tedious notation, let

$$W = \min_{\substack{j \in \Gamma(x) \\ j \neq i}} \{Y_j\} \tag{6.63}$$

so that $P[E' = i] = P[Y_i \leq W]$. To calculate $P[Y_i \leq W]$, we apply the rule of total probability (see also Appendix I):

$$P[Y_i \leq W] = \int_0^\infty P[Y_i \leq W \mid W = y] \cdot f_W(y) dy \tag{6.64}$$

where we condition on the event $[W = y]$, and then integrate over all possible values of y weighted by the pdf of W, denoted by $f_W(y)$. To evaluate the integral (6.64), we proceed in three steps, as described below.

Step 1. Evaluate $P[Y_i \leq W \mid W = y]$.

By the memoryless property **(P2)** of the ith Poisson process, we have $Y_i \sim G_i(t) = 1 - e^{-\lambda_i t}$. Thus, $P[Y_i \leq W \mid W = y]$ is simply the value of $G_i(t)$ at $t = y$:

$$P[Y_i \leq W \mid W = y] = 1 - e^{-\lambda_i y} \tag{6.65}$$

Step 2. Evaluate $f_W(y)$.

Let

$$\Lambda_i = \sum_{\substack{k \in \Gamma(x) \\ k \neq i}} \lambda_k \tag{6.66}$$

and note that $\Lambda(x) = \Lambda_i + \lambda_i$. Then, $P[W \leq y]$ is obtained as in (6.60) with $\Lambda(x)$ replaced by Λ_i, that is,

$$P[W \leq y] = 1 - e^{-\Lambda_i y}$$

and, therefore, the corresponding pdf is the exponential density function

$$f_W(y) = \Lambda_i e^{-\Lambda_i y} \tag{6.67}$$

Step 3. Evaluate the integral in (6.64).

Using (6.65) through (6.67), we have

$$P[Y_i \leq W] = \int_0^\infty [1 - e^{-\lambda_i y}] \cdot \Lambda_i e^{-\Lambda_i y} dy$$
$$= \Lambda_i \int_0^\infty [e^{-\Lambda_i y} - e^{-\Lambda(x)y}] dy$$

where we have used $\Lambda(x) = \Lambda_i + \lambda_i$. The integration yields

$$P[Y_i \leq W] = \Lambda_i \left[\frac{-e^{-\Lambda_i y}}{\Lambda_i} + \frac{e^{-\Lambda(x)y}}{\Lambda(x)} \right]_0^\infty = \left[1 - \frac{\Lambda_i}{\Lambda(x)} \right]$$
$$= \frac{\lambda_i}{\Lambda(x)}$$

Recalling that the event $[Y_i \leq W]$ is identical to $[E' = i]$ given that the state is x, we finally obtain the result:

$$p(i, x) = \frac{\lambda_i}{\Lambda(x)} \tag{6.68}$$

where we see that $p(i, x)$ depends on the current state x only through the feasible event set $\Gamma(x)$ required to define $\Lambda(x)$ in (6.58).

Example 6.4

Consider a single-server queueing system, but now we allow two different "classes" of customers to be served (see Fig. 6.12). Customers of class 1 form an arrival process which we will assume to be Poisson with rate λ_1. Similarly, customers of class 2 form a Poisson arrival process with rate λ_2, and we will assume that the two processes are independent. In addition, we will assume that when a customer gets served the service time is exponentially distributed with parameter μ_1 for class 1 customers, and μ_2 for class 2 customers. We also assume that all service times are mutually independent and independent of the arrival process.

We can see that this is a DES with event set $\mathcal{E} = \{a_1, a_2, d_1, d_2\}$, where a_i, d_i are arrival and departure events respectively for class i customers,

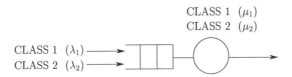

Figure 6.12: Queueing system for Example 6.4.
Customers from each class form a Poisson arrival process and receive exponentially distributed service times.

$i = 1, 2$. Note that the clock sequence for each event in our model is Poisson (the exponentially distributed service times imply a Poisson service process).

To describe the state of this system, we need to specify not only the queue length, but also the class of the customer currently in service (if any), which determines whether d_1 or d_2 is a feasible event at that state. In fact, the state is a vector of the form $[c_1, c_2, \ldots, c_x]$, where c_k is the class of the customer currently in the kth position in the queue (c_1 is the class of the customer in service), and x is the queue length; if $x = 0$, then we use 0 to denote the "empty system" state. This discussion points to the fact that we need a fairly elaborate GSMP model for this system. However, observing that we have a Poisson clock structure, we can exploit some of the properties we have identified in past sections in order to obtain easy answers to some interesting questions.

For instance, suppose we know that the system is empty at some point in time, and ask the question: *How long does it take on the average until the system becomes busy again?* At the empty state, the only feasible events are a_1, a_2. Thus, by (6.60), the interevent time is exponentially distributed with parameter $(\lambda_1 + \lambda_2)$. By the memoryless property, the amount of time already spent at that state is irrelevant, therefore the time until the next event occurs maintains the same distribution. The mean of this distribution is $1/(\lambda_1 + \lambda_2)$, which provides the answer to our question.

Now suppose we are in a state where a class 1 customer is being served, and we pose the question: *What is the probability that the next event is a departure?* The answer is easily obtained by using (6.68). At that state, the total Poisson rate is $(\lambda_1 + \lambda_2 + \mu_1)$, since the feasible events are a_1, a_2 and d_1. Thus, the probability that the next event is d_1 is given by $\mu_1/((\lambda_1 + \lambda_2 + \mu_1))$.

6.8.3   Markov Chains

Recall that the memoryless property is the defining feature of a Markov process, as discussed in Section 6.2.3. Since all of the nice results we derived above for the GSMP driven by a Poisson clock structure are essentially a consequence of this property, it is not surprising that such a GSMP *reduces to a Markov chain*. To see this more clearly, we return to the basic definition of a Markov chain which was given in equation (6.6) and is repeated here:

$$P[X(t_{k+1}) = x_{k+1} \mid X(t_k) = x_k, X(t_{k-1}) = x_{k-1}, \dots, X(t_0) = x_0]$$
$$= P[X(t_{k+1}) = x_{k+1} \mid X(t_k) = x_k]$$

for any $t_0 \le t_1 \le \cdots \le t_k \le t_{k+1}$. As discussed in Section 6.2.3, this definition is equivalent to properties **(M1)** and **(M2)**. Property **(M1)** refers to the fact that no state memory is needed, that is, the value of $X(t_{k+1})$ in the preceding equation depends only on the value of $X(t_k)$. In addition, **(M2)** points to the fact that no state age memory is needed either, that is, if $X(t_k) = x_k$, it is irrelevant how long it has been since the value of the state became x_k.

Now let us consider a GSMP, where, by (6.12), the next state x' is determined based on $p(x'; x, e')$. When the clock structure is Poisson, the triggering event can be probabilistically determined through (6.68), based on knowledge of the current state x and the feasible event set $\Gamma(x)$, needed to specify $\Lambda(x)$. Using this fact, we can also determine x' from knowledge of x alone through a set of transition probabilities

$$p(x', x) = P[X' = x' \mid X = x]$$

determined as follows. We apply the rule of total probability on all possible events $i \in \Gamma(x)$ to get

$$p(x', x) = \sum_{i \in \Gamma(x)} P[X' = x' \mid E' = i, X = x] \cdot P[E' = i \mid X = x]$$

The first term in the sum is the transition probability $p(x'; x, i)$ which is part of the GSMP specification, and the second term is $p(i, x)$ given by (6.68). Thus:

$$p(x', x) = \sum_{i \in \Gamma(x)} p(x'; x, i) \cdot \frac{\lambda_i}{\Lambda(x)} \tag{6.69}$$

and this probability is completely specified by x. In other words, given the current state x, the determination of the next state is completely independent of all past history (as required by **(M1)**), including the amount of time spent at state x (as required by **(M2)**). In fact, in the next chapter, which is devoted to Markov chains, we will derive an explicit expression for $P[X(t_{k+1}) = x_{k+1} \mid X(t_k) = x_k]$ for any two values x_k, x_{k+1} and time instants $t_k \le t_{k+1}$. At this point, it should be clear from (6.69) that the memoryless property is satisfied for any GSMP with a Poisson clock structure. What is particularly attractive about a Markov chain is that its entire stochastic behavior is captured by the transition probabilities $p(x', x)$ specified for all pairs of states (x, x').

The generation of a sample path of a Markov chain is greatly simplified by exploiting the memoryless property. In particular, a set of Poisson parameters

$$G = \{\lambda_1, \ldots, \lambda_m\}$$

is sufficient to describe the entire stochastic clock structure. We can then proceed as follows:

Step 1. With x known, evaluate the feasible event set $\Gamma(x)$.

Step 2. For every event $i \in \Gamma(x)$, sample from $G_i(t) = 1 - e^{-\lambda_i t}$ to obtain a clock value y_i.

Step 3. The triggering event is $e' = \arg \min_{i \in \Gamma(x)} \{y_i\}$.

Step 4. The next state x' is obtained by sampling from $p(x'; x, e')$.

Step 5. The next event time is given by $t' = t + \min_{i \in \Gamma(x)} \{y_i\}$.

Compared with the construction in (6.12) through (6.18), observe that (6.14) is a superfluous step. In other words, no updating of clock values is needed, since the age of feasible events is irrelevant due to the memoryless property. Thus, it suffices to select a completely new set of clock values, regardless of whether they represent new lifetimes or residual lifetimes. Actually, this construction can be further simplified if we assume that the transition probabilities $p(x', x)$ are specified, instead of $p(x'; x, e')$. In this case, it is not even necessary to explicitly determine a triggering event in *Step 3* above, since the next state is directly obtained through $p(x', x)$ alone. We shall have the occasion to further discuss alternative sample path constructions of Markov chains and their implication in later chapters.

6.9 EXTENSIONS OF THE GSMP

When the Generalized Semi-Markov Process (GSMP) was introduced in Section 6.5, it was stated that it covers a broad class of stochastic DES. One aspect of DES not covered by the GSMP is the possible dependence of clock values on the process state. This is because the event clocks that are an integral part of the GSMP model are all defined to tick from an assigned lifetime to zero at some *fixed rate*. We refer to this rate as the *speed* of the clock.

To illustrate this point, consider a DES with a single event. In particular, let this DES be a simple counting process $\{N(t)\}$, not necessarily a Poisson process, where the initial state is 0. A typical sample path is shown in Fig. 6.13, where the stochastic clock structure is a sequence of event lifetimes $\{V_1, V_2, \ldots\}$. In Fig. 6.13 we explicitly show the event clock as a function of time. Initially, the clock is set at $V_1 = v_1$, and it ticks down to 0 at *unit speed*. The first event

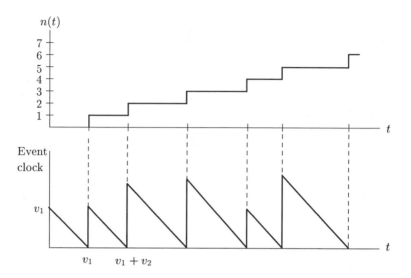

Figure 6.13: A counting process with a fixed-speed event clock.

corresponds to the clock reaching a zero value. At that point, the event clock jumps to the next lifetime value, V_2, and starts ticking down to 0 once again at unit speed. The process then continues in this manner, as shown in Fig. 6.13.

Now let us assume that the process $\{N(t)\}$ is made to be state-dependent in the following sense. When the state is $N(t) = 1$ or 2, the event rate is doubled; and when the state is $N(t) \geq 3$, the event rate is tripled. We can model this change by keeping the clock sequence $\{V_1, V_2, \dots\}$ unchanged, but modifying the event clock speed at different states. This is illustrated in Fig. 6.14 for the sample path of Fig. 6.13. We define the speed of the clock, $r(n)$, to depend on the state $n = 0, 1, 2, \dots$ as follows:

$$r(0) = 1$$
$$r(n) = 2 \quad \text{for } n = 1, 2$$
$$r(n) = 3 \quad \text{for all } n \geq 3$$

In this case, the event clock ticks down at unit speed only initially, when the state is 0. Comparing the event clock of Fig. 6.13 to that of Fig. 6.14, note that in the latter case the clock speed varies with the state. In a more general context, introducing clock speeds allows us to model stochastic DES where event rates are affected by the state of the system.

In general, when a DES consists of two or more events, we define the speed $r_i(x)$ of the ith event clock at state x as a function $r : \mathcal{X} \times \mathcal{E} \to \mathbb{R}^+$, such that $0 < r_i(x) < \infty$ for all $x \in \mathcal{X}, i \in \mathcal{E}$. This causes some modifications to the model defined by (6.12) through (6.18). Specifically, the interevent time Y^* when the state is X is now determined as the smallest clock value scaled by its speed:

$$Y^* = \min_{i \in \Gamma(X)} \left\{ \frac{Y_i}{r_i(X)} \right\} \tag{6.70}$$

Observe that events with higher clock speeds will now tend to occur sooner, since their effective clock values are reduced. Accordingly, the triggering event is given by

$$E' = \arg \min_{i \in \Gamma(X)} \left\{ \frac{Y_i}{r_i(X)} \right\} \tag{6.71}$$

Moreover, when the interevent time is Y^* at the instant an event causes a transition from X to X', the amount of time elapsed since the last event on the ith event clock is given by $r_i(X) \cdot Y^*$, for all $i \in \Gamma(X)$. If i is not the triggering event and it remains feasible in the new state X', then the ith residual lifetime is given by $[Y_i - r_i(X) \cdot Y^*]$. Thus, the new clock value is specified by:

$$Y_i' = \begin{cases} Y_i - r_i(X) \cdot Y^* & \text{if } i \neq E' \text{ and } i \in \Gamma(X) \\ V_{i,N_i+1} & \text{if } i = E' \text{ or } i \notin \Gamma(X) \end{cases} \quad i \in \Gamma(X') \tag{6.72}$$

This leads to an extension of the stochastic timed automaton defined in Section 6.4:

Definition. A *stochastic timed automaton with speeds* is a stochastic timed automaton $(\mathcal{E}, \mathcal{X}, \Gamma, p, p_0, G)$ equipped with a speed function:

$$r : \mathcal{X} \times \mathcal{E} \to \mathbb{R}^+, \text{ s.t. } 0 < r_i(x) < \infty \text{ for all } x \in \mathcal{X}, i \in \mathcal{E}.$$

The automaton operates as specified by (6.12) through (6.18) with (6.13) replaced by (6.71), (6.14) replaced by (6.72), and (6.15) replaced by (6.70). ♦

Definition. A *generalized semi-Markov process with speeds* is a stochastic process $\{X(t)\}$ with state space $\mathcal{X}$, generated by a stochastic timed automaton $(\mathcal{E}, \mathcal{X}, \Gamma, p, p_0, G)$ with speeds $r_i(x), x \in \mathcal{X}, i \in \mathcal{E}$. ♦

In the case of a GSMP with a Poisson clock structure, including speeds allows us to vary the Poisson rates λ_i depending on the process state. Thus, to specify the stochastic clock structure G, we now need to define the state-dependent rates $\lambda_i(x)$ for all $x \in \mathcal{X}$. Note that the event distribution derived in (6.68) now becomes

$$p(i, x) = \frac{\lambda_i(x)}{\Lambda(x)} \tag{6.73}$$

As a result, the transition probability $p(x', x)$ in (6.69) is also affected. However, the Markov property for this GSMP is retained.

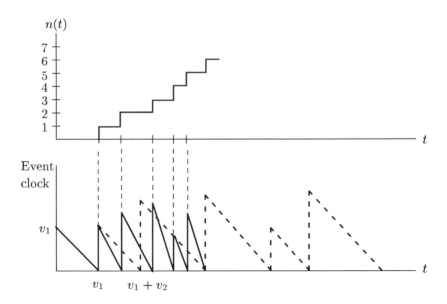

Figure 6.14: A counting process with a variable-speed event clock.

SUMMARY

- A *stochastic timed automaton* is a timed automaton equipped with a stochastic *clock structure*. The stochastic clock structure is a set of distribution functions, one for each event, characterizing the corresponding lifetimes. We assume that all clock sequences are iid and independent of each other.

- A *generalized semi-Markov process* (GSMP) is a stochastic process $\{X(t)\}$ generated by a stochastic timed automaton. The random variable $X(t)$ represents the state of the automaton at time t. A GSMP possesses the Markov property in that the value of the next state at a point where an event takes place is always dependent on the current state only and not the past state history. The interevent times, however, have arbitrary distributions.

- A *Poisson process* is a renewal process $\{N(t)\}$ that counts occurrences of an event over the time interval $(0, t]$. It is a process with *stationary independent* increments, and it constitutes a basic building block for stochastic models of DES.

- The interevent times of a Poisson process are exponentially distributed. This property is preserved when two or more independent Poisson processes are superimposed.

- A unique feature of the exponential distribution is its *memoryless property*. That is, if $P[V \le t] = 1 - e^{-\lambda t}$, for some positive random variable V, then $P[V \le z + t \mid V > z] = P[V \le t]$ for some $z \ge 0$. Thus, if V is an interevent time of a Poisson process, and $Y = V - z$ is a residual lifetime, Y and V are both exponentially distributed with parameter λ. Knowledge of the event age z is irrelevant in determining the residual lifetime.

- A GSMP with a Poisson clock structure reduces to a Markov chain. In this case, it suffices to specify transition probabilities $p(x', x)$ to specify the process, along with an initial cdf $p_0(x)$.

- The standard GSMP construction assumes that event lifetimes are independent of the current state. This can be relaxed by allowing event clocks to tick down to zero at some *clock speed* which can be state-dependent. This allows us to extend the Poisson process to one with state-dependent rates.

PROBLEMS

6.1 The number of messages successfully reaching a receiver at time $k = 0, 1, 2, \ldots$ is denoted by S_k, and it is modeled as an iid sequence with a distribution specified by $S_k = n$ with probability $1/N$, where $n = 1, \ldots, N$ for some given integer N. The cumulative number of successfully received messages is described by the random sequence

$$X_{k+1} = X_k + S_k, \quad S_0 = 0$$

(a) Does $\{X_k\}$ possess the Markov property?

(b) Calculate the *mean* $m_k = E[X_k]$ and *variance* $E[X_k - m_k]^2$ of this random sequence.

6.2 Every day that goes by any given machine deteriorates because of "wear and tear", until its state is so poor that either "preventive maintenance" is performed or it simply fails. Let us model the day-to-day state of a machine by $X_k = A^k, k = 0, 1, \ldots$, where A is a random variable with uniform distribution in $[0, 1]$. We can think of A as representing the random quality of the machine when it is first placed in operation, since $X_1 = A$. Thereafter, X_k continuously decreases as k increases.

(a) Calculate the cdf of the random variable X_k, $F_k(x) = P[X_k \le x]$.

(b) Calculate the joint cdf of X_k, X_{k+1}.

(c) Calculate the *mean* $m_k = E[X_k]$ and *variance* $E[X_k - m_k]^2$ of this random sequence.

6.3 Given an iid sequence $\{X_k\}$, a particularly useful random sequence is defined by collecting N samples from this sequence and defining the *sample mean*:

$$S_N = \frac{1}{N} \sum_{k=1}^{N} X_k$$

Calculate the *mean* $M_N = E[S_N]$ and *variance* $E[S_N - M_N]^2$ of $\{S_N\}$.

6.4 Construct a typical sample path (similar to Fig. 6.4) of the queueing system discussed in Section 6.5.1, with the following characteristics:

(a) All service times are constant and equal to 1 sec. The interarrival time is a random variable given by 0.5 sec with probability 0.5 and 1.0 sec with probability 0.5. The queue length is initially given by $x_0 = 2$. In the long run, what can you say about the probability that an arriving customer will find the server idle?

(b) The same as (a), except that the service time is also a random variable with a uniform distribution in $[0.5, 1.5]$. What can you say about the probability that an arriving customer will find the server idle in this case?

(c) All service times are constant and equal to 1 sec. The interarrival times alternate between 0.8 sec and 1.8 sec, starting with 1.8 sec. Moreover, when the interarrival time is 1.8 sec, two customers arrive simultaneously (this is referred to as a *bulk arrival process*). The queue length is initially given by $x_0 = 0$.

(d) The same as (c), except that the service times are *state-dependent*: Whenever the server is about to start processing a customer, a decision is made as to the service time for that customer; if at the time the number of customers present is ≤ 2, the service time is 1.0 sec, if it is > 2 but ≤ 4 the service time is 0.5 sec, and if it is > 4 the service time is 0.4 sec. The queue length is initially given by $x_0 = 0$.

6.5 Jobs are submitted to a computer system according to a Poisson process with rate λ (jobs/hour). Determine:

(a) The probability that the time interval between two successive job submission events is greater than τ minutes.

(b) The probability that no jobs are lost if the computer goes down for the first a minutes and the last b minutes of an hour (a job is "lost" if it finds the computer down when it is submitted).

(c) The probability that at least two jobs are submitted within the first a minutes.

6.6 Given a Poisson process, suppose that exactly n events are observed in the interval $(0, t]$. Consider a sequence of time instants such that $0 < s_1 < t_1 < s_2 < t_2 < \cdots < s_n < t_n < t$. Let $\tau_i = t_i - s_i, i = 1, \ldots, n$. Show that the probability that exactly one event occurs in each of the intervals $(s_i, t_i], i = 1, \ldots, n$, given that exactly n events occur $(0, t]$ is

$$n! \prod_{i=1}^{n} \frac{\tau_i}{t}$$

6.7 Given a Poisson process $\{N(t)\}$ with parameter λ, suppose that whenever an event occurs it is removed from the process with probability $1 - p$ (this is referred to as *thinning* the original process). The remaining events define a new process $\{N_p(t)\}$. Show that $\{N_p(t)\}$ is also a Poisson process with parameter λp. (*Hint:* Try conditioning the event $[N_p(t) = n]$ on a "convenient" random variable. Also, recall that the binomial distribution for a random variable counting the number of "successes" in k "trials" when the probability of success is p is:

$$P[n \text{ successes in } k \text{ trials}] = \binom{k}{n} p^n (1 - p)^{k-n}, \quad n = 0, 1, \ldots, k)$$

Extend this result to M different processes formed by thinning $\{N(t)\}$, where the ith process consists of events selected from $\{N(t)\}$ with probability $p_i, i = 1, \ldots, M$.

6.8 Fixing $t = 1$ in the expression for the Poisson distribution (6.34) we obtain a probability distribution that characterizes a discrete random variable X:

$$P[X = n] = \frac{\lambda^n}{n!} e^{-\lambda}, \quad n = 0, 1, 2, \ldots$$

Show that the discrete random variable $Y = X_1 + \cdots + X_N$, where every $X_i, i = 1, \ldots, N$ has the above Poisson distribution with parameter λ and all X_i are mutually independent, also has a Poisson distribution with parameter $N\lambda$.

6.9 When a quarter is inserted into a slot machine, the machine returns a dollar with probability p and nothing otherwise. People arrive at the slot machine according to a Poisson process of rate λ. Each person drops one quarter only and then leaves. Let $N(t)$ denote the (total) number of dollars returned by the machine by time t. Determine the probability $P[N(t) = n], n = 0, 1, \ldots$ (*Hint:* Use Problem 6.7).

6.10 This is an extension of Problem 6.9. The slot machine now returns a random number of dollars every time a quarter is inserted, and let this number be a random variable with a Poisson distribution with parameter a (see also Problem 6.8). As before, let $N(t)$ denote the number of dollars returned by the machine by time t. Determine the probability $P[N(t) = n], n = 0, 1, \ldots$

6.11 Consider a simple queueing system where customers arrive according to a Poisson process of rate λ. Assume that service times are exponentially distributed with service rate μ. Let X denote the number of customers that arrive during a service time. Determine $P[X = k], k = 0, 1, 2, \ldots$ The following integral may be of help (but is *not* necessarily needed in obtaining the result):

$$\int_0^\infty x^{k-1} e^{-x} dx = \Gamma(k), \quad k = 0, 1, \ldots$$

where the gamma function $\Gamma(k)$ satisfies $\Gamma(k+1) = k!$.

6.12 A warehouse receives parts which arrive according to a Poisson process with rate λ. Every T minutes (T = constant) a truck arrives and loads all parts present at the time (while loading is going on, the warehouse does not accept any more parts, and we assume that such parts are sent to a different warehouse). The time to load a single part is τ minutes (τ = constant) and the total loading time is proportional to the number of parts loaded. When the truck leaves, the process repeats.

(a) Draw a typical sample path for this queueing model (as in Fig. 6.4), assuming the warehouse is initially empty.

(b) The (random) time interval between two successive points when the warehouse is empty is referred to as a "cycle". The area under the queue length curve over such a cycle measures the "workload" of the system and is denoted by W. Determine the average workload $E[W]$.

6.13 A computer processor is preceded by a queue which can accommodate one job only (so the total capacity of this system is 2 jobs). Two different job types may request service from the processor, and type 1 has a higher priority than type 2. Thus, if a type 1 job arrives and finds a type 2 job in service, it *preempts* it: The type 2 job is returned to the queue, and the type 1 job receives service. If any job arrives and finds the system full, it is rejected and lost. Assume that each job type arrives according to a Poisson process with rate λ_i, $i = 1, 2$, and that its processing time is exponentially distributed with rate μ_i, $i = 1, 2$.

(a) Define a stochastic timed automaton model using four event types: a_i is a type i job arrival, and d_i is a type i job departure, $i = 1, 2$, and draw a state transition diagram.

(b) When the system is processing a type 2 job with no other job waiting for service, what is the probability that the next arriving job will not have to wait to get processed?

(c) Suppose you observe the processor when it is idle. On the average, how long is it going to be before it becomes busy again?

SELECTED REFERENCES

■ *Review of Probability and Stochastic Processes*

  – Clarke, A. B., and R.L. Disney, *Probability and Random Processes*, Wiley, New York, 1985.

  – Gallager, R.G., *Discrete Stochastic Processes*, Kluwer Academic Publishers, Boston, 1996.

  – Hoel, P.G., S.C. Port, and C.J. Stone, *Introduction to Stochastic Processes*, Houghton Mifflin, Boston, 1972.

  – Parzen, E., *Stochastic Processes*, Holden-Day, San Francisco, 1962.

■ *Queueing Models*

  – Kleinrock, L., *Queueing Systems*, Volume I: Theory, Wiley, New York, 1975.

  – Walrand, J., *An Introduction to Queueing Networks*, Prentice-Hall, Englewood Cliffs, 1988.

■ *Stochastic Timed Automata*

  – Glynn, P., "A GSMP Formalism for Discrete Event Systems," *Proceedings of the IEEE*, Vol. 77, No. 1, pp. 14-23, 1989.

  – Ho, Y.C., and X. Cao, *Perturbation Analysis of Discrete Event Dynamic Systems*, Kluwer Academic Publishers, Boston, 1991.

  – Glasserman, P., and D.D. Yao, *Monotone Structure in Discrete-Event Systems*, Wiley, New York, 1994.

Chapter 7

Markov Chains

7.1 INTRODUCTION

In the previous chapter, we presented the Generalized Semi-Markov Process (GSMP) framework as a means of modeling stochastic DES. By allowing event clocks to tick at varying speeds, we also provided an extension to the basic GSMP. In addition, we introduced the Poisson process as a basic building block for a class of stochastic DES which possess the Markov (memoryless) property. Thus, we obtained the class of stochastic processes known as *Markov chains*, which we will study in some detail in this chapter. It should be pointed out that the analysis of Markov chains provides a rich framework for studying many DES of practical interest, ranging from gambling and the stock market to the design of "high-tech" computer systems and communication networks.

The main characteristic of Markov chains is that their stochastic behavior is described by transition probabilities of the form $P[X(t_{k+1}) = x' \mid X(t_k) = x]$ for all state values x, x' and $t_k \le t_{k+1}$. Given these transition probabilities and a distribution for the initial state, it is possible to determine the probability of being at any state at any time instant. Describing precisely how to accomplish this and appreciating the difficulties involved in the process are the main objectives of this chapter.

We will first study *discrete-time* Markov chains. This will allow us to present the main ideas, techniques, and analytical results before proceeding to *continuous-time* Markov chains. In principle, we can always "solve" Markov chains, that is, determine the probability of being at any state at any time instant. However, in practice this task can be a formidable one, as it often involves the solution of complicated differential equations. Even though we would like to obtain general *transient* solutions, in most cases we have to settle for *steady-state* or *stationary* solutions, which describe the probability of being at any state in the long run only (after the system has been in operation for a "sufficiently long" period of time). Even then we often need to resort to numerical techniques. Fortunately, however, explicit closed-form expressions are available for several cases of practical interest. In particular, we will focus attention on the class of *birth-death Markov chains*, of which, as we shall see, the Poisson process turns out to be a special case. Finally, we will present a methodology for converting continuous-time Markov chains into equivalent discrete-time Markov chains. This is known as *uniformization*, and it relies on the fundamental memoryless property of these models.

7.2 DISCRETE-TIME MARKOV CHAINS

Recall that in a discrete-time Markov chain events (and hence state transitions) are constrained to occur at time instants $0, 1, 2, \ldots, k, \ldots$ Thus, we form a stochastic sequence $\{X_1, X_2, \ldots\}$ which is characterized by the Markov (memoryless) property:

$$
\begin{aligned}
&P[X_{k+1} = x_{k+1} \mid X_k = x_k, X_{k-1} = x_{k-1}, \ldots, X_0 = x_0] \\
&= P[X_{k+1} = x_{k+1} \mid X_k = x_k]
\end{aligned}
\tag{7.1}
$$

Given the current state x_k, the value of the next state depends only on x_k and not on any past state history (no state memory). Moreover, the amount of time spent in the current state is irrelevant in determining the next state (no age memory).

In the rest of this section, we first discuss what it means to "obtain a model" of a stochastic DES that qualifies as a discrete-time Markov chain. We then proceed with analyzing such models. One of our main objectives is the determination of probabilities for the chain being at various states at different times.

7.2.1 Model Specification

Thus far, we have been modeling stochastic DES by means of the stochastic timed automaton formalism based on the six-tuple

$$(\mathcal{E}, \mathcal{X}, \Gamma, p, p_0, G)$$

In the framework of the previous chapter, state transitions are driven by events belonging to the set $\mathcal{E}$. Thus, the transition probabilities are expressed as $p(x'; x, e')$ where $e' \in \Gamma(x)$ is the triggering event, and $\Gamma(x)$ is the feasible event set at state x. In Markov chains, however, we will only be concerned with the total probability $p(x', x)$ of making a transition from x to x', regardless of which event actually causes the transition. Thus, similar to (6.68), we apply the rule of total probability to get

$$p(x', x) = \sum_{i \in \Gamma(x)} p(x'; x, i) \cdot p(i, x)$$

where $p(i, x)$ is the probability that event i occurs at state x. This transition probability is an aggregate over all events $i \in \Gamma(x)$ which may cause the transition from x to x'. In general, we will allow transition probabilities to depend on the time instant at which the transition occurs, so we set

$$p_k(x', x) = P[X_{k+1} = x' \mid X_k = x]$$

With these observations in mind, we will henceforth identify events with state transitions in our specification of Markov chain models. Although it is always useful to keep in mind what the underlying event set is, we will omit the specification of the sets $\mathcal{E}$ and $\Gamma(x)$. As we shall see, the clock structure G is implicitly defined by the Markov property in (7.1). Therefore, to specify a Markov chain model we only need to identify:

1. A state space $\mathcal{X}$.

2. An initial state probability $p_0(x) = P[X_0 = x]$, for all $x \in \mathcal{X}$.

3. Transition probabilities $p(x', x)$ where x is the current state and x' is the next state.

Since the state space $\mathcal{X}$ is a countable set, we will subsequently map it onto the set of non-negative integers (or any subset thereof):

$$\mathcal{X} = \{0, 1, 2, \dots\}$$

which will allow us to keep notation simple.

In the next few sections, we provide the basic definitions and notation associated with discrete-time Markov chains.

7.2.2 Transition Probabilities and the Chapman-Kolmogorov Equations

Since we have decided to use the nonnegative integers as our state space, we will use the symbols i, j to denote a typical current and next state respectively. We will also modify our notation and define *transition probabilities* as follows:

$$p_{ij}(k) \equiv P[X_{k+1} = j \mid X_k = i] \tag{7.2}$$

where $i, j \in \mathcal{X}$ and $k = 0, 1, \ldots$ As already pointed out, transition probabilities are allowed to be time-dependent, hence the need to express $p_{ij}(k)$ as a function of the time instant k.

Clearly, $0 \leq p_{ij}(k) \leq 1$. In addition, observe that for any state i and time instant k:

$$\sum_{\text{all } j} p_{ij}(k) = 1 \tag{7.3}$$

since we are summing over all possible mutually exclusive events causing a transition from i to some new state.

The transition probabilities $p_{ij}(k)$ defined in (7.2) refer to state transitions that occur in one step. A natural extension is to consider state transitions that occur over n steps, $n = 1, 2, \ldots$ Thus, we define the *n-step transition probabilities*:

$$p_{ij}(k, k+n) \equiv P[X_{k+n} = j \mid X_k = i] \tag{7.4}$$

Let us now condition the event $[X_{k+n} = j \mid X_k = i]$ above on $[X_u = r]$ for some u such that $k < u \leq k+n$. Using the rule of total probability, (7.4) becomes

$$p_{ij}(k, k+n) = \sum_{\text{all } r} P[X_{k+n} = j \mid X_u = r, X_k = i] \cdot P[X_u = r \mid X_k = i] \tag{7.5}$$

By the memoryless property (7.1),

$$P[X_{k+n} = j \mid X_u = r, X_k = i] = P[X_{k+n} = j \mid X_u = r] = p_{rj}(u, k+n)$$

Moreover, the second term in the sum in (7.5) is simply $p_{ir}(k, u)$. Therefore,

$$p_{ij}(k, k+n) = \sum_{\text{all } r} p_{ir}(k, u) p_{rj}(u, k+n), \qquad k < u \leq k+n \tag{7.6}$$

This is known as the *Chapman-Kolmogorov equation*. It is one of the most general relationships we can derive about discrete-time Markov chains.

It is often convenient to express the Chapman-Kolmogorov equation (7.6) in matrix form. For this purpose, we define the matrix

$$\mathbf{H}(k, k+n) \equiv [p_{ij}(k, k+n)], \qquad i, j = 0, 1, 2, \ldots \tag{7.7}$$

Then, (7.6) can be rewritten as a matrix product:

$$\mathbf{H}(k, k+n) = \mathbf{H}(k, u)\mathbf{H}(u, k+n) \tag{7.8}$$

Choosing $u = k + n - 1$ we get

$$\mathbf{H}(k, k+n) = \mathbf{H}(k, k+n-1)\mathbf{H}(k+n-1, k+n) \tag{7.9}$$

This relationship is known as the *forward Chapman-Kolmogorov equation*. The *backward Chapman-Kolmogorov equation* is obtained by choosing $u = k + 1$, which gives

$$\mathbf{H}(k, k+n) = \mathbf{H}(k, k+1)\mathbf{H}(k+1, k+n) \tag{7.10}$$

7.2.3 Homogeneous Markov Chains

Whenever the transition probability $p_{ij}(k)$ is independent of k for all $i, j \in \mathcal{X}$, we obtain a *homogeneous* Markov chain. In this case, (7.2) is written as

$$p_{ij} = P[X_{k+1} = j \mid X_k = i] \tag{7.11}$$

where p_{ij} is independent of k. In simple terms, a state transition from i to j always occurs with the same probability, regardless of the point in time when it is observed. Note that homogeneity is a form of stationarity which applies to transition probabilities only, but not necessarily to the Markov chain itself. In particular, $P[X_{k+1} = j \mid X_k = i]$ may be independent of k, but the joint probability $P[X_{k+1} = j, X_k = i]$ need not be independent of k. To see this more clearly, observe that

$$\begin{aligned} P[X_{k+1} = j, X_k = i] &= P[X_{k+1} = j \mid X_k = i] \cdot P[X_k = i] \\ &= p_{ij} \cdot P[X_k = i] \end{aligned}$$

where $P[X_k = i]$ need not be independent of k.

Example 7.1
Consider a machine which can be in one of two states, UP or DOWN. We choose a state space $\mathcal{X} = \{0, 1\}$ where 1 denotes UP and 0 denotes DOWN. The state of the machine is checked every hour, and we index hours by $k = 0, 1, \ldots$ Thus, we form the stochastic sequence $\{X_k\}$, where X_k is the state of the machine at the kth hour. Let us further assume that if the machine is UP, it has a probability α of failing during the next hour. If the machine is in the DOWN state, it has a probability β of being repaired during the next hour.

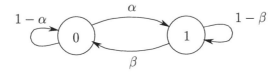

Figure 7.1: State transition diagram for Example 7.1.

We therefore obtain a homogeneous discrete-time Markov chain with transition probabilities:

$$p_{10} = \alpha, \qquad p_{11} = 1 - \alpha, \qquad p_{01} = \beta, \qquad p_{00} = 1 - \beta$$

where $0 \leq \alpha \leq 1$ and $0 \leq \beta \leq 1$. We can also draw a state transition diagram (as we did for automata), but the arcs connecting states now refer to *probabilities* rather than events (see Fig. 7.1).

Alternatively, it is reasonable to expect that as the machine wears out over time it becomes more failure-prone. Suppose we modify the transition probabilities above so that

$$p_{10}(k) = 1 - \gamma^k, \qquad p_{11}(k) = \gamma^k$$

for some $0 < \gamma < 1$ and $k = 0, 1, \ldots$ In this case, the probability of the machine failing increases with k and approaches 1 as $k \to \infty$. The new chain is no longer homogeneous, since $p_{10}(k)$ is time-dependent.

In a homogeneous Markov chain, the n-step transition probability $p_{ij}(k, k + n)$ is also independent of k. In this case, we denote it by p_{ij}^n, and (7.4) is written as

$$p_{ij}^n = P[X_{k+n} = j \mid X_k = i], \qquad n = 1, 2, \ldots \tag{7.12}$$

Then, by setting $u = k + m$ in the Chapman-Kolmogorov equation (7.6), we get

$$p_{ij}^n = \sum_{\text{all } r} p_{ir}^m p_{rj}^{n-m}$$

and by choosing $m = n - 1$:

$$p_{ij}^n = \sum_{\text{all } r} p_{ir}^{n-1} p_{rj} \tag{7.13}$$

or, in the matrix notation of (7.7), where $\mathbf{H}(n) \equiv [p_{ij}^n]$,

$$\mathbf{H}(n) = \mathbf{H}(n - 1)\mathbf{H}(1) \tag{7.14}$$

In simple terms, (7.13) breaks up the process of moving from state i to state j in n steps into two parts. First, we move from i to some intermediate state r in $(n - 1)$ steps, and then we take one more step from r to j. By aggregating over all possible intermediate states r , we obtain all possible paths from i to j in n steps.

From this point on, we will limit ourselves to *homogeneous* Markov chains (unless explicitly stated otherwise).

7.2.4   The Transition Probability Matrix

The transition probability information for a discrete-time Markov chain is conveniently summarized in matrix form. We define the *transition probability matrix* $\mathbf{P}$ to consist of all p_{ij}:

$$\mathbf{P} \equiv [p_{ij}], \qquad i, j = 0, 1, 2, \ldots \tag{7.15}$$

By this definition, in conjunction with (7.3), all elements of the ith row in this matrix, $i = 1, 2, \ldots$, must always sum up to 1. Note that $\mathbf{P} = \mathbf{H}(1)$ in the matrix equation (7.14).

Example 7.2 (A simplified telephone call process)

We describe a simple telephone call process in discrete time. Let the time line consist of small intervals indexed by $k = 0, 1, 2, \ldots$, which are sometimes called *time slots*. The process operates as follows:

- At most one telephone call can occur in a single time slot, and there is a probability α that a telephone call occurs in any one slot.

- If the phone is busy, the call is lost (no state transition); otherwise, the call is processed.

- There is a probability β that a call in process completes in any one time slot.

- If both a call arrival and a call completion occur in the same time slot, the new call will be processed.

We also assume that call arrivals and call completions in any time slot occur independently of each other and of the state of the process.

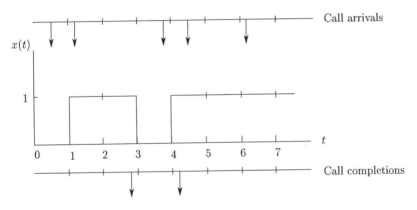

Figure 7.2: Typical sample path for Example 7.2.

Let X_k denote the state of the system at the kth slot, which is either 0 (phone is idle) or 1 (phone is busy). A typical sample path is shown in Fig. 7.2. In this example, $x_0 = 0$ and a call arrives in slot 0, hence $x_1 = 1$. Note that in the slot that starts at $t = 4$ both a new call arrival and a call completion occur; the new call is not lost in this case, and we have $x_5 = 1$.

The transition probabilities are as follows:

$$p_{00} = 1 - \alpha$$ (phone remains idle if no new call occurs in current slot)

$$p_{01} = \alpha$$ (phone becomes busy if new call occurs in current slot)

$$p_{10} = \beta \cdot (1 - \alpha)$$ (phone becomes idle if call completes in current slot *and* no new call occurs)

$$p_{11} = (1 - \beta) + \alpha\beta$$ (phone remains busy if call does not complete in current slot *or* call does complete but a new call also occurs)

A state transition diagram is shown in Fig. 7.3. The probability transition matrix for this Markov chain is

$$\mathbf{P} = \left[\begin{array}{cc} 1 - \alpha & \alpha \\ \beta(1 - \alpha) & (1 - \beta) + \alpha\beta \end{array} \right]$$

where we can easily check that (7.3) is satisfied for both rows. Clearly, once p_{10} is determined, we can simply write $p_{11} = 1 - p_{10}$. Independently evaluating p_{11} serves as a means of double-checking that we have correctly evaluated transition probabilities.

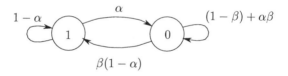

Figure 7.3: State transition diagram for Example 7.2.

Remark. As mentioned earlier, in Markov chain modeling we concentrate on state transition probabilities regardless of which event causes the transition. As a result, the stochastic behavior of a DES is compactly captured in the transition probability matrix $\mathbf{P}$. But this comes at some cost: The *structural information* contained in the event-driven description is lost. This can be seen in Example 7.2 when one considers $p_{11} = (1 - \alpha) + \alpha\beta$. This is just a number telling us the probability of a self-loop transition at state 1, but it does not reveal how this transition might come about (because of no call completion or because of both a call completion and call arrival). Nor does it provide any information on possible call arrivals that were lost during such self-loop transitions. One should be aware of this basic tradeoff: compact representation versus structural information.

Example 7.3 (A two-processor computer system)

Let us consider a computer system consisting of two identical processors working in parallel. Here, we have two "servers" (the processors), as opposed to only one (the phone) in Example 7.2. Once again, let the time line consist of slots indexed by $k = 1, 2, \ldots$ The operation of this system is only slightly more complicated than that of the telephone call process:

- At most one job can be submitted to the system in a single time slot, and such an event occurs with probability α.

- When a job is submitted to the system, it is served by whichever processor is available.

- If both processors are available, the job is given to processor 1.

- If both processors are busy, the job is lost.

- When a processor is busy, there is a probability β that it completes processing in any one time slot.

- If a job is submitted in a slot where both processors are busy and either one of the processors completes in that slot, then the job will be processed.

We also assume that job submissions and completions at either processor in any time slot all occur independently of each other and of the state of the system.

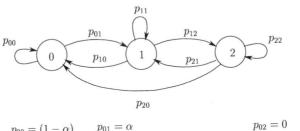

$$p_{00} = (1 - \alpha) \qquad p_{01} = \alpha \qquad\qquad\qquad p_{02} = 0$$

$$p_{10} = \beta \cdot (1 - \alpha) \quad p_{11} = (1 - \beta) \cdot (1 - \alpha) + \beta \cdot \alpha \quad p_{12} = (1 - \beta) \cdot \alpha$$

$$p_{20} = \beta^2 \cdot (1 - \alpha) \quad p_{21} = \beta \cdot (1 - \beta) \cdot (1 - \alpha) + \beta \cdot (1 - \beta) \cdot (1 - \alpha) + \beta^2 \cdot \alpha$$

$$p_{22} = (1 - \beta)^2 + \beta \cdot (1 - \beta) \cdot \alpha + \beta \cdot (1 - \beta) \cdot \alpha$$

Figure 7.4: State transition diagram for Example 7.3.

Let X_k denote the number of jobs being processed by the system at slot k. Thus, the state space is the set $\{0, 1, 2\}$. The transition probability matrix $\mathbf{P}$ is obtained from the state transition diagram shown in Fig. 7.4, where we summarize all $p_{ij}, i, j = 0, 1, 2$. The derivation of each p_{ij} is straightforward. As an example, we explain the derivation of p_{21}:

$$
\begin{aligned}
p_{21} \quad &= \beta \cdot (1 - \beta) \cdot (1 - \alpha) && \text{(proc. 1 completes,} \\
&&& \text{proc. 2 remains busy,} \\
&&& \text{no new job submitted)} \\
&+ \beta \cdot (1 - \beta) \cdot (1 - \alpha) && \text{(proc. 2 completes,} \\
&&& \text{proc. 1 remains busy,} \\
&&& \text{no new job submitted)} \\
&+ \beta^2 \cdot \alpha && \text{(both processors complete,} \\
&&& \text{new job submitted)}
\end{aligned}
$$

Let us write $\mathbf{P}$ for a specific numerical example when $\alpha = 0.5$ and $\beta = 0.7$. Evaluating all p_{ij} shown in Fig. 7.4 for these values, we obtain

$$
\mathbf{P} = \begin{bmatrix}
0.5 & 0.5 & 0 \\
0.35 & 0.5 & 0.15 \\
0.245 & 0.455 & 0.3
\end{bmatrix}
$$

7.2.5 State Holding Times

It might be argued that the specification of a discrete-time Markov chain model should require the specification of probability distributions characterizing the amount of time spent at some state i whenever this state is visited. However, due to the memoryless property, these distributions are automatically determined. We will use $V(i)$ to denote the number of time steps spent at state i when it is visited. This is a discrete random variable referred to as the *state holding time*.

It is a fundamental property of discrete-time Markov chains that *the distribution of the state holding time* $V(i), P[V(i) = n], n = 1, 2, \ldots$, *is geometric with parameter* p_{ii}. This is a direct consequence of the memoryless property, as shown below.

Suppose state i is entered at the kth time step. Note that the event $[V(i) = n]$ is identical to the event $[X_{k+1} = i, X_{k+2} = i, \ldots, X_{k+n-1} = i, X_{k+n} \neq i \mid X_k = i]$, since the chain remains at state i for precisely n time steps. Therefore,

$$
\begin{aligned}
P[V(i) = n] &= P[X_{k+1} = i, \ldots, X_{k+n-1} = i, X_{k+n} \neq i \mid X_k = i] \\
&= P[X_{k+n} \neq i \mid X_{k+n-1} = i, \ldots, X_k = i] \\
&\quad \cdot P[X_{k+n-1} = i, \ldots, X_{k+1} = i \mid X_k = i] \\
&= [1 - P[X_{k+n} = i \mid X_{k+n-1} = i, \ldots, X_k = i]] \\
&\quad \cdot P[X_{k+n-1} = i, \ldots, X_{k+1} = i \mid X_k = i]
\end{aligned}
$$

By the memoryless property (7.1), we have

$$
\begin{aligned}
P[X_{k+n} = i \mid X_{k+n-1} &= i, \ldots, X_k = i] \\
&= P[X_{k+n} = i \mid X_{k+n-1} = i] = p_{ii}
\end{aligned}
$$

and, therefore,

$$P[V(i) = n] = (1 - p_{ii}) \cdot P[X_{k+n-1} = i, \dots, X_{k+1} = i \mid X_k = i] \qquad (7.16)$$

Similarly,

$$\begin{aligned}
P[X_{k+n-1} &= i, \dots, X_{k+1} = i \mid X_k = i] \\
&= P[X_{k+n-1} = i \mid X_{k+n-2} = i, \dots, X_k = i] \\
&\quad \cdot P[X_{k+n-2} = i, \dots, X_{k+1} = i \mid X_k = i] \\
&= P[X_{k+n-1} = i \mid X_{k+n-2} = i] \cdot P[X_{k+n-2} = i, \dots, X_{k+1} = i \mid X_k = i] \\
&= p_{ii} \cdot P[X_{k+n-2} = i, \dots, X_{k+1} = i \mid X_k = i]
\end{aligned}$$

and (7.16) becomes

$$P[V(i) = n] = (1 - p_{ii}) \cdot p_{ii} \cdot P[X_{k+n-2} = i, \dots, X_{k+1} = i \mid X_k = i]$$

Repeating this process finally yields

$$P[V(i) = n] = (1 - p_{ii}) \cdot (p_{ii})^{n-1} \qquad (7.17)$$

This is a geometric distribution with parameter p_{ii}. It is independent of the amount of time already spent in state i, which is a consequence of the memoryless property: At each time step the chain determines its future based only on the fact that the current state is i. Hence, $P[V(i) = n]$ depends only on p_{ii}. The geometric distribution is the discrete-time counterpart of the exponential distribution which we saw in the previous chapter, and which we shall have the opportunity to see once again later in this chapter.

7.2.6 State Probabilities

One of the main objectives of Markov chain analysis is the determination of probabilities of finding the chain at various states at specific time instants. We define *state probabilities* as follows:

$$\pi_j(k) \equiv P[X_k = j] \qquad (7.18)$$

Accordingly, we define the *state probability vector*

$$\boldsymbol{\pi}(k) = [\pi_0(k), \pi_1(k), \dots] \qquad (7.19)$$

This is a row vector whose dimension is specified by the dimension of the state space of the chain. Clearly, it is possible for $\boldsymbol{\pi}(k)$ to be infinite-dimensional.

A discrete-time Markov chain model is completely specified if, in addition to the state space $\mathcal{X}$ and the transition probability matrix $\mathbf{P}$, we also specify an initial state probability vector

$$\boldsymbol{\pi}(0) = [\pi_0(0), \pi_1(0), \dots]$$

which provides the probability distribution of the initial state, X_0, of the chain.

7.2.7 Transient Analysis

Once a model is specified through $\mathcal{X}$, $\mathbf{P}$, and $\pi(0)$, we can start addressing questions such as: What is the probability of moving from state i to state j in n steps? Or: What is the probability of finding the chain at state i at time k? In answering such questions, we limit ourselves to given finite numbers of steps over which the chain is analyzed. This is what we refer to as *transient* analysis, in contrast to the *steady-state* analysis we will discuss in later sections.

Our main tool for transient analysis is provided by the recursive equation (7.13). A similar recursive relationship can also be derived, making explicit use of the transition probability matrix $\mathbf{P}$. Using the definitions of $\pi(k)$ and $\mathbf{P}$ we get

$$\pi(k+1) = \pi(k)\mathbf{P}, \qquad k = 0, 1, \ldots \qquad (7.20)$$

This relationship is easily established as follows. Let $\pi_j(k+1)$ be a typical element of $\pi(k+1)$. Then, by conditioning the event $[X_{k+1} = j]$ on $[X_k = i]$ for all possible i, we get

$$\pi_j(k+1) = P[X_{k+1} = j] = \sum_{\text{all } i} P[X_{k+1} = j \mid X_k = i] \cdot P[X_k = i]$$

$$= \sum_{\text{all } i} p_{ij} \cdot \pi_i(k)$$

which, written in matrix form, is precisely (7.20).

Moreover, by using (7.20) we can obtain an expression for $\pi(k)$ in terms of a given initial state probability vector $\pi(0)$ and the transition probability matrix $\mathbf{P}$. Specifically, for $k = 0$ (7.20) becomes

$$\pi(1) = \pi(0)\mathbf{P}$$

Then, for $k = 1$, we get

$$\pi(2) = \pi(1)\mathbf{P} = \pi(0)\mathbf{P}^2$$

and continuing in the same fashion (or using a formal induction argument) we obtain

$$\pi(k) = \pi(0)\mathbf{P}^k, \, k = 1, 2, \ldots \qquad (7.21)$$

Using (7.13), (7.20), and (7.21) we can completely study the transient behavior of homogeneous discrete-time Markov chains.

Example 7.4 (A simple gambling model)

Many games of chance are stochastic DES and can be modeled through discrete-time Markov chains. Suppose a gambler starts out with a capital of 3 dollars and plays a game of "spin-the-wheel" where he bets 1 dollar at a time. The wheel has 12 numbers, and the gambler bets on only one of them. Thus, his chance of winning in a spin is 1/12. If his number comes up, he receives 3 dollars, hence he has a net gain of 2 dollars. If his number does not come up, his bet is lost.

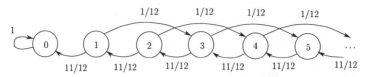

Figure 7.5: State transition diagram for Example 7.4.
In the "spin-the-wheel" game a gambler either wins two dollars per bet with probability 1/12, or loses one dollar with probability 11/12. The game ends if the gambler's capital becomes 0.

This is a Markov chain with state space $\{0, 1, \ldots\}$, where the state X_k represents the gambler's capital after the kth bet, and the initial state is $X_0 = 3$. The state transition diagram is shown in Fig. 7.5. From any state $i > 0$, there are two possible transitions: to state $(i-1)$ when the gambler loses, and to state $(i+2)$ when the gambler wins. At state 0 the gambler has no more capital to bet with; the game is effectively over, which we model by a self-loop transition with probability 1. The corresponding transition probability matrix is

$$\mathbf{P} = \begin{bmatrix} 1 & 0 & 0 & 0 & 0 & \cdots \\ 11/12 & 0 & 0 & 1/12 & 0 & \cdots \\ 0 & 11/12 & 0 & 0 & 1/12 & 0 \\ \vdots & \vdots & \vdots & \vdots & \vdots & \vdots \end{bmatrix}$$

We now use (7.13) to compute the probability that the gambler doubles his capital after 3 bets. Note that since $X_0 = 3$,

$$P[\text{gambler doubles his capital after 3 bets}]$$
$$= P[X_3 = 6 \mid X_0 = 3] = p_{36}^3$$

It follows from (7.13) that

$$p_{36}^3 = \sum_{r=0}^{\infty} p_{3r}^2 \cdot p_{r6}$$

Looking at $\mathbf{P}$ (or the state transition diagram), observe that $p_{r6} = 0$ for all $r \neq 4, 7$. Therefore, we can rewrite the equation above as

$$p_{36}^3 = p_{34}^2 \cdot p_{46} + p_{37}^2 \cdot p_{76} = p_{34}^2 \cdot \frac{1}{12} + p_{37}^2 \cdot \frac{11}{12} \qquad (7.22)$$

Using (7.13) again, we get

$$p_{34}^2 = \sum_{r=0}^{\infty} p_{3r} \cdot p_{r4}$$

where $p_{3r} = 0$ for all $r \neq 2, 5$, therefore,

$$p_{34}^2 = p_{32} \cdot p_{24} + p_{35} \cdot p_{54} = \frac{11}{12} \cdot \frac{1}{12} + \frac{1}{12} \cdot \frac{11}{12} = \frac{22}{(12)^2} \qquad (7.23)$$

Similarly,

$$p_{37}^2 = p_{32} \cdot p_{27} + p_{35} \cdot p_{57} = p_{32} \cdot 0 + \frac{1}{12} \cdot \frac{1}{12} = \frac{1}{(12)^2} \qquad (7.24)$$

Combining (7.22) through (7.24) we get

$$p_{36}^3 = \frac{33}{(12)^3} \approx 0.019$$

In this case, the same result can be obtained by inspection of the state transition diagram in Fig. 7.5. One can see that there are only three possible ways to get from state 3 to state 6 in exactly three steps:

$3 \to 2 \to 4 \to 6$ with probability $\dfrac{11}{12} \cdot \dfrac{1}{12} \cdot \dfrac{1}{12}$

$3 \to 5 \to 4 \to 6$ with probability $\dfrac{1}{12} \cdot \dfrac{11}{12} \cdot \dfrac{1}{12}$

$3 \to 5 \to 7 \to 6$ with probability $\dfrac{1}{12} \cdot \dfrac{1}{12} \cdot \dfrac{11}{12}$

which gives the same answer as before.

Remark. The type of Markov chain described in Example 7.4 is often referred to as a gambler's ruin chain: If the gambler continues to play, he is condemned to eventually enter state 0 and hence be ruined. The reason is that there is always a positive probability to reach that state, no matter how large the initial capital is. This situation is typical of most games of chance. In Example 7.4, note that the expected gain for the gambler after every bet is $2(1/12) + (-1)(11/12) = -9/12 < 0$, so the game hardly seems "fair" to play. In a "fair" game, on the other hand, the expected gain after every bet is 0. The chain describing such a game is called a *martingale*. In particular, a martingale is a sequence of random variables $\{X_k\}$ with the property:

$$E[X_{k+1} \mid X_k = i, X_{k-1} = x_{k-1}, \dots, X_0 = x_0] = i$$

Thus, if in a game of chance X_k represents the gambler's capital at time k, this property requires that the expected capital after the kth bet, given all past capital values, is the same as the present value $X_k = i$. A martingale is not necessarily a Markov chain. If, however, $\{X_k\}$ is a Markov chain, then the property above is equivalent to the requirement that the transition probability p_{ij} satisfy

$$\sum_{\text{all } j} j \cdot p_{ij} = i$$

for all states i.

Example 7.5

Let us return to the two-processor system in Example 7.3. With $\alpha = 0.5$ and $\beta = 0.7$, we obtained the transition probability matrix

$$\mathbf{P} = \begin{bmatrix} 0.5 & 0.5 & 0 \\ 0.35 & 0.5 & 0.15 \\ 0.245 & 0.455 & 0.3 \end{bmatrix}$$

We will now pose three questions regarding this system, and answer them using (7.21). Suppose the system starts out empty, that is,

$$\pi(0) = [1, 0, 0]$$

Then:

1. What is the probability that the system is empty at $k = 3$?

2. What is the probability that no job completion occurs in the third slot?

3. What is the probability that the system remains empty through slots 1 and 2?

To answer the first question, we need to calculate $\pi_0(3)$. Let us determine the entire vector $\pi(3)$ using (7.21). Thus, we first calculate $\mathbf{P}^3$:

$$\mathbf{P}^3 = \begin{bmatrix} 0.425 & 0.5 & 0.075 \\ 0.38675 & 0.49325 & 0.12 \\ 0.35525 & 0.4865 & 0.15825 \end{bmatrix} \begin{bmatrix} 0.5 & 0.5 & 0 \\ 0.35 & 0.5 & 0.15 \\ 0.245 & 0.455 & 0.3 \end{bmatrix}$$

$$= \begin{bmatrix} 0.405875 & 0.496625 & 0.0975 \\ 0.3954125 & 0.4946 & 0.1099875 \\ 0.3866712 & 0.4928788 & 0.12045 \end{bmatrix}$$

and since $\pi(0) = [1, 0, 0]$, (7.21) gives

$$\pi(3) = [0.405875, 0.496625, 0.0975] \tag{7.25}$$

Therefore, $\pi_0(3) = 0.405875$. Note that the same result could also be obtained by calculating the three-step transition probability p_{00}^3 from (7.13).

To answer the second question, note that

$$P[\text{no job completes at } k = 3]$$
$$= \sum_{j=0}^{2} P[\text{no job completes at } k = 3 \mid X_3 = j] \cdot \pi_j(3)$$

We then see that $P[\text{no job completes at } k = 3 \mid X_3 = 0] = 1$, $P[\text{no job completes at } k = 3 \mid X_3 = 1] = (1 - \beta)$, and $P[\text{no job completes at } k = 3 \mid X_3 = 2] = (1 - \beta)^2$. Moreover, the values of $\pi_j(3), j = 0, 1, 2$, were determined in (7.25). We then get

$$P[\text{no job completes at } k = 3] = 0.563612$$

To answer the third question, we need to calculate the probability of the event $[X_1 = 0, X_2 = 0]$. We have

$$P[X_1 = 0, X_2 = 0] = P[X_2 = 0 \mid X_1 = 0] \cdot P[X_1 = 0] \qquad (7.26)$$
$$= p_{00} \cdot \pi_0(1)$$

From the transition probability matrix we have $p_{00} = 0.5$, and from (7.21):

$$\pi(1) = [1, 0, 0]\mathbf{P} = [0.5, 0.5, 0]$$

Therefore,

$$P[X_1 = 0, X_2 = 0] = 0.25$$

As (7.26) clearly indicates (and one should expect), $P[X_1 = 0, X_2 = 0] \neq P[X_1 = 0] \cdot P[X_2 = 0] = \pi_0(1) \cdot \pi_0(2)$.

In general, solving equation (7.20) or (7.21) to obtain $\pi(k)$ for any $k = 1, 2, \ldots$ is not a simple task. As in classical system theory, where we resort to z-transform or Laplace transform techniques in order to obtain transient responses, a similar approach may be used here as well. We will not provide details, but refer the reader to Appendix I. The main result is that $\mathbf{P}^k$, which we need to calculate $\pi(k)$ through (7.21), is the inverse z-transform of the matrix $[\mathbf{I} - z\mathbf{P}]^{-1}$, where $\mathbf{I}$ is the identity matrix.

7.2.8 Classification of States

We will now provide several definitions in order to classify the states of a Markov chain in ways that turn out to be particularly convenient for our purposes, specifically the study of steady-state behavior. We begin with the notion of state reachability, which we have encountered earlier in our study of untimed DES models.

Definition. A state j is said to be *reachable* from a state i if $p_{ij}^n > 0$ for some $n = 1, 2, \ldots$
♦

If we consider a state transition diagram, then finding a path from i to j is tantamount to reachability of j from i.

Next, let S be a subset of the state space $\mathcal{X}$. If there is no feasible transition from any state in S to any state outside S, then S forms a *closed* set:

Definition. A subset S of the state space $\mathcal{X}$ is said to be *closed* if $p_{ij} = 0$ for any $i \in S$, $j \notin S$.
♦

A particularly interesting case of a closed set is one consisting of a single state:

Definition. A state i is said to be *absorbing* if it forms a single-element closed set.
♦

Clearly, by (7.3), if i is an absorbing state we have $p_{ii} = 1$. Another interesting case of a closed set is one consisting of mutually reachable states:

Definition. A closed set of states S is said to be *irreducible* if state j is reachable from state i for any $i, j \in S$.
♦

Definition. A Markov chain is said to be *irreducible* if its state space $\mathcal{X}$ is irreducible.
♦

If a Markov chain is not irreducible, it is called *reducible*. In this case, there must exist at least one closed set of states which prohibits irreducibility.

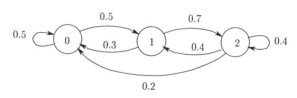

Figure 7.6: State transition diagram for Example 7.6.

Example 7.6

Consider the discrete-time Markov chain with state transition diagram shown in Fig. 7.6. This is a simple case of a three-state irreducible chain, since every state is reachable from any other state. Its transition probability matrix is

$$P = \begin{bmatrix} 0.5 & 0.5 & 0 \\ 0.3 & 0 & 0.7 \\ 0.2 & 0.4 & 0.4 \end{bmatrix}$$

Example 7.7

Consider the five-state Markov chain shown in Fig. 7.7, with transition probability matrix

$$\mathbf{P} = \begin{bmatrix} 0.5 & 0.5 & 0 & 0 & 0 \\ 0.3 & 0 & 0.3 & 0 & 0.4 \\ 0 & 0 & 0 & 1 & 0 \\ 0 & 0 & 0.5 & 0.5 & 0 \\ 0 & 0 & 0 & 0 & 1 \end{bmatrix}$$

This is a reducible Markov chain, since there are states which are not reachable from other states; for instance, state 0 cannot be reached from state 4 or from state 3. There are two closed sets in this chain (both proper subsets of the state space), which are easy to detect by inspection in Fig. 7.7:

$$S_1 = \{4\} \quad \text{and} \quad S_2 = \{2,3\}$$

State 4 is absorbing. The closed set S_2 is irreducible, since states 2 and 3 are mutually reachable. These observations can also be made by inspection of $\mathbf{P}$. Note that $p_{ii} = 1$ indicates that state i is absorbing, as in the case $i = 4$. Also, in the case of $i = 2, 3$, we can see that $p_{ij} = 0$ for all $j \neq 2,3$, indicating that the chain is trapped in these two states forming a closed set.

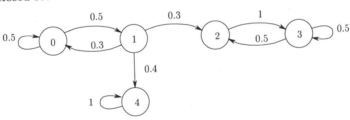

Figure 7.7: State transition diagram for Example 7.7. *This is a reducible chain. State 4 is absorbing. States 2 and 3 form a closed irreducible set.*

Transient and Recurrent States

Suppose a chain is in state i. It is reasonable to ask the question: Will the chain ever return to state i? If the answer is "definitely yes", state i is *recurrent*, otherwise it is *transient*. To formalize this distinction, we introduce the notion of the *hitting time*, T_{ij}, defined as follows:

$$T_{ij} \equiv \min\{k > 0 : X_0 = i, X_k = j\} \tag{7.27}$$

The hitting time represents the *first* time the chain enters state j given that it starts out at state i. If we let $j = i$, then T_{ii} is the first time that the chain returns to state i given that it is currently in state i. We refer to the random variable T_{ii} as the *recurrence time* of state i.

In discrete time, T_{ii} can take on values $1, 2, \ldots$ (including ∞). We define ρ_i^k to be the probability that the recurrence time of state i is k:

$$\rho_i^k \equiv P[T_{ii} = k] \tag{7.28}$$

Then, let ρ_i be the probability of the event [ever return to i | current state is i], which is given by

$$\rho_i = \sum_{k=1}^{\infty} \rho_i^k \tag{7.29}$$

Observe that the event [ever return to i | current state is i] is identical to the event $[T_{ii} < \infty]$. Therefore, we also have

$$\rho_i = P[T_{ii} < \infty] \tag{7.30}$$

Definition. A state i is said to be *recurrent* if $\rho_i = 1$. If $\rho_i < 1$, state i is said to be *transient*.   ◆

Thus, recurrence implies that a state is definitely visited again. On the other hand, a transient state may be visited again, but, with some positive probability $(1 - \rho_i)$, it will not.

Example 7.8
In the Markov chain shown in Fig. 7.8, state 2 is absorbing. Clearly, this is a recurrent state, since the chain will forever be returning to 2, (i.e., $\rho_2 = 1$). On the other hand, if the current state is 0 or 1, there is certainly some positive probability of returning to this state. However, this probability is less than 1, since the chain will eventually be trapped in state 2. For example, if the current state is 0, then with probability $p_{02} = 0.5$ the chain will never return to 0; and if the current state is 1, then with probability $p_{10}p_{02} = 0.25$ the chain will never return to state 1.

Example 7.9
Returning to the gambler's ruin chain of Example 7.4 (see Fig. 7.5), observe that state 0 is absorbing and hence recurrent (the gambler has lost his capital and can no longer play). All remaining states in the countably infinite set $1, 2, \ldots$ are transient. Thus, the gambler wanders around the chain, possibly for a long period of time, but he eventually gets trapped at state 0.

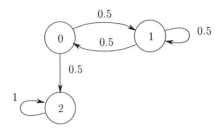

Figure 7.8: State transition diagram for Example 7.8.
In this chain, state 2 is recurrent. States 0 and 1 are transient.

There are several simple facts regarding transient and recurrent states which can be formally proved. We state some of the most important ones without providing proofs (the proofs are left as exercises; the reader is also referred to Chapter 1 of Hoel et al., 1972).

Theorem 7.1 If a Markov chain has a finite state space, then at least one state is recurrent. ◆

Theorem 7.2 If i is a recurrent state and j is reachable from i, then state j is recurrent. ◆

Theorem 7.3 If S is a finite closed irreducible set of states, then every state in S is recurrent. ◆

We can use Fig. 7.7 to illustrate these results. The finite Markov chain in Fig. 7.7 has five states of which three (states 2, 3, 4) are recurrent (Theorem 7.1). Also, state 2 is recurrent and 3 is reachable from 2; as expected, state 3 is also recurrent (Theorem 7.2). Finally, states 2 and 3 form a finite closed irreducible set which consists of recurrent states only (Theorem 7.3).

Null and Positive Recurrent States

Let i be a recurrent state. We denote by M_i the *mean recurrence time* of state i, given by

$$M_i \equiv E[T_{ii}] = \sum_{k=1}^{\infty} k\rho_i^k \tag{7.31}$$

Depending on whether M_i is finite or not we classify state i as *positive recurrent* or *null recurrent*.

Definition. A recurrent state i is said to be *positive* (or *non-null*) *recurrent* if $M_i < \infty$. If $M_i = \infty$, state i is said to be *null recurrent*. ◆

A null recurrent state is not a transient state, because the probability of recurrence is 1; however, the expected recurrence time is infinite. We can view transient states and positive recurrent states as two extremes: Transient states may never be revisited, whereas positive recurrent states are definitely revisited with finite expected recurrence time. Null recurrent states may be viewed as "weakly recurrent" states: They are definitely revisited, but the expected recurrence time is infinite.

A result similar to Theorem 7.2 is the following (see also Chapter 2 of Hoel et al., 1972):

Theorem 7.4 If i is a positive recurrent state and j is reachable from i, then state j is positive recurrent. ◆

By combining Theorems 7.2 and 7.4, we obtain a very useful fact pertaining to irreducible closed sets, and hence also irreducible Markov chains:

Theorem 7.5 If S is a closed irreducible set of states, then every state in S is positive recurrent or every state in S is null recurrent or every state in S is transient. ◆

We can also obtain a stronger version of Theorem 7.3:

Theorem 7.6 If S is a finite closed irreducible set of states, then every state in S is positive recurrent. ◆

Example 7.10 (Discrete-time birth-death chain)

To illustrate the distinctions between transient, positive recurrent and null recurrent states, let us take a close look at the Markov chain of Fig. 7.9. In this model, the state increases by 1 with probability $(1 - p)$ or decreases by 1 with probability p from every state $i > 0$. At $i = 0$, the state remains unchanged with probability p. We often refer to a transition from i to $(i + 1)$ as a "birth", and from i to $(i - 1)$ as a "death". This is a simple version of what is known as a discrete-time *birth-death chain*. We will have the opportunity to explore its continuous-time version in some depth later in this chapter.

Before doing any analysis, let us argue intuitively about the effect the value of p should have on the nature of this chain. Suppose we start the chain at state 0. If $p < 1/2$, the chain tends to drift towards larger and larger values of i, so we expect state 0 to be transient. If $p > 1/2$, on the other hand, then the chain always tends to drift back towards 0, so we should expect state 0 to be recurrent. Moreover, the larger the value of p, the faster we expect a return to state 0, on the average; conversely, as p approaches $1/2$, we expect the mean recurrence time for state 0 to increase. An interesting case is that of $p = 1/2$. Here, we expect that a return to state 0 will occur, but it may take a very long time. In fact, it turns out that the case $p = 1/2$ corresponds to state 0 being null recurrent, whereas if $p > 1/2$ it is positive recurrent.

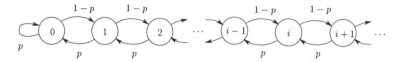

Figure 7.9: State transition diagram for Example 7.10.

Let us now try to verify what intuition suggests. Recalling (7.30), observe that

$$\rho_0 = P[T_{00} < \infty] = p + (1 - p) \cdot P[T_{10} < \infty] \qquad (7.32)$$

In words, starting at state 0, a return to this state can occur in one of two ways: in a single step with probability p, or, with probability $(1 - p)$, in some finite number of steps consisting of a one-step transition to state 1 and then a return to 0 in T_{10} steps. Let us set

$$q_1 = P[T_{10} < \infty] \qquad (7.33)$$

In addition, let us fix some state $m > 1$, and define for any state $i = 1, \ldots, m - 1$,

$$q_i(m) = P[T_{i0} < T_{im}] \quad \text{for some } m > 1 \qquad (7.34)$$

Thus, $q_i(m)$ is the probability that the chain, starting at state i, visits state 0 before it visits state m. We also set $q_m(m) = 0$ and $q_0(m) = 1$. We will now try to evaluate $q_i(m)$ as a function of p, which we will assume to be $0 < p < 1$. This will allow us to obtain $q_1(m)$, from which we will finally obtain q_1, and hence ρ_0.

Taking a good look at the state transition diagram of Fig. 7.9, we observe that

$$q_i(m) = p \cdot q_{i-1}(m) + (1 - p) \cdot q_{i+1}(m) \qquad (7.35)$$

The way to see this is similar to the argument used in (7.32). Starting at state i, a visit to state 0 before state m can occur in one of two ways: from state $(i - 1)$ which is entered next with probability p, or from state $(i + 1)$ which is entered next with probability $(1 - p)$. Then, adding and subtracting the term $(1 - p)q_i(m)$ to the right-hand side of (7.35) above, we get

$$q_{i+1}(m) - q_i(m) = \frac{p}{1 - p}[q_i(m) - q_{i-1}(m)]$$

For convenience, set

$$\beta = \frac{p}{1-p} \tag{7.36}$$

We now see that

$$q_{i+1}(m) - q_i(m) = \beta \cdot \beta[q_{i-1}(m) - q_{i-2}(m)]$$
$$= \ldots = \beta^i[q_1(m) - q_0(m)] \tag{7.37}$$

and by summing over $i = 0, \ldots, m-1$, we get

$$\sum_{i=0}^{m-1} q_{i+1}(m) - \sum_{i=0}^{m-1} q_i(m) = [q_1(m) - q_0(m)] \sum_{i=0}^{m-1} \beta^i$$

which reduces to

$$q_m(m) - q_0(m) = [q_1(m) - q_0(m)] \sum_{i=0}^{m-1} \beta^i \tag{7.38}$$

Recalling that $q_m(m) = 0$ and $q_0(m) = 1$, we immediately get

$$q_1(m) = 1 - \frac{1}{\sum_{i=0}^{m-1} \beta^i} \tag{7.39}$$

We would now like to use this result in order to evaluate q_1 in (7.33). The argument we need requires a little thought. Let us compare the number of steps T_{12} in moving from state 1 to state 2 to the number of steps T_{13}. Note that to get from 1 to 3 we must necessarily go through 2. This implies that $T_{13} > T_{12}$. This observation extends to any T_{1i}, T_{1j} with $j > i$. In addition, since to get from state 1 to 2 requires at least one step, we have

$$1 \le T_{12} < T_{13} < \cdots \tag{7.40}$$

and it follows that $T_{1m} \ge m-1$ for any $m = 2, 3, \ldots$ Therefore, as $m \to \infty$ we have $T_{1m} \to \infty$. Then returning to the definition (7.34) for $i = 1$,

$$\lim_{m \to \infty} q_1(m) = \lim_{m \to \infty} P[T_{10} < T_{1m}] = P[T_{10} < \infty] \tag{7.41}$$

The second equality above is justified by a basic theorem from probability theory (see Appendix I), as long as the events $[T_{10} < T_{1m}]$ form an increasing sequence with $m = 2, 3, \ldots$, which in the limit gives the event $[T_{10} < \infty]$; this is indeed the case by (7.40).

Combining the definition of q_1 in (7.33) with (7.39) and (7.41), we get

$$q_1 = P[T_{10} < \infty] = \lim_{m \to \infty} \left[1 - \frac{1}{\sum_{i=0}^{m-1} \beta^i} \right] = 1 - \frac{1}{\sum_{i=0}^{\infty} \beta^i}$$

Let us now take a closer look at the infinite sum above. If $\beta < 1$, the sum converges and we get

$$\sum_{i=0}^{\infty} \beta^i = \frac{1}{1 - \beta}$$

which gives $q_1 = \beta$. Recall from (7.36) that $\beta = p/(1-p)$. Therefore, this case corresponds to the condition $p < 1 - p$ or $p < 1/2$. If, on the other hand, $\beta \geq 1$, that is, $p \geq 1/2$, we have $\sum_{i=0}^{\infty} \beta^i = \infty$, and obtain $q_1 = 1$.

We can now finally put it all together by using these results in (7.32):

1. If $p < 1/2, q_1 = \beta = p/(1 - p)$, and (7.32) gives

$$\rho_0 = 2p < 1$$

which implies that state 0 is transient as we had originally guessed.

2. If $p \geq 1/2, q_1 = 1$, and (7.32) gives

$$\rho_0 = 1$$

and state 0 is recurrent as expected. We will also later show (see Example 7.13) that when $p = 1/2$ (the point at which ρ_0 switches from 1 to a value less than 1) state 0 is in fact null recurrent.

Observing that the chain of Fig. 7.9 is irreducible (as long as $0 < p < 1$), we can also apply Theorem 7.5 to conclude that in case 1 above all states are transient, and hence the chain is said to be transient. Similarly, in case 2 we can conclude that all states are recurrent, and, if state 0 is null recurrent, then all states are null recurrent.

Periodic and Aperiodic States

Sometimes the structure of a Markov chain is such that visits to some state i are constrained to occur only in a number of steps which is a multiple of an integer $d \geq 2$. Such states are called periodic, and d is called the *period*. In order to provide a formal definition, consider the set of integers

$$\{n > 0 : p_{ii}^n > 0\}$$

and let d be the *greatest common divisor* of this set. Note that if 1 is in this set, then $d = 1$. Otherwise, we may have $d = 1$ or $d > 1$. For example, for the set $\{4, 8, 24\}$ we have $d = 4$.

Definition. A state i is said to be *periodic* if the greatest common divisor d of the set $\{n > 0 : p_{ii}^n > 0\}$ is $d \geq 2$. If $d = 1$, state i is said to be *aperiodic*. ♦

Example 7.11

Consider the Markov chain of Fig. 7.10 with transition probability matrix

$$\mathbf{P} = \begin{bmatrix} 0 & 1 & 0 \\ 0 & 0 & 1 \\ 1 & 0 & 0 \end{bmatrix}$$

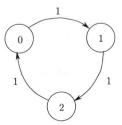

Figure 7.10: Periodic states with $d = 3$.

This is a simple example of a Markov chain whose states are all periodic with period $d = 3$, since the greatest common divisor of the set $\{n > 0 : p_{ii}^n > 0\} = \{3, 6, 9, \dots\}$ is 3 for all $i = 0, 1, 2$.

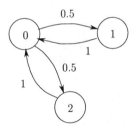

Figure 7.11: Periodic states with $d = 2$.

Another example of a Markov chain whose states are all periodic with $d = 2$ is shown in Fig. 7.11, where

$$\mathbf{P} = \begin{bmatrix} 0 & 0.5 & 0.5 \\ 1 & 0 & 0 \\ 1 & 0 & 0 \end{bmatrix}$$

In this case, $\{n > 0 : p_{ii}^n > 0\} = \{2, 4, 6, \dots\}$ for all $i = 0, 1, 2$, and the greatest common divisor is 2.

Finally, consider the Markov chain of Fig. 7.12, where

$$\mathbf{P} = \begin{bmatrix} 0.5 & 0.5 & 0 \\ 0 & 0.5 & 0.5 \\ 0.5 & 0 & 0.5 \end{bmatrix}$$

In this case, $\{n > 0 : p_{ii}^n > 0\} = \{1, 2, 3, 4, \ldots\}$ for all $i = 0, 1, 2$, and $d = 1$. Thus, all states are aperiodic.

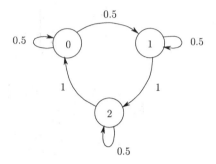

Figure 7.12: Aperiodic states.

It is easy to see that if $p_{ii} > 0$ for some state i of a Markov chain, then the state must be aperiodic, since 1 is an element of the set $\{n > 0 : p_{ii}^n > 0\}$ and hence $d = 1$. This is the case in the chain of Fig. 7.12. It is also possible to show the following (see, for example, Chapter 2 of Hoel et al., 1972):

Theorem 7.7 If a Markov chain is irreducible, then all its states have the same period. ♦

It follows that if $d = 1$ for any state of an irreducible Markov chain, then all states are aperiodic and the chain is said to be aperiodic. On the other hand, if any state has period $d \geq 2$, then all states have the same period and the chain is said to be periodic with period $d \geq 2$.

Summary of State Classifications

There is of course good reason why we have gone through the process of defining new concepts to classify states in a Markov chain. This reason will become clear in the next section, when we derive stationary state probabilities for certain types of chains. We now summarize the state classifications described above as shown in Fig. 7.13. A state i is recurrent if the probability ρ_i of ever returning to i is 1. A recurrent state i is positive recurrent if the expected time required to return to i, M_i, is finite. Finally, state i is aperiodic if the period d_i of the number of steps required to return to i is 1. As we will see, irreducible Markov chains consisting of positive recurrent aperiodic states possess particularly attractive properties when it comes to steady state analysis.

7.2.9   Steady State Analysis

In Section 7.2.7, we saw how to address questions of the type: What is the probability of finding a Markov chain at state i at time k? In transient

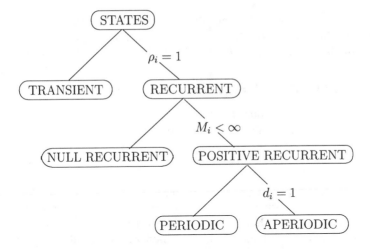

Figure 7.13: Summary of state classifications.

analysis we limited ourselves to given finite numbers of steps over which the chain is observed. In steady state analysis we extend our inquiry to questions such as: What is the probability of finding a Markov chain at state i in the long run? By "long run" we mean that the system we are modeling as a Markov chain is allowed to operate for a sufficiently long period of time so that the state probabilities can reach some fixed values which no longer vary with time. This may or may not be achievable. Our study, therefore, centers around the quantities

$$\pi_j = \lim_{k \to \infty} \pi_j(k) \qquad (7.42)$$

where, as defined in (7.18), $\pi_j(k) = P[X_k = j]$, and the existence of these limits is not always guaranteed. Thus, we need to address three basic questions:

1. Under what conditions do the limits in (7.42) exist?

2. If these limits exist, do they form a legitimate probability distribution, that is, $\sum_j \pi_j = 1$?

3. How do we evaluate π_j?

If π_j exists for some state j, it is referred to as a *steady-state, equilibrium,* or *stationary state probability.* Accordingly, if π_j exists for all states j, then we obtain the *stationary state probability vector*

$$\boldsymbol{\pi} = [\pi_0, \pi_1, \ldots]$$

It is essential to keep in mind that the quantity reaching steady state is a state probability - not a state which of course remains a random variable.

Our study of the steady state behavior of discrete-time Markov chains greatly depends on whether or not we are dealing with an irreducible or a reducible chain. Thus, in the next two sections we consider the two cases separately.

7.2.10 Irreducible Markov Chains

The main objective of our study is the determination of the limits in (7.42) if they exist. Recalling the recursive equation (7.20):

$$\pi(k+1) = \pi(k)\mathbf{P}, \qquad k = 0, 1, \ldots$$

it is often the case that after a long time period (i.e., large values of k) we have $\pi(k+1) \approx \pi(k)$. In other words, as $k \to \infty$ we get $\pi(k) \to \pi$, where π is the stationary state probability vector. This vector, if it exists, also defines the stationary probability distribution of the chain. Then, if indeed in the limit as $k \to \infty$ we get $\pi(k+1) = \pi(k) = \pi$, we should be able to obtain π from (7.20) by solving a system of linear algebraic equations

$$\pi = \pi\mathbf{P}$$

where the elements of π satisfy $\pi_j \geq 0$ and $\sum_j \pi_j = 1$.

The first important observation is that the presence of periodic states in an irreducible Markov chain prevents the existence of the limits in (7.42). We will use the example of the periodic chain of Fig. 7.11 to illustrate this point. Recall that the transition probability matrix for this chain is

$$\mathbf{P} = \begin{bmatrix} 0 & 0.5 & 0.5 \\ 1 & 0 & 0 \\ 1 & 0 & 0 \end{bmatrix}$$

and let $\pi(0) = [1, 0, 0]$, that is, it is known that the chain is initially at state 0. We can calculate $\mathbf{P}^2, \mathbf{P}^3, \ldots$ to get

$$\mathbf{P}^2 = \begin{bmatrix} 1 & 0 & 0 \\ 0 & 0.5 & 0.5 \\ 0 & 0.5 & 0.5 \end{bmatrix} \qquad \mathbf{P}^3 = \begin{bmatrix} 0 & 0.5 & 0.5 \\ 1 & 0 & 0 \\ 1 & 0 & 0 \end{bmatrix}$$

$$\mathbf{P}^4 = \begin{bmatrix} 1 & 0 & 0 \\ 0 & 0.5 & 0.5 \\ 0 & 0.5 & 0.5 \end{bmatrix} \ldots$$

and, in general,

$$\mathbf{P}^{2k} = \begin{bmatrix} 1 & 0 & 0 \\ 0 & 0.5 & 0.5 \\ 0 & 0.5 & 0.5 \end{bmatrix}, \qquad \mathbf{P}^{2k-1} = \begin{bmatrix} 0 & 0.5 & 0.5 \\ 1 & 0 & 0 \\ 1 & 0 & 0 \end{bmatrix}$$

for all $k = 1, 2, \ldots$ Hence, from (7.21) we have

$$\pi(2k) = [1, 0, 0], \qquad \pi(2k - 1) = [0, 0.5, 0.5]$$

Thus, the state probability $\pi_0(k)$ oscillates between 1 and 0 depending on the number of steps k (even or odd). The same is true for the other two state probabilities, which oscillate between 0 and 0.5. It follows that

$$\lim_{k \to \infty} \pi_j(k)$$

cannot exist for $j = 0, 1, 2$.

Fortunately, as long as an irreducible Markov chain is not periodic the limit of $\pi_j(k)$ as $k \to \infty$ always exists. We state without proof this important result (see, for example, Asmussen, 1987).

Theorem 7.8 In an irreducible aperiodic Markov chain the limits

$$\pi_j = \lim_{k \to \infty} \pi_j(k)$$

always exist and they are independent of the initial state probability vector. ◆

Note that Theorem 7.8 guarantees the existence of π_j for all states j, but it does not guarantee that we can form a legitimate stationary state probability distribution satisfying $\sum_j \pi_j = 1$.

Next, let us recall Theorem 7.5, where it was stated that an irreducible Markov chain consists of states which are all positive recurrent or they are all null recurrent or they are all transient. Intuitively, if a state is transient it can only be visited a finite number of times; hence, in the long run, the probability of finding the chain in such a state must be zero. A similar argument holds for null recurrent states. We now present the two fundamental results allowing us to answer the questions we raised earlier: Under what conditions can we guarantee the existence of stationary state probability vectors and how do we determine them? The proofs require some rather elaborate convergence arguments and are omitted (the reader is again referred to Asmussen, 1987).

Theorem 7.9 In an irreducible aperiodic Markov chain consisting of transient states or of null recurrent states

$$\pi_j = \lim_{k \to \infty} \pi_j(k) = 0$$

for all states j, and no stationary probability distribution exists. ◆

Theorem 7.10 In an irreducible aperiodic Markov chain consisting of positive recurrent states a unique stationary state probability vector π exists such that $\pi_j > 0$ and

$$\pi_j = \lim_{k \to \infty} \pi_j(k) = \frac{1}{M_j} \tag{7.43}$$

where M_j is the mean recurrence time of state j defined in (7.31). The vector π is determined by solving

$$\pi = \pi\mathbf{P} \tag{7.44}$$

$$\sum_{\text{all } j} \pi_j = 1 \tag{7.45}$$

$\blacklozenge$

It is clear that aperiodic positive recurrent states are highly desirable in terms of reaching steady state. We term such states *ergodic*. A Markov chain is said to be ergodic if all its states are ergodic.

Note that a combination of Theorems 7.6 and 7.10 leads to the observation that every *finite* irreducible aperiodic Markov chain has a unique stationary state probability vector π determined through (7.44) and (7.45). In this case, obtaining π is simply a matter of solving a set of linear equations. However, solving (7.44) and (7.45) in the case of an infinite state space is certainly not an easy task.

Remark. The fact that $\pi_j = 1/M_j$ in (7.43) has an appealing physical interpretation. The probability π_j represents the fraction of time spent by the chain at state j at steady state. Thus, a short recurrence time for j ought to imply a high probability of finding the chain at j. Conversely, a long recurrence time implies a small state probability. In fact, as M_j increases one can see that π_j approaches 0; in the limit, as $M_j \to \infty$, we see that $\pi_j \to 0$, that is, j behaves like a null recurrent state under Theorem 7.9.

Example 7.12

Let us consider the Markov chain of Example 7.3 shown in Fig. 7.4. Setting $\alpha = 0.5$ and $\beta = 0.7$ we found the transition probability matrix for this chain to be:

$$\mathbf{P} = \begin{bmatrix} 0.5 & 0.5 & 0 \\ 0.35 & 0.5 & 0.15 \\ 0.245 & 0.455 & 0.3 \end{bmatrix}$$

This chain is clearly irreducible. It is also aperiodic, since $p_{ii} > 0$ for all states $i = 0, 1, 2$ (as pointed out earlier, $p_{ii} > 0$ for at least one i is a sufficient condition for aperiodicity). It is also easy to see that the chain contains no transient or null recurrent states, so that Theorem 7.10 can be used to determine the unique stationary state probability vector $\pi = [\pi_0, \pi_1, \pi_2]$. The set of equations (7.44) in this case is the following:

$$\pi_0 = 0.5\pi_0 + 0.35\pi_1 + 0.245\pi_2$$
$$\pi_1 = 0.5\pi_0 + 0.5\pi_1 + 0.455\pi_2$$
$$\pi_2 = 0\pi_0 + 0.15\pi_1 + 0.3\pi_2$$

These equations are not linearly independent: One can easily check that multiplying the first and third equations by -1 and adding them gives the second equation. This is always the case in (7.44), which makes the normalization condition (7.45) necessary in order to solve for π. Keeping the second and third equation above, and combining it with (7.45), we get

$$0.5\pi_0 - 0.5\pi_1 + 0.455\pi_2 = 0$$
$$0.15\pi_1 - 0.7\pi_2 = 0$$
$$\pi_0 + \pi_1 + \pi_2 = 1$$

The solution of this set of equations is:

$$\pi_0 = 0.399, \quad \pi_1 = 0.495, \quad \pi_2 = 0.106$$

It is interesting to compare the stationary state probability vector $\pi = [0.399, 0.495, 0.106]$ obtained above with the transient solution $\pi(3) = [0.405875, 0.496625, 0.0975]$ in (7.16), which was obtained in Example 7.5 with initial state probability vector $\pi(0) = [1, 0, 0]$. We can see that $\pi(3)$ is an approximation of π. This approximation gets better as k increases, and, by Theorem 7.10, we expect $\pi(k) \to \pi$ as $k \to \infty$.

Example 7.13 (Steady-state solution of birth-death chain)

Let us come back to the birth-death chain of Example 7.10. By looking at Fig. 7.9, we can see that the transition probability matrix is

$$\mathbf{P} = \begin{bmatrix} p & 1-p & 0 & 0 & 0 & \cdots \\ p & 0 & 1-p & 0 & 0 & \cdots \\ 0 & p & 0 & 1-p & 0 & 0 \\ 0 & 0 & p & 0 & 1-p & 0 \\ \vdots & \vdots & \vdots & \vdots & \vdots & \vdots \end{bmatrix}$$

Assuming $0 < p < 1$, this chain is irreducible and aperiodic (note that $p_{00} = p > 0$). The system of equations $\pi = \pi\mathbf{P}$ in (7.44) gives

$$\pi_0 = \pi_0 p + \pi_1 p$$
$$\pi_j = \pi_{j-1}(1-p) + \pi_{j+1}p, \quad j = 1, 2, \ldots$$

From the first equation, we get

$$\pi_1 = \frac{1-p}{p}\pi_0$$

From the second set of equations, for $j = 1$ we get

$$\pi_1 = \pi_0(1-p) + \pi_2 p$$

and substituting for π_1 from above we obtain π_2 in terms of π_0:

$$\pi_2 = \left(\frac{1-p}{p}\right)^2 \pi_0$$

Proceeding in similar fashion, we have

$$\pi_j = \left(\frac{1-p}{p}\right)^j \pi_0, \quad j = 1, 2, \ldots \tag{7.46}$$

Summing over $j = 0, 1, \ldots$ and making use of the normalization condition (7.45), we obtain

$$\pi_0 + \sum_{j=1}^{\infty} \pi_j = \pi_0 + \pi_0 \sum_{j=1}^{\infty} \left(\frac{1-p}{p}\right)^j = 1$$

from which we can solve for π_0:

$$\pi_0 = \frac{1}{\sum_{i=0}^{\infty} \left(\frac{1-p}{p}\right)^i}$$

where we have replaced the summation index j by i so that there is no confusion in the following expression which we can now obtain from (7.46):

$$\pi_j = \frac{\left(\frac{1-p}{p}\right)^j}{\sum_{i=0}^{\infty} \left(\frac{1-p}{p}\right)^i}, \quad j = 1, 2, \ldots \tag{7.47}$$

Now let us take a closer look at the infinite sum above. If $(1-p)/p < 1$, or equivalently $p > 1/2$, the sum converges,

$$\sum_{i=0}^{\infty} \left(\frac{1-p}{p}\right)^i = \frac{p}{2p-1}$$

and we have the final result

$$\pi_j = \frac{2p-1}{p} \left(\frac{1-p}{p}\right)^j, \quad j = 0, 1, 2, \ldots \tag{7.48}$$

Now let us relate these results to our findings in Example 7.10:

1. Under the condition $p < 1/2$ we had found the chain to be transient. Under this condition, the sum in (7.47) does not converge, and we get $\pi_j = 0$; this is consistent with Theorem 7.9 for transient states.

2. Under the condition $p \geq 1/2$ we had found the chain to be recurrent. This is consistent with the condition $p > 1/2$ above, which, by (7.48), yields stationary state probabilities such that $0 < \pi_j < 1$.

3. Finally, note in (7.48) that as $p \to 1/2, \pi_j \to 0$. By (7.43), this implies that $M_j \to \infty$. Thus, we see that state 0 is null recurrent for $p = 1/2$. This was precisely our original conjecture in Example 7.10.

From a practical standpoint, Theorem 7.10 allows us to characterize the steady state behavior of many DES modeled as discrete-time Markov chains. The requirements of irreducibility and aperiodicity are not overly restrictive. Most commonly designed systems have these properties. For instance, one would seldom want to design a reducible resource-providing system which inevitably gets trapped into some closed sets of states.[1] Another practical implication of Theorem 7.10 is the following. Suppose that certain states in a DES are designated as "more desirable" than others. Since π_j is the fraction of time spent at j in the long run, it gives us a measure of system performance: Larger values of π_j for more desirable states j imply better performance. In some cases, maximizing (or minimizing) a particular π_j represents an actual design objective for such systems.

Example 7.14
Consider a machine which alternates between an UP and a DOWN state, denoted by 1 and 0 respectively. We would like the machine to spend as little time as possible in the DOWN state, and we can control a single parameter β which affects the probability of making a transition from DOWN to UP. We model this system through a Markov chain as shown in Fig. 7.14, where β ($0 \leq \beta \leq 2$ so that the transition probability 0.5β is in $[0, 1]$) is the design parameter we can select. Our design objective is expressed in terms of the stationary state probability π_0 as follows:

$$\pi_0 < 0.4$$

The transition probability matrix for this chain is

$$\mathbf{P} = \left[\begin{array}{cc} 1 - 0.5\beta & 0.5\beta \\ 0.5 & 0.5 \end{array} \right]$$

Using (7.44) and (7.45) to obtain the stationary state probabilities, we have

$$\pi_0 = (1 - 0.5\beta)\pi_0 + 0.5\pi_1$$
$$\pi_1 = 0.5\beta\pi_0 + 0.5\pi_1$$
$$\pi_0 + \pi_1 = 1$$

---

[1] A supervisory controller S of the type considered in Chapter 3 could be synthesized, if necessary, to ensure that the controlled DES S/G (now modeled as a Markov chain) satisfies these requirements. One would rely upon the notions of marked states and nonblocking supervisor for this purpose.

Once again, the first two equations are linearly dependent. Solving the

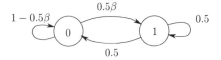

Figure 7.14: Markov chain for Example 7.14.

second and third equations for π_0, π_1 we get

$$\pi_0 = \frac{1}{1+\beta}, \quad \pi_1 = \frac{\beta}{1+\beta}$$

Since the design objective is that $\pi_0 < 0.4$, we obtain the following requirement for the parameter β:

$$\frac{1}{1+\beta} < 0.4 \quad \text{or} \quad \beta > 1.5$$

Thus, given the additional constraint $0 \leq \beta \leq 2$, we can select any β such that $1.5 < \beta \leq 2$.

7.2.11 Reducible Markov Chains

In a reducible Markov chain the steady state behavior is quite predictable: The chain eventually enters some irreducible closed set of states S and remains there forever. If S consists of two or more states, we can analyze the steady state behavior of S as in the previous section. If S consists of a single absorbing state, then the chain simply remains in that state (as is the case with state 0 in the gambler's ruin chain of Fig. 7.5). The only remaining problem arises when the reducible chain contains two or more irreducible closed sets. Then, the question of interest is: *What is the probability that the chain enters a particular set S first?* Clearly, if the chain enters S first it remains in that set forever and none of the other closed sets is ever entered.

Let $\mathcal{T}$ denote the set of transient states in a reducible Markov chain, and let S be some irreducible closed set of states. Let $i \in \mathcal{T}$ be a transient state. We define $\rho_i(S)$ to be the probability of entering the set S given that the chain starts out at state i, that is,

$$\rho_i(S) \equiv P[X_k \in S \text{ for some } k > 0 \mid X_0 = i] \tag{7.49}$$

The event $[X_k \in S \text{ for some } k > 0 \mid X_0 = i]$ can occur in one of two distinct ways:

1. The chain enters S at $k = 1$, that is, the event $[X_1 \in S \mid X_0 = i]$ occurs, or

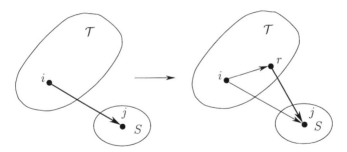

Figure 7.15: Eventual transition from a transient state to an irreducible closed set S.
State j is entered from i in one of two ways: in a single step, or by visiting some state $r \in \mathcal{T}$ first. In the latter case, the process repeats from the new state r.

2. The chain visits some other state $r \in \mathcal{T}$ at $k = 1$ and then eventually enters S, that is, the event $[X_1 \in \mathcal{T}$ and $X_k \in S$ for some $k > 1 \mid X_0 = i]$ occurs.

This is illustrated in Fig. 7.15, where the thinner arrows represent one-step transitions. We can therefore write

$$\rho_i(S) = P[X_1 \in S \mid X_0 = i]$$
$$+ P[X_1 \in \mathcal{T} \ \text{and} \ X_k \in S \text{ for some } k > 1 \mid X_0 = i] \quad (7.50)$$

The first term in (7.50) can be rewritten as the sum of probabilities over all states $j \in S$, that is,

$$P[X_1 \in S \mid X_0 = i] = \sum_{j \in S} P[X_1 = j \mid X_0 = i] = \sum_{j \in S} p_{ij} \quad (7.51)$$

Similarly, the second term in (7.50) can be rewritten as the sum of probabilities over all states $r \in \mathcal{T}$:

$$P[X_1 \in \mathcal{T} \ \text{and} \ X_k \in S \text{ for some } k > 1 \mid X_0 = i]$$
$$= \sum_{r \in \mathcal{T}} P[X_1 = r \ \text{and} \ X_k \in S \text{ for some } k > 1 \mid X_0 = i] \quad (7.52)$$

Each term in this summation can be written as follows:

$$P[X_1 = r \ \text{and} \ X_k \in S \text{ for some } k > 1 \mid X_0 = i]$$
$$= P[X_k \in S \text{ for some } k > 1 \mid X_1 = r, X_0 = i] \cdot P[X_1 = r \mid X_0 = i]$$
$$= P[X_k \in S \text{ for some } k > 1 \mid X_1 = r, X_0 = i] \cdot p_{ir}$$
$$(7.53)$$

By the memoryless property,

$$P[X_k \in S \text{ for some } k > 1 \mid X_1 = r, X_0 = i]$$
$$= P[X_k \in S \text{ for some } k > 1 \mid X_1 = r]$$

and by setting $n = k - 1$, we get

$$P[X_k \in S \text{ for some } k > 1 \mid X_1 = r] = P[X_n \in S \text{ for some } n > 0 \mid X_0 = r]$$
$$= \rho_r(S)$$

Thus, (7.53) becomes

$$P[X_1 = r \text{ and } X_k \in S \text{ for some } k > 1 \mid X_0 = i] = \rho_r(S) \cdot p_{ir}$$

and hence (7.52) becomes

$$P[X_1 \in \mathcal{T} \text{ and } X_k \in S \text{ for some } k > 1 \mid X_0 = i] = \sum_{r \in \mathcal{T}} \rho_r(S) \cdot p_{ir} \qquad (7.54)$$

Finally, using (7.51) and (7.54) in (7.50), we get

$$\rho_i(S) = \sum_{j \in S} p_{ij} + \sum_{r \in \mathcal{T}} \rho_r(S) \cdot p_{ir} \qquad (7.55)$$

In general, the solution of (7.55) for the unknown probabilities $\rho_i(S)$ for all $i \in \mathcal{T}$ is not easy to obtain. Moreover, if the set of transient states $\mathcal{T}$ is infinite, the solution of (7.55) may not even be unique. If, however, the set $\mathcal{T}$ is finite, it can be formally shown that the set of equations in (7.55) has a unique solution.

Example 7.15
The four-state Markov chain of Fig. 7.16 consists of a transient state set $\mathcal{T} = \{1, 2\}$ and the absorbing states 0 and 3. Suppose the initial state is known to be $X_0 = 1$. We now apply (7.55) to determine the probability $\rho_1(0)$ that the chain is absorbed by state 0 rather than state 3. Thus, we obtain the following two equations with two unknown probabilities, $\rho_1(0)$ and $\rho_2(0)$:

$$\rho_1(0) = p_{10} + \sum_{r=1}^{2} \rho_r(0) \cdot p_{1r} = 0.3 + 0.3\rho_1(0) + 0.4\rho_2(0)$$

$$\rho_2(0) = p_{20} + \sum_{r=1}^{2} \rho_r(0) \cdot p_{2r} = 0.2\rho_1(0) + 0.5\rho_2(0)$$

Solving for $\rho_1(0)$ we get

$$\rho_1(0) = 5/9$$

It follows that $\rho_1(3) = 4/9$, which can also be explicitly obtained by solving a similar set of equations in $\rho_1(3)$ and $\rho_2(3)$.

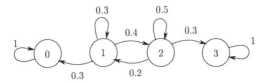

Figure 7.16: Markov chain for Example 7.15.

7.3   CONTINUOUS-TIME MARKOV CHAINS

In the case of a continuous-time Markov chain, the Markov (memoryless) property is expressed as:

$$P[X(t_{k+1}) = x_{k+1} \mid X(t_k) = x_k, X(t_{k-1}) = x_{k-1}, \ldots, X(t_0) = x_0]$$
$$= P[X(t_{k+1}) = x_{k+1} \mid X(t_k) = x_k] \tag{7.56}$$

for any $t_0 \leq t_1 \leq \cdots \leq t_k \leq t_{k+1}$. Thus, if the current state x_k is known, the value taken by $X(t_{k+1})$ depends only on x_k and not on any past state history (no state memory). Moreover, the amount of time spent in the current state is irrelevant in determining the next state (no age memory).

The study of continuous-time Markov chains parallels that of discrete-time chains. We can no longer, however, use the (one-step) transition probability matrix **P** introduced in (7.15), since state transitions are no longer synchronized by a common clock imposing a discrete-time structure. Instead, we must make use of a device measuring the *rates* at which various events (state transitions) take place. This will allow us to specify a model and then proceed with the analysis.

7.3.1   Model Specification

In the discrete-time case, a Markov chain model consisted of a state space $\mathcal{X}$, an initial state probability distribution $p_0(x)$ for all $x \in \mathcal{X}$, and a transition probability matrix **P**. In continuous time, as pointed out above, we cannot use **P**, since state transitions may occur at any time. Instead, we need to specify a matrix $\mathbf{P}(\tau)$ whose (i,j)th entry, $p_{ij}(t)$, is the probability of a transition from i to j within a time interval of duration t for all possible t. Clearly, this is a much more challenging task, since we need to specify all values of the functions $p_{ij}(t)$. As we will see, one solution to this problem is offered by introducing information regarding the *rate* at which various state transitions (events) may occur at any time.

In the next few sections, we introduce the basic definitions and notation leading to the specification and analysis of a continuous-time Markov chain model. In most cases, the definitions are similar to those of the discrete-time case.

7.3.2 Transition Functions

Recall our definition of the n-step transition probability $p_{ij}(k, k+n)$ in (7.4). A special case of $p_{ij}(k, k+n)$ was the one-step transition probability $p_{ij}(k) = p_{ij}(k, k+1)$, based on which the matrix $\mathbf{P}$ was defined.

Our starting point in the continuous-time case is the analog of $p_{ij}(k, k+n)$ defined in (7.4). In particular, we define time-dependent transition probabilities as follows:

$$p_{ij}(s, t) \equiv P[X(t) = j \mid X(s) = i], \qquad s \le t \tag{7.57}$$

We will refer to $p_{ij}(s, t)$ as a *transition function*; it is a function of the time instants s and t. We reserve the term "transition probability" for a different quantity, which will naturally come up in subsequent sections.

The transition functions in (7.57) satisfy a continuous-time version of the Chapman-Kolmogorov equation (7.6). To derive this equation, we proceed by conditioning the event $[X(t) = j | X(s) = i]$ on $[X_u = r]$ for some u such that $s \le u \le t$. Then, using the rule of total probability, (7.57) becomes

$$p_{ij}(s, t) = \sum_{\text{all } r} P[X(t) = j \mid X(u) = r, X(s) = i] \cdot P[X(u) = r \mid X(s) = i] \tag{7.58}$$

By the memoryless property (7.56),

$$P[X(t) = j \mid X(u) = r, X(s) = i] = P[X(t) = j \mid X(u) = r] = p_{rj}(u, t)$$

Moreover, the second term in the sum in (7.58) is simply $p_{ir}(s, u)$. Therefore,

$$p_{ij}(s, t) = \sum_{\text{all } r} p_{ir}(s, u) p_{rj}(u, t), \qquad s \le u \le t \tag{7.59}$$

which is the continuous-time *Chapman-Kolmogorov equation*, the analog of (7.6).

To rewrite (7.59) in matrix form, we define:

$$\mathbf{H}(s, t) \equiv [p_{ij}(s, t)], \qquad i, j = 0, 1, 2, \ldots \tag{7.60}$$

and observe that $\mathbf{H}(s, s) = \mathbf{I}$ (the identity matrix). We then get from (7.59):

$$\mathbf{H}(s, t) = \mathbf{H}(s, u)\mathbf{H}(u, t), \qquad s \le u \le t \tag{7.61}$$

As in the discrete-time case, the Chapman-Kolmogorov equation is one of the most general relationships we can derive.

7.3.3   The Transition Rate Matrix

Let us consider the Chapman-Kolmogorov equation (7.61) for time instants $s \leq t \leq t + \Delta t$, where $\Delta t > 0$. Thus, we have:

$$\mathbf{H}(s, t + \Delta t) = \mathbf{H}(s, t)\mathbf{H}(t, t + \Delta t)$$

Subtracting $\mathbf{H}(s, t)$ from both sides of this equation gives

$$\mathbf{H}(s, t + \Delta t) - \mathbf{H}(s, t) = \mathbf{H}(s, t)[\mathbf{H}(t, t + \Delta t) - \mathbf{I}]$$

where $\mathbf{I}$ is the identity matrix. Dividing by Δt and taking the limit as $\Delta t \to 0$, we get

$$\lim_{\Delta t \to 0} \frac{\mathbf{H}(s, t + \Delta t) - \mathbf{H}(s, t)}{\Delta t} = \mathbf{H}(s, t) \lim_{\Delta t \to 0} \frac{\mathbf{H}(t, t + \Delta t) - \mathbf{I}}{\Delta t} \qquad (7.62)$$

Note that the left-hand side of (7.62) is the partial derivative of $\mathbf{H}(s, t)$ with respect to t, provided of course that the derivatives of the transition functions $p_{ij}(s, t)$ (the elements of $\mathbf{H}(s, t)$) actually exist. Let us also define

$$\mathbf{Q}(t) \equiv \lim_{\Delta t \to 0} \frac{\mathbf{H}(t, t + \Delta t) - \mathbf{I}}{\Delta t} \qquad (7.63)$$

which is called the *Transition Rate Matrix* of the Markov chain, or the *Infinitesimal Generator* driving the transition matrix $\mathbf{H}(s, t)$. Then, (7.62) reduces to the matrix differential equation

$$\frac{\partial \mathbf{H}(s, t)}{\partial t} = \mathbf{H}(s, t)\mathbf{Q}(t), \qquad s \leq t \qquad (7.64)$$

which is also referred to as the *forward Chapman-Kolmogorov equation* in continuous time. In similar fashion, choosing time instants $s \leq s + \Delta s \leq t$ instead of $s \leq t \leq t + \Delta t$, we can obtain the *backward Chapman-Kolmogorov equation*

$$\frac{\partial \mathbf{H}(s, t)}{\partial s} = -\mathbf{Q}(s)\mathbf{H}(s, t) \qquad (7.65)$$

Concentrating on the forward equation (7.64), under certain conditions which the matrix $\mathbf{Q}(s)$ must satisfy, a solution of this equation can be obtained in the form of a matrix exponential function

$$\mathbf{H}(s, t) = exp\left[\int_s^t \mathbf{Q}(\tau)d\tau\right] \qquad (7.66)$$

where $exp[\mathbf{A}t] = e^{\mathbf{A}t} = \mathbf{I} + \mathbf{A}t + \mathbf{A}^2 t^2/2! + \cdots$

As we will soon see, if the transition rate matrix $\mathbf{Q}(t)$ is specified, a complete model of a continuous-time Markov chain is obtained (along with its state space

$\mathcal{X}$ and initial state probability distribution $p_0(x)$ for all $x \in \mathcal{X}$). However, if we are to use $\mathbf{Q}(t)$ as the basis of our models, we should be able to identify the elements $q_{ij}(t)$ of this matrix with actual physically measurable quantities. Thus, we will address the question: What exactly do the entries of $\mathbf{Q}(t)$ represent?

To simplify our discussion, we will (as in the discrete-time case) limit ourselves to homogeneous Markov chains. The definition and implications of homogeneity for continuous-time chains are presented in the next section.

7.3.4 Homogeneous Markov Chains

In a homogeneous discrete-time Markov chain the n-step transition probability $p_{ij}(k, k+n)$ is independent of k. In the continuous-time case we define a chain to be homogeneous if all transition functions $p_{ij}(s, t)$ defined in (7.57) are independent of the absolute time instants s, t, and depend only on the difference $(t - s)$. To make this more clear, we can rewrite (7.57) as

$$p_{ij}(s, s + \tau) = P[X(s + \tau) = j \mid X(s) = i]$$

Then, homogeneity requires that, for any time s, $p_{ij}(s, s + \tau)$ depends only on τ. We therefore denote the transition function by $p_{ij}(\tau)$:

$$p_{ij}(\tau) = P[X(s + \tau) = j \mid X(s) = i] \tag{7.67}$$

It follows that the matrix $\mathbf{H}(s, s+\tau)$ defined in (7.60) is also only dependent on τ. To distinguish this case from the general one, we will use the symbol $\mathbf{P}(\tau)$, and we have

$$\mathbf{P}(\tau) \equiv [p_{ij}(\tau)], \quad i, j = 0, 1, 2, \ldots \tag{7.68}$$

where, similar to (7.3), we have

$$\sum_{\text{all } j} p_{ij}(\tau) = 1 \tag{7.69}$$

Looking at the definition of $\mathbf{Q}(t)$ in (7.63), note that in the homogeneous case we have $\mathbf{H}(t, t + \Delta t) = \mathbf{P}(\Delta t)$, and hence the transition rate matrix is independent of t, that is,

$$\mathbf{Q}(t) = \mathbf{Q} = \text{constant} \tag{7.70}$$

Finally, the forward Chapman-Kolmogorov equation (7.64) becomes

$$\frac{d\mathbf{P}(\tau)}{d\tau} = \mathbf{P}(\tau)\mathbf{Q} \tag{7.71}$$

with the following initial conditions (assuming that a state transition from any i to $j \neq i$ cannot occur in zero time):

$$p_{ij}(0) = \begin{cases} 1 & \text{if } j = i \\ 0 & \text{if } j \neq i \end{cases} \tag{7.72}$$

The solution of (7.71) is of the form

$$\mathbf{P}(\tau) = e^{\mathbf{Q}\tau} \tag{7.73}$$

From this point on, we will limit our discussion to *homogeneous* Markov chains (unless explicitly stated otherwise).

7.3.5   State Holding Times

As in the discrete-time case, let $V(i)$ denote the amount of time spent at state i whenever this state is visited, which we also refer to as the *state holding time*. It is a fundamental property of continuous-time Markov chains that *the distribution of the state holding time* $V(i), P[V(i) \leq t], t \geq 0$, *is exponential.* Thus,

$$P[V(i) \leq t] = 1 - e^{-\Lambda(i)t}, \qquad t \geq 0 \tag{7.74}$$

where $\Lambda(i) > 0$ is a parameter generally dependent on the state i. This is a direct consequence of the memoryless property (7.56) and should not come as a surprise given the results of the previous chapter; in particular, we saw in Section 6.8.3 that a GSMP with a Poisson clock structure reduces to a Markov chain and inherits the memoryless property of the Poisson process.

We will use a simple informal argument to justify (7.74). Suppose the chain enters state i at time T. Further, suppose that at time $T + s \geq T$, no state transition has yet occurred. We consider the conditional probability

$$P[V(i) > s + t \mid V(i) > s]$$

and observe that it can be rewritten as

$$P[V(i) > s + t \mid X(\tau) = i \text{ for all } T \leq \tau \leq T + s]$$

Now, by the memoryless property (7.56), the information "$X(\tau) = i$ for all $T \leq \tau \leq T + s$" can be replaced by "$X(T + s) = i$". That is, the chain behaves as if state i had just been entered at time $T + s$, and the probability of the above event is the same as that of the event $[V(i) > t]$. Therefore,

$$P[V(i) > s + t \mid V(i) > s] = P[V(i) > t]$$

We have already seen in Theorem 6.1 of Chapter 6 that the only probability distribution satisfying this property is the exponential, which immediately implies (7.74).

The interpretation of the parameter $\Lambda(i)$ follows from our discussion of the GSMP with a Poisson clock structure in Chapter 6. In particular, we saw in (6.58) that $\Lambda(i)$ is the sum of the Poisson rates of all active events at state i. In the case of a Markov chain, an "event" is identical to a "state transition", so "interevent times" are identical to "state holding times". Let us, therefore, define e_{ij} to be events generated by a Poisson process with rate λ_{ij} which cause transitions from state i to state $j \neq i$. If two or more underlying "physical" events can cause such a transition, we do not care to distinguish them, and simply refer to any one of them as "e_{ij}". Then, as long as e_{ij} is a feasible event at state i, $\Lambda(i)$ is given by (6.58):

$$\Lambda(i) = \sum_{e_{ij} \in \Gamma(i)} \lambda_{ij} \tag{7.75}$$

where $\Gamma(i)$ is the set of feasible events at state i. In other words, the parameter $\Lambda(i)$, which fully characterizes the holding time distribution for state i, may be thought of as the sum of all Poisson rates corresponding to feasible events at state i.

7.3.6 Physical Interpretation and Properties of the Transition Rate Matrix

We now attempt to gain insight into the meaning of the elements of the transition rate matrix $\mathbf{Q}$. Let us go back to the forward Chapman-Kolmogorov equation (7.71), and write down the individual scalar differential equations:

$$\frac{dp_{ij}(\tau)}{d\tau} = p_{ij}(\tau)q_{jj} + \sum_{r \neq j} p_{ir}(\tau)q_{rj} \tag{7.76}$$

We first concentrate on the case $i = j$. Thus, (7.76) is written as

$$\frac{dp_{ii}(\tau)}{d\tau} = p_{ii}(\tau)q_{ii} + \sum_{r \neq i} p_{ir}(\tau)q_{ri}$$

Setting $\tau = 0$ and using the initial conditions (7.72) we get

$$\left. \frac{dp_{ii}(\tau)}{d\tau} \right|_{\tau=0} = q_{ii} \tag{7.77}$$

which can be rewritten as

$$-q_{ii} = \left. \frac{d}{d\tau}[1 - p_{ii}(\tau)] \right|_{\tau=0}^{c}$$

Here, $[1 - p_{ii}(\tau)]$ is the probability that the chain leaves state i in an interval of length τ. Thus, we see that $-q_{ii}$ is the *instantaneous rate* at which a state transition out of i takes place. It describes the propensity the chain has to leave state i.

Some additional insight into the meaning of $-q_{ii}$ is obtained by returning to (7.74) and attempting to link q_{ii} to $\Lambda(i)$. Suppose the state is i at some time instant t and consider the interval $(t, t + \tau]$, where τ can be made arbitrarily small to guarantee that at most one event (and hence state transition) can take place in $(t, t + \tau]$. Now suppose that no state transition occurs in this interval. Thus,

$$p_{ii}(\tau) = P[V(i) > \tau] = e^{-\Lambda(i)\tau}$$

Taking derivatives of both sides, we get

$$\frac{dp_{ii}(\tau)}{d\tau} = -\Lambda(i)e^{-\Lambda(i)\tau}$$

Evaluating both sides at $\tau = 0$ and using (7.77) yields:

$$-q_{ii} = \Lambda(i) \qquad (7.78)$$

In other words, $-q_{ii}$ represents the *total event rate* characterizing state i. This event rate also captures the propensity of the chain to leave state i, since a large value of $\Lambda(i)$ indicates a greater likelihood of some event occurrence and hence state transition out of i.

Next, returning to (7.76) for the case $j \neq i$, we obtain similar to (7.77):

$$q_{ij} = \left. \frac{dp_{ij}(\tau)}{d\tau} \right|_{\tau=0} \qquad (7.79)$$

Here the interpretation of q_{ij} is the *instantaneous rate* at which a state transition from i to j takes place. Moreover, similar to (7.78), we have

$$q_{ij} = \lambda_{ij} \qquad (7.80)$$

which is the Poisson rate of the event e_{ij} causing transitions from i to j.

From (7.78), since $\Lambda(i) > 0$, it follows that $q_{ii} < 0$. Therefore, all diagonal elements of the matrix $\mathbf{Q}$ must be negative. From (7.80), on the other hand, we see that $q_{ij} > 0$, if the event e_{ij} is feasible in state i; otherwise, $q_{ij} = 0$. Thus, all off-diagonal terms must be nonnegative. In addition, since $\sum_j p_{ij}(\tau) = 1$ in (7.69), differentiating with respect to τ and setting $\tau = 0$ we get

$$\sum_{\text{all } j} q_{ij} = 0 \qquad (7.81)$$

7.3.7   Transition Probabilities

We are now in a position to see how the transition rate matrix $\mathbf{Q}$ specifies the Markov chain model. Let us define the *transition probability*, P_{ij}, of a continuous-time Markov chain as follows. Suppose state transitions occur at random time instants $T_1 < T_2 < \cdots < T_k < \cdots$ The state following the transition at T_k is denoted by X_k. We then set

$$P_{ij} = P[X_{k+1} = j \mid X_k = i] \qquad (7.82)$$

Let us recall (6.67) of Chapter 6, where we derived the probability distribution of events at a given state for a GSMP with Poisson clock structure. In particular, using our Markov chain notation, given that the state is i, the probability that the next event is e_{ij} (therefore, the next state is j) is given by $\lambda_{ij}/\Lambda(i)$. Using (7.78) and (7.80) we now see that the transition probability P_{ij} is expressed in terms of elements of $\mathbf{Q}$:

$$P_{ij} = \frac{q_{ij}}{-q_{ii}}, \qquad j \neq i \tag{7.83}$$

Moreover, summing over all $j \neq i$ above and recalling (7.81), we get $\sum_{j \neq i} P_{ij} = 1$. This implies that $P_{ii} = 0$ as expected, since the only events defined for a Markov chain are those causing actual state transitions.

In summary, once the transition rate matrix $\mathbf{Q}$ is specified, we have at our disposal a full model specification:

- The state transition probabilities P_{ij} are given by (7.83)

- The parameters of the exponential state holding time distributions are given by the corresponding diagonal elements of $\mathbf{Q}$, that is, for state i, $-q_{ii} = \sum_{j \neq i} q_{ij}$.

Conversely, note that a Markov chain can also be specified through:

- Parameters $\Lambda(i)$ for all states i, characterizing the exponential state holding time distributions.

- Transition probabilities P_{ij} for all pairs of states $i \neq j$.

Example 7.16
In this example, we provide the physical description of a simple DES, and obtain a continuous-time Markov chain model.

The DES is a queueing system with a total capacity of two customers (Fig. 7.17). There are two different events that can occur: an arrival a, and a departure d. However, the system is designed so that service is only provided to two customers simultaneously (in manufacturing, for example, this situation arises in *assembly* operations, where a machine combines two or more parts together). Using our standard DES notation from previous chapters, we have a state space

$$\mathcal{X} = \{0, 1, 2\}$$

an event set

$$\mathcal{E} = \{a, d\}$$

and feasible event sets

$$\Gamma(x) = \{a\} \text{ for } x = 0, 1, \qquad \Gamma(2) = \{a, d\}$$

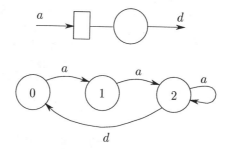

Figure 7.17: Queueing system and state transition diagram for Example 7.16. *Service in this system is provided only when two customers are present, at which time they are combined together.*

The state transition diagram for the system is shown in Fig. 7.17. Departure events are only feasible at state 2 and result in emptying out the system. Arrivals are feasible at state 2, but they cause no state transition, since the capacity of the system is limited to two customers.

By assuming that events a and d both occur according to Poisson processes with rates λ and μ respectively, we obtain a Markov chain model of this system. The transition rate matrix $\mathbf{Q}$ is given by

$$\mathbf{Q} = \begin{bmatrix} -\lambda & \lambda & 0 \\ 0 & -\lambda & \lambda \\ \lambda & 0 & \lambda \end{bmatrix}$$

The Markov chain model can also be specified through the transition probabilities P_{ij} for all $i \neq j, i, j = 0, 1, 2$. By looking at the state transition diagram in Fig. 7.17, we immediately see that

$$P_{01} = P_{12} = P_{20} = 1$$

which is consistent with (7.83). In addition, we can specify the parameters of the exponential state holding time distributions as follows:

$$\Lambda(0) = -q_{00} = \lambda, \quad \Lambda(1) = -q_{11} = \lambda, \quad \Lambda(2) = -q_{22} = \mu$$

Example 7.17

A warehouse keeps in stock one unit of each of two types of products, P_1 and P_2. A truck periodically comes to take away one of these products. If both are in stock, the truck gives preference to P_1. In this DES, let $x_i = 1$ if P_i is present in the warehouse, $i = 1, 2$, and $x_i = 0$ otherwise. Thus, the state space is

$$\mathcal{X} = \{(x_1, x_2) : x_1 = 0, 1, x_2 = 0, 1\}$$

Let us assume that when $x_i = 0$, there is a Poisson rate λ_i with which P_i is replaced. Let us also assume that when a unit of P_1 is present at the warehouse, the truck arrives to pick it up with a Poisson rate μ_1; if only a unit of P_2 is present, then this rate is μ_2. We can now write a transition rate matrix $\mathbf{Q}$ for this model, labeling states $0, 1, 2$ and 3 to correspond to $(0,0), (0,1), (1,0), (1,1)$ respectively:

$$\mathbf{Q} = \begin{bmatrix} -(\lambda_1 + \lambda_2) & \lambda_2 & \lambda_1 & 0 \\ \mu_2 & -(\lambda_1 + \mu_2) & 0 & \lambda_1 \\ \mu_1 & 0 & -(\lambda_2 + \mu_1) & \lambda_2 \\ 0 & \mu_1 & 0 & -\mu_1 \end{bmatrix}$$

Note that $q_{10} = \mu_2$ is due to the fact that the only time the truck picks up a P_2 unit is at state $(0,1)$.

We can now use (7.83) to determine all the transition probabilities. For example,

$$P_{01} = \frac{\lambda_2}{\lambda_1 + \lambda_2}, \quad P_{20} = \frac{\mu_1}{\lambda_2 + \mu_1}, \quad P_{31} = 1, \quad P_{03} = 0$$

In addition, state holding times are characterized by the exponential distribution parameters specified through (7.78). For example, $\Lambda(0) = \lambda_1 + \lambda_2$, and $\Lambda(3) = \mu_1$.

7.3.8 State Probabilities

Similar to the discrete-time case, we define state probabilities as follows:

$$\pi_j(t) \equiv P[X(t) = j] \tag{7.84}$$

Accordingly, we have a *state probability vector*

$$\boldsymbol{\pi}(t) = [\pi_0(t), \pi_1(t), \ldots] \tag{7.85}$$

This is a row vector whose dimension is specified by the dimension of the state space of the chain (not necessarily finite).

A continuous-time Markov chain model is completely specified by the state space $\mathcal{X}$ and the transition matrix $\mathbf{P}(\tau)$, and an initial state probability vector

$$\boldsymbol{\pi}(0) = [\pi_0(0), \pi_1(0), \ldots]$$

which provides the probability distribution of the initial state of the chain $X(0)$. By (7.73), however, $\mathbf{P}(\tau) = e^{\mathbf{Q}\tau}$. Therefore, the specification of $\mathbf{P}(\tau)$ is immediately provided by the transition rate matrix $\mathbf{Q}$.

7.3.9   Transient Analysis

Our main objective is to determine the probability vector $\pi(t)$ given a Markov chain specified by its state space $\mathcal{X}$, transition rate matrix $\mathbf{Q}$, and and initial state probability vector $\pi(0)$. Starting out with the definition of $\pi_j(t)$ in (7.84), let us condition the event $[X(t) = j]$ on the event $[X(0) = i]$, and use the rule of total probability to obtain

$$\pi_j(t) = P[X(t) = j] = \sum_{\text{all } i} P[X(t) = j \mid X(0) = i] \cdot P[X(0) = i]$$
$$= \sum_{\text{all } i} p_{ij}(t)\pi_i(0)$$

Recalling the definition of the transition matrix in (7.68), we can rewrite this relationship in matrix form:

$$\pi(t) = \pi(0)\mathbf{P}(t) \qquad (7.86)$$

and since $\mathbf{P}(t) = e^{\mathbf{Q}t}$, the state probability vector at time t is given by

$$\pi(t) = \pi(0)e^{\mathbf{Q}t} \qquad (7.87)$$

Therefore, in principle, one can always obtain a solution describing the transient behavior of a chain characterized by $\mathbf{Q}$ and an initial condition $\pi(0)$. However, obtaining explicit expressions for the individual state probabilities $\pi_j(t), j = 0, 1, 2, \ldots$ is far from simple, as we shall see later on in this chapter. The time functions $\pi_j(t)$ are fairly complicated, even for the simplest Markov chain models of DES.

Note that by differentiating (7.87) with respect to t, we can obtain the differential equation

$$\frac{d\pi(t)}{dt} = \pi(t)\mathbf{Q} \qquad (7.88)$$

which is of the same form as the forward Chapman-Kolmogorov equation (7.71). In practice, it is often useful to write (7.88) as a set of scalar differential equations

$$\frac{d\pi_j(t)}{dt} = q_{jj}\pi_j(t) + \sum_{i \neq j} q_{ij}\pi_i(t) \qquad (7.89)$$

for all $j = 0, 1, \ldots$ Note that we have separated the $i = j$ element from the sum above in order to emphasize the fact that the summation term is independent of $\pi_j(t)$.

Clearly, if the matrix $\mathbf{Q}$ is specified, then (7.89) can be directly obtained. We have seen, however, that Markov chains are often more efficiently described in terms of state transition diagrams. In such diagrams, nodes represent states and a directed arc from node i to node j represents a transition from i to j. We obtain a state transition rate diagram by associating to an arc (i, j) the transition rate q_{ij}. This is illustrated in Fig. 7.18 for a typical state j. Such a diagram provides a simple graphical device for deriving (7.89) by inspection. This is done as follows.

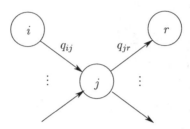

Figure 7.18: Probability flow balance for state j.

We view $\pi_j(t)$ as the level of a "probability fluid" residing in node j and taking on values between 0 (empty) and 1 (full). The transition rate q_{ij} represents the "probability flow rate" from i to j. Hence, the flow from i to j at time t is given by $q_{ij}\pi_i(t)$. We can then write simple flow balance equations for the "probability fluid" level of state j by evaluating the total incoming flow at j and the total outgoing flow from j. Referring to Fig. 7.18, we see that

$$\text{Total flow into state } j = \sum_{i \neq j} q_{ij} \pi_i(t)$$
$$\text{Total flow out of state } j = \sum_{r \neq j} q_{jr} \pi_j(t)$$

Therefore, the net probability flow rate into state j is described by

$$\frac{d\pi_j(t)}{dt} = \sum_{i \neq j} q_{ij} \pi_i(t) - \left[\sum_{r \neq j} q_{jr} \right] \pi_j(t)$$

From (7.81), we have

$$\sum_{r \neq j} q_{jr} = -q_{jj}$$

so that the flow balance equation above becomes

$$\frac{d\pi_j(t)}{dt} = q_{jj} \pi_j(t) + \sum_{i \neq j} q_{ij} \pi_i(t)$$

which is precisely (7.89). Thus, the state transition rate diagram contains the exact same information as the transition rate matrix $\mathbf{Q}$.

Example 7.18

We return to the Markov chain model in Example 7.16. To obtain the state transition rate diagram shown in Fig. 7.19, observe that the probability flow out of state 0 is only due to events a occurring with rate λ, whereas the probability flow into state 0 is only due to events d occurring at state 2 with rate μ. Similarly, at state 1 we have an incoming flow from 1 with rate λ, and an outgoing flow due to a events with rate λ. Finally, state 2 receives a probability flow from state 2 with rate λ, and has an outgoing flow due to d events with rate μ. The state transition rate diagram also corresponds directly to the transition rate matrix $\mathbf{Q}$ derived in Example 7.16:

$$\mathbf{Q} = \begin{bmatrix} -\lambda & \lambda & 0 \\ 0 & -\lambda & \lambda \\ \mu & 0 & -\mu \end{bmatrix}$$

To determine the state probability vector $\boldsymbol{\pi}(t) = [\pi_0(t), \pi_1(t), \pi_2(t)]$ at any time t, we need to obtain the differential equations (7.89). By inspection of the state transition rate diagram in Fig. 7.19, we can easily write down the following probability flow balance equations:

$$\frac{d\pi_0(t)}{dt} = \mu\pi_2(t) - \lambda\pi_0(t)$$
$$\frac{d\pi_1(t)}{dt} = \lambda\pi_0(t) - \lambda\pi_1(t)$$
$$\frac{d\pi_2(t)}{dt} = \lambda\pi_1(t) - \mu\pi_2(t)$$

If an initial condition $\boldsymbol{\pi}(0) = [\pi_0(0), \pi_1(0), \pi_2(0)]$ is specified, these equations can be solved for $\pi_0(t), \pi_1(t)$, and $\pi_2(t)$. However, obtaining an explicit solution, even for this simple example, is not a trivial matter.

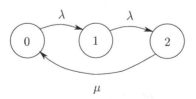

Figure 7.19: State transition rate diagram for Example 7.18.

Example 7.19

We now reconsider the Markov chain model in Example 7.17. First, we obtain a state transition rate diagram as shown in Fig. 7.20, which corresponds to the transition rate matrix $\mathbf{Q}$:

$$\mathbf{Q} = \begin{bmatrix} -(\lambda_1 + \lambda_2) & \lambda_2 & \lambda_1 & 0 \\ \mu_2 & -(\lambda_1 + \mu_2) & 0 & \lambda_1 \\ \mu_1 & 0 & -(\lambda_2 + \mu_1) & \lambda_2 \\ 0 & \mu_1 & 0 & -\mu_1 \end{bmatrix}$$

Recall that states (0,0), (0,1), (1,0), (1,1) are labeled 0,1,2,3 respectively. Also, recall that the truck always prioritizes P_1 units. Thus, from both (1,0) and from (1,1) there are probability flows with rate μ_1 to states (0,0) and (0,1) respectively.

By inspection of Fig. 7.20, we can now write down probability flow balance equations:

$$\frac{d\pi_0(t)}{dt} = \mu_2\pi_1(t) + \mu_1\pi_2(t) - (\lambda_1 + \lambda_2)\pi_0(t)$$
$$\frac{d\pi_1(t)}{dt} = \lambda_2\pi_0(t) + \mu_1\pi_3(t) - (\lambda_1 + \mu_2)\pi_1(t)$$
$$\frac{d\pi_2(t)}{dt} = \lambda_1\pi_0(t) - (\lambda_2 + \mu_1)\pi_2(t)$$
$$\frac{d\pi_3(t)}{dt} = \lambda_1\pi_1(t) + \lambda_2\pi_2(t) - \mu_1\pi_3(t)$$

Once again, we see that obtaining an explicit solution for these equations is not an easy task.

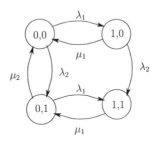

Figure 7.20: State transition rate diagram for Example 7.19.

7.3.10 Steady State Analysis

In the previous section, we saw that the behavior of a continuous-time Markov chain is completely described by the set of differential equations (7.89) or, in matrix form, (7.88). Although the solution to these equations can always be written in the matrix exponential form (7.87), obtaining explicit closed-form expressions for the state probabilities $\pi_0(t), \pi_1(t), \ldots$ is generally a difficult task.

Fortunately, in practice we are often only interested in the steady state behavior of a system; this is much easier to analyze than the transient behavior. In other words, we turn the system on, let it run for some time, and then examine its performance in the "long run". As in the discrete-time case, by "long run" we generally mean that the system has operated sufficiently long to allow all state probabilities to reach some fixed values, no longer varying with time. Of course, there is no guarantee that such values are indeed achievable. Therefore, our study centers around the existence and determination of the limits:

$$\pi_j = \lim_{t \to \infty} \pi_j(t) \tag{7.90}$$

As in the discrete-time case, we need to address three basic questions:

1. Under what conditions do the limits in (7.90) exist?

2. If these limits exist, do they form a legitimate probability distribution, that is, $\sum_j \pi_j = 1$?

3. How do we evaluate π_j?

If π_j exists for some state j, it is referred to as a *steady-state, equilibrium,* or *stationary state probability.* Accordingly, if π_j exists for all states j, then we obtain the *stationary state probability vector*

$$\boldsymbol{\pi} = [\pi_0, \pi_1, \ldots]$$

If the limits in (7.90) exist, it follows that as $t \to \infty$ the derivative $d\boldsymbol{\pi}/dt \to 0$, since $\boldsymbol{\pi}(t)$ no longer depends on t. We therefore expect that at steady state the differential equation (7.88):

$$\frac{d\boldsymbol{\pi}(t)}{dt} = \boldsymbol{\pi}(t)\mathbf{Q}$$

reduces to the algebraic equation

$$\boldsymbol{\pi}\mathbf{Q} = 0$$

The analysis of the steady state behavior of continuous-time Markov chains is very similar to that of the discrete-time case, presented in Sections 7.2.9 through 7.2.11. The concepts of irreducibility and recurrence are still valid, and results paralleling Theorems 7.9 and 7.10 can be formally obtained. Thus, we will limit ourselves here to stating the most important result for our purposes, pertaining to positive recurrent irreducible chains (see also Asmussen, 1987).

Theorem 7.11 In an irreducible continuous-time Markov chain consisting of positive recurrent states, a unique stationary state probability vector $\boldsymbol{\pi}$ exists such that $\pi_j > 0$ and

$$\pi_j = \lim_{t \to \infty} \pi_j(t)$$

These limits are independent of the initial state probability vector. Moreover, the vector $\boldsymbol{\pi}$ is determined by solving:

$$\boldsymbol{\pi} \mathbf{Q} = 0 \tag{7.91}$$

$$\sum_{\text{all } j} \pi_j = 1 \tag{7.92}$$

♦

As in the discrete-time case, it is possible to show that every *finite* irreducible Markov chain has a unique stationary state probability vector $\boldsymbol{\pi}$ determined through (7.91) and (7.92). Then, obtaining $\boldsymbol{\pi}$ is a matter of solving a set of linear algebraic equations. However, in the case of an infinite state space, it is generally a difficult task to solve (7.91) and (7.92). Fortunately, for a large class of Markov chains with a special structure, this task is somewhat simplified. This class of chains, called *birth-death chains*, is the subject of the next section.

Finally, note that (7.91) in scalar form is the set of equations (7.89) with $d\pi_j(t)/dt = 0$. In other words, we can still use the idea of probability flow balancing at steady state, where all the "probability fluid" levels occupying the nodes of the state transition rate diagram have all reached equilibrium. Thus, we have

$$q_{jj}\pi_j + \sum_{i \neq j} q_{ij}\pi_i = 0 \tag{7.93}$$

for all states $j = 0, 1, \dots$ As in the discrete-time case, these equations are not linearly independent. This necessitates including the normalization condition (7.92) in order to obtain the unique stationary state probabilities $\pi_0, \pi_1, \pi_2, \dots$

Example 7.20

Consider the system modeled and analyzed in Examples 7.16 and 7.18, whose state transition rate diagram is shown in Fig. 7.19. By inspection, we write down the probability flow balance equations at steady state, which are precisely the set of equations (7.91) or (7.93):

$$\mu\pi_2 - \lambda\pi_0 = 0$$
$$\lambda\pi_0 - \lambda\pi_1 = 0$$
$$\lambda\pi_1 - \mu\pi_2 = 0$$

The first two equations give

$$\pi_2 = \frac{\lambda}{\mu}\pi_0, \qquad \pi_1 = \pi_0$$

One can then see that the third equation is redundant. The solution is obtained by using the two equations above in conjunction with the normalization condition (7.92):

$$\pi_0 + \pi_1 + \pi_2 = 1$$

which becomes

$$2\pi_0 + \frac{\lambda}{\mu}\pi_0 = 1$$

Hence, we get

$$\pi_0 = \pi_1 = \frac{\mu}{2\mu + \lambda}, \qquad \pi_2 = \frac{\lambda}{2\mu + \lambda}$$

7.4 BIRTH-DEATH CHAINS

Continuous-time Markov chains provide a rich class of models for many stochastic DES of interest. Although we have seen how to analyze both the transient and steady-state behavior of these chains, obtaining explicit expressions for the state probabilities is not an easy task in practice. *Birth-death chains* form a limited class of Markov chains whose special structure facilitates this task, while still providing a sufficiently rich modeling framework.

In simple terms, a birth-death chain is one where transitions are only permitted to/from neighboring states. Since we have adopted a state space $\mathcal{X} = \{0, 1, 2, \ldots\}$, such transitions have a simple appealing interpretation if we think of the state as representing a certain "population" level (e.g., customers, jobs, messages): A transition from state i to state $i+1$ is a *birth*, whereas a transition from i to $i-1$ is a *death*. Obviously, no death can occur at state $i = 0$. Similarly, if the state space is finite, so that $\mathcal{X} = \{0, 1, 2, \ldots, K\}$ for some $K < \infty$, no birth can occur at state $i = K$. A birth-death chain can be defined for discrete or continuous time. A simple version of a discrete-time birth-death chain was considered in Examples 7.10 and 7.13. In what follows, we will limit ourselves to the continuous-time case.

Definition. A *birth-death chain* is a continuous-time Markov chain whose transition rate matrix $\mathbf{Q}$ satisfies

$$q_{ij} = 0 \quad \text{for all } j > i+1 \text{ and } j < i-1 \tag{7.94}$$

♦

It is customary to define a *birth rate*

$$\lambda_j = q_{j,j+1} > 0, \quad j = 0, 1, \ldots \tag{7.95}$$

and a death rate

$$\mu_j = q_{j,j-1} > 0, \quad j = 1, 2, \ldots \tag{7.96}$$

where we normally assume all birth and death parameters above to be strictly positive. Note, however, that we do not necessarily constrain the chain to be homogeneous, since these rates may depend on the states.

By (7.81), we know that all elements in the jth row of $\mathbf{Q}$ must add to 0. Therefore, the diagonal elements q_{jj} are immediately obtained from (7.94) through (7.96) as follows:

$$q_{jj} = -(\lambda_j + \mu_j) \quad \text{for all } j = 1, 2, \ldots$$
$$q_{00} = -\lambda_0$$

The general form of the transition rate matrix $\mathbf{Q}$ for a birth-death chain is as follows:

$$\mathbf{Q} = \begin{bmatrix} -\lambda_0 & \lambda_0 & 0 & 0 & 0 & \cdots \\ \mu_1 & -(\lambda_1 + \mu_1) & \lambda_1 & 0 & 0 & \cdots \\ 0 & \mu_2 & -(\lambda_2 + \mu_2) & \lambda_2 & 0 & 0 \\ 0 & 0 & \mu_3 & -(\lambda_3 + \mu_3) & \lambda_3 & 0 \\ \vdots & \vdots & \vdots & \vdots & \vdots & \vdots \end{bmatrix} \tag{7.97}$$

The special structure of this model is clearly seen in the state transition rate diagram of Fig. 7.21, corresponding to $\mathbf{Q}$ above. We may think of births and deaths as events which can occur with generally different rates at various states. A birth or death at state $j > 0$ causes a transition to $j+1$ or $j-1$ respectively. Similarly, a transition to state $j > 0$ can only occur through a birth at state $j-1$ or a death at state $j+1$. State $j = 0$ represents a "boundary case" where only a birth can occur.

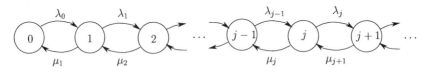

Figure 7.21: State transition rate diagram for a general birth-death chain.

The differential equations characterizing the state probabilities $\pi_0(t), \pi_1(t), \pi_2(t), \ldots$ of a birth-death chain can be obtained from the general case (7.94) by exploiting the special structure of $\mathbf{Q}$. In particular,

$$\frac{d\pi_j(t)}{dt} = -(\lambda_j + \mu_j)\pi_j(t) + \lambda_{j-1}\pi_{j-1}(t) + \mu_{j+1}\pi_{j+1}(t), \quad j = 1, 2, \ldots \tag{7.98}$$

$$\frac{d\pi_0(t)}{dt} = -\lambda_0\pi_0(t) + \mu_1\pi_1(t) \tag{7.99}$$

Even though the solution of (7.98) and (7.99) is easier to obtain than the general case (7.89), it is still an elaborate task. Fortunately, the steady-state analysis of these equations is much simpler to handle, and forms the basis of much of the elementary queueing theory we will develop in the next chapter. In the next few sections, we examine some simple cases for which transient solutions can be obtained and then discuss the steady-state solution of the birth-death chain.

7.4.1   The Pure Birth Chain

In the pure birth chain, we have $\mu_j = 0$ for all $j = 1, 2, \ldots$ Equations (7.98) and (7.99) become

$$\frac{d\pi_j(t)}{dt} = -\lambda_j\pi_j(t) + \lambda_{j-1}\pi_{j-1}(t), \qquad j = 1, 2, \ldots \tag{7.100}$$

$$\frac{d\pi_0(t)}{dt} = -\lambda_0\pi_0(t) \tag{7.101}$$

The solution of (7.101) is easily obtained as an exponential function

$$\pi_0(t) = e^{-\lambda_0 t} \tag{7.102}$$

where we assume that $\pi_0(0) = 1$, that is, the initial state is 0 with probability 1. The solution of (7.100) for $j = 1, 2, \ldots$ is more difficult to obtain. Given (7.101), the following recursive expression allows us to evaluate $\pi_1(t), \pi_2(t), \ldots$:

$$\pi_j(t) = e^{-\lambda_j t}\lambda_{j-1}\int_0^t \pi_{j-1}(\tau)e^{\lambda_j\tau}d\tau, \qquad j = 1, 2, \ldots \tag{7.103}$$

By differentiating (7.103) it is easy to verify that this is indeed the solution of (7.100). It is not a very simple expression, which leads us to suspect (and this is in fact the case) that the general solution of (7.98) and (7.99) is far from simple.

7.4.2   The Poisson Process Revisited

Let us consider the pure birth chain with *constant birth rates*, that is,

$$\lambda_j = \lambda \qquad \text{for all } j = 0, 1, 2, \ldots$$

Then, (7.100) and (7.101) reduce to

$$\frac{d\pi_j(t)}{dt} = -\lambda\pi_j(t) + \lambda\pi_{j-1}(t), \qquad j = 1, 2, \ldots \qquad (7.104)$$

$$\frac{d\pi_0(t)}{dt} = -\lambda\pi_0(t) \qquad (7.105)$$

Returning to our discussion of the Poisson counting process in Section 6.2 in Chapter 6, we can immediately recognize that (7.104) is the same as equation (6.33), which characterizes the counter $N(t)$ taking values in $\{0, 1, 2, \ldots\}$. This is not surprising since counting events with an initial count of 0 is the same as counting births with an initial population of 0. In addition, we have seen that the Poisson process is inherently characterized by exponentially distributed interevent times with parameter λ; so is the birth chain whose only event is a "birth" occurring at rate λ.

In summary, the Poisson counting process may be obtained in one of two ways. From first principles, as a counting process satisfying the basic assumptions (A1) through (A3) of Section 6.2; or as a special case of a Markov chain (pure birth chain with constant birth rate λ). This observation serves to reinforce the strong connection between (a) the memoryless property, (b) the exponential distribution for interevent (or state holding) times, and (c) stationary independent increments in a stochastic process. All three characterize a class of stochastic processes for which the Poisson process is a fundamental building block.

The solution of (7.104) and (7.105) is of course the well-known by now Poisson distribution

$$\pi_j(t) = \frac{(\lambda t)^j}{j!} e^{-\lambda t}, \qquad t \geq 0, \qquad j = 0, 1, 2, \ldots \qquad (7.106)$$

which was also derived in (6.34). Note that for $j = 0$ we obtain $\pi_0(t)$ in (7.102) with $\lambda_0 = \lambda$.

7.4.3 Steady State Analysis of Birth-Death Chains

We will now concentrate on the steady state solution of the birth-death chain, that is, the solution of (7.98) and (7.99) when $d\pi_j(t)/dt = 0$, if it exists:

$$-(\lambda_j + \mu_j)\pi_j + \lambda_{j-1}\pi_{j-1} + \mu_{j+1}\pi_{j+1} = 0, \qquad j = 1, 2, \ldots \qquad (7.107)$$

$$-\lambda_0\pi_0 + \mu_1\pi_1 = 0 \qquad (7.108)$$

Recall that in addition to these equations the normalization condition (7.92) must also be satisfied, that is $\sum_j \pi_j = 1$.

Ignoring for the time being the issue of existence, we can obtain a general solution to these equations by proceeding as follows.

Step 1. From (7.108), we obtain a relationship between π_1 and π_0:

$$\pi_1 = \frac{\lambda_0}{\mu_1} \pi_0$$

Step 2. From (7.107) with $j = 1$, we get

$$\mu_2 \pi_2 = -\lambda_0 \pi_0 + (\lambda_1 + \mu_1)\pi_1$$

and using the result of *Step 1* above we obtain a relationship between π_2 and π_0:

$$\pi_2 = \frac{\lambda_0 \lambda_1}{\mu_1 \mu_2} \pi_0$$

Step 3. Repeat *Step 2* for all $j = 2, 3, \ldots$ to obtain a relationship between π_j and π_0:

$$\pi_j = \left(\frac{\lambda_0 \cdots \lambda_{j-1}}{\mu_1 \cdots \mu_j} \right) \pi_0 \tag{7.109}$$

This is the general solution we are seeking if π_0 can be determined; this is done next.

Step 4. Add all π_j in (7.109) and use the normalization condition (7.92) to obtain π_0:

$$\pi_0 + \pi_0 \sum_{j=1}^{\infty} \left(\frac{\lambda_0 \cdots \lambda_{j-1}}{\mu_1 \cdots \mu_j} \right) = 1$$

which gives

$$\pi_0 = \frac{1}{1 + \sum_{j=1}^{\infty} \left(\frac{\lambda_0 \cdots \lambda_{j-1}}{\mu_1 \cdots \mu_j} \right)} \tag{7.110}$$

This completes the derivation of the steady-state solution, which is provided by the combination of (7.109) and (7.110). These two equations are essential in the study of simple queueing systems and we will make extensive use of them in the next chapter.

The existence of this solution is closely related to the behavior of the sum

$$S_1 = 1 + \sum_{j=1}^{\infty} \left(\frac{\lambda_0 \cdots \lambda_{j-1}}{\mu_1 \cdots \mu_j} \right) \tag{7.111}$$

appearing in the denominator of the expression for π_0 in (7.110). If $S_1 = \infty$, then $\pi_0 = 0$. We can immediately see that in this case the chain cannot be positive recurrent, and Theorem 7.11 does not apply. It can be formally shown that if $S_1 = \infty$ then the chain is either null recurrent or transient. Intuitively, we can see in (7.111) that S_1 tends to blow up when birth rates are larger than death rates, that is, the chain drifts towards ∞ and states near 0 are never revisited (or take infinite expected time to be revisited). Conversely, when death rates are larger than birth rates, then the chain drifts back towards state 0, and the steady-state value π_0 is positive.

Another critical sum, which determines whether the chain is transient or not, is the following:

$$S_2 = \sum_{j=1}^{\infty} \left(\frac{\mu_1 \cdots \mu_j}{\lambda_1 \cdots \lambda_j} \right) \frac{1}{\lambda_j} \tag{7.112}$$

This is a good point to look back at Example 7.10, where we considered a simple discrete-time version of the birth-death chain. In that case, we saw in (7.39) that if a sum of the form of S_2 above is less than ∞ then that chain is transient. Carrying out an analysis similar to that of Example 7.10, it can be shown that $S_2 < \infty$ is a necessary and sufficient condition for our birth-death chain to be transient. Intuitively, we see that S_2 in (7.112) remains finite when birth rates tend to be larger than death rates, and therefore the chain drifts toward ∞.

Most practical DES we care to design and analyze are positive recurrent in nature. Thus, the case of interest for our purposes corresponds to birth and death rates so selected that $S_1 < \infty$, $S_2 = \infty$. It can be seen that this situation arises whenever we have

$$\frac{\lambda_j}{\mu_j} < 1 \tag{7.113}$$

either for all $j = 0, 1, \ldots$, or for all $j > j'$ for some finite positive integer j'. Thus, we generally require that death rates eventually overtake birth rates to reach some form of "stability". In queueing systems, births correspond to customer arrivals requesting service, and deaths correspond to service completions. It naturally follows that service completion rates must exceed arrival rates if our system is to handle customer demand; otherwise, queue lengths tend to grow without bound. Moreover, if the empty state $j = 0$ is never revisited, this implies that the system server is operating above capacity. Therefore, condition (7.113) represents a fundamental "rule of thumb" to remember in designing DES modeled as birth-death chains. We shall have much more to say on this issue when we discuss queueing systems in the next chapter.

In summary, a birth-death chain may be thought of as a stochastic DES with state space $\mathcal{X} = \{0, 1, \ldots\}$ and event set $\mathcal{E} = \{a, d\}$, where a is a birth (arrival) and d is a death (departure). In general, birth and death rates are state-dependent, so the underlying clock uses variable speeds as described in

Section 6.9 of Chapter 6. The interevent times of this DES are always exponentially distributed with parameter $(\lambda_j + \mu_j)$ when the state is $j = 1, 2, \ldots$ and λ_0 when $j = 0$. The framework used in this chapter serves to provide solutions for the state probabilities $\pi_j(t)$. However, as pointed out in Section 7.2.4, the compact transition matrix representation we have used does hide some of the underlying dynamic behavior of an event-driven system. A case in point arises when one compares the state transition diagram in Fig. 7.17 with the state transition rate diagram in Fig. 7.19 for the same DES. The fact that several a events may occur while in state 2 is hidden from the representation of Fig. 7.19. While this has no effect on our effort to determine the state probability vector $\pi(t)$, it does limit us in studying other useful properties of the DES. This is something that one should keep in mind and will be further discussed in later chapters.

7.5   UNIFORMIZATION OF MARKOV CHAINS

Continuous-time Markov chain models allow for state transitions (events) to occur at any (real-valued) time instant, and are therefore viewed as more "realistic" for a large number of applications. On the other hand, there are also applications where discrete-time models are more convenient (e.g., the gambling problem in Example 7.4). Discrete-time models are also often easier to set up and simpler to analyze. It is natural, therefore, to explore how to convert a continuous-time model into a discrete-time model without distorting the information that the former contains. As we will see in Chapter 9, this conversion process is particularly useful in the study of "controlled" Markov chains.

Returning to the transition rate matrix $\mathbf{Q}$, recall that a diagonal element $-q_{ii}$ represents the total probability flow rate out of state i. As in (7.78), let us set $\Lambda(i) = -q_{ii}$. There is of course no reason to expect that $\Lambda(i)$ is the same for all i. Suppose, however, that we pick a *uniform* rate γ such that

$$\gamma \leq \Lambda(i) \qquad \text{for all states } i$$

and replace all $\Lambda(i)$ by γ. We can then argue that the "extra probability flow" $[\gamma - \Lambda(i)]$ at state i corresponds to "fictitious events" occurring at this rate which simply leave state i unchanged. These self-loop transitions should have no effect on the behavior of the chain: The state itself has not been affected, and, by the memoryless property, the time left until the next actual transition out of this state also remains unaffected, that is, it is still exponentially distributed with parameter $\Lambda(i)$. As a result, the new "uniformized" chain should be stochastically indistinguishable from the original one. That is, given a common initial state x_0, the two models should have identical state probabilities $\pi_i(t)$ for all t and states i.

To visualize the uniformization process, consider a state i which can have state transitions to two other states only, j and k, as illustrated in Fig. 7.22. The corresponding transition rates are q_{ij} and q_{ik}, and let $\Lambda(i) = q_{ij} + q_{ik}$. This is the actual total probability flow rate out of state i. We also know that the corresponding transition probabilities are given by

$$P_{ij} = q_{ij}/\Lambda(i) \quad \text{and} \quad P_{ik} = q_{ik}/\Lambda(i)$$

Now, let us define a *uniform* rate γ such that

$$\gamma = q_{ij} + q_{ik} + \nu$$

where $\nu \geq 0$ is an arbitrary rate for the fictitious events we wish to introduce. The uniformized chain is a discrete-time Markov chain where the transition probabilities for state i are given by

$$P_{ij}^U = \left(\frac{\Lambda(i)}{\gamma}\right) P_{ij} = \frac{q_{ij}}{\gamma}, \qquad P_{ik}^U = \left(\frac{\Lambda(i)}{\gamma}\right) P_{ik} = \frac{q_{ik}}{\gamma},$$
$$P_{ii}^U = 1 - \left(\frac{\Lambda(i)}{\gamma}\right) = \frac{\nu}{\gamma}$$

where P_{ii}^U is the self-loop probability in the new chain (see Fig. 7.22). Observe that in the expressions above the uniformized transition probabilities are scaled versions of the original ones with the common scaling factor $\Lambda(i)/\gamma$.

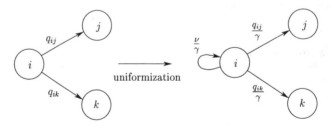

Figure 7.22: Illustrating the uniformization process.
The actual total transition rate out of state i is $\Lambda(i) = q_{ij} + q_{ik}$. We choose a uniform rate $\gamma = q_{ij} + q_{ik} + \nu$, for some $\nu \geq 0$. In the uniformized chain, we allow a self-loop with probability ν/γ, corresponding to fictitious events leaving the state unaffected. The uniformized transition probabilities are the original transition rates scaled by $1/\gamma$.

In summary, given a continuous-time Markov chain with state space $\mathcal{X}$ and transition rate matrix $\mathbf{Q}$, we can construct a stochastically equivalent uniformized discrete-time Markov chain by selecting a uniform transition rate

$$\gamma \geq \max_{i \in \mathcal{X}}\{-q_{ii}\} \tag{7.114}$$

and transition probabilities

$$
P_{ij}^U = \begin{cases} \dfrac{q_{ij}}{\gamma} & \text{if } i \neq j \\[2mm] 1 + \dfrac{q_{ij}}{\gamma} & \text{if } i = j \end{cases} \tag{7.115}
$$

Alternatively, if the Markov chain is specified through transition probabilities $P_{ij}, i \neq j$, and state holding time parameters $\Lambda(i)$, then the uniformization process becomes

$$
\gamma \geq \max_{i \in \mathcal{X}} \{\Lambda(i)\} \tag{7.116}
$$

$$
P_{ij}^U = \begin{cases} \dfrac{\Lambda(i)}{\gamma} P_{ij} & \text{if } i \neq j \\[2mm] 1 - \dfrac{\Lambda(i)}{\gamma} & \text{if } i = j \end{cases} \tag{7.117}
$$

Although we will not explicitly prove that the uniformized chain and the original continuous-time chain are indeed stochastically equivalent, the following example provides a verification of this fact for the steady-state probabilities of homogeneous birth-death chains.

Example 7.21 (Uniformization of a birth-death chain)
Let us consider a homogeneous birth-death chain, as shown in Fig. 7.23. Here, the birth and death rate parameters are fixed at λ and μ respectively. We can now make use of the steady-state solution in equations (7.109) and (7.110). The expression for π_0 becomes

$$
\pi_0 = \frac{1}{1 + \sum_{j=1}^{\infty} \left(\frac{\lambda}{\mu}\right)^j}
$$

The sum in the denominator is a simple geometric series that converges as long as $\lambda/\mu < 1$. Under this assumption, we get

$$
\sum_{j=1}^{\infty} \left(\frac{\lambda}{\mu}\right)^j = \frac{\lambda/\mu}{1 - \lambda/\mu}
$$

and, therefore,

$$
\pi_0 = 1 - \frac{\lambda}{\mu}
$$

Then, from (7.109),

$$
\pi_j = (1 - \lambda/\mu)(\lambda/\mu)^j, \quad j = 0, 1, \ldots \tag{7.118}
$$

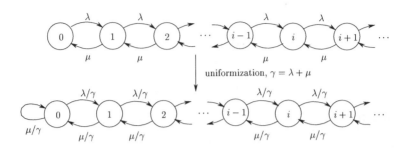

Figure 7.23: Uniformization of birth-death chain of Example 7.21.

Now let us consider a uniformized version of this model. Looking at Fig. 7.23, observe that the total probability flow rate out of every state $i > 0$ is $\Lambda(i) = -q_{ii} = \lambda + \mu$. For state 0, we have $\Lambda(0) = \lambda$. Let us, therefore, select a uniform rate equal to the maximum of all $\Lambda(i)$:

$$\gamma = \lambda + \mu$$

Using equation (7.115), we determine the transition probabilities for the uniformized chain. For all states $i > 0$, we have:

$$P_{i,i+1}^U = \lambda/\gamma, \quad P_{i,i-1}^U = \mu/\gamma, \quad P_{i,i}^U = 0$$

and for state 0:

$$P_{01}^U = \lambda/\gamma, \quad P_{00}^U = \mu/\gamma$$

as shown in Fig. 7.23. We now recognize the discrete-time birth-death chain of Fig. 7.9 that was analyzed in Example 7.13 The steady-state probabilities were obtained in (7.48), and will be denoted here by π_j^U:

$$\pi_j^U = \frac{2p-1}{p} \left(\frac{1-p}{p}\right)^j, \quad j = 0, 1, 2, \dots$$

where p is the probability of a death. In our case, $p = \mu/\gamma$. Thus, substituting for $p = \mu/\gamma$ above, we get

$$\pi_j^U = \frac{2\mu - \gamma}{\mu} \left(\frac{\gamma - \mu}{\mu}\right)^j$$

and since $\gamma = \lambda + \mu$, this becomes

$$p_j^U = (1 - \lambda/\mu)(\lambda/\mu)^j, \quad j = 0, 1, \dots$$

which is precisely what we found in (7.118) to be the steady-state probability of state j in the original chain. We have therefore shown that the steady-state probabilities of the original chain and its uniformized version are indeed identical.

Example 7.22

As a further illustration of the uniformization process, consider the Markov chain of Example 7.19 shown once again in Fig. 7.24. The state transition diagram of the uniformized chain can be obtained by inspection, and is also shown in Fig. 7.24. In this case, a convenient uniform rate is

$$\gamma = \lambda_1 + \lambda_2 + \mu_1 + \mu_2$$

Then, all transition probabilities in the uniform version are obtained by dividing the original transition rates by γ, and by introducing self-loops as indicated in the diagram.

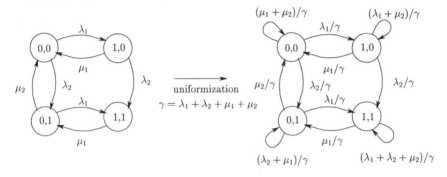

Figure 7.24: Uniformization of chain of Example 7.22.

SUMMARY

- The *n-step transition probabilities* of a discrete-time Markov chain, denoted by $p_{ij}(k, k+n) = P[X_{k+n} = j \mid X_k = i]$, satisfy the *Chapman-Kolmogorov equation*

$$p_{ij}(k, k+n) = \sum_{\text{all } r} p_{ir}(k, u) p_{rj}(u, k+n), \quad k \le u \le k+n$$

- We obtain a *homogeneous* Markov chain whenever the transition probability $p_{ij}(k)$ is independent of k for all i, j. Most of the results in this chapter apply to this case only.

- A consequence of the memoryless property is the fact that state holding times in discrete-time Markov chains are geometrically distributed with parameter p_{ii}.

- A Markov chain is *irreducible* if any state is reachable from any other state by some sequence of transitions. Otherwise, the chain is *reducible*.

If the chain visits a state from which it can never again leave, that state is called absorbing.

- States can be classified as *transient, positive recurrent*, or *null recurrent*. A transient state is revisited with probability less than 1, whereas a recurrent state is revisited with probability 1. The expected recurrence time for a positive recurrent state is finite, whereas for a null recurrent state it is infinite.

- States can also be classified as *periodic* or *aperiodic*. If visits to some state are constrained to occur only in a number of steps which is a multiple of an integer greater than or equal to 2, that state is said to be periodic.

- In an irreducible aperiodic Markov chain consisting of positive recurrent states, a unique stationary state probability vector π exists, which is obtained by solving $\pi = \pi P$ subject to the normalization condition $\sum_j \pi_j = 1$.

- For continuous-time Markov chains, the *Chapman-Kolmogorov equation* applies to *transition functions* $p_{ij}(s,t) = P[X(t) = j \mid X(s) = i]$, $s \geq t$:

$$p_{ij}(s,t) = \sum_{\text{all } r} p_{ir}(s,u)p_{rj}(u,t), \qquad s \leq u \leq t$$

In the homogeneous case, $p_{ij}(s,t)$ depends only on the difference $(t - s)$ for all i, j.

- State holding times in homogeneous continuous-time Markov chains are exponentially distributed.

- A continuous-time Markov chain may be specified either by its transition rate matrix $\mathbf{Q}$, or by defining transition probabilities P_{ij} for all pairs of states $i \neq j$ along with parameters $\Lambda(i)$ for all states i, characterizing the exponential state holding time distributions. Denoting by q_{ij} the (i,j) entry of $\mathbf{Q}$, we have $P_{ij} = -q_{ij}/q_{ii}$, and $\Lambda(i) = -q_{ii}$.

- In an irreducible continuous-time Markov chain consisting of positive recurrent states, a unique stationary state probability vector π exists, which is obtained by solving $\pi \mathbf{Q} = 0$ subject to the normalization condition $\sum_j \pi_j = 1$.

- A *birth-death chain* is a continuous-time Markov chain whose transition rate matrix $\mathbf{Q}$ satisfies $q_{ij} = 0$ for all $j > i+1$ and $j < i-1$. We define the birth rate to be $\lambda_j = q_{j,j+1} > 0$, and the death rate $\mu_j = q_{j,j-1} > 0$. A homogeneous pure birth chain (i.e., when $\mu_j = 0$ and $\lambda_j = \lambda = $ constant) reduces to a Poisson process. Under certain conditions on its parameters, the stationary probabilities of a birth-death chain can be obtained in the closed form (7.109) and (7.110).

- A continuous-time Markov chain can be converted into an equivalent discrete-time chain through the *uniformization* process: A uniform transition rate γ is selected such that $\gamma \geq -q_{ii}$ for all i, and the uniformized transition probabilities are given by q_{ij}/γ for $i \neq j$, $(1 + q_{ii}/\gamma)$ for $i = j$.

PROBLEMS

7.1 The breeding habits of every family in a certain species are as follows: The family has a birth at most once a year; if at the end of a year the family has no children, there is a 50% chance of a birth in the next year; if it has one child, there is a 40% chance of a birth in the next year; if it has two children, there is a 10% chance of a birth in the next year; if it has three or more children, there is never a birth in the next year; when there is a birth, there is a 2% chance of twins, but there are never any triplets, quadruplets, etc. in this species.

Let X_k denote the number of children in such a family at the end of the kth year. Assume that $X_0 = 0$ and that no children ever die for the range of years we are interested in.

(a) What is the state space of this chain?

(b) Determine the transition probabilities $P[X_2 = j \mid X_1 = i]$ for all possible values of i, j. Is this a Markov chain? If so, is it a homogeneous Markov chain?

(c) Compute the probability that the family has one child at the end of the second year.

7.2 An electronic device driven by a clock randomly generates one of three numbers, 0, 1, or 2, with every clock tick according to the following rules: If 0 was generated last, then the next number is 0 again with probability 0.5 or 1 with probability 0.5; if 1 was generated last, then the next number is 1 again with probability 0.4 or 2 with probability 0.6; if 2 was generated last, then the next number is either 0 with probability 0.7 or 1 with probability 0.3. Moreover, before the clock starts ticking, a number is generated from the distribution: 0 with probability 0.3, 1 with probability 0.3, 2 with probability 0.4. The device is attached to a bomb set to explode the first time the sequence {1, 2, 0 } takes place. There is a display on the device showing the most recent number generated.

(a) Draw a state transition diagram for this chain showing all transition probabilities.

(b) Compute the probability that the bomb explodes after exactly two clock ticks.

(c) Compute the probability that the bomb explodes after exactly three clock ticks.

(d) Suppose you are handed the bomb with the number 1 showing. What is the probability that the bomb explodes after exactly two clock ticks? What is the probability that it explodes after exactly three clock ticks?

7.3 The following game is played with a regular 52-card deck. A card is drawn with a player trying to guess its color (red or black). If the player guesses right, he gets 1 dollar, otherwise he gets nothing. After a card is drawn it is set aside (it is not placed back in the deck). Suppose the player decides to adopt the simple policy "guess red all the time", and let X_k denote his total earnings after k cards are drawn, with $X_0 = 0$.

(a) Is $\{X_k\}$ a Markov chain? If so, is it a homogeneous Markov chain?

(b) Compute $P[X_7 = 6 \mid X_6 = 5]$.

(c) Compute $P[X_9 = 4 \mid X_8 = 2]$.

(d) Derive an expression for $P[X_{k+1} = j \mid X_k = i]$ for all $i, j = 0, 1, 2, \ldots$, and $k = 0, 1, \ldots, 51$.

7.4 Repeat Problem 7.3 with the following modification in the game: After a card is drawn from the deck, it is replaced by a card from another deck; this card is chosen to be of the opposite color.

7.5 The following is used to model the genetic evolution of various types of populations. The term "cell" is used to denote a unit in that population. Let X_k be the total number of cells in the kth generation, $k = 1, 2, \ldots$ The ith cell in a particular generation acts independently of all other cells and it either dies or it splits up into a random number of offspring. Let Y_i be an integer-valued non-negative random variable, so that $Y_i = 0$ indicates that the ith cell dies without reproducing, and $Y_i > 0$ indicates the resulting number of cells after reproduction. If $X_k = n$ and $Y_i = 0$ for all cells $i = 1, \ldots, n$ in the kth generation, we get $X_{k+1} = 0$, and we say that the population becomes extinct.

(a) Is $\{X_k\}$ a Markov chain? If so, is it a homogeneous Markov chain?

(b) Suppose $X_0 = 1$ and the random variable Y_i can only take three values, 0, 1, or 2, each with probability 1/3. Draw a state transition diagram for the chain and calculate the transition probabilities $P[X_{k+1} = j \mid X_k = i]$ for $i = 0, 1, 2, 3$.

(c) Compute the probability that the population becomes extinct after four generations.

(d) Compute the probability that the population remains in existence for more than three generations.

7.6 A manufacturing workstation is viewed as a single-server queueing system with a buffer capacity of three parts (including one in process). In a slotted time model (as in Example 7.2) , with probability 0.3 a new part arrives at the workstation in any one slot, and with probability 0.5 a part in process is done (at most one part can arrive or complete processing in a time slot). A part arriving and finding no buffering space is simply rejected and lost. A part completing processing is either removed from the workstation with probability 0.5 or it immediately goes through an inspection process that works as follows: With probability 0.5 a part is good and is removed; with probability 0.3 a part is slightly defective and it is returned to the workstation buffer for rework; and with probability 0.2 a part is hopelessly defective and it is simply scrapped. Let X_k denote the number of parts in the workstation at time $k = 1, 2, \ldots$, and assume that $X_0 = 0$.

(a) Draw a state transition diagram for this chain and compute all transition probabilities involved.

(b) Calculate the probability that there is only one part in the workstation at the second slot.

7.7 Consider the following transition probability matrix:

$$
P = \begin{bmatrix}
0 & 1 & 0 & 0 & 0 & 0 \\
1/4 & 1/4 & 1/2 & 0 & 0 & 0 \\
0 & 0 & 1/2 & 1/2 & 0 & 0 \\
0 & 0 & 1/8 & 0 & 7/8 & 0 \\
0 & 0 & 0 & 4/5 & 1/5 & 0 \\
0 & 0 & 0 & 0 & 1/3 & 2/3
\end{bmatrix}
$$

(a) Is this an irreducible or reducible Markov chain? Is it periodic or aperiodic?

(b) Determine which states are transient and which states are recurrent.

(c) Identify all closed irreducible sets of the chain.

7.8 The weather in some area is classified as either "sunny", "cloudy", or "rainy" on any given day. Let X_k denote the weather state on the kth day, $k = 1, 2, \ldots$, and map these three states into the numbers 0, 1, 2 respectively. Suppose transition probabilities from one day to the next are summarized through the matrix:

$$
P = \begin{bmatrix}
0.4 & 0.4 & 0.2 \\
0.5 & 0.3 & 0.2 \\
0.1 & 0.5 & 0.4
\end{bmatrix}
$$

(a) Draw a state transition diagram for this chain.

(b) Assuming the weather is cloudy today, predict the weather for the next two days.

(c) Find the stationary state probabilities (if they exist) for this chain. If they do not exist, explain why this is so.

(d) If today is a sunny day, find the average number of days we have to wait at steady state until the next sunny day.

7.9 A total of N gas molecules are split between two containers so that at any time step $k = 0, 1, \ldots$ there are X_k molecules in the first container and $(N - X_k)$ molecules in the second one. At each time step, one of the N molecules is randomly selected, independently from all previous selections, and moved from its container to the other container. This type of model is known as the *Ehrenfest chain*.

(a) Determine the transition probability $P[X_{k+1} = j \mid X_k = i]$ for all $i, j = 0, 1, \ldots, N$.

(b) Draw a state transition diagram for this chain when $N = 3$.

(c) Find the stationary state probabilities (if they exist) for this chain. If they do not exist, explain why this is so.

7.10 Repeat Problem 7.9 with the following modification in the molecule movement: After a molecule is selected, it either remains in its container with probability p or it is moved to the other container with probability $1 - p$. Do part (c) by assuming that $p = 1/3$.

7.11 A gambler plays the following game: He tosses two fair dice and if a total of 5 shows he wins one dollar, otherwise he loses one dollar. He has decided that he will quit as soon as he has either won 5 dollars or lost 2 dollars. Calculate the probability that he has won 5 dollars when he quits.

7.12 Recall the definition of the n-step transition probability $p_{ij}^n = P[X_{k+n} = j \mid X_k = i]$ for a homogeneous Markov chain in Section 7.2.3, and the definition of the hitting time $T_{ij} = \min\{k > 0 : X_0 = i, X_k = j\}$ in Section 7.2.8.

(a) Prove that

$$p_{ij}^n = \sum_{m=1}^{n} P[T_{ij} = m \mid X_0 = i] p_{jj}^{n-m}$$

(b) If j is an absorbing state, then show that

$$p_{ij}^n = P[T_{ij} \leq n \mid X_0 = i]$$

7.13 Consider a Markov chain $\{X_k\}, k = 0, 1, \ldots$ with the property

$$\sum_{\text{all } j} j \cdot p_{ij} = i$$

for all states i, and assume that the state space is $\{0, 1, \ldots, N\}$ for some finite integer N. This is an example of a *martingale*, a type of chain that was briefly discussed following Example 7.4.

(a) Show that $E[X_k] = E[X_{k+1}]$ for all $k = 0, 1, \ldots$

(b) Show that states 0 and N are absorbing.

(c) Assume that $X_0 = i$, for some $i \in \{1, 2, \ldots, N - 1\}$. Show that

$$\lim_{k \to \infty} E[X_k] = NP[T_{iN} \leq \infty]$$

(*Note*: You may find the results of Problem 7.12 useful.)

(d) Show that the absorption probability $P[T_{i0} < \infty]$ is given by $(1 - i/N)$.

7.14 A random number of messages A_k is placed in a transmitter at times $k = 0, 1, \ldots$ Each message present at time k acts independently of past arrivals and of other messages, and it is successfully transmitted with probability p. Thus, if X_k denotes the number of messages present at k, and $D_k \leq X_k$ denotes the number of successful transmissions at k, we can write

$$X_{k+1} = X_k - D_k + A_{k+1}$$

Assume that A_k has a Poisson distribution $P[A_k = n] = \lambda^n e^{-\lambda}/n!, n = 0, 1, \ldots$, independent of k.

(a) If the system is at state $X_k = i$, find $P[D_k = m \mid X_k = i], m = 0, 1, \ldots$

(b) Determine the transition probabilities $P[X_{k+1} = j \mid X_k = i], i, j = 0, 1, \ldots$

(c) If $X_0 = 0$, show that X_1 has a Poisson distribution with parameter λ, and that $(X_1 - D_1)$ has a Poisson distribution with parameter $\lambda(1 - p)$. Hence, determine the distribution of X_2.

(d) If X_0 has a Poisson distribution with parameter ν, find the value of ν such that there exists a stationary probability distribution for the Markov chain $\{X_k\}$.

7.15 A DES is specified as a stochastic timed automaton $(\mathcal{E}, \mathcal{X}, \Gamma, f, p_0, G)$ with

$$\mathcal{E} = \{\alpha, \beta, \gamma, \delta\}, \qquad \mathcal{X} = \{0, 1, 2, 3, 4\}$$

$$\begin{aligned}
\Gamma(0) &= \{\alpha, \beta\}, & f(0, \alpha) &= 1, & f(0, \beta) &= 2 \\
\Gamma(1) &= \{\beta, \gamma\}, & f(1, \beta) &= 2, & f(1, \gamma) &= 0 \\
\Gamma(2) &= \{\beta, \gamma, \delta\}, & f(2, \beta) &= 3, & f(2, \gamma) &= 1, \\
& & f(2, d) &= 3 \\
\Gamma(3) &= \{\alpha, \beta, \gamma, \delta\}, & f(3, \alpha) &= 1, & f(3, \beta) &= 4, \\
& & f(3, \gamma) &= 2, & f(3, \delta) &= 3 \\
\Gamma(4) &= \{\gamma, \delta\}, & f(4, \gamma) &= 3, & f(4, \delta) &= 0
\end{aligned}$$

The system is initially at state 0, and the event processes for $\alpha, \beta, \gamma, \delta$ are all Poisson with parameters $\lambda_1, \lambda_2, \lambda_3, \lambda_4$ respectively.

(a) Determine the transition probabilities P_{01}, P_{23}, P_{31}, and P_{42}.

(b) What is the average amount of time spent at state 1? At state 3?

(c) Write down the transition matrix $\mathbf{Q}$ for this Markov chain.

7.16 Consider a Markov chain $\{X(t)\}$ with state space $\mathcal{X} = \{0, 1, 2, 3, 4\}$ and transition rate matrix

$$\mathbf{Q} = \begin{bmatrix}
-\lambda & \lambda & 0 & 0 & 0 \\
0 & -(\lambda + \mu_1) & \lambda & \mu_1 & 0 \\
0 & 0 & -\mu_1 & 0 & \mu_1 \\
\mu_2 & 0 & 0 & -(\lambda + \mu_2) & \lambda \\
0 & \mu_2 & 0 & 0 & -\mu_2
\end{bmatrix}$$

(a) Draw a state transition rate diagram.

(b) Determine the stationary state probabilities (if they exist) for $\lambda = 1, \mu_1 = 3/2, \mu_2 = 7/4$.

(c) If steady state is reached and we know that the state is not 0, what is its expected value?

7.17 One third of all transactions submitted to the local branch of a bank must be handled by the bank's central computer, while all the rest can be handled by the branch's own computer. Transactions are generated according to a Poisson process with rate 1. The central and the branch's computer act independently, they can both buffer an infinite number of transactions, process one transaction at a time, and they are both assumed to require transaction processing times which are exponentially distributed with rates μ_1 and μ_2 respectively. The central computer, however, is also responsible for processing other transactions, which we will assume to be generated by another independent Poisson process with rate ν. Let $\lambda = 10$ transactions per minute, $\mu_1 = 50$ transactions per minute, $\mu_2 = 15$ transactions per minute, and $\nu = 30$ transactions per minute. Finally, let

$X_1(t), X_2(t)$ denote the number of transactions residing in the central and branch computer respectively (see also Problem 6.7).

(a) Do stationary probabilities exist for this process? Why or why not? If stationary probabilities do exist, then at steady state:

(b) Calculate the probability that the branch computer is idle.

(c) Calculate the probability that there are more that 3 transactions *waiting* to be processed by the central computer.

(d) Calculate the probability that both computers have a single transaction in process.

7.18 Consider a pure death chain $\{X(t)\}$ such that in equations (7.98) and (7.99) we have $\lambda_j = 0$ for all $j = 0, 1, \ldots$ and $\mu_j = j\mu, j = 1, 2, \ldots, N$, where N is a given initial state. Show that $\pi_j(t) = P[X(t) = j], j = 1, 2, \ldots, N$, has a binomial distribution with parameters N and $e^{-\mu t}$.

7.19 Use the uniformization process to derive a discrete-time Markov chain model for $\{X(t)\}$ in Problem 7.16, and obtain the stationary probabilities (if they exist). Compare the result to part (b) of Problem 7.16.

SELECTED REFERENCES

- Asmussen, S., *Applied Probability and Queues*, Wiley, New York, 1987.

- Bertsekas, D.P., *Dynamic Programming: Deterministic and Stochastic Models*, Prentice-Hall, Englewood Cliffs, NJ, 1987.

- Bertsekas, D.P., *Dynamic Programming and Optimal Control, Volumes 1 and 2*, Athena Scientific, Belmont, MA, 1995.

- Cinlar, E., *Introduction to Stochastic Processes*, Prentice-Hall, Englewood Cliffs, NJ, 1975.

- Clarke, A.B., and R.L. Disney, *Probability and Random Processes*, Wiley, New York, 1985.

- Gallager, R.G., *Discrete Stochastic Processes*, Kluwer Academic Publishers, Boston, 1996.

- Hoel, P.G., S.C. Port, and C.J. Stone, *Introduction to Stochastic Processes*, Houghton Mifflin, Boston, 1972.

- Kleinrock, L., *Queueing Systems, Volume I: Theory*, Wiley, New York, 1975.

- Parzen, E., *Stochastic Processes*, Holden-Day, San Francisco, 1962.

- Trivedi, K.S., *Probability and Statistics with Reliability, Queuing and Computer Science Applications*, Prentice-Hall, Englewood Cliffs, NJ, 1982.

Chapter 8

Introduction to Queueing Theory

8.1 INTRODUCTION

A simple queueing system was first introduced in Chapter 1 as an example of a DES. We have since repeatedly used it to illustrate many of the ideas and techniques discussed thus far. In this chapter, we will take a more in-depth look at queueing systems.

Queueing theory is a subject to which many books have been devoted. It ranges from the study of simple single-server systems modeled as birth-death chains to the analysis of arbitrarily complex networks of queues. Our main objective here is to present the essential ideas and techniques that are used to analyze simple queueing systems. The word "simple" in queueing theory is often associated with birth-death chain models which we have learned to analyze in the previous chapter. Thus, we will see that several interesting queueing systems can be viewed as special cases of these models. We will also provide some extensions necessary to deal with more complex situations involving event processes which do not satisfy the Markov property and with networks of queues connected in arbitrary ways.

Queueing systems form a very broad and practically useful class of DES, especially when dealing with problems of resource sharing frequently encountered in the design and control of computer, manufacturing, and communication systems, to name just a few examples. By no means, however, should we identify our study of DES with queueing theory. First of all, there are many DES in which queueing may not be the predominant phenomenon. Secondly, traditional queueing theoretic techniques have emphasized the use of stochastic models which do not necessarily reveal some of the basic facets of the dynamic behavior of DES. In addition, queueing theory has as its main goal the determination of a system's performance under certain operating conditions, rather than the determination of the operating policies to be used in order to achieve the best possible performance. Thus, its mission has been largely to develop "descriptive" tools for studying queueing systems, rather than "prescriptive" tools for controlling their behavior in an ever-changing dynamic and uncertain environment (something we will consider in the next chapter). Still, we must stress that queueing theory has made some of the most important contributions to the analysis of stochastic DES where resource contention issues are predominant.

8.2 SPECIFICATION OF QUEUEING MODELS

We begin by establishing the basic definitions and notation required to fully specify a queueing system model. There are normally three components to the "model specification" process:

1. Specification of *stochastic models* for the arrival and service processes.

2. Specification of the *structural parameters* of the system. For example, the storage capacity of a queue, the number of servers, and so forth.

3. Specification of the *operating policies* used. For example, conditions under which arriving customers are accepted, preferential treatment (prioritization) of some types of customers by the server, and so on.

We will discuss these issues in more detail, starting with the simplest cases first. In Fig. 8.1 we show a queueing system with infinite storage space for queueing and a single server. As we have seen, this is a DES with event set $\mathcal{E} = \{a, d\}$, where a is a customer arrival and d is a departure following a service completion.

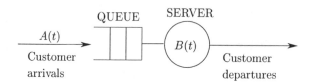

Figure 8.1: Simple queueing system.

8.2.1 Stochastic Models for Arrival and Service Processes

We have seen that in stochastic timed automata we always associate an event with an underlying clock sequence. In a queueing system, we associate with arrival events a a stochastic sequence $\{Y_1, Y_2, \ldots\}$ where Y_k is the kth *interarrival time*, that is, a random variable such that

$$Y_k = \text{time elapsed between the the } (k-1)\text{th and } k\text{th arrival}, k = 1, 2, \ldots$$

For simplicity, we always set $Y_0 = 0$, so that Y_1 is the random variable describing the time of the first arrival. As we know, in general the stochastic behavior of this sequence requires the specification of the joint probability distribution of all events $[Y_k \leq t], k = 1, 2, \ldots$ In most of queueing theory it is assumed that the stochastic sequence $\{Y_k\}$ is iid (i.e., $Y_1, Y_2, \ldots$ are independent and identically distributed). Therefore, a single probability distribution

$$A(t) = P[Y \leq t] \tag{8.1}$$

completely describes the interarrival time sequence. In (8.1), the random variable Y is often thought of as a "generic" interarrival time which does not need to be indexed by k. The mean of the distribution function $A(t), E[Y]$, is particularly important and it is customary to use the notation

$$E[Y] \equiv \frac{1}{\lambda} \tag{8.2}$$

to represent it. Thus, λ is the *average arrival rate* of customers.

Similarly, we associate to the event d a stochastic sequence $\{Z_1, Z_2, \ldots\}$ where Z_k is the kth service time, that is, a random variable such that

$$Z_k = \text{time required for the } k\text{th customer to be served}, k = 1, 2, \ldots$$

If we assume that the stochastic sequence $\{Z_k\}$ is iid, then we define

$$B(t) = P[Z \leq t] \tag{8.3}$$

where Z is a generic service time. Similar to (8.2), we use the following notation for the mean of $B(t)$:

$$E[Z] \equiv \frac{1}{\mu} \tag{8.4}$$

so that μ is the *average service rate* of the server in our system.

8.2.2 Structural Parameters

Typical structural parameters of interest in a queueing system are:

1. The *storage capacity* of the queue, usually denoted by $K = 1, 2, \ldots$ By convention, we normally agree to include in this storage capacity the space provided for customers in service.

2. The *number of servers*, usually denoted by $m = 1, 2, \ldots$.

In the simple system of Fig. 8.1, we have:

- $K = \infty$, that is, infinite storage capacity for queued customers.

- $m = 1$, that is, a single server.

Other structural parameters will be introduced as we consider more complex systems later in this chapter.

8.2.3 Operating Policies

Even for the simple system shown in Fig. 8.1, there are various schemes one can adopt in handling the queueing process. Here are just some of the natural questions that arise about the operation of the system:

- Is the service time distribution the same for all arriving customers?

- Are customers differentiated on the basis of belonging to several different "classes", some of which may have a higher priority than others in requesting service?

- Are all customers admitted to the system?

- Are customers allowed to leave the queue before they get served, and, if so, under what conditions?

- How does the server decide which customer to serve next (if there are more than one in the queue), for example, the first one in queue, anyone at random, and so forth?

- Is the server allowed to preempt a customer in service in order to serve a higher-priority customer that just arrived?

Clearly, operating policies can cover a wide range of possibilities. One can conceive of several imaginative schemes that could be adopted, and some may be better than others depending on the design objectives for a particular system.

In categorizing operating policies, the most common issues we consider are the following:

1. *Number of customer classes.* In the case of a *single-class* system, all customers have the same service requirements and the server treats them all equally. This means that the service time distribution is the same

for all customers. In the case of a multiple-class system, customers are distinguished according to their service requirements and/or the way in which the server treats them.

2. *Scheduling policies.* In a multiple-class system, the server must decide upon a service completion which class to process next. For example, the server may always give priority to a particular class, or it may preempt a customer in process because a higher-priority customer just arrived.

3. *Queueing disciplines.* A queueing discipline describes the order in which the server selects customers to be processed, even if there is only a single class. For example, first-come-first-served (FCFS), last-come-last-served (LCFS), and random order.

4. *Admission policies.* Even if a queue has infinite storage capacity, it may be desirable to deny admission to some arriving customers. In the case of two arriving classes, for instance, higher priority customers may always be admitted, but lower priority customers may only be admitted if the queue is empty or if some amount of time has elapsed since such a customer was admitted.

In the simple case of Fig.8.1, we assume:

- A single-class system.

- All arriving customers are admitted and served.

- The queueing discipline is first-come-first-served (FCFS).

8.2.4 The $A/B/m/K$ Notation

It is customary in queueing theory to employ a particular type of notation in order to succinctly describe a system. This notation is as follows:

$$A/B/m/K$$

where:

A is the interarrival time distribution
B is the service time distribution
m is the number of servers present, $m = 1, 2, \ldots$
K is the storage capacity of the queue, $K = 1, 2, \ldots$

Thus, the infinite queueing capacity single-server system of Fig. 8.1 is described by $A/B/1/\infty$. To simplify the notation, if the K position is omitted it is understood that $K = \infty$. Therefore, in our case, we have $A/B/1$. Furthermore, there is some common notation used to represent the distributions A and B:

> G stands for a *General* distribution when nothing else is known about the arrival/service process.
>
> GI stands for a *General* distribution in a *renewal* arrival/service process (i.e., all interarrival/service times in that process are iid)
>
> D stands for the *Deterministic* case, (i.e., the interarrival/service times are fixed.)
>
> M stands for the *Markovian* case, i.e., the interarrival/service times are exponentially distributed.

These are the most often encountered cases. The reader should be cautioned however that the distinction between G and GI is not always made explicit, that is, many authors use G while implicitly assuming a renewal process. Additional notation for specific distributions will be introduced later as needed. Here are some examples to illustrate the $A/B/m/K$ notation:

> $M/M/1$ A single-server system with infinite storage capacity. The interarrival and service times are both exponentially distributed, that is, $A(t) = 1 - e^{-\lambda t}, B(t) = 1 - e^{-\mu t}$, for some positive parameters λ and μ.
>
> $M/M/1/K$ A single-server system with storage capacity equal to $K < \infty$ (including space at the server). The interarrival and service times are both exponentially distributed.
>
> $M/G/2$ A system with two servers and infinite storage capacity. The interarrival times are exponentially distributed. The service times have an arbitrary (general) distribution.

Note that this notation does not specify the operating policies to be used. It also does not describe the case where two or more customer classes are present, each with different interarrival and service time distributions. When this notation is used and no particular policy is specified, it is to be understood that we are dealing with a single-class system, no admission control (unless no more storage is available), and a FCFS queueing discipline.

Finally, recall that exponentially distributed interevent times imply that the underlying event process is Poisson. In the case of arrival events that are permanently feasible, the statement "the arrival process is Poisson" is the same as "interarrival times are exponentially distributed". In the case of departure events generated by a Poisson process, we use the statement "service times are exponentially distributed", because these events are only feasible when the server is busy.

8.2.5   Open and Closed Queueing Systems

The system of Fig.8.1 is accessible to any customer from the outside world wishing to be served. In other words, the system is open to an infinite population of customers who can request service at any time. We refer to this as an *infinite customer population* or *open* queueing system. In contrast, the two server system shown in Fig. 8.2 is limited to a finite population of customers, usually denoted by $N = 1, 2, \ldots$ In this case, a customer completing service at server 1 always returns for more service and never leaves the system. We refer to this as a *finite customer population* or *closed* queueing system.

In some cases, a single server operates as a finite population system. A typical customer experiences a service time with distribution B, then goes through some delay whose distribution is A before rejoining the queue. Thus, A is intended to emulate an arrival process, but the number of customers served remains fixed at N. We will examine such queueing systems in Section 8.6. To represent them, we extend the $A/B/m/K$ notation above to $A/B/m/K/N$ where N represents the customer population. As in the case of infinite storage, omitting N implies an open system, (i.e., $N = \infty$). Note that if $K = \infty$, but $N < \infty$, we normally write $A/B/m//N$.

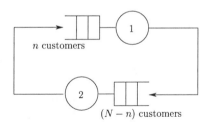

Figure 8.2: A closed queueing system serving N customers.

The customer population N in a closed system should be viewed as a structural parameter along with the storage capacity K and number of servers m. Closed queueing systems, however, should not be thought of as strictly consisting of a particular set of fixed customers. Instead, the population N may indicate a number of resources limiting access to more customers. A typical example arises in modeling a computer system with N terminals. Here, the total number of users is limited to N, but users certainly come and go replacing each other at various terminals. Similarly, in a manufacturing system production parts are often carried in pallets whose number is limited to N. When a finished part leaves the system, it relinquishes its pallet to a new part, so that the effective number of customers in the system is limited to N.

8.3 PERFORMANCE OF A QUEUEING SYSTEM

Let us concentrate on the single-server open queueing system of Fig. 8.1. We have already defined two random variables of interest regarding the kth customer: the *interarrival time* Y_k and the *service time* Z_k. In addition, we define:

A_k is the *arrival time* of kth customer

D_k is the *departure time* of kth customer

W_k is the *waiting time* of kth customer (from arrival instant until beginning of service)

S_k is the *system time* of kth customer (from arrival instant until departure)

Note that

$$S_k = D_k - A_k \tag{8.5}$$

The system time is often also referred to as *response time* or *delay*. It also follows from the definition of S_k that

$$S_k = W_k + Z_k \tag{8.6}$$

and, by combining (8.5) and (8.6), we also have

$$D_k = A_k + W_k + Z_k \tag{8.7}$$

In addition, we define the random variables:

$X(t)$ is the *queue length* at time t, $X(t) \in \{0, 1, 2, \dots\}$

$U(t)$ is the *workload* at time t, that is, the amount of time required to empty the system at t

Note that $X(t)$ is the quantity we normally use to describe the state of a timed queueing system (as in the stochastic timed automaton model of Chapter 6). The workload $U(t)$ is also referred to as the *unfinished work* of the system at t.

The stochastic behavior of the waiting time sequence $\{W_k\}$ provides important information regarding the system's performance. The probability distribution function of $\{W_k\}$, $P[W_k \leq t]$, generally depends on k. We often find, however, that as $k \to \infty$ there exists a stationary distribution, $P[W \leq t]$, independent of k, such that:

$$\lim_{k \to \infty} P[W_k \leq t] = P[W \leq t] \tag{8.8}$$

If this limit indeed exists, the random variable W describes the waiting time of a typical customer at steady state. Intuitively, when the system runs for a sufficiently long period of time (equivalently, the system has processed a sufficiently large number of customers), every new customer experiences a "stochastically identical" waiting process described by $P[W \leq t]$. The mean of this distribution, $E[W]$, represents the *average waiting time* at steady state. Similarly, if a stationary distribution exists for the system time sequence $\{S_k\}$, then its mean, $E[S]$, is the *average system time* at steady state.

The same idea applies to the stochastic processes $\{X(t)\}$ and $\{U(t)\}$. If stationary distributions exist for these processes as $t \to \infty$, then the random variables X and U are used to describe the queue length and workload of the system at steady state. We will use the notation $\pi_n, n = 0, 1, \ldots$, to denote the stationary queue length probability, that is,

$$\pi_n = P[X = n], \quad n = 0, 1, \ldots \tag{8.9}$$

Accordingly, $E[X]$ is the *average queue length* at steady state, and $E[U]$ is the *average workload* at steady state.

In general, we want to design a queueing system so that a typical customer at steady state waits as little as possible. Ideally, we would like every arriving customer to wait zero time. In a stochastic environment this is impossible, since we are not in a position to provide an infinite number of servers or a super-server operating infinitely fast. Thus, we usually settle for the fastest server we can afford. To justify our investment, we strive to serve as many customers as possible by keeping the server as busy as possible, that is, we try to maximize the server's utilization. To do so, we must constantly keep the queue non-empty; in fact, we should make sure there are always a few customers to serve in case several service times in a row turn out to be short. But this is directly contrary to our objective of achieving zero waiting time for customers. This informal argument (which we will soon explicitly quantify) serves to illustrate a fundamental tradeoff in the design and control of queueing systems: To keep a server highly utilized we must be prepared to tolerate long waiting times; conversely, to maintain low waiting times we have to tolerate some server idling.

In most cases we will consider, we assume that steady state can be reached, and so we concentrate on the performance of the queueing system at steady state. This gives rise to performance measures such as:

- the *average waiting time* of customers at steady state, $E[W]$,

- the *average system time* of customers at steady state, $E[S]$,

- the *average queue length* at steady state, $E[X]$,

all of which it is desirable to keep as small as possible. Some additional performance measures are:

- the *utilization* of the system, that is, the fraction of time that the server is busy,

- the *throughput* of the system, that is, the rate at which customers leave the system after service,

which it is desirable to keep as large as possible. Of course, the utilization can never exceed 1. Similarly, the throughput can never be higher than the maximum service rate of the server. Moreover, if a system has reached steady state, the input and output customer flows must be balanced. In this case, the throughput must be identical to the rate of incoming customers.

To gain some more insight on the utilization of a queueing system, let us define the *traffic intensity* as follows:

$$[\text{traffic intensity}] \equiv \frac{[\text{average arrival rate}]}{[\text{average service rate}]}$$

For the single-server system of Fig. 8.1, we use ρ to denote the traffic intensity. Then, by the definitions of λ and μ in (8.2) and (8.4), we have

$$\rho = \frac{\lambda}{\mu} \tag{8.10}$$

In the case of m servers, the average service rate becomes $m\mu$, and, therefore,

$$\rho = \frac{\lambda}{m\mu} \tag{8.11}$$

Now the utilization of a server is the fraction of time that it is busy. In a single-server system at steady state, the probability π_0, defined in (8.9), represents the fraction of time the system is empty, and hence the server is idle. It follows that for a server at steady state:

$$[\text{utilization}] \equiv [\text{fraction of time server is busy}] = 1 - \pi_0$$

Since a server operates at rate μ and the fraction of time that it is actually in operation is $(1 - \pi_0)$, the throughput of a single-server system at steady state is

$$[\text{throughput}] \equiv [\text{departure rate of customers after service}] = \mu(1 - \pi_0)$$

At steady state, the customer flows into and out of the system must be balanced, that is,

$$\lambda = \mu(1 - \pi_0)$$

It then follows from (8.10) that

$$\rho = 1 - \pi_0 \tag{8.12}$$

Thus, the traffic intensity, which is defined by the parameters of the service and interarrival time distributions, also represents the utilization of the system. This relationship holds for any single-server system with infinite storage capacity. If $\pi_0 = 1$, the system is permanently idle because of no customer arrivals ($\lambda = 0$). If $\pi_0 = 0$, the system is permanently busy, which generally leads to instabilities in that the queue length grows to infinity. Thus, as we shall explicitly see later on, the values of ρ must be such that

$$0 \leq \rho < 1$$

The performance measures above are the most common ones we shall encounter. We will introduce additional measures as we look into more complicated systems than that of Fig. 8.1.

A typical design problem for queueing systems is the selection of parameters such as the service rate and number of servers in order to achieve some desirable performance in terms of the measures above. In controlling queueing systems, our task is to select operating policies that help us achieve such performance.

8.4   QUEUEING SYSTEM DYNAMICS

Consider once again the queueing system of Fig. 8.1, operating on a FCFS basis. We have already seen typical sample paths of this queueing system when we modeled it as a stochastic timed automaton in Chapters 5 and 6. Using the notation established in the previous section, a typical sample path is shown once again in Fig. 8.3 (we use lower case letters to represent the values taken by random variables on a sample path).

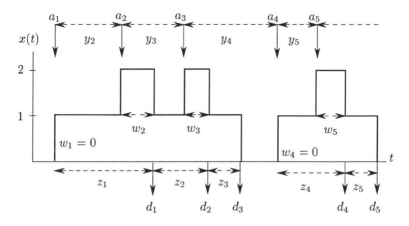

Figure 8.3: A typical queueing system sample path.

In this example, the first arriving customer at time a_1 finds an empty queue.

During the interval starting at a_1 and ending with d_3 the server remains busy. Such an interval is termed a *busy period* of the queueing system. During the interval starting with d_3 and ending with the next arrival at a_4 the server remains idle. We term this an *idle period* of the system. We can see that one way to view this system is as a sequence of alternating cycles each consisting of a busy period followed by an idle period.

Taking a closer look at Fig. 8.3 helps us identify the basic dynamic mechanism of this queueing system. When the kth customer arrives two cases are possible:

Case 1. The system is empty, therefore $W_k = 0$. The system can only be empty when

$$D_{k-1} \leq A_k$$

that is, the previous customer departed before the current customer arrived. Thus:

$$D_{k-1} - A_k \leq 0 \Leftrightarrow W_k = 0 \tag{8.13}$$

This is clearly seen in Fig. 8.3 with the case $w_4 = 0$, which is a result of $d_3 < a_4$.

Case 2. The system is not empty, therefore, $W_k > 0$. In this case, the kth customer is forced to wait until the previous, i.e., $(k-1)$th, customer departs. Thus,

$$D_{k-1} - A_k > 0 \Leftrightarrow W_k = D_{k-1} - A_k \tag{8.14}$$

This situation arises with $w_2 = d_1 - a_2 > 0$ in Fig. 8.3, as well as w_3 and w_5.

Combining (8.13) and (8.14), we obtain:

$$W_k = \begin{cases} 0 & \text{if } D_{k-1} - A_k \leq 0 \\ D_{k-1} - A_k & \text{if } D_{k-1} - A_k > 0 \end{cases} \tag{8.15}$$

which can be conveniently rewritten as

$$W_k = \max\{0, D_{k-1} - A_k\} \tag{8.16}$$

Now using (8.7) for the $(k-1)$th customer on a sample path, and recalling that $A_k - A_{k-1} = Y_k$, we obtain the following recursive expression for waiting times:

$$W_k = \max\{0, W_{k-1} + Z_{k-1} - Y_k\} \tag{8.17}$$

Using (8.6), we can rewrite this relationship for system times:

$$S_k = \max\{0, S_{k-1} - Y_k\} + Z_k \tag{8.18}$$

Finally, we can obtain a similar recursive expression for departure times. From (8.7):

$$W_k = D_k - A_k - Z_k$$

and (8.16) becomes

$$D_k = \max\{A_k, D_{k-1}\} + Z_k \tag{8.19}$$

These relationships capture the essential dynamic characteristics of a queueing system. In (8.17), for instance, we see that if we did not have to account for $W_k = 0$ when $W_{k-1} + Z_{k-1} - Y_k < 0$, waiting times would satisfy a simple linear equation. Thus, waiting time dynamics are linear except for the effect of a customer with interarrival time such that $Y_k > W_{k-1} + Z_{k-1} = S_{k-1}$. Equation (8.17) is referred to as *Lindley's equation*, after D. V. Lindley who studied queueing dynamics in the 1950s (Lindley, 1952).

Also note that these relationships are very general, in the sense that they apply on any sample path of the system regardless of the distributions characterizing the various stochastic processes involved. It is also interesting to observe that equation (8.19) was previously derived in (5.42) from the timed Petri net model of the queueing system. In that case, service times were provided by the clock sequence of the service completion event d.

Finally, returning to the stochastic timed automaton model of a queueing system, note that the stochastic clock structure is equivalent to specifying the probability distributions $A(t)$ and $B(t)$ for the arrival and departure events defining the event set $\mathcal{E} = \{a, d\}$. There are few results we can obtain for queueing systems which are general enough to hold regardless of the nature of these distributions. Aside from the relationships derived above, there is one more general result which we discuss in the next section.

8.5 LITTLE'S LAW

Consider once again the queueing system of Fig. 8.1 with event set $\mathcal{E} = \{a, d\}$. Recall that the event scores $N_a(t)$ and $N_d(t)$ count the number of arrivals and departures respectively in the interval $(0, t]$. Assuming that the system is initially empty, it follows that the queue length $X(t)$ is given by

$$X(t) = N_a(t) - N_d(t) \tag{8.20}$$

Typical sample paths of the score processes $N_a(t)$ and $N_d(t)$ are shown in Fig. 8.4. The difference between them is always positive except during idle periods when it becomes 0.

Let us concentrate on a particular sample path such as that of Fig. 8.4 (again, lower case letters represent the values of the random variables on a sample path). Looking at the shaded area, note that the rectangle of unit height (one customer) and length defined by the first arrival and first departure on the time axis represents the amount of time the first customer spent in the system. A similar rectangle is formed by the second customer, and so on. In general, the shaded area consisting of such rectangles up to time t represents the total amount of time all customers have spent in the system by time t. We denote this area by $u(t)$. Thus:

$$u(t) = \text{total amount of time all customers have spent in the system by time } t$$

Dividing this area by the total number of customers that have arrived in $(0, t]$, $n_a(t)$, we obtain the average system time per customer by time t, denoted by $\bar{s}(t)$:

$$\bar{s}(t) = \frac{u(t)}{n_a(t)} \tag{8.21}$$

Since $u(t)$ has customer-second units and $n_a(t)$ is the number of customers, $\bar{s}(t)$ has units of time (seconds).

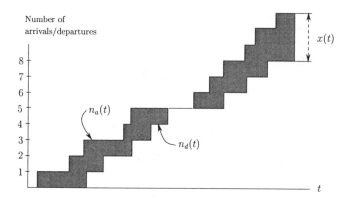

Figure 8.4: Sample paths of arrival and departure score processes, $N_a(t)$ and $N_d(t)$. *The queue length is always given by $x(t) = n_a(t) - n_d(t)$. Observe that each rectangle formed with height $= 1$ customer and length $= (k$th departure time - kth arrival time), $k = 1, 2, \ldots$, represents the amount of time spent by the kth customer in the system.*

Similarly, dividing the area $u(t)$ by t, we obtain the average number of customers present in the system over the interval $(0, t]$, that is, the average queue length along this sample path, $\bar{x}(t)$:

$$\bar{x}(t) = \frac{u(t)}{t} \tag{8.22}$$

Finally, dividing the total number of customers that have arrived in $(0, t]$, $n_a(t)$, by t, we obtain the *average customer arrival rate*, denoted by $\lambda(t)$:

$$\lambda(t) = \frac{n_a(t)}{t} \tag{8.23}$$

Combining (8.21) through (8.23) gives

$$\bar{x}(t) = \lambda(t)\bar{s}(t) \tag{8.24}$$

We now make two crucial assumptions. We assume that as $t \to \infty$, $\lambda(t)$ and $\bar{s}(t)$ both converge to fixed values λ and $\bar{s}$ respectively, that is, the following limits exist:

$$\lim_{t\to\infty} \lambda(t) = \lambda \tag{8.25}$$

$$\lim_{t\to\infty} \bar{s}(t) = \bar{s} \tag{8.26}$$

These values represent the steady-state average arrival rate and system time respectively for a given sample path. If these limits exist, then, by (8.24), $\bar{x}(t)$ must also converge to a fixed value $\bar{x}$. Therefore,

$$\bar{x} = \lambda\bar{s} \tag{8.27}$$

Now (8.27) is true for a particular sample path we selected. Suppose, however, that the limits in (8.25) and (8.26) exist for all possible sample paths and for the same fixed values of λ and $\bar{s}$, and hence $\bar{x}$. In other words, we are assuming that the arrival, system time, and queue length processes are all ergodic (see Appendix I). In this case, $\bar{x}$ is actually the mean queue length $E[X]$ at steady state, and $\bar{s}$ is the mean system time $E[S]$ at steady state. Moreover, λ is the mean arrival rate which was defined in (8.2). We may then rewrite (8.27) as

$$E[X] = \lambda E[S] \tag{8.28}$$

This is known as *Little's Law*. It is a powerful result in that it is independent of the stochastic clock structure of the system, that is, the probability distributions associated with the arrival and departure events. In words, (8.28) states that the mean queue length is proportional to the mean system time with the mean arrival rate λ being the constant of proportionality.

The preceding derivation is hardly a proof of the fundamental relationship (8.28), but it does capture its essence. In fact, this relationship was taken for granted for many years even though it was never formally proved. Formal poofs finally started appearing in the literature in the 1960s (e.g., Little, 1961; Stidham, 1974).

It is important to observe that (8.28) is independent of the operating policies employed in the queueing system under consideration. Moreover, it holds for

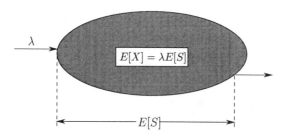

Figure 8.5: Illustrating Little's Law for arbitrary queueing system configurations. *At steady state, the expected queue length is proportional to the expected system time. The arrival rate λ is the constant of proportionality.*

an arbitrary configuration of interconnected queues and servers. Suppose we draw a boundary around part or all of a queueing system and let λ be the total arrival rate of customers crossing the boundary into the system (see Fig. 8.5). Then (8.28) still holds with $E[X]$ being the mean total number of customers in the system, and $E[S]$ the mean total system time (from the point where a customer crosses the boundary into the system to the point where the customer crosses it again to leave the system). This implies that Little's Law holds for a single queue (server not included) as follows:

$$E[X_Q] = \lambda E[W] \qquad (8.29)$$

where $E[X_Q]$ is the mean queue content (without the server) and $E[W]$ is the mean waiting time. Similarly, if our system is defined by a boundary around a single server, we have:

$$E[X_S] = \lambda E[Z] \qquad (8.30)$$

where $E[X_S]$ is the mean server content (between 0 and 1 for a single server) and $E[Z]$ is the mean service time.

Two final words of caution regarding Little's Law. First, it is tempting to observe a queueing system over some period of time, measure the average system time and arrival rate, and then invoke (8.28) to obtain the mean queue length. However, *Little's Law holds only at steady state*, so one must first be certain that the limits in (8.25) and (8.26) are actually reached (unless one wants to use (8.24) instead). Second, the arrival rate λ pertains to customers *actually entering* the system. Thus, if arriving customers are subject to some admission control or are rejected due to finite storage space, one must be careful to apply Little's Law for the rate of customers admitted at steady state.

8.6   SIMPLE MARKOVIAN QUEUEING SYSTEMS

Most interesting performance measures, such as those presented in Section 8.3, can be evaluated from the stationary queue length probability distribution $\pi_n = P[X = n], n = 0, 1, \ldots$, of the system. Obtaining this distribution (if it exists) is therefore a major objective of queueing theory.

We have seen that a simple queueing system, such as the one of Fig. 8.1, may be viewed as a stochastic timed automaton with event set $\mathcal{E} = \{a, d\}$ and state space $\mathcal{X} = \{0, 1, \ldots\}$. The state $X(t)$ represents the queue length, and $\{X(t)\}$ forms a Generalized Semi-Markov Process (GSMP). The evaluation of its stationary state distribution is generally a difficult task that can be tackled in one of two ways:

1. Given a specific stochastic clock structure, that is, some particular probability distributions characterizing the events a and d, we attempt to develop explicit analytical expressions for the stationary state probabilities π_n.

2. By simulating or directly observing sample paths of the system, we estimate the stationary state probabilities π_n. In this case, if the observation period is of length T and the amount of time spent at state n is T_n, then an estimate of π_n is the fraction T_n/T, provided T is "sufficiently long".

We will limit ourselves here to the first approach, leaving the sample-path-based estimation methods to Chapter 10. In particular, we assume that the interarrival and service time distributions are both exponential with parameters λ and μ respectively. As we saw in Chapter 6, the GSMP then reduces to a Markov chain. We have analyzed Markov chains quite extensively in the previous chapter, and queueing theory provides an opportunity to apply some of our results. We refer to queueing systems with exponential interarrival and service time distributions as "Markovian" for obvious reasons.

Recall that the state transition function for our simple queueing system is such that $f(x, a) = x + 1$, and $f(x, d) = x - 1$ as long as $x > 0$. Thus, we may think of an arrival event as a birth and a departure event as a death, and hence model the system as a birth-death chain. This allows us to make immediate use of the results of Section 7.4.3. The key equations, pertaining to the stationary state probabilities $\pi_n, n = 0, 1, \ldots$, are rewritten below:

$$\pi_n = \left(\frac{\lambda_0 \cdots \lambda_{n-1}}{\mu_1 \cdots \mu_n} \right) \pi_0, \qquad n = 1, 2, \ldots \tag{8.31}$$

$$\pi_0 = \frac{1}{1 + \sum_{n=1}^{\infty} \left(\frac{\lambda_0 \cdots \lambda_{n-1}}{\mu_1 \cdots \mu_n} \right)} \tag{8.32}$$

where λ_n and μ_n are the birth and death rates respectively when the state is n.

Before proceeding with applications of this general result to Markovian queueing systems, let us discuss yet another property of the Poisson process, which has significant implications in the performance analysis of some of the systems we will be studying.

Looking once again at the single-server system of Fig. 8.1, consider the event

$$[\text{arriving customer at time } t \text{ finds } X(t) = n]$$

which is to be compared to the event

$$[\text{system state at some time } t \text{ is } X(t) = n]$$

The distinction may be subtle, but it is critical. In the first case, the observation of the system state (queue length) occurs at specific time instants (arrivals) which depend on the nature of the arrival process. In the second case, the observation of the system state occurs at random time instants. In general,

$$P[\text{arriving customer at time } t \text{ finds } X(t) = n]$$
$$\neq P[\text{system state at some time } t \text{ is } X(t) = n]$$

However, equality does hold for a Poisson arrival process (regardless of the service time distribution, as long as the arrival and service processes are independent). This is shown next, following the approach of Kleinrock (1975). It is also known as the PASTA property (Poisson Arrivals See Time Averages). We will use our standard notation:

$$\pi_n(t) = P[\text{system state at some time } t \text{ is } X(t) = n]$$

In addition, let us define

$$\alpha_n(t) \equiv P[\text{arriving customer at time } t \text{ finds } X(t) = n] \qquad (8.33)$$

Theorem 8.1 For a queueing system with a Poisson arrival process independent of the service process, the probability that an arriving customer finds n customers in the system is the same as the probability that the system state is n, that is,

$$\alpha_n(t) = \pi_n(t) \qquad (8.34)$$

Proof. Let $a(t, \Delta t)$ denote the event [arrival occurs in the interval $(t, t + \Delta t]$]. Thus, when $\Delta t \to 0$, the event $a(t, \Delta t)$ becomes the event [arrival at time t]. Now observe that

$$\alpha_n(t) = \lim_{\Delta t \to 0} P[X(t) = n \mid a(t, \Delta t)]$$

which can be rewritten as

$$\alpha_n(t) = \lim_{\Delta t \to 0} \frac{P[X(t) = n \text{ and } a(t, \Delta t)]}{P[a(t, \Delta t)]}$$

We now exploit the nature of the Poisson process, which we know to have stationary increments (Section 6.4), and its independence from the service process. Thus, the number of arrivals in $(t, t + \Delta t]$ is independent of t and of any past arrivals which have contributed to the queue length $X(t)$. Therefore, $a(t, \Delta t)$ is independent of $[X(t) = n]$. It follows that

$$a_n(t) = \lim_{\Delta t \to 0} \frac{P[X(t) = n] \cdot P[a(t, \Delta t)]}{P[a(t, \Delta t)]} = \lim_{\Delta t \to 0} P[X(t) = n]$$
$$= \lim_{\Delta t \to 0} \pi_n(t) = \pi_n(t)$$

and (8.34) is obtained.

Q.E.D.

Clearly, this property extends to steady state (if it exists), for which most of our performance measures are defined. We shall make use of (8.34) in several of the cases we will examine in what follows.

In the next few sections, we analyze several Markovian queueing systems of practical interest. As we shall see, our analysis is essentially a series of direct applications of (8.31)-(8.32) above.

8.6.1   The $M/M/1$ Queueing System

Using the notation we have established, this is a single-server system with infinite storage capacity and exponentially distributed interarrival and service times. It can therefore be modeled as a birth-death chain with birth and death parameters

$$\lambda_n = \lambda \quad \text{for all } n = 0, 1, \dots$$
$$\mu_n = \mu \quad \text{for all } n = 1, 2, \dots$$

where λ is the arrival rate of customers to the system and μ is the service rate. The state transition rate diagram for this system is shown in Fig. 8.6.

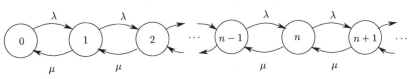

Figure 8.6: State transition rate diagram for the $M/M/1$ system.

It follows from (8.32) that

$$\pi_0 = \frac{1}{1 + \sum_{n=1}^{\infty} \left(\frac{\lambda}{\mu}\right)^n}$$

The sum in the denominator is a simple geometric series that converges as long as $\lambda/\mu < 1$. Under this assumption, we get

$$\sum_{n=1}^{\infty} \left(\frac{\lambda}{\mu}\right)^n = \frac{\lambda/\mu}{1-\lambda/\mu}$$

and, therefore,

$$\pi_0 = 1 - \frac{\lambda}{\mu} \qquad (8.35)$$

Let us set $\rho = \lambda/\mu$, which is the traffic intensity defined in (8.10). Thus, we see that

$$\pi_0 = 1 - \rho \qquad (8.36)$$

which is in agreement with (8.12). Note also that $0 \le \rho < 1$, since (8.35) was derived under the assumption $\lambda/\mu < 1$.

Next, using (8.36) in (8.31), we obtain

$$\pi_n = \left(\frac{\lambda}{\mu}\right)^n (1-\rho)$$

or

$$\pi_n = (1-\rho)\rho^n, \qquad n = 0, 1, \ldots \qquad (8.37)$$

Equation (8.37) gives the stationary probability distribution of the queue length of the $M/M/1$ system.

The condition that ρ be such that $0 \le \rho < 1$ is consistent with the general requirement $S_1 < \infty$, $S_2 = \infty$, discussed in Section 7.4.3, where

$$S_1 = 1 + \sum_{n=1}^{\infty} \left(\frac{\lambda_0 \cdots \lambda_{n-1}}{\mu_1 \cdots \mu_n}\right) = 1 + \sum_{n=1}^{\infty} \left(\frac{\lambda}{\mu}\right)^n$$

and

$$S_2 = \sum_{n=1}^{\infty} \left(\frac{\mu_1 \cdots \mu_n}{\lambda_1 \cdots \lambda_n}\right) \frac{1}{\lambda_n} = \sum_{n=1}^{\infty} \left(\frac{\mu}{\lambda}\right)^n \frac{1}{\lambda}$$

We can then see that $S_1 < \infty$ and $S_2 = \infty$, provided $\rho = \lambda/\mu < 1$, which we refer to as the *stability condition* for the $M/M/1$ system.

We are now in a position to obtain explicit expressions for various performance measures of this system.

Utilization and Throughput

The utilization is immediately given by (8.36), since $1 - \pi_0 = \rho$. The throughput is the departure rate of the server, which is $\mu(1 - \pi_0) = \lambda$. This is to be expected since at steady state the arrival and departure rates are balanced. Thus, for a stable $M/M/1$ system, the throughput is simply the arrival rate λ. On the other hand, if we allow $\lambda > \mu$, then the throughput is simply μ, since the server is constantly operating at rate μ.

Average Queue Length

This is the expectation of the random variable X whose distribution is given by (8.37). Thus,

$$E[X] = \sum_{n=0}^{\infty} n\pi_n = (1 - \rho) \sum_{n=0}^{\infty} n\rho^n \tag{8.38}$$

We can evaluate the preceding sum by observing that

$$\frac{d}{d\rho} \left(\sum_{n=0}^{\infty} \rho^n \right) = \sum_{n=0}^{\infty} n\rho^{n-1} = \frac{1}{\rho} \sum_{n=0}^{\infty} n\rho^n$$

Since

$$\sum_{n=0}^{\infty} \rho^n = \frac{1}{1 - \rho}$$

and

$$\frac{d}{d\rho} \left(\frac{1}{1 - \rho} \right) = \frac{1}{(1 - \rho)^2}$$

we get

$$\sum_{n=0}^{\infty} n\rho^n = \frac{\rho}{(1 - \rho)^2}$$

Then, (8.38) gives

$$E[X] = \frac{\rho}{1 - \rho} \tag{8.39}$$

Note that as $\rho \to 1$, $E[X] \to \infty$, that is, the expected queue length grows to ∞. This clearly reveals the tradeoff we already identified earlier: As we attempt to keep the server as busy as possible by increasing the utilization ρ, the quality of service provided to a typical customer declines, since, on the average, such a customer sees an increasingly longer queue length ahead of him.

Average System Time

Using (8.39) and Little's Law in (8.28), we get:

$$\frac{\rho}{1 - \rho} = \lambda E[S]$$

or, since $\lambda = \rho\mu$,

$$E[S] = \frac{1/\mu}{1 - \rho} \tag{8.40}$$

As $\rho \to 0$, we see that $E[S] \to 1/\mu$, which is the average service time. This is to be expected, since at very low utilizations the only delay experienced by a typical customer is a service time. We also see that as $\rho \to 1, E[S] \to \infty$. Once again, this is a manifestation of the tradeoff between utilization and system time: The higher the utilization (good for the server), the higher the average system time (bad for the customers).

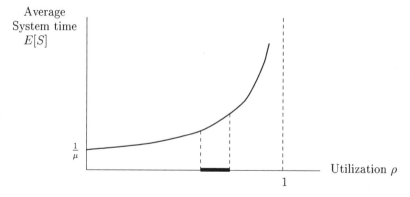

Figure 8.7: Average System Time as a function of utilization in the $M/M/1$ system. *This illustrates the tradeoff between keeping the utilization high (good for the server) and keeping the average system time low (good for customers). As ρ increases, $E[S]$ is initially insensitive to changes (the slope is close to 0). It then suddenly becomes very sensitive (as the slope rapidly approaches ∞). Therefore, a good operating area is the one marked by the shaded interval.*

The system time vs. utilization tradeoff is clearly illustrated in Fig. 8.7, where $E[S]$ in (8.40) is plotted as a function of the utilization (or traffic intensity) ρ. One cannot overemphasize the importance of this relationship. Even though it has been derived only for the $M/M/1$ system, it is representative of most similar queueing systems regardless of models used for the arrival and service processes. Inspection of Fig. 8.7 suggests some simple rules about designing efficient queueing systems. First, note that for a considerable range of values of ρ, the average system time does not significantly vary (the slope of the curve remains small). This, however, tends to change rather suddenly as ρ approaches 1 and $E[S]$ becomes very sensitive to changes in the traffic intensity. Consequently, a good area to operate is *just before the slope of the curve begins to take off* (see, for example, the shaded interval on the utilization axis in Fig. 8.7); this guarantees adequate utilization without excessive customer waiting. Moreover, it prevents the system from being too sensitive to small changes in ρ which can result in very sudden and exceedingly large customer delays. Due to the nonlinear nature of this relationship, the issue of sensitivity is crucial in queueing systems. Our day-to-day life experience (traffic jams, long ticket lines, etc.) suggests that when a queue length starts building up, it tends to build up very fast. This is a result of operating in the range of ρ values where the additional increase in the arrival rate causes drastic increases in $E[S]$.

Average Waiting Time

It follows from (8.6) at steady state that $E[S] = E[W] + E[Z] = E[W] + 1/\mu$. Then, from (8.40) we get

$$E[W] = \frac{1/\mu}{1-\rho} - \frac{1}{\mu}$$

or

$$E[W] = \frac{\rho}{\mu(1-\rho)} \tag{8.41}$$

As expected, we see once again that as $\rho \to 1$, $E[W] \to \infty$, that is, increasing the system utilization towards its maximum value leads to extremely long average waiting times for customers.

Example 8.1
We wish to select the service rate μ of a computer processor providing service to customers submitting jobs at a rate $\lambda = 1$ job/sec so that the average job system time does not exceed 0.5 sec. We assume that job arrivals form a Poisson process and that service times are exponentially distributed. Thus, we can adopt an $M/M/1$ system model.

Using (8.40) with $\rho = 1/\mu$, we must choose μ so that

$$E[S] = \frac{1/\mu}{1 - 1/\mu} < 0.5$$

which implies that

$$0.5\mu > 1.5$$

and, therefore, $\mu > 3$ jobs/sec.

Example 8.2

In a communication system, a transmission line capacity (i.e., transmission speed) is $C = 1200$ bits/sec. Arriving messages to be transmitted form a Poisson process. A message (or packet) consists of L bits, where L is a random variable (the length of the message). We assume that L is exponentially distributed with mean 600 bits. The problem is to determine the maximum arrival rate we can sustain (in messages/sec) in order to guarantee an average message waiting time of less than 1 sec.

Here, the server is the transmission line. We first determine its service rate μ in messages/sec. Note that on the average a message requires L/C sec to be transmitted, therefore $\mu = C/L$ messages/sec or $\mu = 2$ messages/sec.

Using (8.41) with $\rho = \lambda/2$, we have

$$E[W] = \frac{\lambda/2}{2(1 - \lambda/2)} < 1$$

which implies that $\lambda < 4/3$ messages/sec. Under this condition, the system must therefore operate at a utilization of $\rho < 2/3$.

Before leaving the $M/M/1$ system, we briefly discuss the issue of determining the transient solution for the queue length probabilities $\pi_n(t) = P[X(t) = n], n = 0, 1, \ldots$ This requires solving the set of flow balance equations (7.98) and (7.99) with $\lambda_n = \lambda$ and $\mu_n = \mu$. We can easily rederive these equations by inspection of the state transition rate diagram in Fig. 8.6:

$$\frac{d\pi_n(t)}{dt} = -(\lambda + \mu)\pi_n(t) + \lambda\pi_{n-1}(t) + \mu\pi_{n+1}(t), \quad n = 1, 2, \ldots \quad (8.42)$$

$$\frac{d\pi_0(t)}{dt} = -\lambda\pi_0(t) + \mu\pi_1(t) \quad (8.43)$$

Obtaining the solution $\pi_n(t), n = 0, 1, \ldots$, of these equations is a tedious task. We provide the final result below to give the reader an idea of the complexity involved even for the simplest of all interesting queueing systems we can consider:

$$\pi_n(t) = e^{-(\lambda+\mu)t}\left[\rho^{(n-i)/2}J_{n-i}(at) + \rho^{(n-i-1)/2}J_{n+i+1}(at)\right.$$

$$\left. + (1-\rho)\rho^n \sum_{j=n+i+2}^{K}\rho^{-j/2}J_j(at)\right] \quad (8.44)$$

where the initial condition is

$$\pi_i(0) = P[X(0) = i] = 1$$

for some given $i = 0, 1, \ldots,$ and

$$a = 2\mu\rho^{1/2}$$

$$J_n(x) = \sum_{k=0}^{\infty} \frac{(x/2)^{n+2k}}{(n+k)!k!}, \qquad n = -1, 0, 1, \ldots$$

Here, $J_n(x)$ is a modified Bessel function which makes the evaluation of (8.44) particularly complicated.

8.6.2   The $M/M/m$ Queueing System

This is a system with infinite storage capacity and m identical servers, as illustrated in Fig. 8.8. Upon arrival, a customer is served by any one available server. If all servers are busy, the customer is queued until the next departure frees a server. The interarrival times are exponentially distributed with rate λ. The service times at each server are also exponentially distributed with rate μ. Note, however, that the effective service rate varies depending on the state of the system: If there are $n < m$ customers present, then there are n servers busy and the service rate is $n\mu$; if, however, $n \geq m$ customers are present, the service rate is fixed at its maximum value $m\mu$. We can therefore model the system as a birth-death chain with birth and death parameters

$$\lambda_n = \lambda \quad \text{for all } n = 0, 1, \ldots$$
$$\mu_n = \begin{cases} n\mu & \text{if } 0 \leq n < m \\ m\mu & \text{if } n \geq m \end{cases}$$

where λ is the arrival rate of customers to the system and μ is the service rate of each server. The state transition rate diagram for this system is shown in Fig. 8.9.

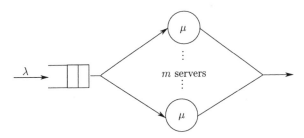

Figure 8.8: The $M/M/m$ queueing system.

To evaluate π_0 from (8.32) note that the infinite sum must be separated into two parts: a finite sum up to $n = m-1$ with varying death rates $\mu, 2\mu, \ldots, (m-$

$1)\mu$, and an infinite sum starting with $n = m$ with fixed death rates $m\mu$. Thus,

$$\pi_0 = \left[1 + \sum_{n=1}^{m-1} \frac{\lambda_n}{(\mu)(2\mu)\cdots(n\mu)} + \frac{\lambda^{m-1}}{(m-1)!\mu^{m-1}} \sum_{n=m}^{\infty} \left(\frac{\lambda}{m\mu}\right)^{n-m+1}\right]^{-1}$$

The infinite sum above is a geometric series that converges as long as $\lambda/m\mu < 1$. It is easy to see that, as in the $M/M/1$ case, this condition is equivalent to the requirement that $S_1 < \infty, S_2 = \infty$ discussed in Section 7.4.3. It is also consistent with the intuition that the arrival rate should not exceed the maximum possible service rate of the system, which in this case is $m\mu$.

Let us now set

$$\rho = \frac{\lambda}{m\mu}$$

where $0 \leq \rho < 1$. The infinite sum in the expression above becomes:

$$\sum_{n=m}^{\infty} \left(\frac{\lambda}{m\mu}\right)^{n-m+1} = \sum_{n=m}^{\infty} \rho^{n-m+1} = \frac{\rho}{1-\rho}$$

and we get

$$\pi_0 = \left[1 + \sum_{n=1}^{m-1} \frac{(m\rho)^n}{n!} + \frac{(m\rho)^{m-1}}{(m-1)!} \frac{\rho}{1-\rho}\right]^{-1}$$

which yields

$$\pi_0 = \left[1 + \sum_{n=1}^{m-1} \frac{(m\rho)^n}{n!} + \frac{(m\rho)^m}{m!} \frac{1}{1-\rho}\right]^{-1} \tag{8.45}$$

In order to obtain π_n for $n = 1, 2, \ldots$, we distinguish two cases in (8.31). First, if $n < m$, we have

$$\pi_n = \left(\frac{\lambda^n}{(\mu)(2\mu)\cdots(n\mu)}\right) \pi_0$$
$$= \frac{(\lambda/\mu)^n}{n!} \pi_0, \quad n = 1, 2, \ldots, m-1 \tag{8.46}$$

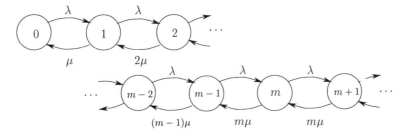

Figure 8.9: State transition rate diagram for the $M/M/m$ system.

and if $n \geq m$,

$$
\begin{aligned}
\pi_n &= \left(\frac{\lambda^{m-1}}{(\mu)(2\mu)\cdots(m-1)\mu} \right) \left(\frac{\lambda^{n-m+1}}{(m\mu)^{n-m+1}} \right) \pi_0 \\
&= \frac{(\lambda/\mu)^{m-1}}{(m-1)!} \left(\frac{\lambda}{m\mu} \right)^{n-m+1} \pi_0 \qquad (8.47) \\
&= \frac{m^m}{m!} \left(\frac{\lambda}{m\mu} \right)^n \pi_0, \qquad n = m, m+1, \ldots
\end{aligned}
$$

Using $\rho = \lambda/m\mu$ and combining (8.46) and (8.47), we get

$$
\pi_n = \begin{cases} \pi_0 \dfrac{(m\rho)^n}{n!} & n = 1, 2, \ldots, m-1 \\[2ex] \pi_0 \dfrac{m^m}{m!} \rho^n & n = m, m+1, \ldots \end{cases} \qquad (8.48)
$$

Equations (8.45) and (8.48) provide the stationary probability distribution of the queue length of the $M/M/m$ system. We can now evaluate several useful performance measures for this system.

Utilization and Throughput

Let B be the random variable denoting the number of busy servers, $B \in \{0, 1, \ldots, m\}$. We calculate the average number of busy servers, $E[B]$, as follows:

$$
E[B] = \sum_{n=0}^{m-1} n\pi_n + mP[X \geq m]
$$

The probability that at least m customers are present, $P[X \geq m]$, is given by

$$
P[X \geq m] = \sum_{n=m}^{\infty} \pi_n = \sum_{n=m}^{\infty} \frac{m^m}{m!} \rho^n \pi_0 = \frac{m^m}{m!} \frac{\rho^m}{1-\rho} \pi_0
$$

Therefore,

$$
\begin{aligned}
E[B] &= \left[\sum_{n=0}^{m-1} n \frac{(m\rho)^n}{n!} + m \frac{(m\rho)^m}{m!} \frac{1}{1-\rho} \right] \pi_0 \\
&= \left[0 + m\rho + \sum_{n=2}^{m-1} \frac{(m\rho)^n}{(n-1)!} + \frac{(m\rho)^m}{m!} \frac{m}{1-\rho} \right] \pi_0 \\
&= m\rho \left[1 + \sum_{n=2}^{m-1} \frac{(m\rho)^{n-1}}{(n-1)!} + \frac{(m\rho)^{m-1}}{m!} \frac{m}{1-\rho} \right] \pi_0
\end{aligned}
$$

Replace the summation index n by $j = n - 1$, and add and subtract the term $(m\rho)^{m-1}/(m-1)!$ to get

$$
\begin{aligned}
E[B] &= m\rho \left[1 + \sum_{j=1}^{m-1} \frac{(m\rho)^j}{j!} + \frac{(m\rho)^{m-1}}{m!} \frac{m}{1-\rho} - \frac{(m\rho)^{m-1}}{(m-1)!} \right] \pi_0 \\
&= m\rho \left[1 + \sum_{j=1}^{m-1} \frac{(m\rho)^j}{j!} + \frac{(m\rho)^m}{m!} \frac{1}{1-\rho} \right] \pi_0
\end{aligned}
$$

Looking at (8.45), note that the term in brackets above is $1/\pi_0$. Therefore, the average number of busy servers in the system is given by:

$$
E[B] = m\rho = \frac{\lambda}{\mu} \tag{8.49}
$$

It follows that an individual server in the system has a utilization given by $E[B]/m = \rho$.

The throughput of the system is once again given by λ, since, at steady state, the arrival and departure rates must be balanced (assuming $\rho < 1$).

Average Queue Length

This is the expected value of the random variable X, that is,

$$
E[X] = \sum_{n=0}^{\infty} n \pi_n
$$

where π_n is given by (8.45) and (8.48). This sum can be evaluated (we omit details) to yield

$$
E[X] = m\rho + \frac{(m\rho)^m}{m!} \frac{\rho}{(1-r)^2} \pi_0 \tag{8.50}
$$

As expected, $E[X] \to 0$ as $\rho \to 0$, and $E[X] \to \infty$ as $\rho \to 1$.

Average System Time

Using (8.50) and Little's Law in (8.28), we get

$$m\rho + \frac{(m\rho)^m}{m!} \frac{\rho}{(1-\rho)^2} \pi_0 = \lambda E[S]$$

or

$$E[S] = \frac{1}{\mu} + \frac{1}{\mu} \frac{(m\rho)^m}{m!} \frac{\pi_0}{m(1-\rho)^2} \tag{8.51}$$

Note that $E[S] \to 1/\mu$ as $\rho \to 0$, since at low traffic intensity values arriving customers always find a server available and their system time is limited to their service time.

Queueing Probability

This is the probability that an arriving customer does not find an idle server, and is therefore forced to wait in the queueing area. We denote this probability by P_Q. It is an important performance measure for this system, which is readily evaluated by taking advantage of our result in (8.34) for Poisson arrivals. Since the probability that an arriving customer finds the system at state n for any $n \geq m$ is simply π_n, we have

$$P_Q = P[X \geq m] = \sum_{n=m}^{\infty} \pi_n$$

which was already obtained earlier in our evaluation of $E[B]$ in (8.49):

$$P_Q = \frac{(m\rho)^m}{m!} \frac{\pi_0}{1-\rho} \tag{8.52}$$

This performance measure finds wide applicability in telephony. The expression in (8.52) is known as the *Erlang C formula*, after the pioneering work of A. K. Erlang in the earlier part of the century. Typically, we are given a call rate λ and the average duration of a telephone call, $1/\mu$. The formula is used to determine the number of required telephone lines (trunks) in order to guarantee that P_Q is less than some specified level p. Thus, we need to determine m such that $P_Q(m) < p$ in (8.52).

8.6.3   The $M/M/\infty$ Queueing System

This may be viewed as a special case of the $M/M/m$ system with $m = \infty$. Proceeding as in the previous section, we adopt a birth-death model with parameters:

$$\lambda_n = \lambda \quad \text{for all } n = 0, 1, \ldots$$
$$\mu_n = n\mu \quad \text{for all } n = 1, 2, \ldots$$

where λ is the arrival rate of customers to the system and μ is the service rate of each server. The state transition rate diagram for this system is shown in Fig. 8.10.

We first evaluate π_0 from (8.32):

$$\pi_0 = \left[1 + \sum_{n=1}^{\infty} \frac{\lambda^n}{(\mu)(2\mu)\cdots(n\mu)}\right]^{-1} = \left[1 + \sum_{n=1}^{\infty} \frac{(\lambda/\mu)^n}{n!}\right]^{-1}$$

Note that the term in brackets above defines the exponential function $e^{\lambda/\mu}$, provided that $\lambda/\mu < \infty$. We now set $\rho = \lambda/\mu$, and get

$$\pi_0 = e^{-\rho} \tag{8.53}$$

Note, however, that ρ is *not* the traffic intensity in this case. Here, we use the symbol ρ only for convenience.

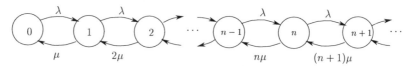

Figure 8.10: State transition rate diagram for the $M/M/\infty$ system.

From (8.53) and (8.31), we obtain

$$\pi_n = e^{-\rho}\frac{\rho^n}{n!}\,, \quad n = 0, 1, \ldots \tag{8.54}$$

which is the stationary queue length distribution for this system. Interestingly, this is simply a Poisson distribution with parameter ρ. We can now evaluate several useful performance measures.

Utilization and Throughput

From (8.53), we immediately obtain the system utilization $1 - \pi_0 = 1 - e^{-\rho}$. As in previous cases, the throughput is simply the arrival rate λ, which in this system is allowed to take on any arbitrarily large finite value.

Average Queue Length

This is the expectation of the random variable X, which we have determined in (8.54) to have a Poisson distribution. Therefore, its mean is known to be

$$E[X] = \rho = \frac{\lambda}{\mu} \tag{8.55}$$

Of course, in this system there is no physical queueing area, since every customer immediately receives service. Thus, X represents the number of busy servers in the system.

Average System Time

Using (8.55) and Little's Law in (8.28), we get

$$\rho = \lambda E[S]$$

and, therefore,

$$E[S] = \frac{1}{\mu} \tag{8.56}$$

This is to be expected, since every arriving customer always finds a server and never has to wait. Hence, the customer's average system time is limited to the average service time $1/\mu$.

Obviously, no real system can have an infinite number of servers. However, the $M/M/\infty$ model is particularly useful in representing physical situations where customers are indeed never queued. This arises, for instance, in transmitting messages over a communication medium. In this case, arriving messages are continuously placed in the medium without delay. If the Poisson arrival rate is λ messages/sec, and the transmission time is exponentially distributed with mean $1/\mu$ sec, then the transmission process can be modeled as an $M/M/\infty$ system. A similar situation arises in some material handling processes in manufacturing, such as perpetually moving belts. In this case, once again, every arriving customer always finds space on the belt without experiencing any waiting. We can see that the $M/M/\infty$ system serves as a "pure delay" modeling element, where every customer simply undergoes some processing without waiting.

8.6.4   The $M/M/1/K$ Queueing System

In this case, the number of customers that can be accommodated in the system is limited to K, as illustrated in Fig. 8.11. An arriving customer that finds the system full is rejected and considered to be lost. This phenomenon is generally referred to as customer *blocking*. Otherwise, the model is similar to the $M/M/1$ case. We can still use a birth-death model to analyze this system by setting the arrival rate $\lambda_n = \lambda$ for all $n = 0, 1, \dots, K - 1$, and $\lambda_n = 0$ for all $n \geq K$. In effect, we shut off the Poisson arrival process while the queue length is K, and turn it back on when it becomes $(K - 1)$. Thus,

$$\lambda_n = \begin{cases} \lambda & \text{if } 0 \leq n < K \\ 0 & \text{if } n \geq K \end{cases}$$
$$\mu_n = \mu \qquad \text{for all } n = 1, 2, \dots, K$$

The state transition rate diagram for this system is shown in Fig. 8.12. The "trick" of shutting off arrivals only works with a Poisson process because of the memoryless property: At the time we turn the process back on, the residual interarrival time of the actual process has the same distribution as the entire interarrival time, and the Markovian structure of the state transition rate diagram in Fig. 8.11 (in particular, the transition from $(K-1)$ to K) is preserved.

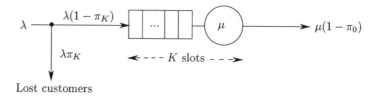

Lost customers

Figure 8.11: The $M/M/1/K$ queueing system.

The probability π_0 is obtained from (8.32):

$$\pi_0 = \left[1 + \sum_{n=1}^{K} \left(\frac{\lambda}{\mu}\right)^n\right]^{-1}$$

The sum in this expression is a finite geometric series which we can evaluate as follows:

$$\sum_{n=1}^{K} \left(\frac{\lambda}{\mu}\right)^n = \frac{(\lambda/\mu)\left(1 - (\lambda/\mu)^K\right)}{1 - \lambda/\mu}$$

In this case, no convergence issue arises since the sum is finite. The ratio λ/μ is therefore unconstrained. Intuitively, if $\lambda > \mu$, then customers are lost but the queue length remains bounded by K and no instability (unbounded queue length growth) can arise.

Once again we set $\rho = \lambda/\mu$, but emphasize that ρ does not stand for the traffic intensity. This is because λ is the rate of customer arrivals, but not the rate of customers actually admitted in the system, since some of them are blocked. We now get

$$\pi_0 = \frac{1 - \rho}{1 - \rho^{K+1}} \tag{8.57}$$

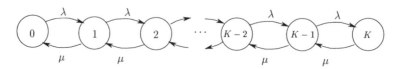

Figure 8.12: State transition rate diagram for the $M/M/1/K$ system.

Next, using (8.57) in (8.31), we obtain:

$$\pi_n = \begin{cases} \dfrac{1-\rho}{1-\rho^{K+1}}\rho^n & \text{if } 0 \le n \le K \\ 0 & \text{if } n > K \end{cases} \tag{8.58}$$

Equation (8.58) provides the stationary probability distribution of the queue length of the $M/M/1/K$ system. We now present some of the interesting performance measures for this system.

Utilization and Throughput

The server utilization is given by $1 - \pi_0$. From (8.57):

$$1 - \pi_0 = \rho\frac{1-\rho^K}{1-\rho^{K+1}} \tag{8.59}$$

Unlike the $M/M/1$ case, the utilization here is dependent on the structural parameter K in addition to λ and μ. Since ρ is allowed to become arbitrarily large, we can see that as $\rho \to \infty$ the utilization approaches 1, as expected. Moreover, if we require that $\rho < 1$, note that as $K \to \infty$ (that is, our model approaches the stable $M/M/1$ case) we get $(1 - \pi_0) \to \rho$, which is precisely the utilization of the $M/M/1$ system.

The throughput is given by the departure rate $\mu(1 - \pi_0)$. Thus,

$$\mu(1 - \pi_0) = \lambda\frac{1-\rho^K}{1-\rho^{K+1}} \tag{8.60}$$

which is smaller than the external arrival rate λ, since some customers are blocked and hence never served.

Blocking Probability

Perhaps the most important performance measure in the case of finite capacity queueing systems is the probability that an arriving customer is *blocked*, that is, the customer is lost because he finds the system full. We denote this probability by P_B. We once again make use of the result in (8.34) to set the probability of an arriving customer finding the queue full equal to the probability of the queue being full (that is, the fraction of time the queue is observed full by an independent observer - not an arriving customer). Thus, from (8.58),

$$P_B = \pi_K = (1 - \rho)\frac{\rho^K}{1-\rho^{K+1}} \tag{8.61}$$

Note once again that if we require that $\rho < 1$ and let $K \to \infty$ our model approaches the stable $M/M/1$ case and we get $P_B \to 0$, as expected for an infinite capacity queueing system. Also, by comparing (8.60) with (8.61), note that $\mu(1 - \pi_0) = \lambda(1 - P_B)$, which simply reflects flow balancing for this system.

In the design of queueing systems, a typical problem is the selection of the storage capacity K so that the blocking probability remains below some desirable level p. In the Markovian case, given arrival and service parameters λ and μ, the solution to this problem is provided through equation (8.61) by determining values of K such that $P_B(K) < p$.

Average Queue Length

This is the expected value of the random variable X, that is,

$$E[X] = \sum_{n=0}^{K} n \pi_n$$

where π_n is given by (8.58). Thus, we have

$$E[X] = \frac{1 - \rho}{1 - \rho^{K+1}} \sum_{n=0}^{K} n \rho^n$$

Let the sum in this expression be denoted by A. Then, observe that

$$A - \rho A = \left[\rho + \rho^2 + \ldots + \rho^K \right] - K \rho^{K+1} = \frac{\rho \left(1 - \rho^K \right)}{1 - \rho} - K \rho^{K+1}$$

from which we obtain

$$A = \frac{\rho \left(1 - \rho^K \right)}{(1 - \rho)^2} - \frac{K \rho^{K+1}}{1 - \rho}$$

and hence,

$$E[X] = \frac{\rho}{1 - \rho^{K+1}} \left[\frac{1 - \rho^K}{1 - \rho} - K \rho^K \right] \tag{8.62}$$

Using Little's Law, we can also obtain from (8.62) the average system time $E[S]$, as well as the average waiting time $E[W] = E[S] - 1/\mu$. We omit these calculations, but must point out that Little's Law can only be applied here as follows:

$$E[X] = \lambda(1 - \pi_K) \cdot E[S] \tag{8.63}$$

where $\lambda(1 - \pi_K)$ is the average arrival rate of *admitted* customers, as opposed to λ which is the arrival rate before some customers are blocked.

Example 8.3

We wish to design a manufacturing cell consisting of a single machine and some finite queueing space for incoming parts. Under Markovian assumptions for the arrival and service processes, we model the cell as an $M/M/1/K$ system. Our design objective is to guarantee that no more than 10% of arriving parts are blocked. The arrival rate is given as 1 part/min. Thus, we need to select the machine processing rate μ and the queue capacity K.

Using (8.61), we require that

$$P_B = (1 - \rho)\frac{\rho^K}{1 - \rho^{K+1}} < 0.1 \tag{8.64}$$

where $\rho = 1/\mu$. We now have some flexibility in selecting the two parameters ρ and K so as to satisfy the inequality above. Suppose that there are three choices for machines:

1. $\mu = 0.5$ parts/min at a cost of $100
2. $\mu = 1.2$ parts/min at a cost of $300
3. $\mu = 2$ parts/min at a cost of $500

In addition, suppose that each queueing space costs $80. Thus, our objective is to choose a combination of μ and K that minimizes the total cost.

1. When $\mu = 0.5$, we have $\rho = 2$ and (8.64) becomes

$$P_B(K) = \frac{2^K}{2^{K+1} - 1} < 0.1$$

For $K = 1$, we get $P_B(1) = 2/3 > 0.1$. In fact, it is easy to see that as K increases, $P_B(K)$ above can never become smaller that $1/2$. It follows that for this machine option there exists no feasible solution to our design problem.

2. When $\mu = 1.2$, we have $\rho = 5/6$ and (8.64) becomes

$$P_B(K) = \frac{5^K}{6^{K+1} - 5^{K+1}} < 0.1$$

For $K = 1$, we get $P_B(1) = 5/9 > 0.1$. Repeating calculations for increasing values of K, we find that

$$P_B(5) \approx 0.1007 \text{ and } P_B(6) \approx 0.0774 < 0.1$$

Choosing $K = 6$, we get a total cost of $780 for this option. On the other hand, if we are willing to compromise just a little bit on performance and choose $K = 5$, we obtain a cost of $700.

3. When $\mu = 2$, we have $\rho = 1/2$ and (8.64) becomes

$$P_B(K) = \frac{1}{2^{K+1} - 1} < 0.1$$

For $K = 1$, we get $P_B(1) = 1/3 > 0.1$. Increasing K we find that

$$P_B(2) = 1/7 > 0.1 \text{ and } P_B(3) = 1/15 \approx 0.0667 < 0.1$$

Therefore, a choice of $K = 3$ results in a total cost of $740 for this option.

In summary, option 3 ($\mu = 2, K = 3$) is optimal. It also provides a lower blocking probability than option 2 ($\mu = 1.2, K = 6$). Still, we see that a slight compromise, that is, a guarantee of no more than 10.07% blocking, would allow us to cut the design cost to $700.

8.6.5 The $M/M/m/m$ Queueing System

This is a system consisting of m identical servers without any queueing space, as shown in Fig. 8.13. Thus, if an arriving customer does not find an idle server, that customer is blocked and lost. As in the $M/M/1/K$ case, the Poisson arrival process is turned off for all queue lengths $n \geq m$. The service rate depends on the number of customers present in the system, that is, $\mu_n = n\mu$ with $n = 1, 2, \ldots, m$. Thus, our birth-death model becomes

$$\lambda_n = \begin{cases} \lambda & \text{if } 0 \leq n < m \\ 0 & \text{if } n \geq m \end{cases}$$
$$\mu_n = n\mu \quad \text{for all } n = 1, 2, \ldots, m$$

The state transition rate diagram for this system is shown in Fig. 8.14.
The probability π_0 is obtained from (8.32):

$$\pi_0 = \left[1 + \sum_{n=1}^{m} \frac{\lambda^n}{(\mu)(2\mu)\cdots(n\mu)}\right]^{-1} = \left[1 + \sum_{n=1}^{m} \left(\frac{\lambda}{\mu}\right)^n \frac{1}{n!}\right]^{-1}$$

Again, let $\rho = \lambda/\mu$, with the reminder that ρ does not represent the traffic intensity in this system. The sum in the expression above is finite, so no convergence issue arises, and ρ may become arbitrarily large. This means that a large number of customers may be blocked as $\lambda > \mu$, but the number of customers in the system is bounded by m. We then get

$$\pi_0 = \left[\sum_{n=0}^{m} \frac{\rho^n}{n!}\right]^{-1} \tag{8.65}$$

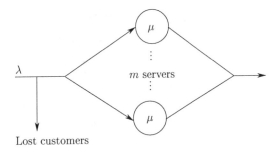

Figure 8.13: The $M/M/m/m$ queueing system.

We now obtain π_n from (8.31) and (8.65):

$$\pi_n = \begin{cases} \dfrac{1}{\sum_{n=0}^{m} \frac{\rho^j}{j!}} \dfrac{\rho^n}{n!} & \text{if } 0 \leq n \leq m \\[2ex] 0 & \text{if } n > m \end{cases} \tag{8.66}$$

Various performance measures can now be calculated as in previous sections. We limit ourselves to the one considered most useful for this system.

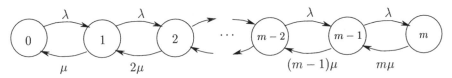

Figure 8.14: State transition rate diagram for the $M/M/m/m$ system.

Blocking Probability

As in the $M/M/1/K$ system, we use P_B to denote the probability that a customer is blocked, that is, the system already contains m customers:

$$P_B = \pi_m = \frac{\rho^m/m!}{\sum_{n=0}^{m} \frac{\rho^j}{j!}} \tag{8.67}$$

Here, we used again (8.34) to equate the probability that an arrival finds m customers in the system with the probability that there are m customers in the system at some random time instant. Note that π_m also represents the fraction of time that the system is fully occupied (all servers busy) at steady state.

Equation (8.67) is also known as the *Erlang B formula*, and it is of wide use in telephony. Given a call rate λ and an average call duration $1/\mu$, we can use (8.67) to determine the number of telephone lines m required to guarantee a blocking probability that is below some desirable level p. Specifically, we determine m such that $P_B(m) < p$.

8.6.6 The $M/M/1//N$ Queueing System

Recalling the notation introduced in Section 8.2.5, the $M/M/1//N$ system consists of a single server with infinite storage capacity, but access to it is limited to a finite population of N customers. The $M/M/1//N$ system is shown in Fig. 8.15. The server has exponentially distributed service times with rate μ. When one of the N customers completes service, he returns to the queue for more service, but only after going through a delay modeled by one of the N servers shown in Fig. 8.15. This delay is exponentially distributed with rate λ. Also, each customer is assumed to act independently of all others.

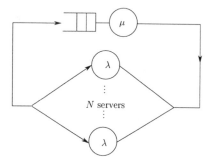

Figure 8.15: The $M/M/1//N$ queueing system.

A delay in returning to the queue arises very naturally in several applications. In a computer system with N terminals, each of the N users is either actively using the system (with a job in queue or in process) or is just "thinking" about what to do next while sitting at a terminal. In a manufacturing process, a machine may be able to process a part only if it is positioned on a special fixture. Suppose there are only N such fixtures and a continuous external supply of parts. In this case, a part completing service relinquishes its fixture to a new part, and the delay represents the time to perform this fixture exchange operation.

The $M/M/1//N$ system can be modeled as a birth-death chain where the state is the number of customers in queue or in process. If this number is n, then there are $(N - n)$ customers that are on their way back to the queue and occupying $(N - n)$ λ-rate servers. We now have the superposition of $(N - n)$ Poisson processes, which we know, from Section 6.7.3, to be a Poisson process with rate $(N - n)\lambda$. Our birth-death model now becomes

$$\lambda_n = \begin{cases} (N - n)\lambda & \text{if } 0 \le n < N \\ 0 & \text{if } n \ge N \end{cases}$$
$$\mu_n = \mu \quad \text{for all } n = 1, 2, \dots, N$$

The state transition rate diagram for this system is shown in Fig. 8.16.

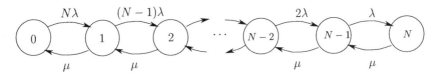

Figure 8.16: State transition rate diagram for the $M/M/1//N$ system.

We evaluate π_0 from (8.32) as follows:

$$\pi_0 = \left[1 + \sum_{n=1}^{N} \frac{[N\lambda][(N-1)\lambda] \cdots [(N-n+1)\lambda]}{\mu^n}\right]^{-1}$$

and since the sum above is finite, no convergence problem arises. The ratio λ/μ is therefore unconstrained. As usual, we set $\rho = \lambda/\mu$, and simplify the expression above to get

$$\pi_0 = \left[\sum_{n=0}^{N} \frac{N!}{(N-n)!}\rho^n\right]^{-1} \tag{8.68}$$

From (8.31), we obtain π_n for $n = 1, 2, \ldots, N$ as follows:

$$\pi_n = \frac{[N\lambda][(N-1)\lambda] \cdots [(N-n+1)\lambda]}{\mu^n}\pi_0 = \frac{N!}{(N-n)!}\rho^n \pi_0$$

and, therefore,

$$\pi_n = \begin{cases} \pi_0 \dfrac{N!}{(N-n)!}\rho^n & \text{if } 0 \le n \le N \\ 0 & \text{if } n > N \end{cases} \tag{8.69}$$

Equations (8.68) and (8.69) provide the stationary probability distribution of the queue length of the $M/M/1//N$ system. Some of the most interesting performance measures for this system are discussed next.

Utilization and Throughput

The server utilization is $1 - \pi_0$, and the throughput is $\mu(1 - \pi_0)$ with π_0 given by (8.68).

Average Response Time

The response time is a random variable R defined by the amount of time elapsing from the instant a customer enters the queue until he completes service. We can calculate $E[R]$ by two applications of Little's Law. First, we apply

Little's Law to the top part of the system in Fig. 8.11 (queue and server) where the total number of customers present is the state X:

$$E[X] = \mu(1 - \pi_0)E[R]$$

Next, note that the number of busy λ-rate servers when the state is X is $(N-X)$. Thus, applying Little's Law to the bottom part of the system in Fig. 8.11 we have

$$E[N - X] = \mu(1 - \pi_0)\frac{1}{\lambda}$$

since the average time spent by a customer in that subsystem is $1/\lambda$. Combining these two equations to eliminate $E[X]$ yields

$$E[R] = \frac{N}{\mu(1 - \pi_0)} - \frac{1}{\lambda} \tag{8.70}$$

This result is particularly useful in modeling computer systems where N is the number of terminals available, μ is the computer processing rate, and $1/\lambda$ is the average "thinking time" of a user at a terminal. Given λ and μ along with a desired value, β, that the average response time should not exceed, (8.70) can be used to determine the number of terminals our system can accommodate, that is, we determine N such that $E[R] < \beta$, or

$$\frac{N}{\mu(1 - \pi_0)} < \beta + \frac{1}{\lambda}$$

In Fig. 8.17, we show a typical plot of $E[R]$ as a function of N. As in our discussion of the average system time characteristics of the $M/M/1$ system in Fig. 8.7, note once again that $E[R]$ is relatively insensitive to increases in N until a point where the slope quickly increases and approaches an asymptote. The asymptote corresponds to the case where $N \to \infty$, and therefore $\pi_0 \to 0$, since the server utilization must approach 1. Thus, from (8.70), we see that the asymptote is given by $E[R] = N/\mu - 1/\lambda$. A good design entails the identification of those values of N which occur just before the curve takes off towards this asymptote.

8.6.7 The $M/M/m/K/N$ Queueing System

This is a system which encompasses all previous cases as we allow the parameters m, K, and N to take on finite or infinite values. It is similar to Fig. 8.15, except that there are m servers instead of one, and the queueing space is limited to K. Of course, we must have $K \geq m$, since queueing space includes the m servers. If $N \geq K$, then customers may be blocked when they try to access the queue and find the queue length to be K (which includes m customers in

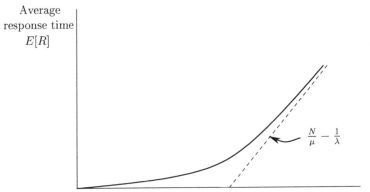

Figure 8.17: Average response time as a function of population size in the $M/M/1//N$ system.

process). In this case, customers return to a λ-rate server and try to enter the queue after another exponentially distributed delay with mean $1/\lambda$.

The analysis of this system becomes rather tedious. We will present the main results without getting into details. First, our birth-death model is now

$$\lambda_n = \begin{cases} (N-n)\lambda & \text{if } 0 \leq n < K \\ 0 & \text{if } n \geq K \end{cases}$$

$$\mu_n = \begin{cases} n\mu & \text{if } 0 \leq n < m \\ m\mu & \text{if } n \geq m \end{cases}$$

Clearly, $\pi_n = 0$ for all $n > K$. Setting $\rho = \lambda/\mu$, the stationary queue length probability distribution, $\pi_n, n = 0, 1, \ldots, K$ is given by

$$\pi_n = \begin{cases} \pi_0 \dbinom{N}{n} \rho^n & n = 1, 2, , \ldots, m-1 \\ \pi_0 \dbinom{N}{n} \dfrac{n!}{m!} m^{m-n} \rho^n & n = m, m+1, \ldots, K \end{cases} \tag{8.71}$$

where π_0 is obtained from (8.32):

$$\pi_0 = \left[1 + \sum_{n=1}^{m-1} \binom{N}{n} \rho^n \right.$$

$$\left. + \binom{N}{m-1} \rho^{m-1} \sum_{n=m}^{K} \frac{(N-m+1)!}{(N-n)!} \left(\frac{\rho}{m}\right)^{n-m+1} \right]^{-1}$$

8.7 MARKOVIAN QUEUEING NETWORKS

The queueing systems we have considered thus far involve customers requesting service from a single service-providing facility (with one or more servers). In some cases, we saw that customers may return to the same facility for additional service. In practice, however, it is common for two or more servers to be connected so that a customer proceeds from one server to the next in some fashion. In communication networks, for instance, messages often go through several switching nodes followed by transmission links before arriving at their destination. In manufacturing, a part must usually proceed through several operations in series before it becomes a finished product. Thus, it is necessary to extend our study of simple queueing systems to *queueing networks*, where multiple servers and queues are interconnected. In such systems, a customer enters at some point and requests service at some server. Upon completion, the customer generally moves to another queue or server for additional service. In the class of *open networks*, arriving customers from the outside world eventually leave the system. In the class of *closed networks*, the number of customers remains fixed, as in some of the systems we have already seen (e.g., the $M/M/1//N$ system).

In the simple systems considered thus far, our objective was to obtain the stationary probability distribution of the state X, where X is the queue length. In networks, we have a system consisting of M interconnected nodes, where the term node is used to describe a set of identical parallel servers along with the queueing space that precedes it. In Fig. 8.18, for example, we show a queueing network with three single-server nodes and one two-server node. In a network environment, we shall refer to X_i as the queue length at the ith node in the system, $i = 1, \ldots, M$. It follows that the state of a queueing network is a vector of random variables

$$\mathbf{X} = [X_1, X_2, \ldots, X_M] \tag{8.72}$$

where X_i takes on values $n_i = 0, 1, \ldots$ just like a simple single-class stand-alone queueing system. The major objective of queueing network analysis is to obtain the stationary probability distribution of X (if it exists), that is, the probabilities

$$\pi(n_1, \ldots, n_M) = P[X_1 = n_1, \ldots, X_M = n_M] \tag{8.73}$$

for all possible values of $n_1, \ldots, n_M, n_i = 0, 1, \ldots$

Our goal in the next few sections is to present the main results pertinent to the analysis of *Markovian* queueing networks. This means that all external arrival events and all departure events at the servers in the system are generated by processes satisfying the Markovian (memoryless) property, and are therefore characterized by exponential distributions. The natural question that arises is

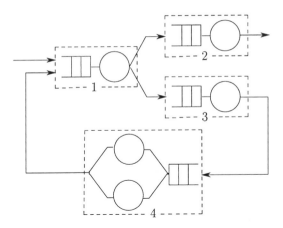

Figure 8.18: Example of a four-node queueing network.

the following: *What about internal arrival processes?* In other words, the arrival process at some queue in the network is usually composed of one or more departure processes from adjacent servers; what are the stochastic characteristics of such processes? The answer to this question is important for much of the classical analysis of queueing networks and is presented in the next section.

8.7.1   The Departure Process of the $M/M/1$ Queueing System

Let us consider an $M/M/1$ queueing system. Recall that Y_k and Z_k denote the interarrival and service time respectively of the kth customer, and that the arrival and service processes are assumed to be independent. Now let us concentrate on the departure times $D_k, k = 1, 2, \ldots$, of customers, and define Ψ_k to be the kth *interdeparture time*, that is, a random variable such that

$$\Psi_k = D_k - D_{k-1}$$
$$= \text{ time elapsed between the } (k-1)\text{th and } k\text{th departure, } k = 1, 2, \ldots$$

where, for simplicity, we set $\Psi_0 = 0$, so that Ψ_1 is the random variable describing the time of the first departure. As $k \to \infty$, we will assume that there exists a stationary probability distribution function such that

$$\lim_{k \to \infty} P[\Psi_k \le t] = P[\Psi \le t]$$

where Ψ describes an interdeparture time at steady state. We will now evaluate the distribution $P[\Psi \le t]$. The result is quite surprising.

Theorem 8.2 The departure process of a stable stationary $M/M/1$ queueing system with arrival rate λ is a Poisson process with rate λ.

Proof. We begin with the fundamental dynamic equation (8.19):

$$D_k = \max\{A_k, D_{k-1}\} + Z_k$$

which we can rewrite as

$$\Psi_k = \max\{A_k - D_{k-1}, 0\} + Z_k \tag{8.74}$$

Thus, there are two cases:

Case 1: If $A_k - D_{k-1} \leq 0$ we have

$$\Psi_k = Z_k$$

that is, the interdeparture process behaves like the service process. This corresponds to the situation where the server remains busy and successive departure times correspond to successive service completion times.

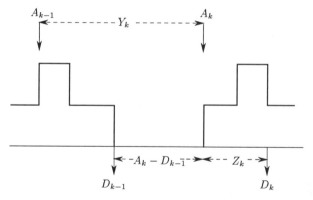

Figure 8.19: Illustrating how an idle period becomes a residual interarrival time when $A_k > D_{k-1}$.

Case 2: If $A_k - D_{k-1} > 0$ we have

$$\Psi_k = A_k - D_{k-1} + Z_k$$

This corresponds to the situation where the server is idle when the kth arrival occurs, as illustrated in Fig. 8.19. Observe that the random variable $(A_k - D_{k-1})$ is the kth residual interarrival time at the $(k-1)$th departure instant. Since the arrival process is Poisson, it follows from the memoryless property that

$$(A_k - D_{k-1}) \sim 1 - e^{-\lambda t} \tag{8.75}$$

Case 1 above occurs when the kth arriving customer finds the queue not empty, that is, $X(A_k) > 0$. Accordingly, Case 2 arises when $X(A_k) = 0$. We would now like to evaluate the probability distribution of the random variable Ψ_k, $P[\Psi_k \leq t]$. Combining these two cases together by conditioning on the state of the queue upon arrival, we have

$$P[\Psi_k \leq t] = P[\Psi_k \leq t \mid X(A_k^-) > 0]P[X(A_k^-) > 0]$$
$$+ P[\Psi_k \leq t \mid X(A_k^-) = 0]P[X(A_k^-) = 0]$$

and using the corresponding expressions for Ψ_k in the two cases above, we get

$$P[\Psi_k \leq t] = P[Z_k \leq t]P[X(A_k^-) > 0]$$
$$+ P[A_k - D_{k-1} + Z_k \leq t]P[X(A_k^-) = 0] \tag{8.76}$$

Let us now look at each term in equation (8.76) at steady state separately.

1. $P[Z_k \leq t]$. For an $M/M/1$ system, we have

$$P[Z_k \leq t] = 1 - e^{-\mu t}$$

for all $k = 1, 2, \ldots$

2. $P[X(A_k^-) = 0]$. Here, we can invoke (8.34), that is, the fact that the probability that Poisson arrivals find the queue at state n is the same as the probability that the state is n. In particular, at steady state,

$$\lim_{k \to \infty} P[X(A_k^-) = 0] = \pi_0 = 1 - \rho$$

since at steady state in a stable (i.e., $\rho < 1$) $M/M/1$ system we have $\pi_0 = 1 - \rho$ by (8.36).

3. $P[X(A_k^-) > 0]$. It follows immediately from above that at steady state

$$\lim_{k \to \infty} P[X(A_k^-) > 0] = 1 - \pi_0 = \rho$$

4. $P[A_k - D_{k-1} + Z_k \leq t]$, with $(A_k - D_{k-1}) \sim (1 - e^{-\lambda t})$ by (8.75). Here, we must determine the probability distribution of the sum of two independent random variables, both of which are exponentially distributed. Specifically, $(A_k - D_{k-1}) \sim (1 - e^{-\lambda t})$ and $Z_k \sim (1 - e^{-\mu t})$. It is well known from elementary probability theory (see also Appendix I) that the pdf of the resulting random variable is the convolution of the individual pdf's. In our case, this convolution involves the two exponential pdf's $\lambda e^{-\lambda t}$ and $\mu e^{-\mu t}, t \geq 0$:

$$\int_0^t \lambda e^{-\lambda \tau} \mu e^{-\mu(t-\tau)} d\tau = \lambda \mu e^{-\mu t} \int_0^t e^{(\mu-\lambda)\tau} d\tau$$
$$= \frac{\lambda \mu e^{-\mu t}}{\mu - \lambda} \left[e^{(\mu-\lambda)t} - 1 \right]$$

This is the pdf of the random variable $A_k - D_{k-1} + Z_k$. The corresponding distribution function is obtained by integrating the expression above, that is,

$$P[A_k - D_{k-1} + Z_k \leq t] = \frac{\lambda\mu}{\mu - \lambda} \int_0^t \left[e^{-\lambda\tau} - e^{-\mu\tau}\right] d\tau$$

$$= \frac{\mu}{\mu - \lambda} \left[1 - e^{-\lambda t}\right] - \frac{\lambda}{\mu - \lambda} \left[1 - e^{-\mu t}\right]$$

which is also clearly the distribution of this random variable at steady state.

We can now combine all the above results into (8.76) at steady state to obtain

$$\lim_{k\to\infty} P[\Psi_k \leq t] = P[\Psi \leq t]$$

$$= \left[1 - e^{-\mu t}\right] \rho + \left\{\frac{\mu}{\mu - \lambda}\left[1 - e^{-\lambda t}\right]\right.$$

$$\left. - \frac{\lambda}{\mu - \lambda}\left[1 - e^{-\mu t}\right]\right\}(1 - \rho)$$

$$= \left[1 - e^{-\mu t}\right]\frac{\lambda}{\mu} + \left\{\frac{\mu}{\mu - \lambda}\left[1 - e^{-\lambda t}\right]\right.$$

$$\left. - \frac{\lambda}{\mu - \lambda}\left[1 - e^{-\mu t}\right]\right\}\frac{\mu - \lambda}{\mu}$$

which, noticing the cancellation of the $(1 - e^{-\mu t})$ terms, yields

$$P[\Psi \leq t] = 1 - e^{-\lambda t} \tag{8.77}$$

This is a Poisson process with parameter λ, and the proof is complete. **Q.E.D.**

Thus, we have established the following fundamental property of the $M/M/1$ queueing system, also known as *Burke's theorem* (Burke, 1956): A Poisson process supplying arrivals to a server with exponentially distributed service times results in a Poisson departure process with the exact same rate. This fact also holds for the departure process of an $M/M/m$ system. Burke's theorem has some critical ramifications when dealing with networks of Markovian queueing systems, because it allows us to treat each component node independently, as long as there are no customer feedback paths (a case we will discuss in the next section). When a node is analyzed independently, the only information required is the number of servers at that node, their service rate, and the arrival rate of customers (from other nodes as well as the outside world).

8.7.2 Open Queueing Networks

We will consider a general open network model consisting of M nodes, each with infinite storage capacity. We will assume that customers form a single class, and that all nodes operate according to a FCFS queueing discipline. Node i, $i = 1, \ldots, M$, consists of m_i servers each with exponentially distributed service times with parameter μ_i. External customers may arrive at node i from the outside world according to a Poisson process with rate r_i. In addition, internal customers arrive from other servers in the network. Upon completing service at node i, a customer is routed to node j with probability p_{ij}; this is referred to as the *routing probability* from i to j. The outside world is usually indexed by 0, so that the fraction of customers leaving the network after service at node i is denoted by p_{i0}. Note that $p_{i0} = 1 - \sum_{j=1}^{M} p_{ij}$.

In this modeling framework, let λ_i denote the total arrival rate at node i. Thus, using the notation above, we have

$$\lambda_i = r_i + \sum_{j=1}^{M} \lambda_j p_{ji}, \qquad i = 1, \ldots, M \qquad (8.78)$$

where the first term represents the external customer flow and the second term represents the aggregate internal customer flow from all other nodes.

Before discussing the general model, let us first consider the simplest possible case, consisting of two single-server nodes in tandem, as shown in Fig. 8.20. In this case, the state of the system is the two-dimensional vector $\mathbf{X} = [X_1, X_2]$, where $X_i, i = 1, 2$, is the queue length of the ith node. The event set of this DES is $\mathcal{E} = \{a, d_1, d_2\}$, where a is an external arrival, and $d_i, i = 1, 2$, is a departure from node i. Since all events are generated by Poisson processes, we can model the system as a Markov chain whose state transition rate diagram is shown in Fig. 8.21.

We can now write down flow balance equations by inspection of the state transition diagram. Let us first look at a state (n_1, n_2) with $n_1 > 0$ and $n_2 > 0$. All possible incoming and outgoing transitions for any such state are shown in Fig. 8.22. We then have

$$\lambda \pi(n_1 - 1, n_2) + \mu_1 \pi(n_1 + 1, n_2 - 1)$$
$$+ \mu_2 \pi(n_1, n_2 + 1) - (\lambda + \mu_1 + \mu_2) \pi(n_1, n_2) = 0 \qquad (8.79)$$

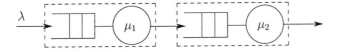

Figure 8.20: A two-node tandem queueing network.

Similarly, for all states $(n_1, 0)$ with $n_1 > 0$ we have

$$\lambda \pi(n_1 - 1, 0) + \mu_2 \pi(n_1, 1) - (\lambda + \mu_1) \pi(n_1, 0) = 0 \qquad (8.80)$$

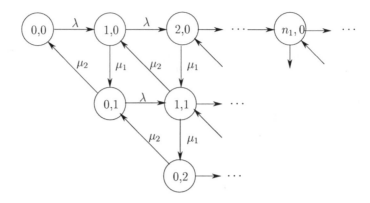

Figure 8.21: State transition rate diagram for a two-node tandem queueing network.

and for states $(0, n_2)$ with $n_2 > 0$:

$$\mu_1\pi(1, n_2 - 1) + \mu_2\pi(0, n_2 + 1) - (\lambda + \mu_2)\pi(0, n_2) = 0 \qquad (8.81)$$

and finally for state $(0, 0)$:

$$\mu_2\pi(0, 1) - \lambda\pi(0, 0) = 0 \qquad (8.82)$$

In addition, the state probabilities must satisfy the normalization condition:

$$\sum_{i=0}^{\infty}\sum_{j=0}^{\infty}\pi(i, j) = 1 \qquad (8.83)$$

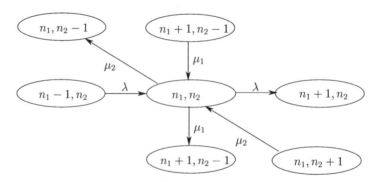

Figure 8.22: State transition rates into and out of state (n_1, n_2) with $n_1 > 0, n_2 > 0$.

The set of equations (8.79) through (8.83) can readily be solved (by proceeding as in the solution of the one-dimensional birth-death chain in Section 7.4.3) to give

$$\pi(n_1, n_2) = (1 - \rho_1)\rho_1^{n_1} \cdot (1 - \rho_2)\rho_2^{n_2} \tag{8.84}$$

where

$$\rho_1 = \frac{\lambda}{\mu_1}, \qquad \rho_2 = \frac{\lambda}{\mu_2}$$

with the usual stability conditions $0 \leq \rho_1 < 1$ and $0 \leq \rho_2 < 1$. Clearly, ρ_1 is the traffic intensity at the first node. Moreover, ρ_2 is the traffic intensity of node 2, since the throughput (departure rate) of node 1, which is λ, is also the arrival rate at node 2.

Observe that if we view each of the two nodes as separate $M/M/1$ systems with stationary state probabilities $\pi_1(n_1)$ and $\pi_2(n_2)$ respectively, we get

$$\pi_1(n_1) = (1 - \rho_1)\rho_1^{n_1}, \qquad \pi_2(n_2) = (1 - \rho_2)\rho_2^{n_2}$$

We see, therefore, that we have a simple *product form solution* for this two-node network:

$$\pi(n_1, n_2) = \pi_1(n_1) \cdot \pi_2(n_2) \tag{8.85}$$

Also note that as λ increases, the node with the smallest service rate will be the first to cause an instability when $\rho_i \geq 1$. This is referred to as the *bottleneck* node of a network.

In fact, the product form in (8.85) is a consequence of Burke's theorem (8.77). It allows us to decouple the two nodes, analyze them separately as individual $M/M/1$ systems, and then combine the results as in (8.85). It is straightforward to extend this solution to any open Markovian network with no customer feedback paths. Let us briefly indicate why *customer feedback* may create a problem by looking at the simple queueing system in Fig. 8.23. The external arrival process is Poisson and service times are exponential. Customers completing service are returned to the queue with probability p, or they depart with probability $(1 - p)$. The difficulty here is that the process formed by customers entering the queue, consisting of the superposition of the external Poisson process and the feedback process, is *not* Poisson (in fact, it can be shown to be characterized by a hyperexponential distribution - we will have more to say about this distribution, shown in equation (8.107), in Section 8.8.1). Remarkably, the departure process of this system is still Poisson. For an insightful explanation of why this is so the reader is referred to Walrand (1988), who discusses the often misunderstood subtleties of this system in some detail.

Returning to the general open network model described above, it was established by Jackson (Jackson, 1957) that a product form solution still exists even if customer feedback is allowed. This type of model is also referred to as a *Jackson network*. What is interesting in this model is that *an individual node*

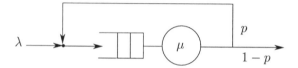

Figure 8.23: A simple queueing system with customer feedback.
*The input process (the superposition of the external arrival process and the feedback process) is **not** Poisson. However, the departure process is Poisson.*

need not have Poisson arrivals (due to the feedback effect), yet it behaves *as if it had Poisson arrivals* and can therefore still be treated as an $M/M/1$ system. Thus, we have the general product form solution:

$$\pi(n_1, n_2, \ldots, n_N) = \pi_1(n_1) \cdot \pi_2(n_2) \cdots \pi_M(n_M) \qquad (8.86)$$

where $\pi_i(n_i)$ is the solution of an $M/M/m_i$ queueing system (see Section 8.6.2) with service rate μ_i and arrival rate given by λ_i in the solution of (8.78). To guarantee the existence of the stationary probability distribution in (8.86), we impose the usual stability condition for each node:

$$\lambda_i = r_i + \sum_{j=1}^{M} \lambda p_{ji} < m_i \mu_i \qquad (8.87)$$

Example 8.4

We will illustrate the use of the product form solution (8.86) for the three-node Jackson network of Fig. 8.24. Note that this includes feedback paths from node 1 to itself and from node 2 to both nodes 1 and 2. There is also a single external arrival process at node 1 with rate r_1. We assume that all routing probabilities $p_{ij}, i, j = 1, 2, 3$, are given, along with the parameters r_1, μ_1, μ_2 and μ_3.

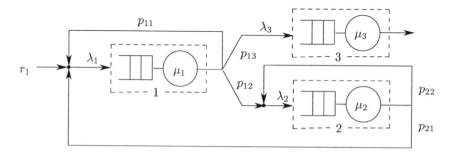

Figure 8.24: Open queueing network for Example 8.4.

The first step is to calculate the arrival rates $\lambda_1, \lambda_2, \lambda_3$ at steady state. Provided all nodes are stable, $\lambda_1, \lambda_2, \lambda_3$ are also the throughputs (departure rates) of nodes 1, 2, 3 respectively. These are obtained by writing down the customer flow balance equations whose general form we saw in (8.78):

$$\lambda_1 = r_1 + p_{11}\lambda_1 + p_{21}\lambda_2$$
$$\lambda_2 = p_{12}\lambda_1 + p_{22}\lambda_2$$
$$\lambda_3 = p_{13}\lambda_1$$

These equations are easily solved to give

$$\lambda_1 = \frac{1}{p_{13}}r_1, \qquad \lambda_2 = \frac{p_{12}}{p_{21}p_{13}}r_1, \qquad \lambda_3 = r_1 \qquad (8.88)$$

Note that the relationship $\lambda_3 = r_1$ could have been directly obtained by the observation that the only customer flow out of the system is λ_3, and it must be balanced by the only incoming flow which is r_1.

The second step is to obtain the utilizations ρ_1, ρ_2, ρ_3. These are immediately given by

$$\rho_1 = \frac{\lambda_1}{\mu_1}, \qquad \rho_2 = \frac{\lambda_2}{\mu_2}, \qquad \rho_3 = \frac{\lambda_3}{\mu_3} \qquad (8.89)$$

where $\lambda_1, \lambda_2, \lambda_3$ are the values obtained in (8.88).

The third step is to solve each node separately as an $M/M/1$ system. Using (8.37), we can write the stationary state probabilities $\pi_i(n_i), i = 1, 2, 3$, as follows:

$$\pi_i(n_i) = (1 - \rho_i)\rho_i^{n_i}, \qquad i = 1, 2, 3$$

Finally, we apply the product form solution (8.86) to get:

$$\pi(n_1, n_2, n_3) = (1 - \rho_1)\rho_1^{n_1} \cdot (1 - \rho_2)\rho_2^{n_2} \cdot (1 - \rho_3)\rho_3^{n_3} \qquad (8.90)$$

Suppose that we set some numerical values for the system as follows:

$$r_1 = 1, \qquad \mu_1 = \mu_2 = 4, \qquad \mu_3 = 2$$
$$p_{12} = 0.2, \qquad p_{13} = 0.5, \qquad p_{21} = 0.2$$

Then, (8.88) gives $\lambda_1 = 2, \lambda_2 = 2, \lambda_3 = 1$. It follows from (8.89) that

$$\rho_1 = 0.5, \qquad \rho_2 = 0.5, \qquad \rho_3 = 0.5$$

which are all less than 1 as required. Finally, (8.90) becomes

$$\pi(n_1, n_2, n_3) = \frac{1}{8}\left(\frac{1}{2}\right)^{n_1}\left(\frac{1}{2}\right)^{n_2}\left(\frac{1}{2}\right)^{n_3}$$

We can now obtain various performance measures of interest. For example, the average number of customers in the system is given by $E[X_1 + X_2 + X_3]$. Since each node behaves like an $M/M/1$ system and all nodes have the same utilization, we have, by (8.39),

$$E[X_i] = \frac{\rho_i}{1 - \rho_i} = 1$$

Therefore, $E[X_1 + X_2 + X_3] = 3$. By Little's Law, we can also calculate the average system time $E[S]$:

$$E[X_1 + X_2 + X_3] = r_1 E[S] = E[S]$$

and hence, $E[S] = 3$.

The open network model we have considered can be extended to include state-dependent arrival and service processes. In particular, Jackson considered the case where an external arrival process depends on the total number of customers $(n_1 + \ldots + n_N)$, and the ith service rate depends on the queue length at node i (Jackson, 1963). Although the analysis becomes considerably more cumbersome, the stationary state probability can still be shown to be of the product form variety.

8.7.3 Closed Queueing Networks

A closed queueing network is one with a finite population of N customers. As already discussed in Section 8.2.5, the constraint in the number of customers may be imposed by a limited number of resources (e.g., terminals in a computer system, part fixtures in a manufacturing system). From a modeling standpoint, a closed network may be obtained from the open network model of the previous section by setting

$$r_i = 0 \quad \text{and} \quad \sum_{j=1}^{M} p_{ij} = 1 \text{ for all } i = 1, \ldots, M$$

In this case, no external arrivals occur and no customers can leave the system. Under these conditions, the state variables of the system, $X_1, \ldots, X_M$, must always satisfy

$$\sum_{i=1}^{M} X_i = N$$

Thus, the state space is finite and corresponds to the number of placements of N customers among M nodes, given by the binomial coefficient

$$\binom{M + N - 1}{M - 1} = \frac{(M + N - 1)!}{(M - 1)!N!}$$

In addition, if we balance customer flows as in (8.78) we get

$$\lambda_i = \sum_{j=1}^{M} \lambda_j p_{ji}, \qquad i = 1, \dots, M \qquad (8.91)$$

There is an important difference between the set of equations (8.78) and that in (8.91). In (8.78), we have M linearly independent equations, from which, in general, a unique solution may be obtained (as in Example 8.4 for $M = 3$). On the other hand, the absence of external arrival rate terms r_i in (8.91) results in $(M - 1)$ linearly independent equations only. Thus, the solution of equation (8.91) for $\lambda_1, \dots, \lambda_M$ involves a free constant. For instance, suppose we choose λ_1 to be this constant. We may then interpret $\lambda_i, i \neq 1$, as the relative throughput of node i with respect to the throughput of node 1.

It turns out that this class of networks also has a product form solution for its stationary state probabilities $\pi(n_1, \dots, n_M)$, with the values of $n_1, \dots, n_M$ constrained to satisfy $\sum_{j=1}^{M} n_i = N$ (Gordon and Newell, 1967). For simplicity, we limit our discussion here to single-server nodes, although the result also applies to the more general case where node i consists of m_i servers.

The starting point is once again the set of probability flow balance equations for the Markov chain model of the closed network. Let us consider a state $(n_1, \dots, n_M)$ and make the following observations:

1. Any transition out of this state is due to a departure from some node i with $n_i > 0$. The rate of such a transition is μ_i. Thus, the total probability flow out of $(n_1, \dots, n_M)$ is given by

$$\sum_{i:n_i>0} \mu_i \pi(n_1, \dots, n_M)$$

2. Any transition into this state is due to a departure from some node j with $n_j > 0$ which is routed to node i. The rate of such a transition is $p_{ji}\mu_j$. Moreover, if the state resulting from such a transition is $(n_1, \dots, n_M)$, then the state when the event with rate $p_{ji}\mu_j$ occurs must be $(n_1, \dots, n_j + 1, \dots, n_i - 1, \dots, n_M)$. In other words, a customer leaves node j and enters node i. The total probability flow into $(n_1, \dots, n_M)$ is, therefore,

$$\sum_{j:n_j>0} \sum_{i} p_{ji}\mu_j \pi(n_1, \dots, n_j + 1, \dots, n_i - 1, \dots, n_M)$$

It follows that the set of flow balance equations is:

$$\sum_{i:n_i>0} \mu_i \pi(n_1, \dots, n_M)$$
$$= \sum_{j:n_j>0} \sum_{i} p_{ji}\mu_j \pi(n_1, \dots, n_j + 1, \dots, n_i - 1, \dots, n_M) \qquad (8.92)$$

for all $n_1, \ldots, n_M$ satisfying $\sum_{j=1}^{M} n_i = N$

The solution to these equations is provided by the following product form (Gordon and Newell, 1967):

$$\pi(n_1, \ldots, n_M) = \frac{1}{C(N)} \rho_1^{n_1} \cdots \rho_M^{n_M} \qquad (8.93)$$

where

> $\rho_i = \lambda_i/\mu_i$ with λ_i obtained from the solution of (8.91) with a free constant (arbitrarily chosen)
>
> $C(N)$ is a constant dependent on the population size N which is obtained from the normalization condition

$$\frac{1}{C(N)} \sum_{n_1, \ldots, n_M} \rho_1^{n_1} \cdots \rho_M^{n_M} = 1 \qquad (8.94)$$

Thus, to obtain the stationary state probability distribution in (8.93) we need to go through several steps of some computational complexity. First, the solution of the linear equations (8.91) to determine the parameters $\lambda_1, \ldots, \lambda_M$ must be obtained. This solution includes a free constant, which must next be selected arbitrarily. This selection will not affect the probabilities in (8.93), but only the values of the parameters ρ_i in the product form. Lastly, we must compute the constant $C(N)$ through (8.94). The latter step, in itself, is usually not a trivial computational task.

Example 8.5

We will illustrate how to obtain the product form solution (8.93) for the three-node closed network of Fig. 8.25. This is a simple version of the central server model for computer systems, with node 1 representing a CPU and nodes 2 and 3 representing I/O devices. We assume that the routing probability $p_{12} = p$ is given, along with the service rate parameters μ_1, μ_2, μ_3. The number of jobs N circulating in the system must also be specified. In our case, we will assume that $N = 2$. It follows that there are 6 possible states in the state space: (0,1,1), (1,0,1), (1,1,0), (2,0,0), (0,2,0), and (0,0,2).

The first step is to write down the customer flow balance equations (8.91). By inspection of the network in Fig. 8.25, we see that:

$$\lambda_1 = \lambda_2 + \lambda_3$$
$$\lambda_2 = p\lambda_1$$
$$\lambda_3 = (1 - p)\lambda_1$$

We immediately see that one of these equations is redundant (adding the second and third together gives the first). Thus, the solution is provided

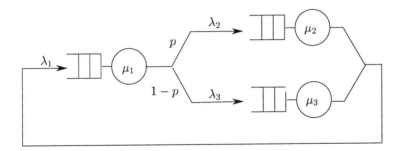

Figure 8.25: Closed queueing network for Example 8.5.

by the last two equations with λ_1 as a free constant. We may now choose any desired value for λ_1. One convenient choice is $\lambda_1 = \mu_1$. We can then write the parameters $\rho_i = \lambda_i/\mu_i$ as follows:

$$\rho_1 = 1, \quad \rho_2 = p\frac{\mu_1}{\mu_2}, \quad \rho_3 = (1-p)\frac{\mu_1}{\mu_3}$$

The next step is the determination of the normalization constant $C(N)$ from (8.94). We let this constant be C, and we get

$$\frac{1}{C}\sum_{n_2,n_3}\left(p\frac{\mu_1}{\mu_2}\right)^{n_2}\left((1-p)\frac{\mu_1}{\mu_3}\right)^{n_3} = 1$$

where n_2, n_3 take on the values allowed in this state space. Specifically, we get

$$C = p(1-p)\frac{\mu_1^2}{\mu_2\mu_3} + (1-p)\frac{\mu_1}{\mu_3} + p\frac{\mu_1}{\mu_2}$$
$$+ 1 + \left(p\frac{\mu_1}{\mu_2}\right)^2 + \left((1-p)\frac{\mu_1}{\mu_3}\right)^2$$

The stationary state probabilities are then given by (8.93) for this value of C:

$$\pi(n_1,n_2,n_3) = \frac{1}{C}\left(p\frac{\mu_1}{\mu_2}\right)^{n_2}\left((1-p)\frac{\mu_1}{\mu_3}\right)^{n_3} \tag{8.95}$$

Note that this is independent of n_1 because of our choice of constant in the determination of the relative throughputs $\lambda_1, \lambda_2, \lambda_3$.

To get some numerical results, let

$$p = 1/2, \quad \mu_1 = 4, \quad \mu_2 = 1, \quad \mu_3 = 2$$

Then, evaluating the expression for C given previously we get $C = 11$. The product form (8.95) becomes

$$\pi(n_1, n_2, n_3) = \frac{1}{11} 2^{n_2}$$

In particular, the six stationary state probabilities are

$$\pi(0, 1, 1) = \pi(1, 1, 0) = 2/11$$
$$\pi(1, 0, 1) = \pi(2, 0, 0) = \pi(0, 0, 2) = 1/11,$$
$$\pi(0, 2, 0) = 4/11$$

Several performance measures of interest can now be evaluated given this probability distribution. As an example, we can easily obtain the utilization of server 1 (the CPU in a central server model) as follows:

$$1 - \sum_{n_1, n_2, n_3 : n_1 = 0} \pi(n_1, n_2, n_3) = 1 - 7/11 = 4/11$$

Finally, it is possible to explicitly obtain the product form solution (8.95) above by directly solving the probability flow balance equations (8.92) for the system of Fig. 8.25. The state transition rate diagram for our model is shown in Fig. 8.26. Writing down the flow balance equations and solving them (or at least verifying that (8.95) satisfies them) is left as an exercise. It is also left as an exercise to check that the selection of a value other than μ_1 for the free constant λ_1 leads to the same solution for the stationary state probabilities $\pi(n_1, n_2, n_3)$.

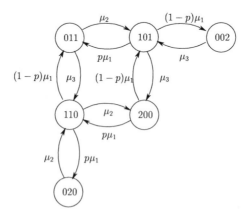

Figure 8.26: State transition rate diagram for Example 8.5.

Computation of the Normalization Constant $C(N)$

As already mentioned, the computation of the constant $C(N)$ through its definition in (8.94) can easily become very tedious, especially when the state space is large. Fortunately, there exist efficient computational algorithms to overcome this difficulty. Moreover, these algorithms allow us to evaluate several useful performance measures directly, without explicitly obtaining all possible values of $\pi(n_1,\ldots,n_M)$. One of the simplest such algorithms is due to Buzen (1973), and it is based on the recursive relationship

$$C_i(k) = C_{i-1}(k) + \rho_i C_i(k-1), \quad i = 2,\ldots,M,\ k = 2,\ldots,N \quad (8.96)$$

with initial conditions:

$$C_1(k) = \rho_1^k, \quad k = 1,\ldots,N$$

and

$$C_i(1) = 1, \quad i = 1,\ldots,M$$

from which $C(N)$ is obtained as

$$C(N) = C_M(N) \quad (8.97)$$

It can also be shown that the utilization of node i when the population size is N is given by

$$\mu_i[1 - \pi_i(0)] = \rho_i \frac{C(N-1)}{C(N)}$$

Expressions for other performance measures can similarly be derived in terms of the parameters ρ_i and $C(k), k = 1,\ldots,N$.

Mean Value Analysis

Suppose we are only interested in obtaining performance measures such as the network throughput and the mean values only of the queue length and system time distributions at all nodes. In this case, the *Mean Value Analysis* (MVA) methodology, developed by Reiser and Lavenberg (1980), bypasses the need for computing the normalization constant $C(N)$. MVA exploits a simple relationship between the average customer system time at a network node and the average queue length at that node. Specifically, consider a customer arriving at node i, and let $\bar{S}_i$ be the average system time the customer experiences at i. Moreover, let $\bar{X}_i$ be the average queue length seen by that arrival. Observe that

$$\bar{S}_i = \frac{1}{\mu_i} + \bar{X}_i \frac{1}{\mu_i}$$

where $1/\mu_i$ is the mean service time at node i. In other words, the customer's system time consists of two parts: his own service time, and the total time required to serve all customers ahead of him. It can be shown that in a closed queueing network with N customers, $\bar{X}_i$ is the same as the average queue length at i in a network with $(N-1)$ customers. Intuitively, what a typical customer sees is the network without that customer. Therefore, if we denote by $\bar{X}_i(N)$ and $\bar{S}_i(N)$ the average queue length and average system time at node i respectively when there are N customers in the network, we obtain the following recursive equation:

$$\bar{S}_i(N) = \frac{1}{\mu_i} \left[1 + \bar{X}_i(N-1) \right], \quad i = 1, \ldots, M \tag{8.98}$$

with initial condition:

$$\bar{X}_i(0) = 0, \quad i = 1, \ldots, M \tag{8.99}$$

In addition, we can use Little's Law (8.28) twice. First, for the whole network, we have

$$N = \Lambda_N \sum_{i=1}^{M} \bar{S}_i(N) \tag{8.100}$$

where Λ_N is the throughput and N is the fixed number of customers in the network. Second, noting that the throughput of all nodes at steady state must be the same, we apply Little's Law to any one node and get

$$\bar{X}_i(N) = \Lambda_N \bar{S}_i(N), \quad i = 1, \ldots, M \tag{8.101}$$

The four equations (8.98) through (8.101) define an algorithm through which $\bar{X}_i(N), \bar{S}_i(N)$, and Λ_N can be evaluated for various values of $N = 1, 2, \ldots$

8.7.4 Product Form Networks

The open and closed queueing network models considered thus far are referred to as *product form* networks. Obviously, this is because the stationary state probabilities $\pi(n_1, \ldots, n_M)$ can be expressed as products of terms as in (8.86) or (8.93). The ith such term is determined by parameters of the ith node only, which makes a decomposition of the network easy and efficient.

One may be tempted to conclude that it is the Markovian nature of the service processes (and, in the case of open networks, arrival processes as well) which lends itself to the elegance and convenience of a product form solution. In addition, our models have been constrained to a single customer class and simple FCFS queueing disciplines. Yet, as it turns out, there exist significantly more complicated types of networks which are also of the product form variety. This suggests that the real reason for product form solutions is not the

Markovian nature of the service or arrival processes, but rather the specific structure of a queueing system that permits some form of decomposition. We will not, however, elaborate further on this topic; the reader is referred to some specialized books on queueing networks listed at the end of this chapter.

The most notable type of product form network is one often referred to as a BCMP network after the initials of the researchers who studied it (Baskett, Chandy, Muntz, and Palacios, 1975). This is a closed network with K different customer classes. Class $k, k = 1, \ldots, K$, is characterized by its own routing probabilities and service rates, that is,

p_{ij}^k is the probability that a class k customer is routed from node i to node j

μ_i^k is the service rate of a class k customer at node i

Four different types of nodes are allowed:

1. A single-server node with exponentially distributed service times and $\mu_i^k = \mu_i$ for all classes. In addition, a FCFS queueing discipline is used. This is the simplest node type which we have used in our previous analysis as well.

2. A single-server node and any service time distribution, possibly different for each customer class, as long as each such distribution is differentiable. The queueing discipline here must be of the *processor-sharing* (PS) type. To briefly explain the PS discipline, consider a policy where each customer in queue receives a fixed "time slice" of service in round-robin fashion and then returns to the queue to wait for more service if necessary. The PS discipline is obtained when this time slice is allowed to become vanishingly small.

3. The same as before, except that the queueing discipline is of the Last-Come-First-Served (LCFS) type with a preemptive resume (PR) capability. This means that a new customer can preempt (i.e., interrupt) the one in service, with the preempted customer resuming service at a later time.

4. A node with an infinite number of servers and any service time distribution, possibly different for each customer class, as long as each such distribution is differentiable.

In this type of network, the state at each node is a vector of the form

$$\mathbf{X}_i = [X_{i1}, X_{i2}, \ldots, X_{iK}]$$

where X_{ik} is the number of class k customers at node i. The actual system state is the vector

$$\mathbf{X} = [\mathbf{X}_1, \mathbf{X}_2, \ldots, \mathbf{X}_K]$$

and, assuming the population of class k is N_k, we must always satisfy the condition $\sum_{j=1}^{M} X_{ik} = N_k$ for all k. We can see that notation can get rather complicated, and that the state space can easily become extremely large. The actual product form solution is cumbersome and is omitted (it can be found in Baskett, Chandy, Muntz, and Palacios, 1975; or several queueing theory textbooks such as Trivedi, 1982). The point to remember, however, is that product form networks need not be Markovian and may involve multiple customer classes and some non-FCFS queueing disciplines.

8.8 NON-MARKOVIAN QUEUEING SYSTEMS

Much of our success in analyzing queueing systems up to this point has rested on our ability to model them as Markov chains, and, in particular, birth-death chains. At the root of this success is our assumption that all events (arrivals, departures) are Markovian. This means that events are generated by Poisson processes (fixed event rate) or by Poisson processes with state-dependent event rates. As we have seen in Chapter 6, this allows us to reduce the Generalized Semi-Markov Process (GSMP) $\{X(t)\}$ representing a queue length into a Markov chain.

We now turn our attention to queueing systems with non-Markovian event processes. Thus, we are back into the realm of GSMPs. The main complication here is that the system state can no longer be specified as the queue length X all by itself. In the absence of the memoryless property, we must also "remember" the residual lifetimes (or the ages) of all events. This point is made all too clear by recalling the mechanism for determining the triggering event E' in a GSMP. Let the observed current state be x, with a feasible event set $\Gamma(x)$. Each event $j \in \Gamma(x)$ has a clock value (residual lifetime) Y_j. Then, the probability that the triggering event is some $i \in \Gamma(x)$ is given by the probability that event i has the smallest clock value among all events in $\Gamma(x)$:

$$P[E' = i] = P\left[Y_i = \min_{j \in \Gamma(x)} \{Y_j\}\right]$$

To determine this probability we generally need information on the random variables $Y_j, j \in \Gamma(x)$. It is only in the case of Markov chains that the memoryless property allowed us to obtain (6.67):

$$P[E' = i] = \frac{\lambda_i}{\Lambda(x)}$$

where λ_i is the Poisson rate of event i and $\Lambda(x) = \sum_{j \in \Gamma(x)} \lambda_j$. In this case, no information on the event clock values Y_j is needed. As an example, for an $M/M/1$ queueing system with arrivals a and departures d, the probability above for all states $x > 0$ becomes

$$P[E' = a] = \frac{\lambda}{\lambda + \mu}, \quad P[E' = d] = \frac{\mu}{\lambda + \mu}$$

To handle non-Markovian event processes in queueing systems, two general approaches can be followed, as described below.

The first approach attempts to build a specific non-Markovian event process by various combinations of Markovian ones. The idea of using simple "building-block-models" to construct more elaborate models is not unusual; it is successfully used in several engineering areas, as in the construction of a complex signal as a Fourier series consisting of sinusoidal "building blocks". In our case, the Poisson process is an obvious convenient "building block".

Let $e_1, e_2, \ldots, e_m$ be a collection of Poisson events with a common fixed rate, which we intend to use as "building blocks". Then, we may define a new event e as occurring only after all events $e_1, e_2, \ldots, e_m$ have occurred in series, one immediately following the other. By adjusting the number m of building blocks used we can expect to generate a variety of events e with different lifetime characteristics. Alternatively, we may define e as occurring only after some event $e_i, i = 1, \ldots, m$, has occurred, where e_i is chosen randomly with probability q_i. Again, by adjusting the probabilities q_i and the number m, we should be able to generate different event processes. In this way, we decompose an event e into stages, each stage represented by a "building-block-event" e_i. The idea here is to preserve the basic Markovian structure (since all building-block-events are Markovian) at the expense of a larger state space (since we will be forced to keep track of the stage the system is at in order to be able to tell when e does in fact occur).

The second approach is to seek structural properties of a system that can be exploited no matter what the nature of the event processes are and without increasing the state space. This is a more difficult task, and we cannot expect to work miracles for completely arbitrary systems. We can, however, identify some special cases for which analytical results of significant generality can be obtained. We will limit our discussion to the $M/G/1$ queueing system, that is, a system with Poisson arrivals, but arbitrary service time distributions.

In what follows, we describe the key elements of the first approach without getting into a great amount of technical detail. We then present one basic result pertaining to the $M/G/1$ queueing system. This result, known as the *Pollaczek-Khinchin mean value formula*, is of great practical importance, and serves to illustrate the second approach. The study of more general queueing systems, including the $G/G/1$ system, can easily become extremely tedious and complex. As already mentioned, however, our goal here is not to engage in the intricacies of advanced queueing theory.

Finally, there have been several efforts to deal with queueing systems of virtually arbitrary generality by means of approximation techniques. These techniques have given rise to a variety of commercially available software packages, intended to capture the complexities of the "real world". Some of these packages are widely used in practice, especially in the design and performance evaluation of actual manufacturing systems, communication networks, and computer systems. We provide a brief overview of these approximate techniques in Section 8.8.3.

8.8.1 The Method of Stages

Let $e_1, e_2, \ldots, e_m$ be a collection of events, each generated by a Poisson process of rate λ. The simplest way to combine these events in order to define a new event e is by requiring them to occur serially: The event e is said to occur after $e_1, e_2, \ldots, e_m$ have all occurred in immediate succession. Thus, if the lifetime of event e_i is denoted by Z_i, then the lifetime of event e is defined by the sum $Z_1 + \ldots + Z_N$. We know that $Z_i \sim 1 - e^{-\lambda t}$, and the question we pose is: *What is the lifetime distribution of event e defined in this manner?*

SERVER

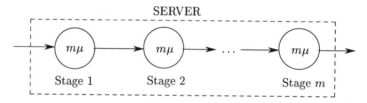

Figure 8.27: A server with m serial (Erlang) stages.

The idea of constructing an event by forming a series of independent Poisson events is due to A. K. Erlang. The construction procedure is best described by considering a server whose average service time is known to be $1/\mu$. However, the service time is not exponentially distributed. We now visualize the server as consisting of m identical and independent stages operating as follows (see Fig. 8.27). Upon arrival, a customer enters stage 1. Stage 1 functions as a server with exponentially distributed service times and rate $m\mu$. When service at stage 1 is complete, the customer proceeds to stage 2, and the process repeats until the customer completes service at stage m and departs. The important observation here is that no new customer can enter the server while this process is ongoing. In other words, *while any one stage is busy, the whole server is considered to be busy.*

Let Z_i be the service time at stage i in this server, and let Z be the actual service time. Clearly,

$$Z = Z_1 + \ldots + Z_m \tag{8.102}$$

Since Z_i is exponentially distributed with rate $m\mu$, our first observation is that

$$E[Z] = m\left(\frac{1}{m\mu}\right) = \frac{1}{\mu} \tag{8.103}$$

Therefore, the average service time in this construction matches the actual average service time, which was known to be $1/\mu$.

We now proceed to determine the distribution of the service time Z. From (8.102) and our assumption that all stages are mutually independent, we can obtain the pdf of Z as the m-fold convolution of the exponential pdf's each with rate $m\mu$ (see Appendix I). Omitting the details of this calculation, the final result is the pdf of Z, $f_Z(t)$:

$$f_Z(t) = \frac{1}{(m-1)!} m\mu (m\mu t)^{m-1} e^{-m\mu t} , \quad t \geq 0 \qquad (8.104)$$

with a corresponding cdf:

$$F_Z(t) = 1 - e^{-m\mu t} \sum_{i=0}^{m-1} \frac{(m\mu t)^i}{i!}, \quad t \geq 0 \qquad (8.105)$$

This is known as the m-stage *Erlang distribution function* and is sometimes denoted by E_m. It is a two-parameter (m and μ) family of distribution functions. However, its mean, which we obtained in (8.103), only depends on μ. The implication is that we can construct a variety of non-exponential service time distributions which have the same fixed mean, but different higher moments. For example, the variance of Z can be calculated to give

$$Var[Z] = \frac{1}{m\mu^2} \qquad (8.106)$$

By varying m, we can generate service time distributions ranging from the pure exponential (when $m = 1$) to a deterministic one (when $m \to \infty$ and hence $Var[Z] \to 0$). Note that increasing m tends to decrease the variance above. Moreover, the variance of the distributions we can construct can never be greater than that of the exponential, $1/\mu^2$.

In summary, the "Erlang server" of Fig. 8.27 is a clever device for modeling a class of non-Markovian events, while preserving a Markovian structure. This is illustrated by considering an $M/E_m/1$ queueing system, where customers arrive according to a Poisson process of rate λ, and get served according to an m-stage Erlang distribution. In this case, the information required to specify the state of the system consists of two parts: the queue length X, $X \in \{0, 1, \ldots\}$, and the current stage K of a customer in service, $K \in \{1, \ldots, m\}$ provided $X > 0$. This state description, however, can be condensed into a single state variable, $\tilde{X}$, defined as

$$\tilde{X} = \text{total number of service stages left in the system}$$

Thus, $\tilde{X} = 0$ if $X = 0$, and $\tilde{X} = (X-1)m + (m - K + 1) = mX - K + 1$ if $X > 0$. The resulting state transition rate diagram is shown in Fig. 8.28.

The model is a Markov chain, but it is no longer a birth-death chain. This is because an arrival causes a transition from state n to state $(n + m)$, that is, it adds m new service stages to be completed. If $m = 1$, however, we obtain the familiar $M/M/1$ state transition diagram of Fig. 8.6.

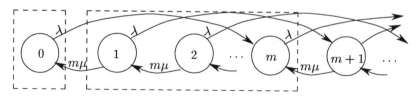

Figure 8.28: State transition rate diagram for the $M/E_m/1$ queueing system.

Finally, it is worth remembering that *the Erlang distribution E_m with mean $1/\mu$ characterizes the sum of m independent exponentially distributed random variables, each with parameter $(m\mu)$.*

As already pointed out, one limitation of the Erlang server with serial stages is that the distributions we can generate can never have a variance greater than $1/\mu^2$. The *coefficient of variation C_X* of a random variable X is defined as the ratio of its standard deviation over its mean (see also Appendix I). It provides a good measure of the uncertainty contained in a physical quantity represented by X ($C_X = 0$ for a deterministic quantity, and $C_X \to \infty$ when $Var[X] \to \infty$). For the service time Z of the m-stage Erlang server, we have

$$C_Z = \left(\frac{1}{m}\right)^{1/2}$$

Thus, we see that this method limits us to distributions such that $0 < C_Z \leq 1$ (as $m \to \infty, C_Z \to 0$).

This limitation may be overcome by a *parallel* - instead of *serial* - configuration of stages, as illustrated in Fig. 8.29. In this case, an arriving customer is routed to stage i with probability $q_i, i = 1, \ldots, m$. Stage i functions as a server with exponentially distributed service times and rate μ_i. When service at the selected stage i is complete, the customer departs. As in the Erlang configuration, no new customer can enter the server while any one of the m stages is busy.

Since the service time at stage i of this server is a random variable with distribution $1 - e^{-\mu_i t}$, and stage i is chosen with probability q_i, we can immediately see that the distribution, $F_Z(t)$, of the service time Z is

$$F_Z(t) = \sum_{i=1}^{m} q_i(1 - e^{-\mu_i t}), \quad \sum_{i=1}^{m} q_i = 1, \quad t \geq 0 \qquad (8.107)$$

This is known as the m-stage *hyperexponential distribution function* and is usually denoted by H_m. Since the mean and variance of each exponentially

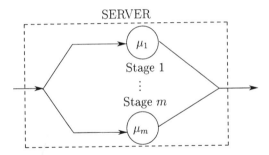

Figure 8.29: A server with m parallel stages.

distributed stage are known to be $1/\mu_i$ and $1/\mu_i^2$ respectively, the mean and variance of the hyperexponential distribution can easily be evaluated:

$$E[Z] = \sum_{i=1}^{m} \frac{q_i}{\mu_i}, \qquad Var[Z] = 2\sum_{i=1}^{m} \frac{q_i}{\mu_i^2} - \left(\sum_{i=1}^{m} \frac{q_i}{\mu_i}\right)^2 \qquad (8.108)$$

It can then be shown that the coefficient of variation is such that $C_Z \geq 1$. Thus, the hyperexponential stage configuration can be used to generate distributions with greater variance than that of a simple exponential, while still maintaining the basic Markovian structure.

Naturally, one can now combine Erlang and hyperexponential distributions to construct mixed series and parallel configurations of considerable generality. The tradeoff in this approach is between the level of generality required and the increase in the state space introduced by each new stage. It is also worth pointing out that in real-time applications it is desirable to observe the state of the system if we are to use such information for control purposes. Whereas a queue length is easy to measure and keep track of, there is no obvious way to know what stage (which is a fictitious device) a real server might be in.

8.8.2   Mean Value Analysis of the $M/G/1$ Queueing System

The simplest non-Markovian queueing system that can be analyzed in detail is the $M/G/1$ system, where customers arrive according to a Poisson process, but the service times form an iid sequence characterized by an arbitrary distribution (strictly speaking, we should employ the notation GI for the service process, but G has been the one traditionally used with the understanding that service times are in fact iid). It is possible to explicitly obtain the probability distribution of the state (queue length) $X, P[X = n], n = 0, 1, \ldots$ (for example, see Kleinrock, 1975). However, we restrict ourselves here to the derivation of the average queue length $E[X]$, which is referred to as "mean value analysis." The result is known as the *Pollaczek-Khinchin formula* (or PK formula) and finds wide use in many applications:

$$E[X] = \frac{\rho}{1-\rho} - \frac{\rho^2}{2(1-\rho)}(1 - \mu^2\sigma^2) \qquad (8.109)$$

where

$1/\mu$ is the average service time
σ^2 is the variance of the service time distribution
$\rho = \lambda/\mu$ is the traffic intensity, as defined in (8.10)
λ is the Poisson arrival rate

We can immediately see that for exponentially distributed service times $\sigma^2 = 1/\mu^2$, and (8.109) reduces to the average queue length of the $M/M/1$ system, $E[X] = \rho/(1-\rho)$, as in (8.39).

The derivation of the PK formula requires some more effort than what was needed to obtain most of the other results of this chapter. To begin with, we need a general result pertaining to single-server queueing systems. This result addresses a question similar to that of Theorem 8.1 in section 8.6. There, we showed that the probability that an arriving customer from a Poisson process finds n customers in the system is the same as the probability that the system state is n at some random time instant. Here, we consider the event

[departing customer at time t leaves $X(t) = n$]

and compare it to the event

[arriving customer at time t finds $X(t) = n$]

for a queueing system with general arrival and service characteristics. In general,

P[arriving customer at time t finds $X(t) = n$]
$\neq P$[departing customer at time t leaves $X(t) = n$]

However, as long as customers arrive and depart one at a time (i.e., the queue length cannot change by more than +1 or -1 at any event time), the following is true: At steady state (if it exists) the two probabilities are equal. This is stated as a theorem below. Recall the notation introduced in (8.33):

$$\alpha_n(t) \equiv P[\text{arriving customer at time } t \text{ finds } X(t) = n]$$

and define similarly:

$$\delta_n(t) \equiv P[\text{departing customer at time } t \text{ leaves } X(t) = n]$$

Theorem 8.3 Consider a queueing system where customers arrive and depart one at a time. If either one of the limits below exists, then:

$$\lim_{t \to \infty} \alpha_n(t) = \lim_{t \to \infty} \delta_n(t) \tag{8.110}$$

Proof. At the kth arrival instant, let X_k^a be the queue length seen by the arriving customer. Similarly, at the kth departure instant, let X_k^d be the queue length left behind by the departing customer. To establish (8.110), we will show that the limiting probability distributions corresponding to $\alpha_n(t)$ and $\delta_n(t)$ are equal:

$$\lim_{k \to \infty} P[X_k^a \leq n] = \lim_{k \to \infty} P[X_k^d \leq n] \tag{8.111}$$

First, consider the kth arrival instant. Let N_k^d be the total number of departures that have taken place prior to that instant. Therefore, just before this arrival, the system has received a total of $(k - 1)$ customers of which X_k^a are in queue and N_k^d have already departed, that is,

$$k - 1 = X_k^a + N_k^d$$

If we assume that $X_k^a \leq n$, for some $n = 0, 1, \ldots$, then it follows that $X_k^a = k - 1 - N_k^d \leq n$, or

$$N_k^d \geq k - 1 - n$$

In other words, the $(k - 1 - n)$th departure must have taken place prior to the kth arrival.

Next, let us consider the $(k - 1 - n)$th departure instant. Thus, X_{k-1-n}^d is the queue length left behind by the departing customer. Let N_{k-1-n}^a be the total number of arrivals by that time, and we have, as before

$$N_{k-1-n}^a = X_{k-1-n}^d + k - 1 - n$$

Since this departure occurs before the kth arrival, at most $(k - 1)$ arrivals could have taken place by this time, that is, $N_{k-1-n}^a \leq k - 1$. It follows that $N_{k-1-n}^a = X_{k-1-n}^d + k - 1 - n \leq k - 1$, or

$$X_{k-1-n}^d \leq n$$

Therefore, we have established that the event $[X_k^a \leq n]$ implies the event $[X_{k-1-n}^d \leq n]$. By a similar argument, we can also show that $[X_{k-1-n}^d \leq n]$ implies $[X_k^a \leq n]$. Since the two events imply each other, we have

$$P[X_k^a \leq n] = P[X_{k-1-n}^d \leq n]$$

It remains to take the limit as $k \to \infty$ (which we assume exists) for any fixed $n = 0, 1, \ldots$ We get

$$\lim_{k \to \infty} P[X_k^a \leq n] = \lim_{k \to \infty} P[X_{k-1-n}^d \leq n] = \lim_{k \to \infty} P[X_k^d \leq n]$$

and (8.111) is obtained. **Q.E.D.**

For our purposes, this theorem has the following implication. In an $M/G/1$ system, the arrival process is Poisson, therefore Theorem 8.1 holds, and we have

$$\alpha_n(t) = \pi_n(t)$$

where $\pi_n(t)$ is the probability of the event $[X(t) = n]$. Combining this with Theorem 8.3, we obtain the steady state result:

$$\lim_{t \to \infty} \delta_n(t) = \lim_{t \to \infty} \pi_n(t) \tag{8.112}$$

We now proceed with the derivation of the PK formula in (8.109), which primarily rests on choosing special points on a sample path of an $M/G/1$ system and establishing a simple relationship for the state changes taking place between two such consecutive points. In particular, we choose to *observe the state at all departure times.* The derivation consists of six steps as detailed below.

Step 1: Observe the state just after departure events and establish a simple recursive equation.

Let $k = 1, 2, \ldots$ index departure events, and let X_k^d be the *queue length just after the kth departure.* Assume the system is initially empty, that is, $X_0 = 0$. As usual, Z_k denotes the kth service time. Let N_k be the *number of customer arrivals during the kth service interval* of length Z_k. We can obtain a simple relationship between X_k^d and X_{k-1}^d as follows:

$$X_k^d = \begin{cases} X_{k-1}^d - 1 + N_k & \text{if } X_{k-1}^d > 0 \\ N_k & \text{if } X_{k-1}^d = 0 \end{cases} \tag{8.113}$$

In the first case, the queue length is not empty prior to the kth service time; the system loses the departing customer and gains the N_k arrivals during the kth service interval. In the second case, the queue is empty prior to the kth service time; a customer arrives during the ensuing idle period and departs, so that the net gain is the N_k arrivals. The two cases are illustrated in Fig. 8.30.

Define the *indicator function* $\mathbf{1}(X_{k-1}^d > 0)$:

$$\mathbf{1}(X_{k-1}^d > 0) = \begin{cases} 1 & \text{if } X_{k-1}^d > 0 \\ 0 & \text{otherwise} \end{cases}$$

and hence rewrite (8.113) as follows:

$$X_k^d = X_{k-1}^d - \mathbf{1}(X_{k-1}^d > 0) + N_k \tag{8.114}$$

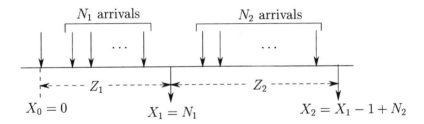

N_1 arrivals N_2 arrivals

$X_0 = 0$ $X_1 = N_1$ $X_2 = X_1 - 1 + N_2$

Figure 8.30: State updates after departures.

Step 2: Take expectations of (8.114) at steady state.

Let $X(t)$ be the state at any time t. We will assume that as $t \to \infty$ there exists a stationary probability distribution

$$\lim_{t \to \infty} P[X(t) = n] = P[X = n], \qquad n = 0, 1, \ldots$$

where, as usual, X is the queue length at steady state. Then, using (8.112), this distribution is the same as that of the queue length left behind by a departing customer at steady state, that is,

$$\lim_{k \to \infty} P[X_k^d = n] = P[X = n], \qquad n = 0, 1, \ldots$$

Therefore,

$$\lim_{k \to \infty} E[X_k^d] = \lim_{k \to \infty} E[X_{k-1}^d] = E[X]$$

and taking expectations in (8.114) at steady state gives

$$E[\mathbf{1}(X > 0)] = E[N] \tag{8.115}$$

Step 3: Square both sides of (8.114), and then take expectations at steady state.

First, squaring both sides of (8.114) and taking expectations gives

$$\begin{aligned}
E\left[(X_k^d)^2\right] = &\, E\left[(X_{k-1}^d)^2\right] + E\left[(\mathbf{1}(X_{k-1}^d > 0))^2\right] \\
&- 2E\left[\mathbf{1}(X_{k-1}^d > 0) \cdot X_{k-1}^d\right] + E\left[N_k^2\right] \\
&+ 2E\left[N_k X_{k-1}^d\right] - 2E\left[\mathbf{1}(X_{k-1}^d > 0) \cdot N_k\right]
\end{aligned}$$

The Poisson arrival process is independent of all past history and of the service process, therefore the random variables X_{k-1}^d and N_k are independent. It follows that $E\left[N_k X_{k-1}^d\right] = E[N_k]E\left[X_{k-1}^d\right]$, and $E\left[\mathbf{1}(X_{k-1}^d > 0) \cdot N_k\right] = E\left[\mathbf{1}(X_{k-1}^d > 0)\right] E[N_k]$. Moreover, note that

$$E\left[\mathbf{1}(X_{k-1}^d > 0) \cdot X_{k-1}^d\right] = E\left[X_{k-1}^d\right]$$

since $\mathbf{1}(X_{k-1}^d > 0) = 0$ for $X_{k-1}^d = 0$ and $\mathbf{1}(X_{k-1}^d > 0) = 1$ otherwise. Using these facts and evaluating the equation above at steady state, as in *Step 2*, we get

$$
\begin{aligned}
0 = & \ E\left[(\mathbf{1}(X > 0))^2\right] - 2E[X] + E[N^2] \\
& + 2E[N]E[X] - 2E[\mathbf{1}(X > 0)]E[N]
\end{aligned}
$$

or

$$
\begin{aligned}
2E[X](1 - E[N]) = & \ E[(\mathbf{1}(X > 0))^2] + E[N^2] \\
& - 2E[\mathbf{1}(X > 0)]E[N]
\end{aligned}
\tag{8.116}
$$

Thus, to obtain $E[X]$, we need to evaluate the first and second moments of the random variables N and $\mathbf{1}(X > 0)$. This is done in the next two steps.

Step 4: Evaluate $E[N]$ and $E\left[N^2\right]$.

The random variable N counts the number of Poisson events in an interval of length Z, also a random variable. If Z were known, say $Z = z$, then the mean of a Poisson distribution is simply $E[N] = \lambda z$. This suggests conditioning on Z, and using the rule of total probability over all possible values of Z. Specifically,

$$
E[N] = \sum_{n=0}^{\infty} n P[N = n]
$$

and by conditioning the event $[N = n]$ on $[Z = z]$ and integrating over all possible z:

$$
P[N = n] = \int_0^{\infty} P[N = n \mid Z = z] f_Z(z) dz
$$

where $f_Z(z)$ is the pdf of the service time Z. Of course, $P[N = n \mid Z = z]$ is simply the familiar Poisson distribution in (6.34):

$$
P[N = n \mid Z = z] = \frac{(\lambda z)^n}{n!} e^{-\lambda z}
$$

We can now evaluate $E[N]$ as follows:

$$
E[N] = \sum_{n=0}^{\infty} n \int_0^{\infty} \frac{(\lambda z)^n}{n!} e^{-\lambda z} f_Z(z) dz = \int_0^{\infty} \left[\sum_{n=0}^{\infty} n \frac{(\lambda z)^n}{n!} e^{-\lambda z} \right] f_Z(z) dz
$$

The sum in brackets is simply the mean of the Poisson distribution with z fixed, that is, $E[N \mid Z = z]$. This was derived in (6.35) and is given by λz. Therefore,

$$
E[N] = \lambda \int_0^{\infty} z f_Z(z) dz = \lambda E[Z]
$$

Using our definition $E[Z] = 1/\mu$ and the definition of the traffic intensity $\rho = \lambda/\mu$, we get

$$E[N] = \rho \tag{8.117}$$

The derivation of $E\left[N^2\right]$ is very similar. Proceeding by conditioning on $[Z = z]$, we get:

$$E\left[N^2\right] = \int_0^\infty \left[\sum_{n=0}^\infty n^2 \frac{(\lambda z)^n}{n!} e^{-\lambda z}\right] f_Z(z)dz$$

where the sum in brackets is the second moment of the Poisson distribution, derived in (6.36), that is, $(\lambda z)^2 + \lambda z$. Thus,

$$E\left[N^2\right] = \int_0^\infty (\lambda^2 z^2 + \lambda z) f_Z(z)dz = \lambda^2 E[Z^2] + \lambda E[Z]$$

The second term of the right-hand side above is once again the traffic intensity ρ. Introducing the variance σ^2 of the service time Z, the first term becomes $\lambda^2(\sigma^2 + 1/\mu^2)$. Therefore,

$$E\left[N^2\right] = \lambda^2 \sigma^2 + \rho^2 + \rho \tag{8.118}$$

Step 5: Evaluate $E[\mathbf{1}(X > 0)]$ and $E\left[(\mathbf{1}(X > 0))^2\right]$.

Combining (8.115) and (8.117), we immediately obtain

$$E[\mathbf{1}(X > 0)] = \rho \tag{8.119}$$

By the definition of the indicator function, it also immediately follows that

$$E[(\mathbf{1}(X > 0))^2] = E[\mathbf{1}(X > 0)] = \rho \tag{8.120}$$

Step 6. Evaluate $E[X]$ from (8.116).

Returning to (8.116) with (8.117) through (8.120), we get

$$2E[X](1 - \rho) = \rho + \lambda^2 \sigma^2 + \rho^2 + \rho - 2\rho^2$$

or

$$E[X] = \frac{\rho}{1 - \rho} - \frac{\rho^2}{2(1 - \rho)}(1 - \mu^2 \sigma^2)$$

which is precisely the PK formula in (8.109).

Remark. It is interesting to compare the $M/M/1$ case with the $M/D/1$ case (deterministic service times). In the $M/M/1$ case, $\sigma^2 = 1/\mu^2$, and (8.109) gives

$$E[X]_{M/M/1} = \frac{\rho}{1-\rho}$$

In the $M/D/1$ case, $\sigma^2 = 0$, and (8.109) gives

$$E[X]_{M/D/1} = \frac{\rho}{1-\rho} - \frac{\rho^2}{2(1-\rho)}) = \frac{\rho}{1-\rho}\left(1 - \frac{\rho}{2}\right)$$

We see that the $M/D/1$ system has a smaller average queue length. This is a manifestation of a point we have already made: Randomness creates more waiting. In this case, the randomness in the exponentially distributed service times causes an additional $\rho^2/2(1-\rho)$ customers to be queued on the average compared to constant service times. Also note that as $\rho \to 1$, the ratio $E[X]_{M/D/1}/E[X]_{M/M/1}$ approaches $1/2$.

From (8.109) we can also easily obtain $E[S]$, the *average system time* for the $M/G/1$ system. Using Little's Law, we have $E[X] = \lambda E[S]$. Thus,

$$E[S] = \frac{1/\mu}{1-\rho} - \frac{\rho/\mu}{2(1-\rho)})(1 - \mu^2\sigma^2) \qquad (8.121)$$

Finally, since $E[S] = E[W] + 1/\mu$, we can also obtain the *average waiting time*:

$$E[W] = \frac{\rho}{2\mu(1-\rho)})(1 + \mu^2\sigma^2) \qquad (8.122)$$

Remark. Comparing once again the $M/M/1$ case ($\sigma^2 = 1/\mu^2$) with the $M/D/1$ case ($\sigma^2 = 0$) in terms of the average waiting time, gives the following interesting relationship:

$$E[W]_{M/D/1} = \frac{\rho}{2\mu(1-\rho)}) = \frac{1}{2}E[W]_{M/M/1} \qquad (8.123)$$

Again, randomness in the service times creates more waiting, in this case by a factor of 2 in the $M/M/1$ system, regardless of the traffic intensity.

8.8.3 Software Tools for the Analysis of General Queueing Networks

We saw that the analysis of Markovian queueing networks entails a considerable amount of computational complexity. When we are interested in the design and performance evaluation of complex DES modeled as queueing networks, we often need to try out many different configurations and parameter settings, in which case repeatedly solving large numbers of equations, even of the simple form (8.78), becomes a very tedious process. The development of software packages for the solution of such networks has proven very helpful in practice, although it is fair to say that it has not yet reached its full potential.

In the case of non-Markovian product-form networks (Section 8.7.4) the computational difficulties are even greater, and the use of computer software becomes almost indispensable. Finally, in the case of more general non-Markovian models for which exact analytical solutions are not available, a number of approximate numerical techniques have been developed, and software packages have become commercially available, often requiring modest processing and storage capabilities on standard personal computers.

In this section, we provide a brief discussion of some of the most popular software packages for the analysis of queueing networks, both exact and approximate, primarily based on a survey article by Snowdon and Ammons (1988). This is clearly not a complete list of available software tools; in fact, with increasing computer processing capabilities and better approximation techniques, new ones are continuously being developed both in research and commercial environments.

CAN-Q (Computer Analysis of Networks of Queues)

CAN-Q was developed by Solberg at Purdue University in the late 1970s. It is based on the analysis of closed Markovian networks presented in Section 8.7.3 (using Buzen's algorithm to calculate the normalization constant $C(N)$), and therefore provides an exact solution for the system analyzed. In the input, a user specifies the means of all the service time distributions, and the output typically provides performance measures such as server utilizations and average queue lengths at the network nodes. Some extensions were later provided to CAN-Q in order to model the presence of multiple customer classes. It should be noted that CAN-Q was developed to address issues of design and performance evaluation of manufacturing systems, so it uses the terminology common in these systems (e.g., machines, parts, pallets, etc.)

RESQ (Research Queueing Package)

RESQ was developed by MacNair at the IBM Watson Research Center, also in the late 1970s. It solves networks that conform to the BCMP model discussed in Section 8.7.4. These networks may be open, closed, or mixed. Since the BCMP model is not limited to exponential distributions, the user here may specify the distributions desired (from the class of allowable ones) and its parameters. The output of RESQ is similar to that of CAN-Q in that it includes server utilizations and average queue lengths. The actual software package includes the capability to simulate a specified queueing network (which can be more general than the BCMP model). Moreover, it can be integrated with RESQME (Research Queueing Package Modeling Environment), which has the appealing feature of allowing a user to graphically specify the model on a screen display.

PANACEA (Package for Analysis of Networks of Asynchronous Computers with Extended Asymptotics)

This is a package developed at the AT&T Bell Laboratories in the early 1980s by Ramakrishnan and Mitra. In contrast to the previous two approaches, PANACEA is aimed at analyzing rather general multiclass open, closed, and mixed networks by utilizing asymptotic expansions of various performance measures of interest. Thus, it generally yields approximate solutions, whose accuracy can be adjusted by trading off between the number of terms included in the asymptotic expansion and the computation time, which is user-controlled. In addition to approximations for the standard performance measures (e.g., server utilizations, mean queue lengths), PANACEA also provides upper and lower bounds for these measures, as well as second moments of the queue length distributions.

QNA (Queueing Network Analyzer)

QNA is a collection of software tools, with different ones invoked depending on the network of interest. It was developed by Segal, Whitt, et al. in the early 1980s at the AT&T Bell Laboratories, and it is intended to provide approximations for various performance measures of fairly general non-Markovian queueing networks. The most common version of QNA to date considers open multiclass networks with arrival and service processes (not necessarily exponential) characterized by their first two moments (which the user specifies as part of the input). Customer routing in the network may either be probabilistic (as in the model of section 8.7.2) or deterministic (i.e., each customer is associated with a specific sequence of servers to be visited). A network node is viewed as a $GI/G/m$ system, and customers may be created or combined at a node (to model, for instance, assembly operations in a manufacturing system). The output of QNA includes approximations of server utilizations, the mean and variance of queue lengths, and the mean and variance of total delays experienced by customers as they go through the network.

It should be noted that another software tool developed at the AT&T Bell Laboratories by Melamed, PAW (Performance Analysis Workstation), provides extensive graphics capabilities through which one can conveniently transform a visual model of a system into input files that either PANACEA or QNA can then read.

MVAQ (Mean Value Analysis of Queues) and PMVA (Priority Mean Value Analysis)

Both of these packages are based on the Mean Value Analysis methodology discussed in Section 8.7.3, including some extensions that were provided in the early 1980s for open and mixed networks. MVAQ was developed by Suri and Hildebrant in the early 1980s as a means to extend CAN-Q to more general manufacturing systems. PMVA was developed around the same time by Shalev-Oren, Schweitzer, et al., and it includes the capability to model priority scheduling of customers at various nodes. Both packages are oriented towards manufacturing applications, and they give approximations of performance measures such as the throughput of each customer class, server utilizations, and average queue lengths.

MPX

MPX was developed by Suri and Diehl at Network Dynamics, Inc. in the mid and late 1980s. MPX models manufacturing systems as general open multiclass networks and uses a number of approximation techniques to obtain performance measures such as the throughput of each class and its average flow time through the system, and server utilizations. However, the mathematical machinery is hidden from the user, due to a simple user-interface. MPX also provides the capability for rapid "what if" analysis: The user specifies one or more desired model changes, and the package evaluates their impact to an existing design in terms of various performance measures. This facilitates decision making and enables a user to quickly compare a large number of alternatives.

One modeling feature that existing software for queueing network analysis does not include is that of customer blocking due to finite queueing capacities (as we saw, for instance, in the $M/M/1/K$ system of Section 8.6.4). Moreover, the effect of complex state-dependent control policies for scheduling customers at servers or routing them through the network are also extremely hard to incorporate. Although this observation does not diminish the value of these software tools, it serves to motivate the use of simulation for the study of systems with such complex features, which we will consider in Chapter 10.

SUMMARY

- The *Lindley equation*, $W_k = \max\{0, W_{k-1} + Z_{k-1} - Y_k\}$, describes the basic dynamics of a single-server queueing system operating under the first-come-first-served (FCFS) discipline.

- *Little's Law*, $E[X] = \lambda E[S]$, provides a simple relationship between the mean system time $E[S]$, mean number of customers in the system $E[X]$, and throughput λ for very general queueing systems at *steady state*.

- Using the steady-state solution of birth-death chains, derived in Chapter 7, we can analyze a variety of simple queueing systems and obtain explicit expressions for a number of performance measures of interest (always at steady state). In Section 8.6, we analyzed the $M/M/1$, $M/M/m$, $M/M/\infty$, $M/M/1/K$, $M/M/m/m$, $M/M/1//N$, and $M/M/m/K/N$ systems.

- In queueing systems, there is a fundamental tradeoff between keeping customer delay low and server utilization high. This tradeoff is nonlinear and such that the mean delay is generally insensitive to changes in utilization when utilization is low. However, this sensitivity rapidly increases and approaches infinity as utilization approaches 1. The tradeoff is typified by the relationship derived for the $M/M/1$ system: $E[S] = (1/\mu)/(1-\rho)$, where $E[S] \to \infty$ as $\rho \to 1$.

- Burke's Theorem asserts that the departure process of an $M/M/1$ system is a Poisson process with the same rate as the input process. Using this fact, open and closed networks of Markovian queues can be analyzed by considering their nodes in isolation, and then combining the results into *product form* solutions. In the case of closed networks, the product form solution involves a normalization constant $C(N)$, dependent on the network population size N.

- Product form solutions are not limited to Markovian networks, but allow for certain non-exponential service time distributions, as well as multiple customer classes, and queueing disciplines other than FCFS.

- One way to analyze non-Markovian queueing systems is to approximate non-exponential service and interarrival time distributions by distributions formed by combining exponential building blocks. This leads to the family of *Erlang* and *hyperexponential* distributions.

- For the $M/G/1$ system, we can obtain an explicit expression for the mean queue length which depends on the first two moments of the service distribution only. The result is known as the *Pollaczek-Khinchin formula*.

PROBLEMS

8.1 Consider an $M/M/1$ queueing system with arrival rate λ and service rate μ. The system starts out empty.

(a) Find the average waiting time of the second customer.

(b) Now suppose the service time becomes fixed at $1/\mu$. Find the average waiting time of the second customer for this case, and compare the result to that of part (a).

8.2 Consider an $M/M/1$ queueing system where customers are "discouraged" from entering the queue in the following fashion: While the queue length is less than or equal to K for some $K > 0$, the arrival rate is fixed at λ; when the queue length is greater than K, the arrival rate becomes $\lambda_n = \lambda/n, n \geq K + 1$. The service rate is fixed at $\mu > \lambda$.

(a) Find the stationary probability distribution of the queue length.

(b) Determine the average queue length and system time at steady state.

(c) How far can you reduce the service rate before the average queue length becomes infinite?

8.3 As a promotional gimmick, a supermarket offers $50 to any of its next 1,000 customers who do not find an available teller when they are ready to check out. It is assumed that customers check out according to a Poisson process of rate $\lambda = 3$ customers per minute, they form a single queue, and that all tellers provide service to customers according to a common exponential distribution with parameter $\mu = 2$ customers per minute. The supermarket manager must decide how many tellers to use during the promotional period so as not to exceed his budget of $1000 for this promotion. Being totally ignorant of queueing theory, he has just picked a number out of a hat: 4 tellers.

(a) What would your recommendation be to the manager (does he need more tellers and if so how many, or is he using too many tellers)?

(b) Suppose that the manager also insists that the chances of all the tellers he uses being idle be kept below 5%. Can this criterion be met with the number of tellers you recommended in part (a)?

(*Note*: Do not hesitate to use the computer if "pencil and paper" calculations get too tedious.)

8.4 A bank is trying to determine how many ATMs to install in a location where it estimates that 40 customers per hour will want to use the facility, and that a transaction takes 1 min on the average to be processed. The bank charges $1 for every transaction, and it assumes that whenever a customer arrives and finds all ATMs busy, the customer leaves (representing a revenue loss for the bank). Assuming Poisson arrivals and exponentially distributed transaction processing times, how many ATMs would you recommend so that the expected revenue loss per hour is kept below $5?

8.5 A manufacturing workstation receives a continuous supply of parts and processes them at a rate of 12 parts per hour with exponentially distributed processing times. However, each part needs to be placed on a special fixture before it can be processed. There are N such fixtures available, and the time it takes to remove a processed part from its fixture and load a new part on it is exponentially distributed with mean 15 minutes. Determine the smallest number of fixtures such that the average waiting time of a part at the workstation queue is less than 2 minutes. Does this number of fixtures satisfy the additional requirement that the utilization of the workstation be at least 80%?

8.6 The time to process transactions in a computer system is exponentially distributed with mean 6 sec. There are two different sources from which transactions originate, both modeled as Poisson processes which are independent of each other and of the state of the system. The first source generates transactions at a rate of 4 per minute, and the second source at a rate of 3 per minute. Up to K transactions (waiting to be processed or in process) can be queued in this system. When a transaction finds no queueing space (i.e., it is *blocked*) it is rejected and lost. Determine the minimum value of K such that the blocking probability for source 1 does not exceed 10%, and that of source 2 does not exceed 2%. Once the value of K is determined, calculate the average response time (system time) for tasks that actually get processed.

8.7 In Problem 7.17, one third of all transactions submitted to the local branch of a bank were handled by the bank's central computer, while all the rest were handled by the branch's own computer.

(a) Determine the mean queue lengths at the central computer and at the local branch computer.

(b) Determine the average system time over all transactions processed by this system.

8.8 Some of the probabilistic calculations encountered in queueing theory are facilitated by the use of *Laplace transforms*. In particular, the Laplace transform of a probability density function (pdf) $f(t)$ is given by

$$\mathcal{F}(s) = \int_0^\infty e^{-st} f(t) dt$$

and the usual properties of Laplace transforms can be used to evaluate moments of the pdf or the convolution of two pdf's.

(a) Show that the Laplace transform of an exponential pdf with parameter μ is $\mu/(s + \mu)$.

(b) Derive the result of Theorem 8.2, that is, equation (8.77), using a Laplace transform approach.

8.9 Using a Laplace transform approach, as in Problem 8.8, and some appropriate conditioning calculations, it is possible to derive the cdf of the system time of customers in an $M/M/1$ queueing system at steady state. Show that the result is

$$P[S \leq t] = 1 - e^{-\mu(1-\rho)t}, \quad t \geq 0$$

where, as usual, $\rho = \lambda/\mu$, λ is the Poisson arrival rate, and μ is the exponential service rate. Similarly, show that the cdf of the waiting time at steady state is

$$P[W \leq t] = 1 - \rho e^{-\mu(1-\rho)t}, \quad t \geq 0$$

8.10 Consider a *bulk arrival* system, where at every event time of a Poisson process with rate λ a fixed number of customers, $m \geq 1$, is generated. Each of these customers is then served in standard fashion (one at a time) by a server with exponentially distributed service times with parameter μ.

(a) Draw a state transition rate diagram for this system and write down the corresponding flow balance equations.

(b) Solve the flow balance equations using the z-transform:

$$P(z) = \sum_{n=0}^{\infty} p_n z^n$$

where p_n is the stationary probability of having n customers in the system, $n = 0, 1, \ldots$ Thus, show that

$$P(z) = \frac{\mu p_0 (1 - z)}{\mu + \lambda z^{m+1} - (\lambda + \mu)z}$$

(c) Argue that $P(1) = 1$, and then show that $p_0 = 1 - \lambda m/\mu \equiv 1 - \rho$, so that

$$P(z) = \frac{\mu(1 - \rho)(1 - z)}{\mu + \lambda z^{m+1} - (\lambda + \mu)z}$$

8.11 A manufacturing process involves operations performed on five machines $M_1, \ldots, M_5$. The processing times at the five machines are modeled through exponential distributions with rates $\mu_1 = 2$ parts/min, $\mu_2 = 1.75$ parts/min, $\mu_3 = 2.2$ parts/min, $\mu_4 = 1$ part/min, and $\mu_5 = 0.5$ parts/min.

All parts first arrive at M_1 according to a Poisson process with rate 1 part/min. After being processed at M_1, a fraction $p_{14} = 0.2$ of all parts is sent to machine M_4, while the rest proceed to M_2. All parts processed at M_4 are sent back to M_1 for one more cycle at that machine. Parts processed at M_2 are all routed to M_3. After being processed at M_3, a fraction $p_{34} = 0.2$ of all parts is sent back to M_4, another fraction $p_{35} = 0.2$ is sent to machine M_5, and all other parts leave the system. Finally, all parts processed at M_5 are sent back to M_2 for more processing.

(a) Obtain expressions for the utilizations $\rho_1, \ldots, \rho_5$ at all five machines in terms of the parameters $\mu_1, \ldots, \mu_5$, λ, and p_{14}, p_{34}, p_{35}.

(b) Calculate the throughput of this system (i.e., the rate of finished part departures from M_3).

(c) Calculate the average system time of parts in the system.

(d) Find the probability that the queue length at machine M_2 is greater than 3 parts.

8.12 A packet-switched communication network is viewed as an open queueing network with each link modeled as an $M/M/1$ queueing system. Packets arrive at various nodes as independent Poisson processes, and the length of every packet is assumed to be exponentially distributed. Let λ_{ij} be the rate of packets to be transmitted over link (i, j), and let μ_{ij} be the transmission rate (packets/sec) of link (i, j).

(a) Show that the average delay of packets (measured from arrival at some node to departure at some other node) is given by

$$E[S] = \frac{1}{\Lambda} \sum_{\text{all } (i,j)} \frac{\lambda_{ij}}{\mu_{ij} - \lambda_{ij}}$$

where Λ is the total arrival rate (packets/sec) into the network.

(b) Consider the 5-node 7-link network shown in Fig. 8.31, where the transmission rate of all links is $\mu = 3$ packets/sec. Packets arrive at nodes 1 and 2 with rates $r_1 = 1$ packet/sec and $r_2 = 3$ packets/sec respectively, and have to be transmitted to nodes 3,4,5. The following routing probabilities are used: $p_{12} = p_{14} = 1/2$, $p_{23} = 3/5$, $p_{24} = p_{25} = 1/5$. Find the average delay of packets in the network, and all the link utilizations.

(c) Suppose that all packets arriving at node 1 are destined for node 5. Find the average delay of these packets.

8.13 Consider the closed queueing network of Example 8.5.

(a) Solve the probability flow balance equations for this model and show that equation (8.95) is satisfied.

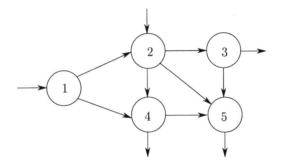

Figure 8.31: Problem 8.12 (b) and (c).

(b) Repeat the solution approach in the example using λ_2 as a free constant and setting $\lambda_2 = p$.

8.14 A closed *cyclic* queueing network consists of M single-server nodes in series with the Mth node connected to the 1st node. In other words, a customer completing service at node $i = 1, \ldots, M - 1$ always proceeds to node $i + 1$, and a customer completing service at node M always proceeds to node 1. There are N customers circulating in this network. For the case $N = 2, M = 3$, let the service times be exponentially distributed with parameters $\mu_1 = 1.5$, $\mu_2 = 1$, $\mu_3 = 1.2$.

(a) Calculate the average time for a customer who just completed service at node 1 to return to the node 1 queue.

(b) Calculate the throughput of the network, i.e., the rate at which customers complete service at node 3.

8.15 Consider the same network as in Problem 8.14 ($N = 2, M = 3$), except that now there is limited queueing capacity at the three nodes. Specifically, suppose that the queueing capacity at node 1 is 2, but at nodes 2 and 3 it is only 1 (i.e., there is only space for a customer in process). Thus, if a customer completes service at node 1 (respectively 2) and node 2 (respectively 3) is busy, then the customer is held at node 1 (respectively 2), and we say that this node is *blocked*.

(a) Draw a state transition rate diagram for this system.

(b) Calculate the probability that any one node in the system is blocked.

(c) Calculate the probability that only one node in the system is actually processing a customer.

8.16 A simple routing problem is illustrated in Fig. 8.32. Customers arrive according to a Poisson process with rate λ. Every arriving customer is routed to queue 1 with probability p and to queue 2 with probability $1-p$. The problem is to determine the value of p such that the average system time of customers in this system is minimized.

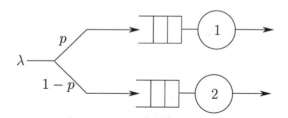

Figure 8.32: Problem 8.16.

(a) Assuming service times at both servers are exponentially distributed with parameters μ_1 and μ_2, find an expression for the value of p that solves this optimization problem. For $\lambda = 2, \mu_1 = 1, \mu_2 = 1.5$, calculate this value of p.

(b) Repeat part (a), except now let server 2 be deterministic, and assume that the service time is fixed at $1/\mu_2$.

(c) For the numerical values given, calculate the average system time under the optimal p in part (a) and in part (b). Compare your results and comment on whether their relative value is as you expected.

SELECTED REFERENCES

- Asmussen, S., *Applied Probability and Queues*, Wiley, New York, 1987.

- Baskett, F., K. M. Chandy, R. R., Muntz, and R. R. Palacios, "Open, Closed, and Mixed Networks with Different Classes of Customers," *Journal of ACM*, Vol. 22, No. 2, pp. 248-260, 1975.

- Bremaud, P., *Point Processes and Queues*, Springer-Verlag, 1981.

- Burke, P. J., "The Output of a Queueing System," *Operations Research*, Vol. 4, pp. 699-704, 1956.

- Buzen, J. P., "Computational Algorithms for Closed Queueing Networks with Exponential Servers," *Communications of ACM*, Vol. 16, No. 9, pp. 527-531, 1973.

- Gordon, W. J., and R. R. Newell, "Closed Queueing Systems with Exponential Servers," *Operations Research*, Vol. 15, No. 2, pp. 254-265, 1967.

- Jackson, J. R., "Jobshop-Like Queueing Systems," *Management Science*, Vol. 10, No. 1, pp. 131-142, 1963.

- Kleinrock, L., *Queueing Systems, Volume I: Theory*, Wiley, New York, 1975.

- Lindley, D. V., "The Theory of Queues with a Single Server," *Proceedings of Cambridge Philosophical Society*, Vol. 48, pp. 277-289, 1952.

- Little, J. D. C., "A Proof of $L = \lambda W$," *Operations Research*, Vol. 9, No. 3, pp. 383-387, 1961.

- Reiser, M., and S. S. Lavenberg, "Mean-Value Analysis of Closed Multichain Queueing Networks," *Journal of ACM*, Vol. 27, No. 2, pp. 313-322, 1980.

- Snowdon, J. L., and J. C. Ammons, "A Survey of Queueing Network Packages for the Analysis of Manufacturing Systems," *Manufacturing Review*, Vol. 1, No. 1, pp. 14-25, 1988.

- Stidham, S., Jr., "A Last Word on $L = \lambda W$," *Operations Research*, Vol. 22, pp. 417-421, 1974.

- Trivedi, K.S., *Probability and Statistics with Reliability, Queuing and Computer Science Applications*, Prentice-Hall, Englewood Cliffs, 1982.

- Walrand, J., *An Introduction to Queueing Networks*, Prentice-Hall, Englewood Cliffs, 1988.

Chapter 9
Controlled Markov Chains

9.1 INTRODUCTION

In Chapter 7 we considered Markov chains as a means to model stochastic DES for which explicit closed-form solutions can be obtained. Then, in Chapter 8, we saw how special classes of Markov chains (mostly, birth-death chains) can be used to model queueing systems. We pointed out, however, that queueing theory is largely "descriptive" in nature; that is, its main objective is to evaluate the behavior of queueing systems operating under a particular set of rules. On the other hand, we are often interested in "prescriptive" techniques, based on which we can make decisions regarding the "best" way to operate a system and ultimately control its performance. In this chapter, we describe some such techniques for Markov chains. Our main objective is to introduce the framework known as *Markov Decision Theory*, and to present some key results and techniques which can be used to control DES modeled as Markov chains. At the heart of these techniques is *dynamic programming*, which has played a critical role in both deterministic and stochastic control theory since the 1960s. The material in this chapter is more advanced than that of previous ones, it involves some results that were published in the research literature fairly recently, and it demands slightly higher mathematical sophistication. The results, however, should be quite gratifying for the reader, as they lead to the solution of some basic problems from everyday life experience, or related to the basic operation of some very familiar DES whose predominant characteristic is resource contention. The last part of this chapter is devoted to some such problems pertaining to queueing systems.

9.2 APPLYING "CONTROL" IN MARKOV CHAINS

Thus far, the Markov chains we considered in previous chapters have had fixed transition probabilities. We have not yet explored the possibility of taking action to control these probabilities in order to affect the evolution of the chain. One simple case of controlling a transition probability was seen in Example 7.14 of Chapter 7, where by selecting a particular parameter value in a two-state chain we were able to adjust the steady state probabilities to meet some desired constraint.

In order to motivate our discussion and illustrate what we mean by "controlling a Markov chain", let us return to a simple gambling problem we first encountered in Chapter 7, where the gambler may now exercise different betting policies.

Example 9.1 (Controlling bets in a simple gambling process)
In the "spin-the-wheel" game of Example 7.4 in Chapter 7, a gambler always bets 1 dollar. He either loses his bet with probability $11/12$, or he wins twice the amount of his bet (2 dollars) with probability $1/12$. Now suppose that he decides to vary the amount of his bet according to some policy (perhaps betting more when his capital increases; or always betting a fixed fraction of his capital; or selecting a fraction of his capital completely at random). In doing so, the gambler is effectively changing the transition probabilities of the Markov chain.

To be more specific, let $u(i)$ be the number of dollars bet when the state (the gambler's capital) is $i = 1, 2, \ldots$ When $i = 0$, the game is obviously over and there is no bet. Thus, $u(i)$ can take any value in the set $\{1, 2, \ldots, i\}$. In Example 7.4 the policy is $u(i) = 1$ for all $i = 1, 2, \ldots$ As a result, the transition probabilities are

$$p_{ij} = \begin{cases} 11/12 & \text{if } j = i - 1 \\ 1/12 & \text{if } j = i + 2 \\ 0 & \text{otherwise} \end{cases}$$

since the bet is lost with probability $11/12$ and the net payoff is $2u(i)$ dollars with probability $1/12$.

Now suppose the gambler adopts the following policy: Bet 1 dollar only if $i \le 2$ and bet 2 dollars otherwise. In this case, the transition probabilities

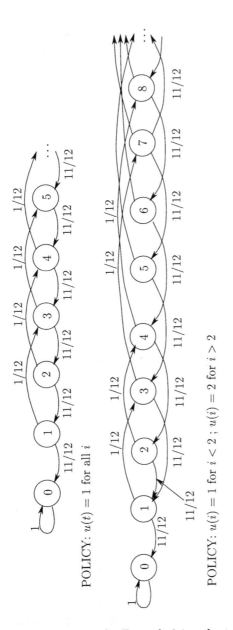

Figure 9.1: State transition diagrams for Example 9.1 under two betting policies. *Under the first policy, the gambler always bets 1 dollar; the state i changes to $(i-1)$ with probability $11/12$, or $(i+2)$ with probability $1/12$. Under the second policy, the gambler increases his bet to 2 dollars if $i > 2$. Thus, for all $i > 2$, the state changes to $(i-2)$ with probability $11/12$, or $(i+4)$ with probability $1/12$.*

become

$$p_{ij} = \begin{cases} 11/12 & \text{if } j = i - 1 \text{ and } i = 1, 2, \text{ or } j = i - 2 \text{ and } i > 2 \\ 1/12 & \text{if } j = i + 2 \text{ and } i = 1, 2, \text{ or } j = i + 4 \text{ and } i > 2 \\ 0 & \text{otherwise} \end{cases}$$

Thus, in general, the Markov chain transition probabilities are controlled by the action taken by the gambler at a particular state. In Fig. 9.1 the state transition diagrams corresponding to the two policies above are shown.

When transition probabilities can be controlled, it is natural to seek to control them in a way that is in some sense "optimal". It is therefore necessary to establish criteria for optimality and a framework within which control problems can be formulated and solved. In the gambling example above, one possible criterion may be to try and prolong the game (i.e., keep the state positive) for as long as possible. Another criterion may be to try and get to a desirable state $i^* > 0$ (where the gambler collects his earnings and quits) before getting to 0.

The theory of controlled Markov chains can become elaborate and mathematically sophisticated, well beyond the level of this book. It is, however, a very exciting area as it allows us to tackle a variety of common problems of resource contention and decision making in dealing with DES. It sometimes enables us to derive results which corroborate our intuition, while occasionally coming up with counter-intuitive ones. Our objective here is limited to a brief introduction to *Markov Decision Processes*, which will allow the reader to formulate a class of control problems and to obtain solutions for at least some simple ones.

9.3 MARKOV DECISION PROCESSES

We begin with a stochastic DES with state space $\mathcal{X}$, and assume that we can observe all state transitions (events). In order to introduce a Markov decision process related to this DES, we need to specify three ingredients: *control actions* taken when a state transition takes place, the *cost* associated with such actions, and *transition probabilities* which may depend on control actions.

When a new state is entered, a *control action* u is selected from a known set of possible control actions, which is denoted by U. There is a cost associated with the selection of a control action u at state i, and we denote this cost by $C(i, u)$. The nature of $C(i, u)$ plays a crucial role in the solution of the problems we will be considering. Most of our discussion will be based on the assumption that this cost is always nonnegative and bounded, that is,

$$0 \le C(i, u) \le K \tag{9.1}$$

for some finite positive real number K, for all controls $u \in U$ and states $i \in \mathcal{X}$. In this case, our cost minimization problems will turn out to be well-defined, and effective methodologies for their solution exist. If $C(i, u)$ is not bounded, the problem of cost minimization becomes significantly more complicated. Nonetheless, unbounded costs are often encountered in practice (see Example 9.2). Fortunately, some of the solution techniques applicable under assumption (9.1) can be extended to the case where $C(i, u)$ cannot be bounded. We will discuss such extensions in Section 9.4.3.

The rule based on which control actions are chosen is called a *policy*. A policy may be quite arbitrary, including, for instance, the possibility of randomly selecting control actions from the set U. We will limit ourselves, however, to a class of policies which is of most interest in practice, characterized by two properties: (a) a control action is not chosen at random, and (b) a control action chosen when the state is i depends only on i. This type of policy is called *stationary*.

Under a stationary policy, a control action is a mapping from the state set $\mathcal{X}$ to the set U, that is, it is a function of the form $u(i)$, $i \in \mathcal{X}$. Sometimes, not all control actions in U may be allowable when the state is i. Thus, we let U_i denote the subset of U containing all control actions *admissible* at state i. We will subsequently use the symbol π to denote a policy.

Now, given that $u(i)$ has been chosen at state i, the next state is selected according to transition probabilities $p_{ij}[u(i)]$, which depend on the value of i only. In addition, for any state i, we assume that the holding time is exponentially distributed with rate parameter $\Lambda(i)$. Recalling the discussion on Markov chain model specifications in Chapter 7 (in particular, Section 7.3.7) , we immediately see that by specifying $p_{ij}[u(i)]$ and $\Lambda(i)$, $i, j \in \mathcal{X}$, we have just specified a Markov chain model. The only thing new here is the fact that the transition probabilities $p_{ij}[u(i)]$ depend on the particular policy π that we wish to adopt.

9.3.1   Cost Criteria

Our objective is to determine policies π which are "optimal" in some sense. Thus, we need to specify a cost criterion that we should seek to minimize. Let the process we have defined above be denoted by $\{X(t)\}$. If at time t the state is $X(t)$ and a specific policy π is given, then a specific action $u(t)$ is taken which depends on $X(t)$, and the resulting cost under π is $C[X(t), u(t)]$.

We now consider several common forms of cost criteria.

- **Total Expected cost over a Finite Horizon.** If we are only interested in the behavior of the chain over a finite time interval $[0, T]$, then we can define the following cost criterion, which measures the total expected cost under policy π:

$$V_\pi(x_0) = E_\pi \left[\int_0^T C[X(t), u(t)]dt \right] \tag{9.2}$$

In this expression, the expectation E_π is evaluated over all states, and the subscript π is there to remind us that a particular policy π has been fixed. The cost $V_\pi(x_0)$ is specified with respect to a given initial state x_0.

■ **Total Expected Discounted Cost over an Infinite Horizon.** If we measure "cost" through a strictly monetary value, then one may question the previous criterion on the following basis: If $C[X(t), u(t)] = 1$ dollar, this cost may mean a lot less at time $t' > t$ than it does at time t. After all, we know from elementary economics that the value of money is discounted over time due to inflation and other factors. It is also generally true that we are more willing to incur a cost in the future than we are willing to incur it right now. Moreover, we have seen in previous chapters that much of our analysis of Markov chains pertains to their steady-state behavior. Thus, if we are interested in the total expected cost accumulated "in the long run" under some policy, the argument regarding discounting becomes even more applicable.

We are therefore led to the introduction of a discount factor $\beta \geq 0$, and we define the following cost criterion which accumulates discounted costs over the infinite future:

$$V_\pi(x_0) = E_\pi \left[\int_0^\infty e^{-\beta t} C[X(t), u(t)] dt \right] \tag{9.3}$$

The exponential function in the integrand causes the cost to be "discounted." Note that if we limit ourselves to a time interval $[0, T]$ and let $\beta = 0$ we obtain the previous total undiscounted expected cost. Once again, the cost is defined with respect to an initial state x_0, which we assume to be given.

■ **Total Expected (Undiscounted) Cost over an Infinite Horizon.** Sometimes, we are interested in minimizing total expected cost without any discounting. This, for instance, may arise when the actual meaning of "cost" cannot be translated into any monetary value. In this case, we set $\beta = 0$ in (9.3), and consider a total expected *undiscounted* cost criterion over an infinite horizon:

$$V_\pi(x_0) = E_\pi \left[\int_0^\infty C[X(t), u(t)] dt \right] \tag{9.4}$$

This criterion, however, only makes sense if it can yield *finite* values for at least some policies and some initial states x_0. This is in fact the case for a number of useful applications, but one must be careful when dealing with this cost criterion: In general, the infinite accumulation of finite costs $C[X(t), u(t)]$ through the integral in (9.4) gives an infinite total cost.

■ **Expected Average Cost.** As pointed out above, if our objective is to minimize a cost criterion, this cost should be a finite quantity under at least some policies. Often, however, this may not be possible. In this case, we can define:

$$V_\pi(x_0) = \lim_{T \to \infty} \frac{1}{T} E_\pi \left[\int_0^T C[X(t), u(t)]dt \right] \tag{9.5}$$

which is often a well-defined finite quantity. Observe that this expression represents the average cost incurred per unit time under policy π. This is in fact a cost criterion which is of interest in many applications, since it addresses the "practical" question: How much does it cost to operate a given system per unit time?

In what follows, we will initially concentrate on cost criteria of the form (9.3). Thus, the problem of interest for us is:

Determine a policy π *to minimize* $V_\pi(x_0) = E_\pi \left[\int_0^\infty e^{-\beta t} C[X(t), u(t)]dt \right]$

where we assume that an initial state for the Markov chain is specified. This is the easiest class of problems that we can analyze. However, it turns out that in many cases the optimal policy for undiscounted criteria can be deduced from the optimal policy of this problem by letting $\beta \to 0$. For several problems of interest, it can in fact be shown that the optimal policy for this problem is of the same form as that for the average cost criterion.

9.3.2   Uniformization

In Section 7.5, we saw that we can convert a continuous-time Markov chain to a discrete-time chain through the process we called "uniformization." To refresh our memory, if a continuous-time Markov chain is specified through transition probabilities $P_{ij}, i \neq j$, and state holding time parameters $\Lambda(i)$, then we choose a uniform rate γ such that

$$\gamma \geq \max_{i \in \mathcal{X}} \{\Lambda(i)\} \tag{9.6}$$

and define a "uniformized" Markov chain through the transition probabilities

$$P_{ij}^U = \begin{cases} \dfrac{\Lambda(i)}{\gamma} P_{ij} & \text{if } i \neq j \\[2mm] 1 - \dfrac{\Lambda(i)}{\gamma} & \text{if } i = j \end{cases} \tag{9.7}$$

where the self-loop transitions corresponding to the rate $(1 - \Lambda(i)/\gamma)$ represent "fictitious events" that we can simply ignore.

The question we address now is: *What happens to the cost criterion* (9.3) *when we uniformize the original Markov chain model?*

Let γ be the uniform rate we have selected. Let the time instants when state transitions occur (including fictitious ones) be $T_0, T_1, \dots, T_k, \dots$, where we set $T_0 = 0$ by convention. We can then rewrite the total expected discounted cost in (9.3) as follows:

$$E_\pi \left[\int_0^\infty e^{-\beta t} C[X(t), u(t)] dt \right] = \sum_{k=0}^\infty E_\pi \left[\int_{T_k}^{T_{k+1}} e^{-\beta t} C[X(t), u(t)] dt \right]$$

Recall that a control action is taken when a new state is entered at time T_k, at which time a cost is incurred. However, the cost remains unaffected for the ensuing time interval of length $(T_{k+1} - T_k)$. We can therefore replace $C[X(t), u(t)]$ for $T_k \le t < T_{k+1}$ by $C(X_k, u_k)$, the cost incurred as soon as the kth transition takes place. Moreover, the value of this cost depends only on the policy π and the state entered at T_k, and not on the actual transition times T_k, T_{k+1} or the interval length $(T_{k+1} - T_k)$. Therefore, the above expression can also be written as

$$E_\pi \left[\int_0^\infty e^{-\beta t} C[X(t), u(t)] dt \right] = \sum_{k=0}^\infty E_\pi \left[\int_{T_k}^{T_{k+1}} e^{-\beta t} dt \right] E_\pi \left[C(X_k, u_k) \right] \quad (9.8)$$

We first consider the case where $\beta > 0$. The first expectation on the right-hand side of (9.8) gives

$$E_\pi \left[\int_{T_k}^{T_{k+1}} e^{-\beta t} dt \right] = -\frac{1}{\beta} \left[E_\pi \left[e^{-\beta T_{k+1}} \right] - E_\pi \left[e^{-\beta T_k} \right] \right]$$

and by writing $T_{k+1} = T_k + V_{k+1}$, where V_{k+1} is the state holding time after the kth transition:

$$E_\pi \left[\int_{T_k}^{T_{k+1}} e^{-\beta t} dt \right] = -\frac{1}{\beta} \left[E_\pi \left[e^{-\beta(T_k + V_{k+1})} \right] - E_\pi \left[e^{-\beta T_k} \right] \right]$$

$$= \frac{1}{\beta} E_\pi \left[e^{-\beta T_k} \right] \left[1 - E_\pi \left[e^{-\beta V_{k+1}} \right] \right]$$

where we have used the fact that V_{k+1} is independent of T_k. Since V_{k+1} is exponentially distributed with parameter γ, we have

$$E_\pi \left[e^{-\beta V_{k+1}} \right] = \int_0^\infty e^{-\beta t} \gamma e^{-\gamma t} dt = \frac{\gamma}{\beta + \gamma}$$

In addition, since $T_k = V_1 + \dots + V_k$, where all the state holding times are mutually independent, we get

$$E_\pi \left[e^{-\beta T_k} \right] = E_\pi \left[e^{-\beta V_1} \right] \cdots E_\pi \left[e^{-\beta V_k} \right] = \left(\frac{\gamma}{\beta + \gamma} \right)^k$$

Setting

$$\alpha = \frac{\gamma}{\beta + \gamma} \tag{9.9}$$

and returning to (9.8) we have

$$E_\pi \left[\int_0^\infty e^{-\beta t} C[X(t), u(t)] dt \right] = \sum_{k=0}^\infty \frac{\alpha^k (1 - \alpha)}{\beta} E_\pi \left[C(X_k, u_k) \right]$$

Finally, note that $(1 - \alpha) = \beta/(\beta + \gamma)$, which gives

$$E_\pi \left[\int_0^\infty e^{-\beta t} C[X(t), u(t)] dt \right] = \frac{1}{\beta + \gamma} E_\pi \left[\sum_{k=0}^\infty \alpha^k C(X_k, u_k) \right] \tag{9.10}$$

In summary, the problem of determining a policy π to minimize the cost criterion (9.3) for a continuous-time Markov chain is converted into the problem of determining a policy π to minimize (9.10) for:

- A discrete-time Markov chain with uniform transition rate γ.

- A discount factor $\alpha = \gamma/(\beta + \gamma)$, where we note that $0 < \alpha < 1$.

- A cost given by $C(i, u)/(\beta + \gamma)$ whenever state i is entered and control action u (dictated by π) is chosen.

Finally, returning to (9.8) and setting $\beta = 0$, we see that the first expectation on the right-hand side of (9.8) simply gives $E_\pi[V_{k+1}] = 1/\gamma$, the average state holding time. It follows that (9.10) is once again obtained, except that now $\alpha = 1$ since $\beta = 0$ in the definition (9.9).

9.3.3 The Basic Markov Decision Problem

Based on the previous discussion, we come to the conclusion that we can limit ourselves to discrete-time Markov chains. If a continuous-time model with discount factor β is of interest, then it is uniformized with rate γ. In the resulting discrete-time model we use a new discount factor $\alpha = \gamma/(\beta + \gamma)$, and replace the original costs $C(i, u)$ by $C(i, u)/(\beta + \gamma)$.

In what follows, we therefore concentrate on a Markov chain $\{X_k\}$, $k = 0, 1, \ldots$ with transition probabilities $p_{ij}[u(i)]$, where the control actions $u(i)$ are determined according to a policy π. The process $\{X_k\}$ is what is commonly referred to as a *Markov decision process*.

In summary, what we shall refer to as the basic *Markov decision problem* is the following. We assume that at every state transition a cost $C(i, u)$ is incurred, where i is the state entered and u is a control action selected from a set of admissible actions U_i. Under a stationary policy π, the control u depends only on the state i. The next state is then determined according to transition probabilities $p_{ij}(u)$. We also assume that a discount factor α, $0 < \alpha < 1$, is given, and that an initial state is specified. We then define the cost criterion

$$V_\pi(i) = E_\pi \left[\sum_{k=0}^{\infty} \alpha^k C(X_k, u_k) \right] \qquad (9.11)$$

which is the total expected discounted cost accumulated over an infinite horizon, given the initial state i. Therefore, strictly speaking, the expectation above is conditioned on $[X_0 = i]$, but we omit this to keep notation simple. Our problem then is: *Determine a policy π to minimize $V_\pi(i)$.*

Example 9.2 (Controlling the operating speed of a machine)

We consider a simple single-server queueing system where customers arrive according to a Poisson process. The server is a machine whose speed can be controlled as follows: There are two settings, a fast one, under which the machine operates at rate μ_1, and a slow one, under which the machine operates at rate $\mu_2 < \mu_1$. We assume that at either setting the service times are exponentially distributed. Therefore, the mean service time at the fast setting is $1/\mu_1 < 1/\mu_2$. The problem is to decide on a machine speed setting every time an event takes place.

We first establish a cost structure. Operating the machine at the fast setting incurs a cost C_1 per unit time (this encompasses the cost of power for running the machine, wear and tear, labor costs for supervising and maintaining the machine, etc.). At the slow rate, the cost is C_2 per unit time, where $C_2 < C_1$ (this makes sense, since more power, more wear and tear, and more supervision are generally needed when the machine runs faster). In addition, whenever a customer is present at this system (in service or in queue) inventory costs are accumulated; we assume that every customer incurs a cost B per unit time.

Let $X(t)$ denote the queue length at time t. This is the state of the system, where $X(t)$ can take values in $\{0, 1, 2, \dots\}$. Now whenever an event takes place (customer arrival or departure), and state i is entered, we observe the following:

1. A cost (Bi) per unit time is incurred for the time interval until the next event occurs.

2. Upon entering state i, we observe the value of i and take one of two control actions: If we choose to operate the machine at rate μ_1, then a cost C_1 per unit time is incurred until the next event; if we choose μ_2 instead, then a cost C_2 per unit time is incurred.

For convenience, define

$$u(i) = \begin{cases} 1 & \text{if } \mu_1 \text{ is chosen at state } i \\ 0 & \text{if } \mu_2 \text{ is chosen at state } i \end{cases}$$

We can then write down a simple expression for the cost $C(i, u)$:

$$C(i, u) = Bi + u(i)C_1 + [1 - u(i)]C_2$$

We will also assume that a discount factor $\beta > 0$ is given. Observe, however, that $C(i, u)$ is not bounded in this case, since the term (Bi) can become infinite as the queue length i grows. We will not let this bother us for now; it is, however, worthwhile pointing out that in queueing systems it is quite common to encounter costs which are proportional to queue lengths, therefore dealing with unbounded costs is often the case.

Next, we invoke uniformization in order to work with a discrete-time model. We choose the uniform rate $\gamma = \lambda + \mu_1$ (which is greater than any one event rate), and proceed as described earlier. Recalling (9.9) and (9.10), we set $\alpha = \gamma/(\beta + \gamma)$ and new cost parameters:

$$b = B/(\beta + \gamma), \quad c_1 = C_1/(\beta + \gamma), \quad c_2 = C_2/(\beta + \gamma)$$

We now have a cost criterion in the form (9.11):

$$V_\pi(i) = E_\pi \left[\sum_{k=0}^{\infty} \alpha^k C(X_k, u_k) \right]$$

where, for any state i, the cost is given by

$$C(i, u) = bi + u(i)c_1 + [1 - u(i)]c_2 \qquad (9.12)$$

Note that when $i = 0$, there is nobody to be served. In this case, we may still want to choose a setting for the machine in anticipation of the next customer.

To specify the transition probabilities for the uniformized model, we set

$$p = \lambda/\gamma, \quad q_1 = \mu_1/\gamma, \quad q_2 = \mu_2/\gamma$$

Since $\gamma = \lambda + \mu_1$, it follows that $p + q_1 = 1$. Note that the transition probability associated with a departure event depends on the choice of service rate and can be written as $q_1 u(i) + q_2[1 - u(i)]$. The probability of a self-loop transition is then given by

$$1 - [p + q_1 u(i) + q_2[1 - u(i)]] = (q_1 - q_2)[1 - u(i)]]$$

We then have for all $i > 0$:

$$p_{ij}(u) = \begin{cases} p & \text{if } j = i + 1 \\ q_1 u(i) + q_2[1 - u(i)] & \text{if } j = i - 1 \\ (q_1 - q_2)[1 - u(i)] & \text{if } j = i \\ 0 & \text{otherwise} \end{cases} \qquad (9.13)$$

and

$$p_{0j} = \begin{cases} p & \text{if } j = 1 \\ 1 - p & \text{if } j = 0 \\ 0 & \text{otherwise} \end{cases} \qquad (9.14)$$

The original Markov chain model and its uniformized version are shown in Fig. 9.2.

We have therefore defined a Markov decision process, where the objective is to determine a stationary policy π that minimizes $V_\pi(i)$ for a given initial state i.

We will return to this problem later in the chapter. At this point, however, we can let intuition guide us as to what might be a "good" policy. The essential tradeoff in choosing between $u = 1$ (fast setting) and $u = 0$ (slow setting) is this: If we choose $u = 0$, we incur less machine cost; but we also operate at a slower rate, which causes arriving customers to build a longer queue, which in turn incurs higher inventory costs proportional to the queue length. Thus, if b is very large, we would probably always want to choose $u = 1$, since inventory costs are dominant. If b is not as large, we can afford to choose $u = 0$ when the queue is empty or relatively small, and only switch to $u = 1$ when the queue becomes large. It follows that we should probably stick to a slow setting as long as we don't see the queue building up, and only switch to the fast (and more expensive) setting when the queue exceeds some *threshold*. If we are correct, then the whole problem reduces to the question: What is the optimal value of this threshold?

As we will see, this simple argument does indeed yield an optimal policy which is of the "threshold" type. For the time being, we only wish to call attention to such threshold-type policies, as they play a special role in many Markov decision processes.

9.4 SOLVING MARKOV DECISION PROBLEMS

A fundamental solution technique for Markov decision problems is based on the concept of *Dynamic Programming* (DP). This is not to say that there are no other approaches for solving these and related problems, but they require mathematical sophistication beyond the level of this book. Thus, we limit ourselves to DP, which is a powerful methodology with applications to a variety of areas, not limited to Markov decision problems. The next section provides a brief review of the DP approach as it pertains to basic optimization problems for the benefit of readers not previously exposed to it. For additional and much more detailed readings on DP, some selected references are provided at the end of this chapter.

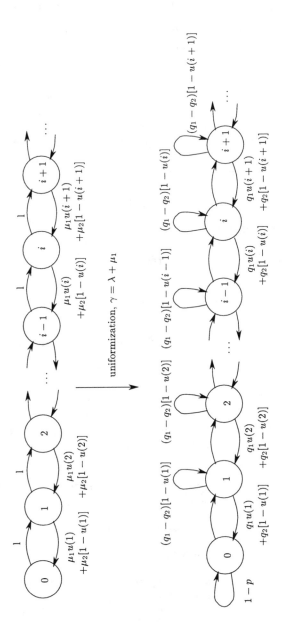

Figure 9.2: Markov chain model and its uniformization for Example 9.2.
The control action $u(i) \in \{0, 1\}$ chooses between the fast (μ_1) and slow (μ_2) server settings at state i. In the uniformized version we set $p = \lambda/\gamma, q_1 = \mu_1/\gamma, q_2 = \mu_2/\gamma$.

9.4.1 The Basic Idea of Dynamic Programming

Dynamic programming, pioneered by R. Bellman in the 1950s, is based on a simple idea known as the *principle of optimality*. To explain this idea, let us consider a discrete-time deterministic setting as follows. At time $k = 0$, the state is x_0. Our time horizon consists of N steps, and when we reach state x_N at time $k = N$, we will incur a terminal cost $C(x_N)$. At each time step $k = 0, 1, \ldots, N - 1$, we can select a control action $u(x_k)$, where x_k is the state at that time. Assume that $u(x_k)$ is chosen from a given set of admissible control actions $U(x_k)$ for that state. Depending on the state x_k and the control action selected $u(x_k)$, we incur a cost $C[x_k, u(x_k)]$. Then, a state transition occurs according to a state equation of the general form

$$x_{k+1} = f_k(x_k, u_k)$$

A policy π is a sequence $\{u_0, u_1, \ldots, u_{N-1}\}$ of control actions over the time horizon N. The optimization problem of interest is to determine a policy $\pi = \{u_0, u_1, \ldots, u_{N-1}\}$ that minimizes the total cost

$$V_\pi(x_0) = C(x_N) + \sum_{k=0}^{N-1} C[x_k, u(x_k)] \tag{9.15}$$

We denote such an optimal policy (which is not necessarily unique) by π^*:

$$\pi^* = \{u_0^*, u_1^*, \ldots, u_{N-1}^*\} \tag{9.16}$$

This is a very common problem arising in a variety of settings and in many different branches of engineering.

Now suppose we have applied an optimal policy π^* and have reached state x_n. We then define the *cost-to-go* $V_\pi(x_n)$:

$$V_\pi(x_n) = C(x_N) + \sum_{k=n}^{N-1} C[x_k, u(x_k)], \quad n = 1, \ldots, N - 1 \tag{9.17}$$

We denote a policy that minimizes $V_\pi(x_n)$ by

$$\pi^o = \{u_n^o, u_{n+1}^o, \ldots, u_{N-1}^o\} \tag{9.18}$$

and pose the question: *What is the optimal policy π^o for this subproblem?* The answer, which certainly agrees with basic intuition, is that π^o is simply the part of the sequence $\pi^* = \{u_0^*, u_1^*, \ldots, u_{N-1}^*\}$ from time step n on, that is,

$$\pi^o = \{u_n^*, u_{n+1}^*, \ldots, u_{N-1}^*\}$$

If this were not so, then the total cost (9.15) for the original problem should be reduced by defining another policy which (a) follows π^* from x_0 to x_n, and then (b) follows π^o from this point on. This, however, would contradict our assertion that π^* is in fact optimal.

A simple way to visualize the principle of optimality is by considering a geometric example, where the problem is to determine the shortest path from one point to another. Thus, looking at Fig. 9.3, suppose the problem is to find a minimum-cost path from point A to point J. The costs associated with each one-step move are shown on the corresponding line segment of this graph. We begin at point A and can choose between two control actions: Move to B or move to C, with corresponding costs $C_{AB} = 3, C_{AC} = 5$. Similarly, from each other point a choice can be made between branches, always moving towards the direction of our final destination J. Sometimes there is no choice (as in point G), and sometimes there are more than two choices (as in point C). For simplicity, we assume that there is no terminal cost in this problem, that is, once point J is reached no additional cost is incurred. If we wish, we may think of this graph as a collection of alternate paths between A and J, and the costs may represent actual travel times along the branches.

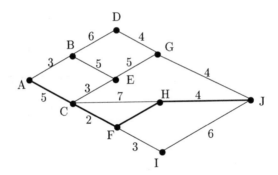

Figure 9.3: A minimum cost path example.
The minimum-cost path from A to J is A-C-F-H-J. The principle of optimality asserts that if a minimum-cost path is sought from point C, that path is C-F-H-J. Similarly, from point F the minimum-cost path is F-H-J.

One can easily verify that the optimal path is given by A-C-F-H-J. The principle of optimality asserts that if one were to start from point C instead of A and seek the optimal path, then surely that path is C-F-H-J. Similarly, the optimal path from F is F-H-J.

Dynamic Programming (DP) refers to an explicit algorithm for determining an optimal policy π^* in (9.16). We will first present the algorithm and then verify it for the problem of Fig. 9.3 (for a detailed proof of the algorithm the reader is referred to Bertsekas (1987), or other related references included at the end of the chapter).

We begin by defining

$$V^*(x_0) = \min_\pi \left[C(x_N) + \sum_{k=0}^{N-1} C[x_k, u(x_k)] \right] \tag{9.19}$$

This is the optimal cost for the problem, and it is obtained under the optimal policy π^*. It is assumed that the minimization is done over all admissible policies, that is, policies such that $u(x_k) \in U(x_k)$ for all states x_k.

Next, set

$$V_k(x_k) = \min_{u_k \in U(x_k)} [(C(x_k, u_k) + V_{k+1}(x_{k+1})], \quad k = 0, \dots, N-1 \tag{9.20}$$

$$V_N(x_N) = C(x_N) \tag{9.21}$$

Thus, $V_k(x_k)$ is the *optimal cost-to-go* from state x_k in the following sense: The controller sees a single-step problem with step cost $C(x_k, u_k)$ and terminal cost $V_{k+1}(x_{k+1})$, and selects the control action that solves that problem. Of course, $V_{k+1}(x_{k+1})$ is not known if one looks at (9.20) by itself. However, $V_N(x_N) = C(x_N)$ is known. Therefore, $V_{N-1}(x_{N-1})$ can be obtained by solving the minimization problem in (9.20) over $U(x_{N-1})$. This in turn allows us to solve another minimization problem to determine $V_{N-2}(x_{N-2})$ and so on. Ultimately, we determine $V_0(x_0)$, and this is precisely the optimal cost in (9.19). In other words,

$$V^*(x_0) = V_0(x_0) \tag{9.22}$$

where $V_0(x_0)$ is obtained by recursively solving the minimization problems (9.20) with initial condition (9.21).

Now let us examine how this scheme can be applied to the minimum-cost path problem shown in Fig. 9.3. This graph is redrawn in Fig. 9.4. In this case, states correspond to the points A,B,...,J. Control actions correspond to the choices of branches out of each point in the graph, or, equivalently, to all admissible next points; for instance, the admissible set of control actions at C is $U_C = \{E,H,F\}$. The time horizon for this problem is $N = 4$. The DP algorithm in this case proceeds as follows.

Step 4. At step $k = N = 4$, there is no terminal cost, so we simply set $V_4(J) = 0$.

Step 3. At step $k = 3$, there are three possible states, G, H, and I. However, there is only one choice to be made from any of these states. Thus, for state G, (9.20) becomes

$$V_3(G) = \min_J [C_{GJ} + V_4(J)] = \min_J[4 + 0] = 4$$

Similarly, we have $V_3(H) = 4, V_3(I) = 6$. For convenience, we mark these costs-to-go on the graph, as shown in Fig. 9.4 (numbers in parentheses).

Step 2. At step $k = 2$, there are three possible states, D, E, and F. Once again, there is only one choice from D or E (which is G), but there are two choices from F. Using (9.20) for D and E, along with the value $V_3(G) = 4$ obtained in the previous step, we get $V_2(D) = 8$, and $V_2(E) = 9$. For state F, we have

$$V_2(F) = \min[C_{FH} + V_3(H), \quad C_{FI} + V_3(I)] = \min[8, 9] = 8$$

Again, we mark these costs-to-go on the graph, as shown in Fig. 9.4.

Step 1. At step $k = 1$, there are two possible states, B and C. The process should now be clear, and we can actually proceed by simple inspection of the graph in Fig. 9.4. From B, the choice D gives a total cost-to-go of $6 + 8 = 14$, and the choice E gives a total cost-to-go of $5 + 9 = 14$. Thus, either action here is optimal, with corresponding optimal cost-to-go $V_1(B) = 14$. Similarly, from C we have to compare three possible choices, E, H, and F. We find the smallest cost-to-go is obtained when selecting F, and we get $V_1(C) = 10$.

Step 0. At the last step, we consider the initial state A. There are two possible choices, leading to B and C respectively. The choice B gives a total cost-to-go of $3 + V_1(B) = 17$, whereas the choice E gives a total cost-to-go of $5 + V_1(C) = 15$. Clearly, the optimal action here is C, with corresponding optimal cost-to-go $V_0(A) = 15$. From (9.22), this also happens to be the optimal cost $V^*(A)$ for the problem. The optimal policy is derived by tracking the optimal control actions selected at each step, starting with point A, and the minimum-cost path A-C-F-H-J is obtained.

Remark. In applying DP there is an implicit assumption that a control action at step k does not affect future admissible control actions at $k' > k$. In Fig. 9.4, for example, choosing to move from A to B will not change the topology of the graph (e.g., by adding or removing some segment of the graph.)

It is important to understand the reason why the DP algorithm provides an advantage over the original problem of determining the minimum-cost path over all possible paths. In the original problem, we have to evaluate the total cost over the entire *policy space*. The policy space consists of all sequences of control actions admissible at different states; in our case, this is the set of all possible paths from A to J. On the other hand, the DP algorithm decomposes the original problem into a number of simpler optimization problems of the

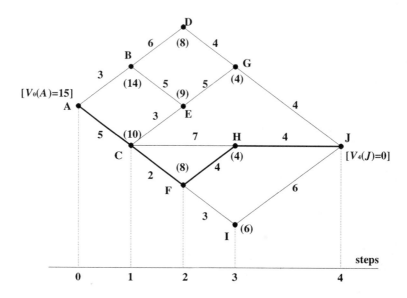

Figure 9.4: Illustrating the Dynamic Programming algorithm.

form in (9.20), one for each time step. Each of these problems requires us to evaluate the cost-to-go from some state over the *control action space* at that state; in our case, this is the number of branches out of that state, which is at most three and sometimes only one.

Despite its attractive features, the DP algorithm often involves prohibitively high computational costs, since the simpler optimization problems in (9.20) still have to be solved for all possible states. Thus, if the state space in our problem is large, the complexity is also very high; this is what is sometimes referred to as the "curse of dimensionality" pertaining to the DP approach. Nonetheless, DP remains a very general methodology for a large class of optimization problems.

It should also be noted that the DP algorithm only gives the value of the optimal cost, and some additional steps are required to derive from that solution an actual optimal policy. Sometimes, a closed-form solution $V^*(x_0)$ may not be possible to obtain; however, the DP algorithm itself may reveal useful structural properties of $V^*(x_0)$ and π^*. For example, we may find that the optimal control action $u^*(x)$ for every state x is of the form $f_\theta(x)$ where $f_\theta(\cdot)$ is a specified family of functions parameterized by θ. The problem then reduces to the determination of the optimal parameter value θ^*, an easier task than searching for an optimal function.

9.4.2 Dynamic Programming and the Optimality Equation

We are now ready to return to the basic problem of determining a stationary policy π to minimize the total expected discounted cost

$$V_\pi(i) = E_\pi \left[\sum_{k=0}^\infty \alpha^k C(X_k, u_k) \right]$$

given an initial state i. We will denote the optimal policy for this problem by π^* and the corresponding optimal cost by $V^*(i)$. We point out that all results derived in this section are based on the cost boundedness assumption (9.1). In Section 9.4.3, this assumption is relaxed, and we will see how some of our results can be extended to this case. In addition, the same extensions pertain to the case where costs are not discounted, so that we can solve problems involving the minimization of the total expected cost criterion (9.4). Finally, in Section 9.4.4, we briefly discuss the case of the average cost criterion in (9.5), and its relationship to (9.3) and (9.4).

We begin by considering a finite-horizon version of our problem, defined by N steps. We will now see how we can place the minimization of expected discounted costs in the framework of the DP algorithm in (9.20). Let us denote the optimal cost-to-go by $V_k'(j)$ when the state is j. Then, comparing with the expression in (9.19) and with (9.20), observe that in our case

- The terminal cost is $V_N'(j) = 0$ for all states j.

- The one-step cost at the kth step when the state is j and a control action u is selected is given by $\alpha^k C(j, u)$.

- The state transition mechanism is probabilistic and governed by given transition probabilities $p_{jr}(u)$ for all states j, r, with $u \in U_j$, an admissible control action at state j.

- The optimal cost-to-go function $V_k'(j)$ in (9.20) consists of the one-step cost and the expected optimal cost at the $(k+1)$th step evaluated over all possible state transitions.

The last observation implies that if control action u is selected at the kth step when the state is j, then the optimal cost-to-go at the $(k+1)$th step is the conditional expectation

$$E\left[V_{k+1}'(X_{k+1}) \mid X_k = j\right] = \sum_{\text{all } r} p_{jr}(u) V_{k+1}'(r)$$

Thus, equations (9.20) and (9.21) become

$$V_k'(j) = \min_{u \in U_j} \left[\alpha^k C(j, u) + \sum_{\text{all } r} p_{jr}(u) V_{k+1}'(r) \right], \quad k = 0, \ldots, N-1$$

$$V_N'(j) = 0 \quad \text{for all } j$$

Next, divide both sides of the equations above by α^k and set

$$V_k'' = \alpha^{-k} V_k'$$

to get

$$V_k''(j) = \min_{u \in U_j} \left[C(j, u) + \alpha \sum_{\text{all } r} p_{jr}(u) V_{k+1}''(r) \right], \quad k = 0, \ldots, N-1$$

$$V_N''(j) = 0 \quad \text{for all } j$$

Finally, it turns out that it is more convenient to associate the "cost-to-go" with the "number-of-steps-to-go" rather than the number of steps already taken. In other words, we prefer to have a terminal cost subscripted by 0 rather than N, and a subscript k indicates that there are k more steps to go (equivalently, $N-k$ steps have been taken). To accomplish this, we make one last transformation by setting

$$V_k = V_{N-k}'', \quad k = 0, 1, \ldots, N$$

and obtain

$$V_0(j) = 0 \quad \text{for all } j \tag{9.23}$$

$$V_{k+1}(j) = \min_{u \in U_j} \left[C(j, u) + \alpha \sum_{\text{all } r} p_{jr}(u) V_k(r) \right], \quad k = 0, \ldots, N-1 \tag{9.24}$$

Then, the optimal cost for this problem is $V_N(i)$, obtained as the solution of (9.23) and (9.24) for a given initial state i.

This discussion, of course, does not guarantee that the DP algorithm above has anything to do with the solution to our basic *infinite-horizon* problem, the minimization of (9.11). However, if the parallels we drew between the DP approach in the previous section and the Markov decision problem of interest here are accurate, then we should suspect some connection between the *finite-horizon* solution $V_N(i)$ and the *infinite-horizon* optimal cost, defined earlier as $V^*(i)$. Thus, our approach is to explore what happens to $V_N(i)$, the solution of (9.24), as $N \to \infty$. Not surprisingly, we will see that $V_N(i)$ converges to $V^*(i)$. Moreover, it turns out that $V^*(i)$ can be obtained as the solution of a single recursive equation, referred to as the *optimality equation* for our problem. We will also see that the optimal policy π^* is obtained through this optimality equation.

Convergence of the Dynamic Programming algorithm

In order to simplify the analysis that follows, we will develop a more compact form for equation (9.24). The starting point is to observe that the right-hand side in (9.24) may be viewed as a special operation on the function $V_k(j)$. Let us consider any function $f(j)$, rather than the specific $V_k(j)$, and introduce the symbol $T[f(j)]$ to represent

$$T[f(j)] = \min_{u \in U_j} \left[C(j,u) + \alpha \sum_{\text{all } r} p_{jr}(u) f(r) \right] \tag{9.25}$$

where $f : \mathcal{X} \to \mathbb{R}$.

What is interesting about $T[f(j)]$ for our purposes is this. Suppose we apply T on the terminal cost $V_0(j)$, where $V_0(j) = 0$ from (9.23). Then, from (9.24), we get

$$V_1(j) = T[V_0(j)]$$

If we now reapply T, again from (9.24) we get

$$V_2(j) = T[T[V_0(j)]]$$

It is clear that if we continue this process k times we get $V_k(j)$. To keep our notation simple and compact, we define, for any $f(j)$:

$$T^k[f(j)] = T[T^{k-1}[f(j)]], \quad k = 2, 3, \ldots \tag{9.26}$$

We now see that we can simply rewrite the DP algorithm in (9.24) as follows:

$$V_0(j) = 0 \quad \text{for all } j \tag{9.27}$$

$$V_k(j) = T^k[V_0(j)], \quad k = 1, 2, \ldots, N \tag{9.28}$$

As pointed out earlier, the recursive equation (9.24) may be thought of as the optimal cost of a one-step problem with step cost $C(j,u)$ and terminal cost $\alpha \sum_r p_{jr}(u) V_k(r)$. Similarly, from the definition in (9.25), $T[f(j)]$ may be thought of as the optimal cost of a one-step problem with step cost $C(j,u)$ and terminal cost $\alpha \sum_r p_{jr}(u) f(j)$.

A useful property of T which follows from the definition (9.25) is the following. Suppose we choose the function $f(j)$ to be of the form

$$f(j) = g(j) + \epsilon$$

for any arbitrary real number ϵ. Then, we see from (9.25) that

$$T[g(j) + \epsilon] = \min_{u \in U_j} \left[C(j,u) + \alpha \sum_{\text{all } r} p_{jr}(u) g(j) + \alpha \epsilon \sum_{\text{all } r} p_{jr}(u) \right]$$

and since the last sum above is simply unity, we get

$$T[g(j) + \epsilon] = T[g(j)] + \alpha \epsilon$$

By applying T once again, we get

$$T^2[g(j) + \epsilon] = \min_{u \in U_j} \left[C(j, u) + \alpha \sum_{\text{all } r} p_{jr}(u) T[g(j)] + \alpha^2 \epsilon \sum_{\text{all } r} p_{jr}(u) \right]$$
$$= T^2[g(j)] + \alpha^2 \epsilon$$

and by repeating this process (or using a formal induction argument) we obtain

$$T^k[g(j) + \epsilon] = T^k[g(j)] + \alpha^k \epsilon, \quad k = 1, 2, \ldots \tag{9.29}$$

We are now ready to establish two results which will guide us toward the "optimality equation" used for the solution of the basic Markov decision problem. The first result (Lemma 9.1) establishes the connection between the solution $V_N(i)$ of the DP algorithm and the optimal cost $V^*(i)$ of the infinite-horizon problem (9.11). In particular, as $N \to \infty$, we will show that $V_N(i)$ converges to $V^*(i)$. The second result (Lemma 9.2) shows that the DP algorithm (9.24), or equivalently (9.28), converges to $V^*(i)$ for any function $f(i)$ we choose as our initial condition in (9.23), as long as this is a bounded function.

Lemma 9.1 Under the assumption $0 \le C(j, u) \le K$ for all states j and control actions $u \in U_j$, the solution $V_N(i)$ of the DP algorithm is such that

$$\lim_{N \to \infty} V_N(i) = V^*(i) \tag{9.30}$$

Proof. We will establish two inequalities relating $V^*(i)$ and $V_N(i)$. The first inequality is obtained by returning to the cost criterion of interest in (9.11):

$$E_\pi \left[\sum_{k=0}^{\infty} \alpha^k C(X_k, u_k) \right]$$

and separating the sum above into two parts as follows:

$$E_\pi \left[\sum_{k=0}^{\infty} \alpha^k C(X_k, u_k) \right] = E_\pi \left[\sum_{k=0}^{N-1} \alpha^k C(X_k, u_k) \right]$$
$$+ E_\pi \left[\sum_{k=N}^{\infty} \alpha^k C(X_k, u_k) \right]$$

Because of our assumption that all one-step costs are bounded by K, we have

$$E_\pi \left[\sum_{k=0}^{\infty} \alpha^k C(X_k, u_k) \right] \le E_\pi \left[\sum_{k=0}^{N-1} \alpha^k C(X_k, u_k) \right] + K \sum_{k=N}^{\infty} \alpha^k$$

Now, let us apply the min operation on both sides of this relationship. The left-hand side is by definition the optimal cost $V^*(i)$. The first term on the right-hand side is the optimal cost of the N-step finite-horizon problem with $V_0(i) = 0$ for all i, which is, by definition, $V_N(i)$. The second term, on the other hand, is simply a geometric series giving

$$K \sum_{k=N}^{\infty} \alpha^k = K \frac{\alpha^N}{1-\alpha}$$

It follows that

$$V^*(i) \leq V_N(i) + K \frac{\alpha^N}{1-\alpha} \tag{9.31}$$

The second inequality is obtained by simply observing that since all one-step costs are non-negative by assumption, the more steps we include in the cost criterion the higher the cost becomes. Under the optimal policy, since $V^*(i)$ is the optimal cost for an infinite horizon, whereas $V_N(i)$ only includes a finite number of steps N, we must have

$$V^*(i) \geq V_N(i) \tag{9.32}$$

Combining (9.31) and (9.32) we get

$$V_N(i) \leq V^*(i) \leq V_N(i) + K \frac{\alpha^N}{1-\alpha}$$

Taking limits as $N \to \infty$ and observing that $\alpha^N \to 0$ since $0 < \alpha < 1$, gives

$$\lim_{N \to \infty} V_N(i) = V^*(i)$$

which is the desired result (9.30).   **Q.E.D.**

We will use Lemma 9.1 in proving the second result, which asserts that the DP algorithm converges to $V^*(i)$ for any bounded function $f(i)$ used as an initial condition.

Lemma 9.2 Under the assumption $0 \leq C(j, u) \leq K$ for all states j and control actions $u \in U_j$, for any bounded function $f(i), f : \mathcal{X} \to \mathbb{R}$, we have

$$\lim_{N \to \infty} T^N[f(i)] = V^*(i) \tag{9.33}$$

Proof. Since the function $f(i)$ is bounded, we can write for some real number ϵ:

$$-\epsilon \leq f(i) \leq \epsilon$$

Recalling from (9.27) that $V_0(i) = 0$ for all states i, we can also write

$$V_0(i) - \epsilon \leq f(i) \leq V_0(i) + \epsilon$$

We now apply T^N to this inequality and use the property of T in (9.29):

$$T^N[V_0(i)] - \alpha^N \epsilon \leq T^N[f(i)] \leq T^N[V_0(i)] + \alpha^N \epsilon$$

From (9.28), $T^N[V_0(i)] = V_N(i)$, and by Lemma 9.1 $V_N(i) \to V^*(i)$ as $N \to \infty$. Therefore, taking limits as $N \to \infty$ above gives

$$V^*(i) - \lim_{N \to \infty} \alpha^N \epsilon \leq \lim_{N \to \infty} T^N[f(i)] \leq V^*(i) + \lim_{N \to \infty} \alpha^N \epsilon$$

and since $\alpha^N \to 0$ as $N \to \infty$, we get

$$V^*(i) \leq \lim_{N \to \infty} T^N[f(i)] \leq V^*(i)$$

yielding the result (9.33). **Q.E.D.**

The importance of the discount factor is evident in deriving these results. As $N \to \infty$, the contribution of one-step costs becomes increasingly smaller, since $\alpha^N \to 0$, provided of course that the costs themselves remain bounded, which is guaranteed by our assumption in (9.1). This gives a hint as to the type of complications introduced when there is no discounting or when costs may not be bounded.

The Optimality Equation

We now state a theorem which provides the main tool for solving infinite-horizon discounted Markov decision problems. We prove that there is a simple "optimality equation" that $V^*(i)$ must satisfy. The optimal policy π^* can then also be obtained through this equation.

Theorem 9.1 Under the assumption $0 \leq C(j, u) \leq K$ for all states j and control actions $u \in U_j$, the optimal cost $V^*(i)$ which minimizes the criterion (9.11) satisfies the *optimality equation*

$$V(i) = \min_{u \in U_i} \left[C(i, u) + \alpha \sum_{\text{all } j} p_{ij}(u) V(j) \right] \tag{9.34}$$

$V^*(i)$ is also the unique bounded solution of this equation. Moreover, a stationary policy π^* is optimal if and only if it gives the minimum value in (9.34) for all states i.

Proof. We will limit our proof to the first part; the second part, pertaining to π^*, is obtained using similar techniques (see also Bertsekas, 1987), and we will omit it.

Following the same steps as in Lemma 9.1, we obtain the two inequalities (9.31) and (9.32) and combine them to get

$$V_N(i) \leq V^*(i) \leq V_N(i) + K\frac{\alpha^N}{1-\alpha}$$

We now apply T to this inequality. The leftmost term becomes $T[V_N(i)] = V_{N+1}(i)$. Also, recalling the property of T in (9.29) to evaluate the rightmost term above, we get

$$V_{N+1}(i) \leq T[V^*(i)] \leq V_{N+1}(i) + \alpha K\frac{\alpha^N}{1-\alpha}$$

Next, we take limits as $N \to \infty$. From Lemma 9.1, $V_{N+1}(i) \to V^*(i)$, and since $\alpha^N \to 0$, we obtain $V^*(i) = T[V^*(i)]$. Then, applying the definition of T in (9.25),

$$V^*(i) = T[V^*(i)] = \min_{u \in U_i} \left[C(i,u) + \alpha \sum_{\text{all } j} p_{ij}(u)V^*(j) \right]$$

Thus, we see that $V^*(i)$ indeed satisfies equation (9.34).

The fact that $V^*(i)$ is the unique bounded solution follows from Lemma 9.2. Suppose that $f(i)$ is some bounded function other than $V^*(i)$ that also satisfies (9.34), that is, $f(i) = T[f(i)]$. It must then be that $T^N[f(i)] \to f(i)$ as $N \to \infty$. But, by Lemma 9.2, $T^N[f(i)] \to V^*(i)$ for any bounded $f(i)$, therefore $f(i) = V^*(i)$. This violates the assumption that $f(i)$ is not the same as $V^*(i)$, and uniqueness is established. **Q.E.D.**

Theorem 9.1 is instrumental in solving the Markov decision problem (9.11) under the cost boundedness assumption (9.1). It should be pointed out, however, that obtaining an explicit closed-form expression for the optimal policy from this theorem is far from a simple task. We often have to resort to numerical techniques involving, for example, the repeated solution of the finite-horizon problem for $N = 1, 2, \ldots$, and then using (9.33) to obtain $V^*(i)$ in the limit as $N \to \infty$.

Example 9.3 (Deciding when to commit a job to a processor)

Suppose we have at our disposal a computer system consisting of M different processors. When a job submitted to this system is assigned to processor $i = 1, 2, \ldots, M$, we assume that a finite cost C_i is incurred, and that the processors are all indexed so that

$$C_1 < C_2 < \cdots < C_M$$

When we submit a job to this system, processor i is assigned to our job with probability p_i. At this point we can (a) decide to go with this processor, or (b) choose to hold the job until a lower-cost processor is assigned.

If the latter decision is made, we assume the system periodically returns to our job and assigns a processor, always with the same probability distribution $p_i, i = 1, \ldots, M$. However, until the next processor assignment is made we have to hold our job and incur a fixed finite cost c. Clearly, the tradeoff in this problem is between the accumulation of holding costs c and the hope that a low-cost processor will ultimately be assigned. The question is: How do we decide to go with the processor currently assigned to our job versus waiting for the next assignment?

To place this problem in the context of Markov decision processes, let the state be i, the currently assigned processor, with $i \in \{1, \ldots, M\}$, and let control actions be defined so that

$$u(i) = \begin{cases} 1 & \text{if } i \text{ is accepted} \\ 0 & \text{if } i \text{ is not accepted} \end{cases}$$

In addition, define a state F representing the "final" state resulting from a control action to accept a processor assignment. This is an absorbing state in the sense that the process terminates when a processor assignment is accepted.

The transition probabilities for this controlled Markov chain are specified as follows. Suppose we are at state $i = 1, \ldots, M, i \neq F$. Then, $p_{ij}(u) = p_j$ for any state $j = 1, \ldots, M, i \neq F$, provided that $u = 0$. In other words, if the decision is to hold the job for another assignment, a transition to a new state j occurs with fixed probability p_j independent of the current state i. On the other hand, if $u = 1$, then $p_{ij}(u) = 0$ for all $j \neq F$, and $p_{iF}(u) = 1$. Moreover, if state F is entered, then $p_{FF} = 1$.

The cost structure is specified as follows. If the state is $i \neq F$ and the control action is $u(i) = 0$, we incur a one-step cost c; if the control action is $u(i) = 1$, we incur a one-step cost C_i. Thus, for all $i = 1, \ldots, M$,

$$C(i, u) = u(i)C_i + [1 - u(i)]c$$

Moreover, if $u(i) = 1$, then we enter state F and the process ends. Therefore, we view state F as one from which no further costs are incurred.

Once the transition probability and cost structures are established, we have at our disposal all the elements necessary to solve the problem using the optimality equation (9.34), assuming a given discount factor $0 < \alpha < 1$ is given. Note that all costs are bounded in this problem, so that Theorem 9.1 can be used.

The solution to this and similar problems usually proceeds in three steps: First, write down the optimality equation; then, try to infer from it a characterization of optimal control actions; finally, try to exploit properties of the optimal cost function and of the optimal policy that can allow us to derive an explicit closed-form expression for the optimal policy.

Step 1. Write down the optimality equation for the problem.

For $i = 1, \ldots, M$, equation (9.34) becomes

$$V(i) = \min_{u \in \{0,1\}} \left[u(i)C_i + [1 - u(i)]c + [1 - u(i)]\alpha \sum_{j=1}^{M} p_j V(j) \right]$$

Here, the first two terms represent the one-step cost: C_i if we accept the processor assignment, and c otherwise. The remaining term is the expected cost resulting from a state transition, which is nonzero only if we do not accept the current assignment (otherwise, state F is entered next and the process ends without incurring any further costs).

We can rewrite this equation as

$$V(i) = \min \left[C_i, \quad c + \alpha \sum_{j=1}^{M} p_j V(j) \right] \tag{9.35}$$

Step 2. Characterize optimal control actions.

Looking at equation (9.35), it is clear that if $C_i < c + \alpha \sum_j p_j V(j)$, the way to minimize $V(i)$ is by choosing the control action $u(i) = 1$, that is, accept the processor assignment i. Since we have indexed processors so that C_i is increasing in i, it follows that there is some critical value i^* such that C_{i^*} is sufficiently small, that is,

$$C_{i^*} < c + \alpha \sum_{j=1}^{M} p_j V(j) \tag{9.36}$$

We immediately see that *the optimal policy is to hold our job until we are assigned any processor* $i \leq i^*$, where i^* is defined by (9.36). The optimal control action at state $i = 1, \ldots, M$, denoted by $u^*(i)$, is therefore

$$u^*(i) = \begin{cases} 1 & \text{if } i \leq i^* \\ 0 & \text{otherwise} \end{cases} \tag{9.37}$$

However, we still need to evaluate $V(i), i = 1, \ldots, M$, in order to apply this optimal policy, since these values are needed for the determination of the critical index i^* in (9.36). Note that this optimal policy is of "threshold type", where the threshold here is an integer index value (as discussed in Example 9.2, threshold-type optimal policies are very common in Markov decision problems).

Step 3. Try to obtain a closed-form expression for the optimal policy.

This is possible in this particular case, by making the following observations. We have already determined that the optimal policy is

$$\pi^* = \text{policy that accepts any processor } i \leq i^*, \quad i^* \in \{1, \ldots, M\}$$

The question now is: What is the value i^*? We start out by defining a policy π_n which accepts any processor such that $i \leq n$. Clearly, $\pi^* = \pi_n$ when $n = i^*$. We will now see that we can explicitly evaluate the total expected discounted cost for this problem under any policy π_n, $n = 1, \ldots, M$.

Let R denote the random number of rejected processor assignments under policy π_n. Then, conditioned on the value of R, the total expected cost under π_n is

$$E_{\pi_n} \left[\sum_{k=0}^{\infty} \alpha^k C(X_k, u_k) \mid R = r \right] \tag{9.38}$$
$$= [c + \alpha c + \ldots + \alpha^{r-1} c] + \alpha^r E[C^i \mid i \leq n]$$

Here, the first r terms in brackets on the right-hand side are the accumulated discounted holding costs for the r rejected processor assignments. The last term is the expected cost incurred by the accepted assigned processor, conditioned on our policy that only a processor $i \leq n$ is accepted.

To evaluate this conditional cost we need two calculations. First, the term in brackets on the right-hand side of (9.38) is a geometric series with r terms; therefore,

$$[c + \alpha c + \ldots + \alpha^{r-1} c] = c \frac{1 - \alpha r}{1 - \alpha} \tag{9.39}$$

Next, to calculate the conditional expectation $E[C_i \mid i \leq n]$, we need an expression for the conditional probability:

$$P[\text{next assignment is } i \mid i \leq n]$$

Since an assignment to processor i occurs with probability p_i, we have

$$P[\text{next assignment is } i \mid i \leq n] = \frac{P[\text{next assignment is } i = 1, \ldots, n]}{P[i \leq n]}$$
$$= \frac{p_i}{\sum_{i=1}^{n} p_i}$$

Therefore,

$$E[C_i \mid i \leq n] = \frac{\sum_{i=1}^{n} C_i p_i}{\sum_{i=1}^{n} p_i} \tag{9.40}$$

Combining (9.39) and (9.40), we get in (9.38):

$$E_{\pi_n}\left[\sum_{k=0}^{\infty}\alpha^k C(X_k, u_k) \mid R = r\right] = c\frac{1-\alpha^r}{1-\alpha} + \alpha^r \frac{\sum_{i=1}^{n} C_i p_i}{\sum_{i=1}^{n} p_i} \qquad (9.41)$$

The last step is to obtain the unconditioned total expected discounted cost under policy π_n. This is given by

$$E_{\pi_n}\left[\sum_{k=0}^{\infty}\alpha^k C(X_k, u_k)\right]$$

$$= \sum_{r=0}^{\infty} E_{\pi_n}\left[\sum_{k=0}^{\infty}\alpha^k C(X_k, u_k) \mid R = r\right] P[R = r]$$

Observe that the random variable R counts the number of rejections before the first acceptance in a process where all assignments occur independently and each rejection occurs with probability

$$P[\text{rejection}] = \sum_{i=n+1}^{M} p_i$$

Therefore, R is geometrically distributed, that is,

$$P[R = r] = \left(\sum_{i=n+1}^{M} p_i\right)^r \left(\sum_{i=1}^{n} p_i\right)$$

and using (9.41) we get

$$E_{\pi_n}\left[\sum_{k=0}^{\infty}\alpha^k C(X_k, u_k)\right]$$

$$= \frac{c}{1-\alpha} - \left[\frac{c}{1-\alpha} - \frac{\sum_{i=1}^{n} C_i p_i}{\sum_{i=1}^{n} p_i}\right]\sum_{r=0}^{\infty}\alpha^r\left(\sum_{i=n+1}^{M} p_i\right)^r\left(\sum_{i=1}^{n} p_i\right)$$

The infinite sum on the right-hand side gives

$$\sum_{r=0}^{\infty}\alpha^r\left(\sum_{i=n+1}^{M} p_i\right)^r\left(\sum_{i=1}^{n} p_i\right) = \left(\sum_{i=1}^{n} p_i\right)\frac{1}{1-\alpha\sum_{i=n+1}^{M} p_i}$$

and we get

$$E_{\pi_n}\left[\sum_{k=0}^{\infty}\alpha^k C(X_k, u_k)\right] = \frac{c}{1-\alpha}\left[1 - \frac{\sum_{i=1}^{n} p_i}{1-\alpha\sum_{i=n+1}^{M} p_i}\right]$$

$$+ \frac{\sum_{i=1}^{n} C_i p_i}{1-\alpha\sum_{i=n+1}^{M} p_i}$$

which, after some algebra, finally reduces to

$$E_{\pi_n}\left[\sum_{k=0}^{\infty} \alpha^k C(X_k, u_k)\right] = \frac{c\sum_{i=n+1}^{M} p_i + \sum_{i=1}^{n} C_i p_i}{1 - \alpha \sum_{i=n+1}^{M} p_i} \qquad (9.42)$$

This expression is a function of $n = 1, \ldots, M$. The value of n that minimizes (9.42) must therefore be the critical index i^* we have been looking for. This minimization is straightforward, since it simply involves comparing M possible values and selecting the smallest one. Once i^* has been determined in this manner, the optimal policy is the one that follows the rule (9.37) for all states $i = 1, \ldots, M$.

Here is a simple numerical example. For $M = 4$ processors, let $p_i = 0.25$ for all $i = 1, \ldots, 4$, $C_i = i$ for all $i = 1, \ldots, 4$, $c = 1$, and $\alpha = 0.9$. Let us denote the expression on the right-hand side of (9.42) by $L(n)$. We can then evaluate $L(n)$ for all $n = 1, \ldots, 4$ to find

$$L(1) = 3.077 \quad L(2) = 2.273$$
$$L(3) = 2.258 \quad L(4) = 2.500$$

We see that the minimum cost occurs at $n = 3$. Therefore, the optimal policy is $\pi_3 =$ accept any one of processors 1, 2, or 3, and reject only 4. Note that even though processor 1 is 3 times less costly than processor 3, waiting for processor 1 to be assigned is clearly not a good idea.

Remark. The preceding example falls in the class of "optimal stopping problems". Here, the basic control action consists of choosing between terminating a particular process or continuing with it, the objective being to minimize a total expected cost criterion (which may be discounted or undiscounted).

9.4.3 Extensions to Unbounded and Undiscounted Costs

In Example 9.2, we saw that one-step costs need not be bounded, and that this is frequently the case in queueing system applications. In the previous section, however, we saw that the cost boundedness assumption (9.1) was critical in the derivation of Lemmas 9.1 and 9.2, and hence Theorem 9.1.

Things get significantly more complicated when $C(i, u)$ cannot be bounded for at least some states i and control actions $u \in U_i$. This is seen in the proof of Lemma 9.1, where we can no longer ensure that $V_\pi(i)$ remains finite. Even with the discounting factor $\alpha, 0 < \alpha < 1$, it is possible for some terms $\alpha^k C(X_k, u_k)$ to be unbounded when $k \to \infty$. Nonetheless, there are several problems of interest where $V_\pi(i)$ is finite for at least some policies and some initial states i, and it therefore makes sense to seek an optimal policy and the corresponding value of the optimal cost.

Fortunately, the optimality equation established in Theorem 9.1 is still applicable when costs may become unbounded or when they are simply not discounted, provided that *all costs are nonnegative* (or all costs are nonpositive).

This is stated in the following theorem, whose proof we omit (the reader is referred to Bertsekas, 1987; or Ross, 1983b).

Theorem 9.2 Under the assumption $C(j, u) \geq 0$ [or $C(j, u) \leq 0$] for all states j and control actions $u \in U_j$, the optimal cost $V^*(i)$ which minimizes the criterion (9.11) satisfies the *optimality equation*

$$V(i) = \min_{u \in U_i} \left[C(i, u) + \alpha \sum_{\text{all } j} p_{ij}(u) V(j) \right] \tag{9.43}$$

where α is no longer constrained to be in $(0, 1)$, but is allowed to take values greater than or equal to 1. ◆

Equation (9.43) may be viewed as a generalization of (9.34), since it allows for unbounded and undiscounted costs. However, unlike Theorem 9.1, we can no longer assert here that $V^*(i)$ is the only solution of the optimality equation, although in most cases of practical interest this does not turn out to pose a serious limitation.

Another difficulty, from a practical standpoint, is that we cannot always ensure that the DP algorithm (9.24) converges to $V^*(i)$, as was shown in Lemma 9.1 for the bounded cost case. Therefore, it may not be possible to obtain $V^*(i)$ by solving the finite horizon problem repeatedly over N and then let $N \to \infty$. Fortunately, however, under certain conditions imposed on the set of admissible control actions U, it is possible to guarantee that the DP algorithm indeed converges to $V^*(i)$. One such simple condition is that U be a finite set. This result, which will suffice for the problems we will be looking at, is stated in the following theorem, whose proof we also omit (see Bertsekas, 1987; or Ross, 1983b).

Theorem 9.3 Assume that the set of control actions U is finite. Then, under the assumption $C(j, u) \geq 0$ for all states j and control actions $u \in U_j$, we have

$$\lim_{N \to \infty} V_N(i) = V^*(i) \tag{9.44}$$

where $V_N(i)$ is the solution of the DP algorithm (9.24). ◆

Thus, if a Markov decision problem with nonnegative costs involves only a finite set of admissible control actions, this theorem can be used to obtain the optimal cost as the limit of the DP algorithm solution letting $N \to \infty$. In addition, the optimal stationary policy π^* can also be obtained in the limit.

Example 9.4 (Controlling the speed of a machine - continued)

After uniformizing the system of Example 9.2, we obtained a discrete-time Markov chain as in Fig. 9.2, with transition probabilities given by

(9.13)-(9.14) and rewritten below:

$$p_{ij}(u) = \begin{cases} p & \text{if } j = i+1 \\ q_1 u(i) + q_2[1 - u(i)] & \text{if } j = i-1 \\ (q_1 - q_2)[1 - u(i)] & \text{if } j = i \\ 0 & \text{otherwise} \end{cases}$$

and

$$p_{0j} = \begin{cases} p & \text{if } j = 1 \\ 1 - p & \text{if } j = 0 \\ 0 & \text{otherwise} \end{cases}$$

where the control action $u(i), i > 0$, is to select between the fast machine setting, corresponding to $u(i) = 1$, and the slow one, corresponding to $u(i) = 0$. This in turn affects the transition probability p_{ij} for $j = i - 1$, corresponding to a departure event: If $u(i) = 1$, then $p_{ij} = q_1$, otherwise $p_{ij} = q_2 < q_1$.

The one-step cost for this problem was defined in (9.12) and is rewritten below:

$$C(i, u) = bi + u(i)c_1 + [1 - u(i)]c_2$$

where b represents the inventory cost for maintaining a customer in the system, and c_1, c_2 are the costs of operating the machine at the fast and slow speeds respectively, with $c_1 > c_2$. As already pointed out, $C(i, u)$ is unbounded due to its linear dependence on the queue length i. Note, however, that $C(i, u) \geq 0$, and we can use the optimality equation (9.43) in Theorem 9.2.

For a given discount factor $0 < \alpha < 1$, we now proceed in three steps, similar to Example 9.3.

Step 1. Write down the optimality equation for the problem.

For $i = 1, 2, \ldots$, (9.43) becomes

$$V(i) = \min_{u \in \{0,1\}} \Big[bi + u(i)c_1 + [1 - u(i)]c_2 +$$
$$\alpha p V(i + 1) + \alpha[q_1 u(i) + q_2[1 - u(i)]]V(i - 1) \qquad (9.45)$$
$$+ \alpha(q_1 - q_2)[1 - u(i)]V(i) \Big]$$

and for $i = 0$,

$$V(0) = \min_{u \in \{0,1\}} \Big[u(0)c_1 + [1 - u(0)]c_2 + \alpha p V(1) + \alpha(1 - p)V(0) \Big] \qquad (9.46)$$

Taking a closer look at (9.45), the first three terms are simply the one-step cost $C(i, u)$ when the state is $i > 0$ and control action $u(i) \in \{0,1\}$ is selected. The remaining terms give the expected optimal cost after a transition takes place, given the transition probabilities: The fourth term corresponds to an arrival, which occurs with probability p and leads to a new state $i + 1$; the fifth term corresponds to a departure, which occurs with probability $[q_1 u(i) + q_2[1 - u(i)]]$ dependent on $u(i)$ and leads to state $i - 1$; and the final term corresponds to the self-loop transition occurring with probability $(q_1 - q_2)[1 - u(i)]$. Similarly, in (9.46), we see that the one-step cost is limited to either c_1 or c_2, depending on our choice of $u(0)$. The remaining terms correspond to the state transition due to an arrival with probability p, and to the self-loop transition occurring with probability $(1 - p)$.

In (9.45), we can separate the terms that depend on $u(i)$ from those that do not:

$$V(i) = \left[bi + c_2 + \alpha pV(i + 1) + \alpha q_2 V(i - 1) + \alpha(q_1 - q_2)V(i)\right]$$
$$+ \min_{u \in \{0,1\}} \left[u(i)(c_1 - c_2) + u(i)\alpha(q_1 - q_2)V(i - 1)\right.$$
$$\left. - u(i)\alpha(q_1 - q_2)V(i)\right]$$

which can also be written as

$$V(i) = \left[bi + c_2 + \alpha pV(i + 1) + \alpha q_2 V(i - 1) + a(q1 - q2)V(i)\right]$$
$$+ \min\left[(c_1 - c_2) + \alpha(q_1 - q_2)[V(i - 1) - V(i)], \ 0\right] \tag{9.47}$$

Similarly, for $i = 0$ we get from (9.46):

$$V(0) = c_2 + \alpha pV(1) + \alpha(1 - p)V(0) + \min_{u \in \{0,1\}} [u(0)(c_1 - c_2)] \tag{9.48}$$

Step 2. Characterize optimal control actions.

Let us concentrate on the min term in equation (9.47). There are clearly two cases:

1. If $(c_1 - c_2) + \alpha(q_1 - q_2)[V(i-1) - V(i)] < 0$, the way to minimize $V(i)$ is by choosing the control action which yields this negative quantity in the min term, that is, $u(i) = 1$.

2. If $(c_1 - c_2) + \alpha(q_1 - q_2)[V(i - 1) - V(i)] \geq 0$, the way to minimize $V(i)$ is by choosing the control action $u(i) = 0$, since this yields 0 in the min term.

We can immediately conclude that the optimal control action at state $i > 0$, denoted by $u^*(i)$, is defined by

$$u^*(i) = \begin{cases} 1 & \text{if } [(c_1 - c_2) + \alpha(q_1 - q_2)[V(i-1) - V(i)]] < 0 \\ 0 & \text{otherwise} \end{cases} \tag{9.49}$$

Note, however, that we still need to evaluate $V(i)$ in order to apply this rule for selecting the optimal control.

Similarly, let us look at the min term in equation (9.48). Here, the optimal choice $u^*(0)$ is clearly defined: Since by assumption $c_1 - c_2 > 0$, the control action that always minimizes $V(0)$ is $u^*(0) = 0$; otherwise, the term in brackets in (9.48) is a positive quantity. This conclusion agrees with the intuition that when there is no customer to serve, and therefore no inventory costs to worry about, the choice of machine speed should be that yielding the smallest possible cost, that is, $c_2 < c_1$, corresponding to $u^*(0) = 0$.

Step 3. Try to obtain a closed-form expression for the optimal policy.

We have already seen that for $i = 0$ it is optimal to keep the machine at its slow speed. In Example 9.2 we argued that it is reasonable to maintain this policy for low queue length values, and then switch to the high speed when the queue length reaches some threshold, which we will denote by i^*. It is now possible to explicitly demonstrate that this is the case.

Looking at (9.49), let us rewrite the condition under which $u^*(i) = 1$ as follows:

$$V(i) - V(i-1) > \frac{c_1 - c_2}{\alpha(q_1 - q_2)}$$

Set

$$D = \frac{c_1 - c_2}{\alpha(q_1 - q_2)} \tag{9.50}$$

and note that $D > 0$ under our assumptions that $c_1 > c_2, q_1 > q_2$, and $0 < \alpha < 1$. Let us also set

$$\Delta V(i) = V(i) - V(i-1), \quad i = 1, 2, \ldots$$

Thus, (9.49) becomes

$$u^*(i) = \begin{cases} 1 & \text{if } \Delta V(i) > D \\ 0 & \text{otherwise} \end{cases} \tag{9.51}$$

Suppose that $\Delta V(i) < D$ for $i = 1$, and that $\Delta V(i)$ is monotonically increasing in i. Then, for small values of i we maintain the slow speed setting, that is, $u^*(i) = 0$, but, for some state i, $\Delta V(i)$ will finally exceed D, at which point we switch to $u^*(i) = 1$. From that point on, $u^*(i) = 1$ for all higher values of i. If this is indeed the case, there is some i^* with the property that

$$\Delta V(i^*) > D, \quad \Delta V(i^* - 1) \le D$$

If we can identify this special value i^*, then the optimal policy reduces to the simple *threshold-based* rule

$$u^*(i) = \left\{ \begin{array}{ll} 1 & \text{if } i \ge i^* \\ 0 & \text{otherwise} \end{array} \right. \tag{9.52}$$

Even though this is one of the simplest Markov decision problems one can solve, the optimality of such threshold-based policies is very common in more complex problems as well. The fact that the optimal policy is of this form rests on the proof that $\Delta V(i)$ is indeed monotonically nondecreasing. Incidentally, if $\Delta V(i)$ is nondecreasing (but not necessarily increasing), it is possible that it never reaches the value D, in which case the optimal policy is to keep the machine at the slow speed setting.

In what follows, we complete our discussion of this problem by providing the details of one way to establish this "structural property" of the optimal cost function $V(i)$. It is worthwhile to go through these details at least once, as this typifies the solution approach to a host of similar Markov decision problems.

Consider a finite-horizon version of the problem, as in (9.24). Therefore,

$$V_0(i) = 0, \text{ for all } i$$

$$V_{k+1}(i) = \min_{u \in \{0,1\}} \Big[bi + u(i)c_1 + [1 - u(i)]c_2 + \alpha p V_k(i + 1)$$
$$+ a[q_1 u(i) + q_2[1 - u(i)]]V_k(i - 1)$$
$$+ \alpha(q_1 - q_2)[1 - u(i)]V_k(i) \Big], \quad \text{for } i > 0$$

$$V_{k+1}(0) = \min_{u \in \{0,1\}} \Big[u(0)c_1 + [1 - u(0)]c_2 + \alpha p V_k(1) + \alpha(1 - p)V_k(0) \Big]$$

for $k = 0, 1, \ldots$ The last two equations can be rewritten in a form analogous to (9.47) and (9.48):

$$V_{k+1}(i) = \Big[bi + c_2 + \alpha p V_k(i + 1) + \alpha q_2 V_k(i - 1) + \alpha(q_1 - q_2)V_k(i) \Big]$$
$$+ \min \Big[(c_1 - c_2) - \alpha(q_1 - q_2)\Delta V_k(i), \ 0 \Big] \tag{9.53}$$

$$V_{k+1}(0) = c_2 + \alpha p V_k(1) + \alpha(1 - p)V_k(0) \tag{9.54}$$

where we have introduced

$$\Delta V_k(i) = V_k(i) - V_k(i-1), \quad i = 1, 2, \ldots$$

Setting $V_k(-1) = V_k(0)$ and $\Delta V_k(0) = 0$, note that (9.53) for $i = 0$ gives

$$
\begin{aligned}
V_{k+1}(0) &= [c_2 + \alpha p V_k(1) + \alpha q_2 V_k(0) + \alpha(q_1 - q_2)V_k(0)) \\
&\quad + \min[(c_1 - c_2), \ 0] \\
&= [c_2 + \alpha p V_k(1) + \alpha q_1 V_k(0)]
\end{aligned}
$$

since $c_1 > c_2$. Moreover, recalling that $p + q_1 = 1$, we have $\alpha q_1 V_k(0) = \alpha(1-p)V_k(0)$, and we see that (9.53) reduces to (9.54). Therefore, with these definitions in mind, we can now use (9.53) for all $i = 0, 1, 2, \ldots$

Since the set of control actions in this problem is limited to $\{0, 1\}$ for all states, Theorem 9.3 can be used, that is, $V_k(i) \to V^*(i)$ as $k \to \infty$, and therefore $\Delta V_k(i) \to \Delta V^*(i)$. Consequently, we can concentrate on proving that $\Delta V_k(i)$ is nondecreasing in i for every $k = 0, 1, \ldots$ This is a good example of the importance of the DP algorithm convergence: Structural properties of the infinite-horizon optimal cost $V^*(i)$ can be inferred from its finite-horizon counterpart, which is often easier to analyze.

It now remains to show that $\Delta V_k(i)$ is nondecreasing in i, which we will accomplish by induction on k. First, when $k = 0$, we have $V_0(i) = 0$ for all i, therefore $\Delta V_0(i) = 0$ which is obviously nondecreasing in i. Next, let us assume that $\Delta V_k(i)$ is nondecreasing in i for some $k > 0$, that is,

$$\Delta V_k(i+1) \geq \Delta V_k(i), \quad i = 0, 1, \ldots \tag{9.55}$$

We shall then prove that $\Delta V_{k+1}(i+1) \geq \Delta V_{k+1}(i)$, $i = 0, 1, \ldots$ Using (9.53), we have

$$
\begin{aligned}
\Delta V_{k+1}(i+1) &= \Big[b(i+1) + c_2 + \alpha p V_k(i+2) \\
&\quad + \alpha q_2 V_k(i) + \alpha(q_1 - q_2)V_k(i+1) \Big] \\
&\quad - \Big[bi + c_2 + \alpha p V_k(i+1) \\
&\quad + \alpha q_2 V_k(i-1) + \alpha(q_1 - q_2)V_k(i) \Big] \\
&\quad + \min\Big[(c_1 - c_2) - \alpha(q_1 - q_2)\Delta V_k(i+1), \ 0\Big] \\
&\quad - \min\Big[(c_1 - c_2) - \alpha(q_1 - q_2)\Delta V_k(i), \ 0\Big] \\
&= \Big[b + \alpha p \Delta V_k(i+2) + \alpha q_2 \Delta V_k(i) + \alpha(q_1 - q_2)\Delta V_k(i+1) \Big] \\
&\quad + \min[(c_1 - c_2) - \alpha(q_1 - q_2)\Delta V_k(i+1), \ 0] \\
&\quad - \min[(c_1 - c_2) - \alpha(q_1 - q_2)\Delta V_k(i), \ 0]
\end{aligned}
\tag{9.56}
$$

Next, we make use of the fact that for any a, b,

$$\min(a, 0) - \min(b, 0) \geq \min(a - b, 0)$$

This can be easily verified by considering all possible cases regarding the relative order of a, b, and 0. Thus, we get

$$\min[(c_1 - c_2) - \alpha(q_1 - q_2)\Delta V_k(i + 1), 0]$$
$$- \min[(c_1 - c_2) - \alpha(q_1 - q_2)\Delta V_k(i), 0]$$
$$\geq \min[-\alpha(q_1 - q_2)[\Delta V_k(i + 1) - \Delta V_k(i)], 0]$$
$$= -\alpha(q_1 - q_2)[\Delta V_k(i + 1) - \Delta V_k(i)]$$

where the last equality follows from the induction hypothesis (9.55).

An expression similar to (9.56) can be obtained for $\Delta V_{k+1}(i)$:

$$\Delta V_{k+1}(i) = \Big[b + \alpha p \Delta V_k(i + 1) + \alpha q_2 \Delta V_k(I - 1) + \alpha(q_1 - q_2)\Delta V_k(i) \Big]$$
$$+ \min[(c_1 - c_2) - \alpha(q_1 - q_2)\Delta V_k(i), \quad 0]$$
$$- \min[(c_1 - c_2) - \alpha(q_1 - q_2)\Delta V_k(I - 1), \quad 0]$$

$$(9.57)$$

Then, using the following fact:

$$\min(a, 0) - \min(b, 0) \leq -\min(b - a, 0)$$

we get

$$\min[(c_1 - c_2) - \alpha(q_1 - q_2)\Delta V_k(i), \quad 0]$$
$$- \min[(c_1 - c_2) - \alpha(q_1 - q_2)\Delta V_k(i - 1), \quad 0]$$
$$\leq -\min[\alpha(q_1 - q_2)[\Delta V_k(i) - \Delta V_k(i - 1), \quad 0] = 0$$

where the last equality again follows from the induction hypothesis (9.55).

Making use of the last two inequalities, we get:

$$\Delta V_{k+1}(i + 1) - \Delta V_{k+1}(i) \geq \alpha p[\Delta V_k(i + 2) - \Delta V_k(i + 1)]$$
$$+ \alpha q_2[\Delta V_k(i) - \Delta V_k(i - 1)]$$
$$+ \alpha(q_1 - q_2)[\Delta V_k(i + 1) - \Delta V_k(i)]$$
$$- \alpha(q_1 - q_2)[\Delta V_k(i + 1) - \Delta V_k(i)]$$

Observing the cancellation of the last two terms and the fact that all other terms are nonnegative by the induction hypothesis (9.55), it follows that $\Delta V_{k+1}(i + 1) - \Delta V_{k+1}(i) \geq 0$, which completes the inductive proof.

The determination of the optimal threshold i^* in (9.52) is illustrated in Fig. 9.5. Since we know that $\Delta V(i)$ is nondecreasing in i, we expect that its value will exceed the critical value D defined in (9.50) for some i.

When this happens, we identify the threshold i^*. In Fig. 9.5, for example, we have $i^* = 4$. Observe, however, that in order to determine i^* explicitly we need knowledge of $\Delta V(i), i = 0, 1, \ldots$, which means that we need to solve the optimality equation. On the other hand, knowing the structure of the optimal policy we may be able to analyze the Markov chain under this policy, parameterized by a threshold, and hence determine the best threshold value i^* (this was the approach followed in Example 9.3). It may also be possible to proceed "experimentally" by repeatedly trying threshold values until the best one is found. This "experimental" procedure typically requires computer simulation or direct trial-and-error for a given system. We will have more to say about this approach in upcoming chapters.

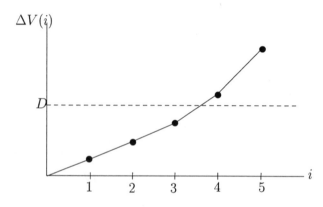

Figure 9.5: Determining the optimal threshold value in Example 9.4.
For the function $\Delta V(i)$ shown here, the optimal threshold is $i^* = 4$, since we have $\Delta V(3) < D$ and $\Delta V(4) > D$.

Example 9.5 (When to commit a job to a processor - no discounting)
For the "optimal stopping" problem in Example 9.3, we found that the policy minimizing a total expected discounted cost criterion is the one that commits the job to any processor $i \le i^*$ for some i^*. We also found that the value of i^* can be determined through (9.42).

Note that the one-step cost for this problem is nonnegative, therefore we may invoke Theorem 9.2, instead of Theorem 9.1, and set $\alpha = 1$ to solve an undiscounted version of this problem. In this case, we may interpret C_i as the time required to process our job when i is assigned, and c as the fixed time we have to wait until the next assignment. Thus, the optimal policy minimizes the total expected time until the job is completed (including possible waiting as well as processing).

Example 9.6 (Some optimal gambling policies)

We return to the "spin-the-wheel" game of Example 9.1. Recall that in this game a bet is lost with probability $11/12$, and the net payoff is twice the bet otherwise. Suppose the gambler has decided to adopt an "all or nothing" approach, whereby he always bets his entire capital i. Thus, either the game immediately ends, or the next state becomes $2i$. The question the gambler poses is: When is it time to quit?

We can place this problem in the context of an undiscounted cost Markov decision process by letting the state represent the gambler's capital, $i = 0, 1, \ldots$ The control actions are defined by

$$u(i) = \begin{cases} 1 & \text{if the gambler quits} \\ 0 & \text{if the gambler bets again} \end{cases}$$

for all $i = 1, 2, \ldots$ For $i = 0$ the game is automatically over.

The transition probabilities are specified as follows:

$$p_{ij} = \begin{cases} 11/12 & \text{if } j = 0 \\ 1/12 & \text{if } j = 2i \\ 0 & \text{otherwise} \end{cases}$$

for all $i = 1, 2, \ldots$ When $i = 0$, we have $p_{00} = 1$.

The gambler's objective is to maximize his capital. Thus, rather than considering a cost minimization formulation, as we have done so far, we consider instead an equivalent reward maximization. Whenever $i > 0$, if the control action is $u(i) = 1$, the gambler simply ends up with his capital i and no further rewards; if the control action is $u(i) = 0$, the gambler's expected future reward is given by $2i$ with probability $1/12$ and 0 otherwise.

Using Theorem 9.2, the optimality equation (9.43) for $i = 1, 2, \ldots$ becomes

$$V(i) = \max_{u \in \{0,1\}} [u(i)i + [1 - u(i)]\frac{1}{12}V(2i)]$$

which can be rewritten as

$$V(i) = \max[i, \ \frac{1}{12}V(2i)] \tag{9.58}$$

It is then immediately clear that the optimal control action is to quit, that is, $u(i) = 1$, whenever

$$i > \frac{1}{12}V(2i)$$

Now observe that at any state $i > 0$ the next state is either 0 and the game ends, or $2i$ and the game simply starts anew. It follows that we can view this problem as a one-step process with a terminal reward $V(2i) = 2i$ when $u(i) = 0$. Thus, the condition for the gambler to quit becomes:

$$i > \frac{1}{12} 2i \qquad 1 > \frac{1}{6}$$

which is always true. This is to be expected, given that this game is so obviously biased against the gambler. As a result, the optimal policy is to quit right away, that is, never play it under an "all or nothing" betting scheme.

Let us now generalize this game as follows. Let the probability of winning be p, and the resulting reward be ri, where i is the gambler's capital and r is a positive real number. The larger the value of r, the higher the winning payoff in the game is. Then, the generalization of the optimality equation (9.58) is

$$V(i) = \max[i, \ pV(ri)]$$

In this case, the condition for quitting the game is

$$i > pri \quad \text{or} \quad pr < 1$$

Thus, the product pr determines whether the game is worth playing, regardless of the gambler's initial capital.

Another interesting problem arises if the gambler decides that he may vary his bet, so that control actions at state i can take on values $\{1, 2, \dots, i\}$ (more generally, any real number in $(0, i]$). In this case, his objective is to play until he either loses his capital or exceeds a given amount M. The question then is: What is the optimal amount to bet each time? It turns out that the "all or nothing" policy discussed above is optimal if the game is biased against the gambler. That is, the gambler either bets his entire capital i if $i + ri < M$, or he bets the amount it takes to reach M if he wins, that is, $(M - i)/r$ (for details, see Bertsekas, 1987).

9.4.4 Optimization of the Average Cost Criterion

The average cost criterion for a continuous-time Markov chain was defined in (9.5). Following uniformization, as described in Section 9.3.2, we can rewrite this cost for a discrete-time model:

$$V_\pi(x_0) = \lim_{N \to \infty} \frac{1}{N} E_\pi \left[\sum_{k=0}^{N-1} C[X_k, u_k] \right] \qquad (9.59)$$

The first observation we can make here is that this criterion should be independent of the initial state x_0, since it represents costs accumulated in the long term, and cost contributions from the first few steps should not influence the overall cost over an infinite horizon. This is indeed the case in almost all problems of practical interest, as expected.

There are two basic difficulties when it comes to minimizing $V_\pi(x_0)$ in (9.59). First, an optimal policy may not exist. In other words, it may be possible to evaluate the smallest possible value of the average cost, and yet not be able to find a policy π that attains that cost. Second, even though we wish to consider stationary policies, an optimal policy that minimizes this criterion need not be stationary. Thus, our effort here must concentrate on identifying conditions under which a stationary optimal policy exists, before even attempting to see how such a policy can be derived.

In what follows, we assume once again that condition (9.1) holds, that is, all one-step costs are bounded, as in Section 9.4.2. In the case of the total expected cost criterion, we had the optimality equation in (9.34) to serve as the basis for solving the minimization problem. In the case of the average cost criterion, there is an analogous (but weaker) result, which is stated below as Theorem 9.4 (its proof may be found in Bertsekas, 1987; or Ross, 1983b).

Theorem 9.4 Suppose there exists a constant ν and a bounded function $h(i)$ which satisfy the following equation:

$$\nu + h(i) = \min_{u \in U_i} \left[C(i, u) + \sum_{\text{all } j} p_{ij}(u)h(j) \right] \tag{9.60}$$

Then, ν is the optimal cost in (9.59), that is,

$$\nu = \min_\pi V_\pi(i) = V^*(i) \tag{9.61}$$

and a stationary policy π^* is optimal if it gives the minimum value in (9.60) for all states i. ◆

Equation (9.60) is the "optimality equation" for the average cost minimization problem, the analog of (9.34) for total expected discounted cost minimization. However, Theorem 9.4 says nothing about conditions guaranteeing the existence of ν and $h(i)$. Interestingly, linking the average cost criterion (9.59) to the associated total expected discounted cost criterion (9.11) provides some additional insight that will help us overcome this difficulty.

Based on Theorem 9.1, we know that the optimal cost for (9.11) satisfies the optimality equation

$$V(i) = \min_{u \in U_i} \left[C(i, u) + \alpha \sum_{\text{all } j} p_{ij}(u)V(j) \right]$$

for some discount factor $\alpha, 0 < \alpha < 1$. Let us denote the optimal cost above by $V_\alpha(i)$ to remind us that it depends on α. Next, fix some state $x \in \mathcal{X}$, and define

$$h_\alpha(i) \equiv V_\alpha(i) - V_\alpha(x) \qquad (9.62)$$

Combining the last two equations, we get

$$V_\alpha(i) = h_\alpha(i) + V_\alpha(x) = \min_{u \in U_i} \left[C(i,u) + \alpha \sum_{\text{all } j} p_{ij}(u)[h_\alpha(j) + V_\alpha(x)] \right]$$

and noting that $\alpha \sum_j p_{ij}(u) V_\alpha(x) = \alpha V_\alpha(x)$ is independent of u, we get

$$h_\alpha(i) + V_\alpha(x) = \alpha V_\alpha(x) + \min_{u \in U_i} \left[C(i,u) + \alpha \sum_{\text{all } j} p_{ij}(u) h_\alpha(j) \right]$$

or

$$(1 - \alpha)V_\alpha(x) + h_\alpha(i) = \min_{u \in U_i} \left[C(i,u) + \alpha \sum_{\text{all } j} p_{ij}(u) h_\alpha(j) \right] \qquad (9.63)$$

This equation is interesting in that it is of the same form as (9.60), but not quite: We can associate ν with $(1 - \alpha)V_\alpha(x)$, and $h_\alpha(i)$ with $h(i)$, but that leaves the coefficient α inside the bracket term in (9.63). This naturally raises the question: What happens as $\alpha \to 1$? Thus, suppose that we construct some sequence $\{\alpha_n\}, n = 1, 2, \ldots$, such that $\alpha_n \to 1$ as $n \to \infty$. Moreover, suppose that the following limits exist:

$$\lim_{n \to \infty} (1 - \alpha_n) V_{\alpha_n}(x) = \nu \qquad (9.64)$$

$$\lim_{n \to \infty} h_{\alpha_n}(i) = h(i) \qquad (9.65)$$

If this is true, then we see that taking limits as $n \to \infty$ in equation (9.63) we can obtain the optimality equation (9.60) for the average cost problem. There are two implications here. First, we have a possible interpretation for the term $h_\alpha(i)$ based on the definition (9.62): The cost difference resulting from starting the process at state i instead of state x. Second, we see that the issue of identifying conditions under which ν and $h(i)$ in (9.60) exist reduces to the issue of existence of the two limits in (9.64) and (9.65), which have to do with the associated *discounted* cost problem.

There are two theorems we will state next which provide relatively simple to check conditions for guaranteeing the existence of ν and $h(i)$ in the optimality equation (9.60). It is actually possible to identify even weaker conditions, but these are not necessary for our purposes.

Theorem 9.5 Assume that the set of control actions U is finite. Suppose that there exists a finite constant L and some state $x \in \mathcal{X}$ such that

$$|V_\alpha(i) - V_\alpha(x)| \leq L \qquad (9.66)$$

for all $i \in \mathcal{X}$ and $\alpha \in (0,1)$. Then, for some sequence $\{\alpha_n\}, n = 1, 2, \ldots$, such that $\alpha_n \to 1$ as $n \to \infty$ and $\alpha_n \in (0,1)$ for all n, the limits (9.64) and (9.65) exist, and ν and $h(i)$ satisfy the optimality equation (9.60).

Proof. Let $\{\alpha_n\}, n = 1, 2, \ldots$, be any sequence that satisfies the assumptions of the theorem. By (9.66), the corresponding sequence $\{V_{\alpha_n}(i) - V_{\alpha_n}(x)\}$ is bounded. It follows that for some subsequence of $\{\alpha_n\}$, say $\{\alpha_m\}, \{V_{\alpha_m}(i) - V_{\alpha_m}(x)\}$ converges to a limit, say $h(i)$.

Next, consider the sequence $\{(1 - \alpha_m)V_{\alpha_m}(x)\}$. Since we are assuming that all one-step costs are bounded, Theorem 9.1 applies, and it follows that $V_{\alpha_m}(x)$ is the unique bounded solution of the optimality equation (9.34) with discount factor α_m. Therefore, the sequence $\{(1 - \alpha_m)V_{\alpha_m}(x)\}$ is bounded. As before, for some subsequence of $\{\alpha_m\}$, say $\{\alpha_r\}, \{V_{\alpha_r}(i) - V_{\alpha_r}(x)\}$ converges to a limit, say ν.

Finally, let us take limits as $n \to \infty$ (and hence $m \to \infty$ and $r \to \infty$) on both sides of equation (9.63). The interchange of the limit and the min operations is allowed because U is assumed finite. Since the left-hand side of (9.63) converges to $\nu + h(i)$, and $\alpha_n \to 1$ as $n \to \infty$, we immediately obtain (9.60). **Q.E.D.**

The next theorem reduces the issue of existence of ν and $h(i)$ in (9.60) to a simpler check of the structural conditions of the Markov chain to be controlled under stationary policies, provided its state space is finite. We omit its proof (see Bertsekas, 1987) which rests on our cost-boundedness assumption, based on which it can be shown that the conditions of Theorem 9.5 are satisfied.

Theorem 9.6 Assume that the set of control actions U and the state space $\mathcal{X}$ are both finite. Suppose that every stationary policy results in an irreducible Markov chain. Then, there exists a constant ν and a bounded function $h(i)$ that satisfy the optimality equation (9.60). Moreover, $\nu = V^*(i)$, and a stationary policy π^* is optimal if it gives the minimum value in (9.60) for all states i. ◆

A key point that emerges from this discussion is that there is a close connection between the expected average cost minimization and the associated expected discounted cost minimization problems, under certain conditions. In particular, suppose we have solved the expected discounted cost minimization problem and obtained $V_\alpha(i)$ for some α. Then, by the definition of $h_\alpha(i)$ in (9.62), if (9.65) is true we have

$$\lim_{\alpha \to 1} [V_\alpha(i) - V_\alpha(x)] = h(i)$$

We see that $h(i)$ inherits the structure of $V_\alpha(i)$. Now, when we determine the optimal stationary policy in the optimality equation (9.34), we look for the actions that minimize the bracket term on the right-hand side. This term is similar to the one on the right-hand side of (9.60), with $h(i)$ behaving just like $V_\alpha(i)$ as $\alpha \to 1$. It follows that when we can satisfy the conditions of Theorem 9.5 or Theorem 9.6 we can *infer the structure of the optimal stationary policy for the average cost criterion from that of the expected discounted cost criterion*. This is of great interest for a number of practical problems where the average cost criterion is a very natural one.

Unfortunately, in the case of many queueing systems (as in Examples 9.2 and 9.4), the conditions of the two theorems above are not satisfied. The main reason is that one-step costs are typically unbounded (e.g., inventory costs are proportional to queue lengths, which may become infinite if the queueing capacity is not limited). Fortunately, there are some alternative conditions which are usually satisfied (see Walrand, 1988), so one can still deduce optimal stationary policies for minimizing the average cost by solving the associated discounted cost problem.

Example 9.7 (Controlling the speed of a machine - average cost)

We return to the system of Examples 9.2 and 9.4, for which we determined that a threshold-type policy minimizes an expected discounted cost criterion. In practice, however, it may be more interesting to determine a stationary policy that minimizes the average cost per unit time of operating this system in the long run, defined as in (9.59). The one-step cost for the problem is the same as in (9.12).

For simplicity, let us assume that the total queueing capacity in the system is limited to $L < \infty$. In this case, the one-step cost in (9.12) is bounded by $(bL + c_1)$. The set of control actions, $U = \{0, 1\}$, and the state space $\mathcal{X} = \{0, 1, \ldots, L\}$ are both finite. Furthermore, the Markov chain is clearly irreducible, as can be seen by the transition diagram of Fig. 9.2. It follows that Theorem 9.6 can be invoked, and the optimal policy for the average cost minimization problem retains the same structure as that of the discounted cost minimization problem, that is, the machine is operated at low speed for all queue lengths $i \leq i^*$, for some threshold value i^*.

9.5 CONTROL OF QUEUEING SYSTEMS

We have already pointed out that our analysis of queueing systems in Chapter 8 was largely "descriptive." We developed models and techniques for evaluating performance measures of particular queueing systems operating under a given set of policies. However, an important component of our effort should be "prescriptive" as well: We should be able to specify the best possible type of queueing system and the best possible operating policies required to achieve some desired level of performance. In general, this is a much more complicated task. Although several techniques and explicit results are currently available, the area of queueing system control (and, more generally, DES control) is still in a state of continuing development. Our discussion in the previous sections provides some of the basic foundations for this effort to date.

There are many aspects of building a queueing system to meet a set of given performance objectives. This includes the configuration or architecture of the system, the selection of parameters (e.g., service rates, queueing capacities), and the operating policies employed (e.g., queueing disciplines, multiple class scheduling).

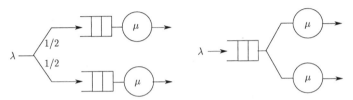

Scheme 1: two distinct queues   Scheme 2: one common queue

Figure 9.6: Comparing two queueing schemes.
Scheme 2, where a customer does not commit to a server until the last possible instant, provides smaller average system time.

We will not dwell here on architecture-related issues. We illustrate, however, the importance of designing proper queue/server configurations to achieve a specific performance objective by means of a simple classic example. Given two identical servers, we wish to minimize the average system time of all customers processed by these servers. The question we pose is: *Which of the two schemes shown in Fig. 9.6 is preferable?* The first scheme forces each customer to join a queue immediately upon arrival; the second scheme allows customers to form a single queue. Assuming a Poisson arrival process with rate λ and exponentially distributed service times with mean $1/\mu$, the answer to this question is readily obtained using the results of the $M/M/1$ and $M/M/m$ system analysis in Chapter 8 (Sections 8.6.1 and 8.6.2).

In scheme 1, a customer is immediately routed to a queue with probability $1/2$. Thus, we have two identical $M/M/1$ systems, each with arrival rate $\lambda/2$ and service rate μ. From (8.40), the corresponding average system time of each $M/M/1$ system is

$$\frac{1/\mu}{1 - \lambda/2\mu} = \frac{2}{2\mu - \lambda}$$

Therefore, the average system time for this scheme, $E[S]_1$, is

$$E[S]_1 = \frac{1}{2}\frac{2}{2\mu - \lambda} + \frac{1}{2}\frac{2}{2\mu - \lambda} = \frac{2}{2\mu - \lambda} \qquad (9.67)$$

In scheme 2, we have an $M/M/2$ system with arrival rate λ and service rate μ. Therefore, with $\rho = \lambda/2\mu$, we get the average system time, $E[S]_2$, from (8.51):

$$E[S]_2 = \frac{1}{\mu} + \frac{1}{\mu}\frac{4\rho^2}{2}\frac{\pi_0}{2(1-\rho)^2}$$

where π_0 is obtained from (8.45):

$$\pi_0 = \left[1 + 2\rho + \frac{4\rho^2}{2}\frac{1}{1-\rho}\right]^{-1} = \frac{1-\rho}{1+\rho}$$

Therefore,

$$E[S]_2 = \frac{1}{\mu} + \frac{\rho^2}{\mu(1-\rho^2)} = \frac{4\mu}{4\mu^2 - \lambda^2} \tag{9.68}$$

Comparing (9.67) to (9.68), we can see that $E[S]_1 > E[S]_2$. This proves that scheme 2, that is, using a common queue to access two servers, is preferable to scheme 1. This is consistent with the intuition that "committing as late as possible" to a server is better than an early commitment arrangement.

The issue of control in queueing systems is closely related to that of *resource contention*. Control is applied to ensure that all customers receive satisfactory service, that all resources are utilized at acceptable levels, and that resource allocation is fair among customers. Many of the fundamental control issues that need to be addressed boil down to questions such as "which resource should a particular customer access next?" or "which customer should a particular resource serve next?" The answers to these questions are usually provided by the operating policies employed by a given system.

The main goal in this section is to broadly categorize the control problems we face in queueing systems, and to formulate some basic such problems. Without getting into excessive technical details, we will also present some of the most well-known solutions that have been obtained to date using the techniques developed in previous sections. We consider three types of dynamic control problems: *admission* to a queueing system, *routing* customers to multiple servers, and *scheduling* customers on a single server. There is a wealth of queueing control problems, which are often variations or extensions of the ones we describe below, and many of them have only recently been treated and solved. One of the most comprehensive up-to-date accounts of the state-of-the-art in this area can be found in Walrand (1988).

9.5.1 The Admission Problem

Consider the simple queueing system of Fig. 9.7. When a customer arrives, we apply "admission control" by deciding whether the customer should be allowed in the queue or be rejected. There are several reasons necessitating admission control. First, we cannot allow the actual admission flow rate λ to exceed the service rate μ; we have seen that this results in building an infinite queue and customers waiting an infinite amount of time. Second, we usually have to ensure that certain acceptable performance levels are achieved. Typically, we wish to keep the average system time of customers as low as possible, while maintaining the server utilized as much as possible. Controlling the incoming customer flow is an obvious way to accomplish this goal. Finally, in the case of customers requesting access to a queueing network (rather than a single server), the effect of too many admissions in the system can manifest itself in a variety of indirect (and sometimes not so obvious) ways within the network.

Let us consider the admission control problem for a simple $M/M/1$ queueing system with arrival rate λ and service rate μ. We assume that queue length information is available whenever a new customer arrives, and the problem is to decide whether to admit or reject such a customer. We can now formulate a Markov decision problem as follows.

First, the state space is defined by $\mathcal{X} = \{0, 1, \dots\}$ and we let $i = 0, 1, \dots$ denote the system state. There are clearly only two control actions: "admit" and "reject". For convenience, set:

$$u(i) = \begin{cases} 1 & \text{if arriving customer is rejected at state } i \\ 0 & \text{if arriving customer is admitted at state } i \end{cases}$$

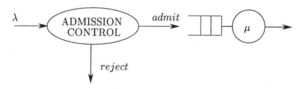

Figure 9.7: Admission control to a queueing system.
Based on observations of the queue length at the time of arrival, a customer may be admitted or rejected. A rejection incurs a fixed cost. On the other hand, admissions lead to queueing costs. The optimal admission policy is threshold-based: Customers are admitted only if they find a queue length $< i^*$, where i^* is the optimal threshold.

To establish a cost structure for the problem, we assume that rejecting a customer incurs a cost R. This cost can only be incurred when an arrival takes place. We therefore define the indicator function

$$\mathbf{1}[\text{arrival occurs at time } t] = \begin{cases} 1 & \text{if an arrival occurs at time } t \\ 0 & \text{otherwise} \end{cases}$$

and express rejection costs at any time by

$$R \cdot \mathbf{1}[\text{arrival occurs at time } t] u[X(t^-)],$$

where $X(t)$ is the system state. We use t^- to indicate that the admission decision is based on the queue length just before the arrival occurs. On the other hand, maintaining a queue for all admitted customers incurs costs, and we assume that there is a cost B per unit time for each admitted customer in queue or in service. We can now define a total expected discounted cost criterion as follows:

$$V_\pi(x_0) = E_\pi \left[\int_0^\infty e^{-\beta t} BX(t)dt \right.$$
$$\left. + \int_0^\infty e^{-\beta t} R\mathbf{1}[\text{arrival occurs at time } t]u[X(t^-)]dt \right]$$

where β is a given discount factor, x_0 is a given initial state, and $u[X(t^-)] = 1$ whenever an arrival finding the system at state $X(t)$ is rejected. Thus, the first integral above measures the cost of queueing customers, while the second one measures rejection costs. A tradeoff arises between these two cost components: To decrease rejection costs we must tolerate larger queues and hence higher queueing costs.

Through uniformization, we obtain a discrete-time model as described in Section 9.3.2 (see also Example 9.2). We choose the uniform rate $\gamma = \lambda + \mu$. We then set $\alpha = \gamma/(\beta + \gamma)$ and new cost parameters $\beta = B/(\beta + \gamma)$, $r = R/(\beta + \gamma)$. Thus, we obtain a cost criterion in the form (9.11):

$$V_\pi(i) = E_\pi \left[\sum_{k=0}^\infty \{\alpha^k bX_k \right.$$
$$\left. + \alpha^k r\mathbf{1}[\text{arrival occurs at } (k+1)\text{th transition}]u_k(X_k)\} \right]$$

Although the mathematical expression above is somewhat awkward, this should not obscure the simplicity of the cost structure: While the state is i, we incur a cost bi; and if an arrival occurs while the state is i, we incur a cost $ru(i)$.

To specify the transition probabilities for the uniformized model (shown in Fig. 9.8), we set $p = \lambda/\gamma$, $q = \mu/\gamma$, so that $p + q = 1$, and obtain the transition probabilities:

$$p_{ij}(u) = \begin{cases} p[1 - u(i)] & \text{if } j = i+1 \\ q & \text{if } j = i-1 \\ pu(i) & \text{if } j = i \\ 0 & \text{otherwise} \end{cases}$$

for $i > 0$, and

$$p_{0j} = \begin{cases} p[1 - u(0)] & \text{if } j = 1 \\ 1 - p + pu(0) & \text{if } j = 0 \\ 0 & \text{otherwise} \end{cases}$$

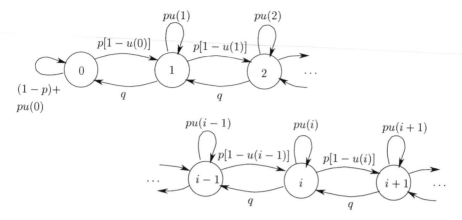

Figure 9.8: Uniformized controlled Markov chain model for admission control.

We can now proceed with the solution of this Markov decision problem in three steps.

Step 1. Write down the optimality equation for the problem.

Observe that costs are unbounded, since the queueing cost is proportional to the queue length, and the queue length is not constrained. However, all costs are non-negative, and we can make use of Theorem 9.2. Thus, the optimality equation (9.43) for our problem becomes, for $i = 1, 2, \ldots$,

$$V(i) = \min_{u \in \{0,1\}} [bi + \alpha p[1 - u(i)]V(i+1) + \alpha pu(i)[r + V(i)] + \alpha q V(i-1)]$$
(9.69)

and for $i = 0$,

$$V(0) = \min_{u \in \{0,1\}} [\alpha p[1 - u(0)]V(1) + pu(0)[r + V(0)] + \alpha(1 - p)V(0)] \quad (9.70)$$

Taking a closer look at (9.69), the first term is the one-step queueing cost when the state is $i > 0$. The second term corresponds to an arrival, which occurs with probability p and leads to a new state $(i+1)$ if the control action $u(i) = 0$ (admit) is taken. The third term also corresponds to an arrival, but this time the control action $u(i) = 1$ (reject) is taken; as a result the state is unchanged and a rejection cost r is incurred. The final term corresponds to a departure, which occurs with probability q. Similarly, in (9.70) we see that when an arrival occurs the state becomes 1 if the customer is admitted, otherwise the state is unchanged and a cost r is incurred. The third term accounts for the fictitious (self-loop) transitions occurring with probability $(1 - p) = q$.

The two equations above can be conveniently combined into one by introducing the notation:

$$[x]^+ = \max\{x, 0\} \tag{9.71}$$

for any real number x. Then, suppose we replace the last term in (9.69) by

$$\alpha q V([i-1]^+) = \begin{cases} \alpha q V(i-1) & \text{if } i = 1, 2, \ldots \\ \alpha q V(0) & \text{if } i = 0 \end{cases}$$

Since $p + q = 1$ in our uniformized model, we see that the last term in (9.70) is identical to that in (9.69) for $i = 0$. Consequently, we combine the two equations and write

$$V(i) = \min_{u \in \{0,1\}} \left[bi + \alpha p[1 - u(i)]V(i+1) + \alpha p u(i)[r + V(i)] + \alpha q V([i-1]^+) \right] \tag{9.72}$$

Removing the terms that do not depend on $u(i)$ from the min bracket, we can rewrite this equation as follows:

$$V(i) = bi + \alpha q V([i-1]^+) + \alpha p \min[V(i+1), r + V(i)] \tag{9.73}$$

Step 2. Characterize optimal control actions.

The structure of the optimal control action at any state $i = 0, 1, \ldots$ is immediately revealed from equation (9.73). If $V(i+1) \leq r + V(i)$, then the min term is minimized by choosing $u(i) = 0$, that is, admit the arriving customer. Thus, denoting the optimal control action at state i, by $u^*(i)$, we have

$$u^*(i) = \begin{cases} 0 & \text{if } V(i+1) \leq r + V(i) \\ 1 & \text{otherwise} \end{cases} \tag{9.74}$$

Step 3. Try to obtain a closed-form expression for the optimal policy.

Set

$$\Delta V(i+1) = V(i+1) - V(i), \quad i = 0, 1, \ldots$$

Thus, (9.73) becomes

$$u^*(i) = \begin{cases} 0 & \text{if } \Delta V(i+1) \leq r \\ 1 & \text{otherwise} \end{cases} \tag{9.75}$$

This structure is the same as that encountered in (9.51) in Example 9.4. Suppose that for $i = 0$ we have $\Delta V(1) \leq r$. Thus, $u^*(0) = 0$. If we can show that $\Delta V(i)$ is monotonically nondecreasing in $i = 0, 1, \ldots$, then $\Delta V(i+1)$ in (9.75) can eventually exceed r. At this point we switch to $u^*(i) = 1$ and maintain it for all higher values of i. It follows that if $\Delta V(i)$ indeed possesses this property, there is some *threshold* i^* such that

$$\Delta V(i^* + 1) > r, \quad \Delta V(i^*) \leq r$$

If we can determine the value of this threshold i^*, then the optimal policy reduces to the simple rule:

$$u^*(i) = \begin{cases} 0 & \text{if } i < i^* \\ 1 & \text{otherwise} \end{cases} \tag{9.76}$$

Thus, the optimal policy is to admit all customers as long as the queue length is less than the threshold i^*, and reject them otherwise. Once again, we see that a threshold-type policy is optimal in a Markov decision problem. It remains, however, to show that $\Delta V(i)$ is indeed *monotonically nondecreasing*. This is accomplished by proceeding exactly as in Example 9.4. We consider a finite-horizon version of the problem and define $\Delta V_k(i+1) = V_k(i+1) - V_k(i)$. Since the set of control actions in our problem is limited to $\{0, 1\}$ for all states, Theorem 9.3 can be used, that is, $V_k(i) \to V^*(i)$ as $k \to \infty$, and therefore $\Delta V_k(i) \to \Delta V^*(i)$. Consequently, it suffices to prove that $\Delta V_k(i)$ is nondecreasing in i for every $k = 0, 1, \ldots$, which is accomplished by induction on k. The details are omitted, but the procedure is identical to that of Example 9.4.

In summary, we have shown that optimal admission control entails the simple threshold-based policy (9.76). The value of the threshold i^* is such that it trades off too many rejections against queue lengths that become too long. Despite the simplicity of the model we have analyzed here, it turns out that this type of admission control is in fact optimal in many practical applications, including communication networks, computer systems, and manufacturing systems (see also Stidham, 1985). An appealing feature of the threshold-based policy is its implementation simplicity, since all it requires is a simple counter of customers, incremented by 1 upon admission and decremented by 1 upon departure from the system. In communication networks (where this problem is also referred to as "flow control"), it is customary to define a fixed number i^* of "tokens" which are kept at the point of admission control. A message or packet is admitted only if a token is present at its time of arrival. The message acquires this token, and relinquishes it only when it leaves the network. We can see that the effect of not finding a token upon arrival is to be rejected. In manufacturing systems, tokens are replaced by fixtures or pallets which are physically required to carry a part through various operations. Again, by choosing the number of available pallets to be the optimal threshold i^*, we effectively exercise threshold-based admission control.

The problem we solved here involved the minimization of the total discounted cost criterion. It is natural to ask whether our results extend to the average cost criterion, which is of more interest in practice. Based on the discussion of Section 9.4.4, this is often the case, provided certain conditions hold as in Theorems 9.5 and 9.6. It can be shown that this is in fact the case for our problem. Moreover, one can determine the value of the optimal threshold i^*, by solving the birth-death chain corresponding to an $M/M/1/i$ queueing system for different values of $i = 1, 2, \ldots$ The value of i^* is the one yielding the minimum average cost, which can be calculated using the stationary probability distribution of the chain (see also Section 8.6.4, where the $M/M/1/i$ system was analyzed).

For more general models, for example a $G/G/1$ system, obtaining the optimal threshold is hard, since we lack analytical expressions for the stationary probability distribution. In this case, one has to resort to numerical techniques, simulation, or some of the sensitivity analysis techniques we will discuss in Chapter 11.

9.5.2 The Routing Problem

Consider the queueing system of Fig. 9.9, where customers can choose between joining queue 1 or queue 2. The selection of a queue is referred to as "routing." The purpose of routing is to balance the use of the two (or more) resources that can be shared among customers.

Let us consider the simplest possible version of a routing problem, where customers arrive according to a Poisson process with rate λ, and must choose between two queues each connected to an exponential server with service rate $\mu_i, i = 1, 2$. In general, $\mu_1 \neq \mu_2$. The arrival and the two service processes are assumed to be mutually independent. As in the admission problem, we assume that queue length information from both queues is available whenever a new customer arrives. We now formulate a Markov decision problem as follows.

Because we have to keep track of two queue lengths, the state space is defined by $\mathcal{X} = \{(i, j) : i, j = 0, 1, \ldots\}$, where i is the length of queue 1, and j the length of queue 2. The set of control actions is defined by $U = \{$route to queue 1, route to queue 2$\}$.

A typical cost structure for this problem involves the total discounted queueing costs. Thus, we assume that there is a cost B_1 per unit time for each customer in queue 1 or in server 1, and a cost B_2 per unit time for each customer in queue 2 or in server 2. We can now define a total expected discounted cost criterion as follows:

$$V_\pi \left(x_0^1, x_0^2 \right) = E_\pi \left[\int_0^\infty e^{-\beta t} \left\{ B_1 X_1(t) + B_2 X_2(t) \right\} dt \right]$$

where β is a given discount factor, (x_0^1, x_0^2) is a given initial state, and $X_i(t)$ is the length of queue i at time t, $i = 1, 2$. Through uniformization, we obtain a discrete-time model by proceeding as described in Section 9.3.2 and similar to the admission control problem of the previous section. In this case, we choose the uniform rate $\gamma = \lambda + \mu_1 + \mu_2$. We then set $\alpha = \gamma/(\beta + \gamma)$ and new cost parameters $b_1 = B_1/(\beta + \gamma), b_2 = B_2/(\beta + \gamma)$. Thus, we obtain a cost criterion of the form (9.11).

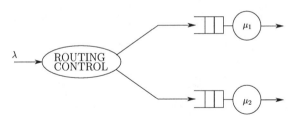

Figure 9.9: Routing between two parallel queues.
*An arriving customer observes the two queue lengths and chooses which queue to join. The objective is to minimize total discounted or average queueing costs. The optimal routing policy is based on a **switching curve** which splits the state space into two parts, one for each possible decision.*

To specify the transition probabilities for the uniformized model, we set $p = \lambda/\gamma$, $q_1 = \mu_1/\gamma$, $q_2 = \mu_2/\gamma$, so that $p + q_1 + q_2 = 1$. The transition probabilities are notationally cumbersome to write down, due to the two-dimensional nature of the system state. In Fig. 9.10, we show a typical state (i, j), $i, j = 1, 2, \ldots$ with all possible transitions resulting from arrivals and departures, with the control action $u(i, j)$ defined by:

$$u(i, j) = \begin{cases} 1 & \text{if arriving customer finding state } (i, j) \text{ is routed to queue 1} \\ 0 & \text{if arriving customer finding state } (i, j) \text{ is routed to queue 2} \end{cases}$$

Note that at state $(0, j)$ we need an additional self-loop transition with probability q_1, corresponding to fictitious departures at queue 1 (which is empty). Similarly, at state $(i, 0)$ we need an additional self-loop transition with probability q_2, corresponding to fictitious departures at queue 2.

We can now proceed with the solution in the usual three steps, as in the admission control problem.

Step 1. Write down the optimality equation for the problem.

As in the admission control problem, the one-step costs are unbounded, since queueing costs are proportional to the queue lengths. Since, however, all costs are nonnegative, we can invoke Theorem 9.2 and use the optimality equation (9.43). In our case, for all $i, j = 0, 1, \ldots$ we have:

$$V(i, j) = \min_{u \in \{0, 1\}} \Big[b_1 i + b_2 j + \alpha p u(i, j) V(i + 1, j)$$
$$+ \alpha p [1 - u(i, j)] V(i, j + 1) \tag{9.77}$$
$$+ \alpha q_1 V([i - 1]^+, j) + \alpha q_2 V(i, [j - 1]^+) \Big]$$

where we have used the notation introduced in (9.71): $[i - 1]^+ = i - 1$ as long as $i \geq 1$, and $[i - 1]^+ = 0$ if $i = 0$. This allows us to account for the self-loop

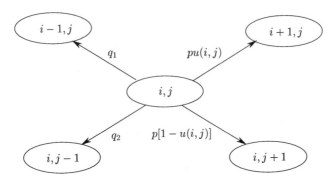

Figure 9.10: Transition probabilities for state $(i, j), i, j > 0$, in the uniformized model.

A customer arrives with probability p, and departs from queue i with probability q_i, $i = 1, 2$. The control action $u(i, j) \in \{0, 1\}$ determines which queue to join. Note that if $i = 0$, the diagram should include a self-loop with probability q_1 (fictitious departure at queue 1), and if $j = 0$, the diagram should include a self-loop with probability q_2 (fictitious departure at queue 2).

transition at states $(0, j)$ which leave the state unchanged. Similarly for the term $[j - 1]^+$ above.

In (9.77), the first two terms represent the one-step queueing costs. The third and fourth terms correspond to arrivals, occurring with probability p: In the third term the arrival is routed to queue 1, whereas in the fourth term it is routed to queue 2. The last two terms correspond to departures from queue 1 and queue 2 respectively.

Separating the terms which are independent of $u(i, j)$ from those that are not, we can rewrite the optimality equation as follows:

$$V(i, j) = b_1 i + b_2 j + \alpha q_1 V([i - 1]^+, j) + \alpha q_2 V(i, [j - 1]^+) \\ + \alpha p \min [V(i + 1, j), V(i, j + 1)] \tag{9.78}$$

Step 2. Characterize optimal control actions.

In equation (9.78), there are two cases regarding the min term:

1. If $V(i+1, j) \leq V(i, j+1)$, this term is minimized by choosing $u(i, j) = 1$, that is, the optimal control action is to route to queue 1.

2. If $V(i+1, j) > V(i, j+1)$, the optimal control action is to route to queue 2.

Thus, denoting the optimal control action at state (i, j) by $u^*(i, j)$, we have

$$u^*(i, j) = \begin{cases} 1 & \text{if } V(i + 1, j) \leq V(i, j + 1) \\ 0 & \text{otherwise} \end{cases} \tag{9.79}$$

Step 3. Try to obtain a closed-form expression for the optimal policy.

This turns out to be quite difficult for this problem. First, we define a subset S of the state space such that all states $(i, j) \in S$ satisfy the condition $V(i+1, j) \leq V(i, j+1)$:

$$S = \{(i, j) : V(i+1, j) \leq V(i, j+1)\} \tag{9.80}$$

The optimal policy is simply: Route to queue 1 if an arrival finds a state $(i, j) \in S$. The problem is that in order to have explicit knowledge of S we need to solve the optimality equation for $V(i, j)$, which is not an easy task.

Fortunately, the set S has some additional structural properties, which can be thought of as the two-dimensional analog of the threshold structure we saw in the admission control problem of Section 9.5.1. The admission control problem is one-dimensional in that the state is described by a single queue length $i = 0, 1, \ldots$ There are also only two possible control actions (admit, reject). The routing problem is two-dimensional, since the state is described by two queue lengths, $i, j = 0, 1, \ldots$ However, there are still only two possible control actions (route to queue 1, route to queue 2). Now, in the one-dimensional problem we saw that one action is optimal as long as the state i is less than a *fixed threshold* i^*. In our two-dimensional problem it turns out that one action is optimal as long as one state variable, say i, is less than a threshold function $s(j)$, $j = 0, 1, \ldots$ This function defines a *switching curve* as shown in Fig. 9.11, which divides the state space into two parts: all i such that $i \leq s(j)$, and all i such that $i > s(j)$. Then, if (i, j) is such that $i \leq s(j)$, the optimal control action is to route to queue 1:

$$u^*(i, j) = \begin{cases} 1 & \text{if } i \leq s(j) \\ 0 & \text{otherwise} \end{cases} \tag{9.81}$$

Intuitively, we route to queue 1 whenever its observed length i is "sufficiently small" compared to the observed queue length j at queue 2. Suppose that for a given state (i, j) the optimal action is to route to queue 1. It then makes sense that we should route to the same queue for all observed $i' < i$, provided j remains the same. This observation is captured by the switching curve in Fig. 9.11. For example, we see that when state $(3, 4)$ is observed, it is optimal to route to queue 1. Then, it is also optimal to route to queue 1 when states $(2, 4)$, $(1, 4)$, and $(0, 4)$ are observed.

To establish the optimal policy structure in (9.81), we return to (9.79) and observe that $u^*(i, j) = 1$ as long as $\Delta V(i, j) \leq 0$, where we define:

$$\Delta V(i, j) = V(i+1, j) - V(i, j+1), \quad i = 0, 1, \ldots$$

Suppose that for some (i, j) we have $\Delta V(i, j) \leq 0$. Then, if $\Delta V(i, j)$ is *monotonically nondecreasing* in i for any fixed j, as i increases we eventually get to some $i^*(j)$ where $\Delta V(i, j)$ switches sign. The state $(i^*(j), j)$ defines a

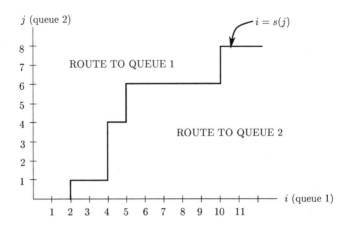

Figure 9.11: A typical switching curve for optimal routing.
When an arriving customer finds a state (i, j), the optimal control action is to route to queue 1 if $i \leq s(j)$, and to queue 2 otherwise.

point on our switching curve $i = s(j)$. Similarly, if i is fixed and $\Delta V(i, j)$ is *monotonically nonincreasing* in j, as j decreases we come to point $(i, j^*(i))$ where $\Delta V(i, j)$ switches sign. Thus, it remains to prove these two properties of $\Delta V(i, j)$. As in the admission control problem and in Example 9.4, this can be accomplished by considering a finite-horizon version of the problem and defining $\Delta V_k(i, j) = V_k(i + 1, j) - V_k(i, j + 1)$. Since the set of control actions in our problem is limited to $\{0, 1\}$ for all states, Theorem 9.3 can be used, that is, $V_k(i, j) \to V^*(i, j)$ as $k \to \infty$, and therefore $\Delta V_k(i, j) \to \Delta V^*(i, j)$. Consequently, it suffices to prove that $\Delta V_k(i, j)$ is nondecreasing in i and nonincreasing in j for every $k = 0, 1, \ldots$ This can be done by induction on k. The details, which can get somewhat tedious, are omitted, but the procedure is identical to that of Example 9.4.

The solution of the optimal routing problem is best summarized by Fig. 9.11. If the switching curve is available, then every routing decision boils down to a simple check on whether the observed state (i, j) belongs to one or the other part of the state space as partitioned by this curve.

One special case of this solution that is worth mentioning arises when $\mu_1 = \mu_2$, that is, the two servers are indistinguishable, and the optimal policy should depend only on the relative values of the two cost parameters β_1, β_2. If, in addition, $\beta_1 = \beta_2$, then the system is completely symmetric. In this case, the switching curve reduces to $i = j$. In other words, the optimal policy is to *route to the shortest queue*. Note, however, that as soon as this symmetry is broken, this intuitive "shortest queue" routing policy need no longer be optimal.

In practice, the routing problem is of particular interest when our objective is to minimize the average cost criterion. As we have seen in Section 9.4.4, it is often possible to derive the optimal policy for this problem from the optimal policy for the total discounted cost criterion by letting the discount factor $\alpha \to 1$. It can be shown that this is the case for the routing problem above, provided that we satisfy the basic stability condition $\lambda < \mu_1 + \mu_2$ (further details are provided in Walrand, 1988, and Hajek, 1984).

There are a number of variations and extensions of the basic routing problem we considered here. For example, if the queues have finite capacities (not necessarily equal), then there is an additional cost consideration (similar to the admission problem) due to customer blocking: If there is no more space in either queue, an arriving customer must be rejected and the system incurs some rejection cost. Thus, when making a decision to route to queue 1 because $i \leq s(j)$ in Fig. 9.11, we must now also anticipate the possibility of future blocking effects, which complicates the problem substantially.

Finally, note that the solution of the basic two-queue routing problem we considered here is not easy to obtain. In fact, we were not able to go beyond identifying the threshold function structure of the optimal policy. Obtaining the explicit form of the switching function is a very tedious process in general. This strongly suggests that the analysis of three and higher-dimensional queueing systems quickly becomes too complex to handle without resorting to numerical techniques or computer simulation.

9.5.3   The Scheduling Problem

Consider the queueing system of Fig. 9.12, where there is a single server providing service to two queues. Here, we distinguish between two "classes" of customers, where customers of class 1 are placed in queue 1 and customers of class 2 are placed in queue 2. Whenever the server is idle, there is a choice to be made as to which queue (customer class) to serve next. This is referred to as a "scheduling" problem. The purpose of scheduling is to provide service to two or more customer classes in a way which is fair and efficient in terms of minimizing an overall cost criterion. Note that the two queues in Fig. 9.12 may be simply "conceptual." Physically, there might be a single queue where both classes are accommodated, but where we are always able to distinguish between customers of each class, and we need not serve them on a purely first-come-first-served basis.

Scheduling may be thought of as a dual of the routing problem: In routing, a customer decides between two servers upon arrival; in scheduling a server decides between two classes of customers upon departure. There are two ways in which we distinguish between customer classes. First, each class has an associated cost parameter $c_i, i = 1, 2$, measuring the queueing cost of a class i customer per unit time. In general, $c_1 \neq c_2$. Second, the service rate of each class is μ_i, $i = 1, 2$, and, in general, $\mu_1 \neq \mu_2$.

The scheduling problem is to choose which queue to serve next, a decision that may, in general, be made at any time. There are two types of policies we may consider: *preemptive* or *non-preemptive*. Under non-preemptive scheduling, once the server commits to a particular customer, no interruption of service is allowed. Thus, the next time a new scheduling decision can be made is only upon this customer's departure. Under preemptive scheduling, the server is allowed to interrupt the current service at any time, and serve instead some other customer in queue. In this case, the server will return to the preempted customer at a later time. The idea here is that one class may be more "important" than the other. Thus, if the server starts serving a customer from the less important class, we want to have the ability to preempt that customer for the benefit of an arrival from the more important class. In what follows, we will assume preemptive scheduling.

Figure 9.12: Scheduling two customer classes on a server.
Each queue corresponds to a customer class with queueing cost per unit time $c_i, i = 1, 2$, and service rate μ_i. The server can observe the two queue lengths and choose which queue (customer class) to serve next. The server may also preempt a customer in service and resume at a later time. The objective is to minimize total discounted queueing costs. The optimal scheduling policy follows the μc-rule: Service is always given to the non-empty queue with the highest $\mu_i c_i$ value. This is a static policy, independent of the system state.

A typical objective of the scheduling problem is the minimization of the total expected discounted queueing costs:

$$V_\pi(x_0^1, x_0^2) = E_\pi \left[\int_0^\infty e^{-\beta t} \{c_1 X_1(t) + c_2 X_2(t)\} dt \right]$$

where β is a given discount factor, (x_0^1, x_0^2) is a given initial state, and $X_i(t)$ is the length of queue i at time t, $i = 1, 2$. To formulate a Markov decision problem as in previous sections, we can assume that customers arrive at each queue according to a Poisson process with rate $\lambda_i, i = 1, 2$, and that service times are exponentially distributed with rate $\mu_i, i = 1, 2$, when a class i customer is served. The arrival and service processes are all assumed to be mutually independent. This approach, however, becomes extremely tedious when one wishes to extend the problem to M queues, where $M > 2$. Moreover, it turns out that the optimal policy is independent of the arrival processes (which we have to assume to be Poisson in the Markov decision process framework).

It can be shown that the optimal scheduling policy is based on the following "μc-rule": *Service is always given to the nonempty queue with the largest $\mu_i c_i$*

value, $i = 1, \ldots, M$. What is particularly interesting here is that that this is a static policy, that is, it is independent of the state of the system (except for knowledge of whether a queue is empty or not). Thus, one can a priori order the queues in descending order of $\mu_i c_i$ values. Then, when a departure occurs, the server simply looks at all nonempty queues in that order and serves the first one it finds. If all queues are empty, the server waits for the next arrival. Preemption manifests itself as follows. Suppose the server starts serving a class j customer while there is at least one empty queue i such that $\mu_i c_i > \mu_j c_j$. Then, during this service interval, suppose that a class i arrival occurs with $\mu_i c_i > \mu_j c_j$. In this case, the server immediately preempts the class j customer and starts serving the class i customer.

The proof of optimality of the μc-rule is based on considering sample paths of the system under specific policies and showing that the policy using the μc-rule is always better in terms of our cost criterion. This is accomplished through a so-called "interchange" argument, which is also used in a number of related problems. We will not go into details of this argument (see Bertsekas, 1987; or Walrand, 1988). We provide, however, a sketch of the proof, which goes as follows.

We consider a discrete-time model of the system, where there are M queues, the arrival processes are arbitrary, and a discount factor α is given. Service times, on the other hand, are assumed to be geometrically distributed with parameters μ_i for class $i = 1, \ldots, M$. Let π^* be the policy that follows the μc-rule. The proof is by induction on the number of steps-to-go $k = 0, 1, \ldots, N$. Thus, if we assume that π^* is indeed optimal for n steps-to-go, we need to establish that it is also optimal for $(n + 1)$ steps-to-go. In other words, π^* is followed at steps $k = 1, 2, \ldots, N$, and the question is: What is the optimal control action at $k = 0$? Suppose queue i is the nonempty queue at $k = 0$ with the largest $\mu_i c_i$ value. Let j be some other nonempty queue, that is, $\mu_i c_i > \mu_j c_j$. Now, define two policies:

$$\pi = (j, i, \pi^*) \quad \text{and} \quad \pi' = (i, j, \pi^*)$$

Under policy π we choose queue j at $k = 0$, and then switch to queue i at time $k = 1$, regardless of whether the class j customer has completed service. Thus, we violate the μc-rule at $k = 0$. Under policy π' we "interchange" the two decisions. Thus, the μc-rule is followed at $k = 0$. After this interchange, both policies simply follow π^*. It remains to show that π' is better than π, that is, that violating the μc-rule always increases cost.

Let x_i and x_j be the given initial queue lengths of queues i and j respectively. Under policy π, the class j customer completes service in one time step with probability μ_j. Ignoring external arrivals for simplicity (which in no way affect the argument), the expected one-step cost after this first step is

$$C(\pi) = \alpha[\mu_j(x_j - 1)c_j + (1 - \mu_j)x_j c_j + x_i c_i] + \alpha \sum_{r \neq i, j} x_r c_r$$

Similarly, under policy π', the class i customer completes service in one time step with probability μ_i, and the expected one-step cost after this first step is:

$$C(\pi') = \alpha[\mu_i(x_i - 1)c_i + (1 - \mu_i)x_i c_i + x_j c_j] + \alpha \sum_{r \neq i,j} x_r c_r$$

Subtracting, we get

$$C(\pi) - C(\pi') = \alpha[\mu_i c_i - \mu_j c_j] > 0$$

since, by assumption, $\mu_i c_i > \mu_j c_j$. Thus, $C(\pi) > C(\pi')$, and it follows that violating the μc-rule at $k = 0$ can only increase cost. Consequently, the μc-rule remains optimal for $(n + 1)$ steps-to-go, which completes the inductive proof.

Problems related to scheduling, be it customers or different types of resources, arise in a variety of settings other than the one shown in Fig. 9.12. It is common, for instance, to encounter applications where scheduling the accessibility of servers to customers waiting at different queues is of critical importance.

Example 9.8 (Scheduling elevators)

Scheduling elevators in large buildings is a major challenge that involves systematically deciding *when* and *where* each elevator car should move, stop, or switch direction based on the current state and available past history. Although the scheduling setting here is different from the one shown in Fig. 9.12, it still fits the framework of Markov decision processes.

The usual performance objective of elevator scheduling is to minimize the average passenger waiting time. Achieving this objective is difficult for a number of reasons, including the need to (a) coordinate multiple cars, (b) satisfy constraints on elevator movement (e.g., a car must stop at a floor where a passenger wants to exit), (c) operate with incomplete state information (e.g., while it is known whether an elevator has been called to a particular floor, it is generally not known how many passengers are waiting at that floor), (d) make decisions in the presence of uncertainty (e.g., passenger arrival times and destinations are uncertain), and (e) handle nonstationary passenger traffic (e.g., for an office building, passenger traffic varies continuously throughout the day, from morning up-traffic, to heavy two-way lunchtime traffic, to evening down-traffic). Even without difficulties (d) and (e), this scheduling problem is combinatorially explosive due to the enormous size of the state space.

A systematic study of the elevator scheduling problem begins by decomposing passenger traffic into four different situations: (a) *uppeak* traffic, (b) *lunchtime* traffic, (c) *downpeak* traffic, and (d) *interfloor* traffic. The uppeak traffic situation arises when all passengers are moving up from the first floor (e.g., the start of the business day in an office building).

Lunchtime traffic is a characterization in which passengers are going to and returning from the first floor (e.g., as they go to and return from lunch in an office building). The downpeak traffic situation is observed when all passengers are moving down to the first floor (e.g., the end of the business day when an office building is emptied). Finally, interfloor traffic is a characterization in which passengers are moving equally likely between floors.

Interestingly, the scheduling problem for the uppeak traffic case is relatively manageable to tackle using the machinery we have developed in this chapter. Assuming the uppeak traffic originates from a single floor (the first floor lobby of a building) and that each elevator serves every floor, we can model the uppeak traffic situation as a single queue of infinite capacity (representing the first floor lobby), served by identical bulk servers (corresponding to identical elevator cars), each with a finite capacity of C passengers. Figure 9.13 illustrates this model, in which passengers arrive one at a time to the queue according to a Poisson process with rate λ. Each passenger arrival generates a "passenger arrival" (pa) event. The passengers are admitted into the cars and the cars are dispatched by the scheduling control. The passengers are served by the cars in batches of size no greater than the car capacity C. The time for a car to serve a batch of passengers is exponentially distributed with parameter μ, a constant, which is the same for each car, and independent of the state. After a car has delivered all of its passengers, it makes an immediate express run returning empty to the first floor lobby to serve more uppeak passengers. The completion of service generates a "car arrival" (ca) event indicating that one of the two elevators has become available for service. Since the elevators are identical, there is no need to distinguish between them.

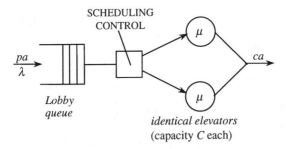

Figure 9.13: Elevator scheduling model for Example 9.8.

The state space for a model limited to two elevator cars is obtained by defining $Y(t) \in \{0, 1, \ldots\}$ to denote the queue length at the first floor lobby at time t and $Z(t) \in \{0, 1, 2\}$ to denote the number of elevators available at the first floor at time t. State transitions in this model are the result of pa and ca event occurrences. Control actions are taken only when any such event occurs and they define a set $U = \{0, 1, 2\}$ where $u = 0$ implies no action (hold all available cars at the first floor), $u = 1$ means loading one elevator car and dispatching it, and $u = 2$ means allowing both cars to load and dispatching them simultaneously. Since cars returning to the lobby are assumed to be empty, each available car can serve up to C passengers from the lobby. Those passengers which would cause a car to overflow will remain at the lobby to wait for another one.

To implement the control action $u = 1$ above when both cars are available, the system must have the ability to load one car before loading the other. This is typically implemented using the so-called "next car" feature: Since returning cars are empty, they do not need to open their doors when they reach the main lobby; thus, to force passengers to load one car at a time, only one car opens its doors and this car is referred to as the "next car" to be dispatched. Observe also that not all actions are admissible at every state. In particular, let $U(y, z)$, a subset of U, denote the set of admissible actions at some state (y, z). Then, $U(y, 0) = \{0\}$, $U(y, 1) = \{0, 1\}$, and $U(y, 2) = \{0, 1, 2\}$.

As in all previous problems we have considered, we proceed by considering a uniformized version of the model. We can choose a uniform rate $\gamma = \lambda + 2\mu$, the total event rate in our two-car model. Without loss of generality, we can also assume the time scale has been normalized so that $\gamma = 1$. Control actions are taken at the beginning of each time step. Let $P_{ij}(u)$ denote the transition probability from state i to state j when the control action taken at the beginning of the current time step is $u \in U(i)$. These state transition probabilities are given by

$$p_{ij}(0) = \begin{cases} \lambda & \text{if } i = (y, 0),\ j = (y + 1, 0) \\ 2\mu & \text{if } i = (y, 0),\ j = (y, 1) \end{cases}$$

$$p_{ij}(0) = \begin{cases} \lambda & \text{if } i = (y, 1),\ j = (y + 1, 1) \\ \mu & \text{if } i = (y, 1),\ j = (y, 2) \\ \mu & \text{if } i = j = (y, 1) \end{cases}$$

$$p_{ij}(0) = \begin{cases} \lambda & \text{if } i = (y, 2),\ j = (y + 1, 2) \\ 2\mu & \text{if } i = j = (y, 2) \end{cases}$$

$$p_{ij}(1) = \begin{cases} \lambda & \text{if } i = (y, 1),\ j = ([y - C]^+ + 1, 0) \\ 2\mu & \text{if } i = (y, 1),\ j = ([y - C]^+, 1) \end{cases}$$

$$p_{ij}(1) = \begin{cases} \lambda & \text{if } i = (y,2), \ j = ([y-C]^+ + 1, 1) \\ \mu & \text{if } i = (y,2), \ j = ([y-C]^+, 2) \\ \mu & \text{if } i = (y,2), \ j = ([y-C]^+, 1) \end{cases}$$

$$p_{ij}(2) = \begin{cases} \lambda & \text{if } i = (y,2), \ j = ([y-2C]^+ + 1, 0) \\ 2\mu & \text{if } i = (y,2), \ j = ([y-2C]^+, 1) \end{cases}$$

where we have used the usual notation $[y - C]^+ = \max\{y - C, 0\}$ and $[y - 2C]^+ = \max\{y - 2C, 0\}$. For each of the preceding equations, the first row corresponds to a state transition induced by a *pa* event. All remaining transitions are induced by a *ca* event, including fictitious *ca* events introduced by uniformization. In the second equation, for example, the last *ca* event is fictitious because one of the cars is available at the lobby and, therefore, cannot generate an actual car arrival event.

To complete the problem formulation, we take the one-step cost to be proportional to the queue length resulting from the control action taken at the beginning of the time step. Letting $C(Y_k, Z_k, u_k)$ denote the cost at the kth time step and b be some given positive and bounded holding cost, we have $C(y, 0, u) = by$ and

$$C(y, 1, u) = \begin{cases} b[y-C]^+ & \text{if } u = 1 \\ by & \text{if } u = 0 \end{cases}$$

$$C(y, 2, u) = \begin{cases} b[y-2C]^+ & \text{if } u = 2 \\ b[y-C]^+ & \text{if } u = 1 \\ by & \text{if } u = 0 \end{cases}$$

Then, our objective is to obtain the optimal stationary policy π^* that minimizes the total discounted cost

$$V_\pi(i) = E_\pi \left[\sum_{k=0}^{\infty} \alpha^k C(Y_k, Z_k, u_k) \right]$$

where α is a discount factor.

It can be shown that the structure of the optimal scheduling policy is *threshold-based* as in other examples we have seen. Specifically, the optimal policy is to dispatch an available elevator car from the first floor *when the number of passengers inside the car reaches or exceeds a threshold value* which depends on λ, μ, α, and b. Details are left as a (challenging) exercise for the reader (see also Pepyne and Cassandras, 1997). It is also worth adding that elevator system scheduling remains an area where DES theory finds several interesting applications.

SUMMARY

- For DES which can be modelled through Markov chains, it is possible to control performance by controlling the chain's transition probabilities. In the Markov decision process framework, whenever a state transition occurs, one may choose a control action which incurs some cost and affects transition probabilities. The rule based on which control actions are chosen is called a *policy*. A policy is *stationary* if a control action is not chosen at random, and if it depends only on the current state.

- The basic Markov decision problem involves the minimization of the total expected discounted cost criterion of equation (9.11). The discount factor α is such that $0 < \alpha < 1$.

- *Dynamic Programming* (DP) offers one of the most general techniques for the solution of Markov decision problems. The DP algorithm transforms an optimization problem requiring searching over all possible policies to a number of simpler problems requiring searching over a set of control actions. Despite its attractive features, the DP algorithm still involves prohibitively high computational costs when the state space of the problem is large.

- When one-step costs are bounded, the solution of the basic Markov decision problem is obtained through the optimality equation (9.34). The DP algorithm can be used to obtain the infinite-horizon optimal cost as the limit of the finite N-step horizon optimal cost as $N \to \infty$.

- When one-step costs are unbounded or undiscounted, the optimality equation is still valid, provided all costs are non-negative (or: all costs are non-positive). In this case, however, the DP algorithm cannot always be used to obtain the infinite-horizon optimal cost as the limit of the finite N-step horizon optimal cost as $N \to \infty$. One condition under which this holds is that the set of control actions is finite.

- In practice, the average cost criterion (9.59) is often of interest. Under certain conditions, the optimal policy that minimizes this criterion is obtained from the optimal policy minimizing the associated discounted cost criterion as $\alpha \to 1$.

- In the control of queueing systems, three basic problems are: *admission control, routing, and scheduling*. In a simple admission control problem, the optimal policy is to admit customers only as long as the queue length is below some *threshold*. In a simple routing problem involving two queues, the optimal policy is defined by a *switching curve*, which divides the state space into two parts as shown in Fig. 9.11. In a basic scheduling problem involving M queues and a single server, different techniques can be used to establish the optimality of the μc-rule: Service is always given to the nonempty queue with the largest $\mu_i c_i$ value, $i = 1, \ldots, M$

- Besides Markov decision processes, the control of DES can be studied through a variety of sample-path-based techniques. It is possible, for example, to compare the performance of two systems operating under two different policies by comparing their respective sample paths under the exact same event sequence (e.g., identical arrival sequences in a queueing system). Sometimes, one can show that a particular policy is always better than any other policy (within a particular class of policies), no matter what sample path one considers, and hence establish optimality results. Some relevant material on these techniques may be found in Ross (1983a), Marshall and Olkin (1979), and Walrand (1988). Some sample-path-based techniques will also be discussed in Chapter 11.

PROBLEMS

9.1 Step 3 of the derivation of an optimal policy for the admission problem in Section 9.5.1 involves a proof of the fact that $\Delta V(i)$ is monotonically nondecreasing, where

$$\Delta V(i + 1) = V(i + 1) - V(i), \quad i = 0, 1, \ldots$$

and

$$V(i) = bi + \alpha q V([i - 1]^+) + \alpha p \min[V(i + 1), r + V(i)]$$

Complete this proof by proceeding as in a similar derivation in Example 9.4.

9.2 Expressions of the form $\min[V(i+1), r+V(i)]$ often appear in DP equations for solving Markov decision problems. Moreover, the structure of the optimal policy frequently relies on the convexity of $V(i)$ (see, for example Problem 9.1). Show that if $V(i)$ is convex in $i = 0, 1, \ldots,$ then the function

$$g(i) = \min[V(i + 1), r + V(i)], \quad r > 0$$

is also convex.

9.3 Step 3 of the derivation of an optimal policy for the routing problem in Section 9.5.2 involves proving some structural properties of $V(i, j)$, where

$$V(i, j) = b_1 i + b_2 j + \alpha q_1 V([i - 1]^+, j) + \alpha q_2 V(i, [j - 1]^+)$$
$$+ \alpha p \min[V(i + 1, j), V(i, j + 1)]$$

Consider a function $V(i, j)$, $i, j = 0, 1, \ldots,$ such that (a) $V(i, j)$ is monotonically nondecreasing in i and j, and (b) $V(i + 1, j) - V(i, j + 1)$ is monotonically nondecreasing in i and nonincreasing in j. Then, show that the function

$$g(i,j) = b_1 i + b_2 j + \alpha q_1 V([i-1]^+, j) + \alpha q_2 V(i, [j-1]^+)$$
$$+ \alpha p \min[V(i+1,j), V(i,j+1)]$$

also satisfies properties (a) and (b).

9.4 Consider a single-server queueing system with an infinite capacity queue as a model of a manufacturing station where parts arrive according to a Poisson process of rate λ and service times are exponentially distributed with rate $\mu > \lambda$, with the arrival and service process being independent. Let us assume that any part completing service is defective with probability p. There is an inspection unit that follows the server and one can decide whether any part completing service should be inspected or not. If a part is inspected and found defective, it is returned to the end of the queue for reprocessing; if not inspected or if inspected and not found defective, the part leaves the system. We assume that there are no errors in the inspection process and that inspection times are negligible. The control variable here is u: If $u = 1$ the decision is to inspect a part completing service, and if $u = 0$ the decision is to skip inspection. We assume that this decision generally depends on the queue length $X(t)$ observed at service completion instants.

Let B denote the cost per unit time for each part in the system, C the inspection cost per part, and D the cost incurred by a defective part allowed to leave the system (B, C, and D are all positive bounded constants). We then define the total expected discounted cost criterion

$$V_\pi(x_0) = E_\pi \left[\int_0^\infty e^{-\beta t} BX(t) dt + \int_0^\infty e^{-\beta t} \{Cu[X(t)] \right.$$
$$\left. + Dp[1 - u[X(t)]]\} \mathbf{1}[\text{service completion occurs at time } t]dt \right]$$

where β is a given discount factor and x_0 is a given initial state.

(a) Using uniformization, derive a discrete time model for this problem and write down the corresponding optimality equation.

(b) Assume that $Dp - C \geq 0$. Show that the optimal policy is threshold-based, that is, the optimal control action is to inspect a part completing service when the queue length is i if $i \leq i^*$, and not to inspect if $i > i^*$, for some i^*.

SELECTED REFERENCES

- Bertsekas, D.P., *Dynamic Programming: Deterministic and Stochastic Models*, Prentice-Hall, Englewood Cliffs, NJ, 1987.

- Bertsekas, D.P., *Dynamic Programming and Optimal Control, Volumes 1 and 2*, Athena Scientific, Belmont, MA, 1995.

- Hajek, B., "Optimal Control of Two Interacting Service Stations," *IEEE Transactions on Automatic Control*, Vol. AC-29, pp. 491-499, 1984.

- Kumar, P.R., and P. Varaiya, *Stochastic Systems: Estimation, Identification, and Adaptive Control*, Prentice-Hall, Englewood Cliffs, NJ, 1986.

- Marshall, A.W., and I. Olkin, *Inequalities: Theory of Majorization and its Applications*, Academic Press, New York, 1979.

- Pepyne, D.L., and C.G. Cassandras, "Optimal Dispatching Control for Elevator Systems During Uppeak Traffic," *IEEE Transactions on Control Systems Technology*, Vol. 5, No. 6, pp. 629-643, 1997.

- Ross, S.M., *Stochastic Processes*, Wiley, New York, 1983a.

- Ross, S.M., *Introduction to Stochastic Dynamic Programming*, Academic Press, New York, 1983b.

- Stidham, S., Jr., "Optimal Control of Admission to a Queueing System," *IEEE Transactions on Automatic Control*, Vol. AC-30, pp. 705-713, 1985.

- Trivedi, K.S., *Probability and Statistics with Reliability, Queuing and Computer Science Applications*, Prentice-Hall, Englewood Cliffs, NJ, 1982.

- Walrand, J., *An Introduction to Queueing Networks*, Prentice-Hall, Englewood Cliffs, NJ, 1988.

Chapter 10

Introduction to Discrete-Event Simulation

10.1 INTRODUCTION

In our study of dynamic systems, our first goal is to obtain a model. For our purposes, a model consists of mathematical equations which describe the behavior of a system. For example, in Chapter 5 we developed the set of equations (5.7)-(5.12) which describe how the state of a DES evolves as a result of event occurrences over time. Our next goal is to use a model in order to obtain explicit mathematical expressions for quantities of interest. For example, in Chapter 7 our model was a Markov chain and the main quantities of interest were the state probabilities $\pi_j(k) = P[X_k = j], j = 0, 1, \ldots$ In some cases, we can indeed obtain such expressions, as we did with birth-death chains at steady state in Section 7.4.3. In general, however, "real world" systems either do not conform to some assumptions we make in order to simplify a model, or they are just too complex to yield *analytical solutions*. Our mathematical model may still be valid; the problem is that we often do not have the tools to solve the equations which make up such a model. *Simulation* is a process through which a system model is evaluated numerically, and the data from this process are used to *estimate* various quantities of interest. As we have repeatedly pointed out in previous chapters, analytical solutions for DES are particularly hard to come by, making simulation a very attractive tool for their study.

It is helpful to think of computer simulation as the electronic equivalent of a good old-fashioned laboratory experiment. The only hardware involved is a computer, and instead of physical devices connected with each other we have software capturing all such interactions. Randomness (noise) is also replaced by appropriate software driven by what we call a "random number generator." Simulation is still not the "real thing", but it is the next best thing we have to actually building some very expensive and complicated systems just to experiment with them. In the case of DES, simulation is widely used in a number of applications. Examples include designing manufacturing systems and evaluating their performance, designing communication networks and testing various protocols for handling messages contending for network resources, and designing airports, road networks, or subways to handle projected traffic loads. Note that these are all highly complex stochastic systems. Thus, building a "laboratory" consisting of highways and cars dedicated to just doing experiments before building the real thing is truly unrealistic. At the same time, it is simply too risky to go ahead and just build such a system based on rough approximations or "gut feeling" alone, given the cost and complexity involved. Finally, one should not exclude the possibility of using approximate solutions for these DES; in this case, simulation becomes a means for testing their accuracy before we can gain any real confidence in them.

Our first objective in this chapter is to present the main components of discrete-event computer simulation and describe its basic functions. In fact, simulation may be viewed as a systematic way of generating sample paths of a DES, something we have already discussed at length in Chapters 5 and 6. Thus, having gone through Chapter 5, and Section 5.2.5 in particular, should make this a relatively easy task. Our second objective is to take a brief look at some existing simulation languages in order to show how one can obtain very general DES models without having to actually write a lot of computer code. Our third objective is to discuss the issue of *random variate generation*, that is, the means by which random numbers are used to obtain samples from various probability distributions of interest. These samples provide the clock structure of the stochastic timed automaton model we introduced in Chapter 6. Finally, we should not forget that the ultimate purpose of simulation is to estimate quantities of interest from the data we generate in our "electronic laboratory". Thus, we will provide a brief introduction to the analysis of simulation output data and estimation techniques and try to point out the dangers of using the results of simulation experiments without carefully assessing their statistical validity.

10.2 THE EVENT SCHEDULING SCHEME

Let us adopt the point of view that simulation is simply a systematic means for generating sample paths of a stochastic DES. Moreover, let us consider the stochastic timed automaton we introduced in Chapter 6 as our model for DES. Before proceeding, let us briefly review how a sample path is generated.

Recall that $\mathcal{E}$ is a countable *event set*, and $\mathcal{X}$ is a countable *state space*. We initialize time at $t = 0$ and assume that an initial state $x_0 \in \mathcal{X}$ is given. For any state x, there is a set of *feasible* events $\Gamma(x) \subseteq \mathcal{E}$. These are the only events which may occur at this state. Given x_0, we have at our disposal $\Gamma(x_0)$. If $i \in \Gamma(x_0)$ is a feasible event, we associate with it a *clock* value, which represents the amount of time required until event i occurs. Initially, the clock value of every feasible event is set to an *event lifetime*. For our purposes, we assume that this is supplied by the computer through some random number generating mechanism. We denote clock values for event i by y_i, and their lifetimes by v_i. Thus, at $t = 0$, the random number regenerating mechanism supplies lifetimes, and we set $y_i = v_i$ for all $i \in \Gamma(x_0)$.

Now given that the current state is x (including the initial state x_0), we look at all clock values y_i where $i \in \Gamma(x)$. The *triggering event* e' is the event which occurs next at that state, that is, the event with the smallest clock value:

$$e' = \arg \min_{i \in \Gamma(x)} \{y_i\}$$

At this point we can update the state based on a given state transition mechanism. Specifically, the new state is x' with probability $p(x'; x, e')$. The actual value of x' is supplied by the computer's random number generating mechanism. If there is only one possible new state when event e' occurs at state x, we have a deterministic state transition mechanism which we represent through the state transition function $f(x, e')$.

The amount of time spent at state x defines the *interevent time*

$$y^* = \min_{i \in \Gamma(x)} \{y_i\}$$

We can then update time by setting:

$$t' = t + y^*$$

We also update clock values for all feasible events in the new state x' as follows. There are two cases to consider. First, if an event $i \in \Gamma(x)$ such that $i \neq e'$ remains feasible in the new state x', the time remaining until its occurrence (the new clock value) is simply given by

$$y_i' = y_i - y^*$$

The second case applies to e' itself if $e' \in \Gamma(x')$ and to all all other events which were not feasible in x, but become feasible in x'. For all such events, we need new lifetimes. These are supplied once again by the computer through the random number generating process.

Finally, recall that the last component of the stochastic timed automaton model is the set of probability distributions $G = \{G_i : i \in \mathcal{E}\}$ characterizing event lifetimes. This information is combined with the computer's *Random Number Generator* (RNG) to supply event lifetime samples as required. We refer to these event lifetime samples as *random variates*, to be distinguished from random variables which have a different mathematical meaning (see also Appendix I).

In Section 5.2.5, we introduced the *event scheduling scheme*. This is precisely the procedure reviewed above with one modification: Whenever an event i is activated at some event time t_n, we *schedule* its next occurrence to be at time $(t_n + v_i)$, where v_i is a lifetime sample supplied by the computer. Thus, instead of maintaining the clock values $y_i, i \in \Gamma(x)$, we maintain a *Scheduled Event List* (SEL)

$$L = \{(e_k, t_k)\}, \quad k = 1, 2, \ldots, m_L$$

where $m_L \leq m$ is the number of feasible events in the current state and m is the number of events in the set $\mathcal{E}$. Moreover, the scheduled event list is *always ordered on a smallest-scheduled-time-first* basis. This implies that the first event in the list, e_1, is always the triggering event. The event scheduling scheme was illustrated in Fig. 5.10. It is shown once again in Fig. 10.1 with only some slight modifications. For simplicity, we assume that the state transition mechanism is fully deterministic, that is, the next state x' when event e' occurs is given by $x' = f(x, e')$.

Referring to Fig. 10.1, the INITIALIZE function sets the STATE to some value x_0, and the simulation TIME to 0 (except in unusual circumstances when TIME may be initially set at some positive value). It is also responsible for providing the RANDOM VARIATE GENERATOR all information regarding the event lifetime distributions. Finally, the RANDOM VARIATE GENERATOR provides event lifetimes for all feasible events at the initial state, and the SCHEDULED EVENT LIST is initialized, with all entries sorted in increasing order of scheduled times. Thus, e_1 is the triggering event at the initial state, and t_1 is its occurrence time.

Subsequently, the simulation procedure consists of continuously repeating the following six steps (also described in Section 5.2.5):

Step 1. Remove the first entry (e_1, t_1) from the SCHEDULED EVENT LIST.

Step 2. Update the simulation TIME by advancing it to the new event time t_1.

Step 3. Update the STATE according to the state transition function, $x' = f(x, e_1)$.

Step 4. Delete from the SCHEDULED EVENT LIST any entries corresponding to infeasible events in the new state, that is, delete all $(e_k, t_k) \in L$ such that $e_k \notin \Gamma(x')$.

Step 5. Add to the SCHEDULED EVENT LIST any feasible event which is not already scheduled (possibly including the triggering event removed in *Step 1*). The scheduled event time for some such i is given by (TIME $+ v_i$), where TIME was set in *Step 2* and v_i is a lifetime obtained from the RANDOM VARIATE GENERATOR.

Step 6. Reorder the updated SCHEDULED EVENT LIST based on a *smallest-scheduled-time-first* scheme.

The procedure then repeats with *Step 1* for the new ordered list.

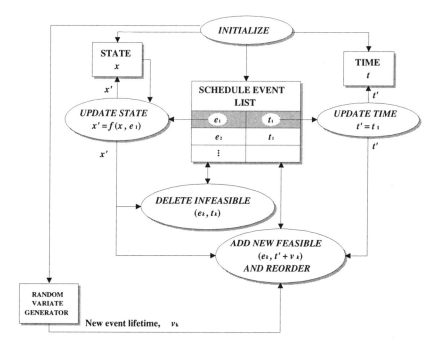

Figure 10.1: The Event Scheduling scheme in computer simulation.

Not included in Fig. 10.1 are the components of the computer simulator responsible for collecting data and for estimating various quantities of interest upon completion of a simulation run. We can now summarize all the components of a simulator as follows:

1. *State*: a list where all state variables are stored.

2. *Time*: a variable where the simulation time is stored.

3. *Scheduled Event List*: a list where all scheduled events are stored along with their occurrence time.

4. *Data Registers*: variables and/or lists where data are stored for estimation purposes.

5. *Initialization Routine*: a routine which initializes all simulation data structures (items 1-4) at the beginning of a simulation run.

6. *Time Update Routine*: a routine which identifies the next event to occur and advances simulation time to the occurrence time of that event.

7. *State Update Routine*: a routine which updates the state based on the next event to occur.

8. *Random Variate Generation Routines*: a collection of routines which transform computer-generated random numbers into random variates according to user-specified event lifetime distributions.

9. *Report Generation Routine*: a routine which computes estimates of various quantities of interest based on all data collected in the course of a simulation run.

10. *Main Program*: this is responsible for the overall coordination of all components. It first calls the initialization routine. It then repeatedly invokes the time and state update routines and updates the Scheduled Event List. It is also responsible for terminating a simulation run based on user-specified criteria, and for invoking the report generation routine.

10.2.1   Simulation of a Simple Queueing System

In this section, we consider a simple single-server queueing system and go through a typical simulation run based on the event scheduling scheme in some detail. Our objective is not only to illustrate the functions described above, but also to show how we can estimate various performance measures of interest by collecting appropriate data in the course of the simulation. Since this, ultimately, is the reason for simulation, it is very important to understand this "estimation" process. In what follows, we consider some of the most common performance measures of interest encountered in queueing systems.

Let us assume that we start the simulation at time zero with the queue empty. Let us also assume that we plan to end the simulation after N customers have completed service, where N is specified in advance. Now before we get started, suppose someone were to ask: How long does it take to serve N customers given that the queue is initially empty? The proper answer is: That depends on the various interarrival times for these customers and their corresponding service times. In short, no specific answer can be given, since this quantity is in fact a random variable. Nonetheless, suppose we carry out one simulation ending after N customers are served, and observe the termination time to be T_N. Then, T_N constitutes a "guess" for answering the question above. This "guess" is what we refer to as an "estimate" of the quantity of interest.

We will now assume that the simulation is performed in order to estimate the following performance measures.

Expected Average System Time

The average system time over the first N customers in the queueing system is in fact a random variable, since every new simulation run will yield a different average system time, depending on the particular interarrival and service time sequences associated with it. Let S_N be the mean of this random variable, that is, the expected average system time. We can estimate S_N from the observations made on our particular simulation run. Thus, suppose we keep track of the time spent in the system by every single customer and denote these observations by $S_1, S_2, \ldots, S_N$. Then, an estimate of S_N, which we denote by $\hat{S}_N$, is given by

$$\hat{S}_N = \frac{1}{N} \sum_{k=1}^{N} S_k \tag{10.1}$$

This is simply the arithmetic mean of the N observed customer system times. Note that to compute this estimate at the end of the simulation we need to maintain data variables which measure the amount of time elapsed from the arrival of any one customer until this customer's service completion.

Probability That Customer System Times Exceed a Given "Deadline"

Suppose we would like to ensure that customers never spend more than some given amount of time, D, in the system, which we refer to as a "deadline." This is often a good measure of customer satisfaction, and becomes a very critical quantity in some applications. We cannot of course guarantee that this deadline is never violated, but we can try and measure the probability P_N^D that the system time of a customer (among the N observed) exceeds the deadline D. Let us define:

$$n_N = \text{number of customers whose system time exceeds } D$$

Then, upon completion of a simulation with N customers, we can obtain an *estimate* of this quantity, denoted by $\hat{P}_N^D$, as follows:

$$\hat{P}_N^D = \frac{n_N}{N} \tag{10.2}$$

Server Utilization

This is the probability that the server is busy during the total time it takes to serve N customers, and we denote it by ρ_N. Equivalently, this probability is the expected fraction of time during which the queue length is positive. In order to estimate this fraction, let

$$T(i) = \text{total observed time during which the queue length is } i, \quad i = 0, 1, \ldots$$

and note that

$$T_N = \sum_{i=0}^{\infty} T(i) \tag{10.3}$$

Thus, an *estimate* of ρ_N, denoted by $\hat{\rho}_N$, is given by

$$\hat{\rho}_N = \frac{\sum_{i=1}^{\infty} T(i)}{T_N}$$

or equivalently, using (10.3),

$$\hat{\rho}_N = 1 - \frac{T(0)}{T_N} \tag{10.4}$$

We therefore see that in order to estimate this performance measure we need to keep track of all times when the queue is empty and store it as a data variable $T(0)$. We also note that this is a *time-average* performance measure (data variables are divided by T_N), in contrast to $\hat{S}_N$ and $\hat{P}_N^D$ which are customer-average measures (data variables are divided by N).

Mean Queue Length

Let Q_N denote the average queue length over the time interval required to serve N customers, and:

$$p_N(i) = \quad \text{probability that queue length is } i \text{ over the time interval required to serve } N \text{ customers, } i = 0, 1, \ldots$$

Then, by definition,

$$Q_N = \sum_{i=0}^{\infty} i p_N(i)$$

In order to obtain an *estimate*, $\hat{Q}_N$, of Q_N we need estimates, $\hat{p}_N(i)$, of $p_N(i), i = 0, 1, \ldots$ Note that we can estimate it as follows:

$$\hat{p}_N(i) = \frac{T(i)}{T_N}$$

since it is the expected fraction of time during which the queue length is i. Therefore,

$$\hat{Q}_N = \frac{1}{T_N} \sum_{i=0}^{\infty} i T(i) \tag{10.5}$$

Thus, we see that in the course of the simulation we need to maintain data variables $T(0), T(1), \ldots$ for all observed queue lengths $i = 0, 1, \ldots$

Let us now proceed with a simulation of this system for $N = 5$. Note that there are two events, an ARRIVAL and a DEPARTURE of a customer. Accordingly, the computer will have to supply random variates for interarrival times and for service times based on the probability distributions specified. For the purposes of this discussion, we will assume that the following random variates are supplied: Interarrival times (sec) are $Y_1 = 0.4$, $Y_2 = 0.3$, $Y_3 = 0.4$, $Y_4 = 1.7$, $Y_5 = 1.7$, $Y_6 = 0.5$, and $Y_7 = 0.9$, while service times (sec) are $Z_1 = 1.6$, $Z_2 = 0.5$, $Z_3 = 1.0$, $Z_4 = 0.3$, $Z_5 = 0.8$.

QUEUE LENGTH

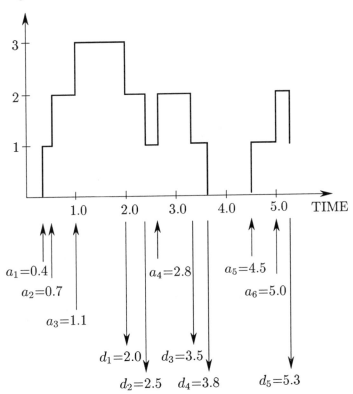

Figure 10.2: Sample path of simulated queueing system. Arrival and departure times are denoted by a_k and d_k respectively.

The resulting sample path is shown in Fig. 10.2. In what follows, we go through the simulation process on an event-by-event basis. For each event occurrence, we follow the six steps outlined in the previous section. In particular, we show how TIME, STATE, and the SCHEDULED EVENT LIST are updated, with random variates supplied by the RANDOM VARIATE GENERATOR as required. In addition, we show how all data variables required for estimating performance measures are stored. Specifically, we store arrival times $a_1, \ldots, a_5$; system times $S_1, \ldots, S_5$; the number of customers, n_N, whose system time exceeds a given deadline value D (which we will take to be 2 sec); and the queue length occupancy times, $T(0), T(1), \ldots$

$t = 0$ *Initialization.* We set TIME = 0.0 STATE = 0. At this state, the only feasible event is an ARRIVAL. Thus, we need a sample from the inter-arrival time distribution. This is supplied by the RANDOM VARIATE GENERATOR: $Y_1 = 0.4$. Thus, an ARRIVAL is scheduled for time 0.4, and the pair (ARRIVAL, 0.4) is entered in the SCHEDULED EVENT LIST, as follows:

RANDOM VARIATE GENERATOR

Time	State
0.0	0

SCHEDULED EVENT LIST	
ARRIVAL	0.4

◄— 0.4 ————

In addition, we initialize all data structures required for estimation purposes, as shown below. The arrival time and system time lists for the five customers we are interested in are left blank, since no data are available yet.

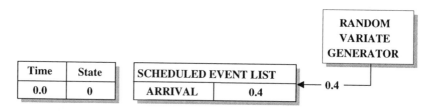

ARRIVAL TIMES	
1.	
2.	
3.	
4.	
5.	

SYSTEM TIMES	
1.	
2.	
3.	
4.	
5.	

n_N	0

QUEUE LENGTH OCCUPANCY TIMES	
T(0)	0.0
T(1)	0.0
T(2)	0.0
T(3)	0.0
T(4)	0.0

$t = 0.4$ *First customer arrives.* This is the only scheduled event from the initialization step. We set TIME = 0.4 and STATE = 1, since the queue length is incremented by 1. The ARRIVAL is removed from the SCHEDULED EVENT LIST. At the new state, both an ARRIVAL and a DEPARTURE event are feasible. Thus, the RANDOM VARIATE GENERATOR is invoked to supply a new interarrival time: $Y_2 = 0.3$, and a service time: $Z_1 = 1.6$. Since the current time is 0.4, the ARRIVAL is scheduled for time $(0.4 + 0.3)$, and the DEPARTURE for $(0.4 + 1.6)$. The two pairs (ARRIVAL, 0.7) and (DEPARTURE, 2.0) are entered in the SCHEDULED EVENT LIST in the order of their scheduled occurrence times as follows:

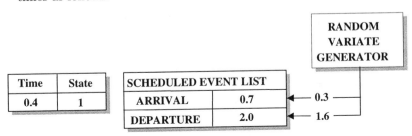

In addition, we update all relevant data as shown below. We record the arrival time of the first customer (this will later help us evaluate the system time of that customer). We can also update $T(0)$, since 0.4 sec have elapsed during which the queue length was 0. No other data are available yet.

ARRIVAL TIMES		SYSTEM TIMES	
1.	0.4	1.	
2.		2.	
3.		3.	
4.		4.	
5.		5.	

n_N	0

QUEUE LENGTH OCCUPANCY TIMES	
T(0)	0.4
T(1)	0.0
T(2)	0.0
T(3)	0.0
T(4)	0.0

$t = 0.7$ *Second customer arrives.* This is the event with the earliest scheduled time in the SCHEDULED EVENT LIST from the previous event occurrence. We set TIME = 0.7 and STATE = 2, since the queue length is incremented once again by 1. The ARRIVAL is removed from the SCHEDULED EVENT LIST. At the new state, the DEPARTURE event is still feasible, so we simply leave it in the list. The RANDOM VARIATE GENERATOR is invoked to supply a new interarrival time: $Y_3 = 0.4$. Since the current time is 0.7, this ARRIVAL is scheduled for time $(0.7 + 0.4)$. Thus, the pair (ARRIVAL, 1.1) is added to the SCHEDULED EVENT LIST. We finally reorder the list on an earliest-scheduled-time-first basis, and end up with the situation shown below:

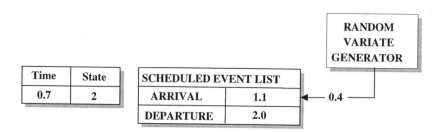

We can also collect some more data for estimation purposes. We record the arrival time of the second customer, and update $T(1)$, since 0.3 sec have elapsed during which the queue length was 1:

ARRIVAL TIMES		SYSTEM TIMES				QUEUE LENGTH OCCUPANCY TIMES	
1.	0.4	1.				$T(0)$	0.4
2.	0.7	2.		n_N	0	$T(1)$	0.3
3.		3.				$T(2)$	0.0
4.		4.				$T(3)$	0.0
5.		5.				$T(4)$	0.0

$t = 1.1$ *Third customer arrives.* Looking at the SCHEDULED EVENT LIST from the previous event occurrence, we see that the next event is once again an ARRIVAL. We set TIME = 1.1 and STATE = 3, since the queue length is incremented by 1. The ARRIVAL is removed from the SCHEDULED EVENT LIST. At the new state, the DEPARTURE event is still feasible and is left in the list. We now need to schedule yet another ARRIVAL, and the RANDOM VARIATE GENERATOR supplies a new interarrival time: $Y_4 = 1.7$. Since the current time is 1.1, this ARRIVAL is scheduled for time (1.1 + 1.7). Thus, the pair (ARRIVAL, 2.8) is added to the SCHEDULED EVENT LIST. We finally reorder the list on an earliest-scheduled-time-first basis as before, and end up with the following situation:

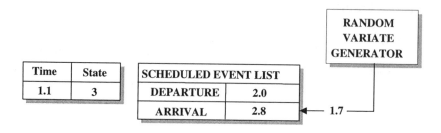

In addition, we record the arrival time of the third customer, and update $T(2)$, since 0.4 sec have elapsed during which the queue length was 2:

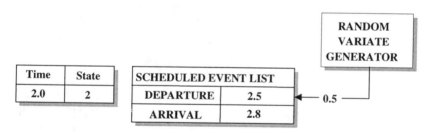

	ARRIVAL TIMES
1.	0.4
2.	0.7
3.	1.1
4.	
5.	

	SYSTEM TIMES
1.	
2.	
3.	
4.	
5.	

n_N	0

QUEUE LENGTH OCCUPANCY TIMES	
$T(0)$	0.4
$T(1)$	0.3
$T(2)$	0.4
$T(3)$	0.0
$T(4)$	0.0

$t = \mathbf{2.0}$ *First customer departs.* Looking at the SCHEDULED EVENT LIST from the previous event occurrence, we see that the next event is now a DEPARTURE. We set TIME = 2.0 and STATE = 2, since the queue length is decremented by 1. The DEPARTURE is removed from the SCHEDULED EVENT LIST. At the new state, a DEPARTURE event is still feasible, so we invoke the RANDOM VARIATE GENERATOR to obtain a new service time: $Z_2 = 0.5$. Since the current time is 2.0, this DEPARTURE is scheduled for time $(2.0 + 0.5)$. Thus, the pair (DEPARTURE, 2.5) is added to the SCHEDULED EVENT LIST. The list is once again reordered, and we have:

RANDOM VARIATE GENERATOR

Time	State
2.0	2

SCHEDULED EVENT LIST	
DEPARTURE	2.5
ARRIVAL	2.8

← 0.5 ─

In terms of data collection, we are now in a position to record the system time of the first customer: Since this customer's departure time is 2.0 and we have in store his arrival time (0.4), the system time is given by $(2.0 - 0.4) = 1.6$. This result is stored in the SYSTEM TIMES list. We can also update n_N. Since the deadline D was given to be 2.0 sec, we see that the first customer's system time is below this value, and $n_N = 0$ remains unchanged. Finally, since 0.9 sec have elapsed during which the queue length was 3, we can also update $T(3)$:

ARRIVAL TIMES		SYSTEM TIMES				QUEUE LENGTH OCCUPANCY TIMES	
1.	0.4	1.	1.6	n_N	0	T(0)	0.4
2.	0.7	2.				T(1)	0.3
3.	1.1	3.				T(2)	0.4
4.		4.				T(3)	0.9
5.		5.				T(4)	0.0

$t = 2.5$ *Second customer departs.* The next scheduled event is another DE-
PARTURE. We set TIME $= 2.5$ and STATE $= 1$, since the queue length is
again decremented by 1. The DEPARTURE is removed from the SCHED-
ULED EVENT LIST. At the new state, a DEPARTURE event is still
feasible, so we invoke the RANDOM VARIATE GENERATOR to obtain
a new service time: $Z_3 = 1.0$. Since the current time is 2.5, this DEPAR-
TURE is scheduled for time $(2.5 + 1.0)$. Thus, the pair (DEPARTURE,
3.5) is added to the SCHEDULED EVENT LIST. The list is reordered,
and we see that the next scheduled event is an ARRIVAL as shown below:

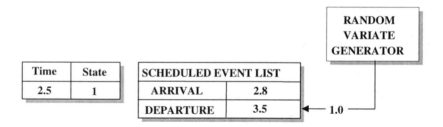

We can now record the system time of the second customer, given by
$(2.5 - 0.7) = 1.8$. This result is stored in the SYSTEM TIMES list. Since
$1.8 < D = 2.0$, $n_N = 0$ remains unchanged. Finally, since 0.5 sec have
elapsed during which the queue length was 2, we can update $T(2)$ to
$(0.4 + 0.5) = 0.9$ sec as shown:

ARRIVAL TIMES		SYSTEM TIMES				QUEUE LENGTH OCCUPANCY TIMES	
1.	0.4	1.	1.6	n_N	0	T(0)	0.4
2.	0.7	2.	1.8			T(1)	0.3
3.	1.1	3.				T(2)	0.9
4.		4.				T(3)	0.9
5.		5.				T(4)	0.0

$t = 2.8$ *Fourth customer arrives.* The next scheduled event is an ARRIVAL. We set TIME = 2.8 and STATE = 2, since the queue length is incremented from 1 to 2. The ARRIVAL is removed from the SCHEDULED EVENT LIST, but since a DEPARTURE event is still feasible at the new state, we leave this event in the list. Next, we invoke the RANDOM VARIATE GENERATOR to obtain a new interarrival time: $Y_5 = 1.7$. Since the current time is 2.8, the next ARRIVAL is scheduled for time $(2.8 + 1.7)$. Thus, the pair (ARRIVAL, 4.5) is added to the SCHEDULED EVENT LIST. The list is reordered, and we now have:

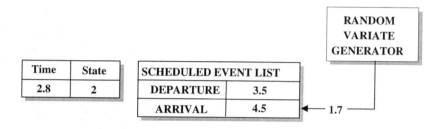

Time	State
2.8	2

SCHEDULED EVENT LIST	
DEPARTURE	3.5
ARRIVAL	4.5

RANDOM VARIATE GENERATOR

1.7

In addition, we record the arrival time of the fourth customer. We also update $T(1)$ to $(0.3 + 0.3) = 0.6$ sec, since 0.3 sec have elapsed during which the queue length was 1.

ARRIVAL TIMES		SYSTEM TIMES				QUEUE LENGTH OCCUPANCY TIMES	
1.	0.4	1.	1.6	n_N	0	$T(0)$	0.4
2.	0.7	2.	1.8			$T(1)$	0.6
3.	1.1	3.				$T(2)$	0.9
4.	2.8	4.				$T(3)$	0.9
5.		5.				$T(4)$	0.0

$t = 3.5$ *Third customer departs.* The next scheduled event is a DEPARTURE. We set TIME = 3.5 and STATE = 1, since the queue length is decremented by 1. The DEPARTURE is removed from the SCHEDULED EVENT LIST. At the new state, a DEPARTURE event is still feasible, so we invoke the RANDOM VARIATE GENERATOR to obtain a new service time: $Z_4 = 0.3$. Since the current time is 3.5, this DEPARTURE is scheduled for time $(3.5 + 0.3)$. Thus, the pair (DEPARTURE, 3.8) is added to the SCHEDULED EVENT LIST. The list is reordered, and we see that the next scheduled event is again a DEPARTURE as shown below:

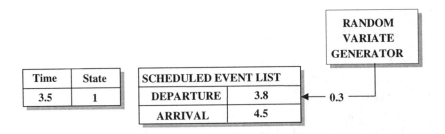

At this point we can record the system time of the third customer, since his arrival time is in store. Thus, $(3.5 - 1.1) = 2.4$ is stored in the SYSTEM TIMES list. Since $2.4 > D = 2.0$, we need to update the count of customers whose system time exceeds the deadline D, so we set $n_N = 1$. Finally, since 0.7 sec have elapsed during which the queue length was 2, we update $T(2)$ to $(0.9 + 0.7) = 1.6$ sec as shown:

ARRIVAL TIMES		SYSTEM TIMES			n_N	0		QUEUE LENGTH OCCUPANCY TIMES	
1.	0.4	1.	1.6					T(0)	0.4
2.	0.7	2.	1.8					T(1)	0.6
3.	1.1	3.	2.4					T(2)	1.6
4.	2.8	4.						T(3)	0.9
5.		5.						T(4)	0.0

$t = 3.8$ *Fourth customer departs.* The next scheduled event is again a DE-PARTURE. We set TIME = 3.8 and STATE = 0, since the queue length is decremented by 1. The DEPARTURE is removed from the SCHED-ULED EVENT LIST. At the new state, only an ARRIVAL event is feasible. Thus, the already scheduled ARRIVAL event is left in the list, and no additional random variate is needed:

Time	State
3.8	1

SCHEDULED EVENT LIST	
ARRIVAL	4.5

We also record the system time of the fourth customer, since his arrival time is in store. Thus, $(3.8 - 2.8) = 1.0$ is stored in the SYSTEM TIMES list. Since $1.0 < D = 2.0$, $n_N = 1$ remains unchanged. Finally, since 0.3 sec have elapsed during which the queue length was 1, we update $T(1)$ to $(0.6 + 0.3) = 0.9$ sec.

ARRIVAL TIMES	
1.	0.4
2.	0.7
3.	1.1
4.	2.8
5.	

SYSTEM TIMES	
1.	1.6
2.	1.8
3.	2.4
4.	1.0
5.	

n_N	1

QUEUE LENGTH OCCUPANCY TIMES	
T(0)	0.4
T(1)	0.9
T(2)	1.6
T(3)	0.9
T(4)	0.0

$t = 4.5$ *Fifth customer arrives.* This is the only scheduled event from the previous step. We set TIME = 4.5 and STATE = 1, since the queue length is incremented by 1. The ARRIVAL is removed from the SCHEDULED EVENT LIST. At the new state, both an ARRIVAL and a DEPARTURE event are feasible, so the RANDOM VARIATE GENERATOR is invoked to supply both a new interarrival time: $Y_6 = 0.5$, and a service time: $Z_5 = 0.8$. Since the current time is 4.5, the ARRIVAL is scheduled for time $(4.5+0.5)$, and the DEPARTURE for $(4.5+0.8)$. The two pairs (ARRIVAL, 5.0) and (DEPARTURE, 5.3) are entered in the SCHEDULED EVENT LIST and ordered as shown below:

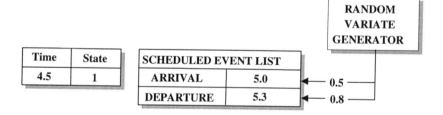

In addition, we record the arrival time. We also update T(0) to $(0.4 + 0.7)$, since 0.7 sec have elapsed during which the queue length was 0:

ARRIVAL TIMES	
1.	0.4
2.	0.7
3.	1.1
4.	2.8
5.	4.5

SYSTEM TIMES	
1.	1.6
2.	1.8
3.	2.4
4.	1.0
5.	

n_N	1

QUEUE LENGTH OCCUPANCY TIMES	
T(0)	1.1
T(1)	0.9
T(2)	1.6
T(3)	0.9
T(4)	0.0

$t = 5.0$ *Sixth customer arrives.* The next scheduled event from the previous step is an ARRIVAL. We set TIME = 5.0 and STATE = 2, since the queue length is incremented by 1. The ARRIVAL is removed from the SCHEDULED EVENT LIST. At the new state, both events are feasible. Thus, the scheduled DEPARTURE is left in the list, while the RANDOM VARIATE GENERATOR is invoked to supply a new interarrival time: $Y_7 = 0.9$. The ARRIVAL is scheduled for time $(5.0 + 0.9)$, and the pair (ARRIVAL, 5.9) is entered in the SCHEDULED EVENT LIST which is reordered as follows:

RANDOM VARIATE GENERATOR	

Time	State
5.0	2

SCHEDULED EVENT LIST	
DEPARTURE	5.3
ARRIVAL	5.9

←— 0.9 —

We can record the arrival time of the sixth customer at this point. But, since in this particular simulation we are interested only in the first five customers, we do not bother to do so. However, we do update $T(1)$ to $(0.9 + 0.5)$, since 0.5 sec have elapsed during which the queue length was 1:

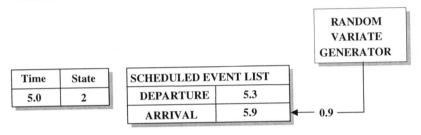

ARRIVAL TIMES	
1.	0.4
2.	0.7
3.	1.1
4.	2.8
5.	4.5

SYSTEM TIMES	
1.	1.6
2.	1.8
3.	2.4
4.	1.0
5.	

n_N	1

QUEUE LENGTH OCCUPANCY TIMES	
$T(0)$	1.1
$T(1)$	1.4
$T(2)$	1.6
$T(3)$	0.9
$T(4)$	0.0

$t = 5.3$ *Fifth customer departs.* This is the last event for this simulation, since we have agreed to use $N = 5$ customer departures as our termination condition. We set TIME $= 5.3$ and STATE $= 1$, since the queue length is decremented by 1. The SCHEDULED EVENT LIST is not updated since this is the end of the simulation, and we are left with:

Time	State
5.3	1

SCHEDULED EVENT LIST	
ARRIVAL	5.9

We record the system time of the fifth and final customer, since his arrival time is in store. Thus, $(5.3 - 4.5) = 0.8$ is stored in the SYSTEM TIMES list. Since $0.8 < D = 2.0$, $n_N = 1$ remains unchanged. Finally, since 0.3 sec have elapsed during which the queue length was 2, we update T(2) to $(1.6 + 0.3) = 1.9$ sec.

ARRIVAL TIMES		SYSTEM TIMES	
1.	0.4	1.	1.6
2.	0.7	2.	1.8
3.	1.1	3.	2.4
4.	2.8	4.	1.0
5.	4.5	5.	0.8

n_N	1

QUEUE LENGTH OCCUPANCY TIMES	
T(0)	1.1
T(1)	1.4
T(2)	1.9
T(3)	0.9
T(4)	0.0

This completes the simulation process for $N = 5$, at time $T_N = 5.3$ sec We are now in a position to compute estimates of the four performance measures of interest:

- **Expected Average System Time.** Using the data from the final SYS-TEM TIMES list above and (10.1) with $N = 5$, we get

$$\hat{S}_5 = \frac{1.6 + 1.8 + 2.4 + 1.0 + 0.8}{5} = 1.52$$

Note that we do not need to continuously maintain a complete list of system times in order to evaluate $\hat{S}_5$; we only need to update the sum of system times which is used in the final calculation above.

- **Probability That Customer System Times Exceed a Deadline D $= 2$.** Using the final value of n_N recorded and (10.2) with $N = 5$, we get

$$\hat{P}_5^D = \frac{1}{5} = 0.2$$

- **Server Utilization.** Using the data from the final QUEUE LENGTH OCCUPANCY TIMES list above and (10.4) with $T_N = 5.3$, we get

$$\hat{\rho}_5 = 1 - \frac{1.1}{5.3} \approx 0.79$$

- **Mean Queue Length.** Using again the data from the final QUEUE LENGTH OCCUPANCY TIMES list above and (10.5), we get

$$\hat{Q}_5 = \frac{1}{5.3} \sum_{i=0}^{\infty} i T(i)$$
$$= \frac{(0 \cdot 1.1) + (1 \cdot 1.4) + (2 \cdot 1.9) + (3 \cdot 0.9) + 0 + \cdots}{5.3} \approx 1.49$$

Note that in this calculation our estimates for all $T(i)$, $i \geq 4$ are 0.

We emphasize again that the quantities we have obtained from this one simulation are only estimates of the actual performance measures. We will have more to say on the basics of estimation theory later in this chapter.

10.3 THE PROCESS-ORIENTED SIMULATION SCHEME

We have seen that a large class of DES consists of resource contention environments, where resources must be shared among many users. Our study of queueing systems in Chapter 8 was largely motivated by this fact. In such environments, it is often convenient to think of "entities" (e.g., customers) as undergoing a *process* as they flow through the DES. This process is a sequence of events separated by time intervals. During such a time interval, an entity is either receiving service at some resource or waiting for service. In the process-oriented simulation scheme, the behavior of the DES is described through several such processes, one for each type of entity of interest.

As an example, the simple single-server queueing system of the previous section may be viewed as follows. There is a single type of entity in this system (a customer) and a single resource (the server). *Every entity* in this system undergoes the following process:

1. It arrives.

2. It enters the queue.

3. It requests service from the server; if the server is idle, the entity "seizes" that resource, otherwise it remains in the queue until the server becomes idle.

4. Once it seizes the server, it remains in service for some period of time corresponding to a service time.

5. When service is complete, it "releases" the server.

6. It leaves the system.

Note that steps 1, 2, 5, and 6 above represent instantaneous actions. In step 3, if the server is not idle, the entity is forced to remain in the queue for an amount of time that depends on the state of the system, that is, the number of other entities already in queue and their corresponding service times, as well as the amount of time left for the entity currently in service. Similarly, in step 4 the entity experiences a time delay corresponding to its service time. It is important to observe, however, that whereas in step 3 the time the entity spends in queue is state-dependent, in step 4 the delay depends only on a specific externally provided (through a random variate generator) service time.

In general, we represent a process as a sequence of *functions*. A function is one of two types:

1. *Logic functions:* Instantaneous actions taken by the entity that triggers this function in its process. For example, checking a condition such as "is the server idle?" or updating a data structure, such as recording the arrival time of the entity.

2. *Time delay functions:* The entity is held by that function for some period of time. There are two types of time delay functions:

   (a) *Specified time:* The delay is fixed, usually determined by a number obtained by the Random Variate Generator used in the simulation. For example, a service time that depends only on some prespecified service time distribution.

   (b) *Unspecified time:* The delay depends on the state of the system. For example, the time spent waiting in a queue until the entity can seize a server.

The process-oriented simulation scheme is particularly suited for queueing systems, where it is natural to think of customers as entities flowing through a network of interconnected queues and servers. From a practical standpoint, the computer code required for simulating such a system through the process-oriented scheme is generally simpler, requiring considerably fewer lines. On the other hand, the event scheduling scheme still provides a very general way to model arbitrary DES including features that the process-oriented scheme is not equipped to handle. Most commercially available simulation languages (discussed in the next section) are primarily process-oriented. Many of them, however, also contain the ability to build simulation models using the more general event scheduling scheme.

In summary, the main components of a process-oriented simulation scheme are the following:

1. *Entities:* objects requesting service (e.g., *parts* in a manufacturing system, *jobs* in a computer system, *packets* in a communication network, *vehicles* in a transportation system). Each *entity type* is characterized by a particular process it undergoes in the DES it enters. For instance, in a computer system there may be two different job types that follow different processes.

2. *Attributes:* information characterizing a particular *individual* entity of any type. Thus, we usually attach a *unique record* to each entity that consists of this entity's attributes (e.g., for a part in a manufacturing system, attributes may be the part's arrival time at the system, its type, and its "due date", i.e., the time by which it is required to leave the system).

3. *Process Functions:* the instantaneous actions or time delays experienced by entities as described above.

4. *Resources:* objects providing service to entities (e.g., *machines* in a manufacturing system, *processors* in a computer system, *switches* in a telephone network). Time delays experienced by an entity are due to either waiting for a particular resource to be made available to that entity, or receiving service at that resource.

5. *Queues:* sets of entities with some common characteristics, usually the fact that they are all waiting for the use of a particular resource. An entity flowing through the system is always located in some queue unless it is in the process of being served by a resource.

10.4 DISCRETE-EVENT SIMULATION LANGUAGES

Using either the event scheduling or the process-oriented scheme, it is certainly possible to build a simulator for a specific DES of interest using standard computer languages such as FORTRAN or C++. On the other hand, by carefully organizing the common components of a simulation (e.g., random variate generation, data collection and processing) it is possible to develop a simulation language which provides building blocks (macros) tailored to DES models. There are several such commercially available simulation languages, and some (but by no means all) are briefly reviewed below. Most of them have built-in extensive capabilities for random variate generation based on a variety of probability distributions, and they provide standard output reports for common performance measures such as resource utilizations and queue length statistics. Moreover, some of them also have animation capabilities: As the simulation unfolds, one can see several icons displayed on a terminal screen differing in shape and/or color, which represent entities moving in the system as they are held in queues or get service at resources (also represented by various icons). Animation is particularly helpful in providing added insight into the event-driven dynamics of a system and in debugging a model. It can be used, for instance, to quickly identify poor design by observing that the number of entities in some queue is continuously increasing. It is not, however, a substitute for proper analysis of a stochastic DES. In fact, it often leads one into the trap of believing that after observing a system evolve over a short period of time a full understanding of its behavior can be gained. This ignores, however, that infrequent events that have not yet occurred can drastically affect the future of the system.

GASP

GASP is a collection of over 30 FORTRAN routines that can be used to simplify the process of building a simulation based on the event scheduling scheme. It is usually available on computer systems that support FORTRAN (see Pritsker, 1974).

GPSS

GPSS (General Purpose Simulation System) was originally developed by Gordon at IBM in the early 1960s. Since then it has evolved, with the most recent version, GPSS/H, developed at Wolverine Software (see Schriber, 1990). GPSS/H is a language based on "blocks" with which one can build a process-oriented simulation model. Entities are referred to as "transactions", and attributes as "parameters." Using standard graphical representations of the more than 60 basic blocks, one can conveniently define a model in terms of a block diagram for most systems of interest. A version of GPSS, GPSS/PC, specifically designed for PCs is also available through Minuteman Software.

SIMAN

SIMAN (SIMulation ANalysis) was developed by Pegden in the early 1980s and is currently available through Systems Modeling Corporation (see Pegden et al., 1990). It allows one to use the process-oriented scheme, the event scheduling scheme, or a combination of both. Like GPSS/H, a process is built in terms of basic blocks which can also be combined through their graphical representations to construct a block diagram. A newer version of SIMAN called ARENA allows one to animate a simulation on screen on most common computer operating systems, including PCs.

SIMSCRIPT

Like SIMAN, SIMSCRIPT has both process-oriented and event scheduling scheme capabilities. However, the former are sufficiently general so as to render the latter unnecessary for most applications. SIMSCRIPT was developed by Markowitz at the Rand Corporation in the early 1960s. The most recent version is available through the CACI Products Company (see Kiviat et al., 1973). The generality and free format style of SIMSCRIPT make it particularly attractive for modeling complex systems which do not need to be characterized exclusively by a queueing structure.

SLAM

SLAM (Simulation Language for Alternative Modeling), like SIMAN and SIMSCRIPT, provides both process-oriented and event scheduling scheme capabilities. SLAM was developed by Pegden and Pritsker in the early 1970s and is available through the Pritsker Corporation (see Pritsker, 1986). In SLAM, the modeling process usually involves a network definition of the system consisting of "nodes" and "branches", which can be done graphically. SLAMSYSTEM and SLAM II/TESS are versions that provide animation and graphical representation of output reports.

EXTEND

EXTEND is an object-oriented simulation package with extensive libraries of "objects" for different applications. It allows the user to "extend" these libraries by building new objects (that is, model building blocks) through simple C-based templates. It also provides the ability to construct hierarchical models and includes a variety of plotting and graphical output analysis tools, as well as animation.

As already mentioned, this is only a brief and partial account of available discrete-event simulation software. There are several additional simulation languages (e.g., SIMPAS, SIMULA, DEMOS, SIM++), and some specifically geared towards particular types of DES, such as manufacturing systems and computer networks. With the recent emergence of object-oriented programming and parallel processing capabilities, it is likely that a new generation of languages or new versions of existing ones will soon be developed; we make no effort here to keep track of all these developments now in progress. The reader is also referred to several books where detailed descriptions and comparisons of simulation languages are provided (Banks and Carson, 1984; Kreutzer, 1986; Law and Kelton, 1991).

10.5 RANDOM NUMBER GENERATION

Returning to Fig. 10.1 where the main components of a discrete-event simulator are shown, we now focus attention on the RANDOM VARIATE GENERATOR. To construct a sample path of a stochastic DES, it is essential to generate *random* numbers which represent, for example, service times or interarrival times with some given probability distribution. Thus, given a probability distribution, our task is to generate samples from that distribution. We refer to these samples as *random variates.*

It turns out, as we will see, that random variates of any distribution can be obtained by transforming random variates from the uniform distribution over the interval [0, 1]; we denote this distribution by U[0,1]. Random variates generated from the U[0,1] distribution are called *random numbers.* Thus, the term "random number generator" refers to a mechanism that generates iid samples from U[0,1].

No matter how hard we try, it is impossible to duplicate mother nature's penchant for randomness. All "random number" generation techniques are inherently systematic, based on some type of algorithm or procedure; therefore, by their very nature, these techniques could not produce true randomness. Consequently, strictly speaking, the term *pseudo-random* number generation should always be used. Nonetheless, we can still do a good job at random number generation, depending on the amount of computational effort we are willing to put in. The key point to be made here is that the *quality of the random number generator used in simulation should never be underestimated.*

By far the most common technique for generating random numbers is based on *linear congruence*, proposed by D.H. Lehmer (1951). We will limit our discussion of random number generation to this technique and refer the reader to Banks and Carson (1984), Bratley et al. (1987), or Law and Kelton (1991) for additional information on this subject.

10.5.1   The Linear Congruential Technique

The linear congruential technique is based on the following recursive relationship:

$$X_{k+1} = (aX_k + c) \bmod m, \quad k = 0, 1, \ldots \tag{10.6}$$

where a is the *multiplier*, c is the *increment*, m is the *modulus*, and X_0 is a given initial value called the *seed*. These are all specified nonnegative integers. The mod operation yields the remainder of the division of $(aX_k + c)$ by m. It follows immediately that

$$0 \leq X_k \leq m - 1 \quad \text{for all } k = 1, 2, \ldots$$

Using (10.6), we can generate a sequence of numbers $\{U_1, U_2, \ldots\}$ constrained to be in the interval $[0, 1]$ by simply setting $U_k = X_k/m$. This is used as the sequence of random numbers in U[0,1] that we are looking for.

Given the values of a, c, m, and X_0, the value of any $X_k, k = 1, 2, \ldots$ is completely specified. So one would be quite justified in wondering where the "randomness" is to be found. In fact, since $0 \leq X_k \leq m - 1$, U_k can only take the following values:

$$0, \quad 1/m, \quad 2/m, \ldots, (m-1)/m$$

It is therefore hard to see how U_k can be treated as a random variable with cdf U[0,1]. For instance, such a random variable can take any value in the interval $[0.5/m, 0.6/m]$ with probability $0.1/m$. However, the probability of that happening with this scheme is 0.

These problems are tackled through careful selection of the parameters of the linear congruential recursion, a, c, m, and X_0. One, for instance, can immediately see that the modulus m should be chosen to be as large as possible for two reasons.

First, by selecting m large we can increase the number of values that U_k can take, $0, 1/m, 2/m, \ldots, (m-1)/m$, and hence make the set of these values as "dense" as desired to approximate covering the entire $[0, 1]$ interval.

Second, it is easy to see that the linear congruential recursion (10.6) has a period which is at most equal to m. For example, let $m = 4, a = 1, c = 5$, and $X_0 = 2$. We then get: $X_1 = 3, X_2 = 0, X_3 = 1, X_4 = 2, X_5 = 3$, and the numbers repeat themselves from this point on. In this example, the period is exactly $m = 4$, which is the best we could hope for. For instance, the period would only be 2 if we had chosen $c = 1$ instead. It turns out that if m and c are not relatively prime (4 and 2 are not, but 4 and 5 are in this example), then the period is less than m. In any case, even if the period attains its maximum value of m, periodicity is hardly compatible with the kind of randomness that we seek. We can see, however, that by selecting m to be very large we can achieve such a large period that in the course of any reasonable simulation a number generated by (10.6) will never occur more than once.

Since we would like to select a value of m which is as large as possible, we look at the characteristics of most computers today: They are generally based on 32-bit words, with the first bit reserved for a sign. The largest integer that can be represented is $(2^{31} - 1)$, a number greater than 2 billion, which should be adequate for most applications. In common practice, m is selected in the form $m = 2^n$ for some integer n. This allows us to exploit modulo arithmetic and the binary representation of numbers on most computers, in order to avoid the division by m required in our procedure. In particular, with $m = 2^n$, the remainder of the division of $(aX_k + c)$ by m is immediately obtained by retaining the n rightmost binary digits of $(aX_k + c)$ and dropping the rest. For example, if $(aX_k + c) = 7$ and $m = 22$, the binary representation of 7 is 111, and the two rightmost digits, 11, give $3 = 7 \bmod 4$.

We shall not get into further details regarding the selection of the remaining parameters in the recursion (10.6). It is only worth mentioning that when we set $c = 0$, we obtain a special case of random number generators called *multiplicative* generators. In contrast, when $c > 0$, we have *mixed* generators. Multiplicative generators have the advantage of not requiring the addition operation in (10.6). On the other hand, because any $m > 0$ and $c = 0$ are not relatively prime, we cannot attain the maximum period possible, (i.e., m). Still, multiplicative generators are widely used, since mixed generators have apparently not provided significantly better performance. If the form $m = 2^n$ is once again adopted, it can be shown that the maximum attainable period in this case is given by 2^{n-2} (Knuth, 1981).

In practice, we would like to associate each stochastic process involved in a DES model with one random number *stream*. We do so by identifying a *stream* of random numbers with a particular seed (thus, by using the same seeds, we can always reproduce our experimental data). As an example, suppose we need two streams, one to model a service time sequence and one to model an interarrival time sequence for a single-server queueing system. We invoke the service time stream by specifying a seed X_0. Let us then assume that we will never use more than N service time samples (say, 10^5 or 10^6). We then invoke the interarrival time stream using X_{N+1} as the seed. In other words, streams are disjoint subsequences of a single random number sequence.

Random number generators are often organized so that they consist of several such disjoint sequences. A user can then invoke any such sequence by specifying its corresponding seed.

10.6   RANDOM VARIATE GENERATION

In the previous section, we saw how to generate random numbers, that is, the special case of random variates from the distribution U[0,1]. In this section, we will assume that a random number sequence is available and concentrate on the issue of how to transform it so as to obtain a sequence of random variates from a specified probability distribution $F(x)$.

There are a number of techniques used for random variate generation. The most common, and in many ways intuitively appealing, approach is based on the *inverse transform* method described in Section 10.6.1. Three additional techniques are also discussed in Sections 10.6.2 through 10.6.4.

10.6.1   The Inverse Transform Technique

Let X be a random variate to be generated from a cdf $F(x)$. Let us first assume that $F(x)$ is continuous and strictly monotonically increasing, as illustrated in Fig. 10.3. Suppose that a random number U is provided from the distribution U[0,1]. Observe that U will always correspond to some value of $F(x) \in [0, 1]$. Therefore, there is a one-to-one mapping between U and some number on the x axis in Fig. 10.3. Next, we show that if we let this number be X, then X is in fact a random variate such that $X \sim F(x)$.

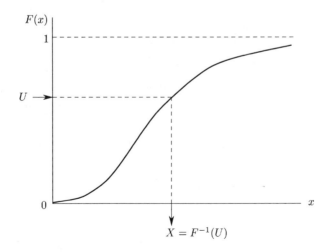

Figure 10.3: Illustrating the inverse transform technique.

Given U, we have $U = F(X)$ for some X. Since the inverse $F^{-1}(U)$ always exists we can write

$$P[X \leq x] = P[F^{-1}(U) \leq x] = P[U \leq F(x)] \tag{10.7}$$

However, $U \sim U[0,1]$, that is, $P[U \leq u] = u$ for all $u \in [0,1]$. Therefore, for any $F(x) \in [0,1]$,

$$P[U \leq F(x)] = F(x)$$

and (10.7) gives

$$P[X \leq x] = F(x)$$

In other words, $X \sim F(x)$ as originally claimed. In short, the inverse transform technique takes a random number U and sets $X = F^{-1}(U)$. The resulting X is a random variate with cdf $F(x)$.

Example 10.1 (Random Variates for the exponential distribution)

Let us generate random variates for the exponential distribution $F(x) = 1 - e^{-\lambda x}$, $x \geq 0$, given random numbers $U \sim U[0,1]$. For any $U \in [0,1]$, we have

$$U = 1 - e^{-\lambda X}$$

for some X. Therefore, solving for X, we get

$$X = \frac{-1}{\lambda} \ln(1 - U) \tag{10.8}$$

which is in fact the most common way for generating samples from the exponential distribution. A side observation here is that since U and $(1 - U)$ have exactly the same distribution, we can also use $X = (-1/\lambda) \ln(U)$. In general, replacing $(1 - U)$ by U does save one arithmetic operation (a subtraction), but it should not be blindly used as a rule; the validity of this replacement depends on the cdf we are inverting (it cannot be applied, for example, when the inverse transform $X = F^{-1}(U)$ is not continuous).

The inverse transform technique is quite general. It can be used, for example, to sample from distributions of *discrete* random variables. This is illustrated in Fig. 10.4 for a random variable X that can take three values $a_1 < a_2 < a_3$:

$$X = \begin{cases} a_1 & \text{with probability } p_1 \\ a_2 & \text{with probability } p_2 \\ a_3 & \text{with probability } p_3 \end{cases}$$

with $p_1 + p_2 + p_3 = 1$. The cdf $F(x)$ in this case is piecewise constant, starting at 0 and having three jumps of size p_1, p_2 and p_3 as shown in Fig. 10.4. Thus,

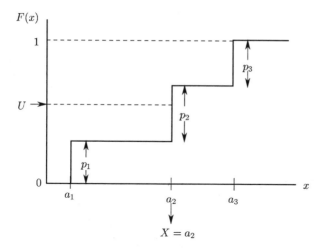

$$X = a_2$$

Figure 10.4: Illustrating the inverse transform technique for discrete random variables.
The random variate X is assigned the smallest value of a_n, $n = 1, 2, 3$, such that $F(a_n) \geq U$.

U is such that it always falls in one of three intervals: $[0, p_1]$, $(p_1, p_1 + p_2]$, or $(p_1 + p_2, 1]$, and X takes the corresponding values a_1, a_2, a_3.

In general, for a discrete random variable X that can take N values $a_1 < \cdots < a_N$ with corresponding probabilities $p_1, \ldots, p_N$, we have

$$X = \begin{cases} a_1 & \text{if } 0 \leq U \leq p_1 \\ a_2 & \text{if } p_1 < U \leq p_1 + p_2 \\ \vdots \\ a_N & \text{if } p_1 + p_2 + \ldots + p_{N-1} < U \leq 1 \end{cases} \tag{10.9}$$

or, equivalently,

$$X = \min\{a_n : F(a_n) \geq U, n = 1, \ldots, N\} \tag{10.10}$$

In Fig. 10.4, for example, the smallest a_n that satisfies $F(a_n) \geq U$ is a_2. Therefore, $X = a_2$.

The obvious generalization to distribution functions with discontinuities (for mixed random variables) is illustrated in Fig. 10.5. Thus, the general form of the inverse transform method is

$$X = \min\{x : F(x) \geq U) \tag{10.11}$$

One drawback of the inverse transform technique is that it may not always be possible to evaluate $F^{-1}(U)$ in closed form. This problem arises with a number

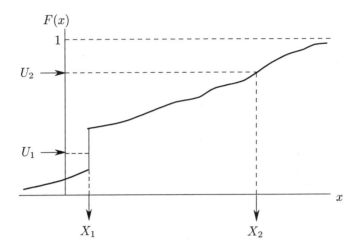

Figure 10.5: Illustrating the generalization of the inverse transform technique.

of common distributions (e.g., the normal distribution). In this case, we resort to various numerical techniques or approximate the function $F(x)$ by some other function whose inverse is easier to obtain. For some special distributions, it is also possible to use alternative special-purpose methods which are more efficient than the inverse transform technique.

One of the main advantages of the inverse transform technique is that it permits the generation of maximally correlated random variates from different distributions. This property has important ramifications in variance reduction techniques (which we will not consider here; see, however, Fishman, 1978; or Law and Kelton, 1991), as well as in perturbation analysis of DES, which we will discuss in Chapter 11. To illustrate the key idea, consider a continuous (for simplicity) cdf $F(x; \lambda)$ where λ is some parameter of the distribution. Suppose we observe a random variate X which we know to be generated from $F(x; \lambda)$. We then ask the question: *What would the random variate have been if the parameter λ had been replaced by a new value λ'?* We can answer this question by arguing as follows. If X had been generated through the inverse transform technique, there would have been some random number U giving rise to X through $X = F^{-1}(U; \lambda)$, that is, $U = F(X; \lambda)$. Now if this same random number U had been used to generate a random variate X' from the new cdf $F(x; \lambda')$, we would get

$$X' = F^{-1}(U; \lambda') = F^{-1}[F(X; \lambda); \lambda'] \qquad (10.12)$$

In other words, we can directly obtain the random variate X' from the observed one, X. This process is shown in Fig. 10.6: First, we identify $U = F(X; \lambda)$, then use this U to obtain the new random variate $X' = F^{-1}(U; \lambda')$. Note that by choosing the same random number for both X and X', we also maximize the correlation between these two random variates.

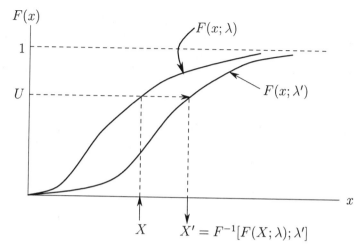

Figure 10.6: Transforming a random variate from $F(x; \lambda)$ into a random variate from $F(x; \lambda')$.

Another interesting observation related to Fig. 10.6 has to do with the comparison of random variables. In general, calling one random variable "larger" or "smaller" than another makes no sense, since random variables are not real numbers subject to standard ordering rules. There are several useful definitions of *stochastic orderings*, however, which allow us to make meaningful comparisons. One such definition is motivated by the following observation. Let X and X' be generated through the inverse transform technique from $F(x; \lambda), F(x; \lambda')$ respectively. Then, if $F(x; \lambda) \geq F(x; \lambda')$ for all x for which $F(\cdot)$ is defined, the two random variates always satisfy $X \leq X'$, as clearly seen in Fig. 10.6. Thus, we say that X' is *stochastically larger* than X if $F(x; \lambda') \leq F(x; \lambda)$.

10.6.2 The Convolution Technique

This technique exploits the fact that a random variable of interest, X, is often defined as the sum of n iid random variables:

$$X = Y_1 + \ldots + Y_n$$

In this case, it is a standard result of basic probability theory (see Appendix I) that the pdf of X can be obtained as the n-fold convolution of the pdf of any one Y_i (since they are iid), which justifies the name of the technique. Assuming that $X \sim F(x)$ and $Y_i \sim G(x)$, we generate a random variate X by proceeding as follows:

1. Generate n random variates, $Y_1, \ldots, Y_n$, from $G(x)$.

2. Set $X = Y_1 + \ldots + Y_n$.

This technique is efficient when the generation of $Y_1, \ldots, Y_n$ is simple. A good example is that of an Erlang cdf with m stages, defined in (8.105), in which case we need to generate m random variates from an exponential distribution, typically using the inverse transform technique. This approach, however, becomes less attractive as m becomes large.

10.6.3 The Composition Technique

The composition technique exploits the fact that the cdf $F(x)$ of a random variable X can sometimes be expressed as a convex combination of n cdf's:

$$F(x) = p_1 F_1(x) + \ldots + p_n F_n(x)$$

with $p_1 + \ldots + p_n = 1$, $p_i \geq 0$ for all $i = 1, \ldots, n$, or, more generally,

$$F(x) = \sum_{i=1}^{\infty} p_i F_i(x) \quad \text{with} \quad \sum_{i=1}^{\infty} p_i = 1, \quad p_i \geq 0 \text{ for all } i = 1, \ldots, n \quad (10.13)$$

Since X can be thought of as a random variate generated from a cdf $F_i(\cdot)$ with probability p_i, the composition technique works as follows:

1. Generate a discrete random variate M defined so that $P[M = i] = p_i, i = 1, 2, \ldots$

2. Generate a random variate X from the cdf $F_M(x)$.

In both steps, the inverse transform technique can be used, as long as this is done independently. It is easy to verify that this approach indeed provides a random variate X from the cdf $F(x)$; we evaluate $P[X \leq x]$ by first conditioning on the random variable M defined in step 1 above:

$$P[X \leq x] = \sum_{i=1}^{\infty} P[X \leq x \mid M = i] P[M = i] = \sum_{i=1}^{\infty} F_i(x) p_i = F(x)$$

An example where this technique is convenient to use is in generating random variates from a hyperexponential cdf with m parallel stages, defined in (8.107). In this case, the inverse transform method can be used in step 2 to generate a random variate from an exponential distribution.

It is worth observing that the convolution and the composition techniques take advantage of some structural properties of the random variate of interest X. We may thus view them as special cases of a more general approach, whereby we first seek such properties of X and then see if we can exploit them in simplifying the process of random variate generation.

10.6.4   The Acceptance-Rejection Technique

This technique is less direct than the ones considered thus far. We assume, as before, that we are interested in generating a random variate X from a cdf $F(x)$ with corresponding pdf $f(x)$. We will describe the technique for continuous variables, but the case of discrete variables is essentially the same.

The first step is to identify some function $g(x)$ with the property: $g(x) \geq f(x)$ for all x. We call $g(x)$ a *majorizing* function, and note that it is clearly not unique. In fact, choosing a convenient majorizing function can enhance the efficiency of this technique (we will return to this point later in this section). Since $g(x)$ is generally not a density function, we determine next a normalization constant, c, which allows us to transform it into a density function. In particular, let

$$c = \int_{-\infty}^{\infty} g(x)dx \tag{10.14}$$

assuming that $g(x)$ is chosen so that $c < \infty$. We can now define a function

$$h(x) = \frac{g(x)}{c} \tag{10.15}$$

which is in fact a density function. The idea now is to generate a random variate Y from the pdf $h(x)$ as efficiently as possible (which is where a "good" choice of $g(x)$ plays a role), and then use a simple rule to decide if Y should be *accepted* as a valid random variate from $f(x)$ or not. The precise procedure is as follows:

1. Select a majorizing function $g(x)$ of $f(x)$, and define $h(x)$ using (10.14) and (10.15).

2. Generate a random variate Y with pdf $h(x)$.

3. Generate a random number U independently from Y.

4. If the condition

$$U \leq \frac{f(Y)}{g(Y)} \tag{10.16}$$

   is satisfied, then set $X = Y$. Otherwise, repeat step 2 for a new Y.

Thus, it is possible that several trials may lead to rejected Y's until the condition in step 4 is finally met. Proving that $X \sim F(x)$ with X obtained from step 4 is based on the observation that the event $[X \leq x]$ is identical to the event $[Y \leq x \mid U \leq f(Y)/g(Y)]$ and it is left as an exercise.

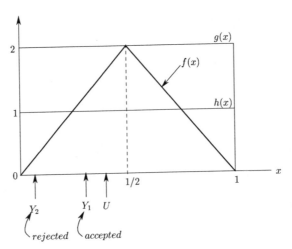

Figure 10.7: Illustrating the acceptance-rejection technique for Example 10.2. *In this example, the acceptance condition is $U \leq 2Y$ if $Y \leq 1/2$, and $U \leq 2 - 2Y$ if $Y > 1/2$. Y_1 is accepted because $U < 2Y_1$, but Y_2 is rejected because it is too small to satisfy $U < 2Y_2$.*

Example 10.2

We will use the acceptance-rejection technique to generate random variates for the triangular pdf

$$f(x) = \begin{cases} 4x & \text{for } 0 \leq x \leq 1/2 \\ 4 - 4x & \text{for } 1/2 < x \leq 1 \\ 0 & \text{otherwise} \end{cases}$$

shown in Fig. 10.7. In this case, we can choose the simple majorizing function

$$g(x) = \begin{cases} 2 & \text{for } 0 \leq x \leq 1 \\ 0 & \text{otherwise} \end{cases}$$

Clearly, in this case $c = 2$, and the function in (10.15) is $h(x) = g(x)/2$, as shown in Fig. 10.7. Therefore, the acceptance condition (10.16) becomes

$$U \leq 2Y \quad \text{if } Y \leq 1/2, \qquad U \leq 2 - 2Y \quad \text{if } Y > 1/2$$

In Fig. 10.7, two values of Y are shown, both such that $Y \leq 1/2 : Y_1$ is accepted because it is clearly such that $U < 2Y_1$ (even though $Y_1 < U$), whereas Y_2 is rejected because it is too small to satisfy $U < 2Y_2$.

Some insight into the acceptance-rejection technique can be gained from Example 10.2 by noticing that Y's that fall near the middle part of the $[0, 1]$

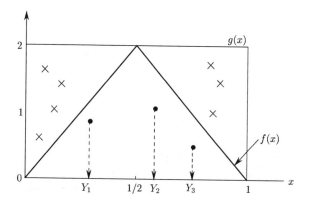

Figure 10.8: Acceptance-rejection based on $Ug(Y) \leq f(Y)$.

interval, where $f(x)$ is high, are more likely to be accepted. On the other hand, Y's that fall very near the edges of the interval (0 or 1) tend to be rejected, as is the case with Y_2 in Fig. 10.7.

Another way of looking at this technique provides even more insight and suggests more general applications of the acceptance-rejection idea. Specifically, the acceptance condition (10.16) can be rewritten as $Ug(Y) \leq f(Y)$. This allows us to visualize the following process: After Y is generated, a point $(Y, Ug(Y))$ is defined in the plane where $f(x)$ is plotted; if this point falls under the curve $f(x)$, then it satisfies $Ug(Y) \leq f(Y)$ and Y is accepted. Thus, the acceptance condition becomes: *Accept all Y such that the point $(Y, Ug(Y))$ falls under the curve defined by the pdf $f(x)$.* This is illustrated in Fig. 10.8 for the triangular pdf of Example 10.2. Of the ten points generated, only three fall under the curve $f(x)$ with corresponding random variates Y_1, Y_2, Y_3.

The Choice of Majorizing Function

We now turn attention to the selection of the majorizing function $g(x)$. There are two issues to be addressed. The first issue is the determination of the normalization constant c in (10.14). To simplify the computation, the selection of $g(x)$ as a piecewise constant function (as in Example 10.2) is usually the best choice, since the integral reduces to summing rectangular areas.

The second issue is that the choice of $g(x)$ affects the probability of acceptance $P[U \leq f(Y)/g(Y)]$, which we would obviously like to maximize. Let us evaluate this probability. We do so by conditioning on the event $[Y = x]$ and applying the rule of total probability:

$$P[U \leq f(Y)/g(Y)] = \int_{-\infty}^{\infty} P[U \leq f(Y)/g(Y) \mid Y = x]h(x)dx \qquad (10.17)$$

Since $U \sim U[0,1]$ and U is independent of Y, we have

$$P[U \leq f(Y)/g(Y) \mid Y = x] = P[U \leq f(x)/g(x)] = f(x)/g(x)$$

and, recalling that $g(x) = ch(x)$, (10.17) yields

$$P[U \leq f(Y)/g(Y)] = \int_{-\infty}^{\infty} \frac{f(x)}{c} dx = \frac{1}{c} \qquad (10.18)$$

since $f(x)$ is a pdf. In other words, the probability of acceptance is inversely proportional to c. It follows that to maximize this probability, c should be made as small as possible. The best we can do is have $c = 1$, which would give us a probability of acceptance equal to 1. This corresponds to the case where we choose $g(x) = f(x)$, since in general $g(x) \geq f(x)$. Thus, the next best thing is to try and determine some function $g(x)$ whose shape envelopes $f(x)$ as closely as possible.

We see, therefore, that while the selection of a simple piecewise constant $g(x)$ is straightforward, it may lead to a large value of c and hence low probability of acceptance. In Example 10.2, we have $c = 2$, which implies that only one out of two Y's is accepted. We could reduce c, and hence increase the probability of acceptance, by choosing a different function $g(x)$ whose shape more closely approximates $f(x)$ while at the same time preserving $g(x) \geq f(x)$. An example is the majorizing function

$$g(x) = \begin{cases} 1 & \text{for } 0 \leq x \leq 1/4 \text{ or } 3/4 < x \leq 1 \\ 2 & \text{for } 1/4 < x \leq 3/4 \\ 0 & \text{otherwise} \end{cases}$$

which gives $c = 1.5$ and therefore increases the probability of acceptance from $1/2$ to $2/3$.

10.7 OUTPUT ANALYSIS

As we have repeatedly emphasized, one of the main objectives of discrete-event simulation is the estimation of quantities of interest from the output data obtained. Returning to the simulation of a simple queueing system in Section 10.2.1, note that output data such as customer system times and queue length occupancy times were used to estimate performance measures of interest such as the average system time and the average queue length of the system over N customers. The average system time over N customers, for example, is a *parameter* (the mean) of a probability distribution characterizing the random variable "average system time over N customers" that was defined in Section 10.2.1. In fact, estimating parameters of various distributions is a major part of output analysis.

Our first task in this section is to characterize different types of simulations based on whether we are interested in the steady-state behavior of a system or a transient finite period of time. This gives rise to the terms *non-terminating* and *terminating* simulations respectively. In either case, output data are used to estimate parameters of various probability distributions, and, if possible, the distribution functions themselves. Concentrating on *parameter estimation*, we will discuss the basic definitions and fundamental principles involved. We will then provide more details for the parameter estimation techniques required for each type of simulation.

10.7.1   Simulation Characterizations

In general, the random variables whose distributions we are interested in are functions of the state of the system. As we saw in Chapters 7 and 8, the state defines a stochastic process $\{X(t)\}$, so that the distribution of $X(t)$ depends on t. In some cases, however, as $t \to \infty$, it is possible that steady state is reached and there are stationary distributions characterizing random variables such as the "queue length at steady state" or the "system time at steady state." Often, therefore, the purpose of simulation is to estimate parameters of these steady-state distributions, and, if possible, these steady-state distribution functions themselves.

A *non-terminating simulation* is a simulation used to study the steady-state behavior of a system and to estimate parameters of various stationary probability distributions. In a non-terminating simulation there is no obvious or natural point to stop, since, in theory, we are interested in the system's behavior as $t \to \infty$. The longer the simulation, the better, but we are obviously limited by practical considerations. The initial conditions of the system should, in theory, not affect the results, but once again this may not be entirely true. Thus, the difficulties in non-terminating simulations involve the determination of initial conditions and of a good stopping rule. Unfortunately, there are few rigorous guidelines to fully resolve these issues, which often depend on the characteristics of the system itself. One must usually resort to simple common-sense (but far from perfect) procedures such as the following. Let the length of the simulation run be initially T and the value of a parameter estimate of interest be $X(T)$. Extend the simulation run length to $2T$ and obtain $X(2T)$. If $|X(2T) - X(T)| < \epsilon$ for some predefined small ϵ, stop. Otherwise, extend the simulation run length to $3T$ and repeat the process for $|X(3T) - X(2T)|$, and so on.

Example 10.3

Suppose a new computer system is installed and we are interested in the mean job response time in the long-run operation of the system. If we assume that the stochastic process representing the submission of jobs is stationary, we can simulate the system starting with no jobs and estimate the mean job response time after T time units, $X(T)$, where T is chosen to be long enough so that the processor doing the simulation can produce output data within a reasonable amount of time (say 5 or 10 minutes). We keep on doubling the length of the run until $|X((n+1)T) - X(nT)| < \epsilon$, for some $n = 1, 2, \ldots$, and for some small predefined ϵ. Since $X(t)$ may fluctuate, this condition does not guarantee that $|X(t_1) - X(t_2)| < \epsilon$ for all $t_1, t_2 \geq nT$. Thus, to doublecheck we may choose to stop only after $|X((n+1)T) - X(nT)| < \epsilon$ is satisfied for some $n = m$ and for $n = m+1$.

There is of course no guarantee that a stationary distribution of the state of the system or any functions of it actually exist. In fact, in many "real-world" systems the operating conditions change so frequently that the system never really gets a chance to reach a point where it is at steady state. Nonetheless, if the time between significant changes is quite long, then the assumption that the system reaches steady state prior to the change can be a very good one.

While studying the steady-state behavior of a system implies that we are interested in what happens "in the long run", there are also many cases where the purpose of simulation is to study a system over some finite period of time. A *terminating simulation* is one whose run length is naturally defined by a specific event. This event may be defined by a prespecified clock value or by a condition satisfied by the state of the system.

Example 10.4
Suppose that the computer system of Example 10.3 is to be turned on every morning at 9:00 am and turned off every night at midnight. Let us also assume that all jobs still running at midnight are aborted. If we are interested in the mean job response time of all jobs actually terminated over this 15-hour period, then we simulate the system starting with no jobs and stop the simulation when the clock reaches 15 hours. In this case, the terminating event is defined by the clock reaching this value.

Example 10.5
Consider a switch in a communication network that can queue up to K messages. We are interested in the mean time it takes for the queue to overflow for the first time, assuming it starts out empty. In this case, the terminating event is specified by the condition $X(t) = K$, where $X(t)$ is the queue length at time t. Note that, in contrast to previous examples, the simulation stopping time here is itself a random variable.

In nonterminating simulations the critical question is: How long is long enough for the simulated system to be at steady state? This is not an issue in terminating simulations. Instead, every simulation defines a "'sample" and estimates are obtained by repeating the simulation under the same initial conditions so that several such samples are obtained. Precisely how such estimates are defined and what their properties happen to be is a topic of continuing interest in the area of discrete-event simulation. The next few sections provide only a limited introduction to parameter estimation theory and its applications to terminating and non-terminating simulations.

10.7.2 Parameter Estimation

The output data of a particular simulation or the data collected over many simulations can be thought of as defining stochastic processes. For example, if we record the system time of customers in a simulated queueing system, we obtain a stochastic sequence $\{S_1, S_2, \dots, S_k, \dots\}$, where S_k is the system time of the kth customer. On the other hand, if we record the queue length at all times t, $X(t)$, we define a stochastic process $\{X(t)\}$. Similarly, suppose a queueing system is simulated over N customers and S_N is the observed average system time over these N customers. If the simulation is repeated with the same initial conditions, a stochastic sequence $\{S_{N,1}, S_{N,2}, \dots, S_{N,k}, \dots\}$ is generated, where $S_{N,k}$ is the average system time over N customers in the kth simulation run.

Let us adopt the following point of view. We collect data from a simulation that are of the form $X_1, X_2, \dots, X_k, \dots$, and we are interested in using them to estimate some fixed quantity of interest θ. For example, θ is the mean of the stationary system time distribution in a queueing system, and X_k is the observed system time of the kth customer. A *point estimate* of θ, denoted by $\hat{\theta}$, is a single number that represents our "best guess" of θ based on collected data $X_1, X_2, \dots, X_n$. An interval estimate of θ, provides a range of numbers defined by an interval $[\hat{\theta} - \alpha_1, \hat{\theta} + \alpha_2]$, where α_1, α_2 are also functions of the data. This interval is usually accompanied by some "level of confidence" expressed as a probability q. The meaning of the interval estimate is: The true value of θ lies within $[\hat{\theta} - \alpha_1, \hat{\theta} + \alpha_2]$ with probability q.

Point Estimation

The simplest and most common estimation problem arises when our collected data $X_1, X_2, \dots, X_n$ form a sequence of iid random variables. In this case, we know (see also Section 6.2.3 in Chapter 6) that there is a single probability distribution function characterizing the sequence. Let θ be the mean of that distribution. What we are interested in is estimating θ through the available data $X_1, X_2, \dots, X_n$. To obtain a point estimate of $\theta, \hat{\theta}_n$, based on n samples, we use the *sample mean*:

$$\hat{\theta}_n = \frac{1}{n} \sum_{i=1}^{n} X_i \tag{10.19}$$

It is important to keep in mind that $\hat{\theta}_n$ is a *random variable*. So, what we are trying to do here is to use a random variable to estimate a real number θ.

The first important property that a good estimator must possess is *unbiasedness*. We say that $\hat{\theta}_n$ is an *unbiased* estimator of θ if it satisfies

$$E[\hat{\theta}_n] = \theta \tag{10.20}$$

In general, however, we have

$$E[\hat{\theta}_n] = \theta + \beta_n \tag{10.21}$$

where β_n is called the *bias* of the estimator.

If the data $X_1, X_2, \ldots, X_n$ form indeed an iid sequence, then the point estimator in (10.19) is unbiased, since

$$E[\hat{\theta}_n] = \frac{1}{n}\sum_{i=1}^{n}E[X_i] = \frac{1}{n}\sum_{i=1}^{n}\theta = \theta$$

In fact, the same is clearly true as long as the n random variables have the same distribution with mean θ, without having to be independent.

In some cases, the data collected form a stochastic process, rather than sequence, $\{X(t)\}$. In this case, the analog of the sample mean in (10.19) over n samples is the *time average* over an interval $[0, T]$:

$$\hat{\theta}_T = \frac{1}{T}\int_0^T X(t)dt \tag{10.22}$$

Returning to (10.19), we already pointed out that the sample mean $\hat{\theta}_n$ is a random variable. It is therefore characterized by a pdf. If the variance of this pdf is large, it is likely that many values of $\hat{\theta}_n$, resulting from different sets of data, do not come very close to the true mean θ. This motivates us to try and get an idea of what the variance of the distribution of $\hat{\theta}_n$ is. We have

$$Var[\hat{\theta}_n] = E[(\hat{\theta}_n - \theta)^2] = E\left[\left(\frac{1}{n}\sum_{i=1}^{n}X_i - \theta\right)^2\right] = \frac{1}{n^2}E\left[\left(\sum_{i=1}^{n}(X_i - \theta)\right)^2\right]$$

$$= \frac{1}{n^2}E\left[\sum_{i=1}^{n}\sum_{j=1}^{n}(X_i - \theta)(X_j - \theta)\right]$$

Since all X_i and X_j, $j \neq i$ are assumed independent, we have, for all $j \neq i$ in the double sum above, $E[(X_i - \theta)(X_j - \theta)] = 0$, and we get

$$Var[\hat{\theta}_n] = \frac{1}{n^2}E\left[\sum_{i=1}^{n}(X_i - \theta)^2\right] = \frac{1}{n^2}\sum_{i=1}^{n}Var[X_i] = \frac{1}{n^2}nVar[X_i]$$

Letting $\sigma^2 = Var[X_i]$, we finally obtain the result

$$Var[\hat{\theta}_n] = \frac{\sigma^2}{n} \tag{10.23}$$

The problem now is that σ^2 is unknown. To estimate it, we resort to the *sample variance* of our collected data, defined by

$$S_n^2 = \frac{1}{n-1} \sum_{i=1}^{n} \left(X_i - \hat{\theta}_n \right)^2 \tag{10.24}$$

which is again a random variable. It is straightforward to evaluate $E[S_n^2]$ and find that $E[S_n^2] = \sigma^2$ (assuming once again that our data form an iid sequence), so that S_n^2 is an unbiased estimator of the actual variance σ^2. We will now replace the unknown σ^2 in (10.23) by its unbiased estimator S_n^2 to define a point estimator of $Var[\hat{\theta}_n]$ which we will denote by $\hat{\sigma}^2(\hat{\theta}_n)$:

$$\hat{\sigma}^2(\hat{\theta}_n) = \frac{S_n^2}{n} = \frac{1}{n(n-1)} \sum_{i=1}^{n} (X_i - \hat{\theta}_n)^2 \tag{10.25}$$

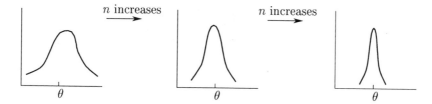

Figure 10.9: Typical behavior of the pdf of the sample mean estimator $\hat{\theta}_n$ as n increases.

In all cases, the mean of the pdf of $\hat{\theta}_n$ is θ. However, the variance of $\hat{\theta}_n$ keeps decreasing as more data are collected, until $\hat{\theta}_n$ is almost always equal to the true value θ which it is trying to estimate.

To understand exactly how estimation works as a function of the number of collected data n, let us take another look at (10.23). Assuming that σ^2 is a finite quantity, as $n \to \infty$ the variance of the point estimate $\hat{\theta}_n$ approaches zero. This reflects the intuitive fact that the more data we collect the better our estimate of θ becomes. A typical situation is shown in Fig. 10.9. For a small value of n, the pdf of the estimate $\hat{\theta}_n$ is rather spread out and $Var[\hat{\theta}_n]$ is likely to be large. As n increases, the shape of the pdf changes so that its mean remains θ, but as $Var[\hat{\theta}_n] = \sigma^2/n$ decreases, it is less likely that our data will yield a value of $\hat{\theta}_n$ which is too far off the true mean θ. In the limit as $n \to \infty$, the pdf essentially collapses to a "spike" at θ, as $\hat{\theta}_n$ comes arbitrarily close to θ. Formally, this asymptotic property of estimators based on sample means of iid sequences is stated as a theorem known as the *Strong Law of Large Numbers*.

Theorem 10.1 Let $X_1, X_2, \ldots, X_n$ form an iid sequence of random variables with mean $\theta < \infty$. Then,

$$\hat{\theta}_n \to \theta \quad \text{with probability 1, as } n \to \infty \tag{10.26}$$

where $\hat{\theta}_n$ is the sample mean of $X_1, X_2, \ldots, X_n$. ◆

The proof of this theorem requires some more advanced material on the theory of stochastic processes, and it is omitted. The statement "with probability 1" always creates some confusion if one has not thoroughly absorbed this advanced material; its meaning is this: There may be just a few data collection experiments where $\hat{\theta}_n$ does not approach θ as $n \to \infty$, but the probability of that happening is zero. In practice, the statement "with probability 1" is normally regarded as equivalent to certainty.

We mentioned earlier that the first desirable property an estimator must possess is unbiasedness. The second desirable property is precisely (10.26), which is referred to as *strong consistency*. Note that an estimator may be biased, but also strongly consistent, in which case it is also called *asymptotically unbiased*. In other words, it is possible that the bias term β_n in (10.21) is such that $\beta_n \neq 0$ for finite values of n, but $\beta_n \to 0$ as $n \to \infty$. Figure 10.10 shows the typical behavior of a strongly consistent point estimator $\hat{\theta}_n$ and its variance $\mathrm{Var}[\hat{\theta}_n]$ as a function of the number of data n. Consistent with (10.23) and Fig. 10.9, we see that $\mathrm{Var}[\hat{\theta}_n]$ approaches 0 with increasing n, and $\hat{\theta}_n$ approaches the true value θ.

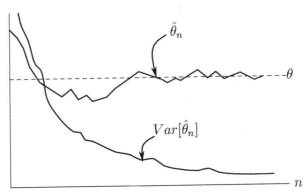

Figure 10.10: Typical behavior of a strongly consistent point estimator $\hat{\theta}_n$ and its variance $\mathrm{Var}[\hat{\theta}_n]$ as a function of the number of data n.

All of the previous discussion was based on the assumption that our data, $X_1, X_2, \ldots, X_n$, form an iid sequence. Unfortunately, in practice (especially when dealing with simulation experiments) one cannot avoid the presence of correlation among data. As we saw earlier, this does not affect the unbiasedness of $\hat{\theta}_n$, but it can definitely affect the unbiasedness of S_n^2 in (10.24). In particular, the derivation of (10.23) depended on the independence assumption so as to set $E[(X_i - \theta)(X_j - \theta)] = 0$ for all $j \neq i$. If, however, X_i and $X_j, j \neq i$, are correlated, one may seriously underestimate (or overestimate) the true variance σ^2 of the estimator. Thus, in general, one must go a step beyond estimating the variance and attempt to estimate correlations as well. This task is not easy, as correlation estimators tend to be biased and to have large variances (see Law and Kelton, 1991; or Fishman, 1978). To bypass this problem in discrete-event simulation, the best approach is to organize output data from different simulations in ways that satisfy iid assumptions. We will further address this point in subsequent sections.

Interval Estimation

We have seen that if our data, $X_1, X_2, \ldots, X_n$, form an iid sequence we can use the sample mean to obtain the unbiased point estimator $\hat{\theta}_n$. Moreover, $\hat{\theta}_n \to \theta$ as $n \to \infty$ (with probability 1). Since we are generally limited to a finite set of data, we would like to develop some systematic way to define an interval that contains $\hat{\theta}_n$ within which we can find the true value θ with some level of "confidence."

The basic tool for accomplishing this task is provided by a classic result of probability theory known as the *Central Limit Theorem*. To state this theorem, we begin by defining the random variable

$$Z_n = \frac{\hat{\theta}_n - E\left[\hat{\theta}_n\right]}{\sqrt{\text{Var}\left[\hat{\theta}_n\right]}} \qquad (10.27)$$

where we have taken the sample mean $\hat{\theta}_n$, subtracted its mean, and normalized the result by its standard deviation. Recalling from (10.20) and (10.23) that $E[\hat{\theta}_n] = \theta$ and $\text{Var}[\hat{\theta}_n] = \sigma^2/n$, Z_n can also be written as

$$Z_n = \frac{\hat{\theta}_n - \theta}{\sqrt{\sigma^2/n}} \qquad (10.28)$$

Let the cdf of Z_n be denoted by $F_n(\cdot)$. Clearly, there is no reason to suspect that F_n has any special properties, since there is nothing special about the cdf of the iid sequence $X_1, X_2, \ldots, X_n$ in the first place. The central limit theorem, however, asserts that as $n \to \infty$, F_n comes arbitrarily close to a normal distribution with mean zero and variance equal to 1. This is also known as the *standard normal distribution*, and is given by

$$\Phi(x) = \frac{1}{\sqrt{2\pi}} \int_{-\infty}^{x} e^{-\tau^2/2} d\tau \qquad (10.29)$$

defined for all real numbers x.

Theorem 10.2 Let $F_n(x)$ be the cdf of the random variable Z_n defined in (10.27). Then,

$$F_n(x) \to \Phi(x) \quad \text{as } n \to \infty \qquad (10.30)$$

where $\Phi(x)$ is the standard normal distribution in (10.29). ◆

Thus, as we increase the number of collected data, the cdf of Z_n approaches the standard normal; equivalently, the cdf of the sample mean $\hat{\theta}_n$ approaches a normal distribution with mean θ and variance σ^2/n.

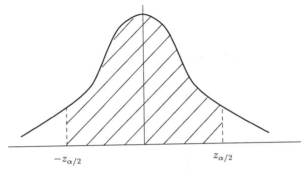

Figure 10.11: The standard normal probability density function. *The points $-z_{\alpha/2}$ and $z_{\alpha/2}$ are such that the shaded area is $(1 - \alpha)$. Thus, for any standard normal random variable $Z : P\left[-z_{\alpha/2} \leq Z \leq z_{\alpha/2}\right] = 1 - \alpha$.*

Now let us see how this theorem can be useful for our purposes of defining an interval within which we can find the true value θ with some "confidence." The standard normal pdf is the well-known symmetric bell-shaped curve shown in Fig. 10.11. We define two points, one negative and one positive, denoted by $-z_{\alpha/2}$ and $z_{\alpha/2}$ respectively, so that the area under the curve between these two points is given by $(1 - \alpha)$. In other words, if Z is a standard normal random variable, then

$$P\left[-z_{\alpha/2} \leq Z \leq z_{\alpha/2}\right] = 1 - \alpha$$

By the central limit theorem, the random variable Z_n in (10.28) behaves like a standard normal random variable as n becomes large. Therefore, *for large enough n*, we have

$$P\left[-z_{\alpha/2} \leq Z_n \leq z_{\alpha/2}\right] \approx 1 - \alpha$$

where we emphasize that the equality represents an approximation which is only valid for large n. Next, recalling the definition of Z_n in (10.28), we get

$$P\left[-z_{\alpha/2} \leq \frac{\hat{\theta}_n - \theta}{\sqrt{\sigma^2/n}} \leq z_{\alpha/2}\right] \approx 1 - \alpha$$

which can be rewritten as

$$P\left[\hat{\theta}_n - z_{\alpha/2}\sqrt{\sigma^2/n} \leq \theta \leq \hat{\theta}_n + z_{\alpha/2}\sqrt{\sigma^2/n}\right] \approx 1 - \alpha \tag{10.31}$$

This defines an interval within which we can (approximately, for large enough n) assert that θ lies with probability $(1 - \alpha)$. This interval is specified by three quantities other than n:

1. The point estimate $\hat{\theta}_n$, which is just the sample mean (10.19).

2. The value of $z_{\alpha/2}$, which can be directly obtained from (10.29) for any given α and is readily available in tabulated form.

3. The variance of the point estimate $\hat{\theta}_n, \sigma^2$, which is unknown.

Fortunately, we can bypass the problem of not knowing σ^2, by replacing it by the sample variance S_n^2 defined in (10.24). It turns out that since S_n^2 does approach σ^2 as $n \to \infty$, the central limit theorem still holds for the random variable

$$T_n = \frac{\hat{\theta}_n - \theta}{\sqrt{S_n^2/n}} \tag{10.32}$$

instead of Z_n. Thus, by making this modification in (10.31), a $(1-\alpha)$ confidence interval for θ is given by

$$P\left[\hat{\theta}_n - z_{\alpha/2}\sqrt{S_n^2/n} \leq \theta \leq \hat{\theta}_n + z_{\alpha/2}\sqrt{S_n^2/n}\right] \approx 1 - \alpha \tag{10.33}$$

In this form, the confidence interval is fully specified by its lower and upper endpoints:

$$L_{n,\alpha} = \hat{\theta}_n - z_{\alpha/2}\sqrt{S_n^2/n} \quad \text{and} \quad U_{n,\alpha} = \hat{\theta}_n + z_{\alpha/2}\sqrt{S_n^2/n}$$

Thus, given a set of data, the confidence interval can be evaluated to give us a better idea of where the true value of θ (which we are trying to estimate) lies. The obvious difficulty here is with the determination of "how large n should be so that it is sufficiently large for the central limit theorem to hold."

To partly overcome the difficulty of deciding how large n should be before the distribution of T_n can be approximated reasonably well by the standard normal distribution, one can proceed as follows. If the iid sequence $X_1, X_2, \ldots, X_n$ were in fact normally distributed, then the distribution of T_n is of a form known as the *Student's t distribution*. One of the parameters of this distribution is the number of collected data $n = 1, 2, \ldots$, which is referred to as the "number of degrees of freedom"; the pdf of T_n has $(n - 1)$ degrees of freedom. Like the standard normal, the t distribution can also be numerically evaluated and tabulated, so that points denoted by $-t_{n-1,\alpha/2}$ and $t_{n-1,\alpha/2}$ can be determined such that

$$P\left[-t_{n-1,\alpha/2} \leq T_n \leq t_{n-1,\alpha/2}\right] = 1 - \alpha$$

for some given α and $n = 2, 3, \ldots$ Thus, we can replace (10.33) by

$$P\left[\hat{\theta}_n - t_{n-1,\alpha/2}\sqrt{S_n^2/n} \leq \theta \leq \hat{\theta}_n + t_{n-1,\alpha/2}\sqrt{S_n^2/n}\right] \approx 1 - \alpha \qquad (10.34)$$

Since the iid sequence $X_1, X_2, \ldots, X_n$ is generally not normally distributed, (10.34) is still an approximation which becomes increasingly better as n becomes larger. It turns out, however, that it is a better approximation than (10.33), and it is the one used in most discrete-event simulation languages discussed in Section 10.4.

In most practical applications, it is customary to seek "95% confidence intervals" for our estimates of θ. The corresponding value of α is then obtained by setting $100(1 - \alpha) = 95$, which yields $\alpha = 0.05$. Most simulation software packages have the built-in capability to provide such confidence intervals for the quantities being estimated, so one does not have to explicitly calculate the endpoints in (10.34) or look up tables for the t distribution. Still, it is important to realize that the information provided by confidence intervals rests on the assumptions that the data form an iid sequence and that n is sufficiently large.

In the next two sections, we will see how we can apply these simple estimation techniques in terminating and non-terminating simulations to obtain estimates of performance measures of interest for DES.

10.7.3 Output Analysis of Terminating Simulations

Suppose we perform a single terminating simulation and collect the data $X_1, X_2, \ldots, X_M$. Recall that in this type of simulation there is a well-defined terminating event that specifies the number of data M. For example, the terminating event could be "stop after exactly M X_i's are collected", in which case M is fixed in advance; or the terminating event could be "stop after the simulation clock reaches a prespecified value t", in which case M is not known in advance and it depends on t and on the characteristics of the system under consideration. In general, therefore, M is a random variable.

Now, except for some special cases, $\{X_1, X_2, \ldots, X_M\}$ is not an iid sequence. For example, suppose that X_k is the system time of the kth customer in a simple single-server queueing system operating in a first-come-first-served manner. Then, X_k and X_{k-1} are generally correlated, as is obvious from the Lindley equation (8.18):

$$X_k = \max\{0, X_{k-1} - Y_k\} + Z_k$$

where Y_k is the kth interarrival time and Z_k is the kth service time. Intuitively, if the $(k - 1)$th customer has a long system time, it is likely that the kth customer also has a long system time, since he has to wait until that customer leaves. Suppose, however, that the data $X_1, X_2, \ldots, X_M$ are used to define some performance measure of interest $L(X_1, X_2, \ldots, X_M)$. Using the example of system times, if we are interested in the average system time over exactly M customers, then

$$L(X_1, X_2, \ldots, X_M) = \frac{1}{M} \sum_{i=1}^{M} X_i$$

which is what we used in (10.1) for the example of Section 10.2.1. Clearly, we can repeat a simulation with the same initial conditions and terminating event as many times as we want, say n times, and collect data $L_1, L_2, \ldots, L_n$. These data are iid, as long as each simulation run is performed independently from each other. This can be accomplished by making sure that each simulation is driven by different random numbers. Therefore, the basic estimation techniques developed in the previous section can be applied to the sequence $\{L_1, L_2, \ldots, L_n\}$, where L_j is a random variable obtained from the jth simulation run.

Let us denote by $X_{i,j}$ the ith random variable observed from the jth simulation run, and let $L_j(X_{1,j}, X_{2,j}, \ldots, X_{M_j,j})$ be some quantity of interest that we obtain from this run, also referred to as a sample function. Observe that the number of data, M_j, is a random variable that generally depends on j and the terminating event that has been defined. Then, $X_{i,j}$ and $X_{k,j}$ are generally not independent for $k \neq i$, but L_j and $L_r, r \neq j$, are independent and identically distributed as long as the jth and rth runs are performed with the same initial conditions and with different random number streams.

A common performance measure of interest is the mean of the distribution characterizing the iid sequence $\{L_1, L_2, \ldots, L_n\}$. Let this mean be θ. We can then estimate θ by using a point estimate of the form (10.19):

$$\hat{\theta}_n = \frac{1}{n} \sum_{j=1}^{n} L_j \qquad (10.35)$$

and a $(1 - \alpha)$ confidence interval of the form (10.34):

$$\left[\hat{\theta}_n - t_{n-1,\alpha/2} \sqrt{S_n^2/n}, \ \hat{\theta}_n + t_{n-1,\alpha/2} \sqrt{S_n^2/n} \right] \qquad (10.36)$$

with the sample variance S_n^2 given by another point estimate of the form (10.24):

$$S_n^2 = \frac{1}{n-1} \sum_{j=1}^{n} (L_j - \hat{\theta}_n)^2 \qquad (10.37)$$

This approach is usually referred to as the method of *independent replications*, since simulation runs are independent from each other and each run is a "'replica" of the system of interest.

Example 10.6

A bank has a single ATM and is interested in estimating the "quality of service" experienced by its customers using the ATM between 9:00 A.M. and 10:00 A.M. every day. In particular, let $D_1, \ldots, D_M$ be the delays (waiting + service) experienced by M customers using the ATM during this time period on a typical day. Then, $(D_1, \ldots, D_M)/M$ is the average delay on a typical day. This is a random variable and the mean of its distribution provides a good measure for the "quality of service."

It is assumed that the queue formed for using the ATM is always empty at 9:00 A.M., which defines a fixed initial condition. Given some information regarding the arrival process between 9:00 A.M. and 10:00 A.M. and the service process at the ATM, we can simulate 1-hour operations of this system and produce n independent replications. In this case, the terminating event for every replication is defined by the clock reaching $t = 1$ hour, assuming that it starts at $t = 0$.

The data collected from the jth replication are $D_{1,j}, \ldots, D_{M_j,j}$, where $D_{i,j}$ is the delay experienced by the ith customer in this replication. Moreover, M_j is the total number of customers that completed service within 1 hour in this replication, which is a random variable. Based on these data, we can define a sample function $L_j(D_{1,j}, \ldots, D_{M_j,j})$ representing the average customer delay in the jth replication:

$$L_j = \frac{1}{M_j} \sum_{i=1}^{M_j} D_{i,j}$$

Now $L_1, L_2, \ldots, L_n$ form an iid sequence of average customer delays for the daily 9:00 A.M. to 10:00 A.M. interval. What we are interested in is the mean of the distribution of these random variables, that is, the expected average delay of customers for this 1-hour interval. If we let this performance measure be D, we obtain a point estimate of $D, \hat{D}_n$, using (10.35):

$$\hat{D}_n = \frac{1}{n} \sum_{j=1}^{n} L_j = \frac{1}{n} \sum_{j=1}^{n} \left(\frac{1}{M_j} \sum_{i=1}^{M_j} D_{i,j} \right)$$

If, in addition, a 95% confidence interval is desired for this estimate, then we can use $\hat{\theta}_n$ in (10.36) and (10.37) with $(1 - \alpha) = 0.95$.

Example 10.7

Consider a manufacturer who is trying to get an estimate of the time it would take to process an order of 100 parts if production were to start under a given set of initial conditions. If a model of the plant is available, then we can perform n independent replications, with each replication starting under the specified initial conditions and ending when 100 parts are produced. Let $T_{i,j}$ denote the time when the ith part is produced in the jth replication, $i = 1, \ldots, 100$. Since we are only interested in the time when the 100th part is done, the obvious sample function is

$$L_j = T_{100,j}$$

and the terminating event is the completion of the 100th part. Then, letting T denote the mean time required to produce 100 parts, a point estimate of T is given by

$$\hat{T}_n = \frac{1}{n}\sum_{j=1}^{n} L_j = \frac{1}{n}\sum_{j=1}^{n} T_{100,j}$$

Example 10.8

Let us return to the ATM system in Example 10.6. We are now interested in getting an idea of the fraction of customers that are delayed for over 1 min. Let R_j be the number of customers in the jth replication whose delay exceeds 1 min. Defining the indicator function

$$\mathbf{1}[D_{i,j} > 1] = \left\{ \begin{array}{ll} 1 & \text{if } D_{i,j} > 1 \\ 0 & \text{otherwise} \end{array} \right.$$

we have

$$R_j = \sum_{i=1}^{M_j} \mathbf{1}[D_{i,j} > 1]$$

We can then define the sample function

$$L_j = \frac{R_j}{M_j} = \frac{1}{M_j}\sum_{i=1}^{M_j} \mathbf{1}[D_{i,j} > 1]$$

which represents the fraction of customers delayed by more that 1 min. in the jth replication. What is interesting in this example is that this performance measure can still be expressed as an average over the data $\mathbf{1}[D_{1,j} > 1]$, $\mathbf{1}[D_{2,j} > 1], \ldots, \mathbf{1}[D_{M,j} > 1]$. Letting B denote the mean of the distribution characterizing the iid sequence $L_1, \ldots, L_n$, i.e., the expected fraction of customers whose delay exceeds 1 min, we can obtain a point estimate similar to (10.35):

$$\hat{B}_n = \frac{1}{n}\sum_{j=1}^{n} L_j = \frac{1}{n}\sum_{j=1}^{n} \frac{R_j}{M_j}$$

The method of independent replications can also be used for performance measures which are not sample means, as defined in (10.19), but rather time averages, as defined in (10.22). This approach is suited to estimating mean queue lengths or server utilizations, as illustrated in the following example.

Example 10.9

In the ATM system of Example 10.6, we are now interested in the mean of the average queue length formed over the daily 9:00 A.M. to 10:00 A.M. interval. We define the sample function

$$L_j = \frac{1}{60} \int_0^{60} X_j(t)dt$$

where $X_j(t)$ is the queue length at time t, and L_j is the time average over 1 hour $= 60$ min (where the simulation clock is set at zero at 9:00 A.M.). Since the queue length changes only when a new customer arrives or when one completes a transaction at the ATM and leaves, it is straightforward to compute L_j with the data collected over the jth simulation run. Letting Q denote the mean of the distribution characterizing the iid sequence $L_1, \ldots, L_n$, that is, the expected time-averaged queue length over a 60 min interval, we can once again obtain a point estimate

$$\hat{Q}_n = \frac{1}{n} \sum_{j=1}^{n} L_j$$

If we are interested in the expected utilization of the ATM over the same interval, we define

$$\mathbf{1}[X_j(t) > 0] = \begin{cases} 1 & \text{if } X_j(t) > 0 \\ 0 & \text{otherwise} \end{cases}$$

and the sample function

$$L_j = \frac{1}{60} \int_0^{60} \mathbf{1}[X_j(t) > 0]dt$$

which represents the fraction of time that the ATM is busy over a time interval $[0, 60]$ min. If U is the expected utilization, We can then obtain the point estimate

$$\hat{U}_n = \frac{1}{n} \sum_{j=1}^{n} L_j$$

10.7.4 Output Analysis of Non-Terminating Simulations

Let $X_1, X_2, \ldots$ be data collected in a non-terminating simulation. In this case, we are interested in estimating parameters of a stationary distribution $F(x) = P[X \leq x]$, where X is a "typical" random variable at steady state. There is of course no guarantee that $F(x)$ exists, but it is often the case that $P[X_k \leq x] \rightarrow P[X \leq x]$ as $k \rightarrow \infty$. We saw, for example, in Section 8.3 of Chapter 8 that the waiting time of a customer in a stable queueing system is expected to behave in this manner, as do several other random variables in such systems. In this section, we will limit ourselves to cases where $F(x)$ exists and try to obtain estimates of performance measures which are expressed as parameters of $F(x)$, usually its mean, which we will denote by θ. Thus, θ may be the mean system time of customers in a queueing system at steady state or the mean queue length at steady state. In general, for a sequence $X_1, X_2, \ldots$,

$$\theta = \lim_{k \to \infty} E[X_k] \tag{10.38}$$

There are a number of techniques that have been proposed for obtaining point and interval estimates of θ. We will limit ourselves here to only three approaches, which are described in the next three sections (for a more extensive discussion of these and other techniques, the reader is referred to Law and Kelton, 1991; or Fishman, 1978).

Replications with Deletions

The main difficulty with estimating steady-state parameters such as θ in (10.38) is that we need to obtain an infinitely long sequence of data $X_1, X_2, \ldots$, a task which is practically infeasible. If we limit ourselves to a finite sequence $X_1, X_2, \ldots, X_m$, then we cannot consider these random variables to be iid: For any finite m, the cdf's of X_i and X_j, $i, j = 1, 2, \ldots, m$, are generally different and not the same as $F(x)$. If we decide to limit ourselves to some "reasonably large" m, then the question is: How do we choose m? On the other hand, since the system becomes more and more representative of its steady-state behavior as m increases, it is reasonable to try and ignore the first part of the sequence $X_1, X_2, \ldots, X_m$, say $X_1, X_2, \ldots, X_r$, for some $r < m$. We can then concentrate on the remaining data $X_{r+1}, X_{r+2}, \ldots, X_m$, which better approximate steady state (see Fig. 10.12). This approach is referred to as *initial data deletion* or as *warming up* the simulation. The key idea is to eliminate the effect of the transient part of the system's behavior as much as possible. We can then proceed with the simple estimation techniques discussed in Section 10.7.2, except that the sample mean (10.19) now becomes a function of both r and m:

$$\hat{\theta}_{m,r} = \frac{1}{m - r} \sum_{i=r+1}^{m} X_i \tag{10.39}$$

Of course, the question now is: How do we choose the length of the warmup interval r? There are a number of techniques that have been suggested for this purpose, which we shall not go into here (the reader is referred to Law and Kelton, 1991).

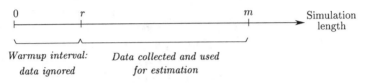

Figure 10.12: Illustrating the initial data deletion approach for non-terminating simulations.

Assuming that a good warmup interval length r and a total simulation run length m have been selected, the most common way to estimate a steady-state performance measure such as the mean θ of a stationary distribution $F(x)$ is once again based on independent replications, as in the previous section. The only difference is that the sample mean obtained from the jth replication is based on (10.39). Thus, the jth sample function is now of the form $L_j(X_{r+1}, X_{r+2}, \ldots, X_m)$, usually the sample mean

$$L_j = \frac{1}{m-r} \sum_{i=r+1}^{m} X_{i,j} \tag{10.40}$$

where $\{X_{r+1,j}, \ldots, X_{r+m,j}\}$ is the data sequence obtained in the jth replication. A point and an interval estimator of θ are given by (10.35) and (10.36) with L_j as in (10.40).

Example 10.10

Consider a $GI/G/1$ queueing system, where we are given an interarrival and a service time distribution. We assume the system is stable, that is, the arrival rate is less than the service rate. Our objective is to estimate the mean customer system time at steady state, denoted by S. We are given, however, a certain simulation budget which only allows us to simulate a total of 100,000 customer departure events. Since we would like to use the method of independent replications, so as to ultimately compute a confidence interval for our estimator, our first decision concerns the number of replications. We choose $n = 10$, which is sometimes sufficiently large for the central limit theorem to give a reasonable approximation. This means that each replication consists of $m = 10,000$ departure events. We now have to decide the length of the warmup period, that is, choose some number r, large enough but considerably smaller than m, which represents the number of customers whose system times will be ignored. Let us choose $r = 1,000$.

Each replication is a simulation of the $GI/G/1$ system starting with an empty queue. Let $S_{i,j}$ be the system time of the ith customer in the jth replication. We then define the sample mean:

$$\hat{S}_j = \frac{1}{9000} \sum_{i=1001}^{10000} S_{i,j}$$

which is a point estimator of S. By repeating this process for all replications $j = 1, \ldots, 10$, we obtain a new point estimator

$$\hat{S} = \frac{1}{10} \sum_{j=1}^{10} \hat{S}_j$$

Now an interval estimate of S can be obtained for some desired confidence $(1 - \alpha)$ by applying (10.34) with $n = 10$.

Suppose that in addition to S we would also like to estimate X, the mean queue length at steady state. One way to proceed is by repeating the "replication with deletions" approach above using a time average over $X_j(t)$, the queue length at time t in the jth replication, after deleting queue length data over some initial warmup period. Alternatively, we can invoke Little's Law in (8.28), $X = \lambda S$, where λ is the arrival rate (which must be given as part of the arrival process specification required to simulate the system). Since Little's Law applies at steady state, we can obtain a point estimate $\hat{X}$ of X, $\hat{X} = \lambda \hat{S}$, where $\hat{S}$ was obtained above. Note that if λ were unknown, we could still estimate it with the techniques of Section 10.7.2, since the arrival process consists of iid interarrival times whose sample mean gives an unbiased and strongly consistent estimate of λ.

Batch Means

The method of batch means is based on a single replication consisting of data $X_1, X_2, \ldots$ which are grouped into *batches*. We can once again use a warmup interval during which no data are collected. Thus, suppose that $X_1, \ldots, X_r$ are the deleted data. Starting with X_{r+1}, we form n batches, each consisting of m data, that is, $\{X_{r+1}, \ldots, X_{r+m}\}$ is the first batch, $\{X_{r+m+1}, \ldots, X_{r+2m}\}$ is the second batch, and so forth. In this way, only a single warmup interval is involved.

The *batch mean* $B_j, j = 1, \ldots, n$ is given by

$$B_j = \frac{1}{m} \sum_{i=r+(j-1)m+1}^{r+jm} X_i \qquad (10.41)$$

and the point estimator of θ is defined to be

$$\hat{\theta}_n = \frac{1}{n} \sum_{j=1}^{n} B_j \qquad (10.42)$$

Proceeding as in the case of replications with deletions, an interval estimator of θ can also be obtained, with batches playing the role of replications.

Regenerative Simulation

This approach is quite different from the two previous ones and leads to different types of point and interval estimators of θ in (10.38). The basic idea is that a stochastic process may be characterized by random points in time when it "regenerates" itself and becomes independent of its past history. Let $R_1, R_2, \ldots$ be such regeneration points in time. Then, $[R_j, R_{j+1}), j = 1, 2, \ldots,$ is called a *regenerative interval*. If such intervals do exist, then the part of the stochastic process defined over one such interval is independent from the part defined over another interval (see Example 10.11).

Consider a simulation of a regenerative stochastic process from which we collect the data $X_1, X_2, \ldots$ Suppose that $\{X_1, \ldots, X_{M_1}\}$ is the portion of the data in the first regenerative interval, $\{X_{M_1+1}, \ldots, X_{M_2}\}$ is the portion of the data in the second regenerative interval, and so on. Then, define

$$L_j = \sum_{i=M_{j-1}+1}^{M_j} X_i \qquad (10.43)$$

with $M_0 = 0$ and $j = 1, 2, \ldots,$ and

$$N_j = M_j - M_{j-1} \qquad (10.44)$$

Observe that N_j is the number of data collected over the jth regenerative interval, but since the length of this interval is random, N_j is a random variable. Because of the regenerative property, $\{L_1, L_2, \ldots\}$ and $\{N_1, N_2, \ldots\}$ are both iid sequences with means $E[L]$ and $E[N]$ respectively. However, L_j is generally not independent of N_j.

A basic property of regenerative processes is that the mean θ defined in (10.38) for the stationary distribution of $\{X_1, X_2, \ldots\}$ is given by

$$\theta = \frac{E[L]}{E[N]} \qquad (10.45)$$

where $E[L]$ is the expected value of data collected over a regenerative interval and $E[N]$ the expected number of data in a regenerative interval, and it is assumed that both quantities are finite. This property is established by considering n regenerative intervals and setting

$$M(n) = \sum_{j=1}^{n} N_j$$

to be the total number of data obtained over all n regenerative intervals. Then, using (10.43), we have

$$\frac{\frac{1}{n} \sum_{j=1}^{n} L_j}{\frac{1}{n} \sum_{j=1}^{n} N_j} = \frac{\sum_{i=1}^{M(n)} X_i}{M(n)}$$

As $n \to \infty$, by the Law of Large Numbers we saw in Theorem 10.1, the numerator of the left-hand side above approaches $E[L]$, and the denominator approaches $E[N]$. Since $M(n) \to \infty$ also as $n \to \infty$, the right-hand side above approaches the mean θ of the iid random variables X_i as defined in (10.38), and (10.45) is obtained.

Based on (10.45), let us now come up with a point estimator of θ using data collected over exactly n regenerative intervals. We denote this estimator by $\hat{\theta}_n$ and define it to be the ratio of two sample means:

$$\hat{\theta}_n = \frac{\frac{1}{n} \sum_{j=1}^{n} L_j}{\frac{1}{n} \sum_{j=1}^{n} N_j} = \frac{\sum_{j=1}^{n} L_j}{\sum_{j=1}^{n} N_j} \qquad (10.46)$$

This is a *ratio estimator* with properties that are quite different from the ones we have seen thus far. In particular, even though the numerator is an unbiased estimator of $E[L]$, and the denominator is an unbiased estimator of $E[N]$, $\hat{\theta}_n$ is *not* an unbiased estimator of θ. It is, however, a strongly consistent estimator of n (as defined in Section 10.7.2), that is, $\hat{\theta}_n \to \theta$ as $n \to \infty$ with probability 1.

Obtaining an interval estimator of θ requires a bit more work, since we now need to estimate the variance of the ratio estimator $\hat{\theta}_n$, which is not as simple as the case of the sample mean estimators we saw in Section 10.7.2. We will not get into this issue (see, however, Law and Kelton, 1991).

Example 10.11 (Regenerative simulation of a $GI/G/1$ system)
Let us return to the $GI/G/1$ queueing system of Example 10.10. Our objective, as before, is to estimate the mean customer system time at steady state, denoted by S. We will attempt to do so using the regenerative approach. For this type of system, the crucial observation is that every arriving customer that finds the system completely empty defines a regeneration point: The behavior of the system over the busy period that follows is independent of the behavior in any other busy period. Thus, for a stable $GI/G/1$ queueing system, a busy period always defines a regenerative interval (see Fig. 10.13).

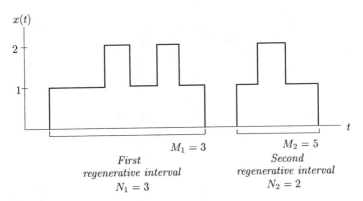

Figure 10.13: Regenerative intervals for a $GI/G/1$ queueing system.
Each busy period defines a regenerative interval. The jth busy period begins with the $(M_{j-1} + 1)$th arrival and contains N_j customers. In this example, three customers are served in the first busy period, and two customers are served in the second busy period.

Let $S_i, i = 1, 2, \ldots$, be the system time of the ith customer in a simulation run. Consider the jth busy period, which starts with the arrival of the $(M_{j-1} + 1)$th customer and consists of $N_j = M_j - M_{j-1}$ customers. The cumulative system time experienced by all N_j customers in this busy period is given by

$$L_j = \sum_{i=M_{j-1}+1}^{M_j} S_i$$

If we observe n busy periods, then an estimator of the mean customer system time at steady state, denoted by $\hat{S}_n$, is obtained from (10.46):

$$\hat{S}_n = \frac{\sum_{j=1}^{n} L_j}{\sum_{j=1}^{n} N_j}$$

Remark. Although it is tempting to extend the idea of regeneration points in the $GI/G/1$ system to all arrivals finding the queue length at some given $x \geq 0$ (not just $x = 0$), this is not true in general. The reason is that a regeneration point must be defined so that the "state" of the system at that point is fixed, where "state" is in the sense discussed in Chapter 1 and considered in the timed automaton model of Section 5.2.3 (what was called the "system state", as opposed to the "process state"). Part of the system state is always the clock value (residual lifetime) of the service completion event corresponding to the customer currently in service. This clock value is a random variable, and it is certainly not fixed at arrival points (unless of course $x = 0$, in which case it is always zero). The only exception is the case of exponentially distributed service times: Because of the memoryless property, the clock values of service completion events seen by arrivals are always identically distributed (the common cdf being the service time distribution). Thus, for a $GI/M/1$ system it is possible to use different definitions of regenerative points.

The main problem with regenerative simulation is that it is difficult to identify regeneration points in practice: Many of the DES of interest are too large and complex for this task. Moreover, even if regeneration points can be identified, the regenerative intervals are usually too long; as a result, a typical simulation run can only include a few such intervals, rendering the accuracy of an estimator of the form (10.46) questionable. As an example, consider a system consisting of N queue-server combinations in series (i.e., a tandem queueing network like the one of Fig. 8.20). In general, the only regeneration points we can identify are defined by external arrivals finding the entire network completely empty, which is a rather rare occurrence. This problem is amplified when the system operates under fairly high utilizations (usually desirable in practice), since the probability of the system becoming empty in this case is very small. In summary, regenerative techniques are attractive for relatively small systems and relatively short regenerative intervals. Otherwise, even if one can define a ratio estimator as in (10.46), the variance of this estimator tends to be large, and hence the estimator tends to be unreliable for most practical purposes.

SUMMARY

- The *event scheduling scheme* (shown in Fig. 10.1) describes a way to construct sample paths of DES for the purpose of simulation. This scheme is consistent with the generation of sample paths of timed automata discussed in Chapter 5. All feasible events at any state are placed in a SCHEDULED EVENT LIST along with their occurrence times in increasing order: The next event is the one with the smallest occurrence time. When a new lifetime is needed for some event, it is obtained from a RANDOM VARIATE GENERATOR, which contains all information regarding the event lifetime distributions.

- An alternative way to simulate DES is the *process-oriented scheme*. This is more suitable for resource-contention environments, where it is convenient to think of "entities" (e.g., customers) undergoing a *process* as they flow through the DES. The behavior of the system is described through several such processes, one for each type of entity of interest.

- In the process-oriented scheme, a process is viewed as a sequence of *functions* triggered by an entity. There are *logic functions*, where the entity triggers instantaneous actions, and *time delay functions*, causing the entity to be held for some period of time, either specified or dependent on the state of the system.

■ Most commercially available discrete-event simulation software packages follow the process-oriented scheme. Examples include GPSS, SIMAN, SIMSCRIPT, and EXTEND.

■ Given probability distributions characterizing the event lifetime processes involved in a DES, an important task of a simulator is to generate samples from these distributions. We refer to these samples as *random variates*. Random variates of any distribution can be obtained by transforming random variates from the uniform distribution over the interval [0, 1], denoted by U[0,1]. Random variates generated from the U[0,1] distribution are called *random numbers*.

■ The most common random number generation technique is based on the *linear congruential* relationship:

$$X_{k+1} = (\alpha X_k + c) \bmod m, \quad k = 0, 1, \ldots,$$

which computes the remainder of the division of $(\alpha X_k + c)$ by m. The quality of the random number sequence generated in this manner largely depends on the choice of the modulus m.

■ The most common way to generate a random variate X from some cdf $F(x)$ given a random number U is based on the *inverse transform* technique: $X = F^{-1}(U)$. Another general-purpose technique employs the *acceptance-rejection* test. Some approaches, such as the convolution and composition techniques, exploit special properties of certain distributions.

■ One of the most important functions of a discrete-event simulator is to collect data based on which *estimates* of various performance measures of interest can be computed.

■ A *non-terminating simulation* is a simulation used to study the steady-state behavior of DES and to estimate parameters of various stationary probability distributions. In a non-terminating simulation there is no obvious or natural point to stop, since, in theory, we are interested in the system's behavior as $t \to \infty$. On the other hand, a *terminating simulation* is one whose run length is naturally defined by a particular event. This event may be defined by a prespecified clock value or by a condition satisfied by the state of the system.

■ If the data collected from simulation form an iid sequence $\{X_1, \ldots, X_n\}$ whose distribution has an unknown mean θ, then the *sample mean* $\hat{\theta}_n =$

$\sum_{i=1}^{n} X_i/n$ is an *unbiased and strongly consistent* point estimator of θ. This means that $E[\hat{\theta}_n] = \theta$ and that $\hat{\theta}_n \to \theta$ with probability 1, as $n \to \infty$ (by the *strong law of large numbers*).

- A $(1 - \alpha)$ *confidence interval* estimate of θ is given by $[L_{n,\alpha}, U_{n,\alpha}]$ where $L_{n,\alpha} = \hat{\theta}_n - z_{\alpha/2}\sqrt{S_n^2/n}$ and $U_{n,\alpha} = \hat{\theta}_n + z_{\alpha/2}\sqrt{S_n^2/n}$. Here, $z_{\alpha/2}$ is defined so that $P[-z_{\alpha/2} \leq Z \leq z_{\alpha/2}] = 1 - \alpha$ for a *standard normal* random variable Z (i.e., having a normal pdf with mean 0 and variance 1). In addition, S_n^2 is the sample variance of $\{X_1, \ldots, X_n\}$, which is given by $\sum_{i=1}^{n}(X_i - \hat{\theta}_n)^2/(n - 1)$.

- For terminating simulations, the method of *independent replications* is the one most often used to obtain point and interval estimates of a performance measure which can be defined as the mean of some distribution.

- For non-terminating simulations, the main problem is to estimate a steady-state expected value of the form $\theta = \lim_{k\to\infty} E[X_k]$. The "replications with deletions" approach is based on obtaining a sample mean over independent simulation runs, but with a certain number of initial data deleted from each simulation to minimize transient effects. Another approach is based on a sample mean over batches of data from a single replication. Finally, if *regenerative points* can be identified in the DES simulated, ratio estimators based on data collected over regenerative intervals of a single replication can also be used.

PROBLEMS

10.1 Write a program to simulate a single-server queueing system (like the one in Section 10.2.1) using the event scheduling scheme. Assume that the system starts out empty and operates on a first-come-first-served basis. Let the interarrival time distribution be uniformly distributed in [1.2, 2.0] and the service time be fixed at 1.0. Simulate the system so as to estimate the average waiting time of the first 10 customers. Then repeat the process for the first 100 and finally 1000 customers and compare your estimates.

To generate random numbers for the interarrival time distribution, use the random number generating function shown below (written in FORTRAN), where the array ISEED is specified through a data statement in your program:

DATA ISEED /1397, 2171, 5171, 7147, 9913/

The function returns a number UNI(I) for some specified integer I. For simplicity, set I=1.

REAL FUNCTION UNI(I)

```
COMMON/SEEDS/ISEED(5)
    m = ISEED(I)/16384
    n = MOD(ISEED(I), 16384)
    l = MOD(13205*m + 74505*n, 16384)
ISEED(I) = INT(AMOD(FLOAT(l*16384 + 13205*n), 268435456.))
    UNI = FLOAT(ISEED(I))/268435456.
RETURN
END
```

10.2 Repeat Problem 10.1 using a process-oriented simulation language such as SIMAN, EXTEND, or SIMSCRIPT. Compare your results with those obtained from Problem 10.1.

10.3 (J. Conway and W. Kreutzer) The "game of life" simulates simple genetic processes driven by births, deaths, and survivals. The game is played on a square board. Each element of the board is itself a square with eight neighbors (except for squares at the edges of the board). Each square is either empty or it contains a member of some species population. The population evolves according to three simple rules: (1) *Birth*: occurs in an empty square which has exactly 3 neighbors, (2) *Death*: occurs when a population member in a square has 4 or more neighbors (and can't stand the overcrowding), (3) *Survival*: occurs when a population member in a square has 2 or 3 neighbors.

Choose dimensions for the board and some appropriate initial conditions for a particular species population. Write a program or use any simulation language at your disposal to simulate the fate of a population for several generations. Note that the game may end by itself if the population becomes extinct.

10.4 (W. Kreutzer) The "dining philosophers" problem is a classic example of resource contention that can lead to deadlocks. In one version of the problem, there are five philosophers seated around a circular table with an infinite supply of noodles within reach of all. There is one fork between each pair of philosophers, but for a philosopher to eat noodles two forks are required, the one on his left and the one on his right. Each philosopher behaves as follows. He spends some time *thinking* (uniformly distributed in [10, 30] min). He then spends some time *sleeping* (uniformly distributed in [10, 180] min). Finally, he spends some time eating, if he can (uniformly distributed in [10, 30] min), and the process repeats.

Write a program or use any simulation language at your disposal to simulate this process for at least 20 hours, making sure to record the average, minimum and maximum waiting time for any philosopher waiting between sleeping and eating. Note that there may be deadlocks, which you have to make sure you avoid. If you cannot avoid them, repeat the simulation several times and estimate the probability of a deadlock if you take no action to avoid one.[1]

10.5 Use any simulation language at your disposal to simulate a manufacturing workstation that operates like a single-server queueing system with finite buffer capacity K (including space at the server), that is, a $GI/G/1/K$ system. Parts are processed on a first-come-first-served basis. If a part arrives and finds the buffer full, it is rejected and lost. Assume that parts arrive according to a Poisson process with rate $\lambda = 1$ part/min, and that processing times are uniformly distributed in [1.0, 1.5] min. In addition, the server (machine) occasionally fails and has to be repaired. Assume that the time between the last repair and the next failure is exponentially distributed with mean 120 min. Repair times are also exponentially distributed with mean 10 min.

Simulate this system starting with an empty workstation and the machine just after a repair. Choose the simulation run length to be such that 1000 parts complete processing. Perform four separate simulations with $K = 2, 3, 10,$ and 50.

(a) Obtain the fraction of rejected parts for each of the four values of K.

(b) For $K = 2$, verify Little's Law. If you think a run length defined by 1000 parts is not long enough to have steady-state conditions, try to increase the run length and comment on the validity of Little's Law as a function of run length.

10.6 Use any simulation language at your disposal to simulate a manufacturing system consisting of three workstations in series, each similar to the one of Problem 10.5. In this case, if a part completing processing at workstation i sees the buffer at workstation $(i+1)$ full, it stays at workstation i causing blocking: No more parts can be processed at this workstation until the part leaves. Assume that parts arrive at workstation 1 according to a Poisson process with rate $\lambda = 1$ part/min. Processing times are defined as follows: At workstation 1 they are uniformly distributed in [1.0, 1.5] min, at workstation 2 they are exponentially distributed with mean 1.5 min., and at workstation 3 they are exponentially distributed with mean 1.2 min. For all three machines, assume that the time between the last repair and the next failure is exponentially distributed with mean 120 min.

---

[1] We considered a two-philospher two-fork version of this problem in Example 2.17 in Chapter 2 and in Example 3.16 in Chapter 3; deadlock avoidance is also discussed in these two examples.

Repair times are exponentially distributed with mean 10 min. Choose buffer capacities to be equal to 5 at all three workstations.

Simulate this system starting with all workstations empty and all machines just after a repair. Choose the simulation run length to be such that 1000 parts complete processing.

(a) Obtain the fraction of rejected parts at workstation 1.

(b) Obtain the average amount of time that each of workstations 2 and 3 spent in a blocked state.

(c) Obtain the flow time for the 1000 parts, that is, the time at the end of the simulation run.

(d) If you are allowed to reallocate the total buffer capacity of 15 in the system, suggest one or more changes that you expect to improve the flow time you obtained in part (c). Then, repeat your simulation to verify (or disprove) your suggestion.

10.7 A computer system has three processors, two fast ones and one slower one. Processors 1 and 2 each take an exponentially distributed amount of time with mean 1 sec to process a job. Processor 3 takes a uniformly distributed amount of time in $[0.5, 2.0]$ sec. to process a job. Jobs arrive according to a Poisson process with rate 2 jobs/min. When a job arrives, it is routed to processor i, $i = 1, 2, 3$, with probability 1/3. If the processor where a job is routed is busy, it joins an infinite-capacity queue at that processor waiting to be processed.

(a) Use any simulation language at your disposal to simulate this system for 1000 total job completions, starting with an empty state. Use your simulation data to estimate the mean system time of customers.

(b) Use the properties of the Poisson process (see Problem 8.7), the analysis of the $M/M/1$ queueing system in Section 8.6.1 of Chapter 8, and the PK formula (8.109) to calculate the average system time at steady state. Compare your result with the simulation result of part (a). If they are not in agreement, try a longer simulation run or a mean system time estimate based on averaging over several independent replications.

10.8 For the system of Problem 10.7, suppose that some alternative routing policies are tried out:

(a) The job routing policy is changed so that the probability of sending a job to each of Processors 1 and 2 is 0.45. Repeat Problem 10.7 and compare your mean system time results from both simulation and analysis. Can you use your analysis to find the optimal routing probability? If so, try to simulate the system under the probabilities you compute.

(b) The job routing policy is changed again so that when jobs arrive they are entered into a single infinite-capacity queue. A job at the head of the queue is sent to the first available processor. If more than one processors are available, first preference is given to processor 1, then 2, and finally 3. Repeat the simulation for this case, and compare your results with those of part (a) and Problem 10.7.

(c) The job routing policy is changed again so that a job is now sent to the processor with the shortest queue at the time of arrival. In case of a tie, first preference is given to processor 1, then 2, and finally 3. Repeat the simulation for this case, and compare your results with those of parts (a) and (b) and Problem 10.7.

10.9 A manufacturing process involves three machines, M_1, M_2, and M_3, and two types of parts, P_1 and P_2. The *part mix*, that is, the ratio of P_1 to P_2 parts arriving at the system, is denoted by m. When P_1 parts arrive, they are always sent to M_1. When P_2 parts arrive, they are sent to M_2 with probability q and to M_3 with probability $(1 - q)$. All parts are inspected upon completion of a machine cycle, and, with probability r, they are immediately reworked at that machine. The process is illustrated in the figure below, in terms of a simple queueing model. All machines operate on a first-come-first-served basis, and all queue capacities are assumed infinite.

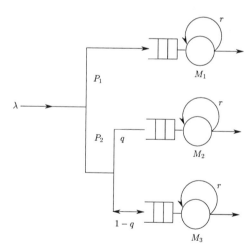

Assume that all parts (either P_1 or P_2) arrive according to a Poisson process of rate λ. Also assume that all service times are exponentially distributed with means s_1, s_2, and s_3 for M_1, M_2, and M_3 respectively. Let the average system time of P_i be denoted by R_i, $i = 1, 2$, and the utilization (fraction of time a machine is not idle) be denoted by $U_i, i =$

1, 2, 3. The parameters of the system are specified as follows: $m = 1, q = 0.5, r = 0.2, 1/\lambda = 1.0$ min, $s_1 = 0.1$ min, $s_2 = 0.4$ min, and $s_3 = 0.4$ min.

(a) Use any simulation language at your disposal to simulate this system in order to estimate R_1, R_2 and U_1, U_2, U_3. The duration of your simulation runs should be determined by the total number of departures observed; let that be denoted by N. Then, obtain results for 7 simulation runs corresponding to $N = 10, 50, 100, 500, 1000, 5000,$ and 10000.

(b) Can you come up with a simple way to get some indication if 10000 defines "long enough" a simulation run to consider the system at steady state? (*Hint*: You know what the average interarrival and service times ought to be.)

(c) Now suppose that all three machines are "flexible", that is, they can process either P_1 or P_2 parts. So we need to establish a routing policy for all arriving parts. Modify the model of part (a) for the following *routing policy*: P_1 parts are sent to the faster machine (M_1) as long as its queue length (including part in process) is less than a threshold value denoted by T; otherwise, send the part to M_2 or M_3, whichever has the shortest queue length at the time (break ties by always going to M_2). P_2 parts are sent to machine M_1, M_2, or M_3 with equal probability.

For $T = 3$ and $N = 10000$, estimate R_1, R_2 and U_1, U_2, U_3.

(d) Suppose your performance objective is to maximize the following function:

$$J = \frac{[0.70 \cdot U_1 + 0.15 \cdot U_2 + 0.15 \cdot U_3]}{[0.6 \cdot R_1 + 0.4 \cdot R_2]}$$

that is, you want to keep your machines as busy as possible, especially M_1, and your average system times as low as possible, especially for P_1. Based on your simulation results, which of (a) and (c) is "best" in this sense ?

10.10 An *assembly* process combines two part types, P_1 and P_2, to produce a third part type P_3. The machine, which performs one assembly operation at a time, is activated only if there are parts of both types present in its input buffer, which is of infinite capacity. P_1 parts arrive according to a Poisson process of rate λ, P_2 parts have an interarrival time which is uniformly distributed in $[a, b]$, and the service time of the machine is a random variable Y given by $Y = c$ with probability p, and $Y = c + X$ with probability $(1 - p)$, where X is an exponentially distributed random variable with parameter m and c is some positive real constant.

Use any simulation language at your disposal to simulate this assembly process for 1000 completed assembly operations, assuming the system starts out empty. Note that the instant P_1 and P_2 parts are combined to initiate a machine cycle, a single new part P_3 may be immediately defined. The parameters of the system are specified as follows: $\lambda = 1$ part/min, $a = 0.5$ min, $b = 1.7$ min, $c = 0.5$ min, $1/\mu = 0.2$ min, $p = 0.4$. Use the simulation data to estimate the utilization of the machine and the mean content of the buffer (including parts of both types).

10.11 A *polling system* is a queueing system where a server moves around to provide service to several queues (e.g., an elevator serving people queueing at floors, a disk controller responding to various user requests, a traveling repairman visiting repair sites). Consider a simple one-server two-queue system as shown in the figure below. In this system, customers arrive at each infinite-capacity queue according to arrival processes A_1 and A_2 respectively. When the server is present, all customers get served (one at a time) according to a process S before the server is allowed to leave, that is, the server leaves only when the queue becomes empty. The server requires T time units to move from Q_1 to Q_2 and vice versa. If the server finds an empty queue when arriving, he immediately leaves and goes to the other queue. Thus, the server is either moving or serving at all times.

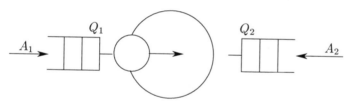

Assume that A_1 is a Poisson process with mean interarrival time τ_1, and A_2 is a deterministic process such that a batch of M customers arrives every τ_2 time units, and $M = i$, $i = 1, 2, 3$, with probability 1/3. In addition, S is a deterministic process with a fixed service time s. Use any simulation language at your disposal to simulate this polling system for 1000 sec with the following initial conditions: The server is positioned at Q_1 and both queues are empty. The parameters of the system are specified as follows: $\tau_1 = 1$ sec, $\tau_2 = 1.5$ sec, $s = 0.1$ sec, $T = 0.2$ sec. Use the simulation data to estimate the utilization of the server (i.e., the fraction of time it is actually serving customers) and the mean waiting time of customers at each of the two queues.

10.12 A common way to exercise flow control in communication networks is known as the *leaky bucket* mechanism and is illustrated in the figure below. Arriving packets form a process $\{A_k\}$, $k = 1, 2, \ldots$, which is fed into a queue of size K (where K may be infinite). The "leaky bucket" is a second queue of size B where "tokens" are stored. The token generation mechanism $\{T_k\}$ is deterministic with period T, that is, every T sec a token is added to the bucket. When a token finds the bucket full, it is simply discarded (the bucket leaks). In order for an arriving packet to proceed, it must find a token in the bucket and remove it. This is the mechanism by which $\{A_k\}$ is converted into a departure process $\{D_k\}$. The goal of this flow control scheme is to decrease the variance of interarrival times (i.e., to "smooth" the incoming traffic) while maintaining the packet queue length (and therefore queueing delays) as small as possible.

Use any simulation language at your disposal to simulate the process of converting $\{A_k\}$ to $\{D_k\}$, assuming that both the packet queue and the bucket are initially empty. Let $\{A_k\}$ be a process that generates packets as follows: With probability 0.6 the next packet interarrival time is fixed at 0.2 sec, and with probability 0.4 the next packet interarrival time is exponentially distributed with mean 1.0 sec. In addition, set $K = 20$, $B = 5$, and $T = 0.3$. If a packet cannot be stored in the queue because there is no more space, it is rejected and lost.

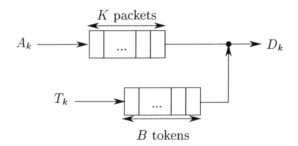

(a) Use simulation data over 1000 departures to estimate the mean packet queue length and the probability of losing packets due to the packet queue capacity being reached.

(b) Use the simulation data to verify that the variance of the interdeparture times is smaller than the variance of the interarrival times.

(c) Vary B and use your simulation model to check whether as B increases the mean packet queue length decreases while the variance of the interdeparture times increases.

10.13 Let N be a geometrically distributed random variable, that is, $P[N = n] = (1 - p)p^n, n = 0, 1, \ldots$, where $0 < p < 1$.

(a) Use the inverse transform technique to describe a procedure for generating random variates from this distribution.

(b) Write a program to implement the procedure of part (a). Use your computer's random number generator or the function given in Problem 10.1.

10.14 Let N be a discrete random variable defined as follows: $N = 0$ with probability 0.2, $N = 1$ with probability 0.4, $N = 2$ with probability 0.1, and $N = $ -1 with probability 0.3.

(a) Use the inverse transform technique to describe a procedure for generating random variates from this distribution.

(b) Write a program to implement the procedure of part (a). Use your computer's random number generator or the function given in Problem 10.1.

10.15 Let X be a gamma-distributed random variable, that is, its pdf is

$$f(x) = \begin{cases} \dfrac{\beta\theta}{\Gamma(\beta)}(\beta\theta x)^{\beta-1}e^{-\beta\theta x} & \text{if } x > 0 \\ 0 & \text{otherwise} \end{cases}$$

where $\Gamma(\beta) = (\beta - 1)!$ if β is an integer.

(a) Use the acceptance-rejection technique to describe a procedure for generating random variates from this distribution.

(b) Write a program to implement the procedure of part (a). Use your computer's random number generator or the function given in Problem 10.1.

10.16 Show that $E[S_n^2] = \sigma^2$ for the sample variance S_n^2 defined in (10.24), where σ^2 is the actual variance of an iid sequence $\{X_1, X_2, \ldots\}$.

10.17 Consider an $M/M/1$ queueing system with $\lambda = 0.6$ and $\mu = 1.0$. Use any simulation language at your disposal or a program as in Problem 10.1 to simulate this system for 1000 customer departures assuming it is initially empty.

(a) Based on the method of independent replications, use 10 simulation runs to obtain a point estimate and a 95% confidence interval for the average system time of these 1000 customers.

(b) Compare your results from part (a) with the steady-state mean system time computed through the analysis of the $M/M/1$ queueing system in Section 8.6.1 of Chapter 8.

(c) Repeat parts (a) and (b) for the server utilization.

10.18 For the $M/M/1$ system of Problem 10.17, we are now interested in estimating the mean system time at steady state. Use the method of replications with deletions, where each of 10 simulation runs consists of 2000 customer departures with the first 200 system times deleted, in order to obtain a point estimate and a 95% confidence interval. Compare your results with the steady-state mean system time computed through the analysis of the $M/M/1$ queueing system in Section 8.6.1 of Chapter 8.

10.19 For the $M/M/1$ system of Problem 10.17, use the regenerative approach to obtain a point estimate of the mean system time at steady state. Use 10 appropriately defined regenerative intervals and compare your results with the steady-state mean system time computed through the analysis of the $M/M/1$ queueing system in Section 8.6.1 of Chapter 8. Increase the number of regenerative intervals to 100 and then 1000 and compare your results.

10.20 For a $GI/G/1$ queueing system, we are interested in estimating various steady-state performance measures using the regenerative approach. We have argued (see Example 10.11) that every arriving customer that finds the system completely empty defines a regeneration point. Are time instants when departures occur leaving a fixed number of customers $x \geq 0$ regeneration points? If so, explain why. If not, why not and are there any special cases where departure instants may define regeneration points?

10.21 In Problem 8.4, a bank was trying to determine the number of ATMs to install in order to keep its revenue loss per hour (due to customers who leave when they do not find an ATM available) below $5. An analytical solution to this problem was possible under the assumption that customers arrive according to a Poisson process with rate 40 customers per hour and that transaction processing times are exponentially distributed with mean 1 min. Now suppose that the bank decides to solve the problem through simulation under a more "realistic" model: Interarrival times are uniformly distributed in $[1.0, 2.0]$ min (which still gives a rate of 40 customers per hour), and transaction processing times have a triangular distribution over $[0.5, 1.5]$ min with the same mean of 1.0 min.

Use any simulation language at your disposal to obtain 95% interval estimates of the expected revenue loss per hour under different numbers of ATM's, and therefore make a final recommendation. If you have already solved Problem 8.4 under Markovian assumptions, compare your results.

10.22 We would like to use simulation to compare two alternative designs for a computer system in terms of mean job response time at steady state. In the first design, we use a single fast processor with rate 2μ, whereas in the second design we use two processors each of rate μ. In both cases, all arriving tasks are placed in a common infinite-capacity queue and are served on a first-come-first-served basis. In the second design, the job at the head of the queue is sent to the first available processor (if both are available, processor 1 is used first).

(a) Suppose the arrival process is Poisson with rate $\lambda = 0.6$ jobs/sec and service times are exponentially distributed with $\mu = 1.0$ jobs/sec. Use any simulation language at your disposal to estimate the mean job response time at steady state and decide which is a better design.

(b) Use the analysis of Sections 8.6.1 and 8.6.2 in Chapter 8 to compare your simulation-based estimates with the analytically obtained mean job response times at steady state.

(c) Repeat part (a) for a case where interarrival times are uniformly distributed in $[0.4, 0.8]$ sec and service times are either equal to 0.5 sec with probability 0.5 or exponentially distributed with mean 1.5 sec with probability 0.5.

10.23 A "mission" is defined as a collection of tasks all of which must be performed before the mission is complete. Some tasks must be performed in sequential order, while others may be performed in parallel. To perform these tasks, we usually have at our disposal a limited number of resources. A common problem is to decide how to allocate resources so as to complete the mission as rapidly as possible.

Consider a mission involving three tasks, T_1, T_2, T_3, such that T_2 must always follow T_1, but T_3 can be performed in parallel with T_1, T_2 (as shown in the figure below). There are five individuals involved in this mission; one of them can only perform T_3, while the remaining four can perform any one task. The objective is to complete the mission in minimum time, where the time to execute task T_i, $i = 1, 2, 3$, is a random variable X_i dependent on the number of individuals working on that task, denoted by n_i. The following information regarding task execution times is available:

- $X_1 \sim U[5.0, 10.0]$ min if $n_1 = 1$, $X_1 \sim U[3.0, 8.0]$ min if $n_1 = 2$, $X_1 \sim U[2.0, 6.0]$ min if $n_1 = 3$, $X_1 \sim U[2.0, 5.0]$ min if $n_1 \geq 4$.
- $X_2 = 3.0$ min for any n_2 with probability 0.6 and $X_2 = (3.0 + Y)$ min with probability 0.4, where $Y \sim U[2.0, 4.0]$ min if $n_2 = 1$ and $Y \sim U[1.0, 3.0]$ min if $n_2 \geq 2$.
- X_3 is an exponentially distributed random variable with mean $\theta = 13.0$ min if $n_3 = 1$, $\theta = 10.0$ min if $n_3 = 2$, $\theta = 9.0$ min if $n_3 = 3$, and $\theta = 8.0$ min if $n_3 \geq 4$.

Consider the following four allocation policies:

P1: Assign four people to T_1 and one person to T_3 (the one who can only perform T_3). If T_1 is completed before T_3, assign three of the four available people to T_2 and the fourth one to T_3.

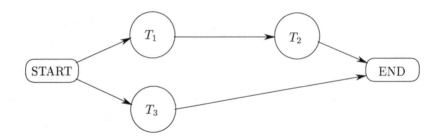

P2: Assign three people to T_1 and two people to T_3 (including the one who can only perform T_3). If T_1 is completed before T_3, assign two of the three available people to T_2 and the third one to T_3.

P3: Assign three people to T_1 and two people to T_3 (including the one who can only perform T_3). If T_1 is completed before T_3, assign two of the three available people to T_3 and the third one to T_2.

P4: Assign two people to T_1 and three people to T_3 (including the one who can only perform T_3). If T_1 is completed before T_3, assign them both to T_2.

In the previous policies, if T_3 is completed before T_1, wait until T_1 is complete and assign all people to T_2 (except the person who can only perform T_3; this person becomes idle). If T_2 is ever completed before T_3, no further action is taken.

(a) Use any simulation language at your disposal to estimate the average mission completion time under each policy, based on 1000 independent replications for each estimate. Obtain 95% confidence intervals and use your results to recommend a policy.

(b) For the best policy found in part (a), estimate the probability that the mission completion time exceeds 10 min.

SELECTED REFERENCES

- Banks, J., and J.S. Carson, *Discrete Event System Simulation*, Prentice-Hall, Englewood Cliffs, NJ, 1984.

- Bratley, P., B.L. Fox, and L.E. Schrage, *A Guide to Simulation*, Springer-Verlag, New York, 1987.

- Fishman, G.S, *Principles of Discrete Event System Simulation*, Wiley, New York, 1978.

- Kiviat, P.J., R. Villanueva, and H.M. Markowitz, *SIMSCRIPT II.5 Programming Language*, (edited by E.C. Russell), CACI Inc., Los Angeles, 1973.

- Knuth, D.E., *The Art of Computer Programming: Vol. 2*, Addison-Wesley, Reading, MA, 1981.

- Kreutzer, W., *System Simulation Programming Styles and Languages*, Addison-Wesley, Reading, MA, 1986.

- Law, A.M., and W.D. Kelton, *Simulation Modeling and Analysis*, McGraw-Hill, New York, 1991.

- Lehmer, D.H., "Mathematical Methods in Large Scale Computing Units," *Annals Comput. Lab. Harvard University*, Vol. 26, pp. 141-146, 1951.

- Pegden, C.D., R.E. Shannon, and R.P. Sadowski, *Introduction to Simulation Using SIMAN*, McGraw-Hill, New York, 1990.

- Pritsker, A.A.B., *The GASP IV Simulation Language*, Wiley, New York, 1974.

- Pritsker, A.A.B., *Introduction to Simulation and SLAM II*, Halsted, New York, 1986.

- Schriber, T.J., *An Introduction to Simulation Using GPSS/H*, Wiley, New York, 1990.

- Zeigler, B.P., *Theory of Modeling and Simulation*, Wiley, New York, 1976.

Chapter 11

Sensitivity Analysis and Concurrent Estimation

11.1 INTRODUCTION

After going through all the previous chapters, it would be natural for readers to conclude that DES are inherently complex and hard to analyze, regardless of the modeling framework adopted. It would also be natural to wonder whether there are any properties at all in DES that we can exploit in our effort to develop mathematical techniques for analysis, design, and control.

In Chapter 1, we saw that systems with event-driven dynamics cannot be modeled through differential (or difference) equations, leaving us with inadequate mathematical tools to analyze their behavior. In Chapters 2 through 5, we developed a number of models, both timed and untimed. Although these models certainly enhance our understanding of DES and give us solid foundations for analysis, one cannot help but notice that they involve rather elaborate notation and sometimes intricate definitions. In fact, the complexity of these models was such that in order to proceed with any manageable analysis of stochastic DES we were forced, in Chapters 7 through 9, to work with the special class of Markov chains. Even within this class, simple formulae in closed form are hard to come by; in Chapter 8, we saw that outside some simple queueing systems at steady state, analysis becomes extremely complicated. This complexity has led to discrete-event simulation as the only universal tool at this point for studying DES, especially stochastic ones. However, as we saw in Chapter 10, simulation has its own limitations. First, it is based on statistical experiments whose validity must always be checked; we saw, for example, that the problem of selecting the length of simulation runs in seeking estimates of steady-state performance measures is far from simple. Second, even with a new generation of very fast computers, simulation remains a slow and costly approach, certainly not suited for real-time applications. A typical problem one faces in the design or control of a system is that of determining how performance is affected as a function of some parameter θ. In the absence of analytical expressions of the form $J(\theta)$, where $J(\theta)$ is a performance measure, one is forced to simulate the system repeatedly, at least once for each value of θ of interest. It is easy to see how tedious a process this becomes for complex systems where θ is in fact a vector of parameters or perhaps a list of alternative designs or control policies to be explored.

Returning to the issue of whether there is anything at all we can exploit in DES to facilitate their analysis, one encouraging possibility lies in the very nature of DES sample paths. In many cases, *observing a sample path under some parameter value θ allows us to predict with relative ease what the sample path would have been like under a different value θ'.* This can be extremely useful in cases where we would like to know the effect of changing θ before actually implementing such a change. This property is due to the inherent nature of DES dynamics, as we will see in this chapter.

There are three types of problems that we wish to consider. First, given a parameter θ and a performance measure $J(\theta)$, we are interested in the *sensitivity* of J with respect to θ, that is, the derivative $dJ/d\theta$. From a control standpoint, if the system is currently operating under $\theta = \theta_0$ and the sensitivity is small, we immediately know that small changes in θ_0 will have little effect on performance. This, in turn, implies that if we are satisfied with $J(\theta_0)$, then it is probably not worthwhile concentrating on fine-tuning the parameter θ; if we are not satisfied, then we know that we should either try large changes in the hope of locating a new value of θ where performance is better, or simply accept the fact that we cannot improve performance by adjusting θ, and probably turn our attention to a different parameter.

The second problem arises when we are interested in the effect of large changes in θ, in which case the derivative $dJ/d\theta$ does not generally provide adequate information. Thus, observing a sample path under θ_0, the question now is: Can we predict what the sample path would have been if θ_0 were changed to different settings $\theta_1, \theta_2, \ldots$?

Finally, there are many situations where θ is a discrete parameter, that is, we can only select a value of θ from a discrete set $\{\theta_1, \theta_2, \ldots, \theta_m\}$. For example, the capacity of a queue can only take on integer values $\{1, 2, \ldots\}$. Once again, we are interested in observing a sample path under one of these values and being able to effectively construct sample paths under all other possible values. Let us think of observing a sample path under θ_1 (e.g., generating a simulation run) as an experiment from which we learn about the behavior of the system under this parameter setting. The techniques we will develop have the property of allowing us to learn about the behavior of the system under all other parameter settings $\{\theta_2, \ldots, \theta_m\}$, without having to perform any additional experiments. Instead, we can infer what the sample paths under these settings would be from direct observations of the one sample path at our disposal (i.e., the original experiment). This justifies the term "rapid learning" that often accompanies such techniques. The ultimate goal of these techniques is to estimate the performance of a DES under $\{\theta_2, \ldots, \theta_m\}$ *concurrently* with that of the system under θ_1; this justifies the term *Concurrent Estimation.*

In this chapter, we will discuss these three problems and present specific techniques that can be used to solve them under various conditions. The basis of these techniques is a theory known as *Perturbation Analysis* (PA), which was developed in the early 1980s and is still a subject of ongoing research. Although the main ideas are generally simple, the material required to formally establish key results in this chapter is more advanced than in previous ones, and will sometimes require a fair amount of mathematical maturity. Readers will find in the next several sections the fundamental principles and concepts on which sensitivity analysis and concurrent estimation techniques for DES are based; they are encouraged to look into some specialized references (for example, Glasserman, 1991; Ho and Cao, 1991; Rubinstein, 1986) to find further details and in-depth proofs.

Finally, because of the variety of approaches and types of problems that come under the heading of this chapter, it is sometimes difficult to keep them all in the proper perspective. Figure 11.44, in the summary section, is an attempt to accomplish this task by presenting a classification of the major techniques, their interconnection, and their applicability to different problems.

11.2 SAMPLE FUNCTIONS AND THEIR DERIVATIVES

Let us consider a parameter θ that generally affects the behavior of a DES. Thus, the state of the DES at time t is written as $X(t, \theta)$ to indicate this dependence on θ. When a sample path of this system is observed, its "performance" is a function of the state history $\{X(t, \theta)\}$. We use the notation $L(\theta, \omega)$ to represent this performance as a function of θ and of ω, a symbol used to denote the particular sample path that was observed (for convenience, we may think of Ω as representing the set of all possible sample paths, and, therefore, $\omega \in \Omega$). For any given θ, since, in general, $\{X(t, \theta)\}$ is a stochastic process, $L(\theta, \omega)$ is a random variable. To emphasize the fact that this performance is sample-path dependent, we refer to $L(\theta, \omega)$ as a *sample performance function* or simply *sample function.*

As an example, consider a single-server queueing system. Let θ be the service rate and $L(\theta, \omega)$ the server utilization observed over a sample path ω. In this case, θ represents the sequences of interarrival and service times that characterize this sample path.

For practical purposes, $L(\theta, \omega)$ gives us an idea of the system's performance, but it only applies to a single sample path; it is entirely possible that the performance observed under a different sample path is drastically different. Thus, what we are interested in is the "average" performance, that is, the expectation over all possible ω:

$$J(\theta) = E[L(\theta, \omega)] \tag{11.1}$$

It is important to keep in mind this distinction between the actual performance measure of interest, $J(\theta)$, and a sample performance function $L(\theta, \omega)$. Recalling our discussion on estimation in Section 10.7.2 of Chapter 10, note that $L(\theta, \omega)$ can also be thought of as one sample in an effort to estimate the true performance $J(\theta)$. If we observe n separate sample paths, denoted by $\omega_1, \dots, \omega_n$, then an estimate of $J(\theta)$ is given by

$$\hat{J}(\theta) = \frac{1}{n} \sum_{i=1}^{n} L(\theta, \omega_i) \tag{11.2}$$

If the sample paths we observe are independent and yield an iid sequence $\{L(\theta, \omega_1), \dots, L(\theta, \omega_n)\}$, then, by the strong law of large numbers (see Section 10.8.2), we get

$$J(\theta) = \lim_{n \to \infty} \left[\frac{1}{n} \sum_{i=1}^{n} L(\theta, \omega_i) \right] \tag{11.3}$$

At this point, it is important to realize that what motivates the material that follows in this chapter is the fact that for many DES *the function $J(\theta)$ is simply unknown.* In Chapter 8, for example, we saw that even under Markovian assumptions for arrival and service processes of queueing systems, there are only limited cases where closed-form expressions for various performance measures can be obtained. This is precisely the reason why we often need to observe several sample paths (e.g., generate simulation runs) from which we can evaluate sample functions $L(\theta, \omega_i)$, $i = 1, 2, \dots$, and hence estimates $\hat{J}(\theta)$.

In practice, what we are interested in is finding a value of θ such that the resulting performance $J(\theta)$ is satisfactory. Ideally, in fact, we would like to determine the value of θ that gives the best possible performance. Since the function $J(\theta)$ is unknown, we usually proceed by trial-and-error, that is, we first set the parameter to some initial value θ_0 and observe a sample path ω_0 in order to get an estimate of $J(\theta_0)$, $\hat{J}(\theta_0) = L(\theta_0, \omega_0)$. We may then change θ_0 to a new value, say θ_1, obtain a new sample path ω_1, and hence a new performance estimate $L(\theta_1, \omega_1)$. The comparison between $L(\theta_0, \omega_0)$ and $L(\theta_1, \omega_1)$ gives us an idea of whether or not (and by how much) we can improve performance by changing θ_0 to θ_1. Obviously, this becomes a very tedious process when the system is complex: The evaluation of $L(\theta_0, \omega_0)$ is time-consuming and there are usually many different parameters to adjust. As a result, this trial-and-error process can take a very long time before even a single action toward improvement can be implemented.

This raises the question: Is there any way in which we could "predict" the effect of a change from θ_0 to θ_1 before actually trying it out? In general, since $J(\theta)$ is unknown and just getting an estimate $\hat{J}(\theta_0)$ at $\theta = \theta_0$ is a challenging task in itself, such predictive capability seems rather hopeless. Yet, as we will see, DES possess some inherent properties one can take advantage of and, in some cases, come up with such "predictive" power.

11.2.1 Performance Sensitivities

As a first step, what would be helpful is if along with $\hat{J}(\theta_0)$, an estimate of $J(\theta)$ at $\theta = \theta_0$, we could also obtain an estimate of the derivative $dJ/d\theta$ evaluated at $\theta = \theta_0$. The derivative $dJ/d\theta$ at $\theta = \theta_0$ is known as the *sensitivity* of J with respect to θ at the point $\theta = \theta_0$. This provides us with at least some local information regarding the effect of θ on performance. For instance, knowing the sign of the derivative $dJ/d\theta$ at $\theta = \theta_0$ immediately tells us the direction in which θ should be changed. Also, if $dJ/d\theta$ is small, we can conclude that J is not very sensitive to changes in θ. Thus, we may concentrate on a different parameter (if possible) for improving performance, rather than work hard at getting minimal improvements from fine-tuning θ.

As already mentioned, a sample path ω (e.g., a simulation run) is used to evaluate $L(\theta, \omega)$, an estimate of $J(\theta)$ for a given value of θ. We can now raise two questions:

1. Can the same sample path be used to evaluate $dL(\theta, \omega)/d\theta$ along with $L(\theta, \omega)$?

   Resolving this issue requires a good understanding of how the parameter θ can affect the order in which events take place, the occurrence times of various events, and the precise way in which changes in event times are transformed into changes in $L(\theta, \omega)$. Based on such understanding, it remains to be seen how $dL(\theta, \omega)/d\theta$ can actually be evaluated from the information contained in the observed sample path.

2. Is $dL(\theta, \omega)/d\theta$ a "good" estimate of $dJ/d\theta$? In the case of the performance measure $J(\theta)$ itself, we use $L(\theta, \omega)$ as an estimate because of (11.1), that is, recalling the discussion on parameter estimation in Section 10.7.2 of Chapter 10, it is an unbiased estimate of $J(\theta)$. The hope then is, by

analogy to (11.1), that

$$\frac{dJ(\theta)}{d\theta} = E\left[\frac{dL(\theta,\omega)}{d\theta}\right] \tag{11.4}$$

However, there is no reason to believe that the equality above holds, or, equivalently, that averaging over all sample derivatives indeed yields the true performance derivative. This may depend on the nature of the function $L(\theta,\omega)$ as well as other factors.

Resolving these issues will be our main concern in the first part of this chapter. We can immediately say a few things, however, regarding the nature of the sample derivative $dL(\theta,\omega)/d\theta$. In particular, suppose we observe a sample path ω that gives rise to $L(\theta,\omega)$, and ask ourselves the question: If θ were perturbed by a small amount $\Delta\theta$ and the same sample path had been generated once again, what would the new result, $L(\theta+\Delta\theta,\omega)$ have been? For example, if the sample path under θ is generated through a simulation experiment, ω corresponds to a specific sequence of random numbers; then, surely we could duplicate the experiment with the exact same sequence of random numbers, but with θ replaced by $\theta+\Delta\theta$. The result of this second experiment is precisely $L(\theta+\Delta\theta,\omega)$. With this interpretation of $L(\theta,\omega)$ and $L(\theta+\Delta\theta,\omega)$, it is reasonable to define the *sample derivative* $dL(\theta,\omega)/d\theta$ in the standard way:

$$\frac{dL(\theta,\omega)}{d\theta} = \lim_{\Delta\theta\to0}\frac{L(\theta+\Delta\theta,\omega)-L(\theta,\omega)}{\Delta\theta} \tag{11.5}$$

Note, however, that this is just a definition of the sample derivative; it gives us no clue as to how we can actually evaluate it for a given θ and ω, since the function $L(\theta,\omega)$ is unknown. The conventional way might be to seek an approximation of it as follows: Obtain $L(\theta,\omega)$ from one sample path generated under θ; consider a "small" change $\Delta\theta$; obtain $L(\theta+\Delta\theta,\omega)$ from a second sample path generated under $\theta+\Delta\theta$; finally, calculate the ratio $[L(\theta+\Delta\theta,\omega)-L(\theta,\omega)]/\Delta\theta$. There are several drawbacks in this approach. First, it is obviously only an approximation of the true derivative $dL(\theta,\omega)/d\theta$. Second, as we make $\Delta\theta$ smaller in an effort to improve the approximation, the difference $[L(\theta+\Delta\theta,\omega)-L(\theta,\omega)]$ also gets smaller, and we end up with the ratio of two small numbers, which leads to a host of numerical problems. And finally, this approach requires the generation of two sample paths, which is time-consuming and contradicts our hope for developing "predictive" power based on a single sample path. Thus, the issue of evaluating $dL(\theta,\omega)/d\theta$ remains open, even though we now have a good idea of what this derivative means based on (11.5).

11.2.2 The Uses of Sensitivity Information

There are a number of areas where sensitivity information, that is, the derivative of a performance measure $J(\theta)$ or some estimate of it, is used for the purpose of analysis and control. In what follows, we single out a few such areas and briefly discuss them.

1. *Local information.* As already pointed out, $dJ/d\theta$ is a good local measure of the effect of θ on performance. We argued earlier, for example, that just knowing the sign of the derivative $dJ/d\theta$ at some point $\theta = \theta_0$ immediately gives us the direction in which θ should be changed. The magnitude of $dJ/d\theta$ also provides useful information in an initial design process: If $dJ/d\theta$ is small, we can conclude that J is not very sensitive to changes in θ and hence concentrate on other parameters that may improve performance. In fact, in the case of multiple parameters $\theta_1, \dots, \theta_m$, the sensitivities $dJ/d\theta_1, \dots, dJ/d\theta_m$ can be used to determine the most and least critical parameters, and guide the designer towards the most promising area for improvement.

   It is, however, important to keep in mind that sensitivities only provide local information regarding the performance measure $J(\theta)$. In Fig. 11.1, for example, the sensitivity at the point $\theta = \theta_0$ shown is a small positive number. One may therefore be led to the conclusion that θ_0 is a good operating point, thus missing the truly optimal point θ^* (and all neighboring points in $[\theta_1, \theta_2]$ which are better that θ_0). Although such situations do arise, it turns out that they are not all that frequent in DES for most performance measures of interest.

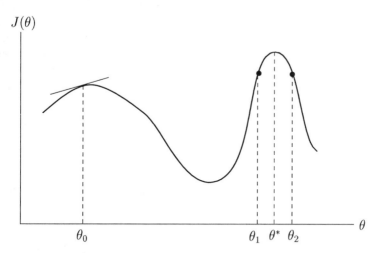

Figure 11.1: An example where sensitivity information near θ_0 misses the optimal point θ^*.

2. *Structural properties.* It is often the case that sensitivity analysis provides not just a numerical value for the sample derivatives in (11.5), but an

expression which captures the nature of the dependence of a performance measure on the parameter θ. The simplest such case arises when $dL/d\theta$ can be seen to be always positive (or always negative) for any sample path; we may not be able to tell what the value of $J(\theta)$ is for some θ, but we can assert nonetheless that $J(\theta)$ is monotonically increasing (or decreasing) in θ. This information in itself is very useful in design and analysis. More generally, the form of $dL/d\theta$ can reveal interesting structural properties of the DES (e.g., monotonicity, convexity). Such properties were exploited, for example, in Chapter 9 in order to determine optimal operating policies for some systems.

3. *Optimization.* The sensitivity $dJ/d\theta$ can be used in conjunction with various optimization algorithms whose function is to gradually adjust θ until a point is reached where $J(\theta)$ is maximized. If no other constraints on θ are imposed, at this point we expect $dJ/d\theta = 0$. A typical such algorithm is described by

$$\theta_{n+1} = \theta_n + \eta_n \left(\frac{dJ}{d\theta} \right)_{\theta=\theta_n} \qquad n = 1, 2, \ldots \qquad (11.6)$$

where the parameter is gradually adjusted from some initial value θ_0 until $dJ/d\theta = 0$ for some θ_n. The amount of adjustment is proportional to the value of the derivative evaluated at $\theta = \theta_n$, with properly selected coefficients η_n, $n = 1, 2, \ldots$ (these coefficients are referred to as *step sizes*, *scaling factors*, or *learning rates*). In our case, the derivative is replaced by an estimate, which makes this optimization process significantly more complicated. Moreover, note that there is no guarantee that the point where $dJ/d\theta = 0$ is indeed optimal. This is illustrated by the example of Fig. 11.1, where $dJ/d\theta = 0$ at three points (one minimum and two maxima), but only θ^* is the true maximum.

4. *Global response curve generation.* Our ultimate goal is often to obtain the function $J(\theta)$, that is, a curve describing how the system responds to different values of θ. Since $J(\theta)$ is unknown, one alternative is to obtain estimates $\hat{J}(\theta)$ of the form (11.2) for as many values of θ as possible. This is clearly a prohibitively difficult task. Derivative information, on the other hand, including not only first, but also higher-order, derivatives $d^2 J/d\theta^2, d^3 J/d\theta^3, \ldots$ can be used to approximate $J(\theta)$. This could be done, for instance, through the well-known Taylor expansion, but also through a number of other schemes. Thus, if such derivative information can be easily and accurately obtained, this task may be accomplished as well.

11.3   PERTURBATION ANALYSIS: SOME KEY IDEAS

Our goal is to develop a framework leading to techniques for evaluating sample derivatives of the form $dL(\theta, \omega)/d\theta$, and for studying their properties as estimates of performance sensitivities $dJ/d\theta$. This is to a large extent the scope of *Perturbation Analysis* (PA). In this section, we present an informal description of the key ideas on which PA is based, before proceeding with a careful development, and returning to the crucial equality in (11.4) to discuss its validity and its limitations.

It is important to emphasize that the ideas on which PA is based are not unique to queueing systems nor do they apply only to stochastic DES. Rather, they are founded on the event-driven nature of sample paths and their intrinsic properties, as the examples used in this section are intended to illustrate.

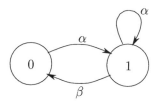

Figure 11.2: A simple DES with event set $E = \{\alpha, \beta\}$ and state space $X = \{0, 1\}$.

Let us consider a simple DES modeled as a timed automaton whose state transition diagram is shown in Fig. 11.2. The event set is $E = \{\alpha, \beta\}$ and the state space is $X = \{0, 1\}$. The event α is always feasible, while β is only feasible at state 1. To complete the specification of the model, as in Chapter 5, let clock sequences $\mathbf{v}_\alpha = \{v_{\alpha,1}, v_{\alpha,2}, \dots\}$ and $\mathbf{v}_\beta = \{v_{\beta,1}, v_{\beta,2}, \dots\}$ be given. For simplicity, suppose that $v_{\beta,k} = T$ for all $k = 1, 2, \dots$, and let T be our parameter of interest. Finally, let us assume that what we would like to control is the total time this system spends at state 0, denoted by $L_0(T)$, which clearly depends on T.

Remark. One may immediately notice that the timed automaton of Fig. 11.2 may be viewed as a $G/D/1/1$ queueing system, that is, a queueing system with fixed service time T and no queueing capacity (other than the customer in service). However, the discussion that follows has nothing to do with this interpretation of the model.

In what follows, we consider a typical sample path of this system and the effects that a perturbation of a particular event time or of a series of event times have on $L_0(T)$. The main observation that will emerge from our discussion is that such effects are easy to predict by simply looking at the sample path at our disposal.

1. A Single "Small" Event Time Perturbation. The top part of Fig. 11.3 shows a typical sample path of the system, where the initial state is 0. The thick line segments correspond to intervals when the system is at state 1, with each such segment being of length T. Note that if an a event occurs while the state is 1, it is simply ignored. We refer to this as the *nominal* sample path of our system.

Now suppose that only the first clock value of the β event is reduced by a "small" amount $\Delta > 0$. This results in a perturbed sample path, as shown in the bottom part of Fig. 11.3. Thus, the first β event, which originally took place at time t_3, will now take place at time $t'_3 = t_3 - \Delta$. Note that in both cases this event is enabled at the same time, t_1. The only difference is that in the nominal path the clock value (lifetime) of β is set to T, whereas in the perturbed path it is set to $(T - \Delta)$. Everything else is identical in both sample paths.

The point to be made here is simple: As long as Δ is "small enough", the only difference between the two sample paths is that the time spent at state 0 is increased by Δ, that is, the change in L_0, denoted by ΔL_0, is

$$\Delta L_0 = \Delta \qquad (11.7)$$

The obvious question is: How small should Δ be? The answer is: Small enough to *prevent an event order change* between α at time t_2 and β at time t_3. As long as this is true, the effect of this change on L_0 is *linear* as indicated by (11.7).

There are two questions that this observation raises. First, what about perturbing T for all clock values $v_{\beta,k}, k = 1, 2, \ldots$, not only $v_{\beta,1}$? And second, what if Δ is not "small enough" to prevent an event order change? We will consider both of these questions next.

2. A "small" perturbation in a parameter. Suppose we now perturb the parameter T by a "small" amount Δ. In other words, *all* clock values of the β event are now reduced by Δ. This results in a *perturbed* sample path, as shown in the bottom part of Fig. 11.4. In this case, every event β occurs Δ time units earlier. Therefore, after K occurrences of the β event, we have

$$\Delta L_0 = K \cdot \Delta, \qquad K = 1, 2, \ldots \qquad (11.8)$$

and the effect of this perturbation on L_0 is once again *linear*.

Once again, the question is: How small should Δ be? The obvious answer is: Small enough to *prevent any event order change* from taking place. In other words, whenever a β event is observed on the nominal path, Δ must be small enough to prevent it from occurring in the perturbed path prior to the α event that precedes it. Here, the difference between perturbing a single event time and a series of event times comes into play. In perturbing $v_{\beta,1}$ only, we can always select Δ to be small enough to prevent an event order change; in Fig. 11.3, we simply make sure that $\Delta < t_3 - t_2$. In perturbing T, on the other hand, we

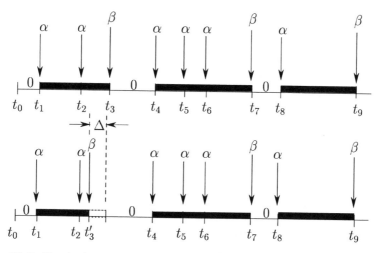

Figure 11.3: Nominal and perturbed sample paths when $v_{\beta,1} = T$ only is perturbed by Δ.

The only effect of this perturbation is to increase the time spent at state 0 by Δ, as long as the order of occurrence of event α at time t_2 and event β at time t_3 remains unaffected.

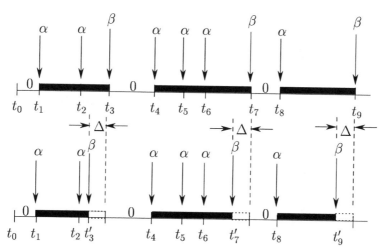

Figure 11.4: Nominal and perturbed sample paths when parameter T is perturbed by Δ.

The effect of this perturbation is to increase the time spent at state 0 by $K \cdot \Delta$, where K is the number of occurrences of the β event. It is assumed that Δ is small enough to prevent any of the K β events from occurring prior to the α event that precedes it.

have to select T so that no event order change ever occurs. We can still do that as long as some finite number K of β event occurrences is specified. However, as $K \to \infty$, sooner or later, for some fixed Δ, there will be a situation where an α event is followed by a β event such that their relative time occurrence is smaller than Δ.

This raises yet another interesting question: What happens as $\Delta \to 0$ *and* $K \to \infty$? This question is at the heart of many of the useful results obtained through PA, since the limiting process $\Delta \to 0$ gives rise to the sensitivities we are interested in. As we will see, for many DES, parameters and performance metrics, it turns out that the effect due to $\Delta \to 0$ "wins out" over the effect due to $K \to \infty$. We leave this issue at that for right now, as we will return to it in subsequent sections. The main point from this example, however, is that for some fixed number of events, K, we can select Δ to be small enough to satisfy the simple linear relationship (11.8). This is already quite an important property of DES sample paths and one that is not commonly found in continuous-variable dynamic systems.

   3. "Large" Perturbations. We now address the question of "large" perturbations, in the sense of causing event order changes with respect to what is observed in the nominal sample path. Thus, as shown in Fig. 11.5, let us reduce the first clock value only of the β event by a "large" amount $\Delta > 0$. This results in the perturbed sample path seen at the bottom part of Fig. 11.5, where the first β event, which originally took place at time t_3, now takes place at time $t_3 - \Delta < t_2$. Thus, in this sample path the first β event precedes the second α event.

   The main consequence of the event order change is the introduction of an interval of length $(t'_3 - t'_2)$ when the state is 0 just after t'_2. In addition, as seen in the perturbed path of Fig. 11.5, subsequent β events are affected. For example, the second β event is now the fifth (instead of seventh) overall event, and it occurs at time $t'_5 < t_7$. At first sight, it appears complicated to predict the effect of this "large" Δ on L_0. A closer look, however, reveals that such a task is not as complicated as it might seem.

   Note that time t_8 in the nominal path corresponds to time t'_9 in the perturbed path, and that beyond this point the two sample paths are once again synchronized, in the sense that Δ can have no further effect on the sample path. As a result, we can focus attention on the interval $[t_0, t_8]$. In the nominal path, the amount of time spent at state 0 in this interval, denoted by $L_0(t_0, t_8)$, is

$$L_0(t_0, t_8) = t_8 - t_0 - 2T$$

since two intervals of length T each when the state is 1 have been observed. Similarly, in the perturbed sample path, the amount of time spent at state 0 in the same interval, denoted by $L'_0(t_0, t_8)$, is

$$L'_0(t_0, t_8) = t_8 - t_0 - 2T - (t'_2 - t_1)$$

since a third interval of length $(t'_2 - t_1)$ when the state is 1 has now been added. Observe, however, that $t'_2 - t_1 = T - \Delta$. Therefore, combining these two equations and recalling that for all $t > t_8$ the two sample paths are identical, we have

$$\Delta L_0 = -(T - \Delta) \tag{11.9}$$

which, interestingly, is less than 0, whereas in the two previous cases it was always greater than 0.

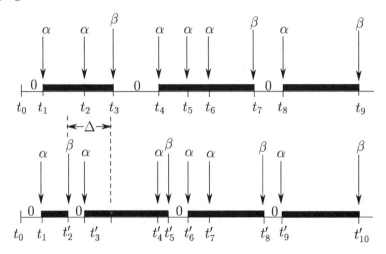

Figure 11.5: Nominal and perturbed sample paths when when $v_{\beta,1} = T$ only is perturbed by some "large" Δ.
The effect of this perturbation is to introduce a fourth interval during which the state is 1 as a result of an event order change.

Therefore, despite the apparent complexity caused by the "large" perturbation, it is still possible to evaluate its effect on L_0 in a simple manner. Although admittedly this is only a simple example, it is true that the same principle applies to many other cases, that is, the effect of a perturbation may not be too difficult to track over an observed sample path of a DES.

The above calculation is based on the assumption that we were able to classify Δ as a "large" perturbation in the first place. This requires us to compare Δ to $(t_3 - t_2)$: If $\Delta > t_3 - t_2$, then we can predict the event order change which is in fact seen in the perturbed path of Fig. 11.5. This is a simple task, however, since both t_2 and t_3 are observable event times on the nominal sample path.

4. Allowable event order changes. Based on our discussion regarding the characterization of a "small" perturbation in the parameter T, one may get the impression that the requirement that no event order change ever occurs is too strong. Actually, it is not necessary to impose such a stringent constraint on the size of Δ. It is not all events that should be prevented from changing their order of occurrence, but only some "critical" ones. Let us illustrate this point by slightly modifying the system of Fig. 11.2 as shown in Fig. 11.6. The event set is the same, but we have now introduced a third state, so that $X = \{0, 1, 2\}$. As before, the event α is always feasible, while β is only feasible at states 1 and 2. The clock sequence $\mathbf{v}_\alpha$ is given, while $v_{\beta,k} = T$ for all $k = 1, 2, \ldots$.

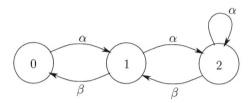

Figure 11.6: A simple DES with event set $E = \{\alpha, \beta\}$ and state space $X = \{0, 1, 2\}$.

Suppose that the first clock value of the β event is reduced by a "large" amount $\Delta > 0$, in the sense that this β event, which originally took place at time t_3, now takes place at time $t_3 - \Delta < t_2$. Thus, in this sample path, the first β event precedes the second α event, as seen in the bottom part of Fig. 11.7.

Let us now evaluate the effect of this perturbation on L_0. First, note that only the interval $[t_1, t_5]$ is affected. In particular, the perturbed sample path includes a new interval $[t_3 - \Delta, t_2]$ when the state is 0, and the interval $[t_4, t_5]$ is increased to $[t_4', t_5]$, as seen in Fig. 11.7. Observe, however, that $t_4' = t_2 + T$, and that $t_4 = t_3 + T$. Therefore, the total change in L_0 is given by

$$\Delta L_0 = [t_2 - (t_3 - \Delta)] + [(t_3 + T) - (t_2 + T)] = \Delta \qquad (11.10)$$

which is the same effect as that of a "small" perturbation in (11.7).

The difference between the "large" perturbation we saw in Fig. 11.5 and the "large" perturbation in this case is that the latter was introduced to an event time occurring at state 2, not state 1. Even though an event order change does result, it is not a "critical" one in the following sense:

- In the nominal path, the event sequence $\{\alpha, \beta\}$ starting at state 1 results in the state sequence $\{1, 2, 1\}$.

- In the perturbed path, the new event sequence $\{\beta, \alpha\}$ starting at state 1 results in the state sequence $\{1, 0, 1\}$.

What is important here is that the final state is the same. This is also illustrated in the diagram of Fig. 11.8. This property of some event order changes preserving the same final state is referred to as the *commuting condition* (as coined by Glasserman; see Glasserman, 1991), and it is a useful and important one. It is particularly useful because it involves purely structural properties of the DES in question, that is, it is independent of the particular clock sequences

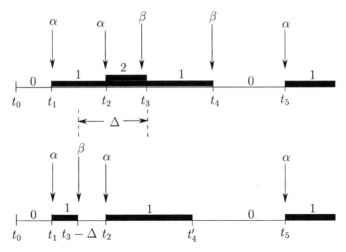

Figure 11.7: Nominal and perturbed sample paths of the DES of Fig. 11.6 when a "large" perturbation in $v_{\beta,1} = T$ is introduced.
Even though this perturbation causes an event order change, this change has no effect on the relationship $\Delta L_0 = \Delta$.

in the sample path, other system parameters, or the precise way in which the perturbation causing the event order change came about.

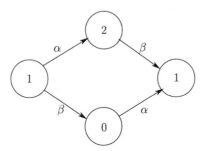

Figure 11.8: The "commuting condition" for the system of Fig. 11.6.
An event order change occurs, but the final state resulting from the interchange of α and β is the same.

It is worthwhile to check that the commuting condition is not satisfied for the system of Fig. 11.2, as we saw in the sample paths of Fig. 11.5. In the nominal path, the event sequence $\{\alpha, \beta\}$ starting at state 1 results in the state sequence $\{1, 1, 0\}$, since the α event at state 1 has no effect; in the perturbed path, the event sequence $\{\beta, \alpha\}$ starting at state 1 results in the state sequence $\{1, 0, 1\}$, leading to a different final state.

11.4 PA OF $GI/G/1$ QUEUEING SYSTEMS

We will now see how the observations made in the previous section can be put to work in the study of DES. We choose to start with the case of $GI/G/1$ queueing systems even though, as already pointed out, the ideas on which PA is based are not limited to queueing systems. Historically, however, the original inception of PA (see Ho et al., 1979) and its first formal developments (Ho and Cassandras, 1983; Ho and Cao, 1983) were motivated by problems involving queueing models of DES. Once the main ideas are presented for the $GI/G/1$ system (see also Suri, 1989), we will generalize them to stochastic timed automata, our basic modeling framework for DES presented in Chapters 5 and 6.

It is important to note that the results we develop will be largely independent of the types of distributions characterizing arrival and service processes. Using the notation established in Chapter 8, let A_k and D_k denote the arrival and departure time respectively of the kth customer, $k = 1, 2, \ldots$, in a single-server queueing system with infinite queueing capacity operating on a first-come-first-served (FCFS) basis. Recall that such a system satisfies the Lindley equation (8.19):

$$D_k = \max\{A_k, D_{k-1}\} + Z_k, \quad k = 1, 2, \ldots \tag{11.11}$$

where Z_k is the kth service time and $D_0 = 0$. In this system (which we will assume to be stable), there are two main parameters of interest: the arrival rate λ and the service rate μ. What we would like to do is investigate the effect of perturbing one of these parameters on the departure times of customers. Let us concentrate on the service rate μ (however, the analysis for perturbations in λ is similar). For simplicity, we set $\theta = 1/\mu$, the mean service time, and consider perturbations $\Delta\theta$.

Now, suppose that we have observed a sample path of this system under θ; we will refer to this as the *nominal* sample path. We would then like to ask the question: What would have happened if instead of θ the same sample path were generated under $\theta' = \theta + \Delta\theta$? Clearly, there would have been a new sequence of departure times, $D_1', D_2', \ldots$ also satisfying the Lindley equation:

$$D_k' = \max\{A_k, D_{k-1}'\} + Z_k', \quad k = 1, 2, \ldots \tag{11.12}$$

We refer to this as the *perturbed* sample path. Here, observe that the arrival times $A_k, k = 1, 2, \ldots$, are the same as in (11.11), since the perturbation in θ does not affect the arrival process. On the other hand, if the parameter θ is perturbed by some amount $\Delta\theta$, we expect all service times $Z_1, Z_2, \ldots$ to be changed by amounts $\Delta Z_1, \Delta Z_2, \ldots$ so that

$$Z_k' = Z_k + \Delta Z_k$$

There are now three questions we need to address:

1. If we observe a service time Z_k for the kth customer and know that the mean θ of the service time distribution has been perturbed by $\Delta\theta$, how can we obtain ΔZ_k? In other words, how does the parameter perturbation $\Delta\theta$ get converted into the sequence of service time perturbations $\Delta Z_1, \Delta Z_2, \ldots$? The process through which this is accomplished is referred to as *perturbation generation*.

2. If we observe the nominal sample path alone and specify some finite perturbation $\Delta\theta$, can we evaluate $D'_k, k = 1, 2, \ldots$, from this information alone?

3. If we observe the nominal sample path alone, can we evaluate the *derivative* of D_k and of other variables or performance measures of interest?

We will provide answers to these questions in the next few sections.

11.4.1   Perturbation Generation

Our first goal is to determine how a perturbation $\Delta\theta$ in the mean of some service time distribution $F(x; \theta)$ affects a sequence of service times $Z_1, Z_2, \ldots$ It should be clear, however, that this discussion applies to any random variables forming an iid sequence with cdf $F(x; \theta)$. There are a few technical assumptions we will make regarding this cdf: (a) $F(x; \theta)$ is continuously differentiable in x and θ, and (b) For any $\Delta\theta > 0$, $F(x; \theta) \geq F(x; \theta + \Delta\theta)$, as illustrated in Fig. 11.9.

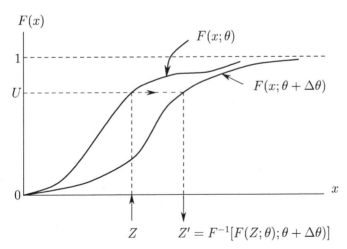

Figure 11.9: Transforming Z into Z' for a given cdf $F(x; \theta)$ and a perturbation $\Delta\theta$. Both Z and Z' are generated by the same underlying random number U. The only difference is due to the fact that $F(x; \theta)$ and $F(x; \theta + \Delta\theta)$ result in different inverse transforms of U.

The simplest way to proceed is to view the service times $Z_1, Z_2, \ldots$, which are observed along the nominal sample path, as random variates generated from $F(x; \theta)$ according to the inverse transform technique discussed in Section 10.6.1 of Chapter 10. Thus, as shown in Fig. 11.9, if Z is such a random variate, we have

$$U = F(Z; \theta) \tag{11.13}$$

for some $U \in [0, 1]$. In other words, there is some random number U, drawn from the uniform cdf over $[0, 1]$, from which Z is obtained through the transformation

$$Z = F^{-1}(U; \theta) \tag{11.14}$$

We now ask the question: *What would the random variate have been if the parameter θ had been replaced by a new value $\theta' = \theta + \Delta\theta$?* We can answer this question by arguing as follows. Since Z is generated through $Z = F^{-1}(U; \theta)$ and the only thing that has changed is θ, the same random number U would have simply resulted in a different random variate Z' given by

$$Z' = F^{-1}(U; \theta + \Delta\theta) = F^{-1}[F(Z; \theta); \theta + \Delta\theta] \tag{11.15}$$

as illustrated in Fig. 11.9. We therefore get (see also Suri, 1983 and 1987):

$$\Delta Z = F^{-1}[F(Z; \theta); \theta + \Delta\theta] - Z \tag{11.16}$$

The importance of this expression is the following: Knowing the cdf $F(x; \theta)$ and the perturbation $\Delta\theta$, and having observed Z, we can evaluate the perturbation ΔZ. Thus, after observing the kth service time on the nominal sample path, we can easily evaluate the kth service time that would have been observed on the perturbed sample path.

Example 11.1

Suppose that service times are exponentially distributed with mean θ, i.e., $F(x; \theta) = 1 - e^{-x/\theta}, x \geq 0$. Thus, setting $U = 1 - e^{-Z/\theta}$ and solving for Z, we get

$$Z = -\theta \ln(1 - U)$$

and, therefore, from (11.16):

$$\Delta Z = -(\theta + \Delta\theta) \ln(1 - U) + \theta \ln(1 - U) = -\Delta\theta \ln(1 - U)$$

Since $\ln(1 - U) = -Z/\theta$ from the previous equation, we finally get

$$\Delta Z = \frac{Z}{\theta} \Delta\theta \tag{11.17}$$

and we see that ΔZ is expressed in terms of the observed random variate Z.

Derivatives of Random Variates

The discussion above allows us to go a step further and define the derivative of Z with respect to the parameter θ, by simply differentiating (11.14), provided of course that the derivative of $F^{-1}(U;\theta)$ in this expression exists. Since determining the explicit form of $F^{-1}(U;\theta)$ may not always be easy, there is a simpler expression we can derive for $dZ/d\theta$ by proceeding as follows.

First, observe that since both Z and $Z' = Z + \Delta Z$ are generated from the same fixed U we have

$$U = F(Z;\theta) = F(Z + \Delta Z; \theta + \Delta\theta) \tag{11.18}$$

Next, since we have assumed that the derivatives of $F(x;\theta)$ with respect to both of its arguments exist, we have

$$
\begin{aligned}
&F(Z + \Delta Z; \theta + \Delta\theta) \\
&= F(Z;\theta) + \left[\frac{\partial F(x;\theta)}{\partial x}\right]_{(Z,\theta)} \Delta Z + \left[\frac{\partial F(x;\theta)}{\partial \theta}\right]_{(Z,\theta)} \Delta\theta + o(\Delta\theta)
\end{aligned}
$$

where $o(\Delta\theta)$ represents all terms of order $(\Delta\theta)^2$ or higher in the expansion (recall that $o(\Delta\theta)$ is defined so that $o(\Delta\theta)/\Delta\theta \to 0$ as $\Delta\theta \to 0$). The notation $[\cdot]_{(Z,\theta)}$ means that the two bracket terms are evaluated at $x = Z$ (the observed service time) and θ (the given mean of the service time cdf). Because of (11.18), this equation reduces to

$$
-\left[\frac{\partial F(x;\theta)}{\partial x}\right]_{(Z,\theta)} \Delta Z = \left[\frac{\partial F(x;\theta)}{\partial \theta}\right]_{(Z,\theta)} \Delta\theta + o(\Delta\theta)
$$

Dividing both sides by $\Delta\theta$ and taking the limit $\Delta\theta \to 0$ yields

$$\frac{dZ}{d\theta} = -\frac{[\partial F(x;\theta)/\partial\theta]_{(Z,\theta)}}{[\partial F(x;\theta)/\partial x]_{(Z,\theta)}} \tag{11.19}$$

which (as some readers may have already noticed) can also be directly obtained from $U = F(Z;\theta)$ in (11.18). This is an important formula, since it enables us to explicitly compute derivatives of random variates (such as service or interarrival times in queueing systems) with respect to a parameter of their cdf. In general, the right-hand side of (11.19) is a function of Z, a quantity which is directly observed on the nominal sample path. This implies that *derivatives of observed service times in the $GI/G/1$ system can be immediately obtained from the service times themselves.* Therefore, it is reasonable to expect that $dL(\theta,\omega)/d\theta$, the derivative of a sample function, can be obtained in terms of $dZ_1/d\theta, dZ_2/d\theta, \ldots$, the derivatives of those observed quantities which are affected by θ. This provides, at least partially, an answer to the first question raised in Section 11.2.1 regarding the feasibility of obtaining $dL(\theta,\omega)/d\theta$ from the observed sample path.

Example 11.2

As in Example 11.1, suppose that service times are exponentially distributed with mean θ, i.e., $F(x; \theta) = 1 - e^{-x/\theta}, x \geq 0$. Thus, we get

$$\frac{\partial F(x; \theta)}{\partial x} = \frac{1}{\theta} e^{-x/\theta}, \qquad \frac{\partial F(x; \theta)}{\partial \theta} = -\frac{x}{\theta^2} e^{-x/\theta}$$

It then follows from (11.19) that

$$\frac{dZ}{d\theta} = \frac{Z}{\theta} \qquad (11.20)$$

Comparing this expression with (11.17), note that we can also write

$$\Delta Z = \frac{dZ}{d\theta} \Delta\theta \qquad (11.21)$$

Scale and Location Parameters

There are two types of parameters which are of particular interest, because they lead to simple expressions in (11.19). First, if $Z \sim F(x; \theta)$, we say that θ is a *scale parameter* of $F(x; \theta)$ if the cdf of Z/θ is independent of θ. This property holds for the mean of the exponential distribution, but also many other useful distribution functions, including the Erlang family (see Section 8.8.1 of Chapter 8). In this case, since the random variable $V = Z/\theta$ has a cdf which is independent of θ, writing $Z = V\theta$ and taking derivatives with respect to θ simply gives

$$\frac{dZ}{d\theta} = \frac{Z}{\theta} \qquad (11.22)$$

which is in agreement with (11.20), since θ is a scale parameter of the exponential distribution of Example 11.2. The second interesting case arises when θ is a *location parameter* of $F(x; \theta)$, which means that if $Z \sim F(x; \theta)$, then the cdf of $(Z - \theta)$ is independent of θ. This property holds for the mean of symmetric distributions such as the normal cdf. Since the random variable $V = Z - \theta$ has a cdf which is independent of θ, writing $Z = V + \theta$ and taking derivatives with respect to θ gives

$$\frac{dZ}{d\theta} = 1 \qquad (11.23)$$

It should be pointed out that distributions with scale or location parameters are often encountered in practice, and can be used to model realistic DES. Thus, (11.22) and (11.23) are very useful in applications of PA. It is also true that if the parameter of interest, θ, is the mean of some *unknown* distribution, then proceeding under the assumption that θ is a scale parameter and using (11.22), since we cannot use (11.19), leads to relatively small errors (see Cassandras et al., 1991).

Parameters of Discrete Distributions

In this case, Z takes on values $v_1, \ldots, v_n$ with corresponding probabilities $p_1, \ldots, p_n$. In general, θ could be a parameter affecting all the values $v_1, \ldots, v_n$ and the probabilities $p_1, \ldots, p_n$. Let us first consider the case where θ affects only one or more of $v_1, \ldots, v_n$. In this case, if we observe $Z = v_i(\theta)$ for some $i = 1, \ldots, n$, then we have $dZ/d\theta = dv_i/d\theta$. For example, consider the deterministic case $Z = \theta$. Clearly, we have $dZ/d\theta = 1$.

If, on the other hand, θ affects one or more of the probabilities $p_1, \ldots, p_n$, things are more complicated. For simplicity, let $n = 2$ and $p_1 = \theta$. Under θ, suppose that a particular U results in v_1 (using the same inverse transform technique as before). If θ changes to $\theta + \Delta\theta$, it is possible that the same U now leads to v_2, that is, the small change $\Delta\theta$ results in a jump from v_1 to v_2 (this is clearly seen in Fig. 10.4). In other words, the inverse transform mapping is no longer continuous in θ. Consequently, sensitivities with respect to these types of parameters cannot be obtained through the analysis that follows. Of course, this does not mean that all hope is lost. There are a number of alternative PA techniques and a variety of tricks one can use to handle these situations, some of which will be discussed later in this chapter (see Section 11.7).

11.4.2   Perturbation Propagation

In the previous section, we addressed the first of the questions raised after equation (11.12). Let us now see how we can address the remaining two questions. Defining

$$\Delta D_k = D'_k - D_k \tag{11.24}$$

we get from equations (11.11) and (11.12):

$$\begin{aligned} \Delta D_k &= \max\{A_k, D'_{k-1}\} - \max\{A_k, D_{k-1}\} + \Delta Z_k = \\ &= \max\{A_k, D_{k-1} + \Delta D_{k-1}\} - \max\{A_k, D_{k-1}\} + \Delta Z_k \end{aligned} \tag{11.25}$$

Thus, we see that the perturbation ΔD_k is due to two sources:

1. The perturbation ΔD_{k-1} that contains the effect of $\Delta\theta$ that has propagated up to the $(k-1)$th customer departure, and

2. The perturbation ΔZ_k generated by the kth customer service.

We have already seen in (11.16) how to evaluate ΔZ_k from a given $\Delta\theta$, so we now concentrate on departure time perturbations propagated through (11.25). There are four cases to be considered.

Case 1.

- Nominal sample path: $A_k \leq D_{k-1}$, i.e., no idle period precedes the kth arrival.

- Perturbed sample path: $A_k \leq D_{k-1} + \Delta D_{k-1}$, i.e., no idle period precedes the kth arrival.

In this case, the perturbation ΔD_{k-1} is either positive or negative, but small enough in magnitude to preserve the order of the two events: kth arrival and $(k-1)$th departure. We then get from (11.25):

$$\Delta D_k = \Delta D_{k-1} + \Delta Z_k \tag{11.26}$$

Case 2.

- Nominal sample path: $A_k > D_{k-1}$, i.e., an idle period of length $(A_k - D_{k-1})$ precedes the kth arrival.
- Perturbed sample path: $A_k > D_{k-1} + \Delta D_{k-1}$, i.e., an idle period of length $(A_k - D_{k-1} - \Delta D_{k-1})$ precedes the kth arrival.

As in Case 1, the perturbation ΔD_{k-1} preserves the order of the two events: $(k-1)$th departure and kth arrival. We then get from (11.25):

$$\Delta D_k = \Delta Z_k \tag{11.27}$$

Observe that the perturbation in D_k here depends only on the current service time perturbation, but no past history.

Case 3.

- Nominal sample path: $A_k \leq D_{k-1}$, i.e., no idle period precedes the kth arrival.
- Perturbed sample path: $A_k > D_{k-1} + \Delta D_{k-1}$, i.e., an idle period of length $(A_k - D_{k-1} - \Delta D_{k-1})$ precedes the kth arrival.

In this case, the perturbation ΔD_{k-1} is negative and sufficiently large to cause the order of the two events: kth arrival and $(k-1)$th departure, to change in the perturbed path. We then get:

$$\Delta D_k = A_k - D_{k-1} + \Delta Z_k \tag{11.28}$$

Case 4.

- Nominal sample path: $A_k > D_{k-1}$, i.e., an idle period of length $(A_k - D_{k-1})$ precedes the kth arrival.
- Perturbed sample path: $A_k \leq D_{k-1} + \Delta D_{k-1}$, i.e., no idle period precedes the kth arrival.

Here, the perturbation ΔD_{k-1} is positive and large enough to cause the order of the two events: $(k-1)$th departure and kth arrival, to change in the perturbed path. We then get

$$\Delta D_k = D_{k-1} + \Delta D_{k-1} - A_k + \Delta Z_k = \Delta D_{k-1} - (A_k - D_{k-1}) + \Delta Z_k \tag{11.29}$$

To illustrate these various cases, let us consider two examples. In the sample path shown in Fig. 11.10, a parameter perturbation $\Delta\theta > 0$ is selected. Therefore, since we have assumed that $F(x;\theta) \geq F(x;\theta+\Delta\theta)$ for $\Delta\theta > 0$, the inverse transform technique in (11.16) implies that every service time is increased by some amount $\Delta Z_k, k = 1, 2, \ldots$ We then see that:

1. $\Delta D_1 = \Delta Z_1$, in accordance with Case 1 above.

2. $\Delta D_2 = \Delta D_1 + \Delta Z_2 = \Delta Z_1 + \Delta Z_2$, again in accordance with Case 1.

3. $\Delta D_3 = \Delta Z_3$, since an idle period of length $(A_3 - D_2)$ precedes the third arrival, and Case 2 applies.

Thus, we see that (a) service time perturbations simply accumulate over a busy period, and (b) when a new busy period starts (e.g., at time A_3), all past is forgotten and a new accumulation of service time perturbations starts. This is true as long as the perturbation ΔD_2 is sufficiently small to preserve the idle period preceding A_3, i.e., prevent any event order change.

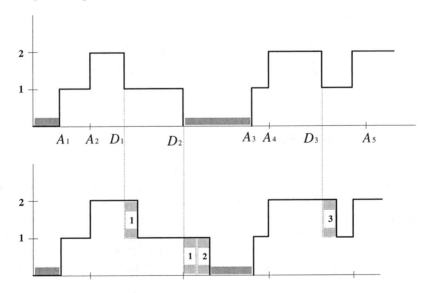

Figure 11.10: Nominal and perturbed sample paths, Cases 1 and 2.

In the sample path shown in Fig. 11.11, on the other hand, we have:

1. $\Delta D_1 = \Delta Z_1$, in accordance with Case 1.

2. $\Delta D_2 = \Delta D_1 + \Delta Z_2 = \Delta Z_1 + \Delta Z_2$, again in accordance with Case 1.

3. $\Delta D_3 = \Delta D_2 + \Delta Z_3 = \Delta Z_1 + \Delta Z_2 + \Delta Z_3$, since Case 1 again applies.

4. $\Delta D_4 = \Delta D_3 - (A_4 - D_3) + \Delta Z_4$, since the nominal path idle period preceding A_4 is eliminated in the perturbed path and Case 4 applies.

In order to give a more meaningful interpretation to the expressions (11.28) and (11.29) arising in Cases 3 and 4, let us set

$$I_k = A_k - D_{k-1} \qquad (11.30)$$

and observe that if $I_k > 0$ it represents the length of an idle period starting with the $(k-1)$th departure and ending with the next, i.e., the kth, arrival. On the other hand, if $I_k < 0$, then $-I_k$ is the waiting time of the kth customer. For instance, in Fig. 11.11 the third customer waits for the period $D_2 - A_3 = -I_3$. Thus, in the case of (11.28) we have

$$\Delta D_k = I_k + \Delta Z_k \qquad (11.31)$$

and in the case of (11.29):

$$\Delta D_k = \Delta D_{k-1} - I_k + \Delta Z_k \qquad (11.32)$$

Thus, combining all four cases, we get:

$$\Delta D_k = \Delta Z_k + \begin{cases} \Delta D_{k-1} & \text{if } I_k \leq 0 \text{ and } \Delta D_{k-1} \geq I_k \\ 0 & \text{if } I_k > 0 \text{ and } \Delta D_{k-1} \leq I_k \\ I_k & \text{if } I_k \leq 0 \text{ and } \Delta D_{k-1} \leq I_k \\ \Delta D_{k-1} - I_k & \text{if } I_k > 0 \text{ and } \Delta D_{k-1} \geq I_k \end{cases} \qquad (11.33)$$

What is important about this equation is that ΔD_k can be evaluated recursively *based on information readily available on the nominal sample path*. In particular, if ΔD_{k-1} is already computed, then the only additional information required in order to evaluate ΔD_k is: (a) I_k, which is either the length of an idle period preceding the kth customer arrival, or the waiting time of the kth customer if no idle period precedes his arrival, and (b) Z_k, the kth service time based on which we can compute ΔZ_k through (11.16). Once these two quantities are obtained, (11.33) can be used to update the departure time perturbation ΔD_k while the nominal sample path evolves. There is no need to generate a separate sample path (e.g., a simulation run) under the new value of θ, (i.e., $\theta + \Delta\theta$)!

Note that ΔD_k in (11.33) consists of two parts. The first part, ΔZ_k, is the *perturbation generation* due to $\Delta\theta$ directly affecting the kth departure time. The second part is due to the *perturbation propagation* from previous departure

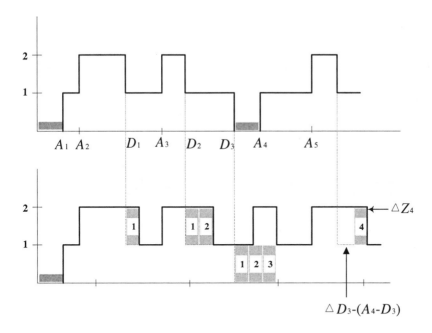

Figure 11.11: Nominal and perturbed sample paths, Cases 1 and 4.

times up to and including the $(k-1)$th one. Finally, note that, since $D'_k = D_k + \Delta D_k$, it is possible to predict the perturbed departure time sequence based on the observed sequence $\{D_k\}, k = 1, 2, \ldots$, in conjunction with (11.33).

Actually, (11.33) can be further simplified once the sign of $\Delta\theta$ is specified. If $\Delta\theta > 0$, then all departure time perturbations are positive (because of assumption (b) in the beginning of Section 11.4.1) and the third case in (11.33) cannot occur, that is, if $I_k \leq 0$ we must have $\Delta D_{k-1} \geq I_k$. Therefore, for $\Delta\theta > 0$:

$$\Delta D_k = \Delta Z_k + \begin{cases} \Delta D_{k-1} & \text{if } I_k \leq 0 \\ 0 & \text{if } I_k > 0 \text{ and } \Delta D_{k-1} \leq I_k \\ \Delta D_{k-1} - I_k & \text{if } I_k > 0 \text{ and } \Delta D_{k-1} \geq I_k \end{cases} \qquad (11.34)$$

A similar simplification takes place when $\Delta\theta < 0$ and the fourth case in (11.33) cannot arise.

At this point, we can either limit ourselves to equation (11.33), certainly a useful mechanism for evaluating departure time perturbations once a specific finite $\Delta\theta$ is selected, or we can ask the natural question: What happens as $\Delta\theta \to 0$, and is it possible to translate the recursive relationship in (11.33) into one which evaluates derivatives of departure times rather than finite differences?

Infinitesimal and Finite Perturbation Analysis

Taking a closer look at (11.33), it is tempting to argue as follows: Let us constrain $\Delta\theta$ to be so "small" that the magnitude of ΔD_k is always such that $\Delta D_{k-1} \leq I_k$ when $I_k > 0$, and $\Delta D_{k-1} \geq I_k$ when $I_k < 0$. Then, (11.33) reduces to

$$\Delta D_k = \Delta Z_k + \begin{cases} \Delta D_{k-1} & \text{if } I_k \leq 0 \\ 0 & \text{otherwise} \end{cases} \qquad (11.35)$$

In other words, all one has to do to keep track of ΔD_k, $k = 1, 2, \ldots$, is to keep on incrementing it by ΔZ_k until an idle period is observed (i.e., $I_k > 0$). At that point, ΔD_k is simply reset to 0. In fact, the value of I_k is irrelevant; it is adequate to detect an idle period, but unnecessary to observe its duration. It is also easy to see that similar simple recursive equations can be derived for other quantities of interest in the $GI/G/1$ system, such as the kth waiting time, W_k, or the kth system time, S_k. For example, since the system time is simply $S_k = D_k - A_k$ and the perturbation $\Delta\theta$ does not affect the arrival process, we have $\Delta S_k = \Delta D_k$ and can write

$$\Delta S_k = \Delta Z_k + \begin{cases} \Delta S_{k-1} & \text{if } I_k \leq 0 \\ 0 & \text{otherwise} \end{cases} \qquad (11.36)$$

In the early development of the PA theory, the term *deterministic similarity* was used (Ho and Cassandras, 1983) to describe nominal and perturbed sample paths satisfying the condition that leads to equation (11.35), i.e., $\Delta D_{k-1} \leq I_k$ when $I_k > 0$, and $\Delta D_{k-1} \geq I_k$ when $I_k < 0$.

The obvious question is: Is it indeed possible to choose such a small $\Delta\theta$ as to satisfy deterministic similarity? Looking at (11.16) we see that we can make ΔZ_k as small as desired by selecting $\Delta\theta$ as small as needed, and therefore force the magnitude of the accumulated perturbations $\Delta D_1, \ldots, \Delta D_{k-1}$ to be less than the magnitude of any specific observed I_k. However, there are several problems with this line of argument. First, no matter how small $\Delta\theta$ is made, sooner or later, as $k \to \infty$, some I_k will be encountered that violates the conditions imposed. And even if this problem were ignored, there is no way to guarantee that a particular choice of $\Delta\theta$ will remain "small enough" over all possible sample paths, hence rendering our analysis over a single sample path practically useless. In short, *deterministic similarity is a condition that cannot be verified* (except for some rather uninteresting situations in view of our effort to estimate $dJ/d\theta$). What we really need in our effort is an approach which is independent of $\Delta\theta$!

Having made this critical point, let us try to draw a distinction between two branches that emerged after the first early work on PA: *Infinitesimal Perturbation Analysis* (IPA) and *Finite Perturbation Analysis* (FPA).

Originally, the term IPA was used to refer to the analysis based on the deterministic similarity condition, e.g., equations of the form (11.35), whereas FPA was used to refer to the more general case, e.g., equation (11.33). Eventually, however, IPA came to refer to the set of techniques that explicitly evaluate *sample derivatives*. As we will see, it often turns out that these techniques can be applied *as if deterministic similarity were valid*, but it is *incorrect* to think that such an assumption is made in the IPA derivative estimation techniques that we will discuss next. In fact, the main advantage of IPA, as will be shown next, is that it is *independent* of $\Delta\theta$.

As for the term FPA, it is currently used to refer to techniques dealing with estimating the effect of finite perturbations $\Delta\theta$, or with estimating derivatives in some complicated situations where IPA cannot be applied so that we resort to $\Delta L/\Delta\theta$ as an approximation of $dL/d\theta$. Although IPA was originally viewed as a special case of FPA (i.e., the case of deterministic similarity), it is again emphasized that IPA explicitly evaluates sample derivatives without any regard to the magnitude of $\Delta\theta$.

In what follows, our analysis will make use of the recursive relationship derived in (11.33). Equations (11.35) and (11.36) cannot be of any use at this point (and will *not* be used), since we have not specified any conditions under which they might hold. It was important, however, to clarify why the step from (11.33) to (11.35) can simply not be taken by blindly invoking an arbitrarily small choice of $\Delta\theta$.

11.4.3   Infinitesimal Perturbation Analysis (IPA)

Let us concentrate on a specific performance measure for the $GI/G/1$ system. In particular, if we observe a sample path consisting of N customer departures (N is fixed and predetermined), we define the mean system time over these N customers as the sample function

$$L_N(\theta, \omega) = \frac{1}{N} \sum_{i=1}^{N} S_i(\theta, \omega) \tag{11.37}$$

where $S_i(\theta, \omega)$ is the system time of the ith customer observed on the sample path ω. The actual performance measure of interest is the mean system time

$$J(\theta) = E[L_N(\theta, \omega)]$$

The sensitivity at some given parameter value θ is the derivative

$$\begin{aligned}
\frac{dJ}{d\theta} &= \lim_{\Delta\theta \to 0} \frac{E[L_N(\theta + \Delta\theta, \omega)] - E[L_N(\theta, \omega)]}{\Delta\theta} \\
&= \lim_{\Delta\theta \to 0} \frac{E[\Delta L_N(\theta, \Delta\theta, \omega)]}{\Delta\theta}
\end{aligned} \tag{11.38}$$

For simplicity, we will drop the argument ω in what follows, but it is always to be understood that the value of $\Delta L_N(\theta, \Delta\theta)$ depends on the nominal sample path observed. We will also limit ourselves to $\Delta\theta > 0$ (the analysis is similar for the case $\Delta\theta < 0$).

Let us assume that the system always starts out empty. Every sample path consists of a series of busy periods separated by idle periods. We will proceed as follows: First, evaluate the change in performance over the first busy period; next, investigate how this change can affect the idle period that follows, and hence customers in the second busy period; then, evaluate the change over the second busy period, and so on.

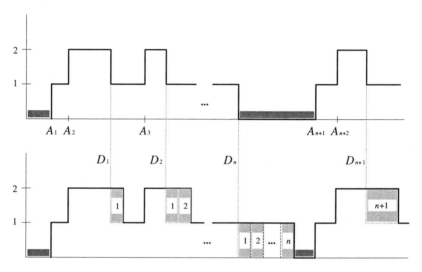

Figure 11.12: Sample path of the $GI/G/1$ system.
In the perturbed path, every service time Z_i is increased by ΔZ_i. These service time perturbations accumulate over the n customers in the first busy period. The last customer's departure time (and hence system time) is perturbed by $\Delta Z_1 + \ldots + \Delta Z_n$.

Let the very first busy period contain n customers, as shown in Fig. 11.12. The rest of the analysis consists of five steps as described next.

Step 1. Evaluate $\Delta L_n(\theta, \Delta\theta)$, the sample function change over the first busy period.

Recall that the system time $S_i(\theta)$ satisfies the recursive equation (8.18):

$$S_i(\theta) = \max\{0, S_{i-1}(\theta) - Y_i\} + Z_i(\theta), \quad i = 1, 2, \ldots \quad (11.39)$$

where Y_i is the interarrival time between the $(i-1)$th and ith customer. In other words, if the ith customer starts a busy period, his system time is limited to his service time $Z_i(\theta)$. Otherwise, he has to wait for an amount given by $[S_{i-1}(\theta) - Y_i]$ before getting served. Note that Y_i is independent of θ, since the mean service time does not affect the arrival process. For consistency in (11.39), set $S_0 = 0$ and $\Delta S_0 = 0$.

The very first customer starts a busy period in both the nominal and the perturbed sample paths. Thus, in the nominal path:

$$S_1(\theta) = Z_1(\theta)$$

and in the perturbed path:

$$S_1'(\theta) = Z_1'(\theta),$$

so that the first customer's system time perturbation is

$$\Delta S_1(\theta) = \Delta Z_1(\theta) > 0$$

since we have taken $\Delta\theta > 0$.

For the second customer, note that $S_1(\theta) = D_1(\theta) - A_1 > Y_2 = A_2 - A_1$ (see also Fig. 11.12), and (11.39) gives for the nominal path:

$$S_2(\theta) = S_1(\theta) - Y_2 + Z_2(\theta)$$

and for the perturbed path:

$$
\begin{aligned}
S_2'(\theta) &= \max\{0, S_1'(\theta) - Y_2\} + Z_2'(\theta) \\
&= \max\{0, S_1(\theta) + \Delta S_1(\theta) - Y_2\} + Z_2'(\theta) \\
&= S_1(\theta) + \Delta S_1(\theta) - Y_2 + Z_2'(\theta)
\end{aligned}
$$

since $\Delta S_1(\theta) > 0$ and $S_1(\theta) - Y_2 > 0$. Thus,

$$\Delta S_2(\theta) = \Delta S_1(\theta) + \Delta Z_2(\theta) = \Delta Z_1(\theta) + \Delta Z_2(\theta)$$

Continuing in a similar way, the perturbation in the ith system time, $i = 1, \ldots, n$, is given by

$$\Delta S_i(\theta) = \sum_{j=1}^{i} \Delta Z_j(\theta) \tag{11.40}$$

Looking at the perturbed path of Fig. 11.12, we can actually immediately see, by inspection, that (11.40) holds, that is, the perturbation in the ith system time is indeed the accumulation of the first i service time perturbations.

Finally, from (11.37), the perturbation in the mean system time of the first n customers is

$$\Delta L_n(\theta) = \frac{1}{n} \sum_{i=1}^{n} \sum_{j=1}^{i} \Delta Z_j(\theta) \tag{11.41}$$

Step 2. Evaluate $\Delta L_{n+m}(\theta, \Delta\theta)$, the sample function change over the first busy period and the first m customers of the second busy period.

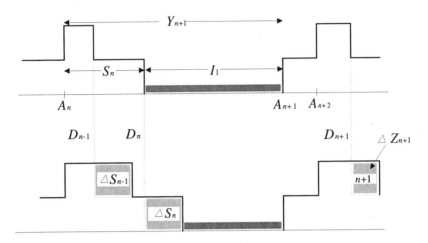

Figure 11.13: $\Delta S_n < I_1$ case.
*The accumulated perturbation over the first busy period is **not** propagated to the second busy period.*

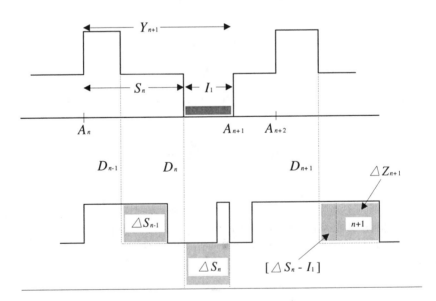

Figure 11.14: $\Delta S_n > I_1$ case.
The accumulated perturbation over the first busy period is propagated causing an additional delay of magnitude $(\Delta S_n - I_1)$ to all system times.

Let the length of the idle period following the first busy period be I_1. First, observe that $Y_{n+1} = A_{n+1} - A_n, S_n = D_n - A_n$, and $I_1 = A_{n+1} - D_n$. Therefore,

$$I_1 = Y_{n+1} - S_n \tag{11.42}$$

as clearly seen in Fig. 11.13 or 11.14.

Next, we consider the $(n + 1)$th customer, who initiates the second busy period. In the nominal path:

$$S_{n+1}(\theta) = Z_{n+1}(\theta)$$

and in the perturbed path:

$$\begin{aligned} S'_{n+1}(\theta) &= \max\{0, S'_n - Y_{n+1}\} + Z'_{n+1}(\theta) \\ &= \max\{0, S_n(\theta) + \Delta S_n(\theta) - Y_{n+1}\} + Z'_{n+1}(\theta) \end{aligned}$$

Thus, using (11.42), the system time perturbation of this customer is

$$\Delta S_{n+1}(\theta) = \max\{0, \Delta S_n(\theta) - I_1\} + \Delta Z_{n+1}(\theta) \tag{11.43}$$

Let us adopt the convenient notation

$$[x]^+ = \max\{x, 0\}$$

which was also used in Chapter 9. Thus, (11.43) is rewritten as

$$\Delta S_{n+1}(\theta) = [\Delta S_n(\theta) - I_1]^+ + \Delta Z_{n+1}(\theta) \tag{11.44}$$

The interpretation here is the following: If the accumulated perturbation $\Delta S_n(\theta)$ is small enough to satisfy $\Delta S_n(\theta) \leq I_1$, then the $(n + 1)$th customer is unaffected by it, and we have $\Delta S_{n+1}(\theta) = \Delta Z_{n+1}(\theta)$, as illustrated in Fig. 11.13. If, on the other hand, $\Delta S_n(\theta) > I_1$, the first and second busy periods coalesce in the perturbed path (see Fig. 11.14) and the system time perturbation of the $(n + 1)$th customer becomes $\Delta S_{n+1}(\theta) = \Delta S_n(\theta) - I_1 + \Delta Z_{n+1}(\theta)$.

For the $(n + 2)$th customer, in the nominal path:

$$S_{n+2}(\theta) = S_{n+1}(\theta) - Y_{n+2} + Z_{n+2}(\theta)$$

and in the perturbed path:

$$\begin{aligned} S'_{n+2}(\theta) &= \max\{0, S'_{n+1} - Y_{n+2}\} + Z'_{n+2}(\theta) \\ &= \max\{0, S_{n+1}(\theta) + \Delta S_{n+1}(\theta) - Y_{n+2}\} + Z'_{n+2}(\theta) \\ &= S_{n+1}(\theta) + \Delta S_{n+1}(\theta) - Y_{n+2} + Z'_{n+2}(\theta) \end{aligned}$$

since $\Delta S_{n+1}(\theta) > 0$, which is clear from (11.44), and $S_{n+1}(\theta) - Y_{n+2} > 0$. Thus, from the preceding two equations and (11.44):

$$\begin{aligned} \Delta S_{n+2}(\theta) &= \Delta S_{n+1}(\theta) + \Delta Z_{n+2}(\theta) \\ &= [\Delta S_n(\theta) - I_1]^+ + \Delta Z_{n+1}(\theta) + \Delta Z_{n+2}(\theta) \end{aligned}$$

Continuing in similar fashion, the perturbation in the $(n + i)$th system time, $i = 1, \ldots, m$, is given by

$$\Delta S_{n+i}(\theta) = [\Delta S_n(\theta) - I_1]^+ + \sum_{j=1}^{i} \Delta Z_{n+j}(\theta) \qquad (11.45)$$

In other words, a portion of the total perturbation, $\Delta S_n(\theta)$, accumulated over the first busy period, propagates to the second busy period only if $\Delta S_n(\theta) > I_1$. Otherwise, there is no such propagation.

We can now combine (11.40), which applies for all customers $i = 1, \ldots, n$, and (11.45) to evaluate the perturbation in the mean system time of the first $(n + m)$ customers from (11.37):

$$\begin{aligned}
\Delta L_{n+m}(\theta) = \frac{1}{n+m} &\left\{ \sum_{i=1}^{n} \sum_{j=1}^{i} \Delta Z_j(\theta) \right. \\
&\left. + \sum_{i=n+1}^{n+m} \left[[\Delta S_n(\theta) - I_1]^+ + \sum_{j=n+1}^{i} \Delta Z_j(\theta) \right] \right\}
\end{aligned} \qquad (11.46)$$

Note that if it were not for the terms contributing $[\Delta S_n(\theta) - I_1]^+$, this expression would be greatly simplified. As we will see, $E[[\Delta S_n(\theta) - I_1]^+]$ is a quantity whose effect approaches 0 as $\Delta\theta \to 0$. This fact will then be exploited when evaluating the limit in (11.38). This will be shown in **Step 4**. First, in **Step 3**, we will extend (11.46) to multiple busy periods.

Step 3. Repeat **Step 2** for subsequent busy periods and generalize (11.46).

Equation (11.46) was derived for a total of $n + m$ customers, the entire n customers forming the first busy period, and the first m customers of the second busy period. Now let us define

$$n_b = \text{index of last customer in } b\text{th busy period}, b = 1, 2, \ldots$$

and agree to set $n_0 = 0$. Thus, with $n_1 = n$ and $n_2 = n + m$, (11.46) can be rewritten as

$$\begin{aligned}
\Delta L_{n_2}(\theta) = \frac{1}{n_2} &\left\{ \sum_{b=1}^{2} \sum_{i=n_{b-1}+1}^{n_b} \sum_{j=n_{b-1}+1}^{i} \Delta Z_j(\theta) \right\} \\
&+ \frac{1}{n_2} \sum_{i=n_1+1}^{n_2} [\Delta S_{n_1}(\theta) - I_1]^+
\end{aligned} \qquad (11.47)$$

The index $(n_{b-1} + 1)$ represents the first customer in the bth busy period, $b = 1, 2$. Thus, the innermost sum in the triple summation accumulates service time perturbations over the first $(i - n_{b-1})$ customers in the bth busy period. The middle sum accumulates system time perturbations over all $(n_b - n_{b-1})$ customers in the bth busy period. The outer sum accumulates perturbations over multiple busy periods. As for the second part of (11.47), it represents the effect that perturbations in the first busy period may have on customers of the second busy period. As we have already seen, this effect is 0 if $\Delta S_{n_1}(\theta) \leq I_1$. Observing that

$$\sum_{i=n_1+1}^{n_2} [\Delta S_{n_1}(\theta) - I_1]^+ = (n_2 - n_1)[\Delta S_{n_1}(\theta) - I_1]^+$$

let us define

$$N_b = n_b - n_{b-1} = \text{number of customers in } b\text{th busy period, } b = 1, 2, \ldots$$

which allows us to rewrite (11.47) as follows:

$$\Delta L_{n_2}(\theta) = \frac{1}{n_2}\left\{ \sum_{b=1}^{2} \sum_{i=n_{b-1}+1}^{n_b} \sum_{j=n_{b-1}+1}^{i} \Delta Z_j(\theta) \right\} + \frac{1}{n_2} N_2 \cdot [\Delta S_{n_1}(\theta) - I_1]^+ \tag{11.48}$$

Adopting this notation, the total system time perturbation accumulated after the first busy period is obtained from (11.40) by setting $i = n_1$:

$$\Delta S_{n_1}(\theta) = \sum_{j=1}^{n_1} \Delta Z_j(\theta) \tag{11.49}$$

Therefore, the system time perturbation of customers in the second busy period, given in (11.45), is written as

$$\Delta S_{n_1+i}(\theta) = [\Delta S_{n_1}(\theta) - I_1]^+ + \sum_{j=n_1+1}^{i} \Delta Z_j(\theta) \tag{11.50}$$

We now move on to the third busy period, which is preceded by an idle period of length I_2. If $[\Delta S_{n_1}(\theta) - I_1]^+ = 0$, then system time perturbations in the third busy period are given by an expression similar to (11.50), that is,

$$\Delta S_{n_2+i}(\theta) = [\Delta S_{n_2}(\theta) - I_2]^+ + \sum_{j=n_2+1}^{i} \Delta Z_j(\theta)$$

On the other hand, if $[\Delta S_{n_1}(\theta) - I_1]^+ = \Delta S_{n_1}(\theta) - I_1 > 0$, then the system times of all customers in the second busy period are perturbed by $(\Delta S_{n_1}(\theta) - I_1)$. In this case, the total system time perturbation at the end of the second period is $(\Delta S_{n_1}(\theta) - I_1) + \Delta S_{n_2}(\theta)$ instead of just $\Delta S_{n_2}(\theta)$, that is,

$$\Delta S_{n_2+i}(\theta) = [\Delta S_{n_1}(\theta) - I_1 + \Delta S_{n_2}(\theta) - I_2]^+ + \sum_{j=n_2+1}^{i} \Delta Z_j(\theta)$$

Combining these two cases, we get

$$\Delta S_{n_2+i}(\theta) = \left[[\Delta S_{n_1}(\theta) - I_1]^+ + \Delta S_{n_2}(\theta) - I_2 \right]^+ + \sum_{j=n_2+1}^{i} \Delta Z_j(\theta)$$

Thus, returning to (11.48), the extension of $\Delta L_{n_2}(\theta)$ to $\Delta L_{n_3}(\theta)$ becomes

$$\Delta L_{n_3}(\theta) = \frac{1}{n_3} \left\{ \sum_{b=1}^{3} \sum_{i=n_{b-1}+1}^{n_b} \sum_{j=n_{b-1}+1}^{i} \Delta Z_j(\theta) \right\}$$
$$+ \frac{1}{n_3} \left\{ N_2 \cdot [\Delta S_{n_1}(\theta) - I_1]^+ \right. \tag{11.51}$$
$$\left. + N_3 \cdot \left[[\Delta S_{n_1}(\theta) - I_1]^+ + \Delta S_{n_2}(\theta) - I_2 \right]^+ \right\}$$

Let R_b be the term that captures the effect of past busy periods that coalesce on customers in the $(b+1)$th busy period, $b = 1, 2, \ldots$, that is, define

$$R_b = [R_{b-1} + \Delta S_{n_b}(\theta) - I_b]^+ \tag{11.52}$$

where $R_0 = 0$. Thus, for B busy periods, (11.51) becomes

$$\Delta L_{n_B}(\theta) = \frac{1}{n_B} \left\{ \sum_{b=1}^{B} \sum_{i=n_{b-1}+1}^{n_b} \sum_{j=n_{b-1}+1}^{i} \Delta Z_j(\theta) \right\} + \frac{1}{n_B} \sum_{b=1}^{B} N_b R_{b-1} \tag{11.53}$$

Now recall that our analysis is based on a sample path containing a fixed number of customers N. Let B be the number of busy periods such that $n_B \leq N < n_{B+1}$. In other words, the Nth customer either coincides with the last customer of the Bth busy period or it belongs to the $(B+1)$th busy period. Thus, (11.53) gives

$$\Delta L_N(\theta) = \frac{1}{N} \left\{ \sum_{b=1}^{B} \sum_{i=n_{b-1}+1}^{n_b} \sum_{j=n_{b-1}+1}^{i} \Delta Z_j(\theta) + \sum_{i=n_B+1}^{N} \sum_{j=n_B+1}^{i} \Delta Z_j(\theta) \right\}$$
$$+ \frac{1}{N} \sum_{b=1}^{B} N_b R_{b-1} + \frac{1}{N}(N - n_B)R_B$$

$$\tag{11.54}$$

Observe that N is a fixed number, whereas B depends on the particular sample path under consideration. Also, note that the second and fourth term in (11.54) correspond to the last $(N-n_B)$ customers belonging to the $(B+1)$th busy period (these terms are 0 if $n_B = N$).

Step 4. Evaluate $E[R_b]$, with R_b defined in (11.52); that is, evaluate the expected effect of the perturbation accumulated over the bth busy period on customers of the $(b+1)$th busy period. Then, show that $E[R_b]/\Delta\theta \to 0$ as $\Delta\theta \to 0$.

Let us concentrate on $b = 1$. The analysis readily extends to all $b > 1$. Thus, looking at (11.48), we see that $\Delta S_{n_1}(\theta)$ is a function of $\Delta Z_j, j = 1, \ldots, n_1$. From (11.16), on the other hand, each ΔZ_j is a function of Z_j, an observed quantity. Thus, with all $Z_j, j = 1, \ldots, n$, observed along the nominal path, $\Delta S_{n_1}(\theta)$ is known. With this in mind, we first rewrite $E[R_1]$ as follows:

$$E[R_1] = E\left[[\Delta S_{n_1}(\theta) - I_1]^+\right] = EE\left[[\Delta S_{n_1}(\theta) - I_1]^+ \mid \Delta S_{n_1}(\theta)\right] \quad (11.55)$$

with the inner expectation obtained by conditioning on $\Delta S_{n_1}(\theta)$. Let $f(x)$ be the pdf of the random variable I_1 (conditioned on all observed $Z_j, j = 1, \ldots, n_1$). Then, considering the inner (conditional) expectation in (11.55), we have

$$E\left[[\Delta S_{n_1}(\theta) - I_1]^+ \mid \Delta S_{n_1}(\theta)\right] = \int_0^{\Delta S_{n_1}(\theta)} [\Delta S_{n_1}(\theta) - x]^+ f(x)dx$$

where we note that $[\Delta S_{n_1}(\theta)-x]^+ = 0$ for all values of x such that $x > \Delta S_{n_1}(\theta)$; therefore, the integral is non-zero only over the interval $[0, \Delta S_{n_1}(\theta)]$.

Let us now assume that $f(x) \le C < \infty$, that is, the pdf $f(x)$ is bounded by some positive constant C. Moreover, note that

$$0 \le [\Delta S_{n_1}(\theta) - x]^+ \le [\Delta S_{n_1}(\theta)]^+ = \Delta S_{n_1}(\theta)$$

where $\Delta S_{n_1}(\theta) > 0$ from (11.49). It follows that

$$\begin{aligned}
E\left[[\Delta S_{n_1}(\theta) - I_1]^+ \mid \Delta S_{n_1}(\theta)\right] &= \int_0^{\Delta S_{n_1}(\theta)} [\Delta S_{n_1}(\theta) - x]^+ f(x)dx \\
&\le \int_0^{\Delta S_{n_1}(\theta)} C \cdot \Delta S_{n_1}(\theta)dx = C\left((\Delta S_{n_1}(\theta))^2\right)
\end{aligned}$$

and returning to (11.55), we get

$$E\left[[\Delta S_{n_1}(\theta) - I_1]^+\right] \le C \cdot E\left[(\Delta S_{n_1}(\theta))^2\right] \quad (11.56)$$

It follows that

$$0 \le \frac{E\left[[\Delta S_{n_1}(\theta) - I_1]^+\right]}{\Delta\theta} \le C \cdot E\left[\frac{(\Delta S_{n_1}(\theta))^2}{\Delta\theta}\right] = C \cdot E\left[\left(\frac{\Delta S_{n_1}(\theta)}{\Delta\theta}\right)^2\right]\Delta\theta$$

and, recalling (11.49), we get

$$0 \le \frac{E\left[[\Delta S_{n_1}(\theta) - I_1]^+\right]}{\Delta\theta} \le C \cdot E \left(\sum_{j=1}^{n_1} \frac{\Delta Z_j}{\Delta\theta}\right)^2 \Delta\theta$$

At this point, we must remember that the number of customers in any busy period depends on the sample path. Since, however, our sample path contains a fixed finite number of customers N, we can provide an upper bound to the sum in this expression as follows:

$$0 \le \frac{E\left[[\Delta S_{n_1}(\theta) - I_1]^+\right]}{\Delta\theta} \le C \cdot E \left(\sum_{j=1}^{n_1} \frac{\Delta Z_j}{\Delta\theta}\right)^2 \Delta\theta \le C \cdot E \left(\sum_{j=1}^{N} \frac{\Delta Z_j}{\Delta\theta}\right)^2 \Delta\theta$$

where the number of terms in the summation is now fixed at N. Next, let us take the limit as $\Delta\theta \to 0$. Under the technical conditions that we have imposed on the service time distribution $F(\cdot; \theta)$ in Section 11.4.1 (essentially, we must require that $|\Delta Z_j|$ be bounded for all j), we can interchange the limit and the expectation in the preceding inequality. Thus, we finally get

$$0 \le \lim_{\Delta\theta \to 0} \frac{E\left[[\Delta S_{n_1}(\theta) - I_1]^+\right]}{\Delta\theta} \le CE \left(\sum_{j=1}^{N} \frac{dZ_j}{d\theta}\right)^2 \left(\lim_{\Delta\theta \to 0} \Delta\theta\right) = 0$$

Recalling (11.55), this establishes the fact that

$$\lim_{\Delta\theta \to 0} \frac{E[R_1]}{\Delta\theta} = 0 \tag{11.57}$$

Another way of seeing this is to notice in (11.56) that $\Delta S_{n_1}(\theta)$ is an expression ultimately involving a sum of $\Delta\theta$ terms, from (11.49) and (11.16), thus its square is $o(\Delta\theta)$. Therefore, $E\left[[\Delta S_{n_1}(\theta) - I_1]^+\right] = o(\Delta\theta)$, which implies (11.57).

We can now repeat this process to show that every $E[R_b], b = 2, 3, \ldots,$ is bounded as in (11.56) by a quantity which is the squared sum of terms in $\Delta\theta$, and hence show, similar to (11.57), that

$$\lim_{\Delta\theta \to 0} \frac{E[R_b]}{\Delta\theta} = 0, \quad b = 2, 3, \ldots \tag{11.58}$$

Step 5. Evaluate $dJ/d\theta$ in (11.38).

Combining (11.38) with (11.54), we have

$$
\frac{dJ}{d\theta} = \frac{1}{N} \lim_{\Delta\theta\to 0} E\left\{ \sum_{b=1}^{B} \sum_{i=n_{b-1}+1}^{n_b} \sum_{j=n_{b-1}+1}^{i} \frac{\Delta Z_j(\theta)}{\Delta\theta} + \sum_{i=n_B+1}^{N} \sum_{j=n_B+1}^{i} \frac{\Delta Z_j(\theta)}{\Delta\theta} \right\}
$$

$$
+ \frac{1}{N} \lim_{\Delta\theta\to 0} E\left\{ \sum_{b=1}^{B} N_b \frac{R_{b-1}}{\Delta\theta} \right\} + \frac{1}{N} \lim_{\Delta\theta\to 0} E\left\{ (N - n_B)\frac{R_B}{\Delta\theta} \right\}
$$

$$(11.59)$$

The first two sums in this expression are of the same form. Under the technical conditions that we have imposed on the service time distribution $F(\cdot; \theta)$ in Section 11.4.1 (essentially, we must require that $|\Delta Z_j|$ be bounded for all j), we can interchange the limit and the expectation to obtain sums of the derivatives $dZ_j/d\theta$:

$$
\frac{dJ}{d\theta} = \frac{1}{N} E\left\{ \sum_{b=1}^{B} \sum_{i=n_{b-1}+1}^{n_b} \sum_{j=n_{b-1}+1}^{i} \frac{dZ_j(\theta)}{d\theta} + \sum_{i=n_B+1}^{N} \sum_{j=n_B+1}^{i} \frac{dZ_j(\theta)}{d\theta} \right\}
$$

$$
+ \frac{1}{N} \lim_{\Delta\theta\to 0} E\left\{ \sum_{b=1}^{B} N_b \frac{R_{b-1}}{\Delta\theta} \right\} + \frac{1}{N} \lim_{\Delta\theta\to 0} E\left\{ (N - n_B)\frac{R_B}{\Delta\theta} \right\}
$$

where the derivative $dZ_j/d\theta$ was given in (11.19). As for the last two terms, we will use the previous step to show that they are 0. Note that B depends on the sample path; however, it is bounded by the fixed number of customers N (it is possible that every one of the B busy periods contains a single customer). The same is true for N_b (i.e., it is also bounded by N). Therefore, for the third term we have

$$
\lim_{\Delta\theta\to 0} E\left\{ \sum_{b=1}^{B} N_b \frac{R_{b-1}}{\Delta\theta} \right\} \leq N \lim_{\Delta\theta\to 0} E\left\{ \sum_{b=1}^{N} \frac{R_{b-1}}{\Delta\theta} \right\} = N \lim_{\Delta\theta\to 0} \sum_{b=1}^{N} \frac{E[R_{b-1}]}{\Delta\theta}
$$

Using (11.57) and (11.58) derived in the previous step, we then get

$$
\lim_{\Delta\theta\to 0} E\left\{ \sum_{b=1}^{B} N_b \frac{R_{b-1}}{\Delta\theta} \right\} \leq N \lim_{\Delta\theta\to 0} \sum_{b=1}^{N} \frac{E[R_{b-1}]}{\Delta\theta} = 0 \qquad (11.60)
$$

Similarly, for the last term

$$
\lim_{\Delta\theta\to 0} E\left\{ (N - n_B)\frac{R_B}{\Delta\theta} \right\} \leq N \lim_{\Delta\theta\to 0} \frac{E[R_B]}{\Delta\theta} = 0
$$

This leads to the final result

$$
\frac{dJ}{d\theta} = \frac{1}{N} E\left\{ \sum_{b=1}^{B} \sum_{i=n_{b-1}+1}^{n_b} \sum_{j=n_{b-1}+1}^{i} \frac{dZ_j(\theta)}{d\theta} + \sum_{i=n_B+1}^{N} \sum_{j=n_B+1}^{i} \frac{dZ_j(\theta)}{d\theta} \right\} \qquad (11.61)
$$

We have thus derived an unbiased estimator of $dJ/d\theta$, defined by

$$\left[\frac{dJ}{d\theta}\right]_{IPA} = \frac{1}{N}\left\{\sum_{b=1}^{B}\sum_{i=n_{b-1}+1}^{n_b}\sum_{j=n_{b-1}+1}^{i}\frac{dZ_j(\theta)}{d\theta} + \sum_{i=n_B+1}^{N}\sum_{j=n_B+1}^{i}\frac{dZ_j(\theta)}{d\theta}\right\}$$

(11.62)

Remark. This derivation of the IPA estimator for the mean system time of N customers in the $GI/G/1$ system is not the most rigorous or the most general one. It combines, however, the intuitive arguments on which IPA is founded with the techniques required to handle the "subtle" problem of establishing (11.57), (11.58), and (11.60). A more rigorous and complete treatment of the same problem is carried out by Zazanis and Suri (1989). We will also see in Section 11.5 that (11.61) becomes a special case of the IPA development for the general class of stochastic timed automata.

Example 11.3 (IPA for a $GI/M/1$ system)
In the case where service times are exponentially distributed with mean θ, we saw in (11.20) that $dZ_i/d\theta = Z_i/\theta, i = 1, 2, \ldots, N$. Thus, the IPA estimator (11.62) becomes

$$\left[\frac{dJ}{d\theta}\right]_{IPA} = \frac{1}{N\theta}\left\{\sum_{b=1}^{B}\sum_{i=n_{b-1}+1}^{n_b}\sum_{j=n_{b-1}+1}^{i}Z_j(\theta) + \sum_{i=n_B+1}^{N}\sum_{j=n_B+1}^{i}Z_j(\theta)\right\}$$

(11.63)

Despite the complexity of the analysis leading to (11.62), the final result has a very appealing simplicity. It requires observations of the service times Z_i and computation of their derivatives through (11.19). These derivatives are simply added up until an idle period is encountered. When the next busy period starts, the process repeats afresh. It is interesting that the recursive equation (11.36) for ΔS_k describes precisely the same procedure. This is one reason why "true" IPA, that is, the derivation of (11.61), and the blind step of eliminating two of the four cases in (11.33) to get (11.35) or (11.36) can become confused. It should be clear, however, that nowhere in the elaborate derivation of this section was there any reference to the magnitude of $\Delta\theta$.

Finally, looking at (11.61), let us carefully distinguish between the *generation* and *propagation* aspects of the overall perturbation process. Each customer experiences a total perturbation given by

[TOTAL PERTURBATION]
= [GENERATION TERM] + [PROPAGATION TERM]

Each departure event contributes a service time perturbation $dZ_j/d\theta$ to the system time of that customer which is *generated* upon occurrence of that event. In addition, the total perturbation from the previous customer propagates to the present customer. When an idle period is encountered, the [PROPAGATION TERM] is reset to 0. Thus, the total perturbation of the first customer in any busy period is given by the [GENERATION TERM] alone due to his own service time perturbation.

11.4.4   Implementation of IPA for the $GI/G/1$ System

Let us now see how the IPA estimator (11.62) can be implemented *on line*, that is, while one observes a sample path of the $GI/G/1$ system under θ. Thus, a sample path may be generated through simulation or it could simply be a collection of data directly observable on a real system. The discussion that follows applies to either case.

Let us first describe the process that the triple sum expression in (11.62) represents. We introduce two "accumulators", A and B, which are initially set to zero. Here, A corresponds to the two innermost sums, and B corresponds to the remaining sum. Whenever a departure event occurs, the following actions are taken:

1. Compute $dZ/d\theta$, the derivative of the service time that just ended, add it to A, and add A to B.

2. Check if a busy period ends (equivalently, check if the queue is empty just after the departure).

Then, depending on the outcome of the check above, if a busy period does end, A is reset to 0. This process stops whenever a given number N of departures is observed. At this point, B/N is precisely the right-hand side in (11.62).

We can therefore implement (11.62) through Algorithm 11.1 as shown. Based on (11.62), note that the Nth departure may not coincide with the end of a busy period. Also, note that the computation of $dZ/d\theta$ at step 2.1 generally requires knowledge of the service time distribution $F(\cdot; \theta)$. However, if θ is known to be a scale parameter, then this step can be omitted and step 2.2 becomes

$$A := A + Z/\theta$$

Similarly, if θ is a location parameter, we have $A := A + 1$ (in which case knowledge of θ is not even needed).

In practice, besides the computation of $dZ/d\theta$, this algorithm requires storing two variables, A and B, and a counter K. It also requires three addition operations in steps 2.2 through 2.4 and a simple check for the idle condition in step 2.5. Thus, the overhead incurred by such an algorithm is indeed very small.

---

1. INITIALIZATION:

$$A := 0, \quad B := 0 \qquad \text{(Perturbation accumulators)}$$
$$K := 0 \qquad \text{(Counter of departure events)}$$

2. WHENEVER DEPARTURE IS OBSERVED WITH
SERVICE TIME Z:

    2.1. Compute $dZ/d\theta$ through (11.19)
    2.2. $A := A + dZ/d\theta$ (accumulate pert's within b.p.)
    2.3. $B := B + A$ (total perturbation)
    2.4. $K := K + 1$ (increment departure counter)

3. STOPPING CONDITION AND OUTPUT:

    If $K = N$, STOP and set $\left[\frac{dJ}{d\theta}\right]_{IPA} = \frac{B}{N}$
    Else, wait for next departure event.

Algorithm 11.1. IPA for estimating $dJ/d\theta$ in $GI/G/1$ system.
($J =$ Mean System Time, $\theta =$ Mean Service Time)

Before leaving the $GI/G/1$ system, it must be pointed out that a similar IPA estimator with respect to the mean interarrival time can be derived by following the same procedure as in Section 11.4.3. Finally, note that one can also estimate $dJ/d\mu$ or $dJ/d\lambda$ by applying (11.19) to the service rate parameter μ in the service time cdf in the first case, or to the arrival rate parameter λ in the interarrival time cdf in the second case. Appendix II provides a self-contained computer program which implements the IPA estimator for $dJ/d\lambda$ in the $M/M/1$ case (this makes it easy to compare IPA estimates with the analytically obtained derivative of the stationary mean system time). The reader is also referred to the Web site http://vita.bu.edu/cgc/IPA/ for an interactive simulator of simple queueing systems that incorporates IPA.

11.5 IPA FOR STOCHASTIC TIMED AUTOMATA

Once again, this is an opportunity to point out that PA is not a technique unique to the $GI/G/1$ system analyzed in the previous section or, for that matter, to queueing systems alone. In fact, the IPA estimator in (11.62) can be rederived as a special case of IPA applied to stochastic timed automata. Recalling the development of this modeling framework for DES in Chapter 6, a stochastic timed automaton generates a stochastic process $\{X(t)\}$ which is referred to as a Generalized Semi-Markov Process (GSMP). Although IPA applied to queueing systems does take advantage of the particular structure of these systems (e.g., the fact that sample paths of the $GI/G/1$ system consist of alternating busy and idle periods), we will see in this section that it is possible to develop a broader theory for stochastic timed automata.

Recall that a stochastic timed automaton, as defined in Section 6.4, is a six-tuple $(\mathcal{E}, \mathcal{X}, \Gamma, p, p_0, G)$, where $\mathcal{E}$ is an event set, $\mathcal{X}$ is a state space, and $\Gamma(x) \subseteq \mathcal{E}$ is the set of feasible events when the state is $\mathcal{X}$, defined for all $x \in \mathcal{X}$. The initial state is drawn from $p_0(x) = P[X_0 = x]$. Subsequently, given that the current state is x, with each feasible event $i \in \Gamma(x)$ we associate a clock value Y_i, which represents the time until event i is to occur. Thus, comparing all such clock values we identify the triggering event

$$E' = \arg \min_{i \in \Gamma(x)} \{Y_i\}$$

where $Y^* = \min_{i \in \Gamma(x)} \{Y_i\}$ is the interevent time (the time elapsed since the last event occurrence). With $E' = e'$ determined, the state transition probabilities $p(x'; x, e')$ are used to specify the next state x'. Finally, the clock values are updated: Y_i is decremented by Y^* for all i (other than the triggering event) which remain feasible in x', while the triggering event (and all other events which are activated upon entering x') are assigned a new lifetime sampled from a distribution G_i. The set $G = \{G_i : i \in \mathcal{E}\}$ defines the stochastic clock structure of the automaton. The resulting state process $\{X(t)\}$ is a GSMP.

In what follows, we will make one simplifying assumption regarding the models to be considered. In particular, we will assume that once an event i is activated, that is, $i \in \Gamma(x)$ for some state x, it cannot be deactivated. In other words, once a clock for some event starts running down, it cannot be interrupted. This is known as the *non-interruption condition* (Glasserman, 1991). For most systems encountered in practice, this is not a very serious limitation (after all, we can often agree to allow an event to occur and to simply ignore it at certain states).

In the discussion that follows, we will use Greek letters (usually α and β) to index events, and reserve standard index letters (such as i, j, k, n and m) for counting event occurrences. We also adopt the following notation:

T_k is the kth event occurrence time, $k = 1, 2, \ldots$ (for any event type)

$T_{\alpha,n}$ is the nth occurrence time of event α

We can now immediately observe the following: When event α takes place at time $T_{\alpha,n}$, it must have been activated at some point in time $T_{\beta,m} < T_{\alpha,n}$ by the occurrence of some event β (note that it is possible that $\beta = \alpha$). In turn, it is also true that event β at $T_{\beta,m}$ must have been activated by some event γ at time $T_{\gamma,k} < T_{\beta,m}$, and so on, all the way back to time $T_0 = 0$. Recalling that $V_{\alpha,k}$ denotes the kth lifetime of event α, since (a) α was activated at time $T_{\beta,m}$, (b) α cannot be deactivated until it occurs (by the non-interruption assumption above), and (c) α finally occurs at time $T_{\alpha,n}$, we have $V_{\alpha,n} = T_{\alpha,n} - T_{\beta,m}$.

Similarly, we have $V_{\beta,m} = T_{\beta,m} - T_{\gamma,k}$, and so on. Therefore, we can always write

$$T_{\alpha,n} = V_{\beta_1,k_1} + \ldots + V_{\beta_s,k_s} \tag{11.64}$$

for some s. Clearly, a similar expression can be written for any event time T_k, since $T_k = T_{\alpha,n}$ for some $a \in \mathcal{E}$ and $n = 1, 2, \ldots$.

The expression in (11.64) can be rewritten in a more convenient form by introducing *triggering indicators*, that is, functions $\eta(\alpha, n; \beta, m)$ taking values in $\{0, 1\}$ as follows:

$$\eta(\alpha, n; \beta, m) = 1 \quad \text{if the } n\text{th occurrence of event } \alpha \text{ is triggered by the } m\text{th occurrence of } \beta$$
$$\eta(\alpha, n; \alpha, n) = 1 \quad \text{for all } \alpha \in \mathcal{E}, n = 1, 2, \ldots$$
$$\eta(\alpha, n; \beta', m') = 1 \quad \text{if } \eta(\alpha, n; \beta, m) = 1 \text{ and } \eta(\beta, m; \beta', m') = 1$$
$$\eta(\alpha, n; \eta, m) = 0 \quad \text{otherwise}$$

Then,

$$T_{\alpha,n} = \sum_{\beta,m} V_{\beta,m} \eta(\alpha, n; \beta, m) \tag{11.65}$$

The sequence of (β, m) pairs that leads to $T_{\alpha,n}$ with $\eta(\alpha, n; \beta, m) = 1$ defines the *triggering sequence* of (α, n).

Example 11.4 (A $GI/G/1$ queueing system)

Consider a typical sample path of a $GI/G/1$ queueing system, as shown in Fig. 11.15. In this case, there are two events, a (arrivals) and d (departures). Focusing on the departure event at $T_9 = T_{d,4}$, note that the fourth departure was activated at time $T_7 = T_{a,4}$ (when the second busy period starts), i.e., $\eta(d, 4; a, 4) = 1$. The fourth arrival, in turn, was activated at $T_4 = T_{a,3}$, the previous arrival, which in turn was activated at $T_2 = T_{a,2}$. Finally, the second arrival was activated at $T_1 = T_{a,1}$, and the first arrival was activated at time 0. We therefore get (see also Fig. 11.15):

$$T_{d,4} = V_{d,4} + V_{a,4} + V_{a,3} + V_{a,2} + V_{a,1}$$

Thus, the triggering sequence of $(d, 4)$ is $\{(a, 1), (a, 2), (a, 3), (a, 4)\}$.

As another example, $T_{d,3}$ is given by

$$T_{d,3} = V_{d,3} + V_{d,2} + V_{d,1} + V_{a,1}$$

since the third departure was activated at time $T_5 = T_{d,2}$ (when the second customer departs and the third one starts service), the second departure was activated at time $T_3 = T_{d,1}$, and the first departure was activated by

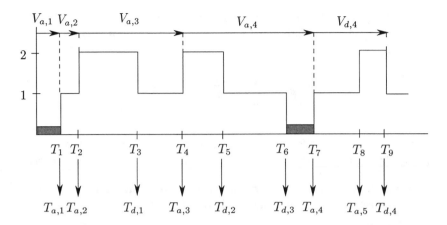

Figure 11.15: Sample path of a $GI/G/1$ system for Example 11.4.

the customer arrival at time $T_1 = T_{a,1}$. The triggering sequence of $(d, 3)$ is $\{(a, 1), (d, 1), (d, 2)\}$.

It is worth making the observation that the kth arrival time is always expressed as the sum of all k interarrival times, $V_{a,1}, \ldots, V_{a,k}$. This is because an a event is always feasible, therefore it is always activated by the previous a event occurrence.

11.5.1   Event Time Derivatives

Let us consider a parameter θ which can only affect one or more of the event lifetime distributions $G_\alpha(x; \theta)$; in particular, θ does not affect the state transition mechanism. We therefore view lifetimes as functions of $\theta, V_{\alpha,k}(\theta)$. The derivative $dV_{\alpha,k}(\theta)/d\theta$ is obtained as in (11.19), that is,

$$\frac{dV_{\alpha,k}}{d\theta} = - \frac{[\partial G_\alpha(x; \theta)/\partial \theta]_{(V_{\alpha,k}, \theta)}}{[\partial G_\alpha(x; \theta)/\partial x]_{(V_{\alpha,k}, \theta)}} \tag{11.66}$$

Since $T_{\alpha,n}$ is expressed as a sum of event lifetimes in (11.65), it is natural to expect that, under certain conditions, the derivative of $T_{\alpha,n}$ can be obtained in terms of the derivatives of these event lifetimes. In particular (see Glasserman, 1991), let us make the following assumptions:

(A1) For all $\alpha \in \mathcal{E}$, $G_\alpha(x; \theta)$ is continuous in θ and $G_\alpha(0; \theta) = 0$.

(A2) For all $\alpha \in \mathcal{E}$ and $k = 1, 2 \ldots, V_{\alpha,k}(\theta)$ is almost surely continuously differentiable in θ.

Under these conditions, we can guarantee that for sufficiently small perturbations of θ, the event sequence leading to time $T_{\alpha,n}$ (assumed finite) remains unchanged. As a result, the sequence of triggering indicators in (11.65) also remains unchanged. Therefore, from (11.65), event time derivatives $dT_{\alpha,n}(\theta)/d\theta$ are given by

$$\frac{dT_{\alpha,n}}{d\theta} = \sum_{\beta,m} \frac{dV_{\beta,m}}{d\theta} \eta(\alpha,n;\beta,m) \qquad (11.67)$$

This expression captures the fact that event time perturbations are *generated* by the effect of θ on event lifetimes $V_{\beta,m}$, and subsequently *propagate* from (β,m) to (α,n) through the events contained in the triggering sequence of (α,n).

Example 11.5 (A $GI/G/1$ queueing system - continued)
Let us return to Example 11.4, where we found that

$$T_{d,4} = V_{d,4} + V_{a,4} + V_{a,3} + V_{a,2} + V_{a,1}$$

and the triggering sequence of $(d,4)$ is $\{(a,1),(a,2),(a,3),(a,4)\}$.

Suppose the parameter θ is the mean of the service time distribution $G_d(x;\theta)$. Thus, all service times (departure event lifetimes) $V_{d,k}$, $k = 1,2,\ldots$, are affected by changes in θ, whereas all interarrival times are not (i.e., $dV_{\alpha,k}/d\theta = 0$ for all $k = 1,2,\ldots$) Therefore,

$$\frac{dT_{d,4}}{d\theta} = \frac{dV_{d,4}}{d\theta} + \frac{dV_{a,4}}{d\theta} + \frac{dV_{a,3}}{d\theta} + \frac{dV_{a,2}}{d\theta} + \frac{dV_{a,1}}{d\theta} = \frac{dV_{d,4}}{d\theta}$$

It is interesting to compare this expression with the analysis of the $GI/G/1$ system in Section 11.4, and equation (11.49) in particular: When the mean service time was perturbed, the departure (and hence system) time of the first customer of every busy period was assigned a perturbation ΔZ_j where j was the index of the first customer in that busy period. We now see above that the fourth customer is the first one in the second busy period in Fig. 11.15, and that his departure time is assigned the perturbation $dV_{d,4}/d\theta$, where $V_{d,4}$ is the first service time in the second busy period.

Similarly, we get

$$\frac{dT_{d,3}}{d\theta} = \frac{dV_{d,3}}{d\theta} + \frac{dV_{d,2}}{d\theta} + \frac{dV_{d,1}}{d\theta} + \frac{dV_{a,1}}{d\theta}$$
$$= \frac{dV_{d,3}}{d\theta} + \frac{dV_{d,2}}{d\theta} + \frac{dV_{d,1}}{d\theta}$$

In this case, three successive perturbations are accumulated, since the first three customers all belong to the same busy period. Note that the presence of $(a,4)$ in the triggering sequence of $(d,4)$ has the effect of "resetting" the perturbations accumulated by the third departure.

The preceding discussion leads to a general-purpose algorithm for evaluating event time derivatives along an observed sample path (see Algorithm 11.2) of a GSMP generated by a stochastic timed automaton. In particular, let us define a perturbation accumulator, Δ_α, for every event $\alpha \in \mathcal{E}$. The accumulator Δ_α is updated at event occurrences in two ways:

1. It is incremented by $dV_\alpha/d\theta$ whenever an event α occurs.

2. It is coupled to an accumulator $\Delta\beta$ whenever an event β (possibly $\beta = \alpha$) occurs that activates an event α.

Note that because of the non-interruption condition, the addition of $dV_\alpha/d\theta$ to Δ_α can be implemented either at the time of occurrence of α or at the time of its activation. In Algorithm 11.2, it is assumed that the system starts out at some state x_0. No particular stopping condition is specified, since this may vary depending on the problem of interest (e.g., stop after a desired total number of event occurrences, stop after a desired number of event type α occurrences).

---

1. INITIALIZATION:

> If event α is feasible at x_0: $\Delta_\alpha := dV_{\alpha,1}/d\theta$
> Else, for all other $\alpha \in \mathcal{E}$: $\Delta_\alpha := 0$

2. WHENEVER EVENT β IS OBSERVED:
   If event α is activated with new lifetime V_α:

> 2.1. Compute $dV_\alpha/d\theta$ through (11.66)
> 2.2. $\Delta_\alpha := \Delta_\beta + dV_\alpha/d\theta$

Algorithm 11.2. Event time derivatives for stochastic timed automata. ($\theta =$ some parameter of the stochastic clock structure)

Example 11.6 (A $GI/G/1$ queueing system - continued)

Returning to Example 11.4, let us apply Algorithm 11.2 in order to evaluate $dT_{d,4}/d\theta$, where θ is the mean of the service time distribution $G_d(x; \theta)$. Assuming the queue is initially empty, we set $\Delta_d = 0$ (because d is not feasible at this state) and $\Delta_a = 0$ (which is always true, since interarrival times are independent of θ). Our goal is to keep track of the d accumulator, Δ_d.

There are two cases to consider in step 2 of Algorithm 11.2:

1. Event d occurs: A new d event is activated as long as d remains feasible in the new state, i.e., as long as $x > 1$ (if $x = 1$, the queue becomes empty and d is no longer feasible). Assuming $dV_d/d\theta$ can

be computed, step 2.2 gives

$$\Delta_d := \Delta_d + dV_d/d\theta$$

2. Event a occurs: A new d event is activated only if $x = 0$, i.e., a new busy period starts. In this case,

$$\Delta_d := \Delta_a + dV_d/d\theta = dV_d/d\theta$$

It is interesting to observe in Example 11.6 that in the case of a $GI/G/1$ system, Algorithm 11.2 recovers the perturbation process described in Section 11.4.3: Service time perturbations (derivatives) are accumulated over a busy period, but the perturbation is reset with every new busy period (after an α event that occurs at state $x = 0$).

11.5.2   Sample Function Derivatives

Since many sample performance functions $L(\theta, \omega)$ of interest can be expressed in terms of event times $T_{\alpha,n}$, we will now see how to use (11.67) and Algorithm 11.2 in order to obtain derivatives of the form $dL(\theta, \omega)/d\theta$. For simplicity, we will omit the argument ω in what follows.

The sample performance functions we will consider are all over a finite horizon and similar to some of those defined in discussing controlled Markov chains in Chapter 9. The main difference is that the quantity we can control here is the parameter θ, as opposed to a sequence of possibly state-dependent control actions. When the state is $X(t, \theta)$, let $C(X(t, \theta))$ be a bounded cost associated with operating the system at that state. We then define (see Glasserman, 1991):

$$L_T(\theta) = \int_0^T C(X(t, \theta))dt \tag{11.68}$$

$$L_M(\theta) = \int_0^{T_M} C(X(t, \theta))dt \tag{11.69}$$

$$L_{\alpha,M}(\theta) = \int_0^{T_{a,M}} C(X(t, \theta))dt \tag{11.70}$$

Here, $L_T(\theta)$ measures the total cost over an interval of time $[0, T]$ for some given finite T. On the other hand, $L_M(\theta)$ is defined so that the total cost is measured over exactly M event occurrences, and $L_{\alpha,M}(\theta)$ over exactly M occurrences of some event type α. These functions cover many (but not all) useful performance measures encountered in practice. In simple queueing systems, $X(t, \theta)$ is usually the queue length. Thus, by setting $C(X(t, \theta)) = X(t, \theta)$ in (11.68), we can obtain the *mean queue length* over $[0, T]$ as $L_T(\theta)/T$. Similarly, by setting $C(X(t, \theta)) = 1$ in (11.70) and choosing α to be the departure event, we get the *throughput* of the system as $M/L_{\alpha,M}(\theta)$. The *mean system time* over M

customers can also be obtained as follows. Recall that the system time of the kth customer is given by $S_k(\theta) = T_{d,k}(\theta) - T_{a,k}(\theta)$, where d is a departure event and a is an arrival event. Thus, setting $C(X(t,\theta)) = 1$ in (11.70), we get $S_k(\theta) = L_{d,k}(\theta) - L_{a,k}(\theta)$. The mean system time over M customers is then given by the sum of M such differences divided by M.

Under the non-interruption condition for all events in our models, and assumptions (A1) and (A2) of the previous section, we can now show that the derivatives of $L_T(\theta), L_M(\theta)$, and $L_{a,M}(\theta)$ exist (with probability 1) and are given by

$$\frac{dL_T}{d\theta} = \sum_{k=1}^{N(T)} \frac{dT_k}{d\theta}[C(X_{k-1}) - C(X_k)] \tag{11.71}$$

$$\frac{dL_M}{d\theta} = \sum_{k=0}^{M-1} C(X_k)\left[\frac{dT_{k+1}}{d\theta} - \frac{dT_k}{d\theta}\right] \tag{11.72}$$

$$\frac{dL_{\alpha,M}}{d\theta} = \sum_{k=0}^{N(T_{\alpha,M})-1} C(X_k)\left[\frac{dT_{k+1}}{d\theta} - \frac{dT_k}{d\theta}\right] \tag{11.73}$$

where $N(T)$ counts the total number of events observed in $[0,T]$. Formal derivations of these expressions are given by Glasserman (1991). The crucial observation is that in DES the state remains unchanged at X_k in any interval $(T_k, T_{k+1}]$. This allows us to rewrite the integrals (11.68) to (11.70) in simpler summation forms. As an example, (11.69) is rewritten as

$$L_M = \sum_{k=0}^{M-1} C(X_k)[T_{k+1} - T_k]$$

In this form, it is not difficult to see that differentiation with respect to θ yields (11.72). We can then see how to use (11.67) in conjunction with (11.71) to (11.73) to obtain sample derivatives: (11.67) (or equivalently Algorithm 11.2) allows us to evaluate the event time derivatives in the expressions above, and hence transform perturbations in event times into perturbations in sample performance functions.

Example 11.7 (A $GI/G/1$ queueing system - continued)
In Example 11.6, we evaluated departure time perturbations of the form $dT_{d,k}/d\theta$ in a $GI/G/1$ system, where θ is the mean of the service time distribution $G_d(x;\theta)$. Let us now see how to use this information to evaluate the derivatives of the mean system time over M customers.

As pointed out earlier, setting $C(X(t,\theta)) = 1$ in (11.70), we get $S_k(\theta) = T_{d,k}(\theta) - T_{a,k}(\theta)$. Thus, the derivative $dS_k/d\theta$ can be obtained directly from the event time perturbations evaluated in Example 11.6. In fact, since $dT_{a,k}/d\theta = 0$ for all arrivals, we have

$$dS_k/d\theta = dT_{d,k}/d\theta$$

where $dT_{d,k}/d\theta$ is given by the value of the accumulator Δ_d of Example 11.6 when the kth departure time occurs. Recall that

$$\Delta_d := \Delta_d + dV_d/d\theta$$

when d occurs and does not end a busy period. In other words, after i customer departures within a busy period, we get

$$\frac{dS_i}{d\theta} = \sum_{j=1}^{i} \frac{dV_{d,j}}{d\theta}$$

which is of the same form as (11.40) obtained in Section 11.4.3 for finite perturbations $\Delta\theta$. It now follows that since the mean system time over a busy period consisting of n customers is

$$L_n(\theta) = \frac{1}{n} \sum_{i=1}^{n} S_i$$

its derivative is obtained by combining the last two equations:

$$\frac{dL_n}{d\theta} = \frac{1}{n} \sum_{i=1}^{n} \sum_{j=1}^{i} \frac{dV_{d,j}}{d\theta} \tag{11.74}$$

which is of the same form as (11.41) obtained for finite perturbations $\Delta\theta$. In addition, recall that in Example 11.6 we found

$$\Delta_d := \Delta_a + dV_d/d\theta = dV_d/d\theta$$

when event a occurs initiating a new busy period. This allows us to evaluate the sample derivative $dL_M/d\theta$ over M customers observed over several busy periods by using (11.74) over separate busy periods. Adopting the same notation as in Section 11.4.3, let n_b be the index of the last customer in the bth busy period, $b = 1, 2 \ldots$, with $n_0 = 0$. Then, the derivative of the mean system time over M customers, $dL_M/d\theta$, is given by

$$\frac{dL_M}{d\theta} = \frac{1}{M} \left\{ \sum_{b=1}^{B} \sum_{i=n_{b-1}+1}^{n_b} \sum_{j=n_{b-1}+1}^{i} \frac{dV_{d,j}}{d\theta} + \sum_{i=n_B+1}^{N} \sum_{j=n_B+1}^{i} \frac{dV_{d,j}}{d\theta} \right\}$$

$$\tag{11.75}$$

which is of the same form as the right-hand side of (11.62), that is, the derivative of the mean system time we derived through different arguments in Section 11.4.3. However, in (11.62) it was established that this expression is indeed an unbiased estimator of the performance measure derivative, $dJ/d\theta$. On the other hand, (11.75) simply provides an expression for the sample derivative; it remains to be shown (see next section) that this is in fact an unbiased estimate of $dJ/d\theta$.

11.5.3   Performance Measure Derivatives

Thus far, we have seen how to evaluate derivatives of sample functions, $dL(\theta,\omega)/d\theta$, when $L(\theta,\omega)$ is of the form (11.68) to (11.70). However, what we are really interested in are derivatives of performance measures $J(\theta) = E[L(\theta,\omega)]$. Since we have developed techniques for evaluating $dL(\theta,\omega)/d\theta$, the hope is that this sample derivative is an unbiased estimate of $dJ/d\theta$, satisfying equation (11.4) rewritten below:

$$\frac{dJ(\theta)}{d\theta} = E\left[\frac{dL(\theta,\omega)}{d\theta}\right]$$

This, in the context of IPA, is the key issue, originally identified by Cao (1985).

This is a good point to confirm our usage of some terminology which, by now, should be quite clear:

- When we say "IPA" we mean the derivation of $dL(\theta,\omega)/d\theta$, a sample derivative based on information obtained from a single observed sample path ω.

- When we say "IPA estimator" we mean $dL(\theta,\omega)/d\theta$, the sample derivative used to estimate the performance measure derivative (or sensitivity) $dJ/d\theta$, that is, we write

$$\left[\frac{dJ}{d\theta}\right]_{IPA} = \frac{dL}{d\theta} \tag{11.76}$$

- When we say "the IPA estimator is unbiased", we mean that equation (11.4) is satisfied.

Before getting into details regarding the issue of unbiasedness, one might immediately suspect that the interchange of expectation and differentiation in (11.4) may be prohibited when $L(\theta,\omega)$ exhibits *discontinuities* in θ. Intuitively, such discontinuities may arise when a change in θ causes event order changes, as described through the simple examples of Section 11.3 and as explicitly seen in the case of the $GI/G/1$ system in equation (11.33). However, as illustrated through the example of Fig. 11.7 in Section 11.3, some event order changes may occur without violating the continuity of $L(\theta,\omega)$.

As we will see, if $L(\theta, \omega)$ is indeed continuous in θ, it can be shown that the interchange of expectation and differentiation in (11.4) does hold. Moreover, there is a simple condition based exclusively on the structure of the DES considered, which can be used to check for continuity. This provides a powerful tool for testing whether IPA can yield unbiased derivative estimators for different classes of DES.

The Commuting Condition

The key to checking for the continuity of sample functions, and hence ultimately establishing the unbiasedness of sample derivative estimators for stochastic timed automata, is the *commuting condition* (due to Glasserman, 1991) that was briefly introduced in Section 11.3 (see Fig. 11.8):

(CC) Let $x, y, z_1 \in \mathcal{X}$ and $\alpha, \beta \in \Gamma(x)$ such that

$$p(z_1; x, \alpha) \cdot p(y; z_1, \beta) > 0.$$

Then, for some $z_2 \in \mathcal{X}$, we have:

$$p(z_2; x, \beta) = p(y; z_1, \beta) \quad \text{and} \quad p(y; z_2, \alpha) = p(z_1; x, \alpha).$$

Moreover, for any $x, z_1, z_2 \in \mathcal{X}$ such that $p(z_1; x, \alpha) = p(z_2; x, \alpha) > 0$, we have: $z_1 = z_2$.

In words, **(CC)** requires that if a sequence of events $\{\alpha, \beta\}$ takes state x to state y, then the sequence $\{\beta, \alpha\}$ must also take x to y. Moreover, this must happen so that every transition triggered by α or β in this process occurs with the same probability. The last part of **(CC)** also requires that if an event α takes place at state x, the next state is unique, unless the transition probabilities to distinct states z_1, z_2 are not equal. The commuting condition is particularly simple to visualize through the diagram of Fig. 11.16.

Remark. The last part of **(CC)** is more of a technicality that has no restrictive impact in applications. In addition, if a model's state transition mechanism is deterministic (as is often the case), then the diagram of Fig. 11.16 tells the whole story: Simply check that $\{\alpha, \beta\}$ applied to state x leads to the same final state as $\{\beta, \alpha\}$ applied to x. What is particularly attractive about **(CC)** is the fact that it is a "structural" condition that makes it very easy to verify (or find that it is not satisfied) as the following few examples illustrate.

Example 11.8 (A $GI/G/1$ queueing system - continued)
Returning to the $GI/G/1$ queueing system of Example 11.4, let us check that the commuting condition indeed holds. Figure 11.17 shows the state transition diagram of this system. Every state other than $x = 0$ has a

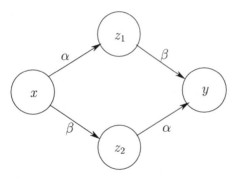

Figure 11.16: The commuting condition (**CC**).
The event sequence $\{\alpha, \beta\}$ applied to state x leads to the same final state y as the sequence $\{\beta, \alpha\}$. Moreover, every transition triggered by α or β in the diagram occurs with the same probability.

feasible event set $\Gamma(x) = \{a, d\}$, and $p(x + 1; x, a) = p(x - 1; x, d) = 1$. It is immediately obvious from Fig. 11.17 that the event sequence $\{a, d\}$ applied to any $x > 0$ results in the state sequence $\{x + 1, x\}$, whereas the event sequence $\{d, a\}$ results in the state sequence $\{x - 1, x\}$. Condition (**CC**) is satisfied, since the final state, x, is always the same.

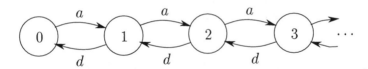

Figure 11.17: State transition diagram for the $GI/G/1$ queueing system in Example 11.8.

Example 11.9 (Checking (CC) for a closed queueing network)
Consider a two-queue network as shown in Fig. 11.18. The network provides service to three customers which cycle around the two servers, both with infinite queueing capacity. The event set is $\mathcal{E} = \{d_1, d_2\}$, where d_i denotes a departure from server $i = 1, 2$. The state transition diagram for this system is shown in Fig. 11.18. In this case, only state $(1, 1)$ has more than a single feasible event. We immediately see that the event sequence $\{d_1, d_2\}$ applied to state $(1, 1)$ results in the state sequence $\{(0, 2), (1, 1)\}$, whereas the event sequence $\{d_2, d_1\}$ results in the state sequence $\{(2, 0), (1, 1)\}$, and (**CC**) once again holds.

Example 11.10 (Checking (CC) for a GI/G/1/K queueing system)
Consider a $GI/G/1/K$ queueing system with state transition diagram as shown in Fig. 11.19. The event set is $\mathcal{E} = \{a, d\}$, as in the $GI/G/1$

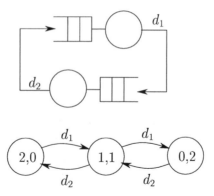

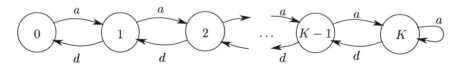

Figure 11.18: Closed queueing network for Example 11.9.

case. The commuting condition is satisfied for all states $x < K$ since the process of changing the order of $\{a, d\}$ in any such state is the same as in the $GI/G/1$ case of Example 11.8. At state K, however, the event sequence $\{a, d\}$ results in the state sequence $\{K, K - 1\}$, whereas the event sequence $\{d, a\}$ results in the state sequence $\{K - 1, K\}$. The final state is now different ($K - 1$ instead of K), therefore, condition **(CC)** is not satisfied. In general, blocking phenomena in queueing systems tend to violate the commuting condition.

Figure 11.19: State transition diagram for the $GI/G/1/K$ queueing system in Example 11.10.

Continuity of Sample Functions

We will now see why the commuting condition plays such a crucial role in IPA. In particular, condition **(CC)** turns out to imply that sample performance functions $L(\theta, \omega)$ of the form (11.68) to (11.70) are continuous in θ. This is stated as a theorem below (the proof may be found in Glasserman, 1991).

Theorem 11.1 Under assumptions **(A1)** and **(A2)** of Section 11.5.1 and condition **(CC)**, the sample functions $L_T(\theta), L_M(\theta)$, and $L_{\alpha,M}(\theta)$ (for finite $T_{\alpha,M}$) defined in (11.68) to (11.70) are (almost surely) continuous in θ. ◆

The derivation of this important result rests on some key properties of sample paths under the commuting condition. We will limit ourselves here to an illustration of these observations through the sample paths shown in Fig. 11.20.

First, let us consider the top part of the left side in Fig. 11.20, and think of it as the sample path of a $GI/G/1$ queueing system. Suppose that the parameter θ affects departure times only. Thus, as θ varies, the d event shown in this sample path approaches the a event shown. As it does so, the total area under the sample path changes smoothly with $\Delta \to 0$, and the shaded rectangle gradually shrinks to zero. At the point where $\Delta = 0$ and the a and d events are about to change order, the total area under the curve is A. Next, looking at the bottom part, we see that as $\Delta \to 0$, the shaded area shown is gradually being added to the total area under the sample path, and when the a and d events are about to change order, the total area is once again A. Thinking of the area as the sample function of interest $L(\theta)$, we see that $L(\theta)$ is continuous in θ, *despite the event order change.* The crucial observation is that the shaded area in the top part becomes zero *precisely at the instant of the event order change.* The role of the commuting condition here is that this event order change does not affect the future state evolution of the system.

Now let us consider the right side of Fig. 11.20, where a sample path of a $GI/G/1/2$ queueing system is shown. Thus, at the top part we see that the a event shown does not affect the sample path, since it represents an arrival that is rejected. Once again, we see that as $\Delta \to 0$, the shaded rectangle shrinks to zero; it becomes precisely zero at the point where the a and d events are about to change order, and the total area under the sample path becomes A, as shown in the figure. Just after the event order change, however, that is, when a occurs after d, the arrival sees a queue length less than 2 and it can be accepted; this arrival then contributes the diagonally shaded area shown as B in Fig. 11.20. Thinking again of the area as the sample function of interest $L(\theta)$, we immediately see that there is a jump at the point where a and d change order and $L(\theta)$ is not continuous in θ at that point. As we saw in Example 11.10, the commuting condition is violated for this queueing system, which is consistent with Theorem 11.1.

Unbiasedness of IPA Estimators

We are now in a position to establish the validity of equation (11.4), that is, the interchange of expectation and differentiation, which is required for IPA estimates (i.e., sample derivatives) to be unbiased. The key idea in Theorem 11.2 below is to identify conditions under which the dominated convergence theorem

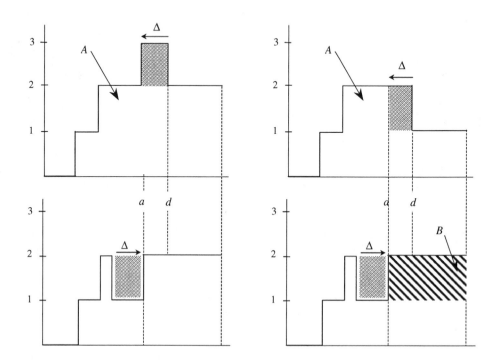

Figure 11.20: Continuous $L(\theta)$ vs. discontinuous $L(\theta)$.
On the left is a typical sample path of a $GI/G/1$ queueing system. As $\Delta \to 0$, the area of the shaded rectangle goes to zero. At the point where a and d change order, the rectangle area becomes exactly 0 and the total area is A.
On the right, a sample path of a $GI/G/1/2$ queueing system is shown. In this case, event a originally contributes no area. However, just after the event order change, a occurs after d and it contributes the rectangular area B, causing a jump in the total area under the sample path.

(see Appendix I) can be used to show that

$$\frac{dJ}{d\theta} = \lim_{\Delta\theta\to 0} E\left[\frac{L(\theta + \Delta\theta, \omega) - L(\theta, \omega)}{\Delta\theta}\right]$$
$$= E\left[\lim_{\Delta\theta\to 0} \frac{L(\theta + \Delta\theta, \omega) - L(\theta, \omega)}{\Delta\theta}\right] = E\left[\frac{dL(\theta, \omega)}{d\theta}\right]$$

In particular, the interchange of the limit and E in the preceding equation is allowed under the dominated convergence theorem if we can determine a bound R, $|([L(\theta + \Delta\theta, \omega) - L(\theta, \omega)]/\Delta\theta| \leq R$, such that $E[R] < \infty$. To this end, we exploit the continuity of $L(\theta, \omega)$ under Theorem 11.1 as follows: For continuous and differentiable functions, the generalized mean value theorem asserts that

$$\left|\frac{L(\theta + \Delta\theta) - L(\theta)}{\Delta\theta}\right| \leq \sup_{\xi\in[a,b]}\left|\frac{dL(\xi)}{d\theta}\right| \tag{11.77}$$

Thus, the only remaining issue is to determine bounds for sample derivatives $dL/d\theta$. This is normally possible to do on a case-by-case basis, exploiting the structure of a particular system. Theorem 11.2 as stated below (Glasserman, 1991), however, provides some more general-purpose conditions:

Theorem 11.2 Suppose assumptions **(A1)** and **(A2)** of Section 11.5.1 hold and condition **(CC)** is satisfied. In addition, assume that $|dV_{\alpha,k}/d\theta| \leq B \cdot (V_{\alpha,k}+1)$ for some $B > 0$. Then, if $E[\sup_\theta N(T)^2] < \infty$, for $L_T(\theta)$ defined in (11.68):

$$\frac{d}{d\theta} E[L_T(\theta)] = E\left[\frac{dL_T(\theta)}{d\theta}\right] \tag{11.78}$$

If $E[\sup_\theta T_M] < \infty$, for $L_M(\theta)$ defined in (11.69):

$$\frac{d}{d\theta} E[L_M(\theta)] = E\left[\frac{dL_M(\theta)}{d\theta}\right] \tag{11.79}$$

If $E[\sup_\theta (T_{\alpha,M})^2] < \infty$, and $E[\sup_\theta (N(T_{\alpha,M}))^2] < \infty$, for $L_{\alpha,M}(\theta)$ defined in (11.70):

$$\frac{d}{d\theta} E[L_{\alpha,M}(\theta)] = E\left[\frac{dL_{\alpha,M}(\theta)}{d\theta}\right] \tag{11.80}$$

◆

Example 11.11 (A $GI/G/1$ queueing system - continued)
Returning to the $GI/G/1$ queueing system of Example 11.4, we already verified in Example 11.8 that condition **(CC)** is satisfied. In addition, we evaluated the sample derivative in Example 11.7 to obtain (11.75):

$$\frac{dL_M}{d\theta} = \frac{1}{M}\left\{\sum_{b=1}^{B}\sum_{i=n_{b-1}+1}^{n_b}\sum_{j=n_{b-1}+1}^{i}\frac{dV_{d,j}}{d\theta} + \sum_{i=n_B+1}^{N}\sum_{j=n_B+1}^{i}\frac{dV_{d,j}}{d\theta}\right\}$$

In order to take the final step and show that this expression, the IPA estimate of the mean system time, is indeed an *unbiased* estimate it remains to invoke Theorem 11.2. Let us, however, see how we can directly apply the generalized mean value theorem by determining a bound in (11.77). Omitting details, the idea is that since we have at our disposal an explicit expression for $dL_M/d\theta$ above, we should be able to determine a simple bound for it, R, by exploiting its structure. First, let us assume that $|dV_{d,j}/d\theta| < C < \infty$, that is, the derivatives evaluated through (11.19) are bounded. Then, the innermost sum in the first term of this expression is bounded by MC; similarly, the innermost sum of the second term is bounded by MC. Thus, since B is also bounded by M, the whole right-hand side is bounded by $R = M^3C/M + M^2C/M$. Then, $E[R] = M^2C + MC < \infty$. This yields the desired bound.

In summary, we have rederived in this example equation (11.61) as a special case of IPA applied to stochastic timed automata.

Remark. We have proceeded all along under **(A1)**, which excludes discontinuities in the event lifetime distributions, and hence excludes discrete cdf's. It should be pointed out that this case can also be handled, as long as θ can affect only the discrete values a lifetime can take, $v_1, \ldots, v_n$, but not the corresponding probabilities, $p_1, \ldots, p_n$ (as already discussed in Section 11.4.1). If, for instance, event α has deterministic lifetimes, θ becomes a location parameter and, by (11.23), $dV_{\alpha,k}/d\theta = 1$ for all k.

Consistency of IPA Estimators

As discussed in Section 10.7.2 of Chapter 10, there are two basic measures of "goodness" for estimators: *unbiasedness* and *consistency*. In the context of estimating $dJ/d\theta$ through the IPA estimator $dL/d\theta$, unbiasedness requires that averaging over all possible sample paths does indeed yield the true performance measure derivative $dJ/d\theta$. Our effort thus far, culminating with Theorem 11.2, has been directed at this problem. We now turn our attention to the issue of consistency.

What is attractive about IPA is the fact that both an estimate of the performance $J(\theta)$ and its sensitivity with respect to $\theta, dJ/d\theta$, can be estimated from a single sample path. If, however, $J(\theta)$ is a steady-state performance measure, we saw in Chapter 10 that the sample path on which our estimate is based must be made "sufficiently long." Thus, if an estimator is based on N samples, the hope is that as $N \to \infty$ the accuracy of the estimator increases. The estimator is "strongly consistent" if in fact it converges (with probability 1) to the quantity being estimated. IPA would be even more attractive if we could assert that, as the length of the observed sample path increases, the IPA estimator, $dL/d\theta$, converges to $dJ/d\theta$ with probability 1 (where $J(\theta)$ is now a steady-state performance measure, no longer based on a finite number, N, of customers as in (11.70) for example). If this is the case, the IPA estimator is not only unbiased, but also strongly consistent.

Let us adopt the notation $J'(\theta) = dJ/d\theta$ to denote the derivative with respect to θ. Similarly, let $L'(\theta) = dL/d\theta$, and we write $L'_N(\theta)$ to represent the fact that an IPA estimator depends on the number of observations $N = 1, 2, \ldots$ on which it is based. For example, the IPA estimator (11.62) depends on the total number N of customer departures observed. Since $L'_N(\theta)$ is our IPA estimate of $dJ/d\theta$, $L'_N(\theta)$ is a strongly consistent estimator of $dJ/d\theta$ if

$$\lim_{N \to \infty} L'_N(\theta) = \frac{dJ}{d\theta} \qquad \text{with probability 1} \qquad (11.81)$$

Of course, we expect this to hold if $L_N(\theta)$ is itself a strongly consistent estimator of $J(\theta)$ in the first place, i.e., $\lim_{N \to \infty} L_N(\theta) = J(\theta)$ with probability 1. Otherwise, it is hard to imagine the sample function derivative having this property when the sample function does not.

During the early developments of PA, it was repeatedly observed through simulation experiments that many estimators derived for single-server queueing systems or more complex queueing networks were in fact strongly consistent. However, formally demonstrating this property and the conditions under which it applies is an arduous task, involving a number of technical conditions and arguments. The first formal proof of strong consistency for an IPA estimator is due to Suri and Zazanis (1988) for the case of derivatives of the steady-state mean system time of an $M/G/1$ queueing system, in which case $dJ/d\theta$ can be analytically evaluated by differentiating the Pollaczek-Khinchin formula (8.109) derived in Chapter 8. But since such explicit expressions for $dJ/d\theta$ are rather few (after all, recall that PA is motivated precisely by the fact that closed-form expressions of performance measures $J(\theta)$ are nonexistent...), this makes it difficult to verify (11.81) above. Nonetheless, the strong consistency of IPA estimators has been established for a number of interesting cases, including $GI/G/1$ queueing systems (Zazanis and Suri, 1989; Hu, 1990) and some classes of queueing networks (Cao, 1987; Wardi and Hu, 1991). In addition, Hu and Strickland (1990) have shown that if (asymptotic) unbiasedness holds and a few technical conditions on the functions $L_N(\theta)$ and $L'_N(\theta)$ can be verified, then (11.81) holds. These technical conditions are usually satisfied by systems characterized by the regenerative structure which was discussed in Section 10.7.4 of Chapter 10 (see also Heidelberger et al., 1988). We will not get into more details on the issue of IPA estimator consistency, but refer the reader to Ho and Cao (1991) for a more complete discussion on the subject and for further references.

11.5.4   IPA Applications

One feature of IPA that should be apparent from the previous sections is that while the analysis leading to IPA estimators can become quite intricate and at times subtle, the final results are usually simple, both to express mathematically and to implement in practice. This is exemplified by the case of the $GI/G/1$ system: The analysis that led to the final estimator in (11.62) was rather elaborate, but the final expression (essentially, a triple sum of service time derivatives) is disarmingly simple, while the implementation of (11.62) is very easily carried out through Algorithm 11.1. Within the framework of stochastic timed automata that we have presented, one has to go through a number of steps: Check whether the sample function $L(\theta)$ is of the type (11.68) to (11.70) and whether it is continuous or not, then derive an expression for $dL/d\theta$ through the event time derivatives obtained in (11.67), then check for the additional technical conditions required for unbiasedness, and finally check for strong consistency. It is therefore helpful to summarize the process of checking whether IPA can be applied as an unbiased and strongly consistent estimator of the performance sensitivity $dJ/d\theta$ of a particular DES as shown in Fig. 11.21. This is a somewhat simplified description of the process, which usually starts by checking whether the commuting condition **(CC)** applies or not. If it does, under generally mild assumptions on the event lifetime distributions $G_\alpha(\cdot; \theta)$, sample functions of the type (11.68) to (11.70) are continuous and one can proceed to derive an expression for $dL/d\theta$. Under some additional conditions, the IPA estimator $dL/d\theta$ of $dJ/d\theta$ is unbiased. Finally, under unbiasedness and some additional conditions, the IPA estimator is also strongly consistent.

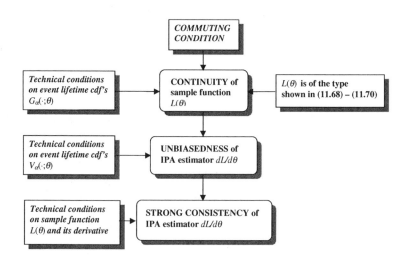

Figure 11.21: A simple roadmap for checking the properties/limitations of IPA.

Sensitivity Analysis of Queueing Networks

One of the areas where IPA has traditionally found most of its applications is in DES modeled as queueing networks. The performance measures considered are such that their sample functions are of the type (11.68) to (11.70), which includes many common metrics like throughput, mean system times or waiting times, and mean queue lengths. For the purpose of sensitivity analysis, we consider parameters of a service time distribution or, in the case of open networks, parameters of some interarrival time distribution. The special structure of queueing networks (i.e., their state transition mechanisms) makes it possible to develop IPA estimators by extending the approach we presented in section (11.4) for the $GI/G/1$ system (for a detailed analysis, see Ho and Cao, 1991, and references therein), without necessarily resorting to the framework of stochastic timed automata presented above.

In some types of networks, it is actually possible to derive equations whose solution provides the *realization probabilities* of perturbations generated at some node in the network (Cao, 1987), through which performance sensitivities can be obtained. The concept of "realization probability" of a perturbation is based on the fact that once a perturbation is generated (e.g., a particular service time at node i of the network is perturbed), then it is either eventually propagated to all nodes (in which case it is "realized") or it is canceled. To see how a perturbation can be canceled, consider the sample path of the $GI/G/1$ system in Fig. 11.10 and suppose that only the first service time was perturbed. This perturbation (marked as a "1" in the figure) propagates to all customers of the first busy period, but it is subsequently canceled by the presence of the idle period that follows; note that it does not appear in the second busy period in Fig. 11.10. A similar, but much more intricate, process takes place in networks, where an idle period at node j terminated by a customer arrival from node i forces the coupling of the two nodes: If node i has a perturbation equal to zero, whereas node j has a positive perturbation, then, after the idle period ends both nodes end up with the same perturbation (in this case, 0).

Viewing queueing networks as special cases of stochastic timed automata makes it fairly straightforward to check the commuting condition (**CC**) as a first step, before attempting to explicitly derive sample derivatives $dL/d\theta$. Although this is readily accomplished on a case-by-case basis, it can be shown (Glasserman, 1991) that all *Jackson-like networks* satisfy (**CC**). This is the class of (open or closed) networks where (a) every node consists of one server with infinite queueing capacity operating under the First-Come-First-Served (FCFS) discipline, (b) the network serves a single class of customers, and (c) routing of customers is purely probabilistic (i.e., a customer completing service at node i is routed to node j with some given probability q_{ij}). Note that no constraints are imposed on the service or interarrival time distributions.

In terms of implementing IPA estimates for queueing network sensitivities, one can always use Algorithm 11.2 to determine event time derivatives. Transforming those into derivatives of specific sample functions is generally not a difficult task (as seen, for instance, in Example 11.7 for the mean system time of a $GI/G/1$ system).

Example 11.12 (IPA for a two-node network)
Consider a simple open network as shown in Fig. 11.22 along with its state transition diagram. The event set is $\mathcal{E} = \{a, d_1, d_2\}$, where a is an external arrival and d_i is a departure from node $i = 1, 2$. The state is represented by (x_1, x_2) where x_i is the number of customers residing at node $i = 1, 2$. Let the parameter of interest be the mean of the first node's service time, θ, and consider throughput as a performance measure. Thus, sample functions are of the form $L(\theta) = N/T_{d_2,N}$, where N is the number of departures observed at node 2, and the sensitivity is easily determined from the sensitivity of the event time $T_{d_2,N}$:

$$\frac{dL}{d\theta} = \frac{-N}{T_{d_2,N}^2} \left(\frac{dT_{d_2,N}}{d\theta} \right)$$

Following the process described in Fig. 11.21, our first task is to check the commuting condition **(CC)**. It is easy to verify by inspection of the state transition diagram that **(CC)** holds. For example, at state $(1,0)$ the sequence $\{a, d_1\}$ leads to the final state $(1,1)$ and so does the sequence $\{d_1, a\}$. Next, assuming that service times at node 1 are such that assumptions **(A1)** and **(A2)** in Section 11.5.1 are satisfied, we proceed to evaluate the event time derivatives $dT_{d_2,N}/d\theta$. This can be accomplished through Algorithm 11.2. Let Δ_a denote the perturbation accumulator for a events, and immediately note that $\Delta_a = 0$ always, since θ cannot affect the arrival process. Let Δ_1 and Δ_2 denote the perturbation accumulators for d_1 and d_2 events respectively. Then, Algorithm 11.2 for this system becomes:

1. INITIALIZATION: $\Delta_a := 0$, $\Delta_1 := 0$, $\Delta_2 := 0$
2. WHENEVER EVENT a IS OBSERVED:

   If $x_1 = 0$: (d_1 activated when a starts a busy period at node 1)
   2.1. Compute $dV_{d_1}/d\theta$ through (11.66)
   2.2. $\Delta_1 := dV_{d_1}/d\theta$

3. WHENEVER EVENT d_1 IS OBSERVED:

   If $x_1 > 1$: (d_1 activated within a busy period)
   3.1. Compute $dV_{d_1}/d\theta$ through (11.66)
   3.2. $\Delta_1 := \Delta_1 + dV_{d_1}/d\theta$

   If $x_2 = 0$: (d_2 activated when d_1 starts busy period at node 2)
   3.3. $\Delta_2 := \Delta_1$

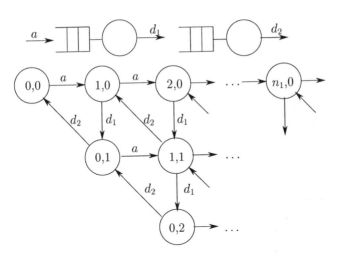

Figure 11.22: Two-node network and its state transition diagram for Example 11.12.

Note that the occurrence of d_2 has no effect on Δ_2, since it can only activate another d_2 event if $x_2 > 1$, resulting in $\Delta_2 := \Delta_2$; nor can it affect Δ_a or Δ_1. This algorithm stops after N events of type d_2 are observed, at which point the IPA estimator of the throughput sensitivity is simply given by $(-N/T_{d_2,N}^2) \cdot \Delta_2$.

It is also interesting to note (recalling also our brief discussion of realization probabilities above) that perturbations generated at node 1 propagate to node 2 when d_1 ends an idle period at node 2 (step 3.3), or they are canceled by a zero perturbation carried by the external arrival process whenever a ends an idle period at node 1 (step 2.2). It is also worth pointing out that only information regarding the service time cdf at node 1 is required; both the arrival process and the service process at node 2 are arbitrary. Finally, we take this opportunity to call once again attention to the fact that the algorithm above (and hence the evaluation of the IPA sensitivity estimator) are easily implemented on line and are based on simple information from a single observed sample path. Moreover, the extension to a system consisting of $n > 2$ nodes in series is straightforward.

Beyond Jackson-like networks, the use of IPA is restricted. For example, when multiple customer classes are present, **(CC)** is satisfied if and only if every node that serves more than one class has a single input node connected to it (Glasserman, 1991). Another serious limitation is imposed by blocking phenomena, as we already saw in Example 11.10. Although in the case of closed networks where each finite-capacity node has a single input node connected to it **(CC)** is still valid, in more general cases, the application of PA techniques to blocking systems requires different approaches which will be presented in subsequent sections.

Regarding sensitivities with respect to (continuous) parameters other than those of service and interarrival time distributions, *routing probabilities* are of particular interest. In the context of controlling DES, recall that routing (discussed in Chapter 9) is one of the main problems encountered in resource contention environments. If we restrict ourselves to IPA techniques, then sensitivities with respect to routing probabilities can be obtained for the class of product form networks. This is because routing probability perturbations in this case can be conveniently transformed into event time perturbations (for details, see Ho and Cao, 1991).

Finally, regarding sample functions other than those in (11.68) to (11.70), it is often the case that they are inherently discontinuous in θ, which makes IPA immediately inapplicable in the sense that it yields biased sensitivity estimates. In order to handle such cases, we will have to resort to more advanced PA techniques discussed later in this chapter.

Performance Optimization

In Section 11.2.2, we pointed out that sensitivity information is essential in using a variety of optimization algorithms, normally of the form shown in (11.6). The fact that IPA provides derivative estimates from a single observed sample path has significant implications in the optimization of DES. When complexity is such that analytical techniques are inadequate, the only feasible approach is to proceed by trial-and-error over the range of a parameter θ until an optimal point is found. On the other hand, for such complex systems, one can often use IPA estimates in (11.6) to make parametric adjustments on line and *without interfering with the regular operation of a system:* One simply evaluates an IPA estimator of the form $dL/d\theta$ "in the background", then adjusts θ and repeats the process, thus providing gradual performance improvements. The fact that few modeling assumptions are required in IPA estimation is also of great importance: It implies that one can control a complex *system without having full knowledge of a model for it.* Recognizing these facts, several applications of IPA to on-line optimization problems have been developed and analyzed (see, for example, Cassandras and Lee, 1988; Suri and Leung, 1989; Fu, 1990; Chong and Ramadge, 1992).

A particularly interesting class of applications involves problems where optimization can be carried out in *decentralized* fashion. This situation often arises in resource allocation problems. To take a specific example of practical interest, consider the problem of routing in packet-switched communication networks: A packet enters the network at some "origin" node a and must be sent to a "destination" node b. Along the way, a packet travels over links connecting various nodes together. Upon leaving a typical node i, a packet can choose among several links connecting i to neighboring nodes. The purpose of such routing decisions is often to minimize the mean delay experienced by all packets over all possible origin-destination pairs over some period of time. It turns out that it is possible to meet this global objective by making decisions in distributed fashion (Gallager, 1977): When a packet leaves node i, its routing decision is made based solely on local information, without any need for an omniscient central routing controller. In particular, the information required at node i is the set of mean delay sensitivities for all outgoing links with respect to the corresponding packet flow rates. In other words, if the packet flow along link (i, j) is λ_{ij} and the corresponding mean delay is D_{ij}, the required information is $dD_{ij}/d\lambda_{ij}$. If links are modeled as $G/G/1$ queueing systems, then what we require is the sensitivity of the mean system time in such a system with respect to its arrival rate (a control parameter in this setting). This problem is ideally suited for IPA, since it only requires the analysis of a single-server system (not an entire network), much like the analysis of Section 11.4. When packet interarrival times and link transmission times have unknown distributions, expressions for $dD_{ij}/d\lambda_{ij}$ are unavailable. On the other hand, IPA requires only knowledge of the arrival process to a link in order to compute link interarrival time derivatives, as required in an IPA estimator of the form (11.62). However, one is often justified in assuming that λ_{ij} is a scale parameter of the interarrival time distribution, in which case (11.22) can be used. Thus, this complex network routing problem can be effectively tackled in practice through the use of simple IPA estimators distributed over links (further details are provided in Cassandras et al., 1990). A more general setting for such decentralized control of DES was recently developed in Vázquez-Abad et al., 1998.

11.6   SENSITIVITY ESTIMATION REVISITED

Let us take a closer look at the basic problem of estimating a sensitivity of the form $dJ/d\theta$ for some given performance measure $J(\theta) = E[L(\theta, \omega)]$. Thus far, we have used the symbol ω to denote a sample path. To be precise, however, a sample path is in fact a sequence $\{X_1(\theta), X_2(\theta), \ldots\}$, so we should write

$$J(\theta) = E\left[L[X_1(\theta), X_2(\theta), \ldots]\right]$$

where the expectation is now over all $X_1(\theta), X_2(\theta), \ldots$ For example, in our $GI/G/1$ queueing system of Section 11.4, the mean system time is evaluated as an expectation over all service and interarrival times that determine a sample path.

To keep notation simple, let us consider the scalar case $J(\theta) = E[L[X(\theta)]]$, and let the cdf of $X(\theta)$ be $F(x; \theta)$. Then, this expectation is given by

$$J(\theta) = \int_{-\infty}^{\infty} L[x(\theta)]dF(x; \theta) \tag{11.82}$$

With this expression as our starting point, we will now see how a differentiation with respect to θ can be carried out in two different ways.

1. The IPA approach. We have already seen in (11.14) that we can view $X(\theta)$ as the inverse transform $X(\theta) = F^{-1}(U;\theta)$, where U is uniformly distributed in $[0,1]$. In other words, the expectation in (11.82) may be taken over $u \in [0,1]$ instead of x, with $u = F(x;\theta)$, that is,

$$J(\theta) = \int_0^1 L[x(\theta, u)]du \qquad (11.83)$$

The interpretation here is that the sample path is really dependent on a uniformly distributed random variable (over which the expectation is taken), which specifies another random variable $X(\theta, U)$, which in turn determines the sample function $L[X(\theta, U)]$.

If we now differentiate (11.83) with respect to θ, we get

$$\frac{dJ}{d\theta} = \int_0^1 \frac{\partial L[x(\theta, u)]}{\partial \theta}du = E\left[\frac{\partial L[X(\theta, U)]}{\partial \theta}\right] \qquad (11.84)$$

provided that the interchange of expectation and differentiation above is allowed. This is the whole basis of IPA: If this interchange is allowed, then the sample derivative $dL/d\theta$ is indeed an unbiased estimator of the performance derivative $dJ/d\theta$ (Cao, 1985). Furthermore, we can see in (11.84) that the sample derivative can be written as follows:

$$\frac{\partial L[X(\theta, U)]}{\partial \theta} = \frac{\partial L[X(\theta, U)]}{\partial X} \cdot \frac{\partial X(\theta, U)}{\partial \theta}$$

which reflects our earlier observations that IPA consists of two parts: perturbation *generation*, that is, how changes in θ introduce changes in $X(\theta)$; and perturbation *propagation*, that is, how a change in $X(\theta)$ ultimately affects the sample function. Observe that in this whole process U (the underlying random number) is kept fixed. This is why this approach is also known as "common random numbers".

2. The Likelihood Ratio approach. Let us now return to (11.82) and proceed in a different way. Suppose that $f(x;\theta)$ is the pdf corresponding to $F(x;\theta)$, so that

$$J(\theta) = \int_{-\infty}^{\infty} L(x)f(x;\theta)dx \qquad (11.85)$$

We have written $L(x)$ here, instead of $L[x(\theta)]$, because we will now view L, a specific realization of the performance metric, as remaining fixed as θ varies (this is why this approach is referred to as "common realizations"). If we differentiate with respect to θ, we now get

$$\frac{dJ}{d\theta} = \int_{-\infty}^{\infty} L(x)\frac{\partial f(x;\theta)}{\partial \theta}dx$$

provided that the interchange of expectation and differentiation in (11.85) *is* allowed. Moreover, observing that

$$\frac{\partial \ln f(x;\theta)}{\partial \theta} = \frac{\partial f(x;\theta)}{\partial \theta} \frac{1}{f(x;\theta)}$$

we get, assuming $f(x;\theta) \neq 0$,

$$\frac{dJ}{d\theta} = \int_{-\infty}^{\infty} L(x) \frac{\partial \ln f(x;\theta)}{\partial \theta} f(x;\theta)dx = E\left[L(X)\frac{\partial \ln f(X;\theta)}{\partial \theta}\right] \qquad (11.86)$$

In this way, we obtain an alternative unbiased estimator of $dJ/d\theta$, given by the term in brackets (11.86). This is known as the *Likelihood Ratio* (LR) estimator (sometimes also referred to as a *Score Function* estimator). We emphasize again that in this case the parameter θ is viewed as affecting the probability distribution of observing a particular value of the sample function, but not that value. This must be contrasted to the IPA approach where we view θ as affecting the sample function itself.

Whereas the IPA approach explicitly seeks to exploit the structure of event-driven sample paths in evaluating $dL/d\theta$, the LR approach does not attempt to do so. In this respect, it is a general-purpose methodology for obtaining derivatives of performance metrics of stochastic processes, not necessarily DES. This, perhaps, explains the fact that it appears to have been used in the 1960s to deal with problems not involving DES (Aleksandrov et al., 1968). In the 1980s, however, with the increasing popularity of discrete-event simulation, this methodology independently emerged as an effective means to obtain sensitivity estimates from a single simulation run (Glynn, 1986; Reiman and Weiss, 1989; Rubinstein, 1989).

A comparison of the IPA and LR approaches reveals several interesting differences. We shall briefly mention a few here, and refer the reader to Ho and Cao (1991) and related references for further details. First of all, both approaches require an interchange of expectation and differentiation in (11.83) and (11.85) respectively. However, the conditions for this interchange in IPA heavily depend on the nature of $L(\theta)$, which is not the case in LR. Therefore, in general, it is easier to satisfy LR unbiasedness conditions. On the other hand, in LR we need the constraint $f(x;\theta) \neq 0$. In fact, if $f(x;\theta)$ has a finite support (that is, x can only take values in some finite interval) which changes with θ, the LR approach cannot work. This is because of the assumption that $L(x)$ remains fixed in (11.85); therefore, the sample space for $X(\theta)$ cannot be altered by a change in θ (this excludes, for example, uniform distributions for event lifetimes).

In terms of information required to implement these estimators, both approaches require knowledge of the cdf (or pdf) in which θ appears as a parameter. In IPA, this is needed to evaluate the event lifetime derivatives in (11.66). In LR, it is needed to compute $\partial \ln f(x;\theta)/\partial \theta$ in (11.86). However, in IPA one may only need to know that θ is a scale or location parameter, without full knowledge of the associated event lifetime cdf (for instance, if θ is the mean of a cdf in the Erlang family, it is a scale parameter, and (11.22) can be used regardless of the actual cdf). Moreover, a scale parameter assumption, when no other information is available, usually leads to small errors in the estimator (see Cassandras et al., 1991). Such "robustness" properties with respect to modeling assumptions are not present in LR, where one must have exact knowledge of $f(x;\theta)$. On the other hand, the nature of θ (which can only affect event lifetime distributions in IPA) may be more general in the LR approach.

In terms of ease of implementation (assuming we have determined expressions for both estimators) IPA and LR are roughly comparable in most cases, although IPA estimators can sometimes get a bit involved. It should be pointed out, however, that one can immediately see how to derive an LR estimator, since $L(\theta)$ is directly observed and $\partial \ln f(x;\theta)/\partial\theta$ is readily evaluated; on the other hand, the evaluation of $dL/d\theta$ in IPA usually requires some analysis and is not immediate. This makes LR attractive when the effort required to derive an expression for $dL/d\theta$ is considerable.

Perhaps the most important criterion for comparison lies in the question of "accuracy" of an estimator, typically measured through its variance. If an estimator is strongly consistent, its variance is gradually reduced over time and ultimately goes to zero. The speed with which this happens can be extremely important: If after a time period T an estimator has variance $V_1(T)$ and a second estimator has variance $V_2(T) = 2V_1(T)$, this implies that it will roughly take T additional time units for the second estimator to reach the same level of accuracy. Since decisions, in practice, normally have to be made in limited time, an estimator whose variance decreases fast is highly desirable. For some simple systems where it is possible to explicitly compute variances of both IPA and LR estimators, it can be shown that the variance of an LR estimator is significantly larger than that of its IPA counterpart (for a two-node cyclic network with a single customer it is about 13 times larger; see Ho and Cao, 1991). In general, when IPA provides unbiased estimators, the variance of these estimators is small. Although it is hard to formally prove that IPA sensitivity estimators have the lowest possible variance, this may very well be the case based on the fact that *IPA fully exploits the structure of DES and their state dynamics* by extracting as much information as possible from the observed sample path.

A major limitation of LR estimators is that their variance not only does not decrease, but it may actually increase with time. It is only in special cases where the system of interest has a regenerative structure (as discussed in Section 10.7.4 of Chapter 10) that LR estimators are strongly consistent. In queueing networks for instance (where IPA is usually applied), determining regenerative intervals is difficult, if at all practically feasible. This imposes a major obstacle for the practical use of LR estimators in systems other than simple ones. It is possible to overcome some of these difficulties through various recently developed techniques, but the added complications make the initial appealing simplicity of the LR approach less attractive.

11.7   EXTENSIONS OF IPA

IPA, as presented up to this point, is far from the whole PA story. When IPA fails (because the commuting condition is violated or a sample function not conforming to the types defined in (11.68) to (11.70) exhibits discontinuities in θ), one can still find ways to derive unbiased performance sensitivity estimates. There are two ways to accomplish this: (a) by modifying the stochastic timed automaton model so that IPA is made to work, and (b) by paying the price of more information collected from the observed sample path, in which case, the same essential PA philosophy can lead to unbiased and strongly consistent estimators, but these are no longer as simple as IPA ones.

In this section, we will concentrate on (a) above, and leave (b) for the next section. The main idea here is that there may be more than one way to construct a GSMP describing a physical process, and while one way leads to discontinuous sample functions $L(\theta)$, another does not. To put it another way: If one model does not suit us vis-a-vis the continuity of $L(\theta)$, let us try to find a more convenient, stochastically equivalent model for the same DES. Since there does not yet appear to be a unified setting with firm guidelines for this approach, we will limit ourselves here to a few examples that illustrate the main "tricks" one can use to overcome the discontinuity problem.

11.7.1   Discontinuities due to Multiple Customer Classes

Consider a $GI/G/1$ queueing system serving two classes of customers. The fraction of customers which are of class 1 is denoted by θ. We assume that when a customer arrives, the class is determined by a probabilistic mechanism so that the customer is assigned to class 1 with probability θ and to class 2 with probability $1 - \theta$. The performance measure of interest is the mean system time over all customers. The system time of the ith customer is denoted by $S_i(\theta)$.

Model 1. With the description of the system above, the following stochastic state automaton model suggests itself. The event set is $\mathcal{E} = \{a, d_1, d_2\}$, where a is an arrival and d_i is a class i departure, $i = 1, 2$. The service time distribution of class i customers is denoted by $G_i(\cdot), i = 1, 2$. The state is described by a vector $[x_1, x_2, \ldots, x_n]$ where n is the queue length and $x_k \in \{1, 2\}$ is the class of the customer in the kth queue position, $k = 1, \ldots, n$. If the system is empty, we use 0 to represent this state.

Clearly, θ does not affect any of the three event lifetime distributions, so the framework of Section 11.5 cannot be applied. In fact, θ affects the state transition mechanism as follows. The state transition probabilities when the state is $[x_1, x_2, \ldots, x_n]$ and a occurs are:

$$p([x_1, x_2, \ldots, x_n, 1]; [x_1, x_2, \ldots, x_n], a) = \theta,$$
$$p([x_1, x_2, \ldots, x_n, 2]; [x_1, x_2, \ldots, x_n], a) = 1 - \theta$$

In other words, $x_{n+1} = 1$ is added to the vector $[x_1, x_2, \ldots, x_n]$ with probability θ. It is important that we also specify the precise mechanism through which the class assignment is done. Let the random variable Y_i denote the class of the ith arriving customer. Thus, Y_i is defined so that $Y_i = 1$ with probability θ and $Y_i = 2$ with probability $1 - \theta$. To generate a random variate Y_i, we adopt the usual inverse transform technique, in this case applied to a discrete cdf. Thus, given some U_i uniformly distributed in $[0, 1]$, we simply define, as in (10.9) of Chapter 10:

$$Y_i = \begin{cases} 1 & \text{if } 0 \leq U_i \leq \theta \\ 2 & \text{if } \theta < U_i \leq 1 \end{cases} \tag{11.87}$$

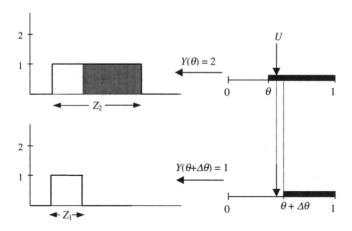

Figure 11.23: Illustrating a sample function discontinuity due to multiple customer classes.

Since $U > \theta$, a customer is assigned the class $Y(\theta) = 2$ and the sample function takes a value given by Z_2. For the same U, but new parameter value $\theta + \Delta\theta$, since $U < \theta + \Delta\theta$, the customer switches class and hence service time. This results in a sample function jump given by $(Z_2 - Z_1)$.

Even though we can no longer use the commuting condition in this model, it is not very difficult to see that the sample function, which is of the form

$$L_N(\theta) = \frac{1}{N} \sum_{i=1}^{N} S_i(\theta)$$

has discontinuities in θ. Intuitively, as θ changes so does the identity of some customers in the queue, and since each class has different service characteristics, $L(\theta)$ experiences a jump when a service time of a class 1 customer is switched to a service time of a class 2 customer (and vice versa). Let us illustrate this through the following simple situation. Suppose the system is empty and the next arrival event takes place. Let Y be the class of the new customer. As shown in Fig. 11.23, under θ we have $Y(\theta) = 2$, since $U > \theta$. The customer immediately starts service and his service time is given by Z_2 (independent of θ). For a perturbation $\Delta\theta$ as shown in Fig. 11.23, we have $\theta < U < \theta + \Delta\theta$. Therefore, under $\theta + \Delta\theta$ we have $Y(\theta + \Delta\theta) = 1$. In this case, it is a class 1 customer who immediately starts service with a service time $Z_1 \neq Z_2$ in general (in Fig. 11.23, $Z_1 < Z_2$). It is clear that as $\Delta\theta$ starts from $\Delta\theta = 0$ and then increases, a discontinuity arises at the point where $\theta + \Delta\theta = U$. In fact, the jump in $L(\theta)$ in this case corresponds to the shaded area shown, or, in other words the service time difference $Z_2 - Z_1$.

Model 2. Observe that even though each customer is assigned a class upon arrival, this class only plays a role when the customer is about to start service. Therefore, we might as well postpone the class assignment in our model until that time. This simplifies the state description as follows. Instead of a vector, it now suffices to describe the state through a pair (x, y) where x is the queue length and y the class of the customer in service. If the system is empty, we still use 0 to represent this state.

In this model, when a departure event (of either type) takes place or an arrival occurs at the empty state, a random number U is used to assign the class of the next customer in service through (11.87). Immediately after $Y(\theta)$ is determined, another random number V is needed to determine the service time through $Z_1 = G_1^{-1}(V)$ if $Y(\theta) = 1$ and $Z_2 = G_2^{-1}(V)$ if $Y(\theta) = 2$. Equivalently, the same random number U may be used to determine the service time as well as the class, as shown in Fig. 11.24. If $U \leq \theta$, then $Y(\theta) = 1$ and the service time is immediately given by

$$Z_1 = G_1^{-1}\left(\frac{U}{\theta}\right)$$

where the scaling factor θ is used here because U is conditioned to be less than or equal to θ, therefore $U \in [0, \theta]$ instead of $[0, 1]$. A quick check can help us verify that Z_1 is indeed drawn from the required distribution. Given that U

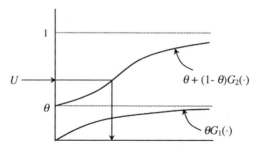

Figure 11.24: Service time generation in **Model 2**.

has already been found to be less than or equal to θ,

$$P[Z_1 \leq z \mid U \leq \theta]$$
$$= P[G_1^{-1}(U/\theta) \leq z \mid U \leq \theta] = P[U \leq \theta G_1(z) \mid U \leq \theta]$$

Observing that $P[U \leq \theta G_1(z) \ \ and \ \ U \leq \theta] = P[U \leq \theta G_1(z)] = \theta G_1(z)$
since U is uniformly distributed in $[0, 1]$, we get

$$P[Z_1 \leq z \mid U \leq \theta] = \frac{P[U \leq \theta G_1(z) \ and \ U \leq \theta]}{P[U \leq \theta]}$$
$$= \frac{P[U \leq \theta G_1(z)]}{P[U \leq \theta]} = G_1(z)$$

as required. Similarly, if $U > \theta$, then the service time is given by

$$Z_2 = G_2^{-1}\left(\frac{U - \theta}{1 - \theta}\right)$$

since $U \in [\theta, 1]$. Figure 11.24 also has the virtue of allowing us to clearly visualize the discontinuity problem: For some fixed $U = u$ imagine θ varying; at $\theta = u^+$, a service time is assigned through $\theta G_1(\cdot)$, whereas at $\theta = u^-$ the service time jumps and is assigned through $\theta + (1 - \theta)G_2(\cdot)$.

Although this model modification does not remedy the discontinuity problem, it does provide a simplification that naturally leads to the third model below.

Model 3. Suppose that the assignment of Z_1 in the previous model is changed to

$$Z_1 = G_1^{-1}\left(1 - \frac{U}{\theta}\right)$$

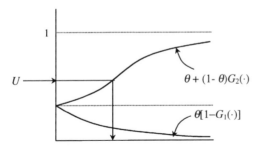

Figure 11.25: Service time generation in **Model 3**.

We can check that this is still a valid construction for the random variate Z_1 as follows:

$$P[Z_1 \leq z \mid U \leq \theta] = P[G_1^{-1}(1 - U/\theta) \leq z \mid U \leq \theta]$$
$$= P[U \geq \theta[1 - G_1(z)] \mid U \leq \theta]$$
$$= \frac{P[\theta[1 - G_1(z)] \leq U \leq \theta]}{P[U \leq \theta]}$$
$$= \frac{\theta - \theta[1 - G_1(z)]}{\theta} = G_1(z)$$

Note that this is equivalent to using the function $\theta[1 - G_1(\cdot)]$ for the inverse transform above, as shown in Fig. 11.25. The interesting thing about this "trick" is that it eliminates the discontinuity, provided that $G_1(0) = G_2(0) = 0$. This is clearly seen in Fig. 11.25 where the two curves are joined at θ as shown (we omit a formal proof; see also Ho and Hu, 1990). In this model, the service time distribution is therefore given by

$$G(z; \theta) = \begin{cases} \theta[1 - G_1(z)] & \text{if } 0 \leq U \leq \theta \\ \theta + (1 - \theta)G_1(z) & \text{if } \theta < U \leq 1 \end{cases} \tag{11.88}$$

Remark. The trick we have used here, that is, using the random number $(1 - U/\theta)$ instead of U/θ was briefly discussed in Example 10.6 of Chapter 10 in the context of improving the efficiency of random variate generation. As mentioned at that point, it cannot be applied to all cdf's. In particular, it may not apply to situations where the inverse transform $X = F^{-1}(U)$ is not continuous (e.g., a triangular distribution). Such cases, however, are excluded by the basic assumptions on which our IPA development is based.

Model 4. We can find yet another way to eliminate the sample function discontinuity, without resorting to the requirement $G_1(0) = G_2(0) = 0$ above, by changing the model once again. The starting point here is to observe that if Z denotes a service time in this system, then

$$Z = \begin{cases} Z_1 & \text{if } 0 \leq U \leq \theta \\ Z_2 & \text{if } \theta < U \leq 1 \end{cases}$$

where $Z_1 \sim G_1(\cdot)$ and $Z_2 \sim G_2(\cdot)$ are random variables corresponding to class 1 and class 2 customers respectively. We then have, by conditioning on the class type,

$$\begin{aligned} P[Z \leq z] &= P[Z \leq z \mid U \leq \theta]P[U \leq \theta] + P[Z \leq z \mid U > \theta]P[U > \theta] \\ &= P[Z_1 \leq z]\theta + P[Z_2 \leq z](1 - \theta) \\ &= \theta G_1(z) + (1 - \theta)G_2(z) \end{aligned}$$

In other words, a service time can be generated from a cdf

$$G(z; \theta) = \theta G_1(z) + (1 - \theta)G_2(z) \tag{11.89}$$

With this observation in mind, let us modify the model by defining an event set $\mathcal{E} = \{a, d\}$, where we do not differentiate between the two types of departures. As long as we are only interested in the mean system time over all customers (not the mean system time by class) it is not necessary that we draw such a distinction anyway. The new event, d, is assigned a lifetime distribution $G(z; \theta)$ as in (11.89) and as shown in Fig. 11.26. Observe that the new service time distribution in (11.89) does depend on θ. It is also easy to see that it is continuous in θ, which makes our task quite simple from this point on, as described next.

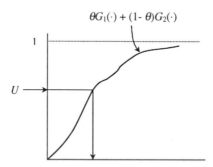

Figure 11.26: Service time generation in **Model 4**.

Either **Model 3** or **Model 4** allows us to treat the original system like a $GI/G/1$ queueing system where the parameter of interest, the class assignment probability θ, affects service times through (11.88) for **Model 3** or (11.89) for **Model 4**. To estimate the mean system time sensitivity, we can therefore resort to the IPA estimator (11.62) or, equivalently, Algorithm 11.1. The service time derivatives in (11.62) are obtained using (11.19) with either (11.88) or (11.89) as the service time distribution.

In the case of **Model 4**, the service time derivatives are obtained from (11.19) once we evaluate:

$$\frac{\partial G}{\partial \theta}\Big|_{(Z,\theta)} = G_1(Z) - G_2(Z) \quad \text{and}$$
$$\frac{\partial G}{\partial z}\Big|_{(Z,\theta)} = \theta \frac{dG_1}{dx}\Big|_{(Z,\theta)} + (1-\theta)\frac{dG_2}{dx}\Big|_{(Z,\theta)}$$

It is interesting to notice that if $G_1(\cdot) = G_2(\cdot)$ above, we get $dZ/d\theta = 0$. This is consistent with the fact that if customers are not distinguished by their service time characteristics, the parameter θ can have no effect on the mean system time.

In short, either **Model 3** or **Model 4** can be used to bypass the discontinuity problem in $L(\theta)$ by finding an alternative stochastic timed automaton that represents the same underlying DES. In fact, **Model 4** reduces the problem of sensitivity estimation to the $GI/G/1$ setting we have already analyzed in Sections 11.4 and 11.5. The basic intuition behind the model modifications that we seek in order to overcome sample function discontinuities can be best summarized as follows: When a parameter introduces inherently large perturbations (e.g., changing service times by large amounts when class switching takes place), *we can often replace one large perturbation by a series of small ones*. In our alternative models, instead of waiting for one "large" change in service time (as shown in Fig. 11.23), we have introduced perturbations to every service time through (11.88) or (11.89).

Remark. It is important to call attention to the following: The fact that the original model is changed does not mean that we are interfering in any way with the actual system. The alternative sample path construction is only a "mind game" we play in order to view real sample paths in the most convenient (and stochastically equivalent) way. The IPA estimator (11.62) that we ultimately use above is still driven by the *real* service times observed, whatever they might be. Our task is simply to find a convenient way for describing the effect of θ on these service times. This observation pertains to the next two problems as well.

11.7.2 Discontinuities due to Routing Decisions

Consider a queueing system with two servers in parallel (see Fig. 11.27), where each server is preceded by an infinite capacity queue. The system handles two customer classes which arrive independently, and each class may be served by either server. The service time distribution for class i customers served at server j is denoted by $G_{ij}(\cdot), i = 1, 2, j = 1, 2$. The four different distributions involved are generally different from each other. When a class i customer arrives in this system, he is routed to server j with probability q_{ij}. Let J_{ij} denote the mean waiting time of class $i = 1, 2$ when routed to server $j = 1, 2$. Thus, in this problem (Vakili and Ho, 1987) there are four performance measures of interest (the problem can of course be extended to N servers and M customer classes).

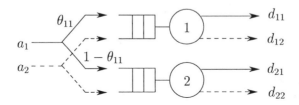

Figure 11.27: A two-class queueing system with routing.

Let us concentrate on a single performance measure, say J_{21}, as a function of a single parameter, say θ_{11}, and try to estimate the sensitivity $dJ_{21}/d\theta_{11}$.

Model 1. For this system, a natural stochastic state automaton model is the following. The event set is $\mathcal{E} = \{a_1, a_2, d_{11}, d_{12}, d_{21}, d_{22}\}$, where a_i is a class i arrival and d_{ij} is a class i departure from server j. The state is described by two vectors: $\mathbf{x} = [x_1, x_2, \ldots, x_n]$ where n is the queue length at server 1 and $x_k \in \{1, 2\}$ is the class of the customer in the kth queue position, $k = 1, \ldots, n$; and similarly $\mathbf{y} = [y_1, y_2, \ldots, y_m]$ for server 2, where m is the queue length at that server. If either queue is empty, we replace the corresponding vector by 0. Thus, we see that this model is very similar to Model 1 of the system in the previous section. The parameter of interest, the routing probability θ_{11}, does not affect any event lifetime distribution, so the framework of Section 11.5 cannot be applied. As in Section 11.7.1, the parameter affects the state transition mechanism as follows. When event a_1 occurs at a state $(\mathbf{x}, \mathbf{y})$, the state transition probabilities are given by

$$p(([x_1, x_2, \ldots, x_n, 1], \mathbf{y}); (\mathbf{x}, \mathbf{y}), a_1) = \theta_{11},$$
$$p((\mathbf{x}, [y_1, y_2, \ldots, y_m, 1]; (\mathbf{x}, \mathbf{y}), a_1) = 1 - \theta_{11}$$

In other words, a class 1 customer is routed to server 1 with probability θ_{11} and to server 2 otherwise. The mechanism through which the routing decision is made is similar to the mechanism through which a class assignment is done in **Model 1** of Section 11.7.1. Specifically, we define R_k to be a random variable representing the routing decision; thus, $R_k = 1$ with probability θ_{11} and $R_k = 2$ with probability $1 - \theta_{11}$. We then use the usual inverse transform technique, as in (11.87), to generate R_k from some random number U_k. By arguing as in the previous section, it is not difficult to see how changes in θ_{11} cause jumps in the sample function

$$L_N(\theta_{11}) = \frac{1}{N} \sum_{k=1}^{N} W_k^{21}(\theta_{11}),$$

where $W_k^{21}(\theta_{11})$ is the waiting time of the kth class 2 customer queued in front of server 1. An example similar to that of Fig. 11.23 shows that as a switch in the routing decision takes place because of a perturbation $\Delta\theta_{11}$, a class 1 customer originally routed to server 2 may now be routed to server 1; the presence of this additional customer in the queue of server 1 contributes an entire additional service time which will affect all subsequent class 2 customers in that queue prior to the next idle period. This constitutes a jump in $L_N(\theta_{11})$ above, and IPA will generally yield biased sensitivity estimates.

Model 2. The discontinuity problem can be resolved by proceeding in a way very similar to that of **Model 2** and then **Model 3** in Section 11.7.1. When a random number U is used to make the routing decision for an arriving class 1 customer, the same number is used to determine the service time of that customer as follows. If $U \leq \theta_{11}$, then the customer is routed to server 1 and assigned a service time from the distribution $\theta_{11}G_{11}(\cdot)$. If $U > \theta_{11}$, then the customer is routed to server 2. We can, however, imagine this customer as also being routed to server 1 with an assigned service time equal to 0. Clearly, this does not affect the waiting time of later customers: When this customer's turn comes to be served, he contributes 0 to all customers behind him. This process is shown on the left side of Fig. 11.28. Note that this is similar to Fig. 11.24, except that now the assignment for $U > \theta_{11}$ is always 0. We have not yet resolved the discontinuity problem, however: As θ_{11} varies on the left side of Fig. 11.28, we can see that there is a jump from 0 to $\theta_{11}G_{11}(\cdot)$.

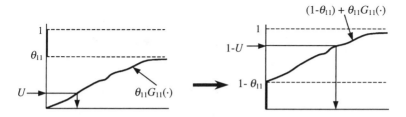

Figure 11.28: Service time assignment in **Model 2** (left) and **Model 3** (right).

Model 3. Having already gone through the trick used in **Model 3** of Section 11.7.1, we can now see how to get to the right side of Fig. 11.28. In this case, the service time distribution for class 1 customers routed to server 1 is given by

$$G(z; \theta_{11}) = \begin{cases} 0 & \text{if } 0 \le U \le 1 - \theta_{11} \\ (1 - \theta_{11}) + \theta_{11} G_{11}(z) & \text{if } 1 - \theta_{11} < U \le 1 \end{cases} \qquad (11.90)$$

Thus, when a "real" class 1 customer is routed to server 1 (i.e., $U > 1 - \theta_{11}$), his service time, Z, is obtained from $U = (1 - \theta_{11}) + \theta_{11} G_{11}(Z)$. Let us verify that Z, generated through this mechanism, is indeed distributed according to the original distribution $G_{11}(z)$:

$$\begin{aligned} P[Z_{11} \le z \mid U > 1 - \theta_{11}] &= P[G_{11}^{-1}((U - 1 + \theta_{11})/\theta_{11}) \le z \mid U > 1 - \theta_{11}] \\ &= P[((U - 1 + \theta_{11})/\theta_{11}) \le G_{11}(z) \mid U > 1 - \theta_{11}] \\ &= \frac{P[1 - \theta_{11} < U \le \theta_{11} G_{11}(z) + 1 - \theta_{11}]}{1 - P[U \le 1 - \theta_{11}]} = G_{11}(z) \end{aligned}$$

as required. The new equivalent service time distribution $G(z)$ in (11.90) is similar to that in (11.89). Once again, we can see that the discontinuity is eliminated. What this alternative model accomplishes is to replace large infrequent perturbations (that is, jumps in the sample function due to occasional switches in routing decisions) by a series of small class 1 service time perturbations at server 1. It remains to evaluate the sample derivative $dL_N(\theta_{11})/d\theta_{11}$ and apply IPA in the usual way (e.g., using Algorithm 11.2), which is left as an exercise.

11.7.3 Discontinuities due to Blocking: IPA with Event Rescheduling (RIPA)

Consider a $GI/G/1/K$ queueing system. We are interested in the mean customer system time as a function of some parameter of the service or of the interarrival time distribution. For simplicity, let θ be the mean of the service time distribution. This system can be studied in the framework of Section 11.5. In fact, we saw in Example 11.10 that the commuting condition **(CC)** is not satisfied; an arrival when the queue length is K followed by a departure leads to a final queue length $K - 1$ (since the arrival is rejected); whereas, a departure followed by an arrival leads to a final queue length K. In Fig. 11.20 we also saw the intuitive reason for the sample path discontinuity causing IPA to fail: An event order change when the queue is at capacity causes an additional customer to be accepted, which introduces a jump in the sample function.

Model 1. Let us start with a natural stochastic state automaton model for this system. The event set is $\mathcal{E} = \{a, d\}$, where a is an arrival and d is a departure. The state space is described by all feasible queue lengths, $\mathcal{X} = \{0, 1, \ldots, K\}$. The feasible event set is $\Gamma(x) = \{a, d\}$ for all $x > 0$, and $\Gamma(0) = \{a\}$. The state transition mechanism, $x' = f(x, e)$ with $x, x' \in \mathcal{X}$ and $e \in \mathcal{E}$, is deterministic: $f(x, a) = x + 1$ for all $x < K$, and $f(K, a) = K$; $f(x, d) = x - 1$ for all

$x > 0$. Finally, let $G_a(\cdot)$ and $G_d(\cdot; \theta)$ denote the interarrival and service time distributions respectively.

We will now limit ourselves to $M/G/1/K$ systems, that is, assume that the arrival process is Poisson with parameter λ: $G_a(t) = 1 - e^{-\lambda t}$. This does not change the fact that the commuting condition is violated. However, it allows us to exploit the memoryless property of the exponential interarrival time distribution as follows. When state $x = K$ is entered, the Poisson process is shut off, and it is turned back on when the next departure event causes a transition to state $K - 1$. This trick is not new; it was used in the analysis of the $M/M/1/K$ system in Section 8.6.4 of Chapter 8. The memoryless property of $G_a(t)$ allows us to treat the new arrival lifetime generated at the point where the process is turned back on the same way as the residual lifetime of the arrival event active when the process is not shut off. Thus, we are led to the following alternative model.

Model 2. The only modification to **Model 1** is the following. Instead of $\Gamma(x) = \{a, d\}$ for all $x > 0$, we set $\Gamma(x) = \{a, d\}$ for all $0 < x < K$, and $\Gamma(K) = \{d\}$. Clearly, the sample paths generated under the two models are stochastically indistinguishable. The general idea here is that we can always deactivate events e with exponential lifetime distributions at states such that $f(x, e) = x$, and *reschedule* them for occurrence when a new state y is entered such that $e \in \Gamma(y)$. This lends its name to the term "IPA with Rescheduling" (RIPA), as coined by Gong et al. (1991) who extended the idea presented in Gong and Glasserman, 1988.

Under **Model 2**, the state transition diagram in Fig. 11.19 is modified so that there is no self-loop transition at state K; the commuting condition is then satisfied (since there is only one feasible event at state K). Therefore, IPA, as developed in Section 11.5, provides unbiased sensitivity estimates for all sample functions of the type (11.68) to (11.70).

The RIPA approach can also be used to overcome discontinuities due to multiple classes when θ is a parameter of an event lifetime distribution (as opposed to the class assignment parameter of Section 11.7.1), as long as all class arrival processes are Poisson (see Gong et al., 1991). The rescheduling trick obviously fails for event processes which are not Poisson, but it can provide reasonable approximations in some cases.

11.8 SMOOTHED PERTURBATION ANALYSIS (SPA)

In order to overcome the problem IPA faces when dealing with discontinuous sample functions (which are often encountered in practice), the methodology of *Smoothed Perturbation Analysis* (SPA) was introduced by Gong and Ho (1987) (with the setting later extended by Glasserman and Gong, 1990, and Fu and Hu, 1997). There are two prices one has to pay in using SPA instead of IPA:

1. *More information extracted from the observed sample path.* In IPA, sensitivity estimators are based on simple observations of some event lifetimes (e.g., customer service times) and some event occurrences (e.g., detecting that a server is idle). In fact, if IPA is used in simulation experiments, event lifetimes have to be generated anyway, so the overhead incurred by sensitivity estimation is truly minimal. When a sample function $L(\theta)$ exhibits discontinuities, however, it is to be expected that more information should be extracted from the observed sample path in order to gain some knowledge about the magnitude of the jumps in $L(\theta)$. In a queueing system, for example, we may have to observe customer waiting times in addition to service times.

2. *More information about the DES model.* In IPA, we have seen that very little information regarding event lifetime distributions is needed. This makes the approach particularly attractive, since, in practice, this type of information may be very difficult to obtain. When $L(\theta)$ is discontinuous, however, we may need some extra knowledge regarding such distributions, which helps us evaluate the probability of occurrence of the jumps in $L(\theta)$.

Based on these observations, one can see that the approach to be presented next lies in the middle ground between sensitivity evaluation derived exclusively from a DES model and sensitivity estimation based exclusively on hard data from simulation or direct observation of a real system. In this context, IPA is but a special case of SPA where (a) minimal modeling information is required, and (b) sensitivity estimation based on data from a single sample path turns out to be particularly efficient.

The main idea of SPA lies in the "smoothing property" of conditional expectation. If we are willing to extract information from a sample path and denote it by $\mathcal{Z}$, then we can evaluate, not just the sample function $L(\theta)$, but also the conditional expectation $E[L(\theta) \mid \mathcal{Z}]$ (provided we have some distributional knowledge based on which this expectation can be evaluated). The result is typically a much smoother function of θ than $L(\theta)$ (in simple terms, expectation is an integration operation on a function, which tends to "smooth out" the function). Precisely how to define $\mathcal{Z}$, which in SPA terminology is called a *characterization* of the observed sample path, may not be an easy task. Every specific problem normally suggests natural quantities for a characterization, but it is difficult to provide general guidelines (except for certain classes of problems).

Before discussing the basic principles of SPA, let us motivate the "smoothing property" of conditional expectation through the following simple example.

Example 11.13

Let X and Y be two random variables, with X exponentially distributed with mean θ (i.e., $X \sim 1 - e^{-x/\theta}$, $\theta > 0$), and Y uniformly distributed in $[1, 2]$ (independent of θ). We are interested in the probability that $X(\theta)$

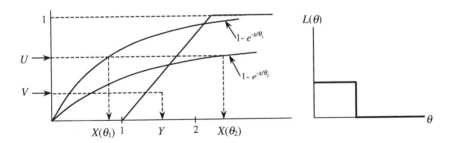

Figure 11.29: Generating a sample function for Example 11.13.
$L(\theta)$ *is piecewise constant.* $L(\theta) = 1$ *for values of* θ *such as* θ_1 *where* $X(\theta_1) < Y$, *and* $L(\theta) = 0$ *for values of* θ *such as* θ_2 *where* $X(\theta_2) > Y$.

is smaller than Y. Thus, we begin by defining the sample function

$$L(\theta) = \mathbf{1}[X(\theta) \leq Y]$$

where $\mathbf{1}[\cdot]$ is the indicator function. Thus, $L(\theta) = 1$ if $X(\theta) \leq Y$, and $L(\theta) = 0$ otherwise. This is a discontinuous function of θ. To see this more clearly, consider a sample path, which in this simple case consists of generating two random variates from the distributions $1 - e^{-x/\theta}$ and $U[1, 2]$, as shown in Fig. 11.29. We use underlying random numbers U and V to generate X and Y respectively. The values of U and V determine the value of the sample function. Note that when $\theta = \theta_1$ we get $X(\theta_1) < Y$, whereas (keeping U fixed) $X(\theta_2) > Y$. Thus, $L(\theta_1) = 1$, whereas $L(\theta_2) = 0$. It is now easy to see that $L(\theta)$ is a piecewise constant function: It is equal to 1 over some interval in which $X(\theta) \leq Y$, and then (letting θ vary) it jumps to 0 as soon as $X(\theta) > Y$. Note that, not only is $L(\theta)$ not continuous, but in fact $dL/d\theta = 0$ always; that is, used as a sensitivity estimate, this sample derivative is always asserting that θ has no effect on the performance measure

$$J(\theta) = E[L(\theta)] = E[\mathbf{1}[X(\theta) \leq Y]] = P[X(\theta) \leq Y]$$

Yet, it is immediately clear that $J(\theta)$ *does depend on* θ: If θ is chosen very small, there is a very high probability that $X(\theta) < Y$ (since $Y \geq 1$ always), while a large value of θ makes the probability that $X(\theta) > Y$ very large (since $Y \leq 2$ always). In other words, $dL/d\theta = 0$ must be a biased estimate of $dJ/d\theta$, and the fact that it is always 0 implies that it contains no information whatsoever about $dJ/d\theta$.

Let us now see how conditioning can help. Suppose that when a sample path is generated we observe the value of Y, say $Y = y$. Then, the

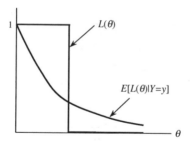

Figure 11.30: Illustrating the smoothing property of the conditional expectation.

conditional expectation of $L(\theta)$ given that $Y = y$ is

$$E[L(\theta) \mid Y = y] = E[\mathbf{1}[X(\theta) \le y]] = \int_0^\infty \mathbf{1}[x \le y] \frac{e^{-x/\theta}}{\theta} dx$$

Recalling the two points made earlier, observe that this process involves (a) making an observation of Y, and (b) knowledge of the pdf of X, based on which the conditional expectation above can be evaluated. Specifically,

$$E[L(\theta) \mid Y = y] = \frac{1}{\theta} \int_0^y e^{-x/\theta} dx = 1 - e^{-y/\theta} \tag{11.91}$$

This is clearly a continuous function of θ, as shown in Fig. 11.30 contrasted to $L(\theta)$. This illustrates the "smoothing" effect that conditioning has on sample functions. Setting $L_y(\theta) = E[L(\theta) \mid Y = y]$, note that $dL_y/d\theta \ne 0$ always, and it does convey useful information about the sensitivity of $P[X(\theta) \le Y]$ with respect to θ. For instance, we see that $dL_y/d\theta$ is small for large θ, which is to be expected since, as already pointed out, the probability that $X(\theta) > Y$ approaches 1 as θ increases; therefore, further increases in θ have a negligible effect on $P[X(\theta) \le Y]$.

For a simple interpretation of this example, we can think of a queueing system at $t = 0$ where a customer has just arrived and is beginning service with random service time $X(\theta)$, where θ is the mean of the service time distribution. The next customer's arrival time is a random variable Y. We are interested in the probability that the next customer will not have to wait, which is given by $P[X(\theta) \le Y] = E[\mathbf{1}[X(\theta) \le Y]$. We have seen that if we are willing to observe the next arrival time (but not the service time), and if we know the service time distribution (but not the interarrival time distribution), we can obtain the conditional expectation (11.91). At this point, we can expect that the unbiasedness property of IPA estimators established in Section 11.5 carries over to sensitivity estimators of the type $dL_y/d\theta$ obtained above. This is because we have

$J(\theta) = E[L_y(\theta)]$, where $L_y(\theta)$ is continuous in θ, and may be thought of as "a sample function that IPA can be applied to."

Of course, in the simple example above, the sample derivative $dL_y/d\theta$ is easily obtained by differentiating (11.91). This is because there is a single perturbation generated, $dX/d\theta$, and no state dynamics through which it might propagate and affect the sample function. In general, even if a characterization $\mathcal{Z}$ is identified and the conditional expectation $E[L(\theta) \mid \mathcal{Z}]$ is determined, the evaluation of $dE[L(\theta) \mid \mathcal{Z}]/d\theta$ remains an issue.

Let us now describe the SPA approach in more detail. Our starting point is equation (11.4):

$$\frac{dJ(\theta)}{d\theta} = E\left[\frac{dL(\theta,\omega)}{d\theta}\right]$$

where, as we have repeatedly seen already, the equality may not hold when $L(\theta,\omega)$ is not continuous in θ (remember that continuity of the sample function is a sufficient condition for the unbiasedness of IPA estimators, provided some additional technical conditions are satisfied). We now rewrite the left-hand side above as shown below, replacing $J(\theta) = E[L(\theta,\omega)]$ by the expectation of a conditional expectation:

$$\frac{dJ(\theta)}{d\theta} = \frac{d}{d\theta}E[(L(\theta,\omega)] = \frac{d}{d\theta}E[E[L(\theta,\omega) \mid \mathcal{Z}]] \qquad (11.92)$$

where the inner expectation is a conditional one, and the conditioning is on the characterization $\mathcal{Z}$. Now, treating $E[L(\theta,\omega) \mid \mathcal{Z}]$ as the new sample function, we expect it to be "smoother" than $L(\theta,\omega)$, and, in particular, continuous in θ. Then, under some additional conditions (comparable to those we made in our development of IPA) we can justify the interchange of differentiation and expectation in (11.92) to get

$$\frac{dJ(\theta)}{d\theta} = E\left[\frac{dE[L(\theta,\omega) \mid \mathcal{Z}]}{d\theta}\right]$$

Letting

$$L_{\mathcal{Z}}(\theta) = E[L(\theta,\omega) \mid \mathcal{Z}] \qquad (11.93)$$

denote this modified sample function, the SPA estimator of $dJ/d\theta$ is

$$\left[\frac{dJ}{d\theta}\right]_{SPA} = \frac{dL_{\mathcal{Z}}(\theta)}{d\theta} \qquad (11.94)$$

This is a general approach for obtaining performance sensitivity estimates when sample functions are not continuous in θ. Naturally, the idea is to minimize the amount of added information represented by $\mathcal{Z}$, since this incurs added costs we would like to avoid. However, there is no specific prescription pertaining to the choice of the characterization $\mathcal{Z}$. This generally depends on the sample function considered and the system under study. In addition, the choice of $\mathcal{Z}$ may be critical: An SPA estimator may be unbiased for some $\mathcal{Z}$, but not for another. A number of applications of SPA to queueing systems has shown that one can obtain unbiased estimators of sensitivities of performance measures that do not conform to the class defined by (11.68) to (11.70). Examples include the average number of customers in a busy period of a $GI/G/1$ queueing system, and the average queue length at a given final time t (see Ho and Cao, 1991; Glasserman, 1991). Note that in these examples the sample functions are event counters; these are common situations that give rise to discontinuities.

Fortunately, there are certain classes of problems where SPA can be applied under specific guidelines and yield very efficient sensitivity estimators (in the sense that little extra information is needed from the sample path, and the actual implementation of the SPA estimators is simple). We will present two such cases in what follows.

11.8.1 Systems with Real-Time Constraints

In our study of IPA, we focused on measures of the average performance of a system (e.g., mean system time, mean queue length), where sample functions are often continuous with respect to interesting parameters we can adjust. Recent applications in DES, on the other hand, have increasingly emphasized the importance of executing tasks "on time." For instance, we may be interested in guaranteeing that the nth occurrence of some event a takes place before a given "deadline" R (commonly used in "timeout" mechanisms, for example). Recalling the notation of Section 11.5, this means that we are interested in guaranteeing that $T_{\alpha,n} \leq R$. In such cases, performance is measured in terms of $P[T_{\alpha,n} > R]$, the probability of violating this requirement, which we seek to minimize. The deadline R is often referred to as a "real-time constraint." Applications of this type arise in communication networks (where we must deliver a message by a certain time, otherwise the information that this message contains becomes useless), computer systems (where a task must be executed within a certain time period), and generally DES intended to provide resources for time-critical operations.

With this motivation in mind, consider a DES and let $X_n(\theta)$ denote the time interval that elapses between two events of given types. For example, $X_n(\theta)$ could be the nth lifetime of an event α, defined as the interval from the instant when α is activated until the time it occurs; or it could be the interval between the nth occurrence of event α and the first subsequent occurrence of another event β. Let us also define $Y_n(\theta)$ in similar fashion. Both $X_n(\theta)$ and $Y_n(\theta)$ generally depend on a parameter θ. Let us further assume that $\{Y_n(\theta)\}$ forms an iid sequence with corresponding cdf $F_Y(\cdot;\theta)$. Finally, let R be some random variable. To simplify matters, however, let us just assume that R is a given constant.

We are interested in events of the type $[X_n(\theta) + Y_n(\theta) > R]$ and their probabilities. A natural sample function for this problem is

$$L_n(\theta) = \mathbf{1}[X_n(\theta) + Y_n(\theta) > R], \quad n = 1, 2, \ldots \qquad (11.95)$$

where $\mathbf{1}[\cdot]$ is the usual indicator function (as in Example 11.13). It is clear that this is a piecewise constant function (like the one we saw in Example 11.13) giving us an IPA sensitivity estimator $dL/d\theta = 0$ (wherever the derivative actually exists). This is of no use, since we generally expect θ to have an effect on the probability of events of the type $[X_n(\theta) + Y_n(\theta) > R]$; for instance, if θ can be adjusted to make all $X_n(\theta)$ larger as it increases, then obviously we expect this probability to increase (unless it is already 1).

This is a typical situation calling for SPA. Suppose that in observing sample paths of this system, we can observe $X_n(\theta)$ for all $n = 1, 2, \ldots$ Then, let us define the characterization $\mathcal{Z}_n = X_n(\theta)$, and hence a conditional expectation as in (11.93):

$$L_{\mathcal{Z}_n}(\theta) = E[\mathbf{1}[X_n(\theta) + Y_n(\theta) > R] \mid \mathcal{Z}_n]$$

which can be rewritten as

$$\begin{aligned} L_{\mathcal{Z}_n}(\theta) &= P[X_n(\theta) + Y_n(\theta) > R \mid X_n(\theta)] \\ &= P[Y_n(\theta) > R - X_n(\theta) \mid X_n(\theta)] \\ &= 1 - F_Y[R - X_n(\theta); \theta] \end{aligned}$$

recalling that $F_Y(\cdot; \theta)$ is the cdf of the iid sequence $\{Y_n(\theta)\}$. Observe that the continuity and differentiability of this modified sample function depend entirely on the properties of $F_Y(\cdot; \theta)$. Then, letting $f_Y(\cdot; \theta)$ denote the pdf obtained from $F_Y(\cdot; \theta)$, we can now obtain the sample derivative $dL_{\mathcal{Z}_n}(\theta)/d\theta$. In doing so, we must be careful to account for the fact that $F_Y(\cdot; \theta)$ generally depends on θ directly, as well as through $X_n(\theta)$. Therefore,

$$\frac{dL_{\mathcal{Z}_n}(\theta)}{d\theta} = f_Y[R - X_n(\theta); \theta] \cdot \frac{dX_n(\theta)}{d\theta} - \left. \frac{\partial F_Y}{\partial \theta} \right|_{[R - X_n(\theta)]} \qquad (11.96)$$

This simple analysis (proposed by Wardi et al., 1992) specifies a family of SPA estimators for this class of problems. Specifically, let us define the "performance" $J(\theta)$ as the expected fraction of events such that $[X_n(\theta) + Y_n(\theta) > R]$ evaluated over N events. Then the SPA estimator is

$$\left[\frac{dJ}{d\theta} \right]_{SPA} = \frac{1}{N} \sum_{n=1}^{N} f_Y[R - X_n(\theta); \theta] \cdot \frac{dX_n(\theta)}{d\theta} - \frac{1}{N} \sum_{n=1}^{N} \left. \frac{\partial F_Y}{\partial \theta} \right|_{[R - X_n(\theta)]}$$

$$(11.97)$$

Note, however, that if $F_Y(\cdot)$ is in fact independent of θ, this reduces to

$$\left[\frac{dJ}{d\theta}\right]_{SPA} = \frac{1}{N}\sum_{n=1}^{N} f_Y[R - X_n(\theta)] \cdot \frac{dX_n(\theta)}{d\theta} \qquad (11.98)$$

There are three requirements involved in this estimator: (a) observations of $X_n(\theta)$ along the sample path, (b) knowledge of the pdf $f_Y(\cdot; \theta)$, and (c) evaluation of the derivatives $dX_n(\theta)/d\theta$. The last requirement is dependent on the system of interest. By its definition, however, $X_n(\theta)$ is a difference of event times, and we generally expect to be able to obtain event time derivatives based on our analysis of Section 11.5.1.

Remark. In the SPA estimators (11.97) and (11.98), we chose $\mathcal{Z}_n = X_n(\theta)$, but we could have also chosen $\mathcal{Z}_n = Y_n(\theta)$. In other words, this approach provides considerable flexibility in terms of choosing the characterization, depending on which is more convenient. Furthermore if R is a third random variable, then there is yet another possibility: We could choose to observe both $X_n(\theta)$ and $Y_n(\theta)$, and express the final result in terms of the distribution of R.

Let us now illustrate this general approach through a specific problem of interest. Consider a stable $GI/G/1$ queueing system with infinite queueing capacity, operating under a FCFS discipline. Let θ be the mean interarrival time. Let $\mathcal{Z}_n$ be the service time of the nth customer, which is independent of θ. Finally, let $S_n(\theta)$ and $W_n(\theta)$ denote the nth customer's system time and waiting time respectively. Now, suppose that every customer is assigned a "deadline" $R > 0$ for completing service, that is, if $S_n(\theta) > R$, the customer is considered useless or lost. Letting $J(\theta)$ denote the probability of violating this constraint at steady state, note that $S_n(\theta) = W_n(\theta) + Z_n$, so we can use the SPA estimator (11.98) by setting $X_n(\theta) = W_n(\theta)$ and $Y_n = Z_n$ in (11.95). Thus, assuming that the service time distribution, $F_Z(\cdot)$, is continuously differentiable (a standard assumption we made in IPA as well), we have

$$\left[\frac{dJ}{d\theta}\right]_{SPA} = \frac{1}{N}\sum_{n=1}^{N} f_Z[R - W_n(\theta)] \cdot \frac{dW_n(\theta)}{d\theta} \qquad (11.99)$$

One can explicitly show that this is indeed an unbiased, as well as strongly consistent, estimator of $dJ/d\theta$ (Wardi et al., 1992). The implementation of this estimator is quite simple. First, with $f_Z(\cdot)$ known, we simply evaluate it at $[R - W_n]$ with every observed waiting time W_n. If $R - W_n \leq 0$, then the expression above gives 0 (since the service time density function is defined to be non-zero over positive values only). If $R - W_n > 0$, we also need to evaluate the derivative $dW_n(\theta)/d\theta$. This is accomplished through standard IPA techniques. In fact, recall the recursive relationship (8.17), $W_i(\theta) = \max\{W_{i-1}(\theta) + Z_{i-1} - Y_i(\theta), 0\}$, where $Y_i(\theta)$ is the interarrival time between the $(i-1)$th and ith customer;

then, for the ith customer within a busy period, we have $W_i(\theta) = W_{i-1}(\theta) + Z_{i-1} - Y_i(\theta), i = 2, 3, \ldots$ Differentiating with respect to θ gives $dW_i(\theta)/d\theta = dW_{i-1}(\theta)/d\theta - dY_i(\theta)/d\theta$, with $dY_i(\theta)/d\theta$ evaluated through (11.19).

11.8.2 Marking and Phantomizing Techniques

The basic routing problem in queueing systems was discussed in Section 9.5.2 of Chapter 9. If customers reach a decision point where they can be routed to one of two servers (with infinite queueing capacity preceding them), we are interested in making the "best" decision in terms of some performance measure, such as the overall mean system time. In Section 9.5.2, we looked at the problem from the point of view of decisions based on some state information (the two queue lengths at the time of a decision). Often, however, such information is not available or too costly to obtain, and one seeks simpler routing mechanisms. One such mechanism is a probabilistic one: route to server 1 with probability θ, and to server 2 with probability $1 - \theta$. Thus, we are interested in the sensitivity of some performance measure, $J(\theta)$, with respect to this routing probability. In fact, we already saw this problem in Section 11.7.2, and identified the key difficulty as being the sample function discontinuities that are caused by changes in θ. In Section 11.7.2, the problem was handled by seeking alternative ways to represent a sample path of the system. Here, we approach the problem from a different perspective, which, in many ways, is more general.

Let us consider a simplified version of the problem, which is closer to admission control, as discussed in Section 9.5.1. As shown in Fig. 11.31, we consider a $G/G/1$ queueing system (the arrival process is completely arbitrary), where a customer is either accepted with probability θ (and joins the queue) or rejected and lost with probability $1 - \theta$. Let $J(\theta) = E[L(\theta, \omega)]$ be some performance measure; we do not care to stipulate any particular type of $L(\theta, \omega)$ at this point.

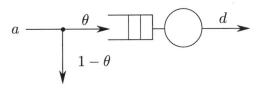

Figure 11.31: Probabilistic admission control (a customer is accepted with probability θ).

The control decisions (accept or reject) can be represented by random variables, $R_i(\theta)$, defined as follows:

$$R_i(\theta) = \begin{cases} 1 & \text{if the } i\text{th customer is accepted} \\ 0 & \text{otherwise} \end{cases} \tag{11.100}$$

Thus, given a particular sample path, we can easily observe the sequence $\{R_i(\theta)\}, i = 1, 2, \ldots$, and use it as the characterization $\mathcal{Z} = \{R_1(\theta), R_2(\theta), \ldots\}$.

Now, in our earlier discussion of SPA, we proceeded to modify the sample function $L(\theta, \omega)$ by considering the conditional expectation $L_{\mathcal{Z}}(\theta) = E[L(\theta, \omega) \mid \mathcal{Z}]$, then defined the SPA estimator as the sample derivative $dL_{\mathcal{Z}}(\theta)/d\theta$. Here, we will take a slightly different approach. Instead of immediately expressing the estimator as a sample derivative, let us instead rewrite (11.94) as

$$\left[\frac{dJ}{d\theta} \right]_{SPA} = \frac{dL_{\mathcal{Z}}(\theta)}{d\theta} = \lim_{\Delta\theta \to 0} \frac{E[L(\theta + \Delta\theta, \omega) \mid \mathcal{Z}] - E[L(\theta, \omega) \mid \mathcal{Z}]}{\Delta\theta}$$

Since $\Delta\theta$ could be positive or negative, let us fix the sign of $\Delta\theta$ to be positive and define

$$\Delta L^+(\theta, \omega) = L(\theta + \Delta\theta, \omega) - L(\theta, \omega) \tag{11.101}$$
$$\Delta L^-(\theta, \omega) = L(\theta, \omega) - L(\theta - \Delta\theta, \omega) \tag{11.102}$$

Therefore, we have two expressions for the SPA estimator above, depending on whether we consider the right or left derivative of $E[L(\theta, \omega) \mid \mathcal{Z}]$. Let us then set

$$\left[\frac{dJ}{d\theta} \right]_{SPA}^+ = \lim_{\Delta\theta \to 0} \frac{E[\Delta L^+(\theta, \omega) \mid \mathcal{Z}]}{\Delta\theta} \tag{11.103}$$

$$\left[\frac{dJ}{d\theta} \right]_{SPA}^- = \lim_{\Delta\theta \to 0} \frac{E[\Delta L^-(\theta, \omega) \mid \mathcal{Z}]}{\Delta\theta} \tag{11.104}$$

It will soon become clear why we have chosen to preserve the limit expression above for our SPA estimator, rather than the derivative expression in (11.94).

Let us now consider the effect of some $\Delta\theta > 0$ on the decision variable $R_i(\theta)$. As usual, we generate $R_i(\theta)$ through the inverse transform technique, exactly as in Section 11.7.2. Thus, for the ith decision, there is an underlying random number U_i such that $R_i(\theta) = 1$ if $0 \le U_i \le \theta$, and $R_i(\theta) = 0$ if $\theta < U_i \le 1$. Let ΔR_i denote the change resulting from $\Delta\theta > 0$. There are two cases to be considered: (a) when θ is changed to $\theta - \Delta\theta$, and (b) when θ is changed to $\theta + \Delta\theta$. Let us concentrate on case (a), since case (b) turns out to be very similar.

As illustrated in Fig. 11.32, the effect of a change from θ to $\theta - \Delta\theta$ (with U_i fixed) can only yield two results: $\Delta R_i = 0$, if $U_i \le \theta - \Delta\theta$, and $\Delta R_i = -1$, if $\theta - \Delta\theta < U_i \le \theta$. In the first case, the decision remains unchanged, in the second the decision is switched: A customer accepted in the nominal path (under θ) would be rejected in the perturbed path (under $\theta - \Delta\theta$). We may think of all such customers, characterized by $\Delta R_i = -1$, as *marked* customers: They are the ones who would be rejected in the perturbed path. In fact, if U_i is known, then (for any given $\Delta\theta$) one could determine all such marked customers while observing the nominal path. On the other hand, if we consider a perturbed path under $\theta + \Delta\theta$, the dual of a marked customer is a *phantom* customer: This is a customer who would be rejected in the nominal path, and would therefore have to be treated like a "phantom" that proceeds through the queueing system without affecting anyone, and without incurring any service time.

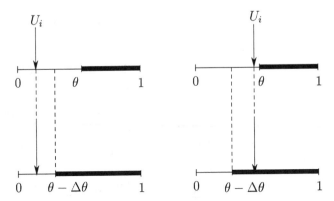

Figure 11.32: Illustrating the effect of $\Delta\theta$ on $R_i(\theta)$: $\Delta R_i = 0$ or $\Delta R_i = -1$.

We now ask the question: What is the probability that a customer is marked, that is, what is $P[\Delta R_i = -1 \mid R_i(\theta) = 1]$? The answer is easy to obtain, since the values of ΔR_i and $R_i(\theta)$ depend entirely on U_i and the given values θ and $\Delta\theta$:

$$P[\Delta R_i = -1 \mid R_i(\theta) = 1] = P[\theta - \Delta\theta < U_i \leq \theta \mid U_i \leq \theta]$$
$$= \frac{P[\theta - \Delta\theta < U_i \leq \theta]}{P[U_i \leq \theta]} = \frac{\Delta\theta}{\theta}$$

since $U_i \sim U[0, 1]$. Similarly, we can answer the more general question: What is the probability that any K customers ($K \geq 1$) are marked? Since all customers are treated independently of each other, the answer is the generalization of the case $K = 1$ above, i.e., $(\Delta\theta/\theta)^K$. Assuming that the observed sample path contains N customer arrivals, let $N_a \leq N$ be the number of accepted arrivals. Then, for any $K \leq N_a$, we have

$P[\Delta R_i = -1$ for any K customers $\mid R_i(\theta) = 1$ for these K customers$]$

$$= \left(\frac{\Delta\theta}{\theta}\right)^K$$

$$(11.105)$$

With this observation in mind, let us evaluate $E[\Delta L^-(\theta, \omega) \mid \mathcal{Z}]$ in (11.104) by further conditioning $\Delta L^-(\theta, \omega)$ on all possible ways to mark customers. To keep the notation manageable, let A be the set of the N_a accepted customer indices (i.e., such that $R_i(\theta) = 1$) and let F and L denote the first and last index respectively in that set. Then

$$
\begin{aligned}
E[\Delta L^-(\theta,\omega) \mid \mathcal{Z}] &= \sum_{i \in A} E[\Delta L^-(\theta,\omega) \mid \mathcal{Z}, \Delta R_i = -1] P[\Delta R_i = -1 \mid \mathcal{Z}] \\
&+ \sum_{i \in A} \sum_{j \in A, j \neq i} E[\Delta L^-(\theta,\omega) \mid \mathcal{Z}, \Delta R_i = -1, \Delta R_j = -1] \\
&\quad \cdot P[\Delta R_i = -1, \Delta R_j = -1 \mid \mathcal{Z}] \\
&+ \ldots + E[\Delta L^-(\theta,\omega) \mid \mathcal{Z}, \Delta R_F = -1, \ldots, \Delta R_L = -1] \\
&\quad \cdot P[\Delta R_F = -1, \ldots, \Delta R_L = -1 \mid \mathcal{Z}]
\end{aligned}
$$

Here, the first term corresponds to marking individual customers one at a time, the second term to marking two distinct customers two at a time, and so on. The last term corresponds to the case where all customers in A are marked. Then, using (11.105), we get

$$
E[\Delta L^-(\theta,\omega) \mid \mathcal{Z}] = \frac{\Delta\theta}{\theta} \sum_{i \in A} E[\Delta L^-(\theta,\omega) \mid \mathcal{Z}, \Delta R_i = -1] + E[H(\Delta\theta, \mathcal{Z})]
$$

where all terms involving $(\Delta\theta)^K$ with $K \geq 2$ are grouped together in $H(\Delta\theta, \mathcal{Z})$. Note that $H(\Delta\theta, \mathcal{Z})$ is an expression of the form $O(\Delta\theta)^2$. Therefore, the ratio $E[H(\Delta\theta, \mathcal{Z})]/\Delta\theta$ vanishes as $\Delta\theta \to 0$, that is,

$$
\lim_{\Delta\theta \to 0} \frac{E[H(\Delta\theta, \mathcal{Z})]}{\Delta\theta} = 0 \tag{11.106}
$$

under some mild technical conditions which we shall not discuss here (see Gong and Schulzrinne, 1992).

Finally, returning to (11.104), we get

$$
\left[\frac{dJ}{d\theta}\right]^-_{SPA} = \lim_{\Delta\theta \to 0} \frac{1}{\Delta\theta} \left\{ \frac{\Delta\theta}{\theta} \sum_{i \in A} E[\Delta L^-(\theta,\omega) \mid \mathcal{Z}, \Delta R_i = -1] \right.
$$
$$
\left. + E[H(\Delta\theta, \mathcal{Z})] \right\}
$$

We can now see the reason for preserving the limit expression in the SPA estimator: The fortuitous cancellation of $\Delta\theta$ in the preceding equation makes the first term in brackets independent of $\Delta\theta$. Finally, taking (11.106) into account and denoting the estimator by $[\cdot]_M$ (to remind ourselves that it corresponds to marked customers), we get

$$
\left[\frac{dJ}{d\theta}\right]^-_{M} = \frac{1}{\theta} \sum_{i \in A} E[\Delta L^-(\theta,\omega) \mid \mathcal{Z}, \Delta R_i = -1] \tag{11.107}
$$

A similar expression can be obtained for (11.103), where customers are phantomized rather than marked. There are several interesting features in this estimator which are worth noting. First, it has been derived under very general conditions on the nature of the arrival and service processes of the system in Fig. 11.31. Second, nothing has been said on the nature of the sample functions themselves. Third, this is the first time that we see an unbiased *derivative* estimator obtained through a *finite difference* which we have to evaluate as we observe a sample path. In particular, what we have to evaluate is the effect that the removal of the ith marked customer has on the sample path, repeat the process for every i, and then average over all i's. Whether or not this can be accomplished may well depend on the nature of $L(\theta, \omega)$, an issue that we shall not get into here (examples may be found in Gong and Schulzrinne, 1992; Cassandras and Julka, 1992). In fact, this observation provides a nice bridge between the sensitivity estimation techniques considered thus far (IPA, SPA) and the techniques used for estimating the effect of finite perturbations considered in the rest of this chapter.

As an example of how (11.107) can be applied to a specific performance measure, let $J(\theta)$ be the mean waiting time of accepted customers in Fig. 11.31. Then, if $W_{N_a}(\theta)$ denotes the cumulative waiting time over N_a customers, the sample function $L_{N_a}(\theta) = W_{N_a}(\theta)/N_a$ is the average waiting time over these N_a customers. Moreover, let $\Delta W_{N_a}(\theta, i^-)$ be the change in cumulative waiting time in the same sample path *when the ith customer is marked* (i.e., removed from the queue). Then, the estimator in (11.107) becomes

$$\left[\frac{dJ}{d\theta}\right]_M^- = \frac{1}{N_a\theta} \sum_{i=1}^{N_a} \Delta W_{N_a}(\theta, i^-) \tag{11.108}$$

As in Section 11.7, the reader must be reminded of the following important fact: Just because we are evaluating the effect of removing (marking) or adding (phantomizing) a customer does not mean that we are interfering in any way with the actual system. The idea is to construct (in our mind only) a perturbed sample path and to evaluate $\Delta W_{N_a}(\theta, i^-)$ based on information which is readily available on the nominal path.

Remark. The idea of marking and phantomizing customers was used by Suri and Cao (1986) in order to evaluate finite changes of the form $\Delta L^+(\theta, \omega)$ or $\Delta L^-(\theta, \omega)$ in closed queueing networks. The same idea later turned up in the context of SPA (Gong and Schulzrinne, 1992). The term *Rare Perturbation Analysis* (RPA) has also been proposed by Brémaud and Vázquez-Abad (1992), on the basis of the fact that marking or phantomizing introduces infrequent (rare) changes in a sample path. Finally, the fact that the marking/phantomizing idea is not limited to customer routing was recognized by Cassandras and Julka (1992): Any discrete "object" which is part of a DES (e.g., a customer, a server) and which may be removed or added with some probability can be treated in the same way. An interesting such example arises in scheduling problems, where access to a resource is based on "time slots" allocated to various competing users; one can mark (or phantomize) such slots in order to evaluate the effect of a removal (or addition) of a slot to an observed sample path (Cassandras and Julka, 1992), and hence be led to some optimal scheduling policies.

11.9 PA FOR FINITE PARAMETER CHANGES

Up to this point, we have concentrated on estimating sensitivities (i.e., derivatives) of performance measures of DES. However, there are also situations where the question of interest becomes: What is the effect of a specific *finite* change $\Delta\theta$ on some performance measure $J(\theta)$? Answering such a question for a "small" $\Delta\theta$ can provide an approximation to the sensitivity itself when other techniques fail. In practice, however, the issue of finite parameter changes is of interest in itself, and it often arises when a set of allowable values of the parameter θ is prespecified. The service rate of a machine in a manufacturing system, for instance, may only be adjustable to some specific settings. It is also possible that a designer needs some estimates of performance over a few extreme values of θ, so as to get a rough idea of the global behavior of the system as a function of θ. Finally, we saw in Section 11.8.2 that some SPA sensitivity estimators actually require the evaluation of finite changes in a sample function caused by marking or phantomizing (e.g., the effect of removing a specific customer from a queueing system on the average waiting time of all other customers in the system).

Answering questions of the type described above is largely the domain of Finite Perturbation Analysis (FPA). FPA was contrasted to IPA in Section 11.4. Equation (11.33), for instance, is a recursive relationship intended to evaluate the effect of a finite change $\Delta\theta$ (where θ is the mean service time) on departure times of the $GI/G/1$ queueing system. Our ability to derive such recursive relationships (that involve *only quantities directly observable on the nominal sample path*) is dependent on the specific DES under study and on the sample function of interest. Since large perturbations generally cause frequent event order changes, one cannot expect the same appealing simplicity and efficiency as IPA in dealing with such problems. Although there have been efforts to categorize FPA estimators in terms of the magnitude of the perturbation $\Delta\theta$ (e.g., Ho et al., 1983), such estimators remain limited to special cases where the specific structure of a DES can be directly exploited. Such is the case, for example, in the evaluation of the change in the mean waiting time of N customers in the sample path of a $G/G/1$ system when one of these customers is removed (Gong and Schulzrinne, 1992). Rather than dwelling on these special cases, we proceed instead by viewing FPA in a broader setting: Given a sample path of a DES operating under some parameter θ, *is it generally possible to construct a sample path of the same DES under $\theta' \neq \theta$?* If not, is it possible to identify the conditions under which it is (similar, perhaps, to the convenient commuting condition (**CC**) that we saw for IPA)? Lastly, are there any general-purpose techniques for constructing a sample path under $\theta' \neq \theta$ short of regenerating complete lifetime sequences and repeating the state update mechanism by "brute force"?

We will discuss these questions in the next section. We will see that it is indeed possible to efficiently construct, not just one sample path under $\theta' \neq \theta$, but actually many sample paths concurrently, all based on the one nominal sample path under θ which is observed. The framework we will present will also allow us to consider parameters which may be either discrete (e.g., buffer capacities, population sizes in closed networks) or continuous (e.g., arrival and service rates).

11.10 CONCURRENT ESTIMATION

Based on the stochastic timed automaton framework developed in Chapters 5 and 6, we have seen that a DES can be viewed as a dynamic system where the input is a set of event lifetime sequences and the output is a sequence of events with their corresponding occurrence times. For example, in a single-server queueing system, the input consists of two event lifetime sequences

$$\mathbf{V}_a = \{V_{a,1}, V_{a,2}, \dots\} \quad \text{and} \quad \mathbf{V}_d = \{V_{d,1}, V_{d,2}, \dots\}$$

corresponding to the two events a (arrivals) and d (departures). The output is a sequence

$$\{(E_1, T_1), (E_2, T_2), \dots\}$$

where (E_k, T_k) denotes the kth event type (a or d) and its occurrence time. Let us now consider a parameter θ that affects the output in one of two basic ways: either through the event lifetimes or through the state transition mechanism. For example, if θ is the mean service time in the single-server queueing system above, then we can write $V_d(\theta)$ to express the fact that θ affects all d event lifetimes. If, on the other hand, θ is the queueing capacity of this system, the output is only affected through the state transition function $f(x, a) = x'$, which gives $x' = x + 1$ for $x < \theta$ and $x' = x$ for $x = \theta$. We can also extend the idea of a "parameter" to describe through θ different operating policies (e.g., $\theta_1 =$ first-come-first-served, $\theta_2 =$ last-come-first-served, $\theta_3 =$ highest-priority-first).

Figure 11.33 describes this model. If the event set $\mathcal{E}$ contains N events, the input consists of the event lifetime sequences $\mathbf{V}_i = \{V_{i,1}, V_{i,2}, \dots\}$, $i = 1, \dots, N$. In addition, we explicitly include θ as an input as well. The output is a sequence $\{(E_k, T_k)\}$, $k = 1, 2, \dots$, which we will denote, for simplicity, by $\xi(\theta)$, emphasizing once again its dependence on θ. Clearly, if we are interested in the state sequence $\{X_k\}$, this can easily be obtained from $x(\theta)$, provided initial conditions have been specified. Also, observe that any sample function of interest can be expressed in terms of $x(\theta)$ as well, which justifies the notation $L[\xi(\theta)]$.

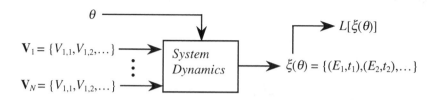

Figure 11.33: Input-Output description of a DES under parameter θ.
Input: lifetime sequences $\mathbf{V}_i = \{V_{i,1}, V_{i,2}, \dots\}, i = 1, \dots, N,$ *and* θ.
Output: $\xi(\theta) = \{(E_1, T_1), (E_2, T_2), \dots\}$ *and* $L[\xi(\theta)]$.

For our purposes, the parameter θ can take values from a finite set $\Theta = \{\theta_0, \theta_1, \dots, \theta_M\}$. Our basic premise, once again, is that the complexity of the DES under study leaves no alternative but to analyze its behavior through its sample paths (e.g., through simulation), from which we can also estimate performance measures of the form $J(\theta) = E[L[\xi(\theta)]]$. Naturally, we would like to be able to obtain such estimates *over all possible values of* θ. The obvious way to accomplish this goal is to go through the process of constructing sample paths under each of $\theta_0, \theta_1, \dots, \theta_M$. This is a rather prohibitive task if we expect to use simulation; it is simply infeasible if we intend to do it by changing parameters on a real system.

11.10.1 The Sample Path Constructability Problem

Following the preceding discussion, in order to obtain performance estimates under parameter values $\theta_0, \theta_1, \dots, \theta_M$, we need to proceed as shown in Fig. 11.34. This raises the question: Is it possible to speed up this process in any way? Since we would like to "learn" the behavior of the system under all possible parameter values without actually interfering with a real system, another critical question is: Can we use the input $V_i = \{V_{i,1}, V_{i,2}, \dots\}, i = 1, \dots, N$, observed under one particular θ, say θ_0, along with information contained in the sample path generated under θ_0, in order to construct all other outputs $\xi(\theta_1), \dots, \xi(\theta_M)$? This is referred to as the *sample path constructability* problem (Cassandras and Strickland, 1989a), and is illustrated in Fig. 11.35. The techniques we will develop to solve this problem come under the heading of *Concurrent Estimation* since their goal is to obtain performance estimates under different parameter settings *concurrently* with an estimate obtained under some given fixed setting.

As we will see in subsequent sections, there are several Concurrent Estimation techniques that address the basic problem of Fig. 11.35. Strictly speaking, these techniques are not directly related to PA as developed thus far, but they have been motivated by the "PA mindset": Extract information from an ob-

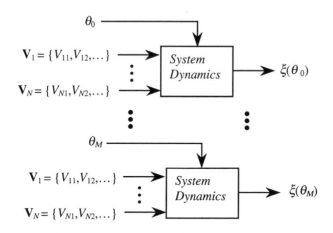

Figure 11.34: Construction of sample paths under all $\theta \in \Theta$.

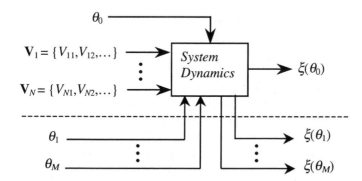

Figure 11.35: The sample path constructability problem.
The goal is to construct $\xi(\theta_1), \ldots, \xi(\theta_M)$ based only on the information contained in the sample path generated under θ_0. This is to be contrasted to Fig. 11.34.

served DES sample path under θ so as to predict its behavior under $\theta' \neq \theta$.

11.10.2 Uses of Concurrent Estimation: "Rapid Learning"

Many of the ideas that motivate sensitivity analysis, as discussed in Section 11.2.2, also apply to Concurrent Estimation. In what follows, we briefly review them and emphasize the fact that $\theta \in \Theta$ where Θ is a finite (possibly very large) set.

1. *Local information*: Even if the goal of constructing all $\xi(\theta_1), \ldots, \xi(\theta_M)$ cannot be met, it is often possible to use the observed sample path, $\xi(\theta_0)$, in order to construct a few sample paths corresponding to neighboring values of θ_0. Information from the resulting performance estimates (e.g., the sign of $L[\xi(\theta_1)] - L[\xi(\theta_0)]$) can provide an indication as to the direction in which θ should be changed.

2. *Structural properties:* As in the case of sensitivity estimates, Concurrent Estimation techniques can not only provide numerical values for the performance measures $L[\xi(\theta_0)], L[\xi(\theta_1)], \ldots$, but they often capture the nature of the dependence of a performance measure on the parameter θ as well. As a result, these techniques can reveal interesting structural properties of a DES.

3. *Optimization:* Clearly, if all $L[\xi(\theta_0)], L[\xi(\theta_1)], \ldots$ can be estimated while a single sample path is observed, then optimization reduces to choosing the largest or smallest performance estimate, depending on whether we wish to maximize or minimize $J(\theta)$. Of course, since $L[\xi(\theta)]$ is only an estimate of $J(\theta)$, this process must be carried out in the context of proper statistical analysis, but the availability of all estimates leads to the direct solution of the optimization problem, at least in principle.

   When Θ is a very large set (as is the case in combinatorial optimization), the availability of many performance estimates simultaneously can be exploited in a different way, based on *ordinal* optimization techniques (Ho et al., 1992; Cassandras et al., 1998). The main idea here is that even when estimates are of poor quality (in terms of accurately predicting $J(\theta)$), their relative order is usually very accurate. In other words, if $L[\xi(\theta_i)] > L[\xi(\theta_j)]$, the actual values $L[\xi(\theta_i)], L[\xi(\theta_j)]$ may be poor estimates of $J(\theta_i), J(\theta_j)$, yet the sign of $\{L[\xi(\theta_i)] - L[\xi(\theta_j)]\}$ may be an excellent estimate of the sign of $\{J(\theta_i) - J(\theta_j)\}$, and hence of the *relative order* of $J(\theta_i), J(\theta_j)$.

4. *Global response curve generation*: Obviously, if all $L[\xi(\theta_0)], L[\xi(\theta_1)], \ldots$ can be estimated while a single sample path is observed, then a global response curve of the system with respect to the parameter θ can immediately be generated. Again, of course, the quality of the estimates

remains a critical issue. From a practical standpoint, however, a designer is often interested in getting a rough idea of how a system behaves over the range of parameter values, and do so as rapidly as possible.

This discussion points to one common element which is at the center of Concurrent Estimation techniques: the ability to obtain many performance estimates simultaneously, hence the ability to *rapidly learn* the behavior of a DES as a function of θ. This "rapid learning" feature of the techniques we will present next is greatly amplified by the increasing availability of parallel processing computing facilities. There is an implicit "parallelism" in the framework of Fig. 11.35, in that the same information (the input and the state evolution of the system under θ_0) is processed M different ways to yield the M desirable sample paths. Thus, in practice, if there are M processors available, the implication is that one can "learn" about the system's behavior under M separate parameter values, $\theta_1, \ldots, \theta_M$, in the same time as learning about a single parameter value, θ_0.

The process of bringing together (a) the constructability framework shown in Fig. 11.35 and related techniques, (b) new ideas on "ordinal optimization", and (c) emerging parallel processing computer architectures, has all the ingredients for an exciting field that seeks to combine new concepts with new technologies. The rest of this chapter is only an attempt to describe the foundations of this developing field of Concurrent Estimation.

11.10.3 Sample Path Constructability Conditions

Once again, our discussion will be based on stochastic timed automata models of DES. In seeking conditions under which the problem depicted in Fig. 11.35 can be solved, we find that there are two aspects of the problem that need to be addressed. First, the structure of the DES itself may or may not be conducive to sample path constructability. Second, the stochastic characteristics of the clock sequences driving the model are critical in ensuring that the sample path we construct is indeed a valid one.

To see these two aspects more clearly, let us go back to the basic mechanism through which a sample path of a timed automaton is constructed. Assuming the current state is x, the key to proceeding with the construction is the determination of the triggering event

$$e' = \arg \min_{i \in \Gamma(x)} \{y_i\}$$

Once $e' = i$ is specified, the next state and next clock values can be updated (see Section 6.4 of Chapter 6). We can now immediately see that the determination of e' above depends (a) on the feasible event set $\Gamma(x)$, and (b) the clock values y_i. While $\Gamma(x)$ depends on the state structure of the system, the clock values mainly depend on the event lifetime distributions characterizing the input to this system.

The first aspect of constructability is addressed by the *observability condition*[1] (as coined by Cassandras and Strickland, 1989a), which is presented next.

The Observability Condition

Let $\theta \in \Theta$ be the parameter value under which an input $\mathbf{V}_i = \{V_{i,1}, V_{i,2}, \dots\}$, $i = 1, \dots, N$, produces an output sequence $x(\theta)$. We refer to this as the nominal sample path, as we did in the development of PA as well. Let $\{X_k(\theta)\}$, $k = 1, 2, \dots$, denote the state sequence observed. Recall that $\Gamma(X_k(\theta))$ is the set of feasible events at state $X_k(\theta)$.

Next, we consider another parameter value $\theta_m \neq \theta$. Suppose this new system is driven by the event sequence generated under θ, giving rise to a new state sequence $\{X_k(\theta_m)\}$, $k = 1, 2, \dots$, and sample path $\xi(\theta_m)$. We then say that $\xi(\theta_m)$ is *observable* with respect to $\xi(\theta)$ if $\Gamma(X_k(\theta_m)) \subseteq \Gamma(X_k(\theta))$ for all $k = 1, 2, \dots$

(OB) Let $\xi(\theta)$ be a sample path under θ, and $\{X_k(\theta)\}, k = 0, 1, \dots$, the corresponding state sequence. For $\theta_m \neq \theta$, let $\xi(\theta_m)$ be the sample path generated by the event sequence in $\xi(\theta)$. Then, for all $k = 0, 1, \dots$, $\Gamma(X_k(\theta_m)) \subseteq \Gamma(X_k(\theta))$.

In simple terms, two sample paths are "coupled" so that the same event sequence drives them both. As the two state sequences subsequently unfold, condition **(OB)** states that every state observed in the nominal sample path is always "richer" in terms of feasible events. In other words, from the point of view of $\xi(\theta_m)$ (the sample path we are trying to construct), all feasible events required at state $X_k(\theta_m)$ are observable in the corresponding state $X_k(\theta)$. As in the case of the commuting condition **(CC)** in Section 11.5.3, the validity of **(OB)** is sometimes easy to check by inspecting the state transition diagrams of the system under θ and θ_m.

Example 11.14 (Checking (OB) for a $G/G/1/K$ system)
Consider a $G/G/1/K$ queueing system, where $K = 1, 2, \dots$ is the parameter of interest. Under $K = 3$, a sample path ξ_3 is obtained. The question is whether a sample path ξ_2, constructed under $K = 2$, is observable with respect to ξ_3. As usual, we adopt a stochastic timed automaton model, with the event set $\mathcal{E} = \{a, d\}$, where a denotes an arrival and d denotes a departure. In what follows, $\{X_k(K)\}, k = 1, 2, \dots$, denotes the state sequence in a sample path generated with queueing capacity $K = 1, 2, \dots$

The two state transition diagrams are shown in Fig. 11.36. Since condition **(OB)** must be tested under the assumption that both sample paths are generated by the exact same input, it is convenient to construct a new

---

[1] Although the same term is used, this notion of observability is different from that presented in Chapter 3 in the context of supervisory control of untimed DES models.

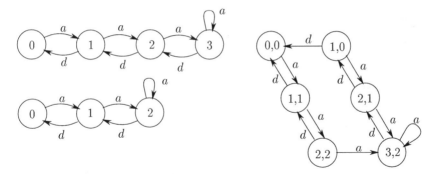

Figure 11.36: State transition diagrams for Example 11.14.
The "Augmented System" state transition diagram on the right describes the joint state (under $K = 3$ and $K = 2$) as a result of a common event sequence input.

system whose state consists of the joint queue lengths of the two original systems. This is referred to as an *Augmented System* (Cassandras and Strickland, 1989b) and plays a crucial role in some of the constructability techniques to be discussed later (it is also related, but not identical, to the parallel composition of automata; see Section 2.3.2 and Problem 11.16). As shown in Fig. 11.36, we assume both sample paths start from a common state (say 0, for simplicity). On the left part of the augmented system state transition diagram, the two states remain the same. In fact, a difference arises only when: (a) $X_k(3) = X_k(2) = 2$ for some k, and (b) event a occurs. At this point, the state transition gives $X_{k+1}(3) = 3$, but $X_{k+1}(2) = 2$. From this point on, the two states satisfy $X_k(3) = X_k(2)+1$, as seen in Fig. 11.36, until state (1,0) is visited. At this point, the only feasible event is a departure, bringing the state back to (0,0).

By inspection of the augmented system state transition diagram, we see that **(OB)** is trivially satisfied for the three states such that $X_k(3) = X_k(2)$. It is also easily seen to be satisfied at (3,2) and (2,1), since $\Gamma(x) = \{a, d\}$ for all $x > 0$. The only remaining state is (1,0), where we see that $\Gamma(X_k(2)) = \Gamma(0) = \{a\} \subset \{a, d\} = \Gamma(1) = \Gamma(X_k(3))$, and **(OB)** is satisfied. Thus, sample paths of this system under $K = 2$ are indeed observable with respect to sample paths under $K = 3$.

Example 11.15 (Checking (OB) for a $G/G/1/K$ system - continued)
Let us take another look at the $G/G/1/K$ queueing system of Example 11.14, except that we now reverse the roles of the two parameter values and corresponding sample paths. Thus, the question is whether a sample path constructed under $K = 3$ is observable with respect to a sample path under $K = 2$. Returning to Fig. 11.36, and state (1,0) in particular, note that $\Gamma(X_k(3)) = \Gamma(1) = \{a, d\} \supset \{a\} = \Gamma(0) = \Gamma(X_k(2))$. Thus, the feasible event d, needed at state $X_k(3) = 1$ is *unobservable* at the

corresponding state $X_k(2) = 0$.

It is interesting to note that condition **(OB)** is not symmetric with respect to the role of the parameter chosen as the "nominal" one.

The Constructability Condition

Observability alone is not sufficient to guarantee a solution to the constructability problem. As already pointed out, while **(OB)** addresses the structural aspect of determining the triggering event at some state $X_k(\theta_m)$ of the constructed path, it does not address the issue of whether the clock values available on the observed path can be used on the constructed path. To see why this is, let us consider sample paths of the $G/G/1/K$ system under $K = 3$ and 2 as in Example 11.14. We have seen that a sample path under $K = 2$ is observable with respect to a sample path under $K = 3$. In Fig. 11.37, let us focus on the joint state $(1,0)$. Thus, the sample path under $K = 2$ is currently at state 0, while the nominal sample path is at state 1. Let $G_d(\cdot)$ denote the lifetime distribution of d events.

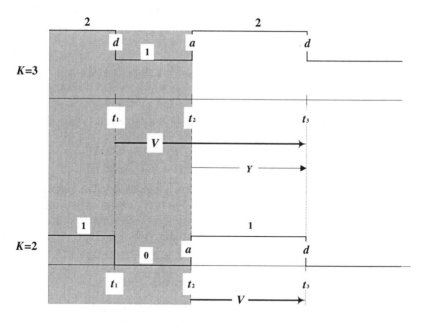

Figure 11.37: Illustrating why observability does not guarantee sample path constructability.
Under $K = 3$, a d event is activated at time t_1 with lifetime V. Under $K = 2$, the d event is only activated at t_2. If Y is used as the lifetime of this event under $K = 2$, its cdf is generally not the same as the correct lifetime cdf of V.

We now make the following observations.

1. When d occurs at time t_1:

   K = 3 path: State 1 entered; a new d event is activated, with lifetime $V \sim G_d(\cdot)$.

   K = 2 path: State 0 entered; no d event is activated, since $d \notin \Gamma(0)$.

2. When a occurs at time t_2:

   K = 3 path: State 2 entered; the clock value of d is $Y < V$.

   K = 2 path: State 1 entered; a new d event must be activated, but the only clock information pertaining to d from the nominal path is Y.

It should now be clear what the problem is at time t_2: d is a feasible event in both sample paths (states 2 and 1), but whereas the $K = 2$ path needs a new lifetime, the only available information from the nominal ($K = 3$) path is a *residual* lifetime Y. If we are to continue the construction, we must ensure that whatever clock value we use in the constructed sample path has the right probability distribution. In this example, we need $Y \sim G_d(\cdot)$, which means that the residual lifetime of d must have the same cdf as the original lifetime. This is only true for exponential random variables, due to their memoryless property. In fact, it can be shown that observability is a sufficient condition for the constructability problem as long as certain event lifetimes (for the d event in our example) are exponentially distributed.

Motivated by this example, let us consider the cdf of the residual lifetime of some event $i \in \mathcal{E}$, which is the conditional distribution $H(t, z)$ defined in (6.42):

$$H(t, z) = P[Y_i \leq t \mid V_i > z]$$

where z is the observed age of the event at the time $H(t, z)$ is evaluated. Given the lifetime distribution $G_i(t)$, and since $V_i = Y_i + z, H(t, z)$ is given by

$$H(t, z) = P[V_i \leq z + t \mid V_i > z] = \frac{P[z < V_i \leq z + t]}{1 - P[V_i \leq z]}$$
$$= \frac{G_i(z + t) - G_i(z)}{1 - G_i(z)}$$

(11.109)

In general, both the distribution $H(t, z)$ and the event age depend on the parameter θ. Thus, when the kth event takes place on a sample path under θ, we write $H(t, z_{i,k}(\theta); \theta)$ to denote the cdf of the clock value of event $i \in \mathcal{E}$, given its age $z_{i,k}(\theta)$. The *constructability condition* is then stated as follows (Cassandras and Strickland, 1989a):

(CO) Let $\xi(\theta)$ be a sample path under θ, and $\{X_k(\theta)\}, k = 0, 1, \ldots$, the corresponding state sequence. For $\theta_m \neq \theta$, let $\xi(\theta_m)$ be the sample path generated by the event sequence in $\xi(\theta)$. Then,

$$\Gamma(X_k(\theta_m)) \subseteq \Gamma(X_k(\theta)) \quad \text{for all } k = 0, 1, \ldots$$

and

$$H(t, z_{i,k}(\theta_m); \theta_m) = H(t, z_{i,k}(\theta); \theta),$$
$$\text{for all } i \in (X_k(\theta_m)), k = 0, 1, \ldots$$

The first part of (CO) is of course the observability condition (OB). The second part imposes a requirement on the event clock distributions.

There are two immediate implications of (CO):

1. For stochastic timed automata with a Poisson clock structure, (OB) implies (CO).

2. If $\Gamma(X_k(\theta_m)) = \Gamma(X_k(\theta))$ for all $k = 0, 1, \ldots$, then (CO) is satisfied.

The first assertion follows from the memoryless property of the exponential distribution, and it reduces the test of constructability to the purely structural condition (OB). As we will see, it enables us to solve the sample path constructability problem for Markov chains even when (OB) is violated. The second assertion simply follows from the fact that if feasible event sets are always the same under θ and θ_m, then all event clock values are also the same. If, for example, a DES consists of events which are always feasible, then constructability immediately follows.

We now come to the issue of developing specific techniques for constructing sample paths $\xi(\theta_1), \ldots, \xi(\theta_m)$, given $\xi(\theta_0)$, as shown in Fig. 11.35. We will describe two approaches in what follows.

11.10.4   The Standard Clock Approach

The *Standard Clock* (SC) approach (Vakili, 1991) applies to stochastic timed automata with a Poisson clock structure. This means that all event lifetime distributions are exponential:

$$G_i(t) = 1 - e^{-\lambda_i t}$$

Therefore, every event process in such systems is characterized by a single Poisson parameter λ_i.

This approach satisfies the constructability condition (CO) as follows:

- Since our scope is limited to a Poisson clock structure, the distributional part of (CO) is immediately satisfied by the memoryless property of the exponentially distributed event lifetimes.

- The observability condition (OB) is forced to be satisfied by choosing a nominal sample path such that $\Gamma(x) = \mathcal{E}$ for all states x. Thus, all events are made permanently feasible, as explained next.

To explain how the SC approach works, let us recall two properties of Poisson processes that were discussed in Chapter 6 and which we shall make use of here. First, when the system enters state x, the distribution of the ensuing interevent time is exponential with rate given by (6.58):

$$\Lambda(x) = \sum_{i \in \Gamma(x)} \lambda_i \qquad (11.110)$$

Second, when the state is x, the distribution of the triggering event is given by (6.68):

$$p(i, x) = \frac{\lambda_i}{\Lambda(x)}, \qquad i \in \Gamma(x) \qquad (11.111)$$

Next, we take advantage of the uniformization of Markov chains, which was discussed in Section 7.5 of Chapter 7. Recall that we can replace the probability flow rate at some state x, $\Lambda(x)$, by a uniform rate $\gamma \geq \Lambda(x)$, common to all x, and simply replace $\Lambda(x)$ by γ in (11.111). The additional probability flow $[\gamma - \Lambda(x)]$ at state x corresponds to "fictitious events" which leave the state unchanged. In our case, we choose this uniform rate to be the maximal rate over all events:

$$\Lambda = \sum_{i \in \mathcal{E}} \lambda_i \qquad (11.112)$$

In the uniformized model, every event is always feasible. However, events i such that $i \notin \Gamma(x)$ for the original model leave the state unchanged, and they are ignored. Moreover, the mechanism through which the triggering event at any state is determined becomes independent of the state. Specifically, in (11.110) we can now set $\Lambda(x) = \Lambda$ for all x. Therefore, the triggering event distribution at any state is simply $p_i = \lambda_i/\Lambda$.

The benefit we derive from uniformization is at the heart of the SC approach. The fact that all events are permanently feasible in the uniformized model, combined with the memoryless property of the exponential event lifetime distributions, immediately implies the constructability condition (CO). In short, the way the SC approach gets around the problem of potentially unobservable events is by forcing all events to occur at all states, and hence satisfy (OB). Although this results in several fictitious events that is, events i such that $i \notin \Gamma(x)$, it also makes for a general-purpose methodology within the realm of Poisson event processes.

Let us now see how the SC approach leads to a specific algorithm for constructing multiple sample paths under different parameter values. Assuming an initial state is specified, there are three steps involved:

Step 1. Generate a sequence of random variates $\{V_1, V_2, \ldots\}$ from an exponential cdf with parameter Λ.

This immediately allows us to define all event times. Assuming the system starts at time zero, we have $T_1 = V_1, T_2 = T_1 + V_2$, and so on.

Step 2. Generate a sequence of discrete random variates corresponding to all events $i \notin \mathcal{E}$ with distribution $p_i = \lambda_i/\Lambda$.

This can be accomplished through the usual inverse transform technique we have invoked on several previous occasions. In particular, assuming an event set $\mathcal{E} = \{1, \ldots, N\}$, the kth event on a sample path, E_k, is determined through a random number U_k as follows:

$$E_k = \begin{cases} 1 & \text{if } 0 \leq U_k \leq \lambda_1/\Lambda \\ 2 & \text{if } \lambda_1/\Lambda < U_k \leq (\lambda_1 + \lambda_2)/\Lambda \\ \vdots \\ N & \text{if } (\lambda_1 + \ldots + \lambda_{N-1})/\Lambda < U \leq 1 \end{cases} \tag{11.113}$$

Step 3. Update the state, based on the model's state transition function $x' = f(x, i)$, where i is the event determined in **Step 2**.

There is one additional simplification we can make to this scheme. If we choose to generate interevent times at **Step 1** from an exponential distribution with parameter 1, the resulting sequence is referred to as the *Standard Clock*. For any specific model with total event rate $\Lambda \neq 1$, we can simply adjust the sequence $\{V_1, V_2, \ldots\}$ obtained by the unity rate standard clock by rescaling, that is, by setting $V_k(\Lambda) = V_k/\Lambda$.

The SC approach is outlined in Algorithm 11.3. This algorithm describes how to construct $M+1$ sample paths, all in parallel, corresponding to the $M+1$ parameter values in Fig. 11.35. In general, each sample path is characterized by a different rate parameter, denoted by $\Lambda_m, m = 0, 1, \ldots, M$. In addition, it is possible for each state transition mechanism to be different; we therefore denote the state transition function by $f_m(\cdot)$.

---

1. CONSTRUCT STANDARD CLOCK:
$$\{V_1, V_2, \ldots\}, \quad V_k \sim 1 - e^{-t}, \quad t > 0$$

For every constructed sample path $m = 0, 1, \ldots, M$:

2. DETERMINE TRIGGERING EVENT E_m THROUGH (11.113)

3. UPDATE STATE X_m: $\qquad\qquad X_m := f_m(X_m, E_m)$

4. RESCALE INTEREVENT TIME V: $\quad V_m = V/\Lambda_m$

Algorithm 11.3. Standard Clock (SC) construction of $M+1$ sample paths concurrently.

Remark. If in step 1 of Algorithm 11.3 the clock sequence has a parameter $\Lambda \neq 1$, then the only modification required is in step 4: $V_m = V \cdot (\Lambda/\Lambda_m)$.

The main advantage of the SC scheme is its simplicity. Even though it only applies to exponentially distributed event lifetimes, it is otherwise very general. It can be used, for instance, for continuous as well as discrete parameters. Its main drawback is the fact that a number of events may be generated at step 2 which are actually wasted, in the sense that they are fictitious and cause no change in the state of one or more of the constructed sample paths.

The SC approach in Algorithm 11.3 is limited to sample paths constructed through simulation. Clearly, an actual system will not generate fictitious events (e.g., we will never see a departure from an empty queueing system). Although this appears to be a limitation of the approach, an extension is possible. With some modifications, the sample path construction can be carried out based on a clock sequence (step 1 of Algorithm 11.3) consisting only of *actual* events as they arise in a nominal sample path (see Cassandras et al., 1991).

Example 11.16 (SC construction for the $M/M/1$ queueing system)
Let us apply Algorithm 11.3 to an $M/M/1$ queueing system with arrival rate λ and service rate μ. In terms of a stochastic timed automaton model, we have an event set $\mathcal{E} = \{a, d\}$, with both lifetimes exponentially distributed; the parameter for an arrival event a is λ, and the parameter for a departure event d is μ. In addition to a nominal (λ, μ) sample path, we are interested in constructing two more sample paths corresponding to parameters $(\lambda + \Delta_1, \mu)$ and $(\lambda + \Delta_2, \mu)$ with $\Delta_2 > \Delta_1$.

Looking at Algorithm 11.3, note that the first step is independent of the system considered. Step 2, on the other hand, becomes

$$E_m = \begin{cases} a & \text{if } U \leq \lambda_m/\Lambda_m \\ d & \text{if } U > \lambda_m/\Lambda_m \end{cases} \qquad (11.114)$$

where U is a random number and $m = 0, 1, 2$, corresponds to the three sample paths. For each m, λ_m is the arrival rate and Λ_m is the maximal Poisson rate. Thus, $\lambda_0 = \lambda$, $\lambda_1 = \lambda + \Delta_1$, $\lambda_2 = \lambda + \Delta_2$, and $\Lambda_0 = \lambda + \mu$, $\Lambda_1 = \lambda + \Delta_1 + \mu$, $\Lambda_2 = \lambda + \Delta_2 + \mu$. Then, at step 3 we set

$$X_m := X_m + 1 \quad \text{if } E_m = a$$
$$X_m := X_m - 1 \quad \text{if } E_m = d \text{ and } X_m > 0$$

If $E_m = d$ and $X_m = 0$, then the state remains unaffected (the d event is ignored). Note that this may happen in all or some of the three constructed sample paths. Finally, at step 4, each interevent time V obtained from step 1 is rescaled, i.e., $V_m = V/\Lambda_m$. Observe that this must be done whether or not E_m is a fictitious event.

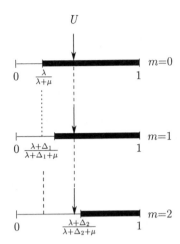

Figure 11.38: Determining triggering events over three concurrently constructed sample paths.

For a concrete example of how the SC scheme operates, suppose all three sample paths start out at state 0. Figure 11.38 illustrates the process of determining the first triggering event for all three sample paths from a single common random number U. For $m = 0$, we see that $U > \lambda/(\lambda+\mu)$, therefore $E_0 = d$. This is a fictitious event, and the state remains at 0. For $m = 1, U > \lambda/(\lambda + \Delta_1 + \mu)$, and once again the state remains at 0. Finally, for $m = 2$, the perturbation Δ_2 is sufficiently large to cause the event to become an arrival, and the next state to become 1.

The potential inefficiency of the SC approach in terms of "wasted random numbers" is visible in Example 11.16: Because of the small value of λ (as seen in Fig. 11.38) a large number of d events is likely to be generated; on the other hand, the system is often empty precisely because λ is small, hence many of these events are fictitious. It is generally the case in queueing systems that under low traffic intensities this situation will arise. The same is true in a system with multiple classes, where a d_i event (corresponding to a departure of class i) may often be wasted because service is provided to some class $j \neq i$ customer.

Example 11.17 (SC construction for the $M/M/1/K$ queueing system)
In this example, we apply Algorithm 11.3 to sample paths defined by different values of a discrete parameter. Consider an $M/M/1/K$ queueing system with arrival rate λ and service rate μ, where K is the parameter of interest. Since K only affects the state transition function in this system, the maximal event rate is always fixed at $\lambda + \mu$. Thus, let us modify step 1 of Algorithm 11.3 to first generate a clock sequence consisting of

random variates from an exponential distribution with parameter $\lambda + \mu$. At step 2, we generate a single triggering event E for all sample paths, which, for convenience, are indexed by $K = 1, 2, \ldots$ This can be done as in (11.114) by simply setting $\lambda_m = \lambda$, and $\Lambda_m = \lambda + \mu$. At step 3, we set for $K = 1, 2, \ldots$:

$$X_K := X_K + 1, \quad \text{if } E = a \text{ and } X_K < K$$
$$X_K := X_K - 1, \quad \text{if } E = d \text{ and } X_K > 0$$

Once again, note that K only affects the sample path construction through the state transition mechanism at step 3. Finally, step 4 is unnecessary, since K does not affect the clock sequence and no rescaling is needed.

The SC approach is well-suited for parallel processing environments, where each processor handles one of the $M + 1$ sample paths to be constructed. In this case, the only factor preventing us from generating a very large number of sample paths in the time normally required to generate a single sample path is the problem of fictitious events mentioned earlier. In order to get a better idea of the tradeoff involved, let C be the "cost" (e.g., CPU time) associated with the generation of a triggering event through the standard inverse transform technique. Suppose that the length of a simulation run is defined by K events (real, not fictitious ones). If $M + 1$ distinct simulation runs are used, the total cost is of the order of $(M + 1)KC$. If $M + 1$ sample paths are *concurrently* constructed through Algorithm 11.3, the total cost is of the order of $(KC + K_f C)$, where K_f is the number of fictitious events that had to be generated in the SC scheme of things. Thus, the computational savings is of the order of $(MK - K_f)C$, which is substantial, unless K_f is a very large number.

A few additional observations are worth making. First, the inverse transform technique for determining event E_k in (11.113) may become inefficient when the number of events in the event set $\mathcal{E}$ is very large (one has to search over all intervals of the form $[(\lambda_1 + \ldots + \lambda_i)/\Lambda, (\lambda_1 + \ldots + \lambda_i + 1/\Lambda]$ in order to locate the one within which U_k lies). It is, however, possible to improve the efficiency of generating discrete random variates of this type through an alternative technique, known as the "Alias Method" (e.g., Bratley and Fox, 1987). What is attractive about this technique is that it is independent of the size of the event set $\mathcal{E}$. Second, note that Algorithm 11.3 constructs sample paths from which any sample function of interest can be evaluated. Thus, unlike PA techniques we saw earlier in this chapter, the nature of the performance measure we are trying to estimate is not critical. Finally, attention must be called to the following fact that makes *Concurrent Simulation* techniques, such as the SC approach, extremely attractive: Performance estimates over many alternatives, $\theta_0, \theta_1, \ldots, \theta_M$, become available quickly; therefore, "good" designs can be distinguished from "bad" designs relatively fast, and emphasis can immediately be placed on improved estimation of the good ones. This brings us back to the point made at the beginning of Section 11.10, that Concurrent Estimation techniques serve the purpose of "rapid learning" for complex DES.

11.10.5 Augmented System Analysis

Augmented System Analysis (ASA) was proposed by Cassandras and Strickland (1989b) as a framework for investigating the sample path constructability problem of Fig. 11.35 for Markov chains. It was later extended to stochastic timed automata where at most one event can be generated by a non-Poisson process (Cassandras and Strickland, 1989c). Unlike the SC approach, which focuses on a simulation-based setting where uniformization can be effectively exploited, the starting point of ASA is a "real" sample path. In other words, we accept the fact that constructability must be satisfied based on whatever events the observed sample path provides. Such, in fact, is the case in most real-time applications.

Let us first limit ourselves once again to exponential event lifetime distributions. This enables us to concentrate on condition **(OB)** alone in trying to satisfy **(CO)**. The setting is the same as before: A sample path $\xi(\theta)$ is directly observed, and we are trying to check whether we can construct $\xi(\theta_m)$ for some $\theta_m \neq \theta$. As the observed state sequence $\{X_k(\theta)\}$ unfolds, we can construct the sequence $\{X_k(\theta_m)\}$ as long as $\Gamma(X_k(\theta_m)) \subseteq \Gamma(X_k(\theta))$. Now, suppose that we encounter a state pair $(X(\theta), X(\theta_m))$ such that $\Gamma(X(\theta_m)) \supset \Gamma(X(\theta))$. This means that there is at least one event, say $i \in (X(\theta_m))$, which is "unobservable" at state $X(\theta)$. At this point, the construction of $\xi(\theta_m)$ must be suspended, since the triggering event at state $X(\theta_m)$ cannot be determined.

Now, the key idea in ASA is the following: Keep the state in $\xi(\theta_m)$ fixed at $X(\theta_m)$ and keep on observing $\xi(\theta)$ *until some new state $X'(\theta)$ is entered that satisfies* $\Gamma(X(\theta_m)) \subseteq \Gamma(X'(\theta))$. The only remaining question is how long $\xi(\theta_m)$ must wait in suspension. Note that we do not require "state matching," that is, we do not require that $X'(\theta) = X(\theta_m)$ in order to continue the construction, but the weaker condition of "event matching": $\Gamma(X(\theta_m)) \subseteq \Gamma(X'(\theta))$.

The validity of the event matching scheme rests on some basic properties of Markov processes which enable us to "cut-and-paste" segments of a given sample path to produce a new sample path which is stochastically equivalent to the original one. This is illustrated in Fig. 11.39 for a sample path of an $M/M/1$ queueing system. The sample path is divided into three segments, as shown. The first segment ends with an arrival at state 2. The third segment starts with a departure from state 2. These two segments are joined to produce another sample path, as shown in the figure. Since a departure event at state 2 is certainly feasible, the only question pertaining to the legitimacy of this new sample path is whether all event clock values have the right distributions. This, however, is ensured by the properties of a Markov chain: The clock values for the arrival and departure events at the "cut" point have the same distribution as the corresponding clock values at the "paste" point, by the memoryless property.

In ASA, the validity of the "cut-and-paste" idea is exploited to solve the constructability problem for *discrete* parameters, which do not alter event times - only state transition mechanisms (see Cassandras and Strickland, 1989b). The "cut-and-paste" idea, however, can also be used for purposes of sensitivity estimation, as in the *Extended Perturbation Analysis* (EPA) approach (Ho and Li, 1988).

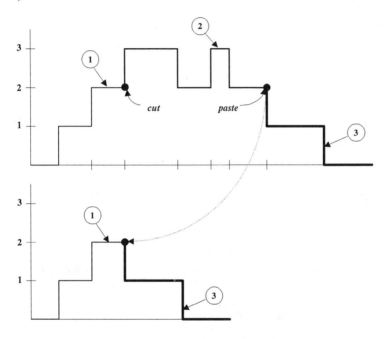

Figure 11.39: Illustrating the "cut-and-paste" idea for Markov chains.

The term "Augmented System" in ASA is due to the construction (illustrated in Fig. 11.36 of Example 11.14) of a system formed by the joint states of $\xi(\theta)$ and $\xi(\theta_m)$ coupled by the fact that both systems are driven by a common event sequence. This provides a convenient means for checking the observability condition **(OB)**, before attempting to construct $\xi(\theta_m)$ from $\xi(\theta)$. It also reflects the idea of treating two (or more) systems, differing only in one parameter setting, as a single system with an "augmented" state of the form $(X(\theta), X(\theta_m))$ with $\theta \neq \theta_m$.

The Event Matching Algorithm.

Let us now see how ASA solves the constructability problem for (a) discrete parameters that affect a system only through its state transition mechanism, and (b) Poisson event processes.

In order to model the process of "suspending" the construction of $\xi(\theta_m)$ when a state of $\xi(\theta)$ is encountered which violates the observability condition **(OB)**, we associate with $\xi(\theta_m)$ an auxiliary variable we shall refer to as the *mode* of $\xi(\theta_m)$, denoted by $\mathcal{M}_m$. The mode can take on either the special value "active", or a value from the state space $\mathcal{X}_m$:

1. Initially, we set $\mathcal{M}_m = $ "active". As long as **(OB)** is satisfied, the construction of $\xi(\theta_m)$ remains active, and the mode is unaffected.

2. If a state $X(\theta)$ is observed where **(OB)** is violated, i.e., $\Gamma(X(\theta_m)) \supset \Gamma(X(\theta))$, we set $\mathcal{M}_m = X_m$. The construction of $\xi(\theta_m)$ is suspended until a new state is encountered where **(OB)** is satisfied.

This procedure for constructing multiple sample paths $m = 1, \ldots, M$, based on an observed sample path $\xi(\theta)$, is outlined in Algorithm 11.4.

---

1. INITIALIZE: For all constructed sample paths $m = 1, \ldots, M$:

$$X_m := X_0 \text{ and } \mathcal{M}_m := \text{ACTIVE}$$

2. WHENEVER ANY EVENT (SAY α) IS OBSERVED
   AND NOMINAL STATE BECOMES X, FOR $m = 1, \ldots, M$:

    2.1. If $\mathcal{M}_m = $ ACTIVE,
        2.1.1. Update state: $X_m := f_m(X_m, \alpha)$
        2.1.2. If $\Gamma(X_m) \supset \Gamma(X)$,   (observability violated)
           set: $\mathcal{M}_m := X_m$ (m constr. suspended)
    2.2. If $\mathcal{M}_m \neq $ ACTIVE, and $\Gamma(\mathcal{M}_m) \subseteq \Gamma(X)$,
        set: $\mathcal{M}_m := $ ACTIVE   (m constr. restarted)
    2.3. If $\mathcal{M}_m \neq $ ACTIVE, and $\Gamma(\mathcal{M}_m) \supset \Gamma(X)$,
        continue   (m remains suspended)

Algorithm 11.4. The Event Matching algorithm.

It is worth pointing out that Algorithm 11.4 is geared towards "real-time" applications (in addition to simulation). Note that it is entirely driven by event and state observations made on an actual system at the beginning of step 2. Thus, at step 2.1.1, it is only the states of constructed sample paths that need to be updated. No stopping condition is specified in Algorithm 11.4, since this is generally application-dependent. Also, as in the case of the SC construction, the result of the Event Matching algorithm is a complete sample path; any desired sample function can subsequently be evaluated from it.

Example 11.18 (ASA construction for the $M/M/1/K$ system)
In this example, we apply Algorithm 11.4 to sample paths defined by

different values of K in an $M/M/1/K$ queueing system with arrival rate λ and service rate μ. This problem was also solved in Example 11.17 using the SC approach. Suppose that we observe a sample path under $K = 2$, and wish to use it in order to construct another sample path under $K = 3$ (for simplicity, we limit ourselves to a single constructed sample path, but the extension to more sample paths is straightforward). Recall that **(OB)** is not satisfied in this case, as shown in Example 11.15, so applying the Event Matching algorithm is of particular interest.

In the application of Algorithm 11.4 to this problem, let X denote the observed state, and X' the constructed state. The mode of the lone sample path to be constructed here is denoted by $\mathcal{M}$. The augmented system state transition diagram shown in Fig. 11.36 is particularly helpful in this case, since it allows us to detect the only possible way that observability can be violated. Specifically, when d occurs, causing a transition into states $X = 0$ (under $K = 2$) and $X' = 1$ (under $K = 3$) respectively, we see that $\Gamma(X') = \{a, d\} \supset \{a\} = G(X)$. Thus, d is unobservable at that joint state, and step 2.1.2 sets $\mathcal{M} = 1$.

The diagram is further helpful in identifying how long the construction needs to remain suspended. Since the only feasible event at state 0 is a, the nominal sample path must enter state 1 next. At this point, we have $X = 1, \mathcal{M} = 1$, and, obviously, $\Gamma(\mathcal{M}) = \Gamma(X)$. In other words, suspension here is limited to an occasional observed idle period, during which the constructed sample path must wait for a new d event to be activated.

With these observations in mind, Algorithm 11.4 reduces to the following procedure in this case:

1. INITIALIZE: $X' := X$ and $m := $ ACTIVE

2. IF EVENT d OBSERVED:

    (a) Update state: $X' := X' - 1$

    (b) If $X = 0$, and $X' = 1$,
        set: $\mathcal{M} := 1$

3. IF EVENT a OBSERVED:

    (a) If $\mathcal{M} = $ ACTIVE, update state: $X' := X' + 1$ if $X' < 3$;
        else: $X' := X'$

    (b) If $\mathcal{M} = 1$, set: $\mathcal{M}_m = $ ACTIVE

A typical example of this construction is shown in Fig. 11.40. When the observed sample path enters state 0 while the constructed sample path is at state 1, the construction is suspended. It is immediately restarted when the next event (always an arrival) takes place. Thus, the length of the suspension interval is given by the length of the idle period shown.

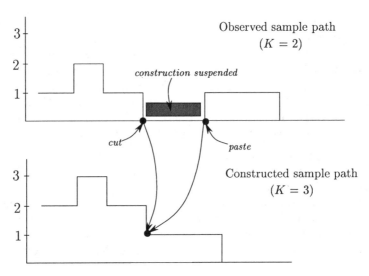

Figure 11.40: Event Matching algorithm for Example 11.18.
Event d is unobservable when the constructed sample path enters state 1 ($\Gamma(1) = \{a,d\}$), while the observed sample path enters state 0 ($\Gamma(0) = \{a\}$). The construction is suspended at state 1 until a new state is observed where d is feasible. This happens when state 1 is entered next.

Example 11.19 (Comparing SC, ASA, and "Brute-Force" schemes)

In this example, we provide explicit numerical data from the simulation of an $M/M/1/K$ system over a range of K values ($K = 1, \ldots, 10$), and compare (a) Concurrent Simulation through the SC scheme (Algorithm 11.3), (b) Concurrent Simulation through the ASA approach (Algorithm 11.4), and (c) "Brute-Force" construction of the sample paths in the form of independently generated simulation runs. There are a number of criteria based on which this comparison should be made, including implementation complexity and variance properties. Here, we limit ourselves to computational efficiency in terms of CPU time measured on a common (serial, not parallel) computer environment (a UNIX-based workstation).

The numerical results shown in Fig. 11.41 are based on a common stopping time defined by 100,000 "real" d events. This means that in the SC scheme the simulation may include a number of fictitious d events. Thus, the comparison in Fig. 11.41 is a "fair" one. The CPU time required for each approach is plotted as a function of the number of constructed sample paths. In the case of ASA, the observed sample path was taken to be the one under $K = 2$. Thus, observability is violated for all sample paths with $K > 2$. Interestingly, and despite the fact that this represents an almost "worst case" for ASA (eight of the nine constructed sample paths have to go through steps 2.3 and 2.4 in Algorithm 11.4), we see that ASA leads to a construction which is virtually insensitive to additional sample paths to be constructed.

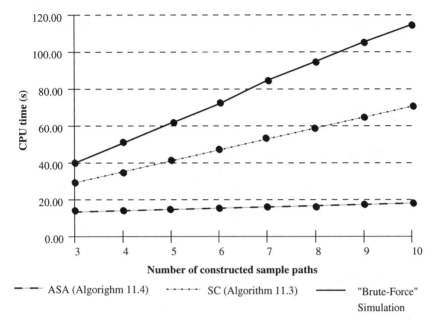

Figure 11.41: Comparison of SC, ASA, and "Brute-Force" sample path construction schemes for an $M/M/1/K$ system.

The idea of "rapid learning" is readily visualized through Fig. 11.41: Instead of taking about 115 sec to "learn" the behavior of the system under 10 different parameter settings by "brute-force" trial-and-error, ASA accomplishes the same task in about 18 sec (about six times faster). The curves in Fig. 11.41 are typical of the improvements gained through the SC and ASA techniques we have discussed. It is also worth pointing out that as the number of constructed sample paths increases, this improvement (in terms of "learning time") grows roughly linearly.

The Age Matching Algorithm

The ASA approach for Concurrent Estimation is not limited to systems with exponential lifetimes for all events. This can be immediately seen in Example 11.18 by focusing on the ages and residual lifetimes of a and d at the "cut" and "paste" points (see Fig. 11.42, where the observed sample path of Fig. 11.40 is redrawn). The memoryless property is needed to ensure that the distributions of the clocks of both events at the "cut" point are the same as those at the "paste" point. This is necessary for the a event, since at the "cut" point its age is some $z_a > 0$, whereas at the "paste" point it is $z_a = 0$ (since a just took place). However, the ages of the d event automatically match: At the "cut" point $z_d = 0$, because d just occurred, while at the "paste" point $z_d = 0$ because d was not active prior to this point. It follows that the lifetime distribution of d need not be exponential. In other words, the Event Matching algorithm in this case works for all $M/G/1/K$ systems.

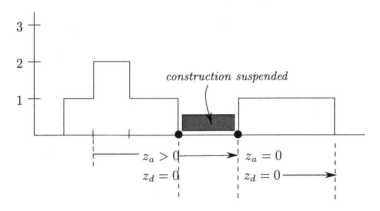

Figure 11.42: Illustrating why Event Matching works for the $M/G/1/K$ system.

This discussion leads to the extension of ASA to more general models of DES. Whenever a state is entered such that the construction of $\xi(\theta_m)$ must be suspended for some m, the ages of all events with non-exponential lifetime distributions are saved. Then, when a state satisfying **(OB)** is entered, the additional requirement is that the construction restarts at a point in time with the following property: The ages of all such events are equal to the saved age values from the "cut" point.

This modification leads to an *Age Matching* sample path construction algorithm. Let us partition the event set $\mathcal{E}$ as follows: $\mathcal{E} = \mathcal{E}_1 \cup \mathcal{E}_2$, where $\mathcal{E}_1$ contains all events with exponentially distributed lifetimes, and $\mathcal{E}_2$ contains all other events. At any point in time, let Z_m^i denote the age of event $i \in \mathcal{E}_2$ on $\xi(\theta_m)$. Algorithm 11.5 describes the "age matching" modification to Algorithm 11.4.

Looking at step 2.2.2 of Algorithm 11.5, we can immediately see that the age matching condition may not always be satisfied. There are four possible situations that can arise:

1. Age matching for $i \in \mathcal{E}_2$ is automatically satisfied at step 2.2.1. This is the case of a events in Example 11.18, as shown in Fig. 11.42. This, however, is a special case that exploits the structure of the $M/G/1/K$ system.

2. $\{\Gamma(X_m) \cap \mathcal{E}_2\}$ contains only one event at step 2.2.1, and $Z_m^i < \text{AGE}(i, m)$. In this case, it is feasible to wait until $Z_m^i = \text{AGE}(i, m)$, and hence satisfy age matching.

3. $\{\Gamma(X_m) \cap \mathcal{E}_2\}$ contains only one event at step 2.2.1, and $Z_m^i > \text{AGE}(i,m)$. In this case, waiting will not help. However, we can maintain the suspension until another event occurs such that step 2.2.1 is revisited.

4. $\{\Gamma(X_m) \cap \mathcal{E}_2\}$ contains more than one event at step 2.2.1. If this situation arises, simultaneous age matching is infeasible. It is only possible to achieve "approximate age matching", if desired. Even this approach, however, becomes increasingly hard to accomplish if the number of events in $\{\Gamma(X_m) \cap \mathcal{E}_2\}$ is large.

The main drawback of ASA (either Event Matching or Age Matching) is the need to suspend a sample path construction for a potentially long time interval. This problem is the counterpart of the drawback we saw for the SC approach, which consists of "wasted random number generations." In either case, valuable time may be wasted in our effort to "rapidly learn" the behavior of a DES under all $\theta \in \Theta$.

---

1. INITIALIZE: For all constructed sample paths $m = 1, \ldots, M$:

$$X_m := X_0 \text{ and } \mathcal{M}_m := \text{ACTIVE}$$

2. WHENEVER ANY EVENT (SAY α) IS OBSERVED
   AND NOMINAL STATE BECOMES X, FOR $m = 1, \ldots, M$:

   2.1. If $\mathcal{M}_m = \text{ACTIVE}$,
        2.1.1. Update state: $X_m := f_m(X_m, \alpha)$
        2.1.2. If $\Gamma(X_m) \supset \Gamma(X)$,   (observability violated)
               2.1.2a. Set: $\mathcal{M}_m := X_m$ (m constr. suspended)
               2.1.2b. For all events $i \in \{\Gamma(X_m) \cap \mathcal{E}_2\}$,
                       set: $\text{AGE}(i,m) := Z_m^i$ (non-Poisson
                                                            event ages saved)
   2.2. If $\mathcal{M}_m \neq \text{ACTIVE}$, and $\Gamma(\mathcal{M}_m) \subseteq \Gamma(X)$,
        2.2.1. If $Z_m^i = \text{AGE}(i,m)$ for all $i \in \{\Gamma(X_m) \cap \mathcal{E}_2\}$,
               set: $\mathcal{M}_m := \text{ACTIVE}$ (m constr. restarted)
               Else, WAIT UNTIL THIS CONDITION
                     IS SATISFIED, AND REPEAT 2.2.1
   2.3. If $\mathcal{M}_m \neq \text{ACTIVE}$, and $\Gamma(\mathcal{M}_m) \supset \Gamma(X)$,
        continue   (m remains suspended)

Algorithm 11.5. The Age Matching algorithm.

One way to overcome this difficulty in ASA is to "inject" information into the constructed sample path at "cut" points, instead of just waiting. For example, in Fig. 11.40, where the construction of an $M/G/1/3$ system is suspended when the d event is unobservable, it is possible to simply "inject" a random variate from the service time distribution $G_d(\cdot)$. This approach, however, does introduce some complications in the coupling of the observed and constructed sample paths, since the injected d event will generally not coincide with any observed event. One interesting special case arises when the d event has a deterministic lifetime, in which case the construction can be quite manageable.

11.10.6 The "Time Warping" Algorithm

The concept of "time warping" is encountered in some software environments where distributed processing takes place, resulting in occasional loss of synchronization across these processes. When this happens, one may "time warp" or "roll back" to a point where data consistency is guaranteed. As it applies to Concurrent Estimation, the idea of "time warping" is the following. We have seen that when an unobservable event is encountered in ASA, we need to suspend the sample path construction at some time instant t_1. While the observed system evolves, several events may occur and their lifetimes recorded. When the unobservable event finally becomes feasible, at some time $t_2 > t_1$, it is possible to utilize much of this information, collected over $[t_1, t_2]$. In particular, we may be able to perform several state updates from time t_1 onwards, possibly constructing an entire piece of the desired sample path all the way up to the current time t_2. This is what we refer to as "time warping" (Cassandras and Panayiotou, 1999).

> **Example 11.20 ("Time warping" for a $GI/G/1/K$ queueing system)**
> Let us return to Example 11.18, except that we will now allow general interarrival and processing time distributions. We observe the system under $K = 2$, and we try to construct a sample path under $K = 3$.
>
> A typical sample path under $K = 2$ is shown in Figure 11.43. As we already saw in Example 11.18, the point where observability is violated is when a d event occurs with the observed state at 1 and the constructed state at 2. At this point, no d event is observable (the first dotted line in the figure). If we wait until the next a event and blindly continue the construction, we are ignoring the possibility that a d event might have occurred first. This uncertainty can be resolved when the next d event is finally observed (the second dotted line in the figure): If we go back to the first dotted line, we now have at our disposal both a sample from the service time distribution (shown as B in Figure 11.43) and a clock value (residual lifetime) for an a event from the point of suspension (shown as A in Figure 11.43). We therefore proceed to determine the triggering event by comparing A and B and find that the next event in this example is d, followed by a (since at state 0 a d event is no longer feasible). At this point, our construction has caught up with the observed sample path, even though the time instants are different. In other words, on the constructed sample path, we "time-warped" from the left dotted line for-

ward by processing as many events as we could based on the information collected while in suspension (shaded area in the figure); in this case, two events were processed instantaneously.

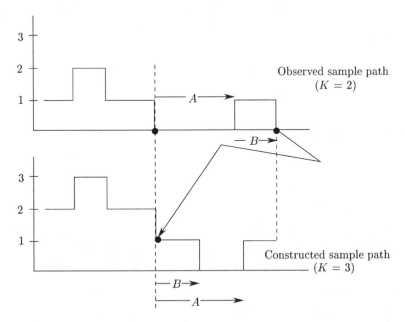

Figure 11.43: Sample path construction for Example11.20.

Time warping is equivalent to delaying the construction of a sample path until adequate information is collected from observed data, in order to correctly update the state, and possibly generate several state updates during the time warp. It should be clear that while this mechanism allows for generality in the system model, it involves additional storage requirements for various observed event lifetimes, as well as some extra data processing.

To gain some more insight on the time-warping mechanism and develop a formal algorithm for general-purpose Concurrent Estimation, let us introduce some notation and basic definitions. First, let $\xi(n,\theta) = \{e_j : j = 1,\dots,n\}$, with $e_j \in \mathcal{E}$, be the sequence of events that constitute the observed sample path up to n total events. Although $\xi(n,\theta)$ is clearly a function of the parameter θ, we will write $\xi(n)$ to refer to the observed sample path and adopt the notation $\hat{\xi}(k) = \{\hat{e}_j : j = 1,\dots,k\}$ for any constructed sample path under a different value of the parameter up to k events in that path. It is important to realize that k is actually a function of n, since the constructed sample path is coupled to the observed sample path through the observed event lifetimes; however, again for the sake of notational simplicity, we will refrain from continuously indicating this dependence.

Next, let $s_i^n = [\xi(n)]_i$ denote the score of event $i \in \mathcal{E}$, that is, the number of instances of event i in the sequence $\xi(n)$. The corresponding score of i in a constructed sample path is denoted by $\hat{s}_i^k = [\hat{\xi}(k)]_i$. In what follows, all quantities with the symbol " $\hat{}$ " refer to a typical constructed sample path.

Associated with every event type $i \in \mathcal{E}$ in $\xi(n)$ is a sequence of s_i^n event lifetimes

$$\mathbf{V}_i(n) = \{v_i(1), \dots, v_i(s_i^n)\} \quad \text{for all } i \in \mathcal{E}$$

The corresponding set of sequences in the constructed sample path is

$$\hat{\mathbf{V}}_i(k) = \{v_i(1), \dots, v_i(\hat{s}_i^k)\} \quad \text{for all } i \in \mathcal{E}$$

which is a subsequence of $\mathbf{V}_i(n)$ with $k \leq n$. In addition, we define the following sequence of lifetimes:

$$\tilde{\mathbf{V}}_i(n, k) = \{v_i(\hat{s}_i^k + 1), \dots, v_i(s_i^n)\} \quad \text{for all } i \in \mathcal{E}$$

which consists of all event lifetimes that are in $\mathbf{V}_i(n)$ but not in $\hat{\mathbf{V}}_i(k)$. Associated with any one of these sequences are the following operations. Given some $\mathbf{W}_i = \{w_i(j), \dots, w_i(r)\}$,

Suffix Addition: $\mathbf{W}_i + \{w_i(r+1)\} = \{w_i(j), \dots, w_i(r), w_i(r+1)\}$.

Prefix Subtraction: $\mathbf{W}_i - \{w_i(j)\} = \{w_i(j+1), \dots, w_i(r)\}$.

Note that the addition and subtraction operations are defined so that a new element is always added as the *last* element (the *suffix*) of a sequence, whereas subtraction always removes the *first* element (the *prefix*) of the sequence.

Next, define the set

$$A(n, k) = \{i : i \in \mathcal{E}, s_i^n > \hat{s}_i^k\} \tag{11.115}$$

which is associated with $\tilde{\mathbf{V}}_i(n, k)$ and consists of all events i whose corresponding sequence $\tilde{\mathbf{V}}_i(n, k)$ contains at least one element. Thus, every $i \in A(n, k)$ is an event that has been observed in $\xi(n)$ and has at least one lifetime that has yet to be used in the coupled sample path $\hat{\xi}(k)$. Hence, $A(n, k)$ should be thought of as the set of *available* events to be used in the construction of the coupled path.

Finally, we define the following set, which is crucial in our approach:

$$M(n, k) = \Gamma(\hat{x}_k) - (\Gamma(\hat{x}_{k-1}) - \{\hat{e}_k\}) \tag{11.116}$$

where, clearly, $M(n, k) \subseteq \mathcal{E}$. Note that $\hat{e}_k$ is the triggering event at the $(k-1)$th state visited in the constructed sample path. Thus, $M(n, k)$ contains all the events that are in the feasible event set $\Gamma(\hat{x}_k)$ but not in $\Gamma(\hat{x}_{k-1})$; in addition,

$\hat{e}_k$ also belongs to $M(n,k)$ if it happens that $\hat{e}_k \in \Gamma(\hat{x}_k)$. Intuitively, $M(n,k)$ consists of all *missing* events from the perspective of the constructed sample path when it enters a new state $\hat{x}_k$: Those events already in $\Gamma(\hat{x}_{k-1})$ which were not the triggering event remain available to be used in the sample path construction as long as they are still feasible; all other events in the set are "missing" as far as residual lifetime information is concerned. This motivates the following definition.

Definition. A constructed sample path is *active* at state $\hat{x}_k$, after an observed event e_n occurs, if, for every $i \in \Gamma(\hat{x}_k)$, $i \in (\Gamma(\hat{x}_{k-1}) - \{\hat{e}_k\}) \cup A(n,k)$.   ◆

Thus, the constructed sample path can only be "active" at state $\hat{x}_k$ if every $i \in \Gamma(\hat{x}_k)$ is such that either $i \in (\Gamma(\hat{x}_{k-1}) - \{\hat{e}_k\})$ (in which case $\hat{y}_i(k)$ is a residual lifetime of an event available from the previous state transition) or $i \in A(n,k)$ (in which case a full lifetime of i is available from the observed sample path).

Let us now see what happens as an observed sample path under θ unfolds and we try to use the information from it in order to concurrently construct another sample path under a different value of the parameter. Upon occurrence of the $(n+1)$th observed event, e_{n+1}, the first step is to update the event lifetime sequences $\tilde{\mathbf{V}}_i(n,k)$ as follows:

$$\tilde{\mathbf{V}}_i(n+1,k) = \begin{cases} \tilde{\mathbf{V}}_i(n,k) + v_i(s_i^n + 1) & \text{if } i = e_{n+1} \\ \tilde{\mathbf{V}}_i(n,k) & \text{otherwise} \end{cases} \qquad (11.117)$$

The addition of a new event lifetime implies that the "available event set" $A(n,k)$ defined in (11.115) may be affected. Therefore, it is updated as follows:

$$A(n+1,k) = A(n,k) \cup \{e_{n+1}\} \qquad (11.118)$$

Finally, note that the "missing event set" $M(n,k)$ defined in (11.116) remains unaffected by the occurrence of observed events:

$$M(n+1,k) = M(n,k) \qquad (11.119)$$

At this point, we are able to decide whether all lifetime information to proceed with a state transition in the constructed sample path is available or not. In particular, the condition

$$M(n+1,k) \subseteq A(n+1,k) \qquad (11.120)$$

may be used to determine whether the constructed sample path is *active* at the current state $\hat{x}_k$ (in the sense of the definition given earlier).

1. INITIALIZATION:

$n := 0$, $k := 0$, $t_n := 0$, $\hat{t}_k := 0$, $x_n := x_0$, $\hat{x}_k = \hat{x}_0$,
$y_i(n) = v_i(1)$ for all $i \in \Gamma(x_n)$, $s_i^n = 0$, $\hat{s}_i^k = 0$ for all $i \in \mathcal{E}$,
$M(0,0) := \Gamma(\hat{x}_0)$, $A(0,0) := \emptyset$

2. WHEN EVENT e_n IS OBSERVED:

2.1. Update e_{n+1}, x_{n+1}, t_{n+1}, $y_i(n+1)$ for all $i \in \Gamma(x_{n+1})$,
    and s_i^{n+1} for all $i \in \mathcal{E}$
2.2. Add the event lifetime of e_{n+1} to $\tilde{V}_i(n+1,k)$ using (11.117)
2.3. Update the available event set $A(n,k)$:
    $A(n+1,k) := A(n,k) \cup \{e_{n+1}\}$
2.4. Update the missing event set $M(n,k)$:
    $M(n+1,k) := M(n,k)$
2.5. If $M(n+1,k) \subseteq A(n+1,k)$ then go to step 3.
    Else, set $n \leftarrow n+1$ and go to step 2.1.

3. TIME WARPING OPERATION:

3.1. Obtain all missing event lifetimes to resume
    sample path construction at state $\hat{x}_k$:
    $$\hat{y}_i(k) = \begin{cases} v_i(\hat{s}_i^k + 1) & \text{for } i \in M(n+1,k) \\ \hat{y}_i(k-1) & \text{otherwise} \end{cases}$$
3.2. Update $\hat{e}_{k+1}$, $\hat{x}_{k+1}$, $\hat{t}_{k+1}$,
    $\hat{y}_i(k+1)$ for all $i \in \Gamma(\hat{x}_{k+1}) \cap (\Gamma(\hat{x}_k) - \{\hat{e}_{k+1}\})$,
    and $\hat{s}_i^{k+1}$ for all $i \in \mathcal{E}$
3.3. Discard all used event lifetimes using (11.121)
3.4. Update the available event set $A(n+1,k)$ using (11.122)
3.5. Update the missing event set $M(n+1,k)$ using (11.123)
3.6. If $M(n+1,k+1) \subseteq A(n+1,k+1)$ then set $k \leftarrow k+1$
    and go to step 3.1.
    Else, set $k \leftarrow k+1$, $n \leftarrow n+1$ and go to step 2.1.

Algorithm 11.6. The Time Warping Algorithm for general-purpose Concurrent Estimation.

Assuming (11.120) is satisfied, we can update the state $\hat{x}_k$ of the constructed sample path in the usual way that applies to any timed state automaton. In so doing, lifetimes $v_i(s_i^k+1)$ for all $i \in M(n+1,k)$ are used from the corresponding sequences $\tilde{V}_i(n+1,k)$. Thus, upon completion of the state update steps, all three variables $\tilde{\mathbf{V}}_i(n, k)$, $A(n, k)$, and $M(n, k)$ need to be updated. In particular,

$$\tilde{\mathbf{V}}_i(n + 1, k + 1) = \left\{ \begin{array}{ll} \tilde{\mathbf{V}}_i(n + 1, k) - v_i(\hat{s}_i^k + 1) & \text{for all } i \in M(n + 1, k) \\ \tilde{\mathbf{V}}_i(n + 1, k) & \text{otherwise} \end{array} \right.$$

$$(11.121)$$

This operation immediately affects the set $A(n + 1, k)$ which is updated as follows:

$$A(n + 1, k + 1) = A(n + 1, k) - \left\{ i : i \in M(n + 1, k), \ \hat{s}_i^{k+1} = s_i^{n+1} \right\} \quad (11.122)$$

Finally, applying (11.116) to the new state $\hat{x}_{k+1}$,

$$M(n + 1, k + 1) = \Gamma(\hat{x}_{k+1}) - (\Gamma(\hat{x}_k) - \{\hat{e}_{k+1}\}) \quad (11.123)$$

Therefore, we are again in a position to check condition (11.120) for the new sets $M(n + 1, k + 1)$ and $A(n + 1, k + 1)$. If it is satisfied, then we can proceed with one more state update on the constructed sample path; otherwise, we wait for the next event on the observed sample path until (11.120) is again satisfied.

We can summarize the preceding process in the form of Algorithm 11.6, which we refer to as the *Time Warping Algorithm* (TWA). If TWA is used as a Concurrent Estimation method *on line*, step 2.1 is taken care of automatically by the actual system. The additional operations involved are steps 2.2 through 2.4 and checking the condition in step 2.5. If the latter is satisfied, then the time warping operation in step 3 is responsible for constructing a segment of the desired sample path for as many events as possible, depending on step 3.6.

Clearly, the computational requirements of TWA are minimal (adding and subtracting elements to sequences, simple arithmetic, and checking condition (11.120)). Rather, it is the storage of additional information that constitutes the major cost of the algorithm.

SUMMARY

- Given a parameter θ and a performance measure $J(\theta)$, the problem of estimating the sensitivity $dJ/d\theta$ can be tackled through *Perturbation Analysis* (PA) or through the *Likelihood Ratio* (LR) (or Score Function (SF)) approach. A comparison of the two approaches is given in Section 11.6.

- PA is based on estimating $dJ/d\theta$ through the *sample derivative $dL/d\theta$*, where $J(\theta) = E[L(\theta, \omega)]$. The evaluation of $dL/d\theta$ is often efficiently done using information obtained from a *single observed sample path* of a DES. This is typically possible because of the event-driven structure of sample paths.

- PA yields unbiased sensitivity estimates if the equality $dE[L(\theta, \omega)]/d\theta = E[dL(\theta, \omega)/d\theta]$ holds. This imposes constraints on the nature of the sample function $L(\theta, \omega)$ and the structure of the DES that can be analyzed through PA.

- For stochastic timed state automata models of DES, *Infinitesimal Perturbation Analysis* (IPA) is a general approach that can efficiently yield unbiased sensitivity estimates for a class of performance measures (described through (11.68) to (11.70) and parameters affecting only the event lifetime distributions in the model. The *commuting condition* (**CC**) is a sufficient condition ensuring the unbiasedness of these estimates (provided some additional technical conditions are satisfied). IPA sensitivity estimates are also often strongly consistent.

- For a $GI/G/1$ queueing system, IPA provides simple expressions for unbiased sensitivity estimates of performance measures such as the mean system time. A typical example is the IPA estimator in (11.61). The reader is also referred to the Web site http://vita.bu.edu/cgc/IPA/ for an interactive simulator of simple queueing systems that incorporates such IPA estimators.

- IPA techniques are independent of event lifetime distributions (except for some technical conditions). Knowledge of the distribution corresponding to the parameter θ is generally required, but no other lifetime distributions in the model need to be known.

- If θ is known to be a scale or location parameter of an event lifetime distribution, then no explicit distribution information is required in IPA. In this case, (11.22) and (11.23) can be used.

- IPA applies to a sizeable class of DES, including Jackson-like queueing networks. IPA sensitivity estimates can be used in conjunction with various parameter adjustment schemes to solve optimization problems of considerable complexity. Such schemes can be applied to simulation models as well as (often more importantly in practice) *on-line control* of DES.

- Modeling features that typically cause IPA to give biased estimates include multiple customer classes, blocking phenomena, and state-dependent control policies.

- IPA fails to provide unbiased estimates when the sample function $L(\theta)$ is discontinuous in θ. In many cases, it is possible to bypass this problem by finding an alternative (stochastically equivalent) model of the system in which $L(\theta)$ is actually continuous. This broadens the applicability of IPA to some models that involve multiple customer classes, or blocking due to finite queueing capacities. It also makes it possible for IPA to be applied in sensitivity estimation with respect to parameters such as routing probabilities in queueing systems.

- *Smoothed Perturbation Analysis* (SPA) extends the domain of sensitivity estimation based on PA techniques. In SPA, the problem of discontinuities in a sample function $L(\theta)$ is overcome at the expense of more information extracted from the observed sample path and more modeling information. SPA sensitivity estimates are of the form

$$\left[\frac{dJ}{d\theta}\right]_{SPA} = \frac{d}{d\theta}E\left[L(\theta,\omega) \mid \mathcal{Z}\right]$$

where z is a characterization of the observed sample path. Conditioning on z has the effect of smoothing out discontinuities in $L(\theta)$.

- SPA estimators for certain types of problems are simple to implement and require limited extra information relative to IPA. Examples include sensitivity estimation problems in systems with real-time constraints (Section 11.8.1) and routing problems (Section 11.8.2).

- When we are interested in the effect of finite changes in some *continuous* parameter, or the effect of changing a *discrete* parameter, *Concurrent Estimation* techniques can be used. Concurrent Estimation techniques are intended to exploit the structure of DES in order to construct multiple sample paths under parameters $\{\theta_1,\ldots,\theta_m\}$, all in parallel, from observations made over a single sample path under a given parameter value θ_0. Solving this constructability problem implies the ability to "rapidly learn" the behavior of a DES over a range of parameter values by limiting observations to a sample path under one parameter value.

- The *constructability condition* (CC) consists of two parts: *observability* (OB) and a requirement for matching event lifetime distributions conditioned on observed event ages.

- The *Standard Clock* (SC) approach solves the sample path constructability problem for models with exponentially distributed event lifetimes by exploiting the uniformization technique for Markov chains. It allows the concurrent construction of multiple sample paths under different (continuous or discrete) parameters, at the expense of some "fictitious" events. This approach is primarily suited for concurrent simulation applications, especially where parallel processing capabilities can be efficiently used.

- *Augmented System Analysis* (ASA) solves the constructability problem for models with at most one non-exponentially distributed event lifetime and for discrete parameters, through the *Event Matching* and *Age Matching* algorithms.

- The *Time Warping Algorithm* (TWA) solves the constructability problem for models with arbitrarily distributed event lifetime and provides the most general method to date for Concurrent Estimation.

- Sensitivity analysis and Concurrent Estimation techniques attempt to exploit the event-driven structure of DES sample paths. They complement analytical and simulation approaches presented in previous chapters. Figure 11.44 is an attempt to classify them in terms of their applicability to continuous and discrete parameters.

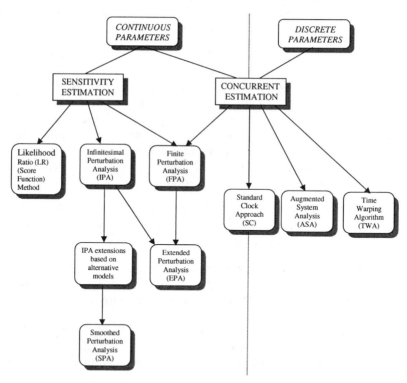

Figure 11.44: Classification of sensitivity analysis and concurrent estimation techniques.

PROBLEMS

11.1 Consider a random variable X uniformly distributed over $[a, b]$, and let θ be its mean.

(a) Suppose θ is perturbed by keeping the length of the interval $[a, b]$ fixed. Is θ a scale parameter for any a, b? Is it a location parameter for any a, b? Find $dX/d\theta$ for this case.

(b) Suppose θ is perturbed by keeping point a fixed and varying b. Is θ a scale parameter for any a, b? Is it a location parameter for any a, b? Find $dX/d\theta$ for this case, and compare your answer to that of part (a).

11.2 Consider a random variable X whose cdf is one of the distributions in the Erlang family:

$$F(t) = 1 - e^{-m\mu t} \sum_{i=0}^{m-1} \frac{(m\mu t)^i}{i!}, \qquad t \geq 0$$

where $m = 1, 2, \ldots$ is the number of stages (see (8.105) in Chapter 8). Find $dX/d\mu$. Is μ a scale parameter? Is it a location parameter?

11.3 Derive an Infinitesimal Perturbation Analysis (IPA) estimator, similar to (11.62), for the mean system time over N customers in a stable $M/G/1$ queueing system with respect to the arrival rate λ. You may use either an approach similar to that of Section 11.4.3, or of Section 11.5. How is your answer different if the interarrival time distribution becomes uniform over $[1.0, 2.0]$? Derive another estimator for the same sensitivity using the Likelihood Ratio (LR) approach.

11.4 Consider a stable $GI/G/1$ queueing system with arrival rate λ and service rate μ. Let $\lambda = 2\theta$ and $\mu = \theta/2$. Derive an Infinitesimal Perturbation Analysis (IPA) estimator for the mean system time over N customers with respect to θ.

11.5 Consider a single-class closed queueing network consisting of two queues, both with infinite capacity, serving three customers. The service time distribution at the first server is exponential with mean $1/\mu$, and at the second server it is Erlang with mean 1 and stage parameter 2. Derive an Infinitesimal Perturbation Analysis (IPA) estimator for the throughput of this system with respect to μ. How will your answer change if the second server's mean service time is doubled?

11.6 Customers arrive to a single-server queueing system with infinite queueing capacity according to a Poisson process. The service time distribution is unknown. Every customer is either admitted to the queue with probability θ or rejected and lost with probability $1 - \theta$. Derive an Infinitesimal Perturbation Analysis (IPA) estimator of the mean system time with respect to θ.

11.7 In Section 11.7.2, we obtained a model for the two-class routing problem of Fig. 11.27 such that the sample function (mean waiting time of class 2 customers at server 1) had no discontinuities in θ_{11}, the probability that a class 1 customer is routed to server 1. Derive an Infinitesimal Perturbation Analysis (IPA) estimator for the sensitivity of this performance measure with respect to θ_{11}.

11.8 Consider a stable $GI/G/1$ queueing system where the performance measure of interest is the mean number of customers in a busy period. Let θ be the mean service time.

(a) Describe how perturbations in θ can cause discontinuities in the sample function for this problem.

(b) Let $i = 1, 2, \ldots$ index busy periods, and define X_i to be the time between the last arrival instant in the ith busy period and the instant when this busy period ends. Suppose we are willing to observe every $X_i, i = 1, 2, \ldots$, along a sample path and record its value, x_i. Let Y_i be the interarrival time that starts with the last arrival in the ith busy period. Let I_i be the length of the idle period that immediately follows the ith busy period. Find an expression for the distribution of I_i conditioned on the fact that $Y_i > x_i$, in terms of the interarrival time cdf, $F(\cdot)$.

(c) Based on the observations $\{x_1, x_2, \ldots\}$ and knowledge of $F(\cdot)$, derive a Smoothed Perturbation Analysis (SPA) estimator for the sensitivity of the mean number of customers in a busy period with respect to θ.

(d) What is the form of your SPA estimator when the arrival process is Poisson?

11.9 In Section 11.8.2, we considered queueing systems with real-time constraints. For a $GI/G/1$ queueing system, let θ be the mean service time. We are interested in the probability that a customer's system time exceeds a given deadline parameter R. Derive an estimator for the sensitivity of this performance measure with respect to θ, assuming that the service time distribution, $F(\cdot)$, is known. What is the expression of your sensitivity estimator when $F(\cdot)$ is exponential?

11.10 Repeat the previous problem with two modifications: (a) the deadline R is an exponentially distributed random variable with mean r, and (b) the service time distribution is *unknown*.

11.11 Repeat Problem 11.9, except that the performance measure of interest is now the probability that a customer's *waiting time* exceeds a given deadline parameter R (*Hint*: Use Lindley's recursive equation for waiting times).

11.12 Check if the observability condition **(OB)** of Section 11.10.2 is satisfied for the following:

(a) A closed queueing network consisting of two queues, both with infinite capacity, serving K customers. Is a sample path under $K = 2$ observable with respect to $K = 3$ (assume that both sample paths start at a state where all customers are at queue 1)? What happens when the role of the two sample paths is reversed? Identify unobservable events at specific states if they ever arise.

(b) Two infinite-capacity single-server queueing systems in parallel, where every arriving customer first attempts to access queue 1 and is routed there if the length of queue 1 is less than K. If the length of queue 1 is greater than or equal to K, the customer is routed to queue 2 provided the length of queue 2 is less than L. If the length of queue 2 is greater than or equal to L, the customer is lost. Is a sample path under $(K = 1, L = 1)$ observable with respect to $(K = 2, L = 1)$ (assume that both sample paths start at a state where both queues are empty)? What happens when the role of the two sample paths is reversed? Identify unobservable events at specific states if they ever arise.

11.13 Use the Standard Clock (SC) approach (Algorithm 11.3) to show how you can construct 10 sample paths in parallel for the system in Problem 11.12, part (a), with $K = 1, \ldots, 10$ (assuming both service times are exponentially distributed with rates μ_1 and μ_2). Write a computer program to implement your algorithm over 1,000 departures at queue 1, using $\mu_1 = 1.0, \mu_2 = 1.5$.

11.14 Repeat the previous problem using the system of Problem 11.12, part (b). In this case, select the 10 sample paths for the following (K, L) pairs: (1,1), (2,1), (1,2), (2,2), (2,3), (3,2), (3,3), (1,3), (3,1), (3,4). Also, assuming both service times are exponentially distributed with rates μ_1, μ_2, and the arrival process is Poisson with rate λ, write a computer program to implement your algorithm over 1,000 total service completions, using $\mu_1 = 1.0$, $\mu_2 = 1.5$, and $\lambda = 2.0$. Use this algorithm to estimate the probability of rejecting customers over the 10 (K, L) pairs above, and the mean system time of the 1,000 accepted customers.

11.15 Use Augmented System Analysis (ASA) to derive an algorithm for constructing 10 sample paths in parallel of an $M/G/1/K$ queueing system over $K = 1, \ldots, 10$ by observing a sample path under $K = 11$. Would you use the Event Matching or the Age Matching algorithm? Next, consider a $GI/M/1/K$ system. Would you use the Event Matching or the Age Matching algorithm in this case? Write a computer program to implement your algorithm over 1,000 total departures, assuming a service rate $\mu = 2.0$, and an interarrival time distribution which is uniformly distributed over $[0.5, 1.0]$.

11.16 Compare the procedure for constructing the augmented system in Section 11.10.3 with the composition operations on automata presented in Chapter 2, Section 2.3.2.

SELECTED REFERENCES

■ *Perturbation Analysis*

– Bao, G., C.G. Cassandras, and M. Zazanis, "First and Second Derivative Estimators for Closed Jackson-Like Queueing Networks Using Perturbation Analysis Techniques," *Journal of Discrete Event Dynamic Systems*, Vol. 7, No. 1, pp. 29-67, 1997.

– Brémaud, P., and F.J. Vázquez-Abad, "On the Pathwise Computation of Derivatives with Respect to the Rate of a Point Process: The Phantom RPA Method," *Queueing Systems*, Vol. 10, pp. 249-270, 1992.

– Cao, X., *Realization Probabilities: The Dynamics of Queueing Systems*, Springer-Verlag, 1994.

– Cao, X., "Convergence of Parameter Sensitivity Estimates in a Stochastic Experiment," *IEEE Transactions on Automatic Control*, Vol. 30, pp. 834-843, 1985.

– Cassandras, C.G., W.B. Gong, and J.I. Lee, "On the Robustness of Perturbation Analysis Estimators for Queueing Systems with Unknown Distributions," *Journal of Optimization Theory and Applications*, Vol. 30, No. 3, pp. 491-519, 1991.

– Cassandras, C.G., and V. Julka, "Scheduling Policies Using Marked/Phantom Slot Algorithms," *Queueing Systems*, Vol. 20, pp. 207-254, 1995.

– Dai, L., and Y.C. Ho, "Structural Infinitesimal Perturbation Analysis (SIPA) for Derivative Estimation of Discrete Event Dynamic Systems," *IEEE Transactions on Automatic Control*, Vol. 40, pp. 1154-1166, 1995.

– Fu, M.C., and J.Q. Hu, "Smoothed Perturbation Analysis for Queues with Finite Buffers," *Queueing Systems*, Vol. 14, pp. 57-78, 1993.

– Fu, M.C., and J.Q. Hu, "Smoothed Perturbation Analysis Derivative Estimation for Markov Chains," *Operations Research Letters*, Vol. 15, pp. 241-251, 1994.

– Fu, M.C., and J.Q. Hu, *Conditional Monte Carlo: Gradient Estimation and Optimization Applications*, Kluwer Academic Publishers, Boston, 1997.

– Glasserman, P., and W.B. Gong, "Smoothed Perturbation Analysis for a Class of Discrete Event Systems," *IEEE Transactions on Automatic Control*, Vol. AC-35, No. 11, pp. 1218-1230, 1990.

– Glasserman, P., *Gradient Estimation via Perturbation Analysis*, Kluwer Academic Publishers, Boston, 1991.

– Gong, W.B., and Y.C. Ho, "Smoothed Perturbation Analysis of Discrete Event Systems," *IEEE Transactions on Automatic Control*, Vol. AC-32, No. 10, pp. 858-866, 1987.

– Gong, W.B., and P. Glasserman, "Perturbation Analysis of the $M/G/1/K$ Queue," *Proceedings of 27th IEEE Conference on Decision and Control*, 1988.

– Gong, W.B., C.G. Cassandras, and J.Pan, "Perturbation Analysis of a Multiclass Queueing System with Admission Control," *IEEE Transactions on Automatic Control*, Vol. AC-36, No. 6, pp. 707-723, 1991.

– Gong, W.B., and H. Schulzrinne, "Perturbation Analysis of the $M/G/1/K$ Queue," *Mathematics and Computers in Simulation*, Vol. 34, pp. 1-19, 1992.

– Heidelberger, P., X. Cao, M. Zazanis, and R. Suri, "Convergence Properties of Infinitesimal Perturbation Analysis Estimates," *Management Science*, Vol. 34, No. 11, pp. 1281-1302, 1988.

– Ho, Y.C., M.A. Eyler, and D.T. Chien, "A Gradient Technique for General Buffer Storage Design in a Serial Production Line," *Intl. Journal of Production Research*, Vol. 17, pp. 557-580, 1979.

– Ho, Y.C., and C.G. Cassandras, "A New Approach to the Analysis of Discrete Event Dynamic Systems," *Automatica*, Vol. 19, No. 2, pp. 149-167, 1983.

– Ho, Y.C., and X. Cao, "Perturbation Analysis and Optimization of Queueing Networks," *Journal of Optimization Theory and Applications*, Vol. 40, No. 4, pp. 559-582, 1983.

– Ho, Y.C., X. Cao, and C.G. Cassandras, "Infinitesimal and Finite Perturbation Analysis for Queueing Networks," *Automatica*, Vol. 19, pp. 439-445, 1983.

– Ho, Y.C., and S. Li, "Extensions of Perturbation Analysis of Discrete Event Dynamic Systems," *IEEE Transactions on Automatic Control*, Vol. AC-33, pp. 427-438, 1988.

– Ho, Y.C., and J.Q. Hu, "An Infinitesimal Perturbation Analysis Algorithm for a Multiclass $G/G/1$ Queue," *Operations Research Letters*, Vol. 9, pp. 35-44, 1990.

– Ho, Y.C., and X. Cao, *Perturbation Analysis of Discrete Event Dynamic Systems*, Kluwer Academic Publishers, Boston, 1991.

– Ho, Y.C., and C.G. Cassandras, "Perturbation Analysis for Control and Optimization of Queueing Systems: An Overview and the State of the Art," in *Frontiers in Queueing* (J. Dshalalow, Ed.), CRC Press, pp. 395-420, 1997.

– Hu, J.Q., and S.G. Strickland, "Strong Consistency of Sample Path Derivative Estimates," *Applied Mathematics Letters*, Vol. 3, No. 4, pp. 55-58, 1990.

– Hu, J.Q., "Convexity of Sample Path Performances and Strong Consistency of Infinitesimal Perturbation Analysis Estimates," *IEEE Transactions on Automatic Control*, Vol. AC-37, No. 2, pp. 258-262, 1992.

– Suri, R., "Implementation of Sensitivity Calculations on a Monte-Carlo Experiment," *Journal of Optimization Theory and Applications*, Vol. 40, No. 4, pp. 625-630, 1983.

– Suri, R., and X. Cao, "The Phantom and Marked Customer Methods for Optimization of Closed Queueing Networks with Blocking and General Service Times," *ACM Performance Evaluation Review*, Vol. 12, No. 3, pp. 234-256, 1986.

– Suri, R., and M. Zazanis, "Perturbation Analysis Gives Strongly Consistent Sensitivity Estimates for the $M/G/1$ Queue," *Management Science*, Vol. 34, pp. 39-64, 1988.

– Suri, R., "Perturbation Analysis: The State of the Art and Research Issues Explained via the $GI/G/1$ Queue," *Proceedings of the IEEE*, Vol. 77, No. 1, pp. 114-137, 1989.

– Vakili, P., and Y.C. Ho, "Infinitesimal Perturbation Analysis of a Multiclass Routing Problem," *Proceedings of 25th Allerton Conference on Communication, Control, and Computing*, pp. 279-286, 1987.

– Wardi, Y., and J.Q. Hu, "Strong Consistency of Infinitesimal Perturbation Analysis for Tandem Queueing Networks," *Journal of Discrete Event Dynamic Systems*, Vol. 1, No. 1, pp. 37-60, 1991.

- Wardi, Y., W.B. Gong, C.G. Cassandras, and M.H. Kallmes, "Smoothed Perturbation Analysis for a Class of Piecewise Constant Sample Performance Functions," *Journal of Discrete Event Dynamic Systems*, Vol. 1, No. 4, pp. 393-414, 1992.

- Zazanis, M., and R. Suri, "Perturbation Analysis of the $GI/G/1$ Queue," Technical Report, IE/MS Dept., Northwestern University, 1989.

■ *Applications of Sensitivity Estimation to Optimization*

- Cassandras, C.G., and J.I. Lee, "Applications of Perturbation Techniques to Optimal Resource Sharing in Discrete Event Systems," *Proceedings of 1988 American Control Conference*, pp. 450-455, 1988.

- Cassandras, C.G., M.V. Abidi, and D. Towsley, "Distributed Routing with On-Line Marginal Delay Estimation," *IEEE Transactions on Communications*, Vol. COM-38, No. 3, pp. 348-359, 1990.

- Chong, E.K.P., and P.J. Ramadge,, "Convergence of Recursive Optimization Algorithms Using Infinitesimal Perturbation Analysis Estimates," *Journal of Discrete Event Dynamic Systems*, Vol. 1, No. 4, pp. 339-372, 1992.

- Fu, M.C., "Convergence of a Stochastic Approximation Algorithm for the $GI/G/1$ Queue Using Infinitesimal Perturbation Analysis," *Journal of Optimization Theory and Applications*, Vol. 65, pp. 149-160, 1990.

- Suri, R., and Y.T. Leung, "Single Run Optimization of Discrete Event Simulations - An Empirical Study using the $M/M/1$ Queue," *IIE Transactions*, Vol. 21, No. 1, pp. 35-49, 1989.

- Vázquez-Abad, F.J., C.G. Cassandras, and V. Julka, "Centralized and Decentralized Asynchronous Optimization of Stochastic Discrete Event Systems," *IEEE Transactions on Automatic Control*, Vol. AC-43, No. 5, pp. 631-655, 1998.

■ *Likelihood Ratio Methodology for Sensitivity Estimation*

- Aleksandrov, V.M., V.J. Sysoyev, and V.V. Shemeneva, "Stochastic Optimization," *Engineering Cybernetics*, Vol. 5, pp. 11-16, 1968.

- Glynn, P.W., "Likelihood Ratio Gradient Estimation: An Overview," *Proceedings of 1987 Winter Simulation Conference*, pp. 366-375, 1987.

- Reiman, M.I., and A. Weiss, "Sensitivity Analysis via Likelihood Ratio," *Operations Research*, Vol. 37, pp. 830-844, 1989.

■ *Concurrent Estimation*

– Cassandras, C.G., and S.G. Strickland, "Sample Path Properties of Timed Discrete Event Systems," *Proceedings of the IEEE*, Vol. 77, No. 1, pp. 59-71, 1989a.

– Cassandras, C.G., and S.G. Strickland, "On-Line Sensitivity Analysis of Markov Chains," *IEEE Transactions on Automatic Control*, Vol. AC-34, No. 1, pp. 76-86, 1989b.

– Cassandras, C.G., and S.G. Strickland, "Observable Augmented Systems for Sensitivity Analysis of Markov and Semi-Markov Processes," *IEEE Transactions on Automatic Control*, Vol. AC-34, No. 10, pp. 1026-1037, 1989c.

– Cassandras, C.G., J.I. Lee, and Y.C. Ho, "Efficient Parametric Analysis of Performance Measures for Communication Networks," *IEEE Journal on Selected Areas in Communications*, Vol. 8, No. 9, pp. 1709-1722, 1990.

– Cassandras, C.G., and C.G. Panayiotou, "Concurrent Sample Path Analysis of Discrete Event Systems," *Journal of Discrete Event Dynamic Systems*, 1999.

– Cassandras, C.G. (Ed.), Special Issue on Parallel Simulation and Optimization of Discrete Event Dynamic Systems, *Journal of Discrete Event Dynamic Systems*, Vol. 5, No. 2/3, 1995.

– Vakili, P., "A Standard Clock Technique for Efficient Simulation," *Operations Research Letters*, Vol. 10, pp. 445-452, 1991.

■ *Miscellaneous*

– Bratley, P.B., B.L. Fox, , and L.E. Schrage, *A Guide to Simulation*, 2nd Edition, Springer-Verlag, New York, 1987.

– Cassandras, C.G., L. Dai, and C.G. Panayiotou, "Ordinal Optimization for Deterministic and Stochastic Resource Allocation," *IEEE Transactions on Automatic Control*, Vol. AC-43, No. 7, pp. 881-900, 1998.

– Ho, Y.C., R. Sreenivas, and P. Vakili, "Ordinal Optimization of Discrete Event Dynamic Systems," *Journal of Discrete Event Dynamic Systems*, Vol. 2, No. 1, pp. 61-88, 1992.

– Gallager, R.G., "A Minimum Delay Routing Algorithm Using Distributed Computation," *IEEE Transactions on Communications*, Vol. COM-23, pp. 73-85, 1977.

– Rubinstein, R.V., *Monte Carlo Optimization, Simulation, and Sensitivity Analysis of Queueing Networks*, Wiley, New York, 1986.

Appendix I

Review of Probability Theory

This Appendix in intended to review the basic definitions, concepts, and techniques encountered in probability theory. It is not meant to be exhaustive. Emphasis is placed on some fundamental notions and on those aspects of probability theory which are most useful in our study of discrete-event systems.

I.1 BASIC CONCEPTS AND DEFINITIONS

A fundamental concept in probability theory is that of a *random experiment*. This is any experiment one can perform whose outcome is uncertain. A simple example is tossing a coin. We know there are two possible outcomes (heads and tails), but we do not know which will be observed until after the coin lands. Another example is observing the number of airplanes that take off from an airport over the period of a particular hour in a day. Again, we know that possible outcomes are $0, 1, 2, \ldots$, but which value will actually be observed when the experiment is performed is unknown. There are three concepts associated with a random experiment, which are discussed next.

First, the set of all possible outcomes of a random experiment is called a *sample space*. We will denote this set by Ω. In the coin tossing experiment, $\Omega = \{\text{heads}, \text{tails}\}$. In the experiment where we count airplanes taking off in a one-hour interval, $\Omega = \{0, 1, 2, \ldots\}$.

The purpose of a random experiment may be thought of as a check of whether a particular statement related to possible outcomes is true or false.

Such a statement may apply to a specific outcome; in the example of counting airplanes, the statement may be "exactly one airplane takes off in a given one-hour interval". Often, however, we are interested in more elaborate statements such as "no more than two airplanes take off in a given one-hour interval" or "the number of airplanes taking off in a given one-hour interval is greater than three but less than ten." We refer to such a statement as an *event*. Choosing the set of events that we wish to consider in performing a random experiment gives rise to an *event space*.

Finally, we would like to provide quantifiable measurements for the likelihood of various events associated with a random experiment. This is accomplished by assigning a *probability* to each event. The simplest way to think of probability is through the so-called "frequency definition." Suppose we toss the same coin N times and count the number of times that we observe heads as the outcome, denoted by N_h. The ratio N_h/N measures the frequency of observing heads. This ratio does vary as N varies. However, as $N \to \infty$ our intuition leads us to believe that N_h/N approaches some fixed number. In fact, if the coin is fair we expect this number to be $1/2$. In general, if A is an event, then the frequency definition of the probability of A is

$$P[A] = \lim_{N \to \infty} \frac{N_A}{N}$$

where N is the number of times the experiment is performed and N_A is the number of times that event A occurs.

Probability space. The preceding discussion is formalized in probability theory by defining a *probability space* $(\Omega, \mathbb{E}, P)$ as follows.

- Ω is a set called the sample space.

- $\mathbb{E}$ is a collection of subsets of Ω called the event space. It is chosen so as to satisfy the following conditions:

  1. If A is an event (i.e., $A \in \mathbb{E}$), then the complement of A, denoted by $\bar{A}$, is also an event (i.e., $\bar{A} \in \mathbb{E}$).

  2. If $A_1, A_2, \ldots$ are all events, then their union is also an event, that is

  $$\bigcup_{i=1}^{\infty} A_i \in \mathbb{E}$$

- P is a function that maps every event $A \in \mathbb{E}$ to a real number $P[A]$ such that:

  1. $P[A] \geq 0$ for all $A \in \mathbb{E}$
  2. $P[\Omega] = 1$

3. If $A_1, A_2, \ldots$ are all disjoint (mutually exclusive) events, then

$$P\left[\bigcup_{i=1}^{\infty} A_i\right] = \sum_{i=1}^{\infty} P[A_i]$$

It is easy to check that $P[A]$ is always such that $0 \leq P[A] \leq 1$ for any $A \in \mathbb{E}$. Since A and $\bar{A}$ are disjoint events, condition 3 implies that $P[A] + P[\bar{A}] = P[A \cup \bar{A}] = P[\Omega]$, with $P[\Omega] = 1$ from condition 2. Since, from condition 1, $P[A]$ and $P[\bar{A}]$ are also non-negative real numbers, it follows that $P[A] \leq 1$.

It is worth pointing out that the choice of an event space may not be unique, and some choices may be better suited than others for the purpose of a particular random experiment. As an example, consider $\Omega = \{\text{heads}, \text{tails}\}$ in the familiar coin tossing experiment. Observe that $\mathbb{E} = \{\Omega, \emptyset, \{\text{heads}\}, \{\text{tails}\}\}$ is an event space (one can easily check that the two requirements in the definition of $\mathbb{E}$ are satisfied). But so is $\mathbb{E} = \{\Omega, \emptyset\}$. The latter, however, is a rather uninteresting event space.

Continuity property. The probability P, defined as a function of events, has a useful continuity property. Suppose we consider an increasing sequence of events $A_0, A_1, \ldots$, that is $A_n \subseteq A_{n+1}$ for all $n = 1, 2, \ldots$ Then, it can be shown that

$$\lim_{n \to \infty} P[A_n] = P\left[\lim_{n \to \infty} A_n\right] \tag{I.1}$$

A similar result holds for a decreasing sequence of events. This property is often useful when evaluating the probability of a complicated event which happens to be of a form expressed as the limit on the right-hand side of (I.1); in this case, evaluating the limit on the left-hand side is often much easier.

Independence. It should be clear that since events are sets, we can apply to them all standard set operations (complementation, union, intersection). For simplicity of notation, it is common to write AB, instead of $A \cap B$, to represent the intersection (logical "and") of A and B. With this notation, two events, A and B, are said to be *independent* if

$$P[AB] = P[A] \cdot P[B] \tag{I.2}$$

I.2   CONDITIONAL PROBABILITY

The concept of "conditioning" in probability theory is intended to capture how information can affect our predictive ability regarding various events and their probabilities. Suppose we are interested in $P[A]$ and know that some other event B has already occurred. If A and B are not independent, we expect the information that B has occurred to affect the probability we should assign to event A. We denote this probability by $P[A|B]$, which is read as "the probability of event A given that event B has occurred." As a simple example, consider rolling a fair die as our random experiment. Thus, $\Omega = \{1, 2, 3, 4, 5, 6\}$. Let A be the event "2 shows up." Without any further information, we can see that $P[A] = 1/6$. Suppose, however, that after the die is rolled - and prior to looking at the number it shows - someone informs us that the number showing is even. Letting B be the event "an even number shows up," we can immediately see that $P[A|B] = 1/3$. In other words, conditioning generally restricts the sample space from Ω to some subset defined by B.

The formal definition of the conditional probability of event A given event B is

$$P[A|B] = \frac{P[AB]}{P[B]} \tag{I.3}$$

provided that $P[B] \neq 0$. We can immediately see that if A and B are independent, $P[A|B] = P[A]$, that is, the initial probability assigned to A is not affected by the information contained in the fact that B has occurred.

Rule of total probability. If $B_1, B_2, \ldots$ are mutually exclusive events such that $\bigcup_i B_i = \Omega$, then the following relationship proves to be extremely useful in a variety of probabilistic calculations:

$$P[A] = \sum_i P[A \mid B_i] \cdot P[B_i] \tag{I.4}$$

This is known as the *rule of total probability*. Note that the collection of events B_i that partitions the sample space Ω may be finite or countable.

Bayes' formula. From (I.4), it is easy to derive another useful result known as *Bayes' formula*:

$$P[B_k \mid A] = \frac{P[A \mid B_k] \cdot P[B_k]}{\sum_i P[A \mid B_i] \cdot P[B_i]} \tag{I.5}$$

I.3   RANDOM VARIABLES

One difficulty we face with our definitions thus far is the fact that the elements of a sample space are abstract objects (e.g., "heads" and "tails") rather than real numbers. Since we would like to develop mathematical techniques based on manipulating real numbers, we introduce a mapping from Ω to the set of real numbers, $X : \omega \to \mathbb{R}$. In other words, we assign a real number to every possible outcome of a random experiment. The function $X(\omega)$, where

$\omega \in \Omega$, is called a random variable. For example, if $\Omega = \{\text{heads}, \text{tails}\}$, we may set $X(\text{heads}) = 0$ and $X(\text{tails}) = 1$.

For purposes of consistency, we also require in the definition of $X(\omega)$ that the set

$$\{\omega : \omega \in \Omega \text{ and } X(\omega) \leq x\}, \text{ for every} x \in \mathbb{R}$$

is an event (i.e., it belongs to the event space $\mathbb{E}$). If, for example, $\Omega = \{0, 1\}, \mathbb{E} = \{\Omega, \varnothing\}$, and we set $X(\omega) = \omega$, then observe that $X(0) = 0 \notin \mathbb{E}$, which violates this requirement. For ease of notation, the event $\{\omega : \omega \in \Omega \text{ and } X(\omega) \leq x\}$ for some $x \in \mathbb{R}$ is written as $[X \leq x]$.

The *cumulative distribution function* (cdf), or just distribution function, of a random variable X, denoted by $F(x)$ is defined as

$$F(x) = P[X \leq x] \tag{I.6}$$

for all $x \in \mathbb{R}$. The most important properties of a cdf are the following. $F(x)$ is a monotonically non-decreasing right-continuous function such that $0 \leq F(x) \leq 1$, with $F(-\infty) = 0$ and $F(\infty) = 1$.

The definition of a cdf extends to random vectors $\mathbf{X} = [X_1, \ldots, X_n]$ by setting

$$F(x_1, \ldots, x_n) = P[X_1 \leq x_1, \ldots, X_n \leq x_n]$$

Given this *joint* cdf $F(x_1, \ldots, x_n)$, the marginal cdf $F_i(x_i)$ of X_i is obtained by evaluating $F(x_1, \ldots, x_n)$ with $x_j = \infty$ for all $j \neq i$.

The random variables $X_1, \ldots, X_n$ are said to be independent if

$$F(x_1, \ldots, x_n) = F_1(x_1) \ldots F_n(x_n) \tag{I.7}$$

for all $x_1, \ldots, x_n$.

For discrete random variables (i.e., the sample space Ω is a discrete set, as in $\Omega = \{\text{heads}, \text{tails}\}$) the *probability mass function* (pmf), $p(x)$, of a (scalar) random variable X is defined as

$$p(x) = P[X = x] \tag{I.8}$$

where x takes on values associated with the discrete set over which the random variable X is defined. In this case, the cdf and pmf are related as follows:

$$F(x) = \sum_{y \leq x} p(y) \tag{I.9}$$

and we have $F(\infty) = \sum_{\text{all } y} p(y) = 1$.

A commonly encountered and useful for our purposes pmf is the geometric one. It characterizes a random variable X that can take values in the set $\{1, 2, \ldots\}$ and it is defined as

$$p(n) = P[X = n] = (1 - p)p^{n-1} \tag{I.10}$$

where p is a parameter such that $0 < p < 1$.

For continuous random variables, the probability density function (pdf), $f(x)$, is defined as

$$f(x) = \frac{dF(x)}{dx} \tag{I.11}$$

provided that this derivative exists everywhere, except possibly a finite number of points. In this case,

$$F(x) = \int_{-\infty}^{x} f(\tau)d\tau \tag{I.12}$$

The probability of an event of the form $[a < X \leq b]$ is then given by

$$P[a < X \leq b] = F(a) - F(b) = \int_{a}^{b} f(\tau)d\tau$$

Since $F(x)$ is a non-decreasing function, (I.11) implies that $f(x) \geq 0$. Moreover, $\int_{-\infty}^{\infty} f(\tau)d\tau = 1$.

An example of a pdf that we make ample use of in this book is the *exponential* one. It characterizes a random variable X that can only take non-negative real values, and it is defined as

$$f(x) = \lambda e^{-\lambda x}, \quad x \geq 0 \tag{I.13}$$

where $\lambda > 0$. Here, it is to be understood that $f(x) = 0$ for all $x < 0$. The corresponding cdf is obtained by integrating $f(x)$ to yield

$$F(x) = 1 - e^{-\lambda x}, \quad x \geq 0 \tag{I.14}$$

I.4 CONDITIONAL DISTRIBUTIONS

We can return to the definition of conditional probability in (I.3) and revisit it for events of the form $[X \leq x]$ and $[Y \leq y]$. In particular, we can define the *conditional distribution function of X given that $Y \leq y$* as follows:

$$H(x,y) = P[X \leq x \mid Y \leq y] \tag{I.15}$$

Of particular interest is the case where Y is a continuous random variable and we would like to condition on the event $[Y = y]$. The problem here is that $P[Y = y] = 0$ and the ratio in (I.3) is undefined. Thus, in order to evaluate the probability $P[X \leq x \mid Y = y]$, we define the *conditional density function*

$$f(x \mid y) = \frac{f(x,y)}{f_Y(y)} \tag{I.16}$$

where we are careful to call attention to the fact that the pdf in the denominator is the marginal pdf of Y, written as $f_Y(y)$. Here, $f(x \mid y)$ is defined only if $f_Y(y) \neq 0$. Accordingly, the *conditional distribution function function of X given that $Y = y$* is obtained by integrating $f(x \mid y)$:

$$F(x \mid y) = P[X \leq x \mid Y = y] = \int_{-\infty}^{x} f(\tau \mid y)d\tau \qquad (\text{I}.17)$$

In this setting, we can also extend the rule of total probability in (I.4) as follows:

$$P[X \leq x] = \int_{-\infty}^{\infty} P[X \leq x \mid Y = y] \cdot f_Y(y)dy \qquad (\text{I}.18)$$

where $f_Y(y)$ is the marginal pdf of the continuous random variable Y.

I.5   FUNCTIONS OF RANDOM VARIABLES

Suppose two random variables X and Y are related through $Y = g(X)$ and the cdf of X is known to be $F_X(x)$. In order to obtain the cdf of $Y, F_Y(y)$, observe that

$$F_Y(y) = P[Y \leq y] = P[g(X) \leq y]$$

At this point, the nature of the function $g(\cdot)$ often determines the way to proceed. For example, if $Y = aX + b$, where $a > 0$, we have

$$F_Y(y) = P[aX + b \leq y] = P\left[X \leq \frac{y-b}{a}\right] = F_X\left(\frac{y-b}{a}\right) \qquad (\text{I}.19)$$

A useful formula that allows us to determine the pdf of $Y, f_Y(y)$, given the pdf of $X, f_X(x)$, is the following:

$$f_Y(y) = \sum_i \frac{f_X(x_i)}{|g'(x_i)|} \qquad (\text{I}.20)$$

where $x_1, x_2, \ldots$ are the roots of the equation $y = g(x)$ and $g'(x_i)$ denotes the derivative $dg(x)/dx$ evaluated at the point $x = x_i$. As an example, if $Y = X^2$, then the two roots of $y = x^2$ are $x_1 = +\sqrt{y}$ and $x_2 = -\sqrt{y}$ for $y \geq 0$. Then, (I.20) yields

$$f_Y(y) = \frac{f_X(\sqrt{y})}{|2\sqrt{y}|} + \frac{f_X(-\sqrt{y})}{|-2\sqrt{y}|} = \frac{f_X(\sqrt{y}) + f_X(-\sqrt{y})}{2\sqrt{y}}$$

In general, it is possible to obtain $F_Y(y)$ from the pdf of $X, f_X(x)$, by evaluating the integral

$$F_Y(y) = \int_{S(y)} f_X(x)dx \qquad (\text{I}.21)$$

where $S(y)$ is the set of points $\{x : g(x) \leq y\}$. The complexity of this integral depends on the complexity of $S(y)$, which in turn depends on the complexity of $g(\cdot)$.

In the case where Y is a function of two or more random variables, there is a natural extension of (I.21). In particular, if $Z = g(X, Y)$, then

$$F_Z(z) = \int \int_{S(z)} f(x, y) dx dy \tag{I.22}$$

where $f(x, y)$ is the joint density of X and Y and $S(z) = \{(x, y) : g(x, y) \leq z\}$.

If X and Y are independent random variables, that is, $f(x, y) = f_X(x) \cdot f_Y(y)$, and $Z = X + Y$, then (I.22) can be used to evaluate $F_Z(z)$. In this case, it turns out that $f_Z(z) = dF_Z(z)/dz$ is given by the convolution of the two marginal density functions $f_X(x)$ and $f_Y(y)$:

$$f_Z(z) = \int_{-\infty}^{\infty} f_X(\tau) \cdot f_Y(z - \tau) d\tau \tag{I.23}$$

This result extends to the sum of n independent random variables, $Z = X_1 + \ldots + X_n$, in which case $f_Z(z)$ is the n-fold convolution of the marginal density functions of $X_1, \ldots, X_n$.

As an example, let $X_i, i = 1, 2$, be independent exponentially distributed random variables with the same parameter λ, as in (I.13). Then, if $Z = X_1 + X_2$, (I.23) yields

$$f_Z(z) = \int_{-\infty}^{\infty} f_X(\tau) \cdot f_Y(z - \tau) d\tau = \int_0^z \lambda e^{-\lambda \tau} \cdot \lambda e^{-\lambda(z-\tau)} d\tau$$
$$= \lambda^2 e^{-\lambda z} \int_0^z d\tau = \lambda(\lambda z) e^{-\lambda z} \tag{I.24}$$

This is known as the *Erlang* pdf with two stages. It generalizes to n stages by defining $Z = X_1 + \ldots + X_n$, where each X_i is exponentially distributed with a common parameter λ.

I.6 EXPECTATION

For a discrete random variable X, the *expectation* of X (also called *expected value* or *mean*) is defined as

$$E[X] = \sum_{\text{all } x} x p(x) \tag{I.25}$$

where $p(x)$ is the pmf of X.

For a continuous random variable, we define

$$E[X] = \int_{-\infty}^{\infty} x f(x) dx \tag{I.26}$$

where $f(x)$ is the pdf of X. It can be shown that $E[X]$ can also be expressed in terms of the cdf $F(x)$ as follows:

$$E[X] = \int_0^\infty [1 - F(x)]dx - \int_{-\infty}^0 F(x)dx \tag{I.27}$$

If X happens to be a non-negative random variable, then we get

$$E[X] = \int_0^\infty [1 - F(x)]dx, \quad \text{if } X \geq 0 \tag{I.28}$$

If a random variable Y is defined as a function of another random variable X through $Y = g(X)$, then

$$E[Y] = E[g(X)] = \int_{-\infty}^\infty g(x)f(x)dx \tag{I.29}$$

where $f(x)$ is the pdf of X. A similar relationship holds for discrete random variables. This is known as the *fundamental theorem of expectation*.

Moments. The nth *moment* of a random variable X is defined as $E[X_n], n = 1, 2, \ldots$ This can be evaluated through (I.29), or its discrete analog, by setting $g(x) = x^n$.

Variance, standard deviation, and coefficient of variation. The variance of a random variable X is defined as

$$Var[X] = E[(X - E[X])^2] \tag{I.30}$$

Since $Var[X] = E[X^2 - 2X \cdot E[X] + (E[X])^2]$ we also get

$$Var[X] = E[X_2] - (E[X])^2 \tag{I.31}$$

The variance of X is sometimes also denoted by σ_X^2. The square root of the variance, σ_X, is called the *standard deviation*.

The variance is a measure of the "spread" of the pdf around the mean. If there were no uncertainty whatsoever, then the random variable X would always equal its mean $E[X]$. Since in general this is not the case, then the larger the variance is the more uncertain we are about the values that X may take relative to its mean $E[X]$. Another measure of this uncertainty is given by the *coefficient of variation* of X, C_X, defined as

$$C_X = \frac{\sigma_X}{E[X]} \tag{I.32}$$

For an exponentially distributed random variable X, using the definition (I.13) we can determine the expectation $E[X]$ as follows:

$$E[X] = \lambda \int_0^\infty xe^{-\lambda x} dx = -[xe^{-\lambda x}]_0^\infty + \int_0^\infty e^{-\lambda x} dx$$
$$= \frac{-1}{\lambda}[e^{-\lambda x}]_0^\infty = \frac{1}{\lambda} \tag{I.33}$$

Similarly, we can determine the second moment

$$E[X_2] = \lambda \int_0^\infty x_2 e^{-\lambda x} dx = \frac{2}{\lambda^2} \tag{I.34}$$

and hence the variance using (I.31):

$$Var[X] = \frac{1}{\lambda^2} \tag{I.35}$$

as well as the coefficient of variation from (I.29):

$$C_X = 1 \tag{I.36}$$

I.7 CHARACTERISTIC FUNCTIONS

The characteristic function of a *continuous* random variable X, denoted by $\Phi(\omega)$, is defined as

$$\Phi(\omega) = E[e^{j\omega X}] = \int_{-\infty}^\infty e^{j\omega x} f(x) dx \tag{I.37}$$

where $j = \sqrt{-1}$ and $f(x)$ is the pdf of X. A basic property of $\Phi(\omega)$ is that $|\Phi(\omega)| \leq 1$.

The importance of $\Phi(\omega)$ becomes apparent in the *moment generating theorem*, which is obtained by differentiating (I.37):

$$\frac{d\Phi(\omega)}{d\omega} = j \int_{-\infty}^\infty xe^{j\omega x} f(x) dx$$

and noticing that at $\omega = 0$ we get

$$\frac{d\Phi(\omega)}{d\omega}\bigg|_{\omega=0} = j \int_{-\infty}^\infty xf(x) dx = jE[X]$$

This process can be repeated for higher moments of X, providing the general formula

$$E[X^n] = (-j)^n \frac{d^n \Phi(\omega)}{d\omega^n}\bigg|_{\omega=0} \tag{I.38}$$

Let us denote the nth moment of X by m_n. Expanding $\Phi(\omega)$ around the point $\omega = 0$, we then obtain

$$\Phi(\omega) = 1 + (j\omega)m_1 + \frac{(j\omega)^2}{2!}m_2 + \cdots \qquad (\text{I.39})$$

This allows us to approximate $\Phi(\omega)$ by evaluating several moments. Moreover, realizing from (I.37) that $\Phi(\omega)$ and $f(x)$ form a Fourier transform pair, we can see that a pdf can be obtained by evaluating all of its moments (or approximated by evaluating as many moments as possible).

For *discrete* random variables, the characteristic function becomes

$$\Phi(\omega) = E[e^{j\omega X}] = \sum_{\text{all } x} e^{j\omega x} p(x) \qquad (\text{I.40})$$

Moment generating functions. Sometimes, it is more convenient to work with the *moment generating function*, denoted by $M(u)$, instead of the characteristic function. This is defined as

$$M(u) = E[e^{uX}] = \int_{-\infty}^{\infty} e^{ux} f(x)dx \qquad (\text{I.41})$$

It should be clear that results similar to (I.38) and (I.39) can be obtained for $M(u)$.

For *discrete* random variables, the moment generating function becomes

$$M(u) = E[e^{uX}] = \sum_{\text{all } x} e^{ux} p(x) \qquad (\text{I.42})$$

Laplace transforms and probability generating functions. The Laplace transform of the pdf $f(x)$ of a continuous random variable X is defined as

$$\mathcal{F}(s) = E[e^{-sX}] = \int_{-\infty}^{\infty} e^{-sx} f(x)dx \qquad (\text{I.43})$$

It should be clear that the three functions $\Phi(\omega)$, $M(u)$, and $\mathcal{F}(s)$, all associated with a random variable X, are closely related. In particular,

$$\mathcal{F}(s) = \Phi(js) = M(-s) \qquad (\text{I.44})$$

As an example, consider an exponentially distributed random variable X, whose pdf was defined in (I.13). Then,

$$\mathcal{F}(s) = \int_0^{\infty} e^{-sx} \lambda e^{-\lambda x} dx = \lambda \int_0^{\infty} e^{-(s+\lambda)x} dx = \frac{\lambda}{s+\lambda} \qquad (\text{I.45})$$

and, from (I.44), we get $\Phi(\omega) = \lambda/(-j\omega + \lambda)$ and $M(u) = \lambda/(-u + \lambda)$.

For discrete random variables, we define the *probability generating function* of the pmf $p(x)$ as the z-transform

$$\mathcal{G}(z) = E[z^X] = \sum_{\text{all } x} z^x p(x) \tag{I.46}$$

The probability generating function is useful in solving equations of the form

$$\pi(k+1) = \pi(k)\mathbf{P}, \quad k = 0, 1, \ldots \tag{I.47}$$

with $\pi(0)$ given. This is equivalent to

$$\pi(k) = \pi(0)\mathbf{P}^k, \quad k = 1, 2, \ldots \tag{I.48}$$

which involves the matrix multiplication $\mathbf{P}^k$. We encountered (I.48), for example, in the transient analysis of discrete-time Markov chains (see Chapter 7). Multiplying both sides of (I.47) by z^k and summing we get

$$\sum_{k=0}^{\infty} z^k \pi(k+1) = \sum_{k=0}^{\infty} z^k \pi(k)\mathbf{P}$$

Changing the index of the sum on the left-hand-side to $n = k + 1$, we get

$$\sum_{n=1}^{\infty} z^{n-1} \pi(n) = \sum_{k=0}^{\infty} z^k \pi(k)\mathbf{P}$$

which we rewrite as follows:

$$\frac{1}{z} \sum_{n=1}^{\infty} z^n \pi(n) = \sum_{k=0}^{\infty} z^k \pi(k)\mathbf{P}$$

Using the probability generating function

$$\mathbf{G}(z) = \sum_{k=0}^{\infty} z^k \pi(k)$$

note that the sum on the right-hand side is precisely $\mathbf{G}(z)$, whereas the sum on the left-hand side is $\mathbf{G}(z) - \pi(0)$. Thus,

$$\frac{1}{z}[\mathbf{G}(z) - \pi(0)] = \mathbf{G}(z)\mathbf{P}$$

Solving for $\mathbf{G}(z)$, we obtain

$$\mathbf{G}(z) = \pi(0)[\mathbf{I} - z\mathbf{P}]^{-1} \tag{I.49}$$

Using (I.48) and the definition of $\mathbf{G}(z)$, we can rewrite (I.49) as follows:

$$\pi(0) \sum_{k=0}^{\infty} z^k \mathbf{P}^k = \pi(0)[\mathbf{I} - z\mathbf{P}]^{-1}$$

It follows that $\mathbf{P}^k, k = 0, 1, \ldots$, and $[\mathbf{I} - z\mathbf{P}]^{-1}$ form a z-transform pair. Therefore, instead of performing the matrix multiplication $\mathbf{P}^k$, we can first evaluate the inverse $[\mathbf{I} - z\mathbf{P}]^{-1}$ and then obtain $\mathbf{P}^k$ as its inverse z-transform (sometimes an easier task than the computation of $\mathbf{P}^k$).

I.8   RANDOM SEQUENCES AND RANDOM PROCESSES

An introduction to *random processes* (or *stochastic processes*) was provided in Section 6.2. Recall that a random process is a collection of random variables indexed by a variable t, with the random variables defined over a common probability space. If t is defined to take values in some set $\mathcal{T} \subseteq \mathbb{R}$, we denote this continuous-time random process by $\{X(t)\}$. If t is defined over a discrete set, then we use the notation $\{X_t\}$, $t = 0, 1, \ldots$, for this discrete-time process, which we refer to as a *random sequence*. When dealing with a random sequence, it is common to use n or k, instead of t, as the discrete time index.

Our discussion here is limited to some issues related to the convergence of random sequences and processes. The main question centers around the behavior of a random sequence $\{X_n\}, n = 0, 1, \ldots$, as $n \to \infty$. There are several forms of convergence that can be defined for random sequences. Two of them were encountered in Chapter 10 in stating the Central Limit Theorem and the Strong Law of Large Numbers.

Convergence in distribution. One way of qualifying the behavior of $\{X_n\}$, $n = 0, 1, \ldots$, as $n \to \infty$ is by investigating whether there exists a random variable, X, such that the sequence of cdf's of the random variables $X_n, F_{X_n}(x)$, converges to the cdf of $X, F_X(x)$, as $n \to \infty$. Formally, we say that $X_n \to X$ *in distribution* if $F_n(x) \to F_X(x)$ as $n \to \infty$ for every x such that $F_X(x)$ is continuous. Thus, the Central Limit Theorem is a special case of this form of convergence, where the random variables $X_1, X_2, \ldots, X_n$ form an iid sequence and $F_X(x)$ is a normal distribution (see Section 10.7.2).

Convergence with probability 1. A much stronger form of convergence of $\{X_n\}, n = 0, 1, \ldots$, as $n \to \infty$, is what is referred to as "convergence with probability 1." We say that $X_n \to X$ *with probability* 1 (or *almost surely* or *almost everywhere*) as $n \to \infty$ if

$$P\left[\left\{\omega : \lim_{n \to \infty} |X_n(\omega) - X(\omega)| = 0\right\}\right] = 1$$

Here, $P[\cdot]$ is applied to every possible sample path ω that gives rise to a sequence

$\{X_n(\omega)\}$. It is then asserted that every such sequence converges to a random variable $X(\omega)$. It is possible for a few sequences not to converge, but the probability that such sequences arise is 0, hence of no practical concern. The Strong Law of Large Numbers is a special case of this type of convergence that applies to the sample mean

$$\frac{1}{n}\sum_{i=1}^{n}X_i$$

of random variables $X_1, X_2, \ldots, X_n$ forming an iid sequence. In this case, if $E[X_i]$ is the common mean of these random variables, then

$$\frac{1}{n}\sum_{i=1}^{n}X_i \to E[X_i] \text{ with probability 1, as } n \to \infty$$

Interchanging limit and expectation operations. In Chapter 10 we introduced point estimation and the idea of unbiasedness. Specifically, if $\hat{\theta}_n$ is an estimator of an unknown parameter θ, we say that $\hat{\theta}_n$ is an unbiased estimator of θ if

$$E[\hat{\theta}_n] = \theta$$

Now, $\hat{\theta}_n$ is a random variable and $\{\hat{\theta}_n\}$ defines a random sequence for $n = 1, 2, \ldots$ Thus, as $n \to \infty$, we are often interested in the expectation of the random variable $[\lim_{n\to\infty} \hat{\theta}_n]$ if it exists, that is, we are interested in $E[\lim_{n\to\infty} \hat{\theta}_n]$. An important question is whether the expectation and limit operations can be interchanged. In general, given a random sequence $\{X_n\}, n = 0, 1, \ldots$, we would like to have criteria for checking the validity of the following relationship:

$$E[\lim_{n\to\infty} X_n] = \lim_{n\to\infty} E[X_n] \tag{I.50}$$

There are two useful theorems that help us establish the validity of (I.50), which are stated next.

Monotone convergence theorem. Let $\{X_n\}, n = 0, 1, \ldots$, be an increasing sequence of non-negative random variables (i.e., $X_n \geq X_{n-1} \geq 0$ for all $n = 0, 1, \ldots$) with $X_n \to X$ with probability 1. Then, (I.50) is satisfied.

Dominated convergence theorem. Let $\{X_n\}, n = 0, 1, \ldots$, be a sequence of random variables with $X_n \to X$ with probability 1. If there exists a random variable Y such that (a) $E[Y] < \infty$, and (b) $|X_n(\omega)| \leq Y(\omega)$ for all $n = 0, 1, \ldots$ and $\omega \in \Omega$, then (I.50) is satisfied.

If X_n in (I.50) happens to be the derivative of a sample function $L(\theta, \omega)$ of a DES (as discussed in Chapter 11), then the form of (I.50) to which we are interested in applying these two theorems becomes

$$E\left[\lim_{\Delta\theta\to 0} \frac{L(\theta + \Delta\theta, \omega) - L(\theta, \omega)}{\Delta\theta}\right] = \lim_{\Delta\theta\to 0} E\left[\frac{L(\theta + \Delta\theta, \omega) - L(\theta, \omega)}{\Delta\theta}\right]$$

Ergodicity. The last topic related to random sequences and random processes that we briefly discuss is that of *ergodicity*. In general, ergodicity is associated with two types of averaging when dealing with a random process (or random sequence). First, we have the *ensemble average* $E[X(t)]$, that is, the expected value of the random variable $X(t)$ over all possible sample paths. Second, we have the *time average*

$$\bar{X} = \lim_{T \to \infty} \frac{1}{T} \int_0^T X(t, \omega) dt \qquad (I.51)$$

which is evaluated over a specific sample path ω. If we are interested in using $\bar{X}$ as an estimate of $E[X(t)]$, the first observation we can make is that the former is a constant, whereas the latter depends on t. Suppose, however, that $E[X(t)] = \eta$, a particular constant independent of t. Then, the question of whether $\bar{X} = \eta$ is a valid and extremely interesting one, since it allows us to infer something about the behavior of a random process based on observing a single sample path.

Ergodicity is linked to questions such as "under what conditions does the limit in (I.51) actually exist?" and "under what conditions can we claim that $\bar{X} = \eta$?" Although a random process is often defined to be ergodic if indeed $\bar{X} = \eta$, the notion of ergodicity is broader; we will not, however, discuss it here.

In the context of Markov chains (see Chapters 7 through 9), a state of the chain is said to be *ergodic* if it is aperiodic and positive recurrent (see also Section 7.2.8). Then, if all states of the chain are ergodic, the chain itself is said to be ergodic. In an ergodic Markov chain, the state probabilities $\pi_j(k)$ converge to the limiting distribution π_j for all states j. In such a chain, we can obtain a stationary state probability π_j through a time average of the form (I.51), where the integrand is the time spent at state j over a particular sample path. In the limit, as $T \to \infty$, this time average provides the fraction of time spent at state j, which is precisely π_j.

Appendix II

IPA Estimator

This Appendix provides a self-contained computer program (written in C) which implements the IPA estimator for the sensitivity of the mean system time (sys_time) with respect to the arrival rate (alambda) in the $M/M/1$ case. A random number generator, referred to as UNI(i), is included, from which random variates for the interarrival and service time distributions are obtained. The length of the simulation run is determined by the user, who is prompted to specify a desired number of arrivals (N_a).

The output includes the IPA sensitivity estimate (DERI) and the steady-state derivative (DERI_theo) evaluated from the analytical expression (8.40). A relative error is also provided. In addition, an estimate of the mean system time and the analytical value from (8.40), along with a relative error, are also output.

Note that the $M/M/1$ case is convenient for the purpose of comparing estimates with values obtained from analytical expressions such as (8.40). The program can be easily modified to implement IPA estimators for a system with non-exponential interarrival and/or service time distributions.

The reader is also referred to the Web site http://vita.bu.edu/cgc/IPA/ for an interactive simulator of a more general class of simple queueing systems that incorporates IPA.

```
/*--------------------------------------------------------------*/

#include <stdlib.h>
#include <stdio.h>
#include <math.h>

#define ARR 0
#define DEP 0

double  time, time_a, time_d, x, y, alambda, amu;
long int number_a, number_bp, length_queue;
double  sys_time;
long int N_a, N_bp, T_sp;
long int ISEED[5]={1397, 2171, 5171, 7147, 9913};
double  PA_L, PA_G, DERI;
double  UNI();
double  U;

main()
{
    INITIAL();
    while (number_a < N_a){
/*
 *  or you can choose the following two options
 *               to 'truncate' the sample path:
 *  while (number_bp < N_bp){
 *  while (time < T_sp){
 */
        if (time_a < time_d)
            ARRIVAL();
        else
            DEPTURE();
    };

            OUTPUT();
}
/*--------------------------------------------------------------*/
INITIAL()
{
  int ic;
        printf("\n");
        printf("\n");
```

```
        printf(" PROGRAM IPA_M/M/1\n");
        printf("\n");
        printf("\n");
            printf("PLEASE ENTER SYSTEM PARAMETERS\n");
            printf("ARRIVAL RATE=\n");
            scanf("%lf", &alambda);
            printf("SERVICE RATE=\n");
            scanf("%lf", &amu);
            printf("SPECIFY THE TOTAL NUMBER OF ARRIVALS");
                printf(" FOR THE SAMPLE PATH\n");
            scanf("%d", &N_a);

        time = 0.0;
   sys_time = 0.0;
            PA_L = 0.0;
            PA_G = 0.0;
                    U = UNI(ARR);
                x = - log((double)(1 - U))/alambda;
/*
*    x = f(U; alambda) is the inverse function of the
*            exponential c.d.f for interarrival times
*/
                time_a = x;
                    U = UNI(DEP);
                y = - log((double)(1-U))/amu;
/*
*  y = g(U; amu) is the inverse function of the
*            exponential c.d.f for service times
*/
                time_d = time_a + y;
        length_queue = 0;
        number_a = 0;
        number_bp = 0;
}
/*----------------------------------------------------------------
   The main IPA algorithm is contained in the following
   subroutine. Whenever an arrival occurs, we reset PA_L
   to zero if the arrival starts a busy period; otherwise,
   we compute the sample derivative of the interarrival
   time and update PA_L and PA_G accordingly.
   ----------------------------------------------------------*/

ARRIVAL()
```

```
{
        if (length_queue == 0)
            PA_L = 0.0;
        else{
            PA_L = PA_L + x/alambda;
            PA_G = PA_G + PA_L;
    sys_time = sys_time + (time_a - time) * length_queue;
    };

        time = time_a;
                    U = UNI(ARR);
                x = - log((double)(1 - U))/alambda;
                time_a = time + x;
        length_queue = length_queue + 1;
        number_a = number_a + 1;
}
/*------------------------------------------------------------*/

DEPTURE()
{
    sys_time = sys_time + (time_d - time) * length_queue;
        time = time_d;
        length_queue = length_queue - 1;
                    U = UNI(DEP);
                y = - log((double)(1 - U))/amu;

        if (length_queue == 0){
            time_d = time_a + y;
            number_bp = number_bp + 1;
                }
        else
            time_d = time + y;
}
/*------------------------------------------------------------*/

OUTPUT()
{
  double DERI_theo, sys_t_theo;
            DERI = PA_G/(number_a-1.0);
        printf("\n");
        printf("lambda=%f mu=%f\n", alambda, amu);
        printf("number of arrivals in the simulation=%d\n", N_a);
        printf("\n");
```

```
         DERI_theo = 1.0/(amu-alambda)/(amu-alambda);
       printf("DERI(ipa)=%f  DERI(theo)=%f\n", DERI, DERI_theo);
       printf("          RELA. ERROR=");
     printf("%f %c\n", (DERI-DERI_theo)*100.0/DERI_theo,37);
             sys_time = sys_time/(number_a-1.0);
             sys_t_theo = 1.0/(amu-alambda);
       printf("\n");
       printf("SYS. TIME(simu)=%f", sys_time);
     printf("    SYS. TIME(theo)=%f\n", sys_t_theo);
       printf("          RELA. ERROR=");
     printf("%f %c\n",(sys_time-sys_t_theo)*100.0/sys_t_theo,37);
}
/*-------------------------------------------------------------*/

double UNI( i )
int i;
{
  long int a,b;

  a = ISEED[i]/16384;
  b = ISEED[i]%16384;
  ISEED[i] = ((( 13205*a+74505*b )%16384 )
            *16384 + 13205*b )%268435456;
  return( (double)ISEED[i]/268435456.0 );
}
```

Index

About the Authors

Christos G. Cassandras is Professor of Manufacturing Engineering and Professor of Electrical and Computer Engineering at Boston University. He received a B.S degree from Yale University, M.S.E.E degree from Stanford University, and S.M and Ph.D degrees from Harvard University. Prior to joining Boston University, he worked on automated manufacturing systems with ITP Boston (1982-84), and was Professor of Electrical and Computer Engineering at the University of Massachusetts at Amherst until 1997. Besides discrete event and hybrid systems, his interests include stochastic control and optimization, computer simulation, manufacturing systems, and computer networks. He has published over 150 technical papers in these areas. In addition to his academic activities, he has worked extensively with industrial organizations on various projects involving system integration of manufacturing facilities and the development of decision-support software for discrete event systems.

His prior book, *Discrete Event Systems: Modeling and Performance Analysis*, on which a substantial portion of the present book is based, received the 1999 Harold Chestnut Prize, awarded by the International Federation of Automatic Control for best control engineering textbook.

Dr. Cassandras is Editor-in-Chief of the *IEEE Transactions on Automatic Control* and a member of several other journal editorial boards. He is a Fellow of the IEEE and a recipient of a 1991 Fellowship from the Lilly Foundation.

Stéphane Lafortune is Professor of Electrical Engineering and Computer Science at the University of Michigan. He received his B.Eng. degree from the École Polytechnique de Montréal, M.Eng. degree from McGill University, and Ph.D. degree from the University of California at Berkeley. He joined the University of Michigan in 1986. The principal focus of his research has been the area of discrete event systems, in particular supervisory control, optimal control, failure diagnosis, and more recently distributed and decentralized control and diagnostics.

Dr. Lafortune is a Fellow of the IEEE. He received the Presidential Young Investigator Award from the National Science Foundation in 1990 and the George S. Axelby Outstanding Paper Award from the Control Systems Society of the IEEE in 1994 (for a paper co-authored with S. L. Chung and F. Lin).